Commercial, Industrial, and Institutional Refrigeration

Design, Installation, and Troubleshooting

Commercial, Industrial, and Institutional Refrigeration

Design, Installation, and Troubleshooting

WILLIAM B. COOPER
Consulting Engineer

Professor Emeritus
Division of Technology
Macomb County Community College
Warren, Michigan

PRENTICE-HALL, INC. Englewood Cliffs, New Jersey 07632

Library of Congress Cataloging-in-Publication Data

COOPER, WILLIAM B.
Commercial, industrial, and institutional refrigeration.

Includes index.
1. Refrigeration and refrigerating machinery.
I. Title.
TP492.C686 1986 621.5'6 86-12279
ISBN 0-13-152018-0

Editorial/production supervision
and interior design: *Theresa A. Soler*
Cover design: *Whitman Studios, Inc.*
Manufacturing buyer: *Rhett Conklin*

Printed in the United States of America

10 9 8 7 6 5 4 3 2 1

ISBN 0-13-152018-0 025

PRENTICE-HALL INTERNATIONAL (UK) LIMITED, *London*
PRENTICE-HALL OF AUSTRALIA PTY. LIMITED, *Sydney*
PRENTICE-HALL CANADA, INC., *Toronto*
PRENTICE-HALL HISPANOAMERICANA, S.A., *Mexico*
PRENTICE-HALL OF INDIA PRIVATE LIMITED, *New Delhi*
PRENTICE-HALL OF JAPAN, INC., *Tokyo*
PRENTICE-HALL OF SOUTHEAST ASIA, PTE. LTD., *Singapore*
EDITORA PRENTICE-HALL DO BRASIL, LTDA., *Rio de Janeiro*

I wish to dedicate this book to my wife

. . .and partner. . .

Mary Anne

Contents

PREFACE xix

1 THE REFRIGERATION PROCESS 1

What Is Refrigeration? 1
Making Things Cold 1
How Heat Is Moved 3
How the Refrigerant Produces Refrigeration 4
Understanding the Two Types of Heat 7
Relating Latent Heat to Refrigeration 7
Specific Heat 8
Saturated Vapor 8
Superheat and Subcooling 8
Components of a Refrigeration System 10
Changes During the Refrigeration Cycle 11
Changes in State, 11
Changes in Pressure, 12
Changes in Temperature, 12
Exchanges of Heat, 12
Why Measurement Is Important 13
Review Exercise 15

2 TOOLS AND SUPPLIES 16

Good Tools for Quality Work 16
Caring for Tools 16

Piping 17
Refrigeration Service Tubing (soft copper), 17
Nitrogenized ACR copper tubing (hard drawn), 18

Fittings for Pipe and Tubing 18
Flare Fittings, 19
Wrought Copper Fittings, 19
Special Service Fittings, 19

Cutting tools 22
Tube-bending tools 22
Swaging Tools 23
Flare Connections for Copper Tubing 23
Soft Soldering 24
Hard Soldering and Silver Brazing 25
Brazing Equipment, 25
Flowing the Silver Alloy, 26

Wrenches 26
Socket Wrenches, 26
Box and Flare Nut Wrenches, 26
Open-end Wrenches, 26
Adjustable Wrenches, 27
Special-purpose Wrenches, 27

Thermometers 27
Gauges 28
Refrigeration Lubricating Oils 29
Refrigerants 30
Grouping and Classifying Refrigerants, 31
Cryogenic Fluids, 31
Refrigerant Containers, 31

Portable Charging Cylinders 32
Solvents and Cleaning 32
Steam Cleaning, 32
Solvent Sprays, 32
Liquid Acids, 33

Review Exercise 34

3 COMPRESSORS **36**

Types of Compressors 36
Reciprocating Compressors 36
Hermetically Sealed Compressors, 37
Semihermetic Compressors, 38
Open-type Compressors, 39

Condensing Units 40
Capacity Data, 42
Using Air-cooled and Water-cooled Condensers, 48

Determining which Refrigerant Is Best 48
Determining Performance from the Manufacturers' Literature 49
Compressor Unloaders 49
Advantages of the Rotary Compressor 51
Rotary Booster Compressors 51
Helical Rotary Compressors 52
Ammonia Compressors 52
Centrifugal Compressors 57
Review Exercise 62

4 EVAPORATORS 64

Classification of Evaporators 64
Evaporator Construction, 65
Classification by Use, 66

Evaporator Defrosting 67
Air Defrosting, 67
Electric Defrosting, 67
Hot Gas Defrosting, 69
Glycol Defrosting, 72

Heat Exchangers for Evaporators 72
Installation of Unit Coolers 72
Selecting a Large Product Cooler 73
Selection Procedure, 74
Selection Example, 74

Selecting a Small Unit Cooler 78
Matching Coils and Compressors 78
Evaporator Performance 79
Proper Temperature Difference 79
Fin Spacing 80
Liquid Chilling Evaporators 80
Shell-and-tube Chillers, 81
Shell-and-coil Chillers, 81
Tank Chillers, 82
Double-pipe Chillers, 82
Baudelot Chillers, 83

Secondary Refrigerants 83
Selecting a Liquid Cooler 83
Selecting a Cooler for a Liquid Other Than Water, 84

Review Exercises 85

5 CONDENSERS AND WATER ECONOMIZERS 87

Heat Transfer 87
Selection Based on Total Heat Rejection 87

Air-cooled Condensers 88

Remote Air-cooled Condensers, 89
Head Pressure Control, 92

Water-cooled Condensers 94

Series and Parallel Flow, 95
Determining the Capacity, 96
Effect of Noncondensables, 96
Shell-and-tube, 96
Shell-and-coil, 97
Double-tube, 98
Atmospheric, 99

Evaporative Condensers 99

Cooling Towers, 101
Closed-circuit Evaporative Water Coolers, 107

Water Regulating Valves 107

Example, 110
Three-way Water Regulating Valves, 110

Condenser Controls 111
Condenser Maintenance 112
Review Exercise 113

REFRIGERANT METERING DEVICES AND CONTROLS **115**

Types of Metering Devices 115
Hand Expansion Valves 116
Automatic Expansion Valves 116
Thermostatic Expansion Valves 117

Setting the Superheat, 121
Distributors, 122
Pressure-limiting Valves, 122
Thermostatic Expansion Valve Charges, 122
Expansion Valve and Bulb Location, 124
TEV Hunting, 124

Capillary Tubes 124
Low-side Floats 126
High-side Floats 127
Solenoid Valves 128
Pressure Regulating Valves 129

Evaporator Pressure Regulating Valves, 130
Head Pressure Controls, 130
Crankcase Pressure Regulating Valves, 131
Hot-gas Bypass Valves, 131

Ammonia Valves 131

Ammonia Expansion Valves, 133
Float Valves, 133

Ammonia Solenoid Valves, 134
Ammonia Pressure Regulators, 134

Review Exercise 1 137
Review Exercise 2 138

7 REFRIGERANT PIPING AND ACCESSORIES 140

Piping Functions 140
Materials 140
Joints 141
Supports 141
Pipe Sizing 142
System Practices for Halocarbon Refrigerants 148
Suction Lines, 148
Hot-gas Lines, 150
Condenser-to-receiver Lines, 151
Liquid Lines, 152
Installation, 152

Multiple-system Practices for Halocarbon Refrigerants 152
Hot-gas Lines, 153
Suction Lines, 153
Evaporator Piping, 154
Condenser Piping, 156

Accessories 157
Vibration Eliminators, 157
Liquid Refrigerant Indicators, 158
Pressure Relief Valves, 158
Oil Separators, 158
Heat Exchangers, 159
Receiver Tank Valves, 160
Compressor Service Valves, 160
Manual Valves, 160
Dehydrators, 161

System Practices for Ammonia Refrigeration 161
Compressor Piping, 161
Check Valves, 162
Condenser and Receiver Piping, 162
Parallel Condensers, 163
Evaporative Condensers, 163
Air Blowers, 163
Suction Traps, 163

Review Exercise 168

8 REFRIGERATION SYSTEMS 170

Classification of Systems 170
Basic Systems 171

Automatic Defrost Systems 173
Automatic Ice-making Systems 174
Supermarket Systems 177
Ultralow (Cryogenic) Temperature Systems 178
Compound Systems, 180
Cascade Systems, 181
Heat Pumps 184
Secondary Refrigerant (Brine) Systems 186
Review Exercise 189

9 ELECTRICAL THEORY 191

Volts-amperes-ohms 191
Simple Circuits 191
Ohm's Law 192
Power 192
Power Factor 193
Horsepower 193
Direct and Alternating Current 194
Electromagnetism 194
Conductance 195
Types of Circuits 195
Voltages 197
Phase 198
Circuit Conditions 198
Motor Nameplates 198
Manufacturers' Information 199
Electrical Meters 200
Ohmmeters, 200
Voltmeters, 201
Clamp-on Ammeter, 201
Wattmeters, 202
Capacitor-checker Meters, 202
Review Exercises 203

10 ELECTRICAL COMPONENTS AND WIRING DIAGRAM SYMBOLS 205

Electrical Equipment 205
Load Devices 207
Motors, 207
Solenoids, 208
Transformers, 216
Lights, 219
Heaters, 219
Switches 219
Switching Actions, 219

Pressurestats and Thermostats, 220
Time Clocks, 225
Oil Lubrication Safety Switches, 226
Fuses and Circuit Breakers, 228
Thermal Overloads, 229

Combination Devices 229
Relays, 229
Dual-circuit Overloads, 233

Capacitors 233
Review Exercise 235

11 ELECTRIC MOTORS AND STARTERS **237**

Alternating Current Motors 237
Wye and Delta Power Systems 237
Single-phase Motors 238
Split-phase Motors, 239
Capacitor-start Motors, 239
Capacitor-start, Capacitor-run Motors, 240
Permanent Split Capacitor Motors, 240
Shaded-pole Motors, 241
Wound-rotor Motors, 241

Three-phase Motors 242
Squirrel-cage Motors, 242
Wound-rotor Motors, 242
Synchronous Motors, 242

Motor Speed 242
Hermetic Compressor Motors (Single Phase) 243
Thermal Overload Protection, 244

Motor Starters 245
Wye-delta Starters, 245
Part-winding Starters, 247
Primary-resistance Starters, 247
Autotransformer Magnetic Starters, 247

Oil Lubrication Safety Controls 251
Interlocking Circuits 251
Review Exercise 252

12 REFRIGERATION EQUIPMENT **254**

Classification of Equipment 254
Drinking Water Coolers 255
Automatic Ice Dispensers 259
Ice Cube Makers, 260
Flake Ice Machines, 262

Large Ice-making Machines, 264
Tube Ice Machines, 267

Ice Builders 270
Prefabricated Refrigerating Buildings 273
Food Store Equipment 273
Walk-in Coolers and Freezers, 274
Dairy Product Display Cases, 275
Frozen Food and Ice Cream Display Cases, 277
Meat Display Cases, 278
Delicatessen Display Cases, 279
Produce Display Cases, 279
Rack-mounted Condensing Units, 281
Multiple-compressor Units, 282
Reach-in Refrigerators, 290

Packaged Liquid Coolers 292
Large Industrial Packaged Chillers 295
Transport Refrigeration Units 296
Review Exercise 303

13 **WIRING DIAGRAMS AND CONTROL CIRCUITS** **305**

Wiring Diagram Practices 305
Types of Wiring Diagrams 305
Typical Wiring Diagrams 306
Drinking Water Coolers, 306
Electric Defrost Systems for Walk-in Freezers, 307
Hot Gas Defrost Systems for Walk-in Freezers, 308
Ice Cube Makers, 309
Ice Dispensers, 310
Flake Ice Machines, 311
Large-capacity Flake Ice Equipment, 312
Tube Ice Machines, 315
Refrigeration Compressors, 315
Supermarket Systems, 317
Transport Refrigeration Units, 320

Review Exercise 1 324
Review Exercise 2 325

14 **INSTALLATION AND OPERATION** **326**

Installation Arrangement 326
Selecting a Location for the Equipment 327
Ventilation, 327
Weather Conditions, 327
Space for Installation and Service, 327
Isolation of Noise and Vibration, 327
Availability of Utility Services, 328

Refrigerant Piping 329
Electrical Connections 330
Dehydration and Evacuation 330
Evacuation, 331
Charging the System with Refrigerant, 333
Purging Noncondensables, 335

Check, Test, and Start 336
Testing for Leaks, 336
Loosening the Hold-down Bolts, 337
Positioning the Valves, 337
Checking the Lubrication, 337
Checking the Compressor Oil, 337
Belt Tension, 338
Checking the Water System, 338
Checking the Electrical System, 338
Supplying Customer Information, 339

Review Exercise 340

15 SERVICE 342

The Service Business 342
Using the Gauge Manifold 342
Connecting the Gauge Manifold, 344

Using and Caring for Service Valves 345
Servicing Compressors and Condensing Units 346
Testing Open-type Compressors, 346
Removing Compressors, 346
Replacing Crankshaft Seals, 347
Replacing Valve Plates, 347

Servicing Burned-out Hermetics 348
Servicing Condensers 348
Purging Noncondensables, 349
Correcting Undercharge or Overcharge, 349
Cleaning Condensers, 349

Servicing Evaporators 350
Servicing Refrigerant Controls 350
Thermal Expansion Valves, 351
Low-side Float Systems, 352
Solenoid Valves, 353
Pressure Regulators, 353

Using Filters and Driers 353
Removing the Refrigerant 354
Cleaning the System with Liquid Refrigerant 354
Summary of Halocarbon Refrigeration System Service 355
Review Exercise 356

16 TROUBLESHOOTING 358

Troubleshooting Skills 358
Eight Steps of Troubleshooting Procedure 358
Interviewing the Operator, 358
Verifying the Symptoms, 359
Attempting a Quick-fix, 359
Reviewing Troubleshooting Aids, 359
Following a Step-by-Step Search, 359
Correcting the Problem, 360
Making Final Checks, 360
Completing the Project, 360
Troubleshooting Aids 360
Examples of Quick-fix Procedures 360
Troubleshooting Symptoms, Problems, and Solutions 361
Review Exercises 366

17 APPLICATIONS 368

Classification of Applications 368
Food Preservation 369
Meats, 370
Poultry, 372
Seafood, 373
Fruits and Vegetables, 373
Dairy Products and Eggs, 374
Frozen Foods, 376
Transport Refrigeration 383
Long-haul Systems, 383
Local Delivery Equipment, 384
Mechanical Refrigeration, 384
Low-temperature Applications 385
Making Ice Cream, 385
Bakery Refrigeration, 386
Ice Rinks, 387
Review Exercise 390

18 CALCULATING THE REFRIGERATION LOAD AND SELECTING THE EQUIPMENT 392

Refrigeration Load Calculation Is Unique 392
The Survey 393
Guidelines for Room Conditions, 405
Calculating the Refrigeration Load 406
Transmission Load, 406
Infiltration Load, 406
Product Load, 410

Miscellaneous Loads, 422
Safety Factor, 423

Selecting the Equipment 423
Equipment Selection Example, 423
Checking the Balancing-out Point, 426

Review Exercise 428

19 USING TABLES AND CHARTS **429**

Reading the Refrigerant Table 429
Adding Heat to the Refrigerant, 430
Calculating from the Tables and Charts, 431

Volumetric Efficiency 432
Operating Conditions that Increase the Capacity of the System 433
Plotting System Performance on a Pressure-enthalpy Diagram, 434
Reading the P-E Chart, 436
Pressure-Enthalpy Chart in SI Units, 442
Summary, 446

Review Exercise 447

APPENDIX I. SI Units and Conversion Factors **448**

APPENDIX II. Psychometric Charts **457**

APPENDIX III. Safety Code of Mechanical Refrigeration **460**

APPENDIX IV. Pressure-Enthalpy Diagrams and Tables **480**

ANSWERS TO REVIEW QUESTIONS **490**

INDEX **499**

Preface

This book has been written for service people, maintenance people, installers, and operators of commercial refrigeration equipment.

Few people realize the millions of dollars that are spent on the commercial refrigeration jobs throughout our industrial society. It has been estimated that 83% of the total cost of these systems over a period of 25 years is in operating costs and maintenance, compared to 17% for the original installation.

The traditional work of service people has been to repair a breakdown and to get a system back into operation in the shortest possible time. It is the desire of the writer to impress upon contemporary service people the urgent need to assume additional responsibility and to recognize opportunities for whole-system efficiencies. Ideally, the service person would look for ways to save energy and maintain the commercial refrigeration system at its design level of performance.

The byword of the day is energy conservation and, since commercial refrigeration jobs are highly energy intensive, it is a very important function of the technical experts to look for ways to save energy.

Let's take a supermarket, for example. How can energy be saved in the course of replacing a burned-out compressor, repairing malfunctioning solenoid valves, cleaning condensers, calibrating controls, and so on? A survey of the system could indicate the following "opportunities":

1. Installing a heat reclaimer for the refrigeration condenser to be used to heat the building in winter.
2. Lowering the relative humidity of the air conditioning system to 40% to reduce frosting on the evaporators and eliminate the need for antisweat heaters.

3. Recovering the cooling that is dumped on the floor from open display cases.

Each type of system has its own energy-saving opportunities and, in most instances, it is the refrigeration technician who is in the right place at the right time to initiate the action for putting these conservation measures into effect.

Another area where the refrigeration service person can exercise responsibility is the maintenance of commercial refrigeration equipment at its original level of performance. This means preventive surveillance which has a proven record of payback for the owner. If, due to wear or improved technology, replacement or retrofit is desirable, service people looking after the equipment can initiate the necessary action.

Unfortunately, it is the exception rather than the rule to find this kind of maintenance. The writer has had the occasion numerous times to inspect equipment that has been in operation for ten years or more and to find maintenance neglected and essential controls inoperative.

The responsibility and dedication of service people is therefore essential to the ultimately effective operation and continued reliable use of commercial refrigeration systems. This book presents information to achieve these objectives.

In teaching refrigeration, I have always found a rather large group of students who desire to specialize in commercial refrigeration. Some of the advantages claimed for this branch of the business are:

Less competition
Better pay
More repeat business
Greater challenge

In reviewing the current texts it is evident that there is a need for more information on this subject. Much of the material is available from manufacturers; however, the information needs to be assembled, edited, and interpreted to make it useful to the student. For one thing, it is difficult to find a complete listing of commercial refrigeration products. In numerous instances throughout my experience in the industry, I have been asked to service new types of units without having had previous experience with the equipment and have been provided with inadequate manuals. This text is, therefore, designed to include comprehensive information on most of the commonly used commercial refrigeration products that the service person is most likely to encounter.

The text is also designed as a stand-alone handbook. It includes the basics necessary for understanding the advanced information in the text and for reference. Many of the larger commercial systems are complex and the student will need to develop his or her knowledge gradually through experience to fully appreciate the techniques associated with large sophisticated systems.

The instructor as well as the students will find the Objectives at the beginning of each chapter and the Review Exercises at the end extremely useful. The answers to the Review exercises are given at the end of the book.

The first chapter introduces four important aspects of refrigeration systems which are basic to a full understanding of the refrigeration cycle. The second chapter addresses another fundamental need by supplying information on the correct use of refrigeration tools and supplies.

In Chapters 3, 4, 5, and 6 particularly, examples are given of the selection procedure for major items of equipment. These procedures indicate the parameters that affect the capacity and performance of the system. The service person should be able to distinguish the difference between the need to replace or repair a defective element and the need to correct the original poor design or selection of equipment.

Chapter 3 covers the wide range of compressors available, from small hermetics to the large centrifugals. The section on the effect of using various refrigerants on compressor performance should be particularly helpful. Obtaining the proper balance between the compressor and evaporator is explained in Chapter 4. Special attention to the winter operation of condensers is given in Chapter 5. The special features of ammonia metering devices is included in Chapter 6. Chapter 7 stresses the importance of good piping practices for both halocarbon and ammonia systems.

The refrigeration system is developed in Chapter 8. This is an extremely important chapter since it assembles the separate components into the various commonly used commercial systems.

A great many of the problems on refrigeration systems are electrical. Chapters 9, 10, 11, and 13 deal with the electrical elements of commercial refrigeration equipment: first the basics, in Chapter 9, then the electrical components and their symbols, in Chapter 10. Motors and controls are covered in Chapter 11. Finally, in Chapter 13, wiring diagrams for many of the common refrigeration systems are shown along with a description of the sequence of operations. Sandwiched in between Chapters 11 and 13 is Chapter 12, giving a comprehensive review of the many types of commercial refrigeration equipment.

Some of the most important chapters in the book are Chapters 14, 15, and 16, covering installation, service, and troubleshooting. The text approaches these subjects from a systems standpoint, yet supplies details for solving problems associated with individual components. The chapter on troubleshooting is unique to refrigeration procedures and offers an advanced approach for teaching this critical subject. The term "quick fix" refers to a repair that can solve the problem at hand in a minimum amount of time. It is a technique used by expert service people.

Chapters 17, 18, and 19 are optional chapters. Some schools desire to teach the student how equipment is used (Chapter 17), how to calculate a refrigeration load (Chapter 18), and how to use refrigeration tables and charts (Chapter 19).

Chapter 17 supplements Chapter 12 on equipment, supplying information on additional refrigeration products used for special applications. Some refrigeration licenses require the applicant to be able to calculate loads and read refrigeration tables. This information is covered in Chapters 18 and 19.

One of the important items included in the Appendix is the Standard Safety Code for mechanical refrigeration, which has been adopted by most localities.

I particularly wish to thank the many manufacturers who have generously contributed information and illustrations to make this text complete and "user-friendly." Special appreciation goes to William Walrep for his expert work in making many of the drawings included in the text.

William B. Cooper

1 The Refrigeration Process

OBJECTIVES

After studying this chapter, the reader will be able to:

- Define common terms used in refrigeration.
- Explain operation of the vapor-compression refrigeration cycle.
- Explain the proper function of the basic components of a system.

WHAT IS REFRIGERATION?

Refrigeration makes it possible to remove heat from an area or a substance where it is not wanted to some other space or material where it is not objectionable. Heat is a form of energy related to the action of atoms and molecules. The greater the heat, the more rapid the action of these minute particles of matter. It is believed that all motion stops at absolute zero, about $-460\,^{\circ}$F. This low temperature has never been reached, but scientists have plotted the decreased action of the atoms due to reduction in temperature and have arrived at a base for "absolute" zero temperature. Another way to define refrigeration is "making things cold." So what is this condition we call "cold"?

MAKING THINGS COLD

The best definition of cold is the absence of heat. Thus, to produce the cooling effect, heat is removed. To calculate the size of equipment to perform refrigeration work, heat needs to be defined in terms of some unit that can be measured. In the United States the most common unit of heat is the British thermal unit (Btu). One Btu is the amount of heat required to raise 1 pound of water 1 degree Fahrenheit ($^{\circ}$F). Thus, if we raise the temperature of 10 lb of water 10 °F, it takes 100 Btu to do it. Or, if it is desired to *cool* 10 lb of water 10 °F, 100 Btu must be removed.

In most tables that give the heat content of substances such as refrigerants, the base for heat measurements is −40 °F. This is an accepted standard, so that various measurements of heating or cooling can be related.

Usually, heat is measured indirectly. The actual measurement is temperature, measured on an indicating thermometer of some type. Metric thermometers measure in Celsius degrees. English thermometers measure in Fahrenheit degrees. The two thermometers can be related as shown in Figure 1–1. Electronic thermometers are being used more commonly, as shown in Figure 1–2.

In refrigeration the main concern is the amount of heat that is removed. Thus, if a refrigeration coil removes heat at the rate of 100 Btu per hour, this would be equivalent to reducing the temperature of 10 lb of water 10 °F.

In order to relate the heat removed from various substances, the ability of the substance to absorb heat must be known. This quality of the substance is called *specific heat.* Specific heat is the amount of heat required to raise the temperature of 1 pound

Figure 1–1 Comparison of Fahrenheit and Celsius temperature scales.

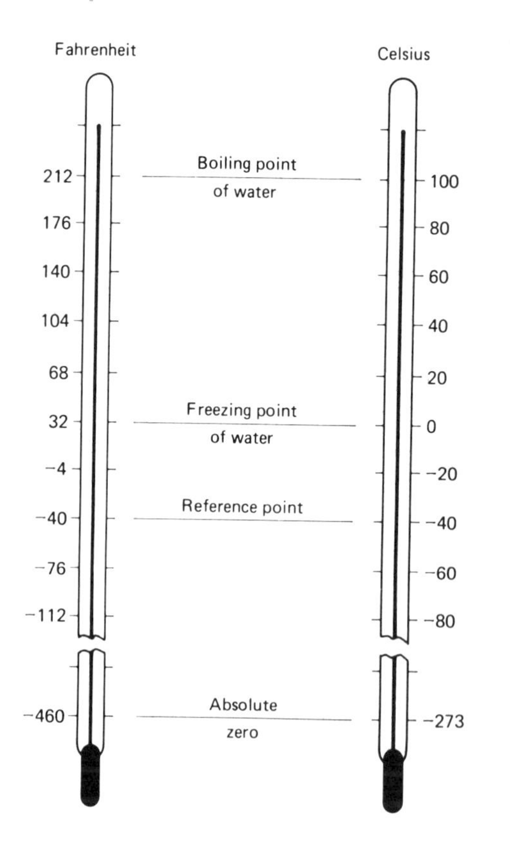

Figure 1–2 Electronic thermometer. (By permission of IMC Instruments, Inc.)

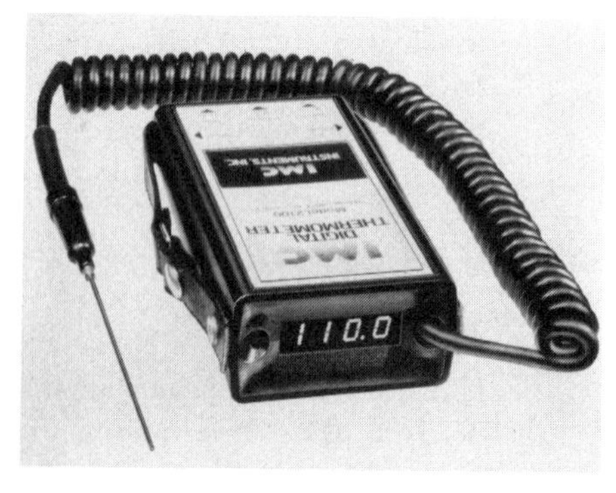

SPECIFIC HEAT VALUES

MATERIAL	SPECIFIC HEAT BTU/LB./DEG F (KCAL/KG/DEG C)
WATER	1.00
ICE	0.50
AIR (DRY)	0.24
STEAM	0.48
ALUMINUM	0.22
BRICK	0.20
CONCRETE	0.16
COPPER	0.09
GLASS	0.20
IRON	0.10
WOOD (HARD)	0.45
WOOD (PINE)	0.67

THESE VALUES MAY BE USED FOR COMPUTATIONS WHICH INVOLVE NO CHANGE OF STATE.

Figure 1-3 Specific heat values for some common substances.

of a substance or material 1°F. A table giving some common specific heats is shown in Figure 1-3.

The amount of heat removed from a substance is determined by the following formula:

$$H \text{ (heat removed)} = W \text{ (weight of material)} \times \text{sp. ht. (specific heat)} \times T \text{ (temperature rise)}$$

For example, since water has a specific heat of 1, the heat removed by cooling 10 lb of water at 10°F is

$$10 \text{ lb} \times 1 \text{ sp. ht.} \times 10°\text{F} = 100 \text{ Btu}$$

The rate at which heat is removed is important. The unit of time is generally considered as 1 hour. Therefore, if 100 Btu of heat is removed in 1 hour the rate would be 100 Btuh (British thermal units per hour).

HOW HEAT IS MOVED

The process of moving heat is called *refrigeration.* The medium or substance used to move heat is a *refrigerant.* The refrigerant absorbs heat from inside a refrigerator and transfers it to the air outside the cabinet. Many substances can be used as refrigerants.

Figure 1–4 Reach-in refrigerator, 65 ft^3 size. Aircooled refrigerating condenser unit on top. Corrosion resistant. Self-closing door. Urethane foamed-in-place insulation. Automatic electric evaporating condensate system. (By permission of Howard Refrigeration Company, Inc.)

Some are better for one purpose, others are better for another application. The most commonly used refrigerants today derive their cooling effect primarily from the process of evaporation.

Heat moves from a warmer area to a colder area. The refrigerant is the medium of exchange. Observe the simple refrigeration system shown in Figure 1–4. Heat from inside of the refrigerator is picked up by the cold refrigerant. The refrigeration system utilizes the compressor to raise the temperature of the refrigerant so that the warm refrigerant can lose heat to the colder air outside the cabinet. The refrigeration system is so designed that the refrigerant continues to circulate from inside the cabinet to the outside and outside to inside, carrying heat away from the products to be cooled.

HOW THE REFRIGERANT PRODUCES REFRIGERATION

The refrigeration cycle in modern refrigerators is usually a closed system. That is, the refrigerant is reused. The principle of physics that is useful in understanding this action of the refrigerant is the transfer of energy by *change of state.* Whenever a substance changes from a liquid to a vapor, or a solid to liquid, or the reverse of any of these processes, there is an exchange of heat.

Reference to Figure 1–5 shows that Refrigerant-12 (R-12) has the heat content of liquid at 40°F of 17.273 Btu/lb. R-12 as vapor at the same temperature has a heat content of 81.436 Btu/lb. The difference is 64.163 Btu/lb, resulting from a change of state. In the refrigeration cycle the evaporating temperature can be controlled. The heat absorbed by the refrigerant inside the cabinet is used to vaporize or boil the refrigerant at 40°F, thus picking up 64.163 Btu/lb.

Temp F	Pressure psia	Pressure psig	Volume cu ft/lb Vapor v_g	Density lb/cu ft Liquid $1/v_f$	Enthalpy** Btu/lb Liquid h_f	Enthalpy** Btu/lb Vapor h_g	Entropy** Btu/lb · R Liquid s_f	Entropy** Btu/lb · R Vapor s_g
−150	0.15359	29.60849*	178.65	104.36	−22.697	60.837	−0.062619	0.20711
−140	0.25623	29.39951*	110.46	103.54	−20.652	61.896	−0.056123	0.20208
−130	0.41224	29.08187*	70.730	102.71	−18.609	62.968	−0.049830	0.19760
−120	0.64190	28.61429*	46.741	101.87	−16.565	64.052	−0.043723	0.19359
−110	0.97034	27.94558*	31.777	101.02	−14.518	65.145	−0.037786	0.19002
−100	1.4280	27.0138*	22.164	100.15	−12.466	66.248	−0.032005	0.18683
− 90	2.0509	25.7456*	15.821	99.274	−10.409	67.355	−0.026367	0.18398
− 80	2.8807	24.0560*	11.533	98.382	− 8.3451	68.467	−0.020862	0.18143
− 70	3.9651	21.8482*	8.5687	97.475	− 6.2730	69.580	−0.015481	0.17916
−60	5.3575	19.0133*	6.4774	96.553	− 4.1919	70.693	−0.010214	0.17714
− 50	7.1168	15.4313*	4.9742	95.616	− 2.1011	71.805	−0.005056	0.17533
− 40	9.3076	10.9709*	3.8750	94.661	0.0000	72.913	0.000000	0.17373
− 30	11.999	5.490*	3.0585	93.690	2.1120	74.015	0.004961	0.17229
− 20	15.267	0.571	2.4429	92.699	4.2357	75.110	0.009831	0.17102
− 10	19.189	4.493	1.9727	91.689	6.3716	76.196	0.014617	0.16989
0	23.849	9.153	1.6089	90.659	8.5207	77.271	0.019323	0.16888
†5	26.483	11.787	1.4580	90.135	9.6005	77.805	0.021647	0.16842
10	29.335	14.639	1.3241	89.606	10.684	78.335	0.023954	0.16798
12	30.539	15.843	1.2748	89.392	11.118	78.546	0.024871	0.16782
14	31.780	17.084	1.2278	89.178	11.554	78.757	0.025786	0.16765
16	33.060	18.364	1.1828	88.962	11.989	78.966	0.026699	0.16750
18	34.378	19.682	1.1399	88.746	12.426	79.176	0.027608	0.16734
20	35.736	21.040	1.0988	88.529	12.863	79.385	0.028515	0.16719
22	37.135	22.439	1.0596	88.310	13.300	79.593	0.029420	0.16704
24	38.574	23.878	1.0220	88.091	13.739	79.800	0.030322	0.16690
26	40.056	25.360	0.98612	87.870	14.178	80.007	0.031221	0.16676
28	41.580	26.884	0.95173	87.649	14.618	80.214	0.032118	0.16662
30	43.148	28.452	0.91880	87.426	15.058	80.419	0.033013	0.16648
32	44.760	30.064	0.88725	87.202	15.500	80.624	0.033905	0.16635
34	46.417	31.721	0.85702	86.977	15.942	80.828	0.034796	0.16622
36	48.120	33.424	0.82803	86.751	16.384	81.031	0.035683	0.16610
38	49.870	35.174	0.80023	86.524	16.828	81.234	0.036569	0.16598
40	51.667	36.971	0.77357	86.296	17.273	81.436	0.037453	0.16586
42	53.513	38.817	0.74798	86.066	17.718	81.637	0.038334	0.16574
44	55.407	40.711	0.72341	85.836	18.164	81.837	0.039213	0.16562
46	57.352	42.656	0.69982	85.604	18.611	82.037	0.040091	0.16551
48	59.347	44.651	0.67715	85.371	19.059	82.236	0.040966	0.16540
50	61.394	46.698	0.65537	85.136	19.507	82.433	0.041839	0.16530
52	63.494	48.798	0.63444	84.900	19.957	82.630	0.042711	0.16519
54	65.646	50.950	0.61431	84.663	20.408	82.826	0.043581	0.16509
56	67.853	53.157	0.59495	84.425	20.859	83.021	0.044449	0.16499
58	70.115	55.419	0.57632	84.185	21.312	83.215	0.045316	0.16489
60	72.433	57.737	0.55839	83.944	21.766	83.409	0.046180	0.16479
62	74.807	60.111	0.54112	83.701	22.221	83.601	0.047044	0.16470
64	77.239	62.543	0.52450	83.457	22.676	83.792	0.047905	0.16460
66	79.729	65.033	0.50848	83.212	23.133	83.982	0.048765	0.16451
68	82.279	67.583	0.49305	82.965	23.591	84.171	0.049624	0.16442
70	84.888	70.192	0.47818	82.717	24.050	84.359	0.050482	0.16434
72	87.559	72.863	0.46383	82.467	24.511	84.546	0.051338	0.16425
74	90.292	75.596	0.45000	82.215	24.973	84.732	0.052193	0.16417
76	93.087	78.391	0.43666	81.962	25.435	84.916	0.053047	0.16408
78	95.946	81.250	0.42378	81.707	25.899	85.100	0.053900	0.16400
80	98.870	84.174	0.41135	81.450	26.365	85.282	0.054751	0.16392
82	101.86	87.16	0.39935	81.192	26.832	85.463	0.055602	0.16384
84	104.92	90.22	0.38776	80.932	27.300	85.643	0.056452	0.16376
†86	108.04	93.34	0.37657	80.671	27.769	85.821	0.057301	0.16368
88	111.23	96.53	0.36575	80.407	28.241	85.998	0.058149	0.16360
90	114.49	99.79	0.35529	80.142	28.713	86.174	0.058997	0.16353
92	117.82	103.12	0.34518	79.874	29.187	86.348	0.059844	0.16345
94	121.22	106.52	0.33540	79.605	29.663	86.521	0.060690	0.16338
96	124.70	110.00	0.32594	79.334	30.140	86.691	0.061536	0.16330
98	128.24	113.54	0.31679	79.061	30.619	86.861	0.062381	0.16323
100	131.86	117.16	0.30794	78.785	31.100	87.029	0.063227	0.16315
102	135.56	120.86	0.29937	78.508	31.583	87.196	0.064072	0.16308
104	139.33	124.63	0.29106	78.228	32.067	87.360	0.064916	0.16301
106	143.18	128.48	0.28303	77.946	32.553	87.523	0.065761	0.16293
108	147.11	132.41	0.27524	77.662	33.041	87.684	0.066606	0.16286
110	151.11	136.41	0.26769	77.376	33.531	87.844	0.067451	0.16279
112	155.19	140.49	0.26037	77.087	34.023	88.001	0.068296	0.16271
114	159.36	144.66	0.25328	76.795	34.517	88.156	0.069141	0.16264
116	163.61	148.91	0.24641	76.501	35.014	88.310	0.069987	0.16256
118	167.94	153.24	0.23974	76.205	35.512	88.461	0.070833	0.16249
120	172.35	157.65	0.23326	75.906	36.013	88.610	0.071680	0.16241
130	195.71	181.01	0.20364	74.367	38.553	89.321	0.075927	0.16202
140	221.32	206.62	0.17799	72.748	41.162	89.967	0.080205	0.16159
160	279.82	265.12	0.13604	69.209	46.633	91.006	0.088927	0.16053
180	349.00	334.30	0.10330	65.102	52.562	91.561	0.098039	0.15900
200	430.09	415.39	0.076728	60.026	59.203	91.278	0.10789	0.15651
220	524.43	509.73	0.053140	52.670	67.246	89.036	0.11943	0.15149
233.6 (Critical)	596.9	582.2	0.02870	34.84	78.86	78.86	0.1359	0.1359

aFrom published data (1955 and 1956) of E. I. du Pont de Nemours & Co., Inc. Used by permission.
*Inches of mercury below one standard atmosphere.

**Based on 0 for the saturated liquid at −40 F.
†Standard cycle temperatures.

Figure 1–5 Refrigerant 12: properties of liquid and saturated vapor. (By permission of *ASHRAE Handbook.*)

Temp, F	Viscosity, $lb_m/ft \cdot h$			Thermal Conductivity, $Btu/h \cdot ft \cdot F$			Specific Heat, c_p, $Btu/lb_m \cdot F$				Temp, F
	Sat. Liquid	Sat. Vapor	Gas, $P = 1$ atm $\times 10^{-2}$‡	Sat. Liquid	Sat. Vapor	Gas, $P = 1$ atm $\times 10^{-3}$‡	Sat. Liquid	Sat. Vapor	Gas $(c_p)_{0\,atm}$	Gas $(c_p)_{1\,atm}$	
−140	2.47			0.0655			0.199		0.1085		−140
−120	1.97			0.0631			0.202		0.1123		−120
−100	1.612			0.0608			0.204		0.1160		−100
− 80	1.347			0.0585			0.207		0.1196		− 80
− 60	1.146			0.0561			0.209	0.126	0.1230		− 60
− 40	0.990			0.0538			0.212	0.133	0.1264		− 40
− 20	0.866	0.0249	2.49	0.0514	0.0040	4.00	0.214	0.139	0.1296		− 20
0	0.767	0.0265	2.61	0.0490	0.0043	4.31	0.217	0.145	0.1327		0
20	0.687	0.0279	2.72	0.0467	0.0046	4.63	0.220	0.150	0.1356		20
40	0.620	0.0291	2.83	0.0443	0.0050	4.95	0.224	0.157	0.1385		40
60	0.564	0.0301	2.94	0.0420	0.0053	5.28	0.229	0.164	0.1413		60
80	0.517	0.0311	3.05	0.0397	0.0056	5.61	0.234	0.174	0.1439		80
100	0.477	0.0324	3.15	0.0373	0.0060	5.94	0.240	0.185	0.1465		100
120	0.441	0.0339	3.26	0.0350	0.0064	6.27	0.251	0.199	0.1490		120
140	0.409	0.0359	3.36	0.0326	0.0068	6.60	0.266	0.216	0.1513		140
160	0.370	0.0384	3.47	0.0302	0.0072	6.94	0.288	0.235	0.1536		160
180	0.329	0.0417	3.57	0.0276	0.0076	7.28	0.317	0.290	0.1558		180
200	0.273	0.0458	3.67	0.0246	0.0083	7.63	0.356	0.362	0.1579		200
220	0.200	0.051	3.77	0.0204	0.0093	7.98	0.406		0.1599		220
230	0.149	0.060	3.82	0.0161	0.0107	8.16			0.1609		230
234*	0.075	0.075	3.84	0.0130	0.0130	8.23			0.1612		234*
240			3.87			8.34			0.1618		240
260			3.96			8.71			0.1637		260
280			4.06			9.08			0.1654		280
300			4.15			9.45			0.1671		300
320			4.25			9.82			0.1687		320
340			4.34			10.1			0.1703		340
360			4.43			10.5			0.1718		360
380			4.53			10.8			0.1732		380
400			4.62			11.2			0.1746		400

*Critical Temperature. Tabulated properties ignore critical region effects.
‡Actual value = (Table value) × (Indicated multiplier).

Figure 1–5 (continued)

The compressor in the cycle compresses the 40 °F vapor, raising its temperature higher than the air temperature outside the cabinet. As a result, the refrigerant loses its heat to the air and condenses back to liquid, ready for reuse.

Refer again to Figure 1–5, which shows that under a saturation condition where the liquid is at the same temperature as the vapor, for every pressure exerted on the refrigerant there is a corresponding temperature. By controlling the pressure exerted by the compressor, the refrigerant boiling temperature and the refrigerant condensing temperature can be controlled.

The melting of ice, another example of a change of state, has produced a standard for refrigeration capacity known as the *ton* of refrigeration. It takes 144 Btu to melt 1 lb of ice. If a ton of ice, 2000 lb, is melted in 1 day (from heat absorbed inside a refrigerator), the heat required is

$$\begin{aligned} 1 \text{ ton refrigeration} &= 2000 \text{ lb} \times 144 \text{ Btu/lb} \\ &= 288{,}000 \text{ Btu/day} \end{aligned}$$

The more useful figure is one on an hourly basis. Thus

$$\begin{aligned} 1 \text{ ton refrigeration} &= \frac{288{,}000 \text{ Btu/day}}{24 \text{ h}} \\ &= 12{,}000 \text{ Btu/h} \end{aligned}$$

The ton is also expressed on a per minute basis, which is

$$\begin{aligned} 1 \text{ ton refrigeration} &= \frac{12{,}000 \text{ Btu/h}}{60 \text{ min}} \\ &= 200 \text{ Btu/min} \end{aligned}$$

UNDERSTANDING THE TWO TYPES OF HEAT

It is often useful to separate the two types of heat. One of these is called *sensible heat,* which is the type of heat that changes the temperature of a material, with no change in state. The other type of heat is called *latent heat.* This type of heat changes the state but does not change the temperature of a material. For example, if 1 lb of water is heated to the boiling point of water, 212 °F at atmospheric pressure, the amount of heat added as indicated on an ordinary thermometer is sensible heat. At the boiling point, an amount of heat—approximately 980 Btu/lb—needs to be added to vaporize the water without changing the temperature of 212 °F. This is latent heat. In this particular example, the heat is also called *heat of vaporization.*

When ice turns to liquid, 144 Btu/lb needs to be added, without a change of temperature. This heat is also called *heat of fusion.*

RELATING LATENT HEAT TO REFRIGERATION

When liquid refrigerant enters the evaporator it has a relatively low heat content. When it vaporizes it picks up heat from inside the refrigerator. The heat that is added to vaporize the refrigerant is latent heat or heat of vaporization. The difference between the total heat of the vapor and the original heat of the liquid is the amount of cooling or the amount of refrigeration performed by the refrigerant.

The heat content of the refrigerants is given in the refrigerant tables (see Figure 1-5 for data on R-12 refrigerant). The refrigerant tables show total heat content as *enthalpy*. Enthalpy is expressed in Btu per pound. Assuming that the evaporating temperature is 0°F and referring to Figure 1-5, we find the following:

1. *Enthalpy of saturated vapor:* 77.271 Btu/lb
2. *Enthalpy of saturated liquid:* 8.521 Btu/lb

By subtracting line 2 from line 1, the difference is 68.750 Btu/lb, which is the heat of vaporization for this refrigerant under these conditions.

To accurately calculate the total "refrigerating effect," this determination needs to be refined somewhat. However, the heat of vaporization is a major factor in producing refrigeration.

SPECIFIC HEAT

Specific heat is a measure of the amount of heat required to raise 1 pound of a substance 1 degree in temperature. It is fairly constant unless the product goes through a change of state. For example, the specific heat of water is 1 Btu/lb per °F, and for ice, 0.5 Btu/lb per °F. The specific heat of fresh lean beef above freezing temperature is 0.77 Btu/lb per °F. Below freezing temperature, its specific heat is 0.47 Btu/lb per °F (see Figure 18-15).

SATURATED VAPOR

One of the very useful properties of a refrigerant is the pressure–temperature relationship of saturated vapor. A refrigeration vapor is said to be saturated whenever both liquid and vapor are present in the same container. Under these conditions there is a fixed relationship between the temperature of the refrigerant in the container and the pressure.

It is customary to use absolute pressures in most refrigeration tables and charts. Most changes take place in relation to absolute pressure. Absolute pressure is found by adding gauge pressure to atmospheric pressure at sea level and is usually considered to be 14.7 psi.

Referring to the refrigerant chart in Figure 1-5, R-12 at 70°F has an absolute pressure of 84.9 psi. For any given pressure there is an equivalent temperature. Handy pocket-sized charts (see Figure 1-6) show the saturated gauge pressure for various common refrigerants for a useful temperature range. These charts have continuous value to the service person in determining the condition of the refrigerant in a system.

SUPERHEAT AND SUBCOOLING

If additional heat is added to a refrigerant vapor in a saturated condition, with no liquid present, the additional heat is called *superheat* and the temperature–pressure relationship shown in the saturated refrigerant chart no longer applies. This feature is useful in the

Vacuum-Inches of Mercury
Italic Figures

Pressure-Pounds Per Square Inch
Bold Figures

TEMPERATURE °F.	REFRIGERANT — CODE				
	12-F	22-V	500-D	502-R	717-A
−60	*19.0*	*12.0*	*17.0*	*7.2*	*18.6*
−55	*17.3*	*9.2*	*15.0*	*3.8*	*16.6*
−50	*15.4*	*6.2*	*12.8*	*0.2*	*14.3*
−45	*13.3*	*2.7*	*10.4*	1.9	*11.7*
−40	*11.0*	0.5	*7.6*	4.1	*8.7*
−35	*8.4*	2.6	*4.6*	6.5	*5.4*
−30	*5.5*	4.9	*1.2*	9.2	*1.6*
−25	*2.3*	7.4	1.2	12.1	1.3
−20	0.6	10.1	3.2	15.3	3.6
−18	1.3	11.3	4.1	16.7	4.6
−16	2.0	12.5	5.0	18.1	5.6
−14	2.8	13.8	5.9	19.5	6.7
−12	3.6	15.1	6.8	21.0	7.9
−10	4.5	16.5	7.8	22.6	9.0
−8	5.4	17.9	8.8	24.2	10.3
−6	6.3	19.3	9.9	25.8	11.6
−4	7.2	20.8	11.0	27.5	12.9
−2	8.2	22.4	12.1	29.3	14.3
0	9.2	24.0	13.3	31.1	15.7
1	9.7	24.8	13.9	32.0	16.5
2	10.2	25.6	14.5	32.9	17.2
3	10.7	26.4	15.1	33.9	18.0
4	11.2	27.3	15.7	34.9	18.8
5	11.8	28.2	16.4	35.8	19.6
6	12.3	29.1	17.0	36.8	20.4
7	12.9	30.0	17.7	37.9	21.2
8	13.5	30.9	18.4	38.9	22.1
9	14.0	31.8	19.0	39.9	22.9
10	14.6	32.8	19.7	41.0	23.8
11	15.2	33.7	20.4	42.1	24.7

TEMPERATURE °F.	REFRIGERANT — CODE				
	12-F	22-V	500-D	502-R	717-A
12	15.8	34.7	21.2	43.2	25.6
13	16.4	35.7	21.9	44.3	26.5
14	17.1	36.7	22.6	45.4	27.5
15	17.7	37.7	23.4	46.5	28.4
16	18.4	38.7	24.1	47.7	29.4
17	19.0	39.8	24.9	48.8	30.4
18	19.7	40.8	25.7	50.0	31.4
19	20.4	41.9	26.5	51.2	32.5
20	21.0	43.0	27.3	52.4	33.5
21	21.7	44.1	28.1	53.7	34.6
22	22.4	45.3	28.9	54.9	35.7
23	23.2	46.4	29.8	56.2	36.8
24	23.9	47.6	30.6	57.5	37.9
25	24.6	48.8	31.5	58.8	39.0
26	25.4	49.9	32.4	60.1	40.2
27	26.1	51.2	33.2	61.5	41.4
28	26.9	52.4	34.2	62.8	42.6
29	27.7	53.6	35.1	64.2	43.8
30	28.4	54.9	36.0	65.6	45.0
31	29.2	56.2	36.9	67.0	46.3
32	30.1	57.5	37.9	68.4	47.6
33	30.9	58.8	38.9	69.9	48.9
34	31.7	60.1	39.9	71.3	50.2
35	32.6	61.5	40.9	72.8	51.6
36	33.4	62.8	41.9	74.3	52.9
37	34.3	64.2	42.9	75.8	54.3
38	35.2	65.6	43.9	77.4	55.7
39	36.1	67.1	45.0	79.0	57.2
40	37.0	68.5	46.1	80.5	58.6
41	37.9	70.0	47.1	82.1	60.1

TEMPERATURE °F.	REFRIGERANT — CODE				
	12-F	22-V	500-D	502-R	717-A
42	38.8	71.4	48.2	83.8	61.6
43	39.8	73.0	49.4	85.4	63.1
44	40.7	74.5	50.5	87.0	64.7
45	41.7	76.0	51.6	88.7	66.3
46	42.6	77.6	52.8	90.4	67.9
47	43.6	79.2	54.0	92.1	69.5
48	44.6	80.8	55.1	93.9	71.1
49	45.7	82.4	56.3	95.6	72.8
50	46.7	84.0	57.6	97.4	74.5
55	52.0	92.6	63.9	106.6	83.4
60	57.7	101.6	70.6	116.4	92.9
65	63.8	111.2	77.8	126.7	103.1
70	70.2	121.4	85.4	137.6	114.1
75	77.0	132.2	93.5	149.1	125.8
80	84.2	143.6	102.0	161.2	138.3
85	91.8	155.7	111.0	174.0	151.7
90	99.8	168.4	120.6	187.4	165.9
95	108.2	181.8	130.6	201.4	181.1
100	117.2	195.9	141.2	216.2	197.2
105	126.6	210.8	152.4	231.7	214.2
110	136.4	226.4	164.1	247.9	232.3
115	146.8	242.7	176.5	264.9	251.5
120	157.6	259.9	189.4	282.7	271.7
125	169.1	277.9	203.0	301.4	293.1
130	181.0	296.8	217.2	320.8	—
135	193.5	316.6	232.1	341.2	—
140	206.6	337.2	247.7	362.6	—
145	220.3	358.9	264.0	385.0	—
150	234.6	381.5	281.1	408.4	—
155	249.5	405.1	298.9	432.9	—

Figure 1–6 Temperature–pressure chart. (By permission of Sporlan Valve Company.)

refrigeration process, as, for example, for refrigerant entering a compressor. The compressor is a vapor pump and most compressors will not tolerate liquid. In order to be certain that only vapor reaches the compressor, it is useful to add at least 10 °F of superheat to the saturated vapor. Even if the vapor is cooled a few degrees in its path to the compressor, the refrigerant will remain in a vapor state and will not harm the compressor. There are a number of other uses of superheat, which will be described later.

Subcooling is the condition of lowering the temperature of liquid refrigerant below the saturated temperature. Only liquid must be present to make this possible. Subcooling increases the "refrigerating effect" since the temperature of the liquid is precooled before entering the evaporator.

Both superheating and subcooling can be measured by a thermometer. These measurements are easily determined by an electric temperature-measuring instrument. Both superheating and subcooling need to be controlled since they represent changes in the performance of the refrigeration cycle.

COMPONENTS OF A REFRIGERATION SYSTEM

The discussion of refrigeration in this text centers primarily around the vapor-compression refrigeration cycle. The basic requirements are a compressor, a condenser, a refrigerant metering device, and an evaporator. The receiver is optional, depending on the type of system (see Figure 1–7).

The compressor is a mechanical device for pumping vapor from a low-pressure area to a high-pressure area. Since pressure, temperature, and the volume of a gas are related, a change in pressure will also affect other variables in the formula:

$$PV = WRT$$

where P = absolute pressure, psia
V = volume of the vapor, ft^3
W = weight of the gas, lb
R = a gas constant
T = absolute temperature, °F

When a low-temperature, low-pressure vapor refrigerant is compressed, two important actions occur: (1) the volume is reduced, and (2) the temperature of the refrigerant is raised.

In the condenser, the high-pressure, high-temperature vapor loses heat to the surrounding medium (usually water or air). When the temperature of the vapor reaches saturation temperature, additional heat removed causes condensation of the refrigerant, producing liquid refrigerant.

The metering device is a separation between the high-pressure and low-pressure parts of the system. High-pressure, medium-temperature liquid at the metering device enters a low-pressure, low temperature area (the evaporator). The metering device controls the flow of refrigerant into the evaporator. Two actions occur in the metering device: (1) the refrigerant liquid is cooled to evaporation temperature by actual evaporation of some of the liquid refrigerant, and (2) the pressure of the refrigerant is reduced to a pressure corresponding to the evaporating temperature at the saturated condition.

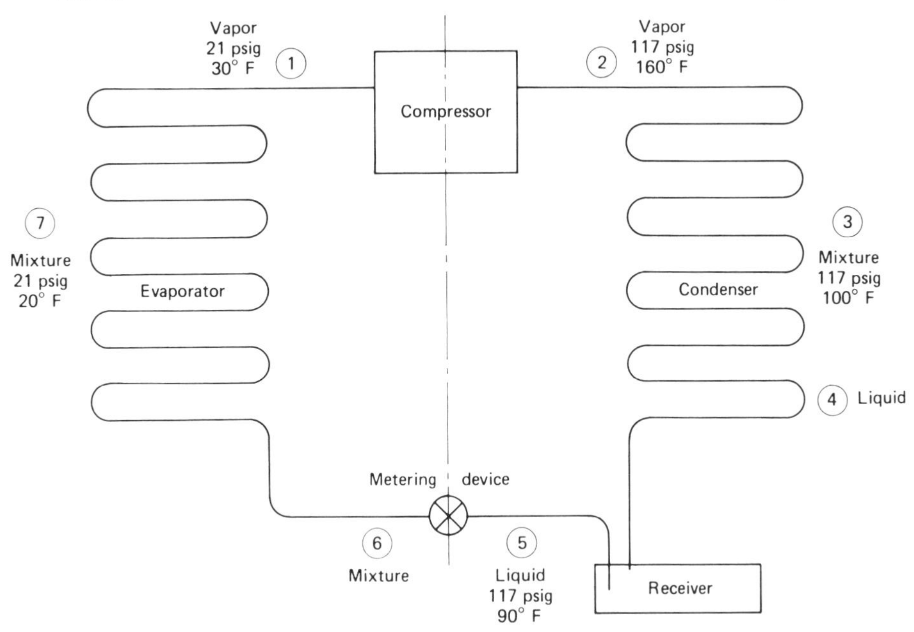

Figure 1–7 Changes that take place in the refrigerant during the refrigeration cycle.

In the evaporator the saturated refrigerant absorbs heat from its surroundings and boils into a low-pressure vapor. Some superheating of the vapor takes place before the suction gas reaches the compressor.

The receiver is really an accessory in the system providing additional space to store the liquid refrigerant. The receiver is on the high side of the system with its refrigerant at approximately the same pressure as in the condenser. Some subcooling may take place in the lower part of the condenser thus providing a lower energy level for the refrigerant in storage.

CHANGES DURING THE REFRIGERATION CYCLE

The various changes that take place in the refrigeration cyc fall into four groups (see Figure 1–7).

Changes in State

Position 1. The refrigerant enters the compressor as a low-pressure, low-temperature vapor.

Position 2. The refrigerant leaves the compressor as a high-pressure, high-temperature vapor.

Position 3. The refrigerant temperature is reduced to saturated conditions and the vapor condenses. At this point the condition of the refrigerant is a mixture of vapor and liquid.

Position 4. At the lower portion of the condenser the refrigerant has been completely condensed and is in liquid form.

Position 5. Same refrigerant condition as position 4. Refrigerant is all liquid. However, some subcooling has taken place.

Position 6. In passing through the metering device some refrigerant is evaporated, reducing the temperature to the evaporating temperature. The refrigerant is a mixture at this point.

Position 7. The remaining liquid in the refrigerant evaporates to absorb the load. The refrigerant is a mixture, as in position 6.

Position 1. The refrigerant has picked up some superheat. The refrigerant enters the compressor as a superheated vapor.

Changes in Pressure

Positions 6, 7, and 1 represent the low side of the system. Positions 2, 3, 4, and 5 represent the high side of the system. For all practical purposes these are the two pressures in the system: the low-side pressure and the high-side pressure.

Changes in Temperature

Positions 6 and 7. From the metering valve through the central area of the evaporator the temperature of the refrigerant is at the evaporating temperature.

Position 1. The refrigerant has picked up some superheat in the latter part of the evaporator and in the suction lines to the compressor. The superheat is indicated by a temperature higher than saturated conditions.

Position 2. Superheat in large quantities has been added to the refrigerant vapor, raising the discharge temperature.

Position 3. The superheat is removed by the first part of the condenser, bringing the temperature of the refrigerant down to condensing temperature.

Positions 4 and 5. Some subcooling is accomplished in the lower part of the condenser as well as in the receiver and liquid line to the metering device. The temperature of the liquid is below that of saturated conditions.

Exchanges of Heat

Positions 6 and 7. Heat is added to the refrigerant in absorbing the product load in the evaporator. This constitutes primarily the net refrigerating effect plus a small gain that occurs in the piping up to position 1, where the refrigerant enters the compressor.

Between positions 1 and 2. The compressor adds heat to the refrigerant in a sizable quantity. This is equivalent to the work done in compressing the refrigerant.

Position 3. The heat that has been added in the evaporator and by the compressor is removed in the condenser. Some relatively small additional losses occur in the receiver and liquid-line piping up to the metering device in positions 4 and 5.

WHY MEASUREMENT IS IMPORTANT

In analyzing a refrigeration system to determine its performance or nonperformance, measurement of the operating conditions is important. Suitable instruments are used for obtaining operating values, such as:

1. Temperature (°F)
2. Pressure (psig or psia)
3. Amount of refrigerant in the system (lb)
4. Power characteristics (volts, amperes, ohms, watts)
5. Revolutions per minute of a fan or compressor (rpm)
6. Cubic feet of air per minute of a fan (cfm)
7. Relative humidity of a space (RH)
8. Capacity of the system (Btuh or tons)

These measurements, and many others, make it much easier to determine performance accurately. Remedial action or improvement of performance can be planned based on knowing these values.

The measurement of temperature was described earlier. The measurement of pressure is usually performed by the use of a gauge manifold, described in Chapter 15. The measurement of the refrigerant in the system is explained in Chapter 14. Electrical measurements are outlined in Chapter 9. To measure the rpm value of a fan or compressor, various types of rpm meters are available. Air velocities can be measured with an anemometer or velometer and the quantity of airflow calculated knowing the area of the air outlet. Use the formula:

$$\text{cfm} = \text{fpm} \times A$$

where
cfm = cubic feet of air per minute
fpm = velocity of air, feet per minute
A = area of coil, ft^2

Relative humidity is measured with a psychrometer.

The measurement of the capacity of a system is usually performed indirectly. Knowing the quantity of air (CFM) handled by the fan and the change in temperature and humidity conditions of the air in passing through the coil, the total heat removed can be calculated. For example, assume the following conditions:

Dry-bulb temperature of air entering coil	40 °F DB
Wet-bulb temperature of air entering coil	37 °F WB
Dry-bulb temperature of air leaving coil	30 °F DB
Wet-bulb temperature of air leaving coil	29 °F WB
Air quantity, cfm	1000 cfm

The formula for *total heat* is

$$\text{Btuh} = \text{cfm} \times 4.5 \times (TH_2 - TH_3)$$

where
CFM = cubic feet of air per minute
TH_2 = total heat of leaving air
TH_3 = total heat of entering air

Applying the formula gives us

$$\text{Btuh} = 1000 \text{ cfm} \times 4.5 \times (13.9 - 10.5)$$

Note that total heat values are obtained from the psychrometric chart.

$$\text{Total heat removed} = 15.300 \text{ Btuh or } 1.28 \text{ tons}$$

If the medium is a liquid, by knowing the flow rate in gallons per minute, the temperature drop, and the specific heat of the liquid, the capacity of the system can be calculated. For example, assume the following conditions:

Entering water temperature	50 °F
Leaving water temperature	40 °F
Flow, gpm	10 gpm
Specific heat of water	1.0

The following formula applies:

$$\text{Btuh} = \text{gpm} \times 500 \times \text{sp. ht.} \times (T_2 - T_1)$$

where GPM = gallons per minute
sp. ht. = specific heat of liquid
T_2 = entering temperature of liquid, °F
T_1 = leaving temperature of liquid, °F

Applying the formula gives us

$$\begin{aligned}\text{Btu} &= 100 \times 500 \times 1.0 \times (50 - 40) \\ &= 10 \times 500 \times 1.0 \times 10 \\ &= 50{,}000 \text{ Btuh or } 4.17 \text{ tons}\end{aligned}$$

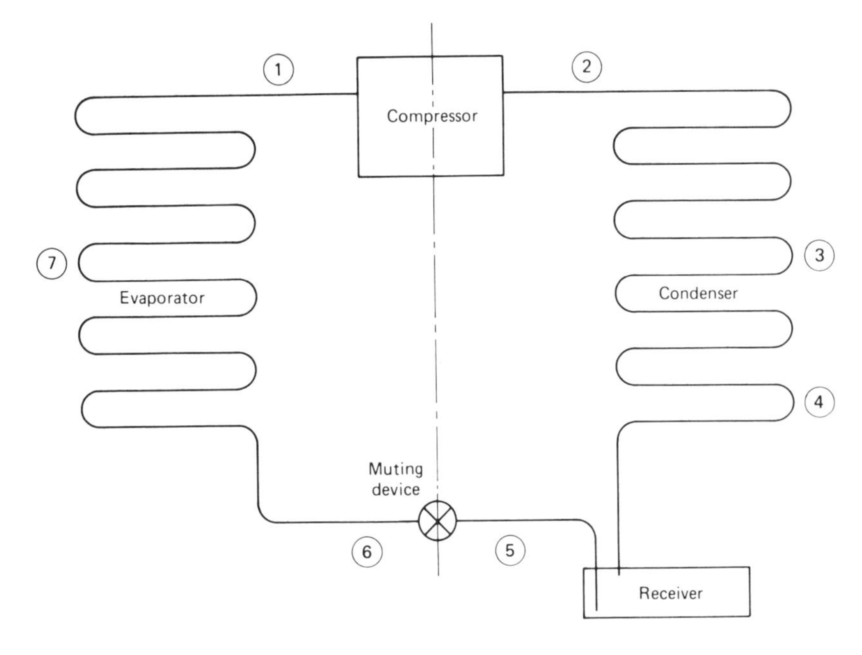

Figure 1–8 Refrigeration cycle for Review Exercise.

REVIEW EXERCISE

Directions: Using the information given in this unit, fill in the missing blanks in the following exercise. The numbers indicate positions in Figure 1–8.

1. Pressure 24.6 psig
   Temperature 55 °F
   Refrigerant ______
2. Pressure 126.6 psig
   Temperature 160 °F
   Refrigerant ______
3. Pressure ______
   Temperature ______
   Refrigerant mixture
4. Pressure ______
   Temperature 95 °F
   Refrigerant ______
5. Pressure ______
   Temperature ______
   Refrigerant ______
6. Pressure ______
   Temperature ______
   Refrigerant ______
7. Pressure ______
   Temperature ______
   Refrigerant ______

Note: For R-12 refrigerant.

2 Tools and Supplies

OBJECTIVES

After studying this chapter, the reader will be able to:

- Make a list of the essential tools and supplies needed by commercial refrigeration mechanics.
- Describe the care required to keep refrigeration tools in proper operating order.
- Demonstrate the proper use of essential tools and supplies.

GOOD TOOLS FOR QUALITY WORK

The commercial refrigeration mechanic should have the correct tools and the necessary supplies to perform high-quality work. The tools used most often can be placed in pockets in a belt worn by the service person. The next most important tools are placed in one or more portable tool boxes. The least used tools are stored in the truck or in the shop. Some of the newer models of service trucks have side-opening panels for easy access to tools, parts, and other supplies.

CARING FOR TOOLS

Tools must be kept in good working order. All movable parts should be lubricated. Tools that cannot be repaired should be replaced. A supply of replaceable blades should be readily available for cutting tools. Replacement washers should be on hand to repair damaged hose connections.

Gauges and instruments must be kept in calibration to retain their usefulness. Tools should be arranged in organized containers so that they are easy to find. The use of tool boards for truck or shop storage is highly recommended.

Missing wrenches should be replaced. Spares should be carried for essential tools that may be lost or broken.

Above all, tools should be kept clean. A refrigeration mechanic must always guard against possible contamination of the refrigeration system.

PIPING

The most common material for piping for refrigeration work is copper. Some aluminum is used and some steel pipe. Steel pipe is required for ammonia (R-717) work. Copper cannot be used due to its reactivity with ammonia refrigerant.

There are two commonly used types of refrigeration tubing: refrigeration service tubing and nitrogenized ACR copper tubing.

Refrigeration Service Tubing (Soft Copper)

This tubing is dehydrated and nitrogen purged and sealed at the ends. It is manufactured in accordance with ASTM B280 and ANSI B9.1 refrigeration industry standards. Figure 2-1 shows one coil of this tubing together with sizes available and safe working pressures.

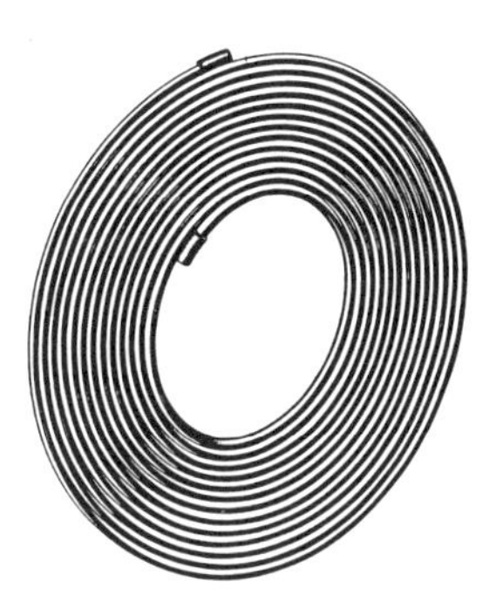

Figure 2-1 Soft copper refrigerant tube is dehydrated and sealed at both ends. (By permission of Mueller Brass Company.)

Computed allowable stress for annealed copper tube at indicated temperature

SAFE WORKING INTERNAL PRESSURES — ANSI-B31.5

Tube OD	Wall Thickness	Wt. Per Foot	150°F PSI	250°F PSI	350°F PSI	400°F PSI
1/8	.030	.0347	2660	2450	2080	1570
3/16	.030	.0575	1690	1560	1320	1000
1/4	.030	.0804	1230	1130	970	720
5/16	.032	.109	1040	960	820	610
3/8	.032	.134	860	790	670	500
1/2	.032	.182	630	580	490	370
5/8	.035	.251	540	500	430	320
3/4	.035	.305	440	400	350	260
7/8	.045	.455	500	460	390	300
1 1/8	.050	.655	430	400	340	250
1 3/8	.055	.884	390	360	300	230
1 5/8	.060	1.140	370	340	280	220

Nitrogenized ACR Copper Tube (Hard Drawn)

This tubing is cleaned, purged, and pressurized with nitrogen to protect against harmful oxides during brazing. Supplied in 20-ft lengths with reusable plugs, nitrogenized tubing is available in sizes from ⅜ to 3⅛ in. O.D. Larger tubing, 3⅝ to 6⅛ in. O.D., is cleaned and capped. ACR tubing is manufactured in accordance with ASTM B88 type L and cleaned in accordance with ASTM B280. Figure 2–2 shows wall thicknesses for various sizes and safe working pressures.

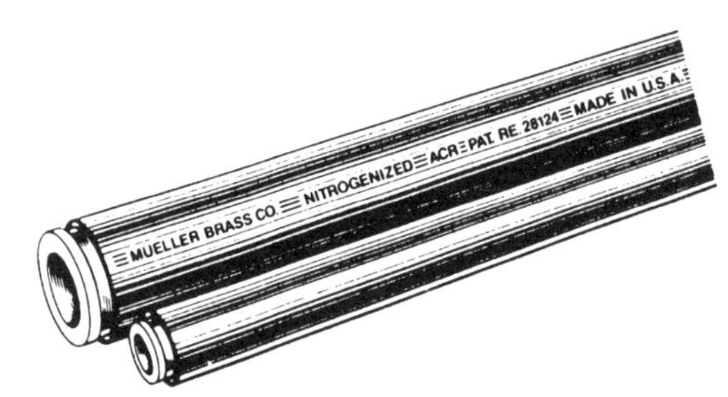

Figure 2–2 ACR hard-tempered copper tube comes in 20-ft lengths, is cleaned and capped. (By permission of Mueller Brass Company.)

Computed allowable stress for annealed copper tube at indicated temperatures
SAFE WORKING INTERNAL PRESSURES — ANSI-B31.5-1974

Tube OD	Wall Thickness	Wt. Per Foot	Lengths Per Bundle	150°F PSI	250°F PSI	350°F PSI	400°F PSI
⅜	.030	.126	100	760	700	600	460
½	.035	.198	25	680	620	520	400
⅝	.040	.285	25	630	580	490	370
¾	.042	.362	10	550	510	430	320
⅞	.045	.455	10	500	460	390	300
1⅛	.050	.655	5	430	400	340	250
1⅜	.055	.884	5	390	360	300	230
1⅝	.060	1.14	5	370	340	280	220
2⅛	.070	1.75	3	310	290	240	180
2⅝	.080	2.48	2	300	280	230	170
3⅛	.090	3.33	1	280	260	220	170
3⅝	.100	4.29	1	270	250	210	160
4⅛	.110	5.38	1	250	230	200	140
5⅛	.123	7.61	1	240	220	190	140
6⅛	.140	10.20	1	220	200	170	130

**U.S. Patent 3,200,984 **U.S. Patent RE. 28124, Canadian Patents — 723,463 and 751,099*

Soft copper tubing may become brittle through continual bending or hammering. It can be made flexible again by heating to a blue surface color and allowed to cool slowly to ambient temperature.

FITTINGS FOR PIPE AND TUBING

Various types of fittings are used by the refrigeration mechanic, depending on the requirements of the job.

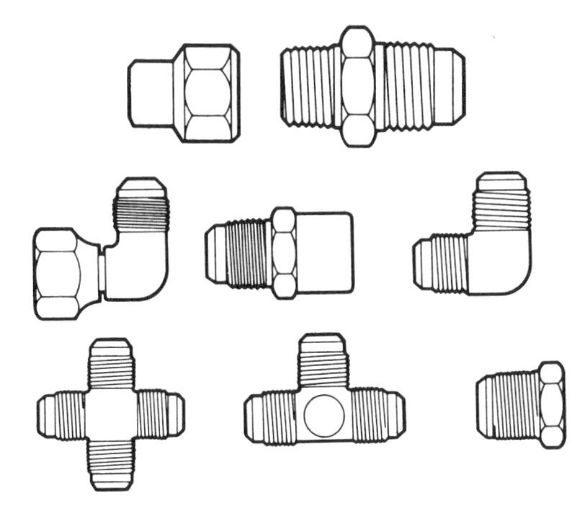

Figure 2–3 Flare fittings have coded numbers to indicate type and size. (By permission of Mueller Brass Company.)

Flare Fittings

Flare fittings are used with soft copper tubing. The size of the fitting is the same as the connecting tube. A wide variety of fittings are available (Figure 2–3). It is important to know the various configurations because often one fitting can be used in place of two, thus effecting a considerable savings.

In ordering fittings for an installation it is important to know the use of each connection. For example, fittings are made flare-to-flare, flare-to-pipe, and flare-to-sweat. Much time and money can be saved by having the right fitting available. When connections are made to the pipe, the pipe dimensions are given in inside dimensions (I.D.) as opposed to the outside dimensions (O.D.) used for copper tubing.

To prevent errors in ordering, each fitting is given a number. For example, a $\frac{3}{16}$ in. male flare to $\frac{1}{8}$ in. male iron pipe size (I.P.S.) 90° elbow is No. E1–3A.

Wrought Copper Fittings

Wrought copper fittings are also known as *sweat* fittings since they are connected to tubing by soldering or brazing. These fittings are also made in transition types to connect to flare or pipe connections. Generally speaking, these fittings are ordered by description. The various sizes and types are shown in Figure 2–4.

In sizing a tee fitting, the inlet size is first, the inline outlet is second, and the tee connection size is last. For example, a solder × solder × solder tee $\frac{5}{8}$ in. × $\frac{3}{8}$ in. × $\frac{1}{2}$ in. has an inlet of $\frac{5}{8}$ in., an outlet of $\frac{3}{8}$ in. and a tee connection of $\frac{1}{2}$ in. All sizes refer to the O.D. copper tube connection.

Special Service Fittings

Access fittings are in the special service category, as shown in Figure 2–5. These fittings are placed in the piping to provide easy access for servicing on connecting pressure gauges or controls. Some of these are equipped with Shrader valves, similar to those used on pneumatic tires, for connecting gauge manifolds.

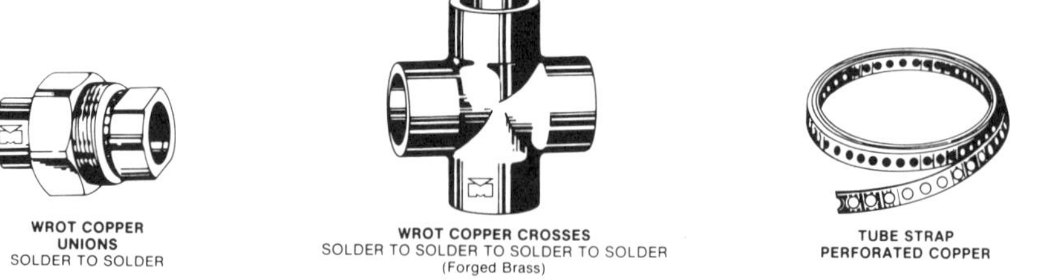

Figure 2-4 Wrought copper fittings are used for soldered or brazed connections. (By permission of Mueller Brass Company.)

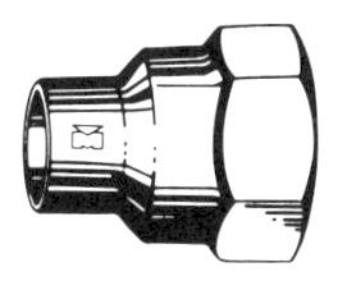

WROT COPPER
FEMALE ADAPTERS
SOLDER TO FPT

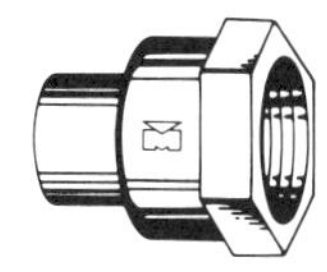

WROT COPPER
FEMALE FITTING ADAPTERS
FITTING TO FPT

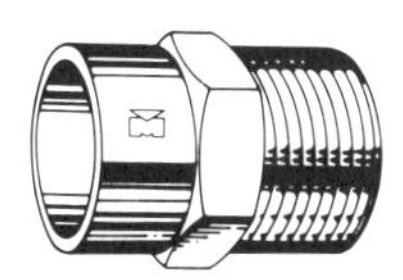

WROT COPPER
MALE ADAPTERS
SOLDER TO MPT

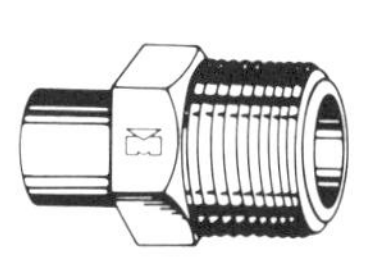

WROT COPPER
MALE FITTING ADAPTERS
FITTING TO MPT

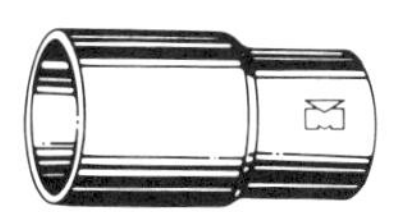

WROT COPPER
FITTING REDUCERS EXTENDED
FITTING TO SOLDER

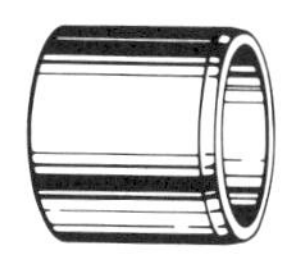

WROT COPPER
FLUSH BUSHINGS
FITTING TO SOLDER

WROT COPPER
FLUSH BUSHINGS
FITTING TO FPT

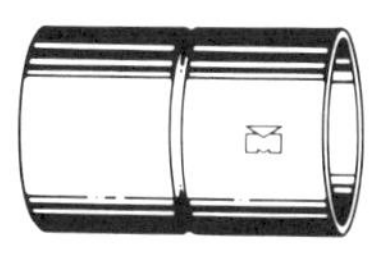

WROT COPPER
COUPLINGS ROLLED STOP
SOLDER TO SOLDER

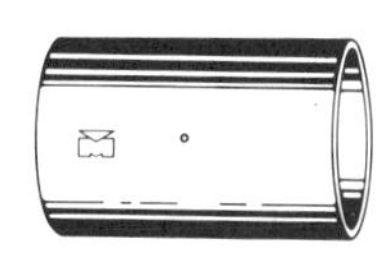

WROT COPPER
COUPLINGS STAKED STOP
SOLDER TO SOLDER

WROT COPPER
COUPLINGS WITHOUT STOP
SOLDER TO SOLDER

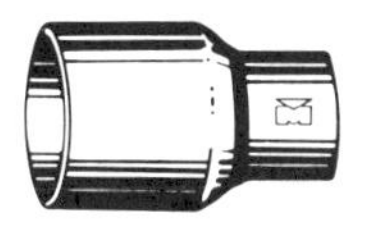

WROT COPPER
REDUCING COUPLINGS
SOLDER TO SOLDER

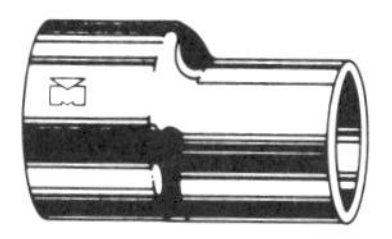

WROT COPPER
ECCENTRIC COUPLINGS
SOLDER TO SOLDER

WROT COPPER
45° ELLS
SOLDER TO SOLDER

WROT COPPER
45° FITTING ELLS
FITTING TO SOLDER

WROT COPPER
45° FITTING ELLS
FITTING TO FITTING

Figure 2-4 Continued

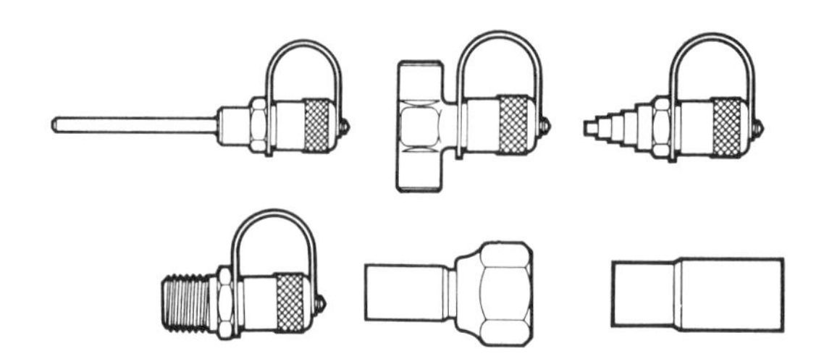

Figure 2–5 Special access fittings are used to facilitate entrance into the system. (By permission of Mueller Brass Company.)

CUTTING TOOLS

Cutting tools for copper tubing are made in various sizes to accommodate the size of tube being cut. There is a mini-size cutter that is useful when space is limited. It can be used for ⅛ to ⅞ in. O.D. tubing. Standard tube cutters come in two sizes: one for ⅛ to 1⅛ in. O.D. tubes and the other for ⅜ to 2⅜ in. O.D. tubes. The cutter can be replaced by a restricter, which is useful in reducing the size of a tube for connecting to smaller tubing. Extra cutter wheels should be kept on hand for replacements in case one becomes damaged or excessively worn.

Heavy-duty cutters are available for use with iron pipe. Two sizes are made, one for ¾ to 2 in. O.D. tubing and one for 2 to 4 in. O.D. tubing. Blades on these cutters are made of steel.

The advantage of a wheel-type cutter is the clean cut it makes. In some work it is more convenient to use a hacksaw. A sawing fixture is often used for holding the tube to assure a square cut. Care must be taken to keep the inside of the tube clean and free from burrs. All tubes should be reamed after being cut to assure a full opening size. Any rough edges should be removed with sanding material or a suitable file. Figure 2–6 illustrates the cutting tool.

Figure 2–6 Tubing cutters are used to cut copper, aluminum, or stainless steel tubing. (By permission of Robinaire Division of Sealed Power Corp.)

TUBE-BENDING TOOLS

The easiest tube to bend is one made of soft copper. Soft copper is often used for final connections to a fixture or compressor where flexibility is important. Spring-type benders are usually used for soft copper. The spring may be placed outside or inside the tube. The advantage of using a spring is that it can easily be twisted out for removal after the bend is made. Springs are normally used only on small tubes of ⅝ in. O.D. or less.

Mechanical bending tools are available for bending hard-drawn copper tubing. Bending tools can be used for tubing up to 1⅜ in. O.D. provided that the wall thickness meets the manufacturer's requirements. Figure 2–7 illustrates some of the types of bending tools available.

Figure 2–7 Tube benders can be used for soft or hard brown copper tube. (By permission of Robinaire Division of Sealed Power Corp.)

Tubing should be bent so that no strain is put on the fittings after assembly. Also, no reduction should be made in the cross-sectional area of the tube. The minimum bending radius should not be less than five times the diameter of the tube.

There are many situations where a bending tool can be used to form an elbow, thus avoiding the need for a special fitting and thereby reducing the material cost.

SWAGING TOOLS

Another tool that is useful in reducing fitting costs for a job is the swaging tool shown in Figure 2–8. These are only used for relatively small soft copper tubes. A set of six punches is available in sizes from $\frac{3}{16}$ to $\frac{5}{8}$ in. with a clamping bar. Two tubes of the same size can be joined by swaging one of the tubes and inserting the original tube to form an overlap. The amount of swaging can be controlled to make the overlap equal to or greater than the diameter of the tube.

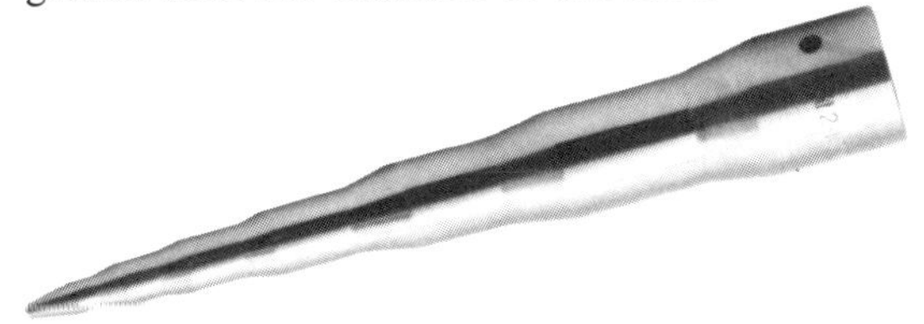

Figure 2–8 A swaging tool can be used to enlarge an end of a tube to join two tubes of similar size. (By permission of Robinaire Division of Sealed Power Corp.)

FLARE CONNECTIONS FOR COPPER TUBING

Tubing is prepared for flaring by providing a square cut, cleaning the tubing if required, and removing the burr that is formed during the cutting operation. Many cutting tools incorporate a reaming blade for removing the burr. Before flaring, the flare nut must be slid onto the tube to provide the final holding arrangement.

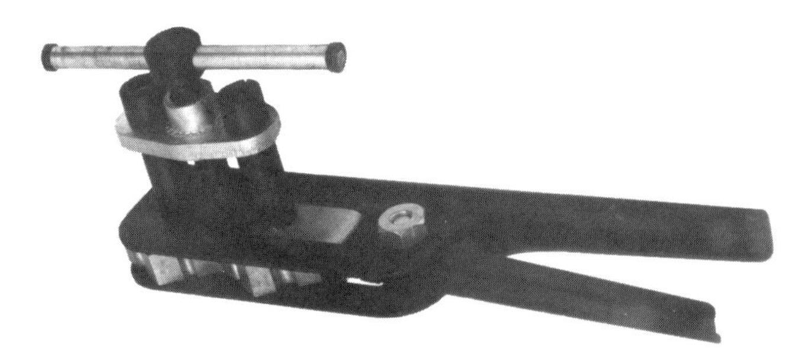

Figure 2–9 Flaring tools are used to prepare tubing for flare connections. (By permission of Robinaire Division of Sealed Power Corp.)

When soft copper tubing is used for refrigerant piping it is usually good practice to provide flared connections. A flaring tool such as the one shown in Figure 2–9 is used to construct the flare. The flaring tool consists of a split block with openings for various sized tubes, usually $\frac{3}{16}$, $\frac{1}{4}$, $\frac{5}{16}$, $\frac{3}{8}$, $\frac{1}{2}$, and $\frac{5}{8}$ in. O.D., and a screw-type spreading tool. The tubing is inserted in the block with the end extended about one-third the width of the tube and held securely by the clamping arrangement. The flare point of the spreading tool is forced into the end of the tube, shaping it at a 45° angle to fit the pattern in the block. Sufficient pressure is exerted to form the end of the tube but not to work-harden the material. A few drops of refrigerant oil are used to lubricate the expanding tube.

SOFT SOLDERING

Soft soldering refers to the heat bonding of metals using a solder material that melts below 500°F. This technique is suitable for water pipes or drain lines. Most code requirements require hard solder (melting point in the 1100°F range) for refrigeration piping.

A commonly used soft solder is composed of half tin and half lead (called "fifty–fifty") which starts to melt at 360°F and becomes fluid at 450°F. Another type of soft solder is composed of 95% tin and 5% antimony (called "ninety five–five"). This material starts to melt at 450°F and becomes fluid at 465°F. A suitable flux must be used with the soft solder.

The preparations are about the same for both soft and hard soldering:

1. Joints must be properly fitted with a minimum of clearance.
2. The metals must be clean. Sandpaper is the preferred abrasive; avoid emery cloth because of its oil content.
3. The joint is thoroughly fluxed before soldering or brazing. Do not use an excess of flux.
4. The joint is assembled and properly supported to prevent strain that might separate the joint when it is heated.
5. The joint with the flux applied is heated to the proper temperature. Excessive temperatures must be avoided.
6. The solder flows into the joint by capillary action. Excess solder is removed and the joint allowed to cool by natural action.
7. The final joint is cleaned to remove all traces of flux and to prevent corrosion.

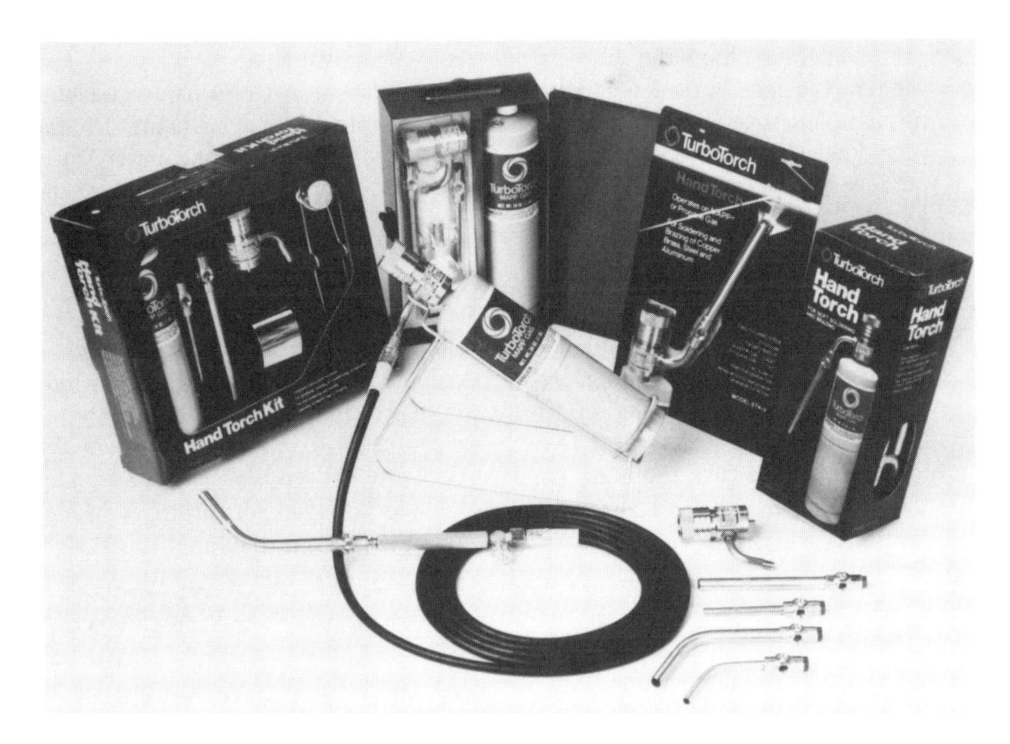

Figure 2–10 A propane torch is used for soft solder connections. (By permission of Wingaersheek, Division of Victor Equipment Co.)

Heat is usually applied to the shoulder of the joint first. As this is the heaviest portion of the connection, it requires the most heat. Heat should be distributed uniformly to both the fitting and tube by moving the torch during the heating process.

On small piping, a propane torch can be used. Where more heat is required, acetylene torches are advisable. Torches should be equipped with a "spark" lighter. A typical propane torch is shown in Figure 2–10.

HARD SOLDERING AND SILVER BRAZING

Brazing Equipment

For silver brazing a hot flame is required, produced by burning a combination of oxygen and acetylene. The two tanks are usually carried in a portable stand or rack known as an "MG" outfit (see Figure 14–2).

In addition to the carrying stand, the outfit includes a welding torch, suitable welding tips, goggles, friction lighter, regulators, and hose. Each tank has a separate regulator and two gauges, one for tank pressure and one for torch pressure. Valves on the torch handle are fully opened when the outfit is in use. The mixture is controlled by the adjustable pressure regulators.

To light the torch the acetylene valve is opened one-quarter turn and the oxygen valve cracked open. A spark lighter is used to ignite the flame. After the flame is established two gases are supplied in the correct proportion to produce a good flame. The flame should have a luminous blue cone with a slightly reddish tip. Too much acetylene will produce a greenish flame, which produces soot. Too much oxygen will produce an oxidizing flame, which will pit the metal.

Flowing the Silver Alloy

The general method of preparation for silver alloy is the same as that for soft soldering: cutting, fitting, cleaning, fluxing, and supporting. Both the flux and the brazing rod are selected for the brazing operation. Heat is applied first to the tube and then to the joint. During the heating process the flame is moved constantly, to provide even heating. When the flux becomes clear the flame is pulled back and the rod is touched to the joint.

Carbon dioxide or nitrogen should be used to flow through the tubing during the brazing operation to prevent contamination on the inside of the tube.

WRENCHES

A service person needs many types of wrenches. It is important that the wrench be of good quality, using good alloy steel properly heat treated. The wrench should fit to rest snugly and hold its shape during hard service use. By design, some types of wrenches fit better than others. Some of the common wrenches, in order of their fitting capability, are: (1) socket, (2) box, (3) open-end, and (4) adjustable.

Socket Wrenches

A set of socket wrenches is essential. The square shank varies in size: ¼, ⅜, and ½ in. square shanks are available. The smallest sockets usually have the smallest shank. The wrench points come with 6, 8, and 12 points, the higher the number of points, the larger the wrenches. Wrench sizes are available from 3⁄16 to 1 in., in 1⁄16 in. steps.

Box and Flare Nut Wrenches

Although the socket wrench is considered the safest type for maintaining a grip on the head of the bolt, the box wrench is next best. The handles of these wrenches are curved to permit better access to the bolt head.

The flare nut wrench is similar to the box type except it has an open end, to permit fitting over the tubing, and the handles are straight.

Open-End Wrenches

The open end wrench is probably the most commonly used wrench. It can access a bolt from the side and fit into close quarters. It is used for compressor head bolts and flare nuts. It comes in a wider range of sizes than does the socket wrench. If the opening spreads, it should be discarded.

Adjustable Wrenches

Adjustable wrenches come in sizes ranging from ½ to 2½ in. maximum opening. This type of wrench should be pulled to put the greatest stress on the fixed portion of the wrench.

Special-Purpose Wrenches

Special-purposes wrenches include: (1) service valve, (2) torque, (3) hex, and (4) pipe.

The service valve wrench is an extremely useful tool. It is important to use it when adjusting a valve stem on a compressor. The fixed end is used for "cracking" the valve and the ratchet end is used for opening or closing the valve quickly.

The torque wrench is used to tighten compressor head bolts, using a torque strength specified by the manufacturer.

Hex wrenches usually come in sets for tightening and loosening setscrews on pulley wheels.

The pipe wrench has a self-clamping feature for tightening and loosening pipe fittings, valves, and unions.

THERMOMETERS

A service person needs a good supply of thermometers. The most commonly used portable type in use today is the digital thermometer, illustrated in Figure 1-2. There are times, however, when this instrument may not be operative for one reason or another and then glass pocket thermometers like the one shown in Figure 2-11 are handy. The mercury-filled type is more accurate, although more expensive. A range of −40 to 120 °F is the most useful for refrigeration purposes. A disk thermometer (Figure 2-12) is also a favorite of many service people.

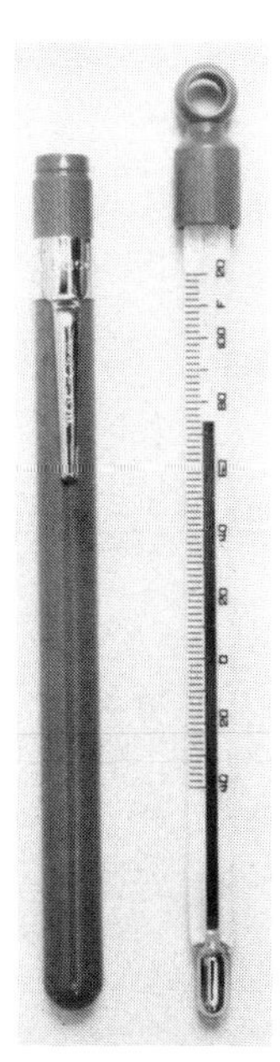

Figure 2-11 Pocket thermometers are red liquid or mercury fitted and are available in various temperature ranges. (By permission of Taylor Scientific Instruments, Sybron Corp.)

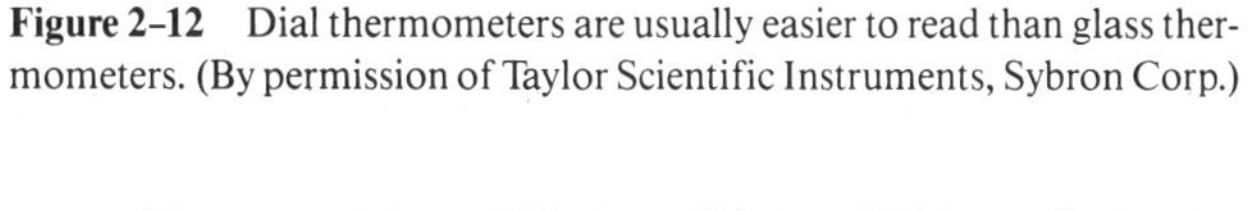

Figure 2-12 Dial thermometers are usually easier to read than glass thermometers. (By permission of Taylor Scientific Instruments, Sybron Corp.)

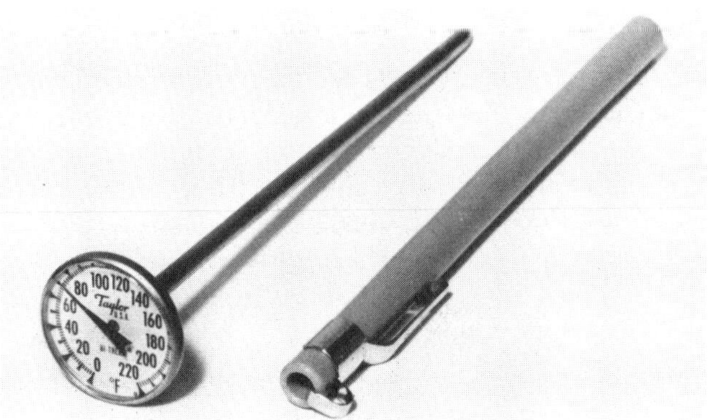

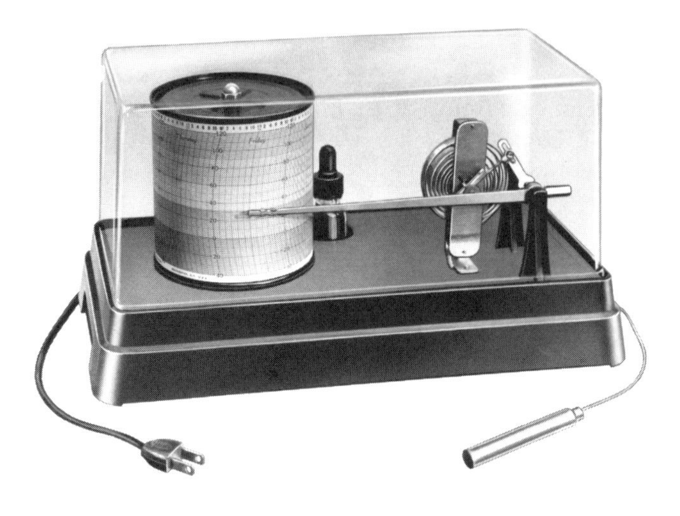

Figure 2-13 Recording thermometers are available with 24-hour or 7-day charts. (By permission of Taylor Scientific Instruments, Sybron Corp.)

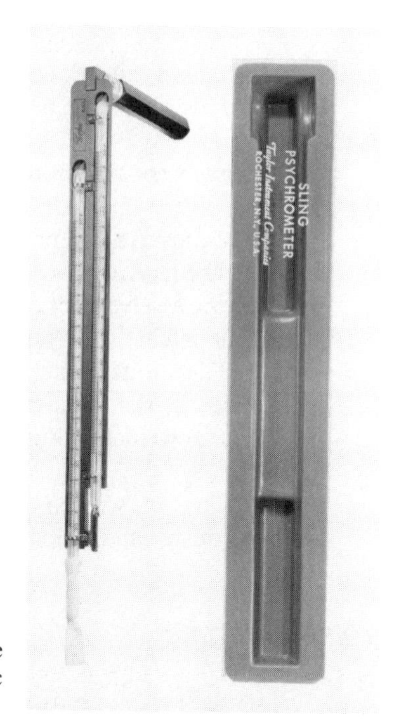

Figure 2-14 Sling psychrometers are used to measure relative humidity. (By permission of Taylor Scientific Instruments, Sybron Corp.)

Occasionally, it is advisable to use a portable recording thermometer (Figure 2-13) to monitor the temperature in a refrigerated room over a period of time. These will record the on/off cycle of the compressors as well as the temperatures. Recording thermometers are available within 24-hour or 7-day charts. Some of them have clocks that are spring wound, whereas others are electrically driven.

A sling psychrometer (Figure 2-14) is a valuable instrument for measuring humidity when the moisture content of the air is important. These units have two thermometers, for reading wet- and dry-bulb temperatures. With these readings the relative humidity can be determined by using a sliding scale which is part of the instrument.

GAUGES

Pressure gauges are one of the most useful instruments carried by service people. In most cases the gauges are carried on a "gauge manifold" with valve and hose connections. The use of these manifolds is discussed fully in Chapter 15.

Pressure and compound gauges are illustrated in Figure 2-15. The compound gauge reads from 30 in. of vacuum to 120 psi, and the pressure gauge reads from 0 to 250 psi. A restriction screw is usually supplied to reduce pulsation. The gauge has a ⅛-in. MPT connection. An equivalent saturated vapor temperature scale for R-12, R-22, and R-502 is printed on the face of the instrument. Most standard gauges are 2½ in. in diameter. Extra hoses, gauges, and gaskets should be carried by all service people.

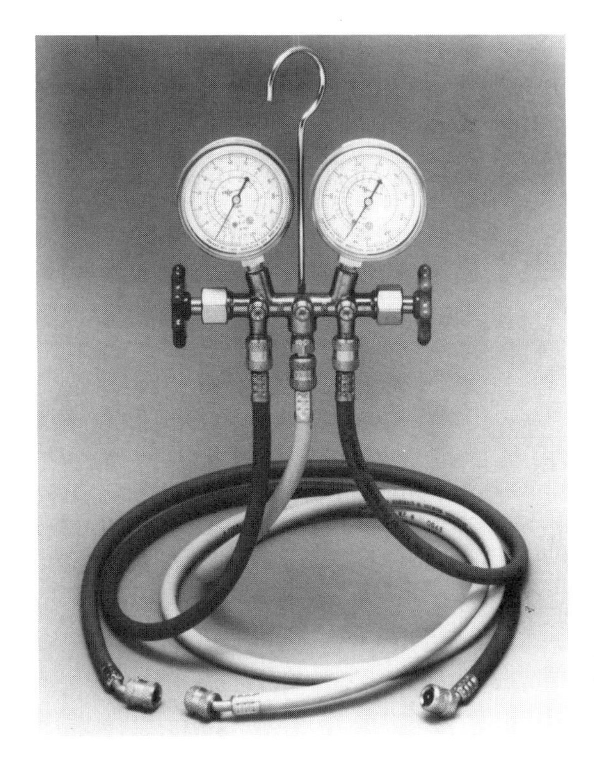

Figure 2-15 Pressure and compound gauges are assembled in a charging and testing manifold. (By permission of Robinaire Division of Sealed Power Corp.)

Hoses that attach to the gauge manifold are available in 12 to 120 in. lengths. Connections are usually for ¼ in. flare, but connections for ⅜ in. flare are also available. Hoses are usually color coded, along with the gauges: blue for the compound gauge system and red for the pressure gauge system. Hoses for most uses are built to withstand 2500 psi burst pressure and 500 psi working pressure.

REFRIGERATION LUBRICATING OILS

Under this heading, reference is being made to oils for the compressor, which, in most systems, partially flow through the refrigeration piping with the refrigerant. These special-purpose oils are refined for this specific purpose and are not suitable for other lubrication uses, such as for motors. The type of oil used for compressors is a naphenic-base oil and contains low quantities of wax. It has a lower viscosity than that of the paraffin-base oils recommended for electric motors. A good refrigeration oil has a high lubrication ability, high stability, low moisture, low wax, and nonfoaming characteristics.

Most wholesalers carry one or two brands of good compressor oil. The mechanic selects the viscosity needed. The viscosity refers to the thickness of the oil: the lower the viscosity number, the thinner the oil.

The selection of viscosity is as follows:

1. Use 300 viscosity down to $-20\,^{\circ}$F, 150 viscosity for lower temperatures for group 1 refrigerants.

2. Use 150 viscosity for refrigerants in groups 2 and 3. Refrigerants in each group are defined in the Safety Code included as Appendix III.

Refrigerants are classified in three types in relation to their reaction with oil:

1. Refrigerants that mix with all proportions of oil in the operating range, such as R-12, R-113, and methyl chloride.
2. Refrigerants that separate into two phases in the operating range, such as R-22, R-502, and R-114.
3. Refrigerants that are insoluble in oil, such as R-13, sulfur dioxide, ammonia, and carbon dioxide.

REFRIGERANTS

The refrigerant is the medium used to transfer heat from one location to another. Thus heat is absorbed by the refrigerant in the evaporator and exhausted from the system in the condenser. For changes that take place in the refrigerant during the cycle, refer to Chapter 1.

If a perfect refrigerant were produced, it would have some of the following qualities:

1. Nonpoisonous, nonexplosive, noncorrosive, and nonflammable
2. Leaks easy to detect and locate
3. Low operating pressures for vaporizing and for condensing
4. Chemically stable liquid and vapor (not easily decomposed)
5. Small volume of vapor displacement for the compressor
6. Small difference between vaporizing pressure and condensing pressure
7. Large amount of heat absorbed by vaporizing a pound of liquid
8. Low price per pound

Efforts are still being made to produce a perfect refrigerant. Actually, depending on the application, refrigerants are selected based on high priorities for certain qualities. For example, some years back, there was a major switch to the so-called "safe refrigerants," such as R-12, R-22, and R-502—the halocarbon refrigerants. Safety was a primary concern. Some of the desirable characteristics, such as "refrigeration effect per pound," were sacrificed for safety features.

In the refrigeration business, the use of ammonia has persisted due to some of its excellent performance characteristics. One of these is the heat-absorbing rate of boiling a pound of liquid. At 5 °F evaporating temperatures the heat absorbed by each of the common refrigerants, in Btu per pound, is as follows:

R-12	R-22	R-502	R-717
68.2	93.2	68.9	565.0

From this comparison it is evident why ammonia is still in use in large refrigeration systems, where codes permit it, even with its explosive characteristics.

Grouping and Classifying Refrigerants

Refrigerants may be identified by their chemical name, their chemical symbols, or by number. The numbers are most commonly used. The *Fundamental Handbook of ASHRAE* lists 86 different refrigerants. Only a few of these are commonly used. The ones most frequently referred to in this text are the following:

Number	Chemical name	Symbol
R-12	Dichlorodifluoromethane	CCl_2F_2
R-22	Monochlorodifluoromethane	$CHClF_2$
R-502	Azeotropic mixture of 48.8% of R-22 and 51.2% of R-115	$CHClF_2/CClF_2CF_3$
R-717	Ammonia	NH_3

The National Refrigeration Safety Code classifies refrigerants in these groups:

Group 1: *Safest.* Refrigerants included in this group are R-744 (carbon dioxide), R-12, R-22, R-502, R-40 (methyl chloride), etc.

Group 2: *Toxic and somewhat flammable.* Refrigerants included in this group are R-717, R-764 (sulfur dioxide), etc.

Group 3: *Flammable.* Refrigerants included in this group are R-600 (butane), R-240 (propane), etc.

The American Standard Safety Code (see Appendix III) indicates the conditions under which each of these groups can be used.

Cryogenic Fluids

The temperature range between −250°F and absolute zero (−454.69°F) is called the *cryogenic* range. Some refrigeration systems described in this text use refrigerants that boil at these low temperatures at atmospheric pressure; for example, R-728 (nitrogen). Containers for these refrigerants must be capable of maintaining their strength at these low temperatures. These containers are heavily insulated and not sealed. The temperature of the liquid is maintained at atmospheric pressure or nearly atmospheric pressure by boiling off some of the liquid. If the vessel were sealed, extreme pressures would develop as the refrigerant is warmed and would probably burst the container.

Refrigerant Containers

Most service people use refrigerant stored in disposable cylinders for R-12, R-22, and R-502. These containers are made in sizes to hold 15-, 30-, and 50-lb quantities. Larger, reusable cylinders are available for big installations. Ammonia refrigerant for large installations is usually delivered by tank truck. Containers for the commonly used refrigerants are color-coded as follows:

Refrigerant	Cylinder color
R-12	White
R-22	Green
R-502	Orchid

Cylinders of all types must have an Interstate Commerce Commission (ICC) stamp indicating their approval of the container.

It is extremely important that cylinders not be filled more than 80 percent of their total volume. Expansion of the liquid will occur due to temperature rise. There must be space in the cylinder for the increased volume of the liquid or the cylinder will rupture.

PORTABLE CHARGING CYLINDERS

One of the most useful devices for a service person supplying refrigerant to a critically charged system is the portable charging cylinder. This device can be used for a number of refrigerants and measures the amount added to the system to the nearest ¼ oz. These are made in a variety of sizes to hold 2.5, 5, and 10 lb of refrigerant. They can be used for vapor or liquid charging. Heating elements are available to vaporize the refrigerant for vapor charging.

SOLVENTS AND CLEANING

One extremely important operation performed by the service person is cleaning. All heat transfer equipment must be cleaned periodically if it is to function at a high level. This applies particularly to condensers and evaporator coils. Good solvents are available from most refrigeration wholesalers. Proper cleaning often requires a great deal of ingenuity on the part of the service person, using a number of methods. Some of the approved methods of cleaning are as follows:

Steam Cleaning

This method melts the grease and applies steam pressure to float off the grease and dirt at the same time. This method has been used for years to clean engines and applies equally well to cleaning condensing units. Safety precautions need to be used to prevent scalding.

Solvent Sprays

A hand sprayer can be used to clean condensers. The solvent is sprayed into the finned surfaces to remove dirt and grease. Brushing the surface will remove lint and other fibers that have collected. Carbon dioxide or nitrogen under pressure can follow the use of the solvent to assist in removing the loose particles.

Liquid Acids

These fluids are used to remove scales that form on the outside of a wetted surface, such as on a bare tube coil. Acid is also used to clean the interior surface of water-cooled condensers. Special precautions need to be used to remove all traces of acid before the condenser is put back into service. Tests should be made with litmus paper. Handling acid requires the use of rubber gloves to protect the skin. Plenty of fresh air needs to be available to prevent injury due to vapor.

REVIEW EXERCISE

Select the letter representing your choice of the correct answer (there may be more than one).

2-1. What is the most common material used for refrigeration piping?

_______ a. Aluminum

_______ b. Copper

_______ c. Iron

_______ d. Brass

2-2. What special treatment is given to refrigeration tubing?

_______ a. It is annealed and case hardened

_______ b. It is stretcher leveled

_______ c. It is cleaned, dehydrated, and sealed

_______ d. It is polished

2-3. Type K tubing refers to:

_______ a. The wall thickness

_______ b. The manufacturer's name

_______ c. The length

_______ d. The diameter

2-4. A flare fitting has what type of connection?

_______ a. Threaded

_______ b. Soldered

_______ c. Sleeve

_______ d. Butt joint

2-5. All copper tubing for refrigeration work is measured by:

_______ a. The I.D. dimension

_______ b. The O.D. dimension

2-6. The largest copper tubing that is normally cut with a tube cutter is:

_______ a. ⅞ in. I.D.

_______ b. ⅞ in. O.D.

_______ c. 1⅛ in. I.D.

_______ d. 1⅝ in. O.D.

2-7. Bending tools can be used up to what size tubing?

_______ a. 1⅛ in. O.D.

_______ b. 1⅝ in. O..D.

_______ c. 2⅛ in. O.D.

_______ d. 2⅝ in. O.D.

2-8. A swaging tool is used to:

_______ a. Increase the size of a tube

_______ b. Crimp a tube

_______ c. Bind a tube

_______ d. Straighten a tube

2-9. The highest melting point for soft solder is:

________ a. 300°F

________ b. 500°F

________ c. 700°F

________ d. 900°F

2-10. Silver brazing requires the use of what combination of gases?

________ a. Nitrogen and oxygen

________ b. R-12 and R-22

________ c. Helium and oxygen

________ d. Oxygen and acetylene

3 Compressors

OBJECTIVES

After studying this chapter, the reader will be able to:

- Describe the identifying features of various types of refrigeration compressors.
- Select the type of compressor or condensing unit that best fits the application.
- Determine compressor performance from manufacturers' data.

TYPES OF COMPRESSORS

The use of a compressor in a refrigeration system makes possible the reuse of refrigerant. There are two general types of compressors: positive-displacement units and centrifugal units. The positive-displacement compressors include:

1. Reciprocating
2. Rotary
3. Helical rotary (screw)

The fourth type of compressor is a continuously compressing unit, much like a high-speed fan:

4. Centrifugal

There are a number of other classifications of compressors that influence their selection. Compressors may also be of open type, semihermetic, or hermetic in design.

Compressors may be made specifically for certain evaporator temperature ranges—low, medium, and high. In addition, the most suitable refrigerant must be selected for the compressor and the system in which it operates.

RECIPROCATING COMPRESSORS

Reciprocating compressors are made in three types: (1) hermetically sealed, (2) semi-hermetic, and (3) open type.

Hermetically Sealed Compressors

The *hermetically sealed units* are used in relatively small packaged units, such as water coolers and reach-in refrigerators. The compressor and motor are mounted on a common shaft. Some units have internal spring mounting, others external. Piping and electrical connections protrude through the welded steel shell. One of the advantages of this type of unit is that it has a refrigerant gas-cooled motor. One of the disadvantages is that in case of motor burnout the refrigerant is contaminated.

Probably the greatest advantage of the hermetic compressor is that using a capillary-type metering device, it can be welded into a hermetically sealed system. The enclosed motor-compressor eliminates shaft seals, and belts and pulleys, which are a source of maintenance. These machines are ruggedly built and have many excellent features.

Commercial-sized hermetic compressors are made in sizes up to 30 hp, as shown in Figure 3–1 and in the cutaway view in Figure 3–2.

Figure 3–1 Exterior view of hermetically sealed compressor. Note suction and discharge valves welded to the shell. Note enclosed electrical access and oil level sight glass. (By permission of The Trane Company.)

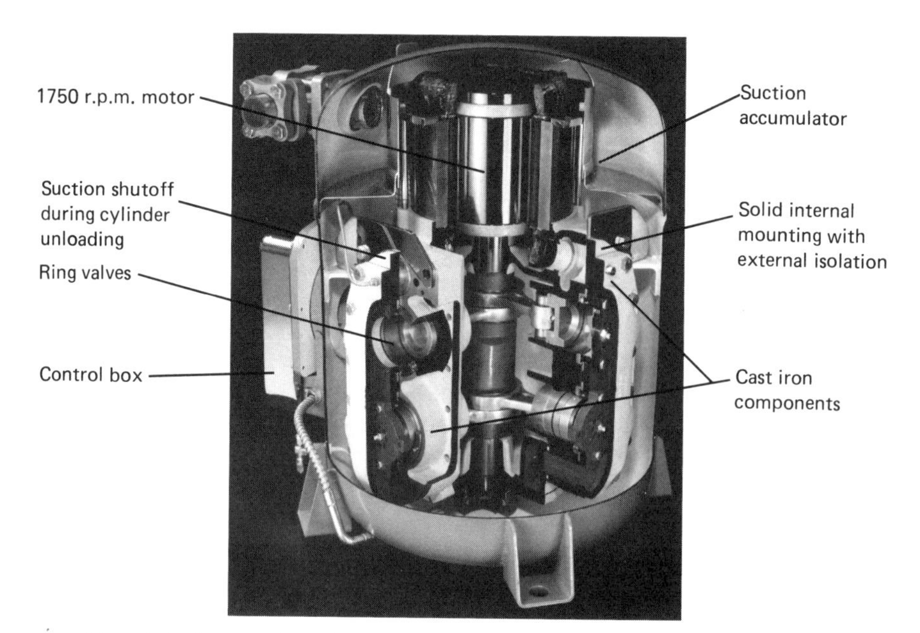

Figure 3-2 Cutaway view of hermetic compressor, four cylinder model, 1750 rpm motor. It is provided with unloaders. Electrical access includes high/low pressure control and crankcase heater. (By permission of The Trane Company.)

Semihermetic Compressors

This unit is popular in the medium-capacity range. Figure 3-3 shows an exterior view and Figure 3-4 shows a cutaway view of a semihermetic compressor. It differs from the full hermetic model in that the valve plate is removable, which is an advantage from a service standpoint. In fact, all the internal parts of the compressor can be removed by loosening appropriate bolts. Two important features that it has retained from the hermetic design are the gas-cooled motor and the elimination of the shaft seal. Machines of this type range in size from ½ to 50 or 60 hp.

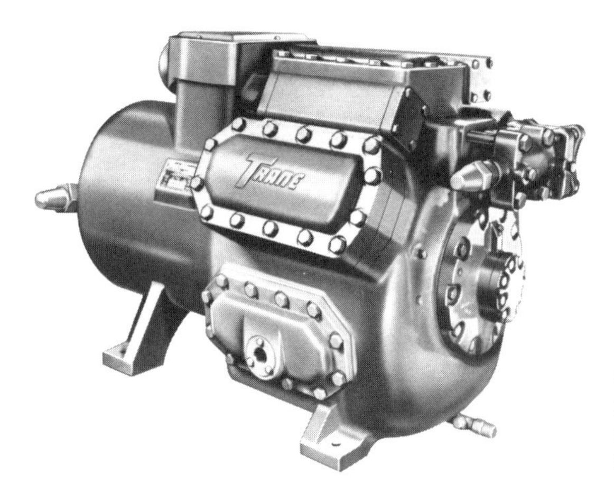

Figure 3-3 Exterior view of semi-hermetic compressor. Note accessability for servicing compressor. (By permission of The Trane Company.)

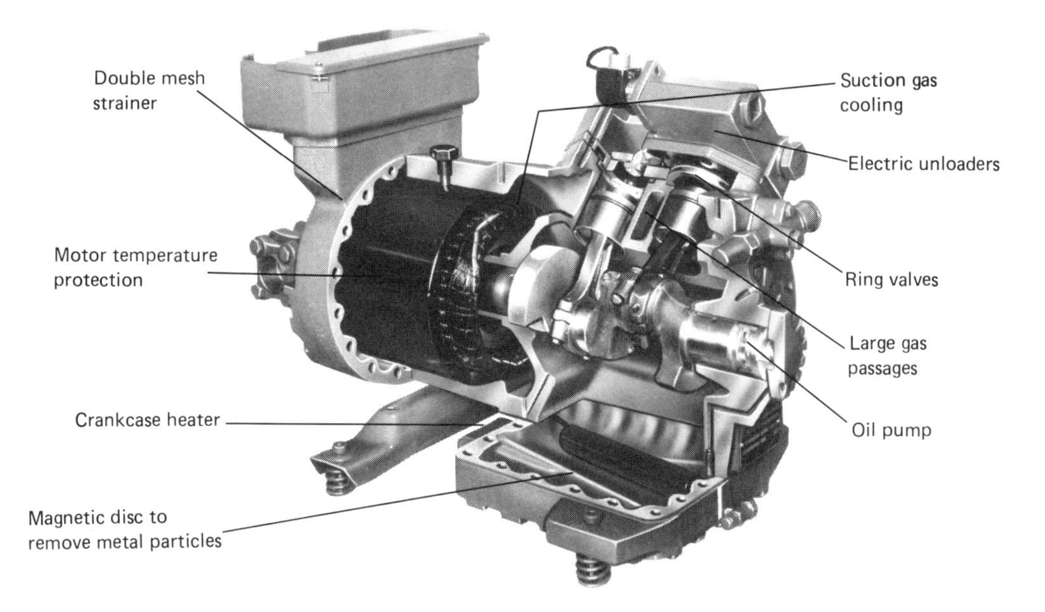

Figure 3–4 Cutaway view of semi-hermetic compressor. It uses double mesh suction strainer. The motor has three sensors in the motor windings to detect excessive temperatures. Crankcase heat, electric unloaders, and oil pump are important features. (By permission of The Trane Company.)

Open-Type Compressors

A complete unit with a direct drive is shown in Figure 3–5 and a cutaway view is shown in Figure 3–6. The open-type compressor is more flexible in application in view of the opportunity it presents to change the speed to match the load, by altering the pulley-drive arrangement. Many industry people prefer this unit because a burned-out motor does not contaminate the refrigerant. Shaft seals are now made with greater reliability, thus minimizing this potential problem.

Figure 3–5 Exterior view of open type compressor. Motor and controls are factory installed on a suitable base, depending on application. (By permission of The Trane Company.)

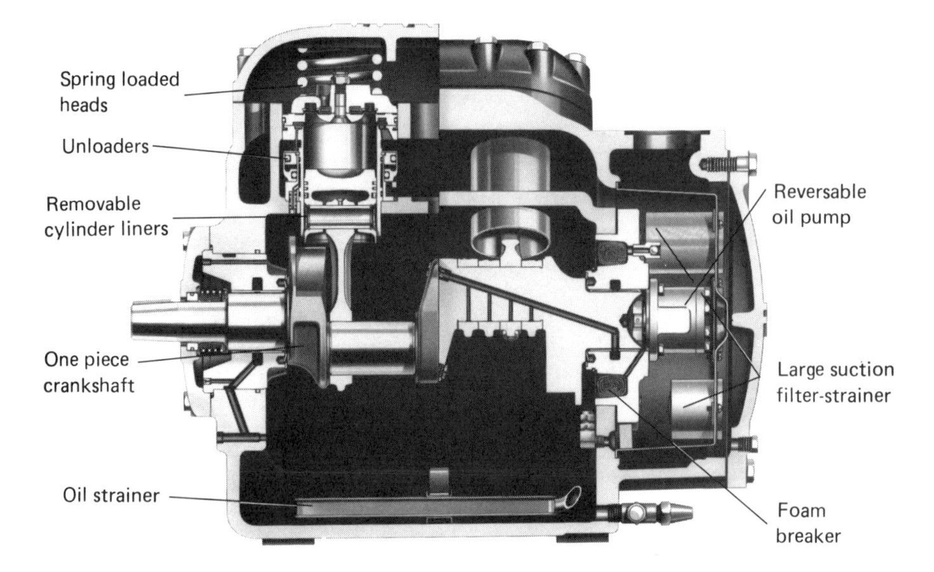

Figure 3-6 Cutaway view of open-type compressor. Note the large suction gas filter, foam breaker in the crankcase, cylinder unloader, removable cylinder liners, and crankcase oil strainer. (By permission of The Trane Company.)

CONDENSING UNITS

Condensing units ½ to 20 hp in size, semihermetic, are supplied in four configurations:

1. Air cooled (Figure 3-7)
2. Water cooled (Figure 3-8)
3. Air/water cooled (Figure 3-9)
4. Compressor/receiver type for use with a remote condenser (Figure 3-10)

Air-cooled units can be mounted on the roof in a well-ventilated penthouse or on double tier racks in the machine room. Water costs are eliminated. For outdoor use rain-proof enclosures are provided.

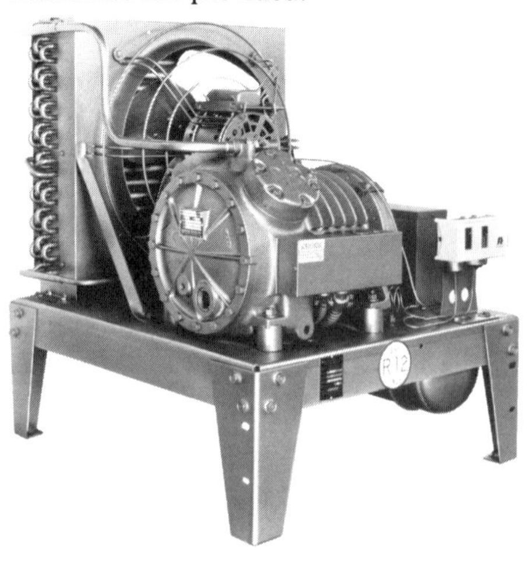

Figure 3-7 Air-cooled condensing units are designed to use ambient air to condense the refrigerant. These units are often placed on racks, two high, to conserve floor space. When used in an equipment room the ambient air temperature must be controlled within efficient limits. (By permission of Dunham-Bush, Inc.)

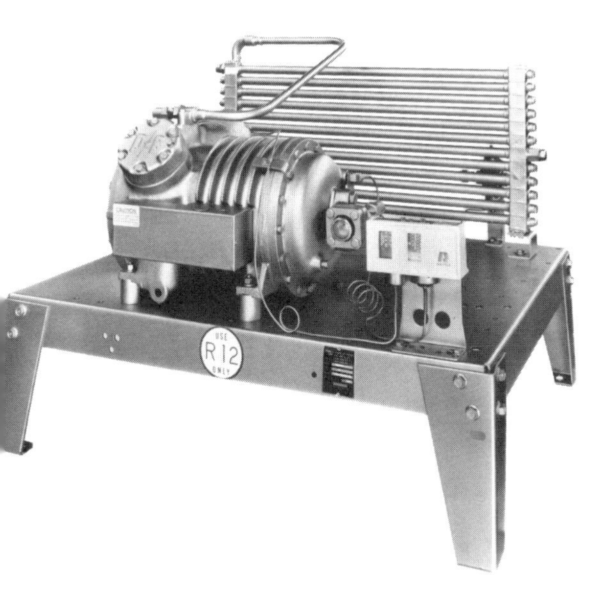

Figure 3–8 Water-cooled condensing units in this size range are designed primarily for use with city water but can be used for cooling-tower application. A minimum amount of ventilation air is required. (By permission of Dunham-Bush, Inc.)

Figure 3–9 These units use both air and water for condensing refrigerants. For normal ambient temperatures the air-cooled condenser only is used. When the load becomes heavy or when the ambient temperature is excessive the water valve will open, permitting the use of both condensers. (By permission of Dunham-Bush, Inc.)

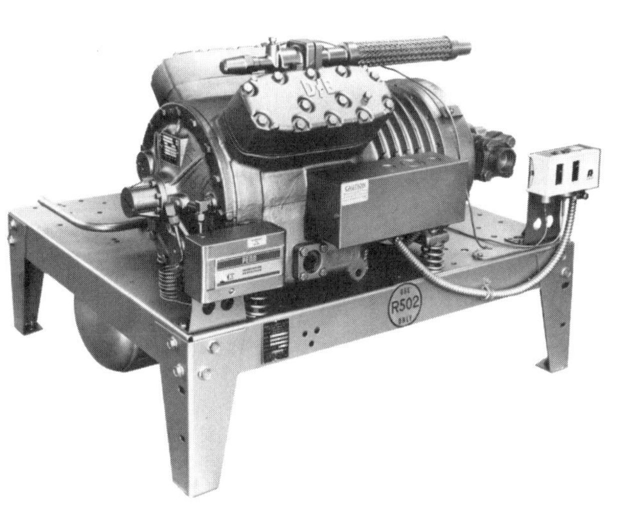

Figure 3–10 Compressor–receiver units are designed to be used with remote-mounted condensers and/or when a multicircuited air-cooled condenser is used for a series of these units. Units are equipped with suction, discharge, and receiver outlet valves. (By permission of Dunham-Bush, Inc.)

Water-cooled units are applied when a good supply of water is available or where proper ventilation for air-cooled units is not available. Units can use city water or cooling-tower water. All units are equipped with water regulating valves.

Air/water units normally use air-cooled condensers. However, if excessive head pressures are reached, the water-cooled condenser will automatically operate to reduce the condensing temperature. The water-cooled condenser is cleanable. A water regulating valve is standard equipment.

The compressor–receiver unit is used with a remote air-cooled condenser. These

HP	60 HZ. AH/AWH	50 HZ. AH/AWH	Ambient Air Temp. °F	LOW TEMP. -40°F	-30°F	-20°F	-10°F	COMMERCIAL TEMP. 0°F	10°F	20°F	25°F	HIGH TEMP. 30°F	40°F	50°F
				SATURATED SUCTION TEMPERATURE °F										
				SUCTION PRESSURE - PSIG OR *VACUUM INCHES OF MERCURY										
				10.9*	5.5*	0.6	4.5	9.2	14.64	21.04	24.61	28.45	36.97	46.70
				BTUH	BTUH	BTUH	BTUH	BTUH	BTUH	BTUH	BTUH	BTUH	BTUH	BTUH
½	△ 5HCL		90	-	1,370	1,900	2,600	3,350	4,150	5,000	5,400	6,000	6,900	7,800
			100	-	1,250	1,750	2,350	3,050	3,800	4,650	5,100	5,800	6,700	7,600
¾	△ 8HCL		90	-	1,550	2,200	3,000	3,800	4,700	5,700	6,250	6,700	7,900	9,500
			100	-	1,400	2,000	2,700	3,500	4,400	5,400	5,950	6,500	7,700	9,000
	△ 9HCL		90	-	1,950	2,750	3,700	4,700	5,800	7,050	7,800	8,400	9,700	11,100
			100	-	1,750	2,500	3,400	4,350	5,400	6,500	7,200	7,700	8,900	10,200
1	△ 10HCL		90	-	2,250	3,200	4,350	5,550	6,900	8,300	9,150	9,800	11,400	13,000
			100	-	2,050	2,850	3,900	5,100	6,450	7,850	8,700	9,400	11,000	12,600
	△ 11HCL**		90	2,070	2,900	4,150	5,600	7,100	8,800	10,600	11,700	12,500	14,600	16,600
			100	1,850	2,650	3,800	5,100	6,600	8,400	10,000	11,100	12,000	14,100	16,100
1½	15H	15H5	90	-	-	-	-	-	11,800	12,200	13,400	14,600	16,800	18,900
			100	-	-	-	-	-	9,700	11,600	12,600	13,800	15,900	18,000
	15C	15C5	90	-	-	-	7,100	9,300	11,500	14,500	16,000	-	-	-
			100	-	-	-	6,500	8,700	10,900	13,600	15,300	-	-	-
	15L	15L5	90	2,800	4,200	5,800	7,600	9,600	-	-	-	-	-	-
			100	2,500	3,700	5,400	7,000	8,900	-	-	-	-	-	-
2	20H	20H5	90	-	-	-	-	-	14,500	16,500	17,500	19,000	22,000	26,000
			100	-	-	-	-	-	12,000	14,500	15,500	17,000	20,000	23,000
	20C	20C5	90	-	-	-	8,600	11,300	15,400	17,500	19,400	-	-	-
			100	-	-	-	8,100	10,600	14,000	16,200	18,000	-	-	-
	20L	22L5	90	3,600	5,500	7,500	9,800	12,200	-	-	-	-	-	-
			100	3,400	5,000	7,000	9,100	11,400	-	-	-	-	-	-
3	30H	32H5	90	-	-	-	-	-	18,000	23,000	25,500	28,000	32,500	38,000
			100	-	-	-	-	-	17,000	21,500	24,000	26,000	30,500	35,400
	30C	30C5	90	-	-	-	-	18,600	23,000	28,700	32,000	-	-	-
			100	-	-	-	-	17,300	21,500	27,000	30,000	-	-	-
	32C		90	-	-	-	12,500	16,200	19,900	25,400	28,200	-	-	-
			100	-	-	-	11,500	15,000	18,500	24,000	26,400	-	-	-
	30L	30L5	90	6,000	8,500	12,000	16,000	20,500	-	-	-	-	-	-
			100	5,400	8,000	11,000	15,000	19,000	-	-	-	-	-	-
		41PL5	90	6,560	9,700	13,900	18,900	24,100	-	-		-	-	-
			100	6,070	9,000	12,400	17,500	22,500	-	-		-	-	-
5	50H	50H5	90	-	-	-	-	-	30,000	37,50[illegible]	[illegible]1,000	45,000	52,500	61,500
			100	-	-	-	-	-	29,300	36,00[illegible]	[illegible]9,000	42,700	50,000	58,500
	50C	51PC5	90	-	-	-	21,300	27,400	33,500	41,80[illegible]	[illegible]6,300	-	-	-
			100	-	-	-	18,600	24,600	30,600	38,300	[illegible]3,000	-	-	-
	51PL		90	8,000	11,800	17,000	23,000	29,400	-	-	-	-	-	-
			100	7,400	11,000	15,100	21,300	27,400	-	-	-	-	-	-
		51PH5	90	-	-	-	-	-	37,200	46,500	50,800	55,800	65,100	76,300
			100	-	-	-	-	-	36,300	44,600	48,400	53,000	62,000	72,500
		62PC5	90	-	-	-	26,200	33,700	41,200	51,400	57,000	-	-	-
			100	-	-	-	22,900	30,200	37,600	47,100	52,900	-	-	-
		61PL5	90	9,800	14,600	20,900	28,300	36,200	-	-	-	-	-	-
			100	9,100	13,500	18,600	26,200	33,700	-	-	-	-	-	-
7½	76PH/C		90	-	-	-	26,500	34,000	41,500	54,000	60,000	67,000	84,500	105,000
			100	-	-	-	23,900	32,400	38,000	50,500	56,000	62,500	78,500	96,000
	77PC	77PC5	90	-	-	-	32,000	41,000	54,000	67,000	75,000	-	-	-
			100	-	-	-	30,000	39,000	50,000	63,500	70,000	-	-	-
	76PL	75PL5	90	10,700	16,600	25,400	35,400	47,100	-	-	-	-	-	-
			100	9,700	15,500	23,300	32,600	43,500	-	-	-	-	-	-
		77PH5	90	-	-	-	-	-	50,400	66,400	73,800	82,400	104,000	129,000
			100	-	-	-	-	-	46,700	62,100	68,900	76,900	96,600	118,000
10	101PH	101PH5	90	-	-	-	-	-	68,900	90,600	100,700	112,300	141,700	176,100
			100	-	-	-	-	-	64,600	84,700	93,900	104,800	131,700	161,000
	101PC	101PC5	90	-	-	-	40,000	51,300	67,500	83,800	93,800	-	-	-
			100	-	-	-	37,500	48,800	62,500	79,400	87,500	-	-	-
	104PL		90	12,500	20,000	30,000	46,000	61,000	-	-	-	-	-	-
			100	11,750	18,800	28,200	43,300	57,400	-	-	-	-	-	-
15	D154PH		90	-	-	-	-	-	74,800	101,200	106,400	119,000	147,600	182,000
			100	-	-	-	-	-	70,600	92,700	99,300	111,000	137,000	166,200
	154PC†		90	-	-	-	42,700	56,800	71,200	95,900	-	-	-	-
			100	-	-	-	39,400	52,500	67,200	89,600	-	-	-	-

NOTES: † This unit will not operate at suction temperatures greater than 20°F.

△ AH units only.

* Inches of mercury below one atmosphere.

** Not available for +30°F through +50°F SST 208V/1φ applications.

Figure 3–11 Capacity data for typical *air*-cooled condensing units, ½ to 15 hp, using R-12. (By permission of Dunham-Bush, Inc.)

condensers often have a number of circuits, each usable for a separate compressor–receiver unit. Units have a minimum height, which makes it possible to rack them two or three units high. Units are supplied with compressor and receiver service valves.

Capacity Data

Typical capacity data for air- and water-cooled condensing units are shown in Figures 3–11 through 3–16 using R-12, R-22, and R-502 refrigerants. The range of size, using semihermetic compressors is ½ to 20 hp.

HP	60 HZ.	50 HZ.	Saturated Discharge Temp. °F	LOW TEMP.				COMMERCIAL TEMP.				HIGH TEMP.		
				SATURATED SUCTION TEMPERATURE °F										
				-40°F	-30°F	-20°F	-10°F	0°F	10°F	20°F	25°F	30°F	40°F	50°F
				SUCTION PRESSURE - PSIG OR * VACUUM INCHES OF MERCURY										
	EH/WH	EH/WH		10.9*	5.5*	0.6	4.5	9.2	14.64	21.04	24.61	28.45	36.97	46.70
HP				BTUH	BTUH	BTUH	BTUH	BTUH	BTUH	BTUH	BTUH	BTUH	BTUH	BTUH
½	5HCL		105	-	1,225	1,800	2,530	3,410	4,450	5,640	6,310	8,000	9,900	11,800
			115	-	1,120	1,660	2,355	3,200	4,200	5,360	6,000	7,500	9,200	11,000
			125	-	1,015	1,520	2,180	2,990	3,960	5,080	5,680	6,900	8,500	10,200
¾	8HCL		105	-	1,400	2,055	2,890	3,900	5,080	6,440	7,210	8,690	11,400	13,900
			115	-	1,280	1,900	2,690	3,660	4,800	6,120	6,850	7,950	10,250	12,700
			125	-	1,160	1,740	2,490	3,420	4,520	5,800	6,490	7,540	9,500	11,800
	9HCL		105	-	1,750	2,570	3,610	4,870	6,350	8,050	9,010	10,600	13,900	16,000
			115	-	1,600	2,370	3,360	4,570	6,000	7,650	8,560	9,700	12,500	15,000
			125	-	1,450	2,170	3,110	4,270	5,650	7,250	8,110	9,200	11,600	14,000
1	10HCL		105	-	2,060	3,030	4,260	5,740	7,480	9,490	10,620	11,500	14,200	17,000
			115	-	1,890	2,800	3,960	5,390	7,070	9,010	10,100	10,500	13,000	15,500
			125	-	1,710	2,560	3,670	5,030	6,660	8,540	9,560	9,800	11,800	14,500
	11HCL**		105	1,725	2,625	3,855	5,420	7,310	8,400	11,000	12,400	14,000	17,800	19,600
			115	1,525	2,400	3,555	5,040	6,860	7,600	10,000	11,500	13,000	16,600	18,400
			125	1,425	2,175	3,255	4,670	6,410	6,800	9,200	10,500	12,000	15,200	17,000
1½	15H	15H5	105	-	-	-	-	-	9,800	13,200	14,900	16,600	20,600	24,800
			115	-	-	-	-	-	8,900	12,200	13,700	15,500	19,400	23,000
			125	-	-	-	-	-	8,200	11,300	12,700	14,400	17,800	21,500
	15C	15C5	105	-	-	-	6,400	9,600	12,800	16,800	19,000	-	-	-
			115	-	-	-	6,200	9,000	11,800	15,600	17,600	-	-	-
			125	-	-	-	5,700	8,300	10,900	14,400	16,200	-	-	-
	15L	15L5	105	2,700	4,000	5,900	8,300	10,800	-	-	-	-	-	-
			115	2,400	3,600	5,400	7,500	9,900	-	-	-	-	-	-
			125	2,100	3,200	4,900	6,800	9,000	-	-	-	-	-	-
2	20H	20H5	105	-	-	-	-	-	12,800	17,400	19,500	22,000	27,500	33,000
			115	-	-	-	-	-	12,500	16,500	18,500	20,500	25,500	31,000
			125	-	-	-	-	-	11,200	15,300	17,000	19,400	24,000	29,000
	20C	20C5	105	-	-	-	8,100	11,800	15,500	20,500	23,200	-	-	-
			115	-	-	-	7,500	11,000	14,500	19,200	21,700	-	-	-
			125	-	-	-	7,100	10,200	13,300	18,000	20,500	-	-	-
	20L	22L5	105	3,200	4,800	7,100	10,000	13,400	-	-	-	-	-	-
			115	2,900	4,000	6,600	9,300	12,400	-	-	-	-	-	-
			125	2,700	4,000	6,000	8,400	11,400	-	-	-	-	-	-
3	30H	32H5	105	-	-	-	-	-	18,000	25,000	28,500	32,000	39,700	48,000
			115	-	-	-	-	-	17,500	23,000	26,000	29,500	37,000	44,500
			125	-	-	-	-	-	15,000	21,000	23,700	27,000	34,000	41,000
	30C	30C5	105	-	-	-	-	19,500	25,500	33,500	38,000	-	-	-
			115	-	-	-	-	17,800	23,300	31,000	35,000	-	-	-
			125	-	-	-	-	16,000	21,000	28,200	32,000	-	-	-
	32C		105	-	-	-	10,800	16,100	21,400	27,400	31,700	-	-	-
			115	-	-	-	9,600	14,900	20,200	25,300	26,400	-	-	-
			125	-	-	-	8,300	13,600	18,900	23,100	24,100	-	-	-
	30L	30L5	105	5,200	7,500	11,000	15,700	21,000	-	-	-	-	-	-
			115	4,800	6,500	9,800	14,200	19,200	-	-	-	-	-	-
			125	4,000	5,800	8,800	13,000	17,500	-	-	-	-	-	-
		41PL5	105	7,000	9,800	14,400	19,800	27,000	-	-	-	-	-	-
			115	6,400	9,000	13,100	18,500	24,600	-	-	-	-	-	-
			125	5,700	8,000	11,600	16,600	22,100	-	-	-	-	-	-
5	50H	50H5	105	-	-	-	-	-	32,000	41,000	46,000	52,000	65,000	79,000
			115	-	-	-	-	-	29,000	37,500	43,000	48,500	61,000	74,000
			125	-	-	-	-	-	26,000	34,500	39,300	45,000	57,000	-
	50C	51PC5	105	-	-	-	19,900	28,500	37,100	49,000	55,500	-	-	-
			115	-	-	-	18,500	26,500	34,500	45,000	51,000	-	-	-
			125	-	-	-	16,700	24,200	31,700	41,200	46,200	-	-	-
	51PL		105	8,500	12,000	17,500	24,200	33,000	-	-	-	-	-	-
			115	7,800	11,000	16,000	22,500	30,000	-	-	-	-	-	-
			125	7,000	9,800	14,200	20,300	27,000	-	-	-	-	-	-
		51PH5	105	-	-	-	-	-	39,700	50,800	57,000	64,500	80,600	98,000
			115	-	-	-	-	-	36,000	46,500	53,300	60,100	75,600	91,800
			125	-	-	-	-	-	32,000	42,800	48,700	55,800	70,700	-
		62PC5	105	-	-	-	24,500	35,000	45,600	60,300	68,300	-	-	-
			115	-	-	-	22,800	32,600	42,400	55,400	62,700	-	-	-
			125	-	-	-	20,500	29,800	39,000	50,700	56,800	-	-	-
		61PL5	105	10,500	14,800	21,500	29,800	40,600	-	-	-	-	-	-
			115	9,600	13,500	19,700	27,700	36,900	-	-	-	-	-	-
			125	8,600	12,000	17,500	25,000	33,200	-	-	-	-	-	-
7½	76PHC		105	-	-	-	23,600	35,000	46,500	61,000	68,900	78,700	98,000	120,000
			115	-	-	-	21,600	32,400	43,200	56,700	65,200	73,500	92,500	113,000
			125	-	-	-	19,100	29,500	39,400	52,700	60,600	68,600	87,000	105,000
	77PC	77PC5	105	-	-	-	28,400	42,200	54,500	75,600	85,500	-	-	-
			115	-	-	-	26,000	39,000	50,600	70,500	80,000	-	-	-
			125	-	-	-	23,000	35,500	46,000	65,200	74,500	-	-	-
	76PL	75PL5	105	11,900	17,900	26,700	38,300	51,300	-	-	-	-	-	-
			115	10,400	16,400	24,700	35,200	47,400	-	-	-	-	-	-
			125	9,100	14,500	22,600	32,100	43,500	-	-	-	-	-	-
		77PH5	105	-	-	-	-	-	56,000	73,500	83,000	96,800	120,500	147,600
			115	-	-	-	-	-	52,000	68,300	78,500	90,400	113,800	139,000
			125	-	-	-	-	-	47,500	63,500	73,000	84,400	107,000	129,000
10	101PHC	101PHC5	105	-	-	-	35,500	52,800	68,100	94,500	106,900	117,000	145,000	176,000
			115	-	-	-	32,500	48,800	63,300	88,100	100,000	110,000	137,000	168,000
			125	-	-	-	28,800	44,400	57,500	81,500	93,100	102,000	128,000	156,000
	104PL		105	14,800	21,500	31,800	40,000	64,000	-	-	-	-	-	-
			115	13,600	19,800	29,300	42,400	58,900	-	-	-	-	-	-
			125	12,500	18,200	26,900	39,000	54,200	-	-	-	-	-	-
15	154PHC		105	-	-	-	46,200	65,500	87,500	111,600	125,000	140,600	173,400	211,000
			115	-	-	-	42,500	61,200	81,200	103,100	116,200	131,200	163,600	198,900
			125	-	-	-	38,700	55,700	75,400	95,000	108,100	122,500	153,100	187,900

* Inches of mercury below one atmosphere.

** Not available for +30°F through +50°F SST 208V/1Φ application

Figure 3–12 Capacity data for typical *water*-cooled condensing units, ½ to 15 hp, using R-12. (By permission of Dunham-Bush, Inc.)

				COMMERCIAL TEMP.					HIGH TEMP.	
				SATURATED SUCTION TEMPERATURE °F						
				0°F	10°F	20°F	25°F	30°F	40°F	50°F
			Ambient	SUCTION PRESSURE PSIG						
	60 HZ.	50 HZ.	Air	24.09	32.73	43.28	47.04	55.25	69.02	84.70
HP	AH/AWH	AH/AWH	Temp °F	BTUH	BTUH	BTUH	BTUH	BTUH	BTUH	BTUH
1½	15CF	15CF5	90	10,400	13,100	16,200	17,800	–	–	
			100	9,000	12,000	15,000	16,500	–	–	–
		15HF5	90	–	7,850	11,200	13,000	14,550	18,250	22,400
			100	–	7,300	10,200	11,600	13,100	16,100	19,400
2	20HF	20HF5	90	–	10,500	15,000	17,400	19,500	24,500	30,000
			100	–	9,900	13,700	15,500	17,500	21,500	26,000
	20CF	20LF5	90	13,400	17,800	23,000	26,000	–	–	
			100	12,000	16,400	21,500	24,200	–	–	–
3	30HF	30HF5	90	–	17,800	22,000	23,900	26,100	31,300	38,600
			100	–	16,000	19,800	21,600	23,500	28,200	34,700
	30CF	30CF5	90	16,700	21,400	27,200	30,500	–	–	–
			100	15,000	19,600	25,100	28,200		–	
5	52HF	52HF5	90	–	32,500	40,000	43,500	47,500	57,000	–
			100	–	29,900	36,200	39,000	42,500	–	
	50CF	50CF5	90	29,000	35,000	43,000	48,000	–	–	
			100	26,000	32,000	40,000	44,500	–	–	
7½	75HF	75HF5	90	–	42,900	52,700	57,300	62,500	75,000	–
			100	–	38,800	47,400	51,400	56,000	–	–
	75CF	75CF5	90	42,300	51,000	62,700	70,000	–	–	–
			100	37,900	46,600	58,300	66,000	–	–	–
10	100PHF	101PHF5	90	–	52,600	64,700	72,200	75,000	90,000	111,000
			100	–	48,100	60,200	68,100	67,200	80,500	98,800
		101PCF5	90	43,700	51,400	63,200	68,800	–	–	–
			100	39,100	46,600	56,900	61,700	–	–	–
	111PHF	111PHF5	90	–	63,700	77,400	85,300	93,000	111,600	137,600
			100	–	57,800	70,500	76,500	83,300	99,800	122,500
	111PCF	111PCF5	90	52,900	63,800	78,400	87,500	–	–	–
			100	47,400	58,200	72,900	80,500	–	–	–
15	D151PHF	D151PHF5	90	–	75,800	93,200	101,400	110,600	132,700	163,700
			100	–	68,800	83,900	91,000	99,100	118,700	145,700
	D151PCF	D151PCF5	90	62,400	75,300	92,500	103,200	–	–	–
			100	55,900	68,700	86,000	95,000	–	–	–
17½	D201PHF	D201PHF5	90	–	104,600	115,900	128,700	137,500	165,000	203,600
			100	–	98,400	107,200	118,400	123,200	147,600	181,200
	D201PCF	D201PCF5	90	77,800	93,900	115,300	126,200	–	–	–
			100	69,700	85,600	104,500	113,200	–	–	
20	D204PHF		90	–	115,000	138,000	152,000	164,000	198,000	240,000
			100	–	108,000	129,000	143,000	154,000	186,000	225,000
	D204PCF		90	78,000	105,000	129,000	141,000	–	–	–
			100	71,000	98,800	121,000	132,500	–	–	–

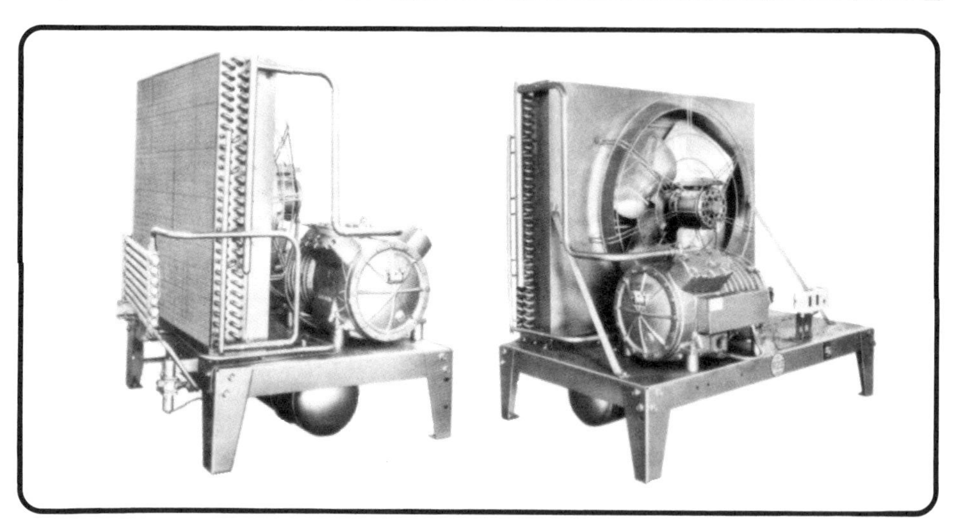

Figure 3-13 Capacity data for typical *air*-cooled (AH) and air/water (AWH) condensing units 1½ to 20 hp, using R-22. (By permission of Dunham-Bush, Inc.)

HP	60 Hz. WH/EH	50 Hz. WH/EH	Saturated Discharge Temp. °F	COMMERCIAL TEMP.				HIGH TEMP.		
				SATURATED SUCTION TEMPERATURE °F						
				0°F	10°F	20°F	25°F	30°F	40°F	50°F
				SUCTION PRESSURE PSIG						
				24.09	32.73	43.28	47.04	55.25	69.02	84.70
				BTUH	BTUH	BTUH	BTUH	BTUH	BTUH	BTUH
1½	15CF	15CF5	105	10,200	12,700	16,300	19,000	–	–	–
			115	9,300	11,500	14,900	18,100	–	–	–
			125	8,400	10,500	13,600	16,000	–	–	–
		15HF5	105	–	8,600	12,200	14,200	15,800	20,900	26,500
			115	–	7,900	11,100	12,700	14,300	18,700	23,500
			125	–	7,700	10,200	11,200	12,700	16,700	20,500
2	20HF	20HF5	105	–	11,400	16,300	19,000	21,200	28,000	35,500
			115	–	10,600	14,900	17,000	19,200	25,000	31,500
			125	–	10,200	13,600	15,000	17,000	22,300	27,500
	20CF	20CF5	105	13,600	18,200	23,800	27,900	–	–	–
			115	12,400	16,500	22,000	26,000	–		–
			125	11,100	15,000	19,900	23,300	–	–	–
3	30HF	30HF5	105	–	16,600	23,800	27,400	31,000	39,000	48,300
			115	–	15,600	21,800	25,000	28,000	35,500	43,500
			125	–	14,200	19,600	22,500	25,000	31,500	38,500
	30CF	30CF5	105	16,500	22,000	30,000	34,500	–	–	
			115	15,000	20,000	27,000	31,500	–	–	–
			125	13,500	18,000	24,500	28,000	–	–	
5	52HF	52HF5	105	–	32,500	43,500	48,500	54,500	69,000	87,000
			115	–	30,000	40,500	45,000	51,000	64,000	81,000
			125	–	28,000	37,500	42,000	47,000	60,000	74,000
	50CF	50CF5	105	28,500	39,000	50,000	56,000	–	–	–
			115	25,500	35,500	46,500	52,000	–	–	–
			125	23,000	32,500	42,500	48,300	–	–	–
7½	75HF	75HF5	105	–	47,000	62,000	69,500	77,000	94,000	111,000
			115	–	41,000	56,000	63,500	71,000	87,000	–
			125	–	36,000	50,000	57,000	64,000	–	–
	75CF		105	42,000	57,500	74,000	83,000	–	–	–
			115	38,000	52,500	68,000	77,500		–	–
			125	34,000	48,500	64,000	72,500	–	–	–
10	100PHF		105	–	54,000	73,500	83,000	93,000	116,000	138,000
			115	–	50,500	69,000	77,500	87,500	107,000	128,000
			125	–	48,500	64,500	72,000	80,500	99,000	118,000
		101PHF5	105	–	59,300	76,400	86,000	96,300	120,000	143,000
			115	–	54,200	71,500	80,300	90,600	110,800	132,600
			125	–	52,000	66,800	77,600	83,400	102,600	122,200
		101PCF5	105	43,300	55,900	76,100	85,700	–	–	–
			115	39,200	52,300	70,200	80,000	–	–	–
			125	35,100	50,100	66,000	74,800	–	–	–
	111PHF/CF	111PH5/CF	105	–	67,000	91,100	102,900	115,300	143,800	171,100
			115	–	62,600	85,600	96,100	108,500	132,200	158,700
			125	–	60,100	80,000	89,300	99,800	122,800	146,300
15	151PHF/CF		105	–	82,600	111,000	124,800	139,600	172,000	205,400
			115	–	77,700	104,200	116,000	130,700	161,200	193,700
			125	–	77,000	96,300	109,100	121,900	151,400	182,800
		151PHF5	105	–	87,000	117,000	131,500	147,100	181,300	216,500
			115	–	81,800	109,800	122,200	137,800	169,900	204,000
			125	–	74,600	101,500	115,000	128,500	159,500	192,700
17½	201PHF/CF	201PHF5/CF5	105	–	104,600	139,900	158,800	175,200	213,500	257,000
			115	–	98,500	131,600	149,300	164,800	202,000	242,500
			125	–	93,200	123,300	137,800	153,300	188,600	227,900
20	204PHF*/CF		105	90,000	120,000	155,000	176,000	194,000	237,000	285,000
			115	82,700	110,000	146,000	166,000	183,000	224,000	269,000
			125	76,000	101,700	137,000	153,000	170,000	209,000	253,000

NOTES: * WH204PHF available on special order only. Contact factory Application Dept. for additional information.

Figure 3–14 Capacity data for typical *water*-cooled (WH) and remote (EH) condensing units, 1½ to 20 hp, using R-22. (By permission of Dunham-Bush, Inc.)

HP	60 HZ. AH/AWH	50 HZ. AH/AWH	Ambient Air Temp.°F	LOW TEMP. -40°F	-30°F	-20°F	-10°F	COMMERCIAL TEMP. 0°F	10°F	20°F	30°F
				SATURATED SUCTION TEMPERATURE °F							
				SUCTION PRESSURE PSIG							
				4.28	9.40	15.52	22.76	31.24	41.10	52.50	65.40
				BTUH	BTUH	BTUH	BTUH	BTUH	BTUH	BTUH	BTUH
1½	15LF	15LF5	90	-	-	6,700	9,000	11,600	-	-	-
			100	-	-	6,200	8,300	10,700	-	-	-
	15ELF	15ELF5	90	4,600	6,400	8,800	-	-	-	-	-
			100	4,200	5,800	8,200	-	-	-	-	-
		15HF5	90		-	-	-	8,200	11,200	12,700	15,000
			100	-	-	-	-	7,500	10,000	12,000	14,200
2	20HF	20HF5	90	-	-	-	-	11,000	15,000	17,000	20,000
			100	-	-	-	-	10,000	13,500	16,000	19,000
	20LF	20HF5	90	4,600	6,400	8,800	11,800	15,000	-	-	-
			100	4,200	5,800	8,200	10,800	13,900	-	-	-
	20ELF	20ELF5	90	5,600	7,800	10,800	14,300	18,300	-	-	-
			100	5,000	7,200	9,900	13,200	17,000	-	-	-
3	30HF	30HF5	90	-	-	-	-	16,500	19,800	23,000	27,000
			100	-	-	-	-	15,000	18,000	21,500	25,000
	30LF	32LF5	90	6,600	9,300	12,700	16,700	21,000	-	-	-
			100	6,000	8,400	11,700	15,400	19,400	-	-	-
	32ELF		90	8,100	11,100	15,200	19,900	25,400	-	-	-
			100	7,400	10,200	14,000	18,400	23,400	-	-	-
	30ELF	30ELF5	90	9,000	12,700	17,400	23,000	29,000	-	-	-
			100	8,200	11,700	16,000	21,000	26,700	-	-	-
5	52HF	52HF5	90	-	-	-	-	30,000	36,000	42,000	49,000
			100	-	-	-	-	27,500	33,000	39,000	46,000
	50LF	50LF5	90	10,800	15,300	20,700	26,600	32,600	-	-	-
			100	9,800	14,200	19,100	24,600	30,200	-	-	-
	50ELF		90	13,000	18,500	25,000	32,000	39,000	-	-	-
			100	11,900	17,000	23,000	29,500	36,500	-	-	-
		51PLF5	90	13,200	18,600	24,800	31,600	38,700	-	-	-
			100	11,900	17,100	23,000	29,300	35,800	-	-	-
		62PELF5	90	15,400	21,900	29,600	37,800	46,100	-	-	-
			100	14,100	20,100	27,200	34,900	43,100	-	-	-
7½	75HF	75HF5	90	-	-	-	-	40,000	48,000	56,000	65,000
			100	-	-	-	-	37,000	44,000	52,000	61,000
	76PLF		90	16,000	22,600	30,200	38,500	47,200	-	-	-
			100	14,400	20,700	28,000	35,700	43,600	-	-	-
	77PELF	77PELF5	90	18,200	26,000	34,800	44,400	54,000	-	-	-
			100	16,500	23,800	32,000	41,200	50,000	-	-	-
10	100PHF		90	-	-	-	-	48,000	58,000	68,000	79,000
			100	-	-	-	-	44,000	53,000	62,000	73,000
		101PHF5	90	-	-	-	-	49,700	60,100	70,500	81,800
			100	-	-	-	-	45,600	55,000	64,200	75,600
	111PHF	111PHF5	90	-	-	-	-	60,000	71,000	83,000	97,000
			100	-	-	-	-	54,000	66,000	77,000	91,000
	101PELF	101PELF5	90	22,400	32,000	42,800	56,600	70,200	-	-	-
			100	20,200	24,100	39,300	52,200	65,300	-	-	-
15	D151PHF	151PHF5	90	-	-	-	-	72,000	85,000	100,000	116,000
			100	-	-	-	-	65,000	78,000	92,000	108,000
	154PELF		90	27,000	38,600	51,600	68,200	84,800	-	-	-
			100	24,000	35,400	47,400	63,000	78,800	-	-	-
17½	D201PHF	201PHF5	90	-	-	-	-	89,000	108,000	124,000	145,000
			100	-	-	-	-	81,000	97,000	115,000	136,000
20	D204PHF		90	-	-	-	-	106,000	128,000	149,000	174,000
			100	-	-	-	-	98,000	117,000	136,000	162,000

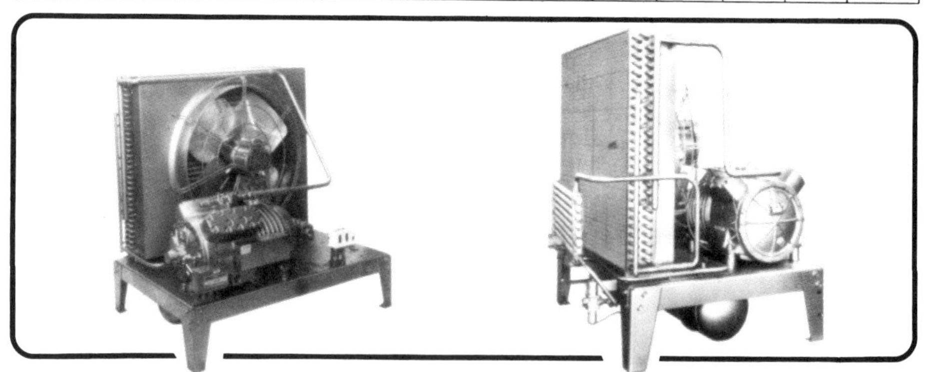

Figure 3-15 Capacity data for typical *air*-cooled (AH) and air/water (AWH) condensing units, 1½ to 20 hp, using R-502. (By permission of Dunham-Bush, Inc.)

HP	60 HZ.	50 HZ.	Saturated Discharge Temp. °F	LOW TEMP.				COMMERCIAL TEMP.			
				SATURATED SUCTION TEMPERATURE °F							
				-40°F	-30°F	-20°F	-10°F	0°F	10°F	20°F	30°F
	WH/EH	WH/EH		SUCTION PRESSURE PSIG							
				4.28	9.40	15.52	22.76	31,24	41.10	52.50	65.40
				BTUH	BTUH	BTUH	BTUH	BTUH	BTUH	BTUH	BTUH
1½	15LF	15LF5	105	-	-	6,600	9,100	12,000	-	-	-
			115	-	-	5,700	8,100	10,800	-	-	-
			125	-	-	4,900	7,100	9,700	-	-	-
	15ELF	15ELF5	105	4,400	6,300	8,800	-	-	-	-	-
			115	3,400	5,200	7,600	-	-	-	-	-
			125	2,400	4,100	6,500	-	-	-	-	-
		15HF5	105	-	-	-	-	9,350	11,600	14,000	16,500
			115	-	-	-	-	8,450	10,500	12,900	15,400
			125	-	-	-	-	7,600	9,550	12,000	14,500
2	20HF	20HF5	105	-	-	-	-	12,500	15,500	18,800	22,100
			115	-	-	-	-	11,300	14,100	17,300	20,700
			125	-	-	-	-	10,200	12,800	16,000	19,400
	20LF	20LF5	105	4,400	6,300	8,800	12,100	16,000	-	-	-
			115	3,400	5,200	7,600	10,600	14,600	-	-	-
			125	2,400	4,100	6,500	9,400	13,000	-	-	-
	20ELF	20ELF5	105	5,400	7,600	10,800	14,700	19,500	-	-	-
			115	4,200	6,400	9,400	13,200	17,600	-	-	-
			125	3,100	5,200	8,000	11,400	15,800	-	-	-
3	30HF	30HF5	105	-	-	-	-	16,800	20,800	25,200	29,500
			115	-	-	-	-	15,200	19,000	23,100	28,000
			125	-	-	-	-	13,800	17,200	21,500	26,000
	30LF	32LF5	105	6,400	9,200	12,900	17,500	23,400	-	-	-
			115	5,000	7,500	11,200	15,600	21,200	-	-	-
			125	3,600	6,200	9,400	13,800	19,000	-	-	-
	32ELF		105	7,800	11,000	15,400	20,900	27,900	-	-	-
			115	6,400	9,300	13,700	18,600	25,400	-	-	-
			125	5,000	7,900	11,900	16,400	23,000	-	-	-
	30ELF	30ELF5	105	8,800	12,600	17,500	24,200	32,000	-	-	-
			115	6,800	10,300	14,800	20,800	28,400	-	-	-
			125	4,800	7,700	12,000	17,600	24,800	-	-	-
5	52HF	52HF5	105	-	-	-		30,400	37,800	45,500	54,000
			115	-	-	-	-	27,500	32,300	42,000	50,500
			125	-	-	-	-	25,000	31,700	39,000	47,800
	50LF	50LF5	105	10,800	15,000	21,600	29,000	38,800	-	-	-
			115	8,300	12,600	18,700	26,000	34,800	-	-	-
			125	6,000	10,000	15,500	22,600	30,200	-	-	-
	50ELF		105	12,800	18,500	26,000	35,600	47,600	-	-	-
			115	10,000	15,400	22,800	31,800	42,000	-	-	-
			125	7,200	12,400	19,100	27,600	37,000	-	-	-
		51PLF5	105	14,900	21,600	29,400	38,600	49,700	-	-	-
			115	12,850	19,000	26,800	35,000	45,500	-	-	-
			125	10,800	16,450	23,700	31,400	41,300	-	-	-
		62PELF5	105	15,800	22,800	31,700	43,500	57,500	-	-	-
			115	12,400	18,800	28,200	39,100	52,000	-	-	-
			125	8,400	15,300	23,600	34,200	46,500	-	-	-
7½	75HF	75HF5	105	-	-	-	-	40,200	50,500	60,500	72,000
			115	-	-	-	-	37,000	46,000	56,200	67,500
			125	-	-	-	-	33,000	42,000	52,000	63,300
	76PLF		105	16,000	23,000	32,000	44,000	58,000	-	-	-
			115	12,500	19,000	28,500	39,500	52,500	-	-	-
			125	8,500	15,500	23,800	34,500	47,000	-	-	-
	77PELF	77PELF5	105	18,800	27,100	37,800	51,900	68,600	-	-	-
			115	14,700	22,100	32,800	46,600	61,200	-	-	-
			125	10,700	17,600	27,900	41,200	56,400	-	-	-
10	100PHF		105	-	-	-	-	49,000	60,000	73,000	87,000
			115	-	-	-	-	44,500	56,000	68,000	81,000
			125	-	-	-	-	40,200	51,000	63,000	76,000
		101PHF5	105	-	-	-	-	50,470	61,800	75,200	89,600
			115	-	-	-	-	45,800	57,700	70,000	83,400
			125	-	-	-	-	41,400	52,500	64,900	78,300
	111PHF	111PHF5	105	-	-	-	-	61,000	75,000	90,000	117,000
			115	-	-	-	-	55,000	69,000	84,000	100,000
			125	-	-	-	-	49,500	63,000	77,000	94,000
	101PELF	101PELF5	105	23,800	33,800	47,400	64,900	87,000	-	-	-
			115	17,800	27,800	41,200	58,000	77,700	-	-	-
			125	13,000	21,900	34,200	49,700	69,400	-	-	-
15	151PHF	151PHF5	105	-	-	-	-	72,000	89,000	107,000	127,000
			115	-		-	-	66,000	82,000	100,000	120,000
			125	-	-	-	-	59,000	75,000	92,000	112,000
	154PELF		105	28,800	40,800	57,300	78,400	105,000	-	-	-
			115	21,500	33,500	49,800	70,000	93,800	-	-	-
			125	15,600	26,500	41,300	60,000	83,800	-	-	-
17½	201PHF	201PHF5	105			-	-	90,000	112,000	134,000	160,000
			115	-	-	-	-	82,000	102,000	124,000	150,000
			125	-	-	-	-	74,000	93,000	114,000	140,000
20	204PHF		105	-	-	-	-	107,000	134,000	162,000	191,000
			115	-	-	-	-	98,000	123,000	150,000	179,000
			125	-	-	-	-	88,000	112,000	138,000	168,000

Figure 3–16 Capacity data for typical *water*-cooled condensing units, 1½ to 20 hp, using R-502. (By permission of Dunham-Bush, Inc.)

Note that each chart is divided into "low temperature," "commercial temperature," and "high temperature" applications. A single machine may span one or more of these temperature ranges. Capacities of air-cooled units are given in Btuh for 90 and 100 °F ambient. Capacities of water-cooled units are given in Btuh for condensing temperatures of 105, 115, and 125 °F.

These charts are useful when selecting a compressor, where it is necessary to select the most suitable refrigerant. For example, if the requirements are as follows:

Capacity required	40,000 Btuh
Evaporating temperature	−10 °F
Ambient temperature	90 °F

If R-12 refrigerant is used (see Figure 3–11), a 10-hp unit is required. However, if R-502 refrigerant is used (see Figure 3–15), a 7½-hp unit can be used.

Using Air-Cooled and Water-Cooled Condensers

In most cases, particularly on smaller units, the type of condensing arrangement is selected at the same time as the compressor. Units are purchased as condensing units rather than as compressors only.

For most compressors the efficiency (horsepower per ton) increases by using a water-cooled condensing unit rather than an air-cooled unit. This is due to the fact that at design conditions the condensing temperature is lower for water-cooled units. For most water-cooled units, even those using cooling-tower water, 105 °F condensing temperature can easily be maintained. With air-cooled condensing units the condensing temperature is usually 30 °F above design ambient temperature. Based on a design ambient of 95 °F, the condensing temperature would be 125 °F.

DETERMINING WHICH REFRIGERANT IS BEST

In common field-assembled systems and in small and medium-sized jobs, the choice of refrigerants is usually R-12, R-22, or R-502.

Usually the type of service, the size of the job, the availability of equipment and/or the preferences of the purchaser are deciding factors. Having given the Btu per hour capacity requirements, the evaporating temperature and preference for air cooled versus water cooled, the manufacturers' literature can be consulted and a selection made on the basis of capacity required and cost.

Generally speaking, R-502 is used for low-temperature applications, R-12 for the full range, and R-22 for high-temperature systems. The reason for the use of R-22 for high-temperature installations is not so much the boiling point of the refrigerant but the fact that R-22 requires less compressor displacement per ton and therefore provides a lower-cost system.

DETERMINING PERFORMANCE FROM THE MANUFACTURERS' LITERATURE

Typical condensing unit performance data for semihermetic compressors is illustrated in Figures 3-11 and 3-16. This data includes capacities for air-cooled and water-cooled condensing units in the range of ½ to 20 hp. for refrigerants R-12, R-22 and R-502. The use of these tables provides flexibility for making the best selection. The method for using these tables is as follows:

1. Choose the most desirable refrigerant. This will reduce the selection to two pages.
2. Choose air-cooled or water-cooled units. This will reduce the selection to one page.
3. Select the required evaporating temperature at the top of the page and drop down the column to the required capacity in Btuh.
4. Trace horizontally to the left and select the required ambient temperatures for air cooled units or the required saturated discharge temperature (condensing temperature) for water-cooled units.
5. Read the unit model number at the left of the selection made in item No. 4 above.

The model number itself if helpful since in most cases it indicates the size, range, refrigerant, air cooled or water cooled, and so on. The model number is also important information for ordering parts as well as for ascertaining the operating characteristics of the machine.

To illustrate the use of the table in Figure 3-11, assume that an air-cooled condensing unit was needed to produce 40,000 Btuh at a 25 °F saturated suction temperature, 90 °F ambient. From the table, two compressors meet this requirement, a medium- or commercial-temperature 5-hp machine and a high-temperature 5-hp machine. It would be good practice to select the high-temperature machine in this case since it would have an extra capacity range at higher suction temperatures to handle "pull-down" loads. Pull-down refers to the original lowering of temperatures during startup of a system. During this period the loads are increased and extra compressor capacity is required.

COMPRESSOR UNLOADERS

Capacity control equipment is usually standard on the larger high-temperature machines, optional on commercial (medium)-temperature machines, and not available on low-temperature machines. When unloaders are furnished, one or more cylinders are left operating under all conditions to provide adequate refrigerant mass flow. Usually, a three-cylinder machine has two steps of unloading to provide 66⅔ and 33⅓% full-load operation. A four-cylinder machine has two steps of unloading to provide 75 and 50% full-load opera-

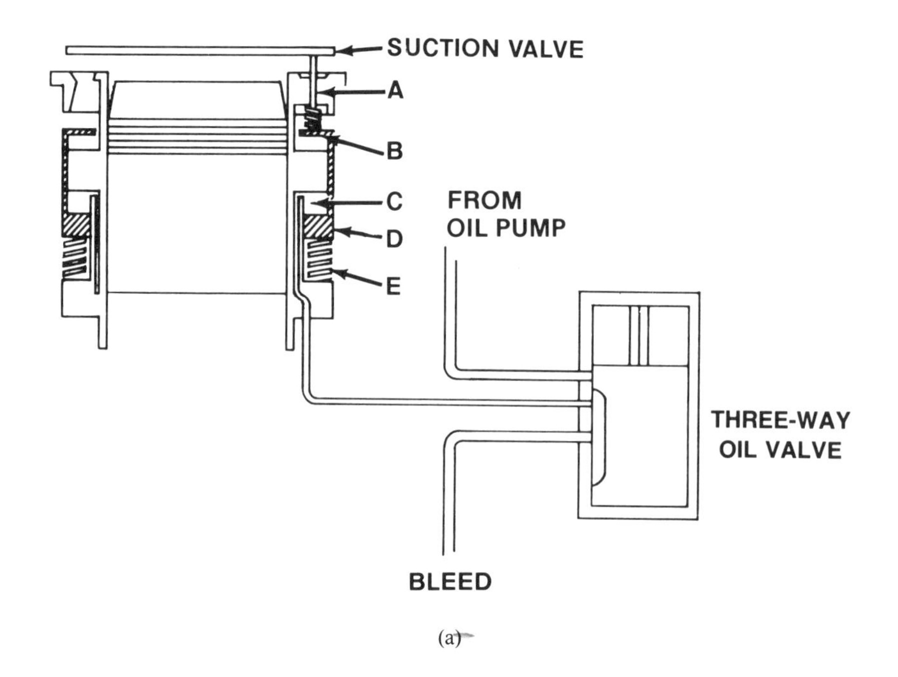

(a)

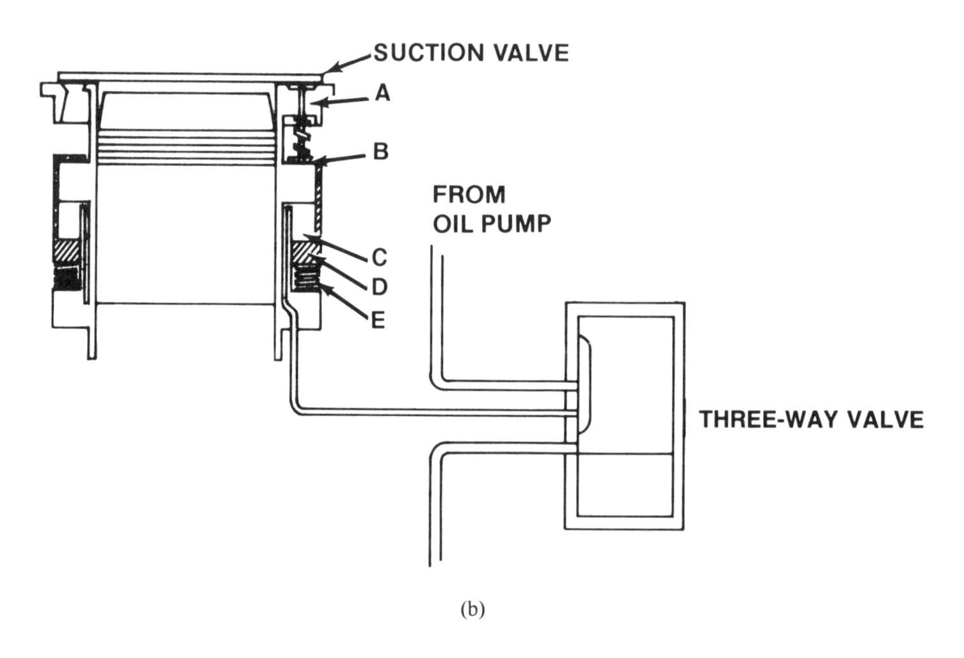

(b)

Figure 3-17 Unloader system: (a) in unloaded position; (b) in loaded position. (By permission of The Trane Company.)

tions. Oil pressure or high-side gas pressure is used to activate the unloader mechanism. A typical unloader mechanism is shown in Figure 3–17. Figure 3–17a shows the unloaded condition and Figure 3–17b shows the loaded condition. Note that this compressor uses ring-type valves. In the unloaded position of the suction valve, the piston lift pins raise the valve from its seat. In the loaded position the oil pressure causes the take-up ring and the piston unloader pin to be drawn downward, allowing the valve to seat normally.

ADVANTAGES OF THE ROTARY COMPRESSOR

The advantage of the small rotary compressor is the high efficiency, low sound level, and compact design. The smaller units are used for packed equipment such as household refrigerators and freezers.

There are two types of design: one with a stationary blade and the other with a rotating blade. The stationary blade type is illustrated in Figure 3–18 and the rotating blade type in Figure 3–19. In either type design a cylinder revolves on an eccentric shaft to compress the vapor refrigerant into the discharge port. A number of ports are constructed at close tolerance. A check valve is required in the exhaust port to prevent exhaust vapors from backing up into the pump.

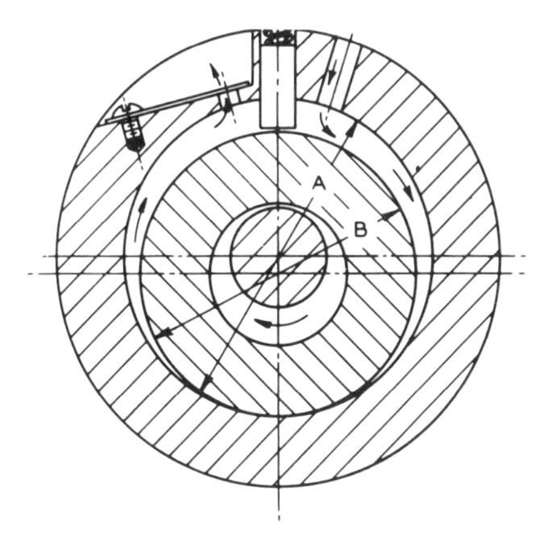

Figure 3–18 Stationary-blade-type rotary compressor. This design is used in domestic refrigerators and small air-conditioning units. In this design the rolling piston is attached to the stator. The compression is produced by the action of the concentric rotor. (By permission of *ASHRAE Handbook*.)

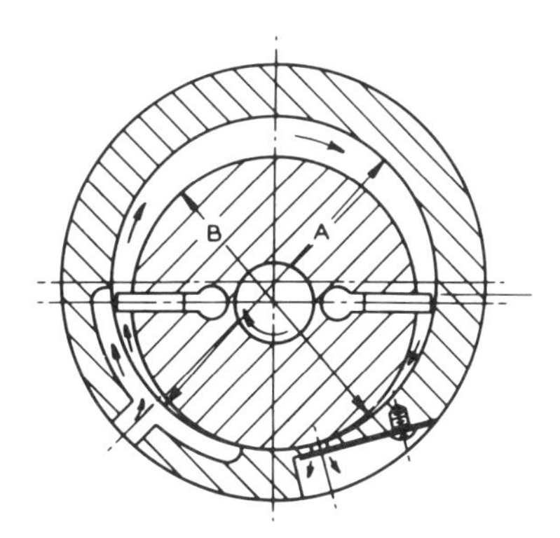

Figure 3–19 Rotating-vane-type rotary compressor. Two rotating vanes operate out of the rotor. Compression is produced by concentric action of the rotor. Motors are usually direct drive. (By permission of *ASHRAE Handbook*.)

ROTARY BOOSTER COMPRESSORS

A rotary booster compressor is shown in Figure 3–20. These machines are used for low-temperature systems when a large volume of low-pressure vapor is compressed. Suction temperatures range from -125 to $-5\,°F$ and use refrigerants R-12, R-22, and R-717. These

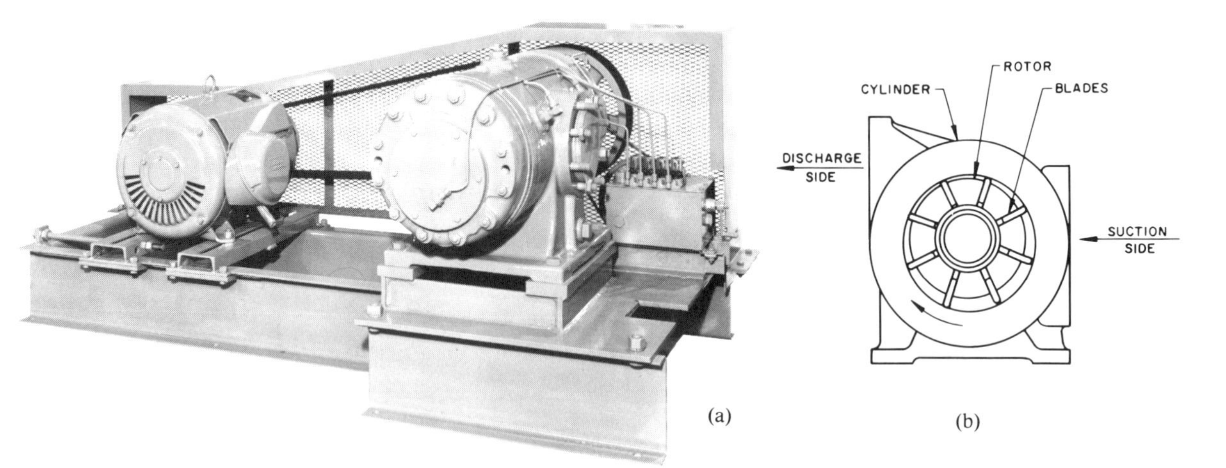

Figure 3–20(a) Rotary booster compressors are available either for V-built drive or direct drive. Sizes range from 5 to 350 tons of refrigeration capacity. (By permission of Vilter Manufacturing Corp.). **(b)** Large rotary compressors; schematic view. (By permission of *ASHRAE Handbook*.)

range in power from 10 to 600 hp. These compressors can start unloaded and have capacity from 100 to 20% during normal operation.

HELICAL ROTARY COMPRESSORS

The helical rotary compressor is constructed as shown in Figure 3–21. It is sometimes called a "screw" compressor since it has two meshing helically designed gears that rotate in opposite directions. The advantage of this equipment is that liquid in the refrigerant will not damage it. Modern machines have oil injection, permitting wide compression ratios, reduced discharge temperature, and lower noise levels. These machines can use R-12, R-22, R-502, and R-717 refrigerants.

An oil separator is required to remove oil from the high-pressure refrigerant. Figure 3–22 shows a helical rotary compressor system with an oil separator. Capacity reduction is possible to nearly zero, using either constant or variable speeds.

AMMONIA COMPRESSORS

Another group of medium-sized compressors are the ammonia units. This equipment is designed exclusively for the ammonia refrigerant, R-717. The equipment is made principally of steel and cast iron and is built for long, rugged service. These units operate at 1200 rpm rather than 1800 rpm as do the halocarbon machines. The brake horsepower (bhp) range is usually in the area between 20 and 250 bhp. Units can be either direct drive or belt driven.

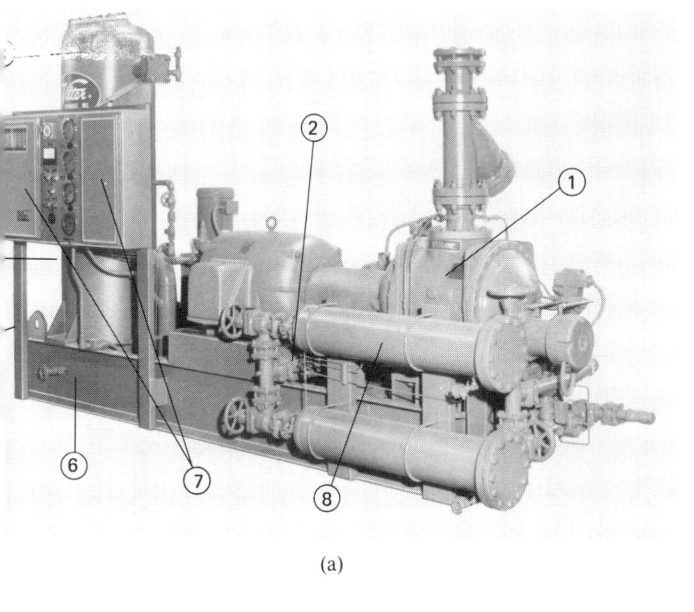

(a)

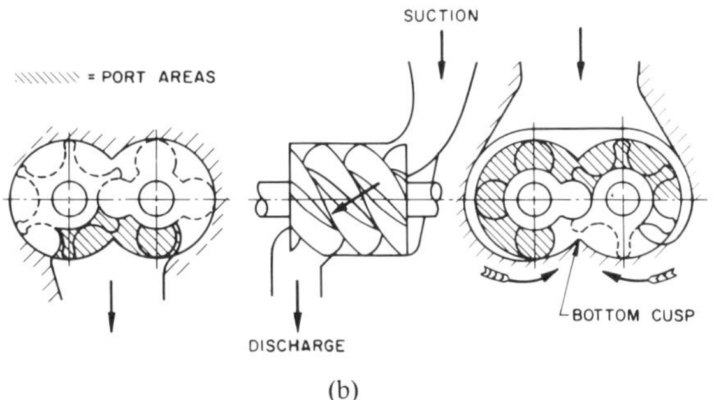

(b)

Figure 3–21(a) Helical rotary (screw) compressor. 1, Compressor; 2, discharge check valve; 3, vertical oil separator; 4, oil pump (not visible); 5, oil cooler (not visible); 6, steel base; 7, control panel; 8, micronic oil filter. (By permission of Vilter Manufacturing Company.) **(b)** Helical rotary (screw) compressor consists of two mating, helical-grooved rotors. Gas flow through the rotors is both radial and axial. Compression obtained by direct volume reduction. Oil-flooded helical rotary compressors are suitable for most applications using R-12, R-22, R-502, and R-717. Compressors are manufactured from 20 to 1500 hp. (By permission of *ASHRAE Handbook*.)

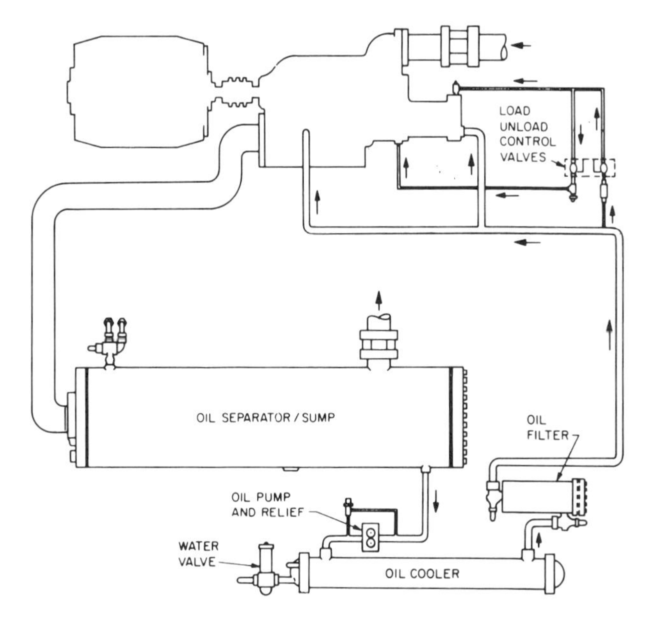

Figure 3–22 Typical oil system for the helical rotary compressor. Components include: oil separator, oil sump, oil heater, pump inlet screen, oil pump, oil cooler, filter pressure regulator, distribution piping, and control valves. (By permission of *ASHRAE Handbook*.)

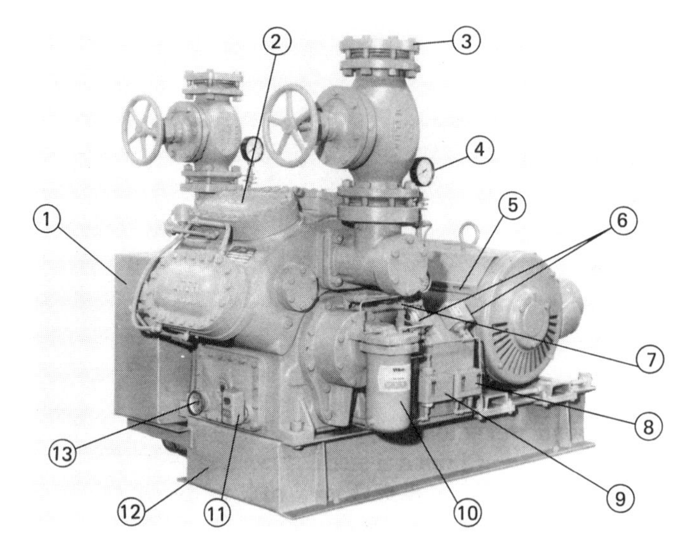

Figure 3–23 Ammonia reciprocating compressor. 1, Compressor drive; 2, water-cooled cylinder covers; 3, service valves; 4, gauges; 5, water-cooled oil cooler; 6, capacity reduction switches; 7, manual valve to check pressure drop across filter; 8, high- and low-pressure cutout; 9, oil failure switch, trimicro oil filter; 11, crankcase heater; 12, steel base; and 13, oil thermometer. (By permission of Vilter Manufacturing Company.)

450 VMC COMPRESSOR RATINGS FOR REFRIGERANT-717 (AMMONIA) AT 1200 rpm[a]

Condensing pressure (psig) and corresponding temperature (°F)	Suction temperature and pressure		452		454		456		458		4512	
	°F	psig	Tons	bhp	Tons	bhp	Tons	bhp	Tons	bhp	Tons	bhp
175 93.0	−15°	6.2	11.3	25.9	22.6	49.3	35.9	75.5	45.0	95.1	67.6	139.8
	−10°	9.0	13.7	27.8	27.3	53.0	42.7	80.3	53.3	103.7	80.0	152.6
	−5°	12.2	16.3	29.7	32.6	56.6	50.7	85.0	63.1	112.0	94.8	164.8
	0°	15.7	19.2	31.6	38.3	60.1	59.3	91.7	74.4	119.9	111.8	176.3
	5°	19.6	22.4	33.2	44.7	63.3	68.7	97.0	86.9	127.4	130.4	187.4
	10°	23.8	25.7	26.9	51.3	66.3	79.2	102.0	101.3	134.4	152.0	197.9
	15°	28.4	29.2	36.3	58.4	69.2	90.6	106.5	116.5	140.8	174.8	207.0
	20°	33.5	33.3	37.5	66.6	71.5	103.2	110.4	133.0	146.5	199.4	215.5
	25°	39.0	37.7	38.7	75.3	73.7	116.3	113.6	150.0	151.5	224.8	222.9
	30°	45.0	42.4	39.6	84.8	75.5	130.2	115.9	168.5	156.0	252.8	229.4
185 96.2	−15°	6.2	11.1	26.3	22.1	50.0	34.3	75.8	44.2	96.3	66.2	141.7
	−10°	9.0	13.4	28.2	26.8	53.8	41.7	82.2	52.6	106.0	79.0	155.9
	−5°	12.2	16.1	30.2	32.1	57.5	49.6	88.5	62.4	114.8	93.4	168.8
	0°	15.7	18.9	32.0	37.8	61.0	58.2	94.3	73.5	124.1	110.2	182.6
	5°	19.6	22.1	33.8	44.1	64.3	67.5	99.9	85.8	130.1	128.8	192.3
	10°	23.8	25.3	35.4	50.6	67.4	77.9	105.1	99.3	138.0	149.0	203.0
	15°	28.4	28.7	37.0	57.4	70.4	89.7	109.6	114.3	145.0	171.4	213.2
	20°	33.5	32.8	38.4	65.5	73.2	102.0	113.7	130.2	151.8	195.4	223.3
	25°	39.0	37.1	39.7	74.2	75.6	114.9	117.3	147.0	158.1	220.4	232.4
	30°	45.0	41.9	40.7	83.7	77.6	128.5	120.2	164.9	163.5	247.4	240.5
205 102.3	5°	19.6	21.0	36.2	41.9	68.9	65.0	103.6	83.3	135.9	125.0	200.0
	10°	23.8	24.2	37.6	48.4	71.6	74.9	109.5	96.7	145.4	145.2	213.9
	15°	28.4	27.8	45.2	55.5	74.9	86.1	114.9	110.9	154.0	166.4	226.5
	20°	33.5	31.7	51.7	63.3	78.0	98.5	119.9	125.8	161.3	188.6	237.2
	25°	39.0	35.9	42.5	71.8	80.9	111.3	124.3	141.3	167.6	212.0	246.4
	30°	45.0	40.8	43.8	81.5	83.4	125.1	128.4	158.4	172.3	237.6	253.5
225 108.0	20°	33.5	30.4	44.7	60.8	85.2	94.0	125.3	121.5	166.8	182.2	245.2
	25°	39.0	34.7	46.6	69.3	88.7	107.5	130.6	136.2	174.3	204.4	256.3
	30°	45.0	39.3	47.7	78.6	90.9	121.4	135.6	151.0	180.7	227.8	265.9
	35°	51.6	44.4	48.7	88.7	92.8	136.5	140.0	168.9	186.2	253.4	273.9

[a]Ratings above the line are for extrapolation only.

Ratings are based on saturated suction and 5 °F liquid subcooling. Reduce tonnage 2½% for no liquid subcooling. Ratings do not include belt losses. Add 3% to bhp for V-belt-driven units.

(a)

Maximum limits:

Suction superheat (R-717)	25 °F
Compressor ratio (R-717)	8:1
Pressure differential	175 psig
Discharge temperature	300 °F
Discharge pressure	300 psig
Suction pressure	150 psig
Direct-drive motor hp	300 hp

Crankcase oil temperature operating range 110 to 130 °F. If oil temperature exceeds 130 °F, premature wear and/or failure may result.

450 VMC V-BELT-DRIVEN HORSEPOWER LIMITATION

Compressor sizes	rpm	Maximum bhp
452 and 454	1200	115
	1130	115
	1000	115
	900	110
	810	90
456, 458, and 4512	1200	140
	1130	135
	1000	125
	900	120
	810	100

(b)

Figure 3–24 Typical compressor performance shows capacity in tons and horsepower for various sizes and operating conditions. (By permission of Vilter Manufacturing Company.)

Figure 3–23 shows an exterior view of a typical ammonia reciprocating compressor with a description of the principal parts. Note the water-cooled oil cooler, the pressure-activated capacity reduction switches, the full-flow oil filter, the oil thermometer, and the water-cooled cylinder covers. The entire machine is built for rugged service and long life.

A typical capacity table is shown in Figure 3–24. These single-stage machines are recommended down to −15 °F suction temperature. Note that the bhp per ton improves with a rise in suction temperature. For low suction temperatures it is advisable to use a two-stage unit.

Figure 3–25 shows a two-stage compressor model. Note the interstage discharge gas cooler suspended at the top of the machine. On a six-cylinder machine, four cylinders are

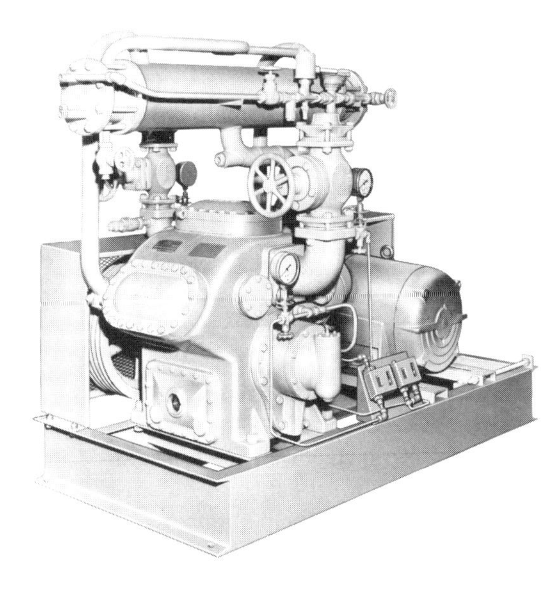

Figure 3–25 Two-stage reciprocating compressor, suitable for R-12, R-22, R-502, and R-717. This type will operate at suction temperatures of −60 to −80 °F, depending on size of machine. Capacities of 5 to 50 tons. Horsepower, 25 to 150. Sizes, 6-cylinder and 12-cylinder. One-third of the cylinders operate on high state. Note intercooler on top side of unit. (By permission of Vilter Manufacturing Company.)

usually used on the low stage and two cylinders on the high stage. Typical two-stage compressors with either 6 or 12 cylinders develop capacities from 6 to 60 tons.

Ammonia compressors are limited to areas where group 2 refrigerants can be used. This limits them chiefly to industrial applications and areas exclusive of public buildings. Ammonia refrigerants have several advantages. For example, Figure 3–26 shows a comparison of certain characteristics of a number of industrial refrigerants. Note the large latent heat effect of 1 lb of ammonia compared with other refrigerants: ammonia, 565.0 Btu/lb; R-12, 69.2 Btu/lb; R-22, 93.2 Btu/lb; and R-502, 68.9 Btu/lb. Thus, less ammonia refrigerant is used in the system at a lower cost per pound, effecting considerable savings.

	Refrigerant			
	R-717	R-12	R-22	R-502
Latent heat (Btu/lb)	565.0	68.2	93.2	68.9
Mass flow rate required to produce 1 ton of refrigeration (lb/min/ton)	0.422	4.0	2.88	4.33
Displacement (cfm/ton)	3.41	5.83	3.55	3.61
Coefficient of performance at standard conditions	4.76	4.70	4.66	4.37
Theoretical power requirement at standard conditions (bhp/ton)	0.989	1.002	1.011	1.079

COMPRESSOR HORSEPOWER REQUIREMENTS FOR TYPICAL SINGLE- AND TWO-STATE SYSTEMS

Refrigeration capacities and bhp based on 1200 rpm	R-717	R-12	R-22	R-502
At 30°F evap. and 100°F cond.[a]				
Tons	90.2	52.0	84.1	76.8
bhp	89.0	59.0	89.5	107.0
bhp ton	.99	1.14	1.06	1.39
At −30°F evap. and 100°F cond.[b]				
tons	15.74	10.90	18.30	19.73
bhp	45.01	36.70	54.40	65.35
bhp ton	2.86	3.37	2.97	3.31

[a]Single-stage six-cylinder compressor.
[b]Integral two-stage six-cylinder compressor with desuperheating and subcooling.

Figure 3–26 Characteristics of four commonly used industrial refrigerants. The primary performance difference between ammonia and the other industrial refrigerants is the latent heat. A much greater amount of heat can be transferred per pound of ammonia refrigerant than by R-12, R-22, or R-502. This means small pipe sizes and less refrigerant circulated. (By permission of Vilter Manufacturing Company.)

There are many other advantages of using ammonia. Oil entrainment in the refrigerant is less of a problem. Compressors pump less oil with ammonia and less gets past the oil separator. Oil is easier to recover. Moisture in the refrigerant is less of a problem. All-steel piping is used (not copper), which is not subject to corrosion.

CENTRIFUGAL COMPRESSORS

Centrifugal compressors are limited primarily to liquid chillers. Where large equipment is required, the centrifugal machine has no substitute. Centrifugals are available in from 1 to 10 stages, depending on the leaving temperature of the fluid being cooled. The impellers are designed like a centrifugal pump (Figure 3–27). Because centrifugal units are designed to cool fluids, the units are made up in packages (Figure 3–28).

Figure 3–27 Centrifugal compressors are available with as many as 9 or 10 stages. The illustration shows a four-stage impeller. (By permission of Carrier Corporation.)

Figure 3–28 Centrifugal compressor package, showing compressor, drive motor, chiller, condenser, receiver, control panel, and accessories completely assembled. (By permission of Carrier Corporation.)

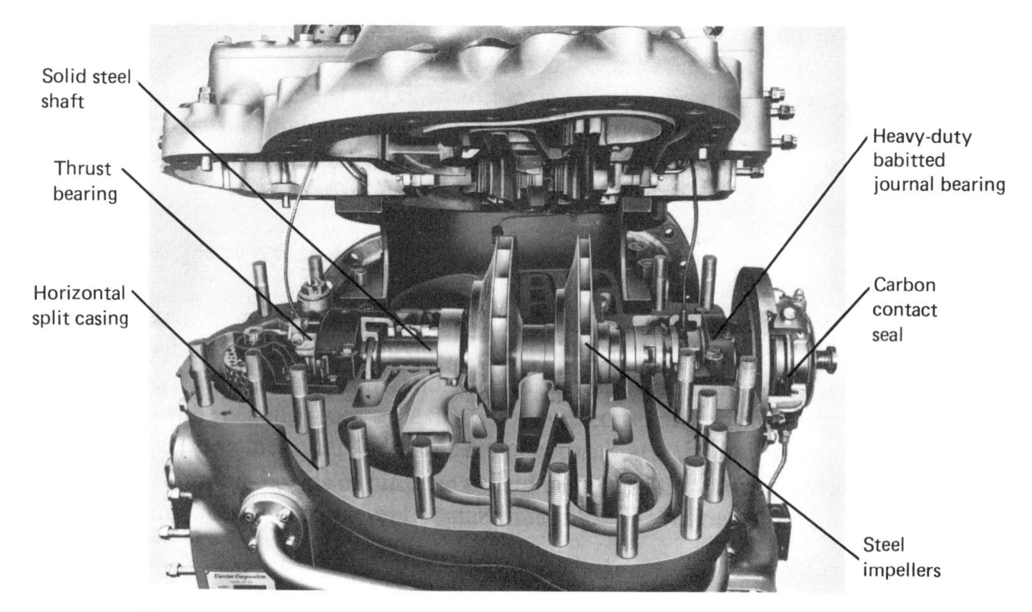

Figure 3–29 Two-stage centrifugal compressor with horizontal split casing showing gas passages and internal assembly. (By permission of Carrier Corporation.)

Centrifugals are made to use the following refrigerants: R-12, R-22, R-114, R-500, and R-502. An interior view of a two-stage centrifugal is shown in Figure 3–29. Note the carbon contact seal used on the open-type machine.

The lubrication system is extremely important since most of these machines run at high rpm levels. A typical lubrication system is shown in Figure 3–30. An auxiliary oil pump package can be furnished, as shown in Figure 3–31. Details of a typical seal are shown in Figure 3–32. A special monitoring system is provided to indicate internal pressures to permit control of oil foaming and loss of compressor oil.

Figure 3–30 Centrifugal compressor oil pump for forced lubrication system and pump drive gear. Pump is driven directly from the main compressor shaft through a worm-and-gear arrangement.

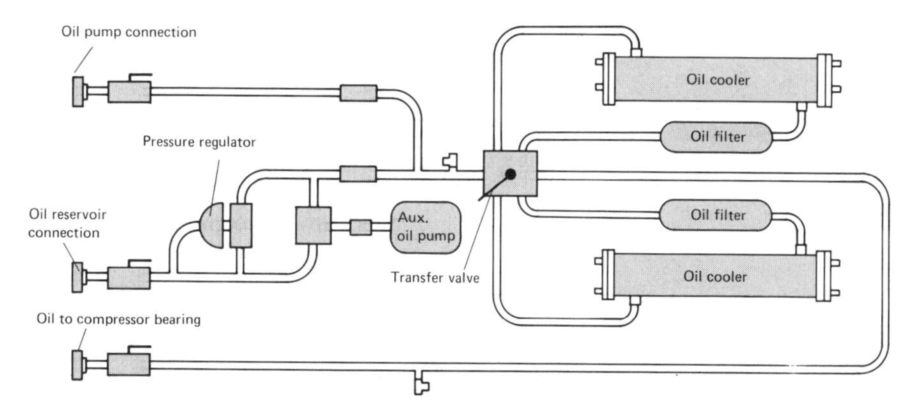

Figure 3–31 Auxiliary lubrication system is available for severe operating conditions using a single or dual oil cooler. (By permission of Carrier Corporation.)

Figure 3–32 Cross section of contact seal assembly. The carbon ring contact seal consists essentially of a rotating contact ring, a stationary contact sleeve, and two finely matched carbon rings. (By permission of Carrier Corporation.)

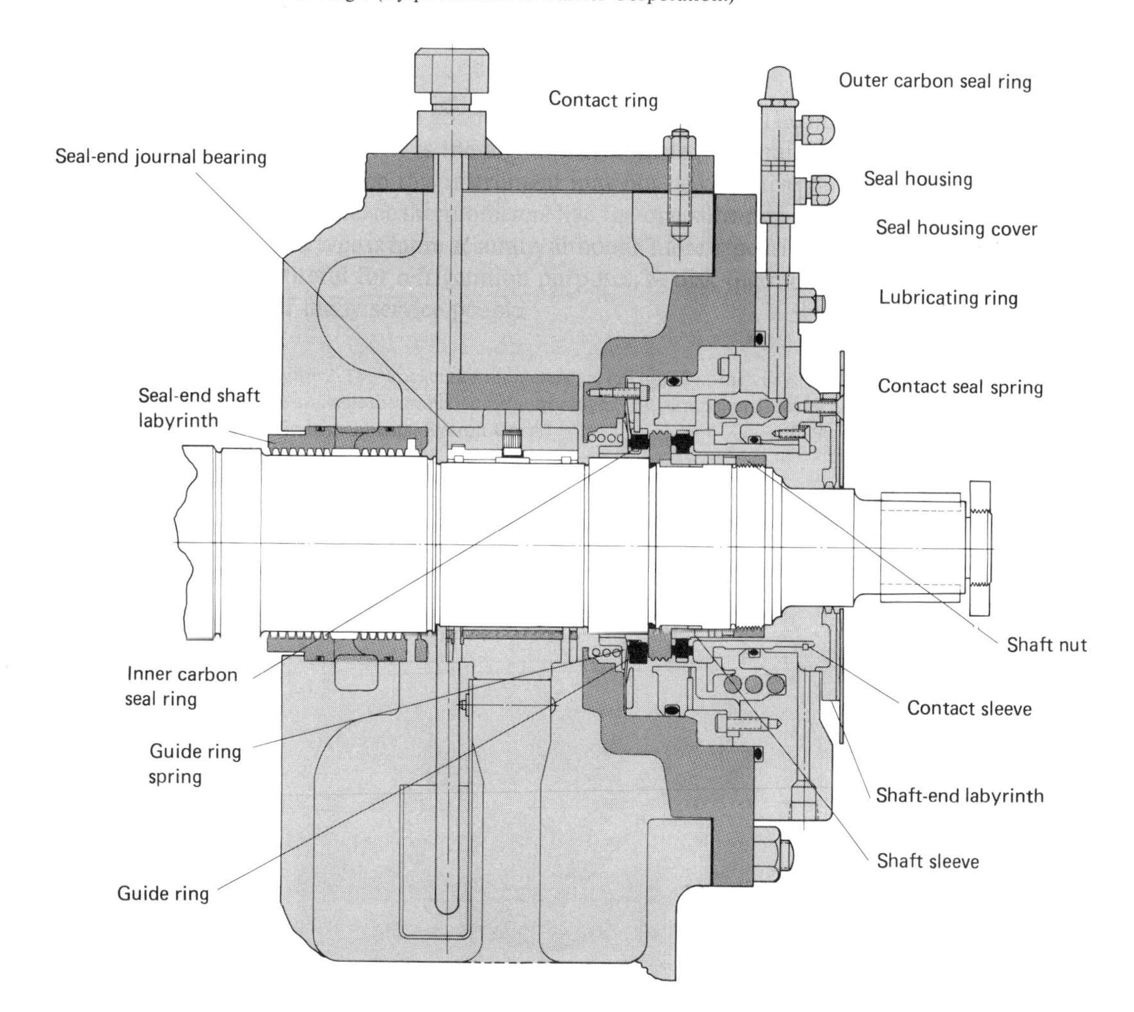

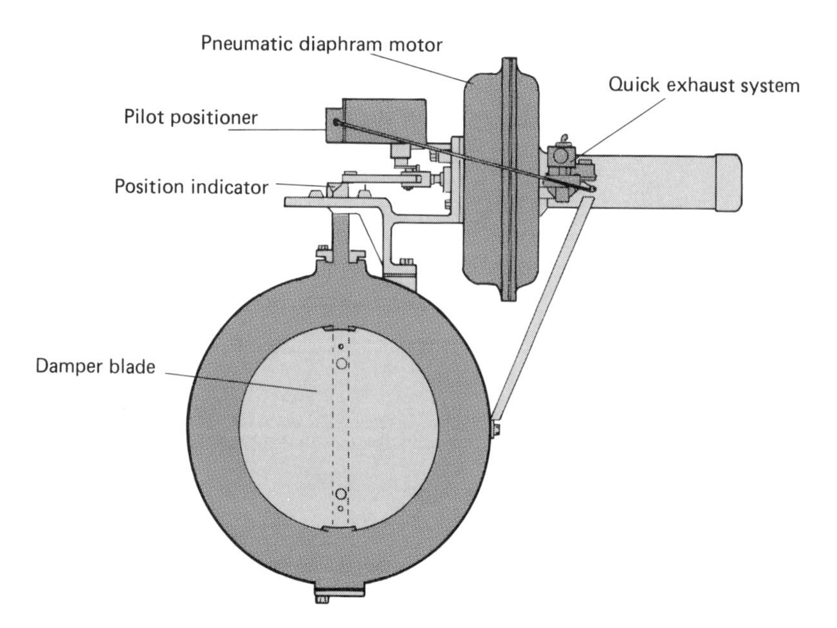

Figure 3–33 Pneumatic suction damper. This capacity control system provides regulation down to 30% of design load. Damper controls volume of refrigerant gas that can enter the compressor. On startup, damper remains closed until compressor is at full speed. (By permission of Carrier Corporation.)

Figure 3–34 Hot-gas bypass can prevent unstable condition at low load. Hot gas is desuperheated by bubbling through liquid refrigerant. With hot-gas bypass, equipment can operate down to zero load. (By permission of Carrier Corporation.)

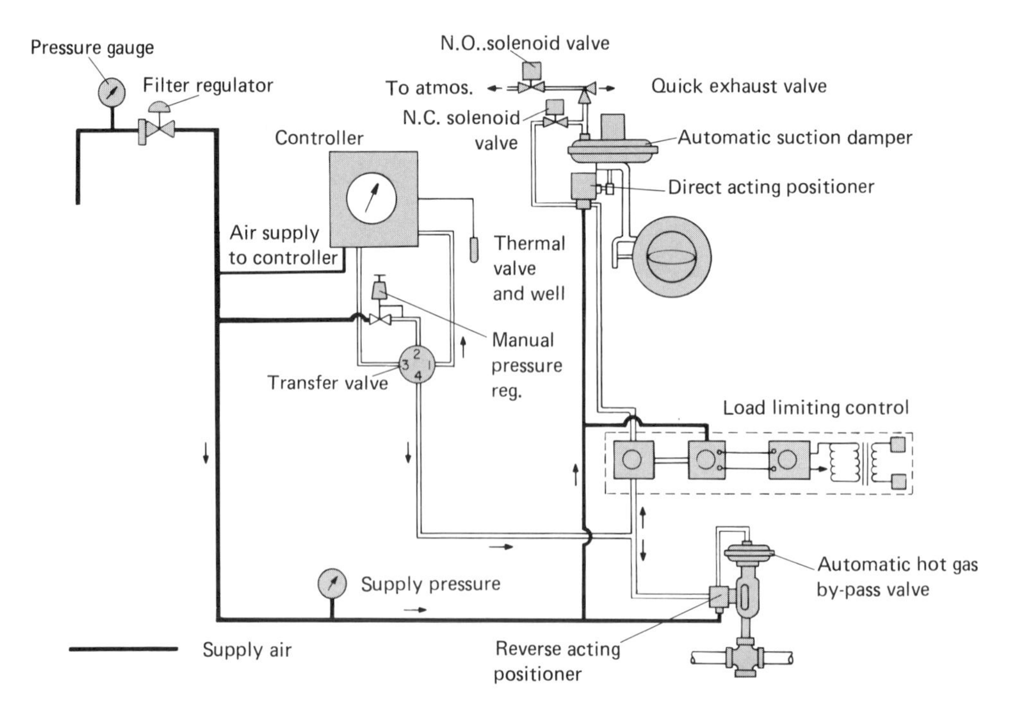

Capacity control is regulated by a suction-line damper, usually pneumatically controlled (Figure 3–33). The suction damper regulates capacity down to about 30 percent of full load. For greater reduction a hot-gas bypass can be used (Figure 3–34).

Centrifugal compressors are available with up to 10 stages of compression. A typical rotor assembly is shown in Figure 3–35. Gas flows from economizers and auxiliary coolers are generally introduced between impellers (Figure 3–36).

With the standard refrigerants, R-12 and R-22, capacities up to 3500 tons at 40 °F suction temperature and 500 tons at −50 °F suction temperature can be reached. Special refrigerants such as propylene (R-1270) and propane (R-290) can be used. The lower temperature limit is about −100 °F suction temperature.

Figure 3–35 Nine-stage impeller for special applications. (By permission of Carrier Corporation.)

Figure 3–36 Side load arrangement. Side loads are used for gas from economizers or auxiliary coolers. (By permission of Carrier Corporation.)

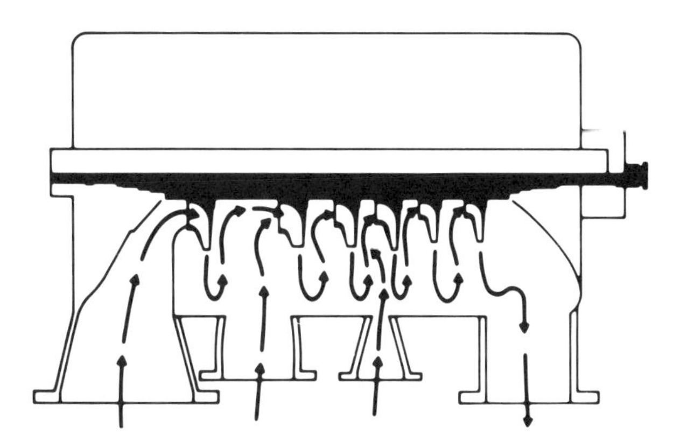

REVIEW EXERCISE

Select the letter representing your choice of the correct answer (there may be more than one).

3–1. Which types of compressors are positive displacement?

________ a. Reciprocating

________ b. Rotary

________ c. Screw

________ d. Centrifugal

3–2. Which type of compressor has a suction gas-cooled motor?

________ a. Centrifugal

________ b. Hermetically sealed

________ c. Helical rotary

________ d. Open type

3–3. What type of compressor uses a "ring"-type valve?

________ a. Medium-sized reciprocating

________ b. Centrifugal

________ c. Rotary

________ d. 4 × 4 ammonia compressor

3–4. Which type of condensing unit is more efficient?

________ a. Air cooled

________ b. Water cooled

3–5. What is usually the normal condensing temperature for water-cooled units?

________ a. 85°F

________ b. 95°F

________ c. 105°F

________ d. 115°F

3–6. Which refrigerant is specifically designed for low-temperature applications?

________ a. R-717

________ b. R-12

________ c. R-22

________ d. R-502

3–7. What is the suction temperature range for commercial-temperature compressors?

________ a. 0 to 25°F

________ b. −40 to −10°F

________ c. +30 to +50°F

________ d. −40 to +50°F

3–8. If a refrigeration load required 25°F suction temperatures, 57,000 Btu, 90°F ambient, and R-22, what horsepower compressor would be selected?

________ a. 5

________ b. 7½

________ c. 10

________ d. 15

3-9. What is the normal suction temperature range of booster compressors?

________ a. −125 to −5°F

________ b. 0 to 32°F

________ c. −87.2 to −20.6°C

________ d. −10 to +40°C

3-10. Which type of positive-displacement compressor cannot be damaged by a liquid slug?

________ a. Fully hermetic reciprocating

________ b. Semihermetic reciprocating

________ c. Open reciprocating

________ d. Helical rotary

4 Evaporators

OBJECTIVES

After studying this chapter, the reader will be able to:

- Identify the various types of evaporators.
- Select the type of evaporator that best fits the applications.
- Improve the performance of an evaporator.

CLASSIFICATION OF EVAPORATORS

The evaporator is one of the essential parts of the refrigeration system, where heat is absorbed from the product or space being cooled and transferred to other parts of the system, where heat is removed.

Evaporators may be classified in many ways: (1) by the methods of refrigerant circulation and control, (2) by construction, and (3) by use.

Refrigerant Circulation and Control

The two principal types of refrigerant controls are: flooded and dry expansion. The flooded type is illustrated in Figure 6–14 and the dry expansion type is illustrated in Figure 4–1.

In the *flooded* type the evaporator is mostly filled with liquid refrigerant, allowing a small portion at the top of the evaporator or flash chamber for the collection of evaporated refrigerant vapor. The advantage of this type of evaporator is the excellent heat transfer rate that is accomplished by having liquid refrigerant in direct contact with the coil tubes. The disadvantage of the flooded evaporator is the large amount of refrigerant that is required to fill the evaporator.

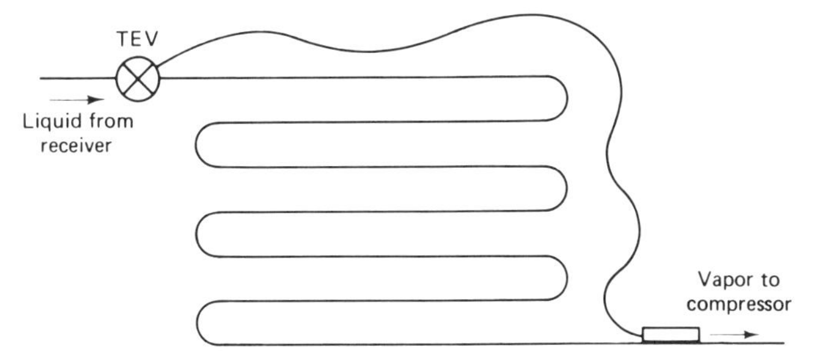

Figure 4–1 Dry-expansion-type evaporator. Liquid refrigerant is metered into the coil as required to match the load. All refrigerant is vaporized before it leaves the evaporator.

The *dry expansion* arrangement uses a metering device that feeds refrigerant into the coil at the rate it is evaporated by the load. The refrigerant must be completely vaporized when it reaches the end of the coil.

Evaporator Construction

From a construction standpoint, evaporator coils can be classified as:

1. Bare tubes
2. Plate surface
3. Finned surface

Bare tube evaporators are shown in Figure 4–2a, plate surface evaporators in Figure 17–21, and exterior finned surface evaporators in Figure 4–2b.

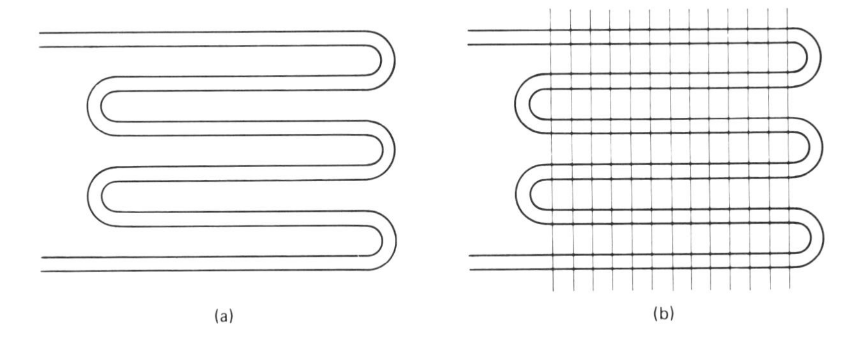

Figure 4–2 Two types of evaporator surface: (a) bare tube; (b) finned tube.

Bare tube evaporators are made of steel or copper tubing. Steel tubing is used for ammonia and copper for most of the other refrigerants. Bare tubc evaporators are easier than finned tube evaporators to defrost and clean. However, for the same capacity, finned tubes occupy less space and have better heat transfer.

Plate surfaces have numerous usages. They are particularly helpful as freezer shelves for contact cooling. The eutectic types are usually made with refrigerant tubes within a metal enclosure containing the eutectic solution. The solution has a freezing point lower than water and provides holdover capacity. Delivery trucks cool down the plates at night, with connected refrigeration equipment, in the storage garage. This holdover capacity is sufficient to provide the necessary refrigeration during the day without further compressor operation.

Finned surface evaporators are constructed with metal sheets firmly attached to the tubes to increase the surface. Interior fins can be added inside the tubes to increase the contact area with the refrigerant. Exterior fins add to the external surface to increase the contact surface with air. Where defrosting is necessary, fins are widely spaced, as few as two or three fins per inch. For above-freezing applications the fins may be as close as 14 or more fins to the inch. Gravity air circulation coils have wider fin spacing than is found on forced-air circulation coils, to prevent high resistance to the airflow.

Classification by Use

Evaporators can be used to cool air, water, or some other fluid, such as brine. Air, water, or brine become the transfer agents to distribute the cooling to the location or product area where it is needed. Air evaporators are required for reach-in refrigerator and walk-in coolers. Liquid coolers are used in industrial plants. Brine coolers are used in food-processing operations.

An example of a reach-in refrigerator evaporator is shown in Figure 4-3. An example of a walk-in refrigerator evaporator is shown in Figure 4-4. Capacities for air evaporators are usually given in terms of 10 °TD (difference between refrigerator box temperature and suction temperature of the refrigerant). Reach-in evaporator capacities vary from 650 to about 4000 Btuh and from 135 to about 540 cfm. Walk-in refrigerator evaporators go up as high as 140,000 Btuh with 22,000 cfm.

A typical industrial product cooler is shown in Figure 4-5. These range from 4000 through 30,000 cfm and 11,000 through 222,000 Btuh. They can be equipped with a pro-

Figure 4-3 Compact evaporators used in reach-in refrigerators. Air is radially discharged from the top (By permission of Dunham-Bush, Inc.)

Figure 4-4 Forced-air-type finned evaporator coil for ceiling mounting. Ratings vary from 12,000 to 72,000 Btuh at 10 °F TD. This equipment is used for above-freezing room temperatures (By permission of Dunham-Bush, Inc.)

Figure 4–5 Industrial-type product cooler. Available with propeller or centrifugal fan, V-belt drive, and standard defrost systems. (By permission of Dunham-Bush, Inc.)

peller or a centrifugal fan. Centrifugal fan units are used where ductwork is required for air distribution.

EVAPORATOR DEFROSTING

The type of defrost cycle is an important factor in making an evaporator selection. The more popular types of defrosting are: (1) air, (2) electric, (3) hot gas, and (4) glycol.

Air Defrosting

The air defrost system generally has its application in refrigerating systems where the box temperature is above 35 °F. Usually, a low-pressure control initiates the defrost cycle. As ice builds up on the coil, the suction pressure drops. When the suction pressure drops to the setting of the low-pressure control, the compressor is turned off and the evaporator fan continues to run. The air, above freezing temperature in the box, defrosts the coil. When defrosting is completed, the suction pressure rises and the compressor is started.

There is another use of air defrost that is becoming more common, particularly to save energy. This application involves the defrosting of coils in open cases in supermarkets. During the defrost cycle, the supply of refrigerant is stopped and air from the store, usually between 65 and 75 °F is picked up by the case fan and blown over the evaporator coil to defrost it. Defrost times are regulated by a time clock.

Electric Defrosting

Electric defrosting consists of resistance heaters that are used to melt the accumulated ice at intervals usually regulated by a time clock. The electric defrost cycle is initiated by closing the refrigerant liquid-line solenoid. The compressor pumps down and shuts off on the low-pressure control (LPC). The heaters are turned on and the fan is turned off. In many systems the defrost cycle is terminated by a defrost thermostat that senses that the ice has been melted. At the end of the defrost period the heaters are turned off, the solenoid valve opens the refrigeration liquid line, and the compressor starts. The fan may come on immediately or after a period of time delay to prevent moisture remaining on the coil from blowing into the refrigerated space.

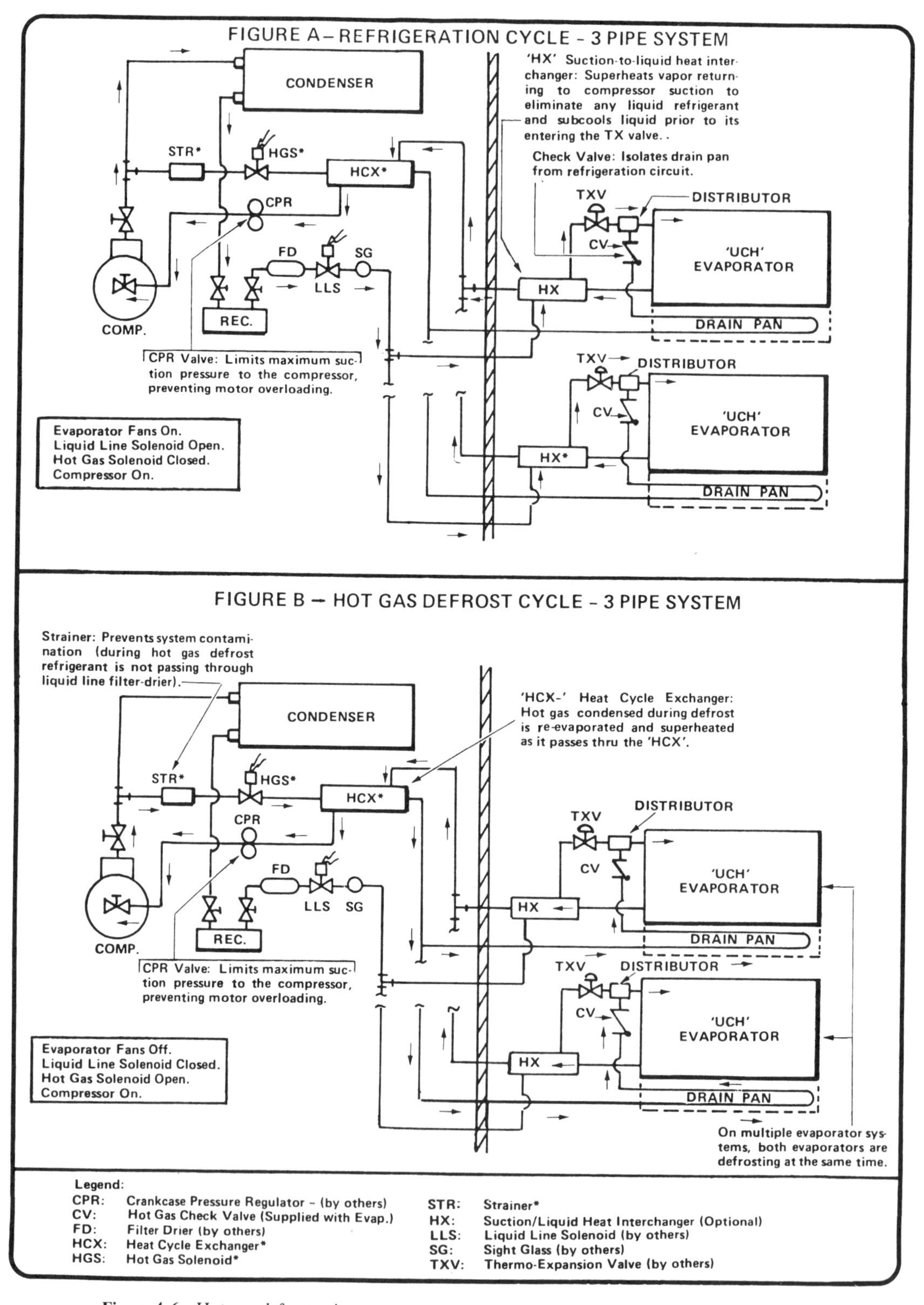

Figure 4–6 Hot-gas defrost using a reevaporating arrangement. (By permission of Dunham-Bush, Inc.)

Hot Gas Defrosting

There are a number of methods used for hot-gas defrosting:

1. Reevaporation
2. Reuse of refrigerant
3. Reverse cycle
4. Thermobank method

Reevaporation can be performed in a number of ways. The important thing is to vaporize the condensed liquid before it reaches the compressor. One method is shown in Figure 4-6, using a special heat exchanger to vaporize the refrigerant.

A system that *reuses the condensed liquid refrigerant* and includes multiple evaporators is shown in Figure 4-7. During normal operation, valves D, 1, 2, 3, and 4

Figure 4-7 Hot-gas defrost using multiple evaporators.

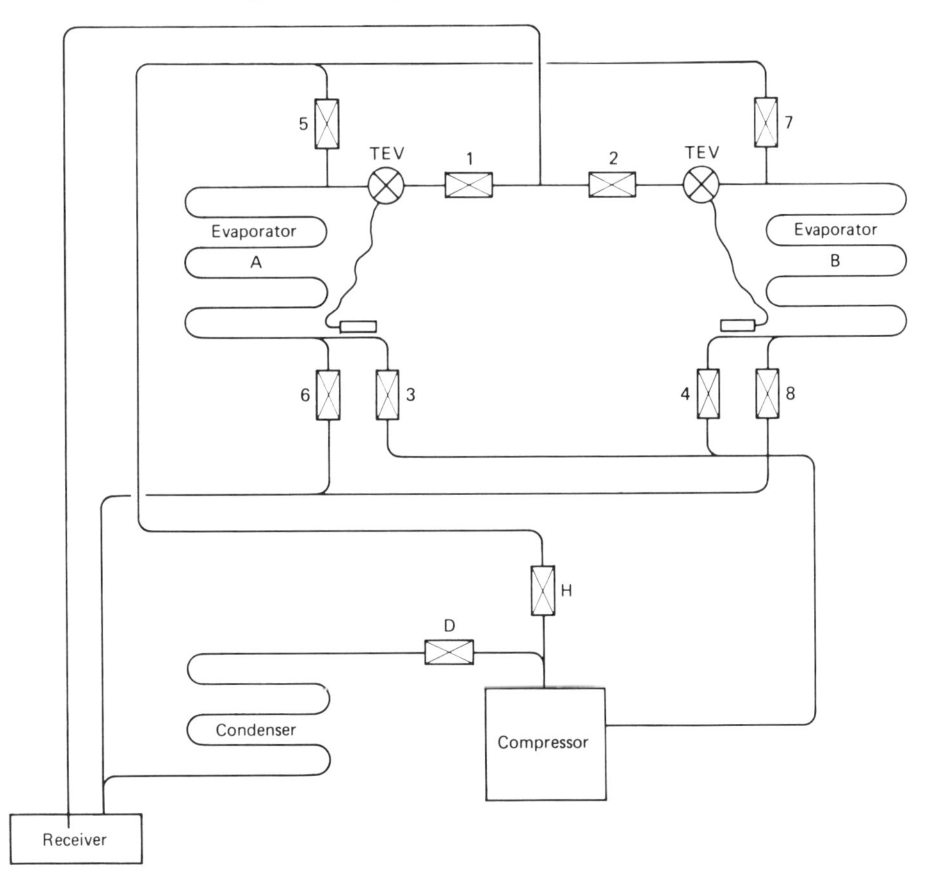

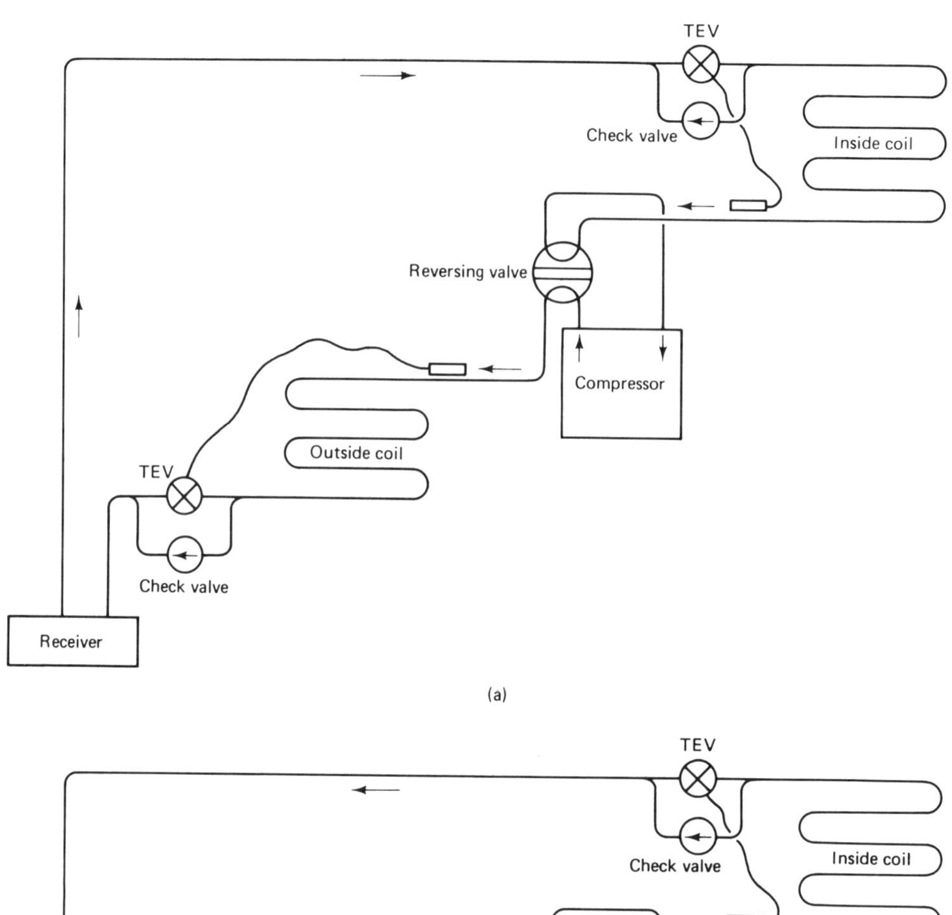

(a)

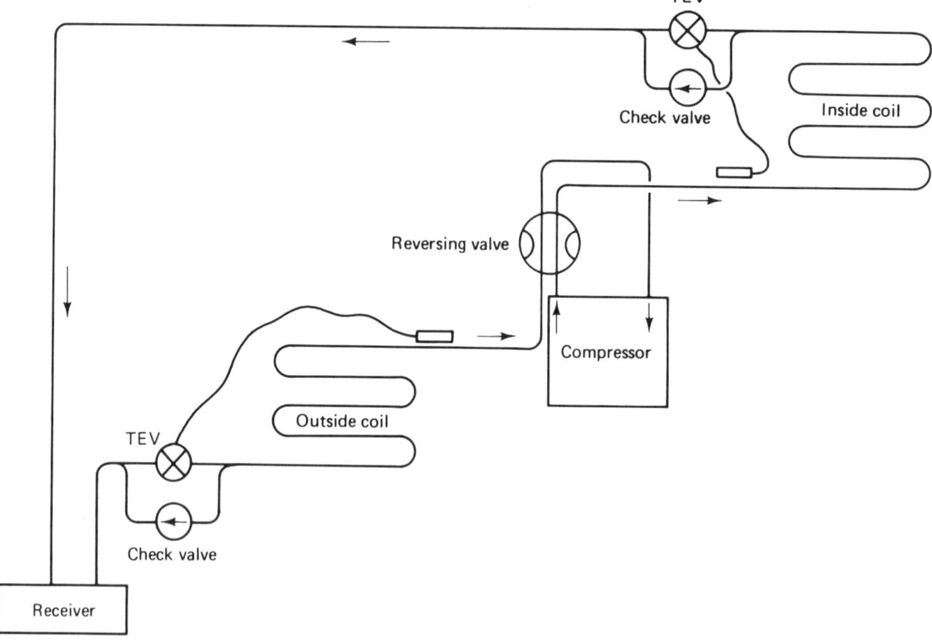

(b)

Figure 4-8 Reverse-cycle defrost: (a) normal cycle; (b) hot-gas defrost.

are open and valves H, 5, 6, 7, and 8 are closed. To defrost the left-hand coil, valves H, 2, 4, 5, and 6 are open and valves, D, 1, 3, 7, and 8 are closed. To defrost the right-hand coil, valves H, 1, 3, 7, and 8 are open and valves D, 2, 4, 5, and 6 are closed. This type of defrost arrangement is commonly applied in supermarket installations, where the defrost times are staggered to allow only a few units to defrost at one time.

The *reverse-cycle defrosting* system is illustrated in Figure 4–8. During the *normal cycle* (Figure 4–8a) the reversing valve directs the discharge gas from the compressor to the outside coil (condenser) and the standard arrangement of components permits cooling to be produced in the inside coil (evaporator). During the *defrost cycle* (Figure 4–8b) the reversing valve directs the discharge gas from the compressor to the inside coil, defrosting it by condensing the refrigerant. The liquid refrigerant then goes through the expansion valve to the outside coil, picking up heat and evaporating the refrigerant before entering the compressor.

Two expansion valves are used in the reverse-cycle system, each of which has a bypass arrangement with a check valve. This permits the refrigerant to flow in the opposite direction without passing through the expansion valve.

On a single evaporator system that accumulates a considerable quantity of ice, this system defrosts slowly. The available heat for defrost is primarily the heat of compression. Other types of hot-gas defrosting are faster, as, for example, the Thermobank (a registered trademark of Kramer Trenton Company) system, which provides storage of heat.

The Thermobank system is illustrated in Figure 4-9. During the *normal cycle,* refrigerant flows through the water "bank" (A), through the air-coded condenser (B),

Figure 4–9 Thermobank hot-gas defrost system uses heat bank to store heating capacity during the normal cycle for use during defrost. (By permission of Kramer Trenton Company.)

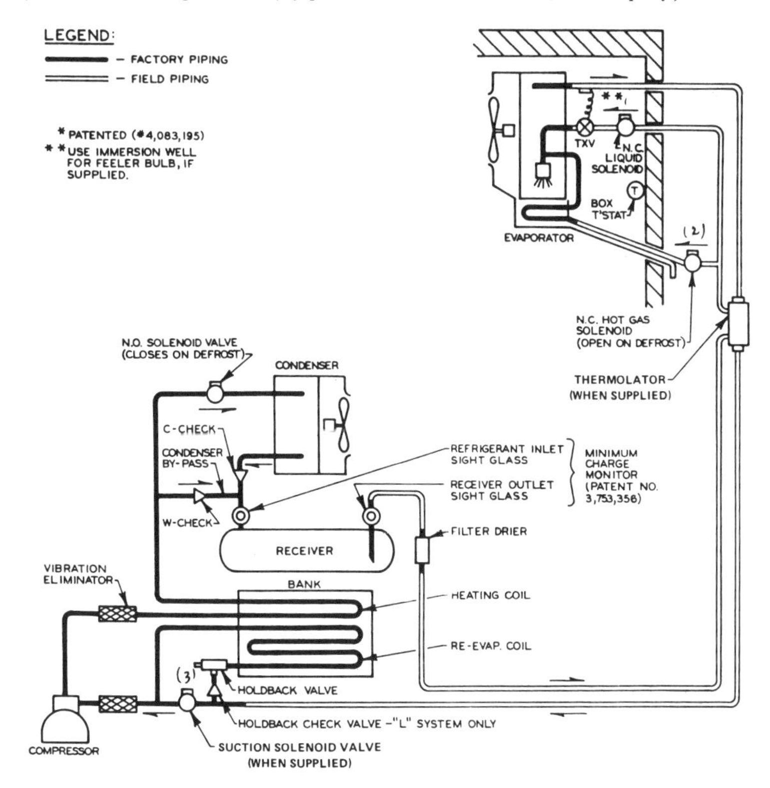

and to the receiver (C). The refrigerant then goes through the heat exchanger (D), to the thermal expansion valve (E), through the evaporator (F), and back to the compressor (R). The hot-gas line to the bank has a built-in bypass which prevents the water in the tank from overheating.

During the defrost cycle, the hot gas from the compressor (R) goes directly to the receiver (C), forcing the liquid in the receiver to enter the evaporator (F), followed by the hot gas from the compressor. The hot gas condenses in the evaporator and enters the bank (A). The bank becomes an evaporator, furnishing vapor to the suction of the compressor. The holdback valve (P) controls the flow of refrigerant and the pressure in the bank (A).

The defrost cycle is terminated by a pressure switch, causing the system to enter into the post-defrost period. Then the discharge solenoid opens and the check valve (M) closes, restoring the condenser in the discharge gas circuit. The hot-gas solenoid valve (J) is closed, stopping hot-gas flow into the evaporator. The suction solenoid (L) and the liquid solenoid (K) remain closed until there is liquid in the receiver and adequate pressure on the high side of the system. A pressure switch restores the system to normal operation and starts the evaporation fans (H).

Glycol Defrosting

The glycol arrangement requires special apparatus to heat the glycol, remove the excess water, and maintain the proper concentration of glycol in the solution.

HEAT EXCHANGERS FOR EVAPORATORS

Suction-to-liquid heat exchangers are installed to superheat the suction gas and subcool the liquid. They are useful in protecting the compressor and preventing the formation of flash gas in the liquid line. They should be used on all low-temperature systems. A heat exchanger often reduces lubrication oil problems. Heat exchangers are selected based on the temperature of the entering liquid, the suction gas temperature, the capacity of the system, and the type of refrigerant used.

INSTALLATION OF UNIT COOLERS

Unit coolers should be mounted level. Installation accessories usually provide means of sloping the drain pan to the drain outlet, based on a level mounting of the unit. Motors should be at least 12 in. from the wall. Piping should be properly sized. Heat exchangers should be used where necessary. A pump-down cycle is recommended, which requires a solenoid valve in the liquid line. The drain line should be the full size of the connection and sloped 4 in. to the foot. Heater cable, 5 ft to the running foot, should be installed around the drain line and insulated. Drain lines must be trapped.

SELECTING A LARGE PRODUCT COOLER

Note that the typical rating table (Figure 4–10) shows the capacity rating of a product cooler with various rows deep of the coil (four, six and eight fins per inch). Capacities are given in Btuh per 1°TD to assist in calculating the capacity for any required TD. Correction factors are given in Figures 4–11 and 4–12.

RU RUE RUH MODEL NO.	ROWS DEEP	CFM*	COOLER CAPACITY DATA - 60 AND 50 HERTZ ** BTUH/1°F TD — ELECTRIC DEFROST AND HOT GAS DEFROST: RUE/RUH(1) 3 FPI	RUE/RUH(1) 4 FPI	RUE/RUH(1) 6 FPI	OFF-CYCLE AMBIENT DEFROST: RU(2) 8 FPI
400-4	4		1100	1220	1410	1860
-6	6	3840	1540	1680	1900	2510
-8	8		1880	2040	2260	2980
600-4	4		1660	1830	2110	2790
-6	6	5760	2300	2520	2850	3760
-8	8		2830	3050	3390	4470
800-4	4		2210	2440	2820	3720
-6	6	7680	3070	3350	3800	5010
-8	8		3770	4070	4520	5960
1000-4	4		2760	3050	3520	4640
-6	6	9600	3840	4190	4750	6260
-8	8		4710	5090	5660	7450
1200-4	4		3450	3810	4400	5800
-6	6	12,000	4800	5240	5940	7830
-8	8		5890	6360	7070	9320
1500-4	4		4140	4570	5280	6960
-6	6	14,400	5760	6290	7130	9400
-8	8		7070	7630	8480	11180
1800-4	4		5180	5720	6600	8700
-6	6	18,000	7200	7860	8910	11750
-8	8		8840	9540	10610	13980
2200-4	4		6210	6860	7920	10440
-6	6	21,600	8640	9430	10690	14090
-8	8		10600	11450	12730	16770
3000-4	4		8280	9140	10560	13920
-6	6	28,800	11520	12580	14260	18790
-8	8		14140	15260	16970	22360

NOTES:

1. Capacity @ -10°F Evaporating, R502
2. Capacity @ 25°F Evaporating, R22 or R12
3. For elevations above 3000′ consult Harrisonburg Applications for unit capacities.

FPI = Fins Per Inch

TD = Temperature Difference, Entering Air Temp. Minus Refrigerant Evaporating Temp.

* = CFM rated @ face velocity of 600 fpm

** 50 Hertz models provide full 60 Hertz capacity and CFM

Figure 4–10 Product cooler capacity data. (By permission of Dunham-Bush, Inc.

CAPACITY CORRECTION FACTORS FOR VARIOUS EVAPORATING TEMPERATURES RUE & RUH ONLY							
SUCTION TEMP. °F:	20°	10°	0°	-10°	-20°	-30°	-40°
FACTOR	1.10	1.04	1.01	1.00	.96	.91	.85

Figure 4–11 Capacity data correction factors for various evaporating temperatures. (By permission of Dunham-Bush, Inc.

CAPACITY CORRECTION FACTORS FOR VARIOUS FACE VELOCITIES ALL RU/RUE/RUH				
ROW DEPTH DIRECTION OF AIR FLOW	COIL FACE VELOCITY, FPM			
	400	500	600	700
4	.82	.91	1.0	1.07
6	.80	.90	1.0	1.08
8	.79	.89	1.0	1.09

Base rating is at 600 FPM

Figure 4–12 Capacity data correction factors for various coil face velocities. (By permission of Dunham-Bush, Inc.)

Selection Procedure

To select a product cooler properly, the following information must be known:

1. Load requirements (Btuh)
2. Refrigerant evaporating temperature (°F) (note that the pressure drop in the suction line will affect the design suction pressure of the compressor but not the evaporator).
3. Box temperature (°F)
4. Refrigerant
5. Electrical power characteristics
6. Type of defrost
7. Fin spacing of evaporator
8. External static pressure (ESP) required

Since most large product coolers require a centrifugal fan, a centrifugal fan performance chart is supplied in Figure 4–13 to match the capacity data in Figure 4–10. The internal static pressure is determined from Figures 4–14 through 4–16. The total static pressure (TSP) is the sum of these three plus the external static pressure. The motor heat equivalent is given in Figure 4–17.

Selection Example

The data are as follows:

Refrigeration load	65,000 Btuh
Refrigerant evaporating temperature	−20°F
Desired space temperature	−10°F
Coil face velocity	500 fpm
External static	1.0 in. H_2O
Fin spacing	4 fpi
Electrical power	460 V, three-phase, 60 Hz
Type of defrost	Hot gas

MODEL HRU OR VRU	CFM	COIL FPM	CENTRIFUGAL FAN PERFORMANCE — TOTAL STATIC PRESSURE - INCHES, WATER GAUGE														
			¼" TSP		½" TSP		¾" TSP		1" TSP		1½" TSP		2" TSP		2½" TSP	
			RPM	BHP	RPM	BHP	RPM	BHP	RPM	BHP	RPM	BHP	RPM	BHP	RPM	BHP
400	2560	400	530	.4	630	.5	730	.7	830	.8	990	1.0	1150	1.4	1300	1.7
	3200	500	610	.7	700	.8	780	.9	860	1.0	1010	1.4	1150	1.7	1275	2.2
	3840	600	710	.9	780	1.1	840	1.3	910	1.5	1050	1.8	1170	2.3	1290	2.8
	4480	700	820	1.3	870	1.6	905	1.9	965	2.1	1095	2.3	1195	3.0	1310	3.6
600	3840	400	500	.5	610	.7	710	.9	810	1.2	1000	1.5	1190	2.2	1330	2.8
	4800	500	530	.8	630	1.0	730	1.2	820	1.4	990	1.8	1160	2.6	1310	3.4
	5760	600	550	1.1	650	1.3	760	1.5	840	1.8	1000	2.4	1140	3.0	1280	3.8
	6720	700	575	1.6	675	1.9	795	2.2	865	2.5	1010	3.1	1120	3.9	1250	4.9
800	5120	400	530	.8	630	1.0	730	1.4	830	1.6	990	2.0	1150	2.8	1300	3.4
	6400	500	610	1.4	700	1.6	780	1.8	860	2.0	1010	2.8	1150	3.4	1275	4.4
	7680	600	710	1.8	780	2.2	840	2.6	910	3.0	1050	3.6	1170	4.6	1290	5.6
	8960	700	820	2.7	870	3.3	905	3.9	965	4.2	1095	4.6	1195	6.0	1310	7.3
1000	6400	400	440	1.0	520	1.2	600	1.4	670	1.6	820	2.6	970	3.0	1090	4.0
	8000	500	470	1.5	550	1.7	630	2.0	695	2.6	820	3.2	950	4.0	1070	5.2
	9600	600	520	2.0	590	2.4	660	2.8	720	3.2	840	4.2	960	5.2	1070	6.2
	11200	700	575	3.0	635	3.6	695	4.2	750	4.5	865	5.3	970	6.8	1075	8.1
1200	8000	400	470	1.2	550	1.6	630	2.0	695	2.6	820	3.2	950	4.0	1070	5.2
	10000	500	540	2.2	605	2.6	670	3.2	735	3.6	850	4.6	960	5.6	1070	6.6
	12000	600	600	3.4	665	4.0	730	4.6	790	5.2	895	6.2	1000	7.8	1090	8.8
	14000	700	665	5.1	730	6.0	795	6.9	850	7.3	945	8.1	1045	10.1	1115	11.4
1500	9600	400	370	1.4	440	1.8	510	2.2	590	2.8	720	3.8	830	5.0	930	6.4
	12000	500	410	2.4	475	2.8	540	3.2	600	4.0	720	5.2	820	6.4	915	7.8
	14400	600	460	3.6	505	4.2	570	4.8	630	5.4	740	6.8	840	8.1	920	9.6
	16800	700	515	5.4	540	6.3	605	7.2	665	7.6	765	8.8	865	10.5	925	12.5
1800	12000	400	410	2.2	475	2.7	540	3.2	600	4.0	720	5.2	820	6.4	915	7.8
	15000	500	450	3.8	515	4.5	580	5.2	640	5.8	740	7.2	840	8.6	930	10.0
	18000	600	510	6.0	570	6.7	630	7.8	680	8.4	780	9.8	870	11.6	950	13.6
	21000	700	575	9.0	630	10.0	685	11.7	725	11.8	825	12.7	905	15.1	975	17.7
2200	14400	400	300	2.2	380	2.8	460	3.4	510	4.2	610	5.4	705	7.0	800	9.0
	18000	500	380	4.0	430	4.6	480	5.2	530	5.8	630	7.6	720	9.6	790	11.2
	21600	600	410	6.1	470	6.8	530	7.8	570	8.8	650	10.6	730	12.8	800	14.4
	25200	700	445	9.1	515	10.2	585	11.7	615	12.3	675	13.8	740	16.6	810	18.7
3000	19200	400	320	3.6	380	4.4	440	5.2	480	6.0	580	8.2	660	10.2	750	13.0
	24000	500	375	6.8	425	7.6	470	8.4	520	9.8	595	11.6	675	14.0	750	16.8
	28800	600	430	10.2	475	11.6	520	13.0	550	14.0	630	17.0	690	19.2	760	22.2
	33600	700	440	15.1	530	16.4	575	19.5	585	19.6	670	22.1	705	25.0	770	28.9

NOTE: All RU/RUE/RUH units are belt drive requiring one motor.

Figure 4-13 Product cooler centrifugal fan performance. (By permission of Dunham-Bush Inc.)

AIRSIDE PRESSURE DROP* - COILS ONLY

ROWS DEEP	TYPE OF COIL	COIL FACE VELOCITY															
		400 FPM				500 FPM				600 FPM				700 FPM			
		FIN SPACING - FPI															
		3	4	6	8	3	4	6	8	3	4	6	8	3	4	6	8
4	DX	.055	.088	.133	.182	.073	.118	.178	.242	.092	.148	.232	.302	.133	.197	.302	.382
6	DX	.077	.126	.189	.259	.104	.169	.254	.345	.132	.210	.330	.432	.189	.281	.431	.546
8	DX	.098	.161	.243	.333	.133	.216	.325	.441	.169	.269	.422	.553	.242	.360	.552	.698
1	HEAT	–	–	.038	–	–	–	.051	–	–	–	.066	–	–	–	.086	–
2	HEAT	–	–	.057	–	–	–	.076	–	–	–	.099	–	–	–	.129	–

* Pressure drop measured in inches of water.

NOTE: For wet or medium frosted D-X coils, multiply pressure drop by 1.25.

Figure 4-14 Product cooler air-side pressure drop for coils. (By permission of Dunham-Bush, Inc.)

AIRSIDE PRESSURE DROP* - CABINET ONLY

CABINET	COIL FACE VELOCITY			
	400 FPM	500 FPM	600 FPM	700 FPM
HRU, HRUE, HRUH - NO COIL	.07	.10	.15	.20
VRU, VRUE, VRUH - NO COIL	.10	.15	.22	.30

* Pressure drop measured in inches of water.

Figure 4-15 Product cooler air-side pressure drop for cabinets. (By permission of Dunham-Bush, Inc.)

Model HRU or VRU	Face Velocity	Outlet Velocity
400	400	1474
	500	1842
	600	2211
	700	2580
600	400	1106
	500	1382
	600	1658
	700	1935
800	400	1474
	500	1842
	600	2211
	700	2580
1000	400	1244
	500	1555
	600	1866
	700	2177
1200	400	1555
	500	1944
	600	2333
	700	2721
1500	400	1256
	500	1570
	600	1884
	700	2198
1800	400	1570
	500	1962
	600	2354
	700	2748
2200	400	1467
	500	1834
	600	2201
	700	2567
3000	400	1571
	500	1964
	600	2357
	700	2749

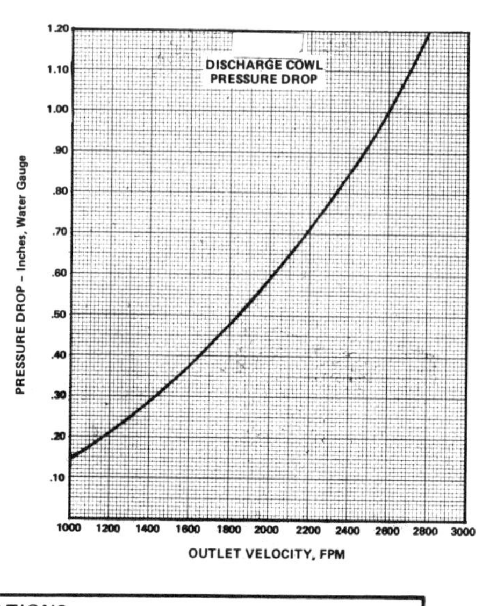

RU, RUE, RUH SPECIFICATIONS

RU RUE RUH MODEL NO.	ROWS DEEP	CFM*	FACE AREA SQ. FT.	INTERNAL VOLUME CU. FT.	NO. OF FANS	RUE MODELS ELECTRIC DEFROST 230/3/60 kW	AMPS	APPROXIMATE** SHIPPING WEIGHTS, LBS. HORIZONTAL MODELS HRU	HRUE HRUH	VERTICAL MODELS VRU	VRUE VRUH
400-4	4	3840	6.4	.31	1	7.0	16	575	585	630	640
-6	6			.47		8.5	19	615	625	670	680
-8	8			.62		12.5	29	655	665	710	720
600-4	4	5760	9.6	.47	2	9.2	21	790	810	860	880
-6	6			.71		11.4	26	855	875	925	945
-8	8			.94		16.2	37	920	940	990	1010
800-4	4	7680	12.8	.63	2	13.6	31	940	970	1020	1050
-6	6			.95		16.4	37	1015	1045	1095	1125
-8	8			1.26		24.4	56	1095	1125	1175	1205
1000-4	4	9600	16.0	.78	2	16.0	36	985	1025	1100	1140
-6	6			1.17		22.2	51	1075	1115	1190	1230
-8	8			1.56		22.2	51	1170	1210	1285	1325
1200-4	4	12000	20.0	.98	2	22.2	51	1380	1430	1510	1560
-6	6			1.47		25.0	58	1500	1550	1630	1680
-8	8			1.96		25.0	58	1625	1675	1755	1805
1500-4	4	14,400	24.0	1.18	2	23.6	55	1610	1670	1875	1935
-6	6			1.77		29.4	68	1750	1810	2015	2075
-8	8			2.36		34.2	79	1895	1955	2160	2220
1800-4	4	18000	30.0	1.47	2	28.4	65	1890	1960	2215	2285
-6	6			2.21		34.2	79	2060	2130	2385	2455
-8	8			2.94		39.0	89	2230	2300	2555	2625
2200-4	4	21600	36.0	1.76	2	33.2	76	2290	2370	2675	2755
-6	6			2.64		42.8	99	2500	2580	2885	2965
-8	8			3.52		48.4	112	2715	2795	3100	3180
3000-4	4	28800	48.0	2.35	2	44.4	109	2930	3030	3430	3530
-6	6			3.53		55.4	128	3200	3300	3700	3800
-8	8			4.70		58.6	136	3475	3575	3975	4075

NOTES: * = CFM Rated @ Face velocity of 600 FPM
** = Approximate shipping weight does not include fan motors.

MATERIALS:
Case and drain pan 16 gauge
Blower end panels: 14 gauge thru Model 1200
11 gauge Model 1500 thru 3000

Figure 4-16 Product cooler air-side pressure drop for cowl. (By permission of Dunham-Bush, Inc.)

MOTOR HEAT EQUIVALENT

FAN MOTOR HORSEPOWER	BTUH INTO REFRIGERATED SPACE
½ HP	1,850
¾ HP	2,700
1 HP	3,500
1½ HP	5,100
2 HP	6,600
3 HP	9,600
5 HP	15,500
7½ HP	22,875
10 HP	30,000
15 HP	45,000
20 HP	60,000
25 HP	75,000

Figure 4–17 Product cooler, motor heat equivalent. (By permission of Dunham-Bush, Inc.)

The procedure is as follows:

1. Determine the temperature difference:

$$-10\,°\mathrm{F} \text{ minus } -20\,°\mathrm{F} = 10\,°\mathrm{TD}$$

2. Select a cooler from the capacity table. For RUH 2200-8 with 4 fpi producing a capacity of 11,450 Btuh per 1°F and 28,400 cfm:

$$\text{capacity} = \frac{11{,}450 \text{ Btuh}}{1\,°\mathrm{F}} \times 10\,°\mathrm{F TD} = 114{,}500 \text{ Btuh}$$

3. Adjust the cooler capacity for −20°F (Figure 4–11) and for a velocity of 500 fpm (Figure 4–12):

$$\text{adjusted capacity} = 114{,}500 \times 0.89 = 97{,}829 \text{ Btuh}$$

4. Determine the interior static pressure:

Coil (Figure 4–14)	0.441
Cabinet (Figure 4–15)	0.15
Cowl (Figure 4–16)	0.500
	1.091 in. H_2O

5. Determine the total static pressure:

$$1.091 \text{ in.} + 1.0 \text{ in.} = 2.091 \text{ in.}$$

6. Determine the motor size (Figure 4–13):

10 hp

7. Adjusted net refrigerating capacity:

Cooler capacity	=	97,829 Btuh
10-hp motor (Figure 4–17)	=	− 30,000 Btuh
Adjusted cooler capacity	=	67,829 Btuh
Less 4% safety factor	=	65,000 Btuh

SELECTING A SMALL UNIT COOLER

Evaporator ratings are supplied by manufacturers giving the Btuh value at a certain temperature difference (TD). The TD is defined as the difference between the temperature of the air entering the coil and the refrigerant-saturated suction temperature. For example, if a walk-in box carries a temperature of 40 °F and the evaporating temperature of the refrigerant is 30 °F, the TD is 10 °F.

It should be noted that the temperature difference between the evaporating temperature and the box temperature (TD) has a great influence on the capacity of a coil. The capacity of a coil can be doubled by doubling the TD. The limiting factor, however, is the condition of the product. As the TD increases the relative humidity in the box decreases. Most fresh produce requires high humidities. Therefore, a large number of evaporators are required to operate at a low TD.

MATCHING COILS AND COMPRESSORS

Usually, evaporator coils and compressors are selected on the basis of load calculation. In actual performance the load will probably be somewhat different than calculated. If the load is reduced on the evaporator, the suction pressure drops and the compressor

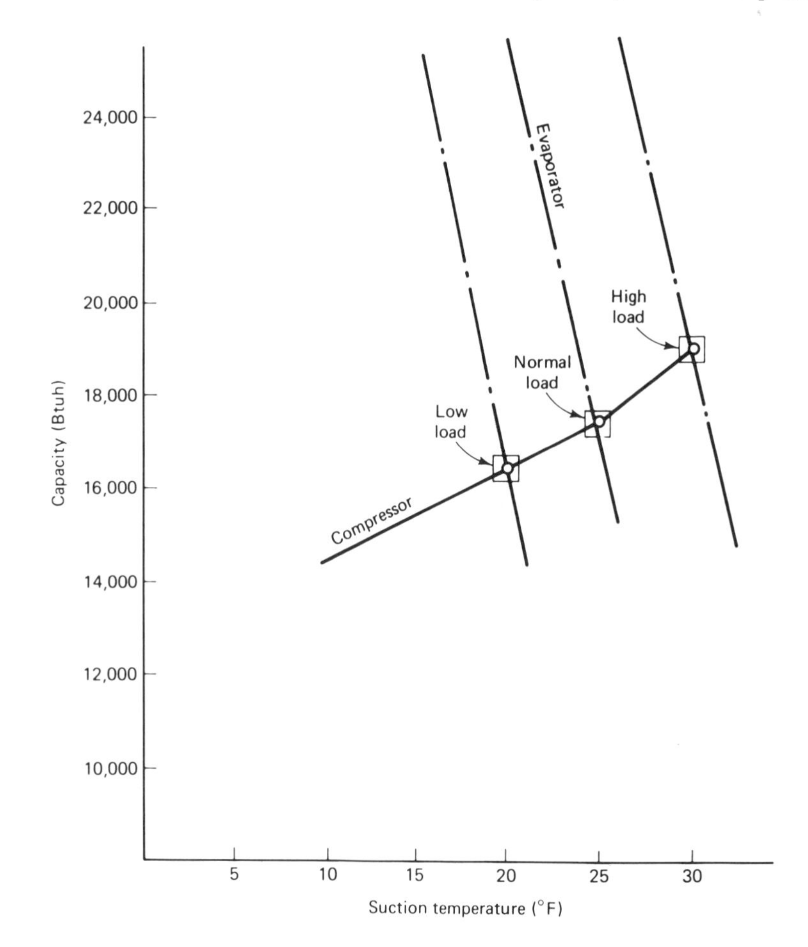

Figure 4-18 Effect of load changes on the combined performance of the compressor and evaporator.

capacity decreases. Figure 4–18 shows the combined performance of the two pieces of equipment. When the load drops off, the TD increases, thus increasing the amount of dehumidification. Limits need to be established to cut off the equipment when excessive dehumidification occurs.

The same problem in performance may be experienced if the equipment selected is too large for the job. To prevent short cycling it may be necessary to change the coil or compressor or both in order to maintain desired conditions. This is particularly important in box temperatures in the range 35 to 40 °F.

EVAPORATOR PERFORMANCE

One way to increase the capacity of an evaporator, exclusive of changing the TD (temperature difference between room and refrigerant), is to increase the air velocities. There are, of course, limitations of pressure drop, noise, and the effect on the product using higher velocities of air. The following information is a guide to the normal performance of above-freezing and below-freezing evaporators.

For a 1-ton-capacity commercial refrigeration coil, operating at 15 °F temperature difference between room temperature and refrigerant temperature, the fan should deliver about 1400 cfm per ton of cooling. On this basis, the temperature difference between air leaving the coil and the refrigerant temperature should be 7 or 8 °F. These conditions would represent the required performance of a unit cooler operating in a 35 °F room.

For a freezer application where the design is based on 10 °F TD, the fan should deliver about 2000 cfm/ton and produce an air temperature leaving the coil of 4 to 5 °F higher than the refrigerant temperature.

PROPER TEMPERATURE DIFFERENCE

There are many factors that affect the relative humidity in a refrigerated space. Probably most important is the TD. The following is a guide to the TD that should be provided for various desired relative humidities.

Condition and product	Approximate relative humidity (%)	Recommended TD °F
Very high: butter, fresh meat, flowers	90	8 to 12
High: fresh fruits, vegetables, fish, general food storage	80 to 85	12 to 16
Medium: dried meats, dried fruits, dried vegetables, beer, milk	75	16 to 22

For low-temperature storage at +10 °F or lower, a 10 °F TD is usually considered satisfactory, mainly to facilitate defrosting.

FIN SPACING

The temperature level is more important than the transfer rate in determining fin spacing. For freezers, the fins should not be closer than four fins per inch to permit defrosting in a reasonable time. For coils in boxes with a medium-temperature range (35 to 40 °F), the fins should not be closer than six fins per inch to permit air defrost.

LIQUID CHILLING EVAPORATORS

Classified according to construction and configuration, there are a number of liquid chiller designs: (1) shell-and tube, (2) shell-and coil, (3) tank types, (4) double-pipes, and (5) Baudelot. The construction of chillers has much in common with that of liquid-cooled condensers. The same configuration can be used for both. They differ in the method of handling the flow of the fluids and, of course, in their position in the refrigeration system.

Refrigerant is distributed in liquid chillers by the dry expansion or the flooded method. Both chillers in Figures 4–19 and 4–20 are flooded chillers. The two chillers shown in Figure 4–21 are dry expansion chillers.

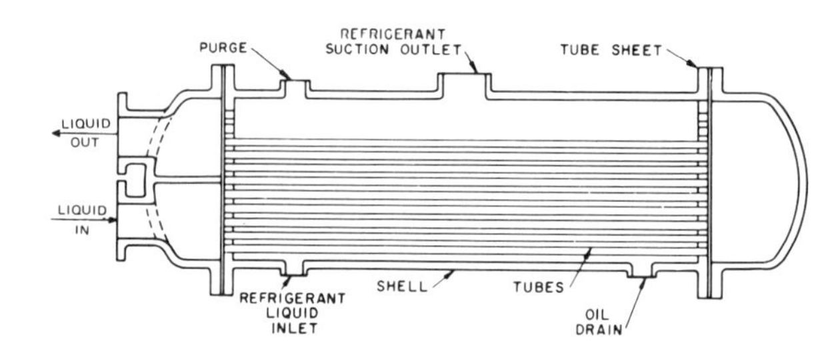

Figure 4–19 Flooded shell-and-tube liquid chiller. Note space provided above tubes for vapor refrigerant. Most chillers use some type of level control (LSF) to regulate the height of the liquid refrigerant in the shell. (By permission of *ASHRAE Handbook.*)

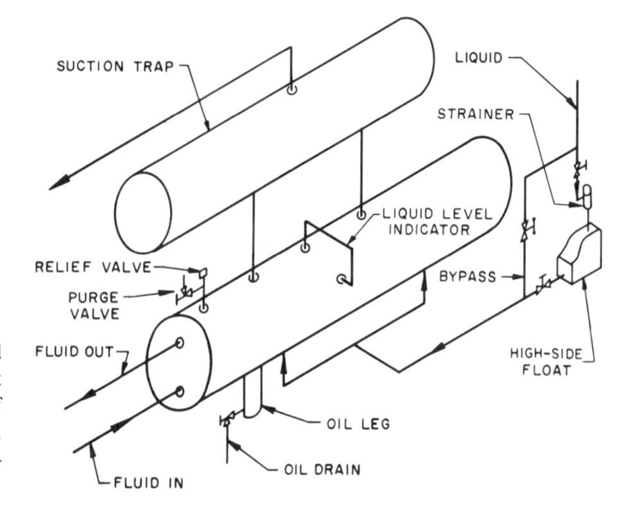

Figure 4–20 Arrangement with horizontal ammonia liquid chiller and high-side float using separate suction trap. The amount of ammonia in this type of system is critical. Freeze protection must be provided. (By permission of *ASHRAE Handbook*.)

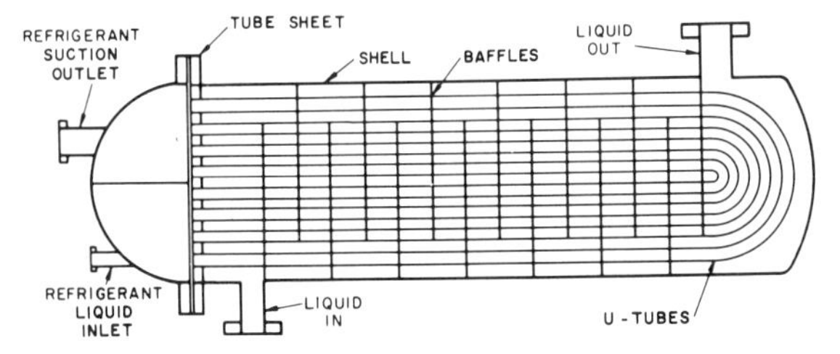

(a) U-tube design.

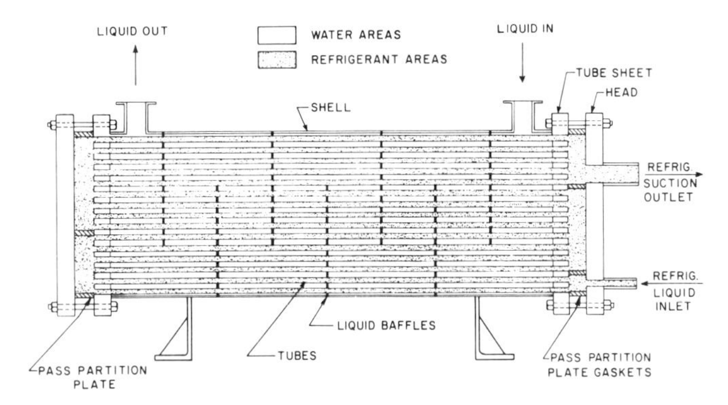

(b) Straight tube design.

Figure 4–21(a) and (b) Direct expansion chillers using thermal expansion valve. This type of chiller gives better protection against freeze up since liquid is in the shell. Will stand some freezing without damage. (By permission of *ASHRAE Handbook*.)

The *dry expansion* type uses a thermal expansion valve for a metering device and the evaporator is filled with vaporizing liquid refrigerant. Due to the fact that much of the coil surface is in contact with vapor, it is termed dry expansion.

In a *flooded* system the tubes of the chiller are immersed in liquid refrigerant. A space is left at the top of the evaporator for vapor. In Figure 4–19 the vapor space is provided in the chiller shell. In Figure 4–20 the vapor space is provided in a separate vessel connected to the top of the chiller, called a *suction trap.*

Shell-and-Tube Chillers

The chillers shown in Figures 4–19 and 4–21 are shell-and-tube chillers. Since the chiller shown in Figure 4–19 is a flooded chiller, the refrigerant is in the shell. The dry expansion chillers in Figure 4–21 have the refrigerant in the tubes.

Shell and tube chillers are constructed with a tube sheet holding the open ends of the tubes and a manifold space to provide access to the tube openings. In the design of the dry expansion chiller, baffles are placed in the shell to increase the transfer rate of the liquid being cooled. The gpm flow of the liquid and the spacing of the baffles affect the liquid-side pressure drop and thus the size of the liquid pump required.

Shell-and Coil Chillers

A shell-and-coil chiller is shown in Figure 4–22. The chiller tubing coil is placed inside a shell. Since, in Figure 4–22, the refrigerant is in the tubes, this is a dry expansion cooler.

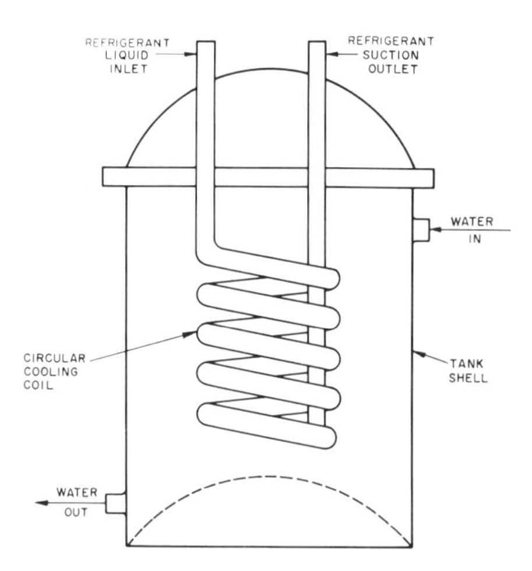

Figure 4–22 Shell and coil chiller. Used for small equipment such as water coolers. Has the advantage of storage capacity to meet peak conditions. (By permission of *ASHRAE Handbook.*)

This type of cooler is limited to small sizes since the refrigerant is evaporated in one continuous tube and the pressure drop must be maintained (on the refrigerant side) within reasonable limits. The unit is compact and is usable on chillers constructed for cooling drinking water, water for photographic laboratories, and water for certain types of bakery units.

Tank Chillers

A tank-type chiller is illustrated in Figure 4–23. This is a flooded chiller using a low-side float to maintain the liquid level in the surge drum. This unit has a heavy concentration of refrigerant tubes, to produce rapid cooling. Some type of liquid agitation should be provided to increase the transfer rate. Coolers of this type were used for many years in making ice.

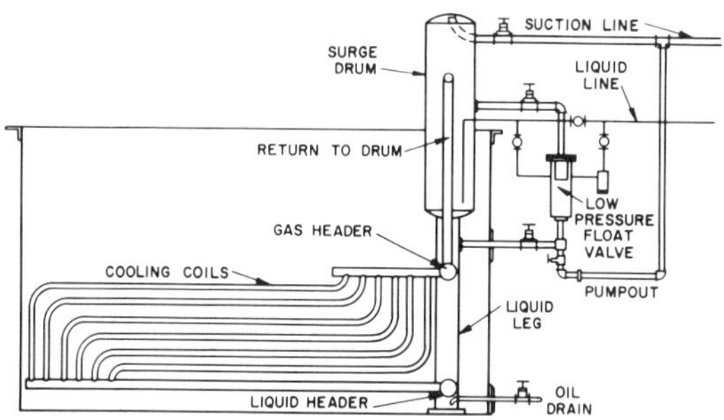

Figure 4–23 Tank type chiller. Wide usage for large systems where sanitation is not critical. The use of an agitator improves efficiency. Typical application is in ice-making. (By permission of *ASHRAE Handbook.*)

Double-Pipe Chillers

Double-pipe chillers are constructed similar to double-tube condensers, using a tube-within-tube arrangement. The flow of refrigerant and liquid being cooled follow counterflow principles. The most common practice is to use the inner tube for the liquid being cooled, since many of these coolers are constructed to permit cleaning the inner tube by removing convenient headers.

Baudelot Chillers

For many years these coolers have been used by dairy processors to cool milk quickly as the final stage of pasteurizing. The milk drops down over the cold surface in thin sheets and quickly approaches the cooler surface temperatures. A diagram of a Baudelot cooler is shown in Figure 4–24. By carefully controlling the temperature of the cooler, the liquid being cooled can be lowered quickly to near-freezing temperatures.

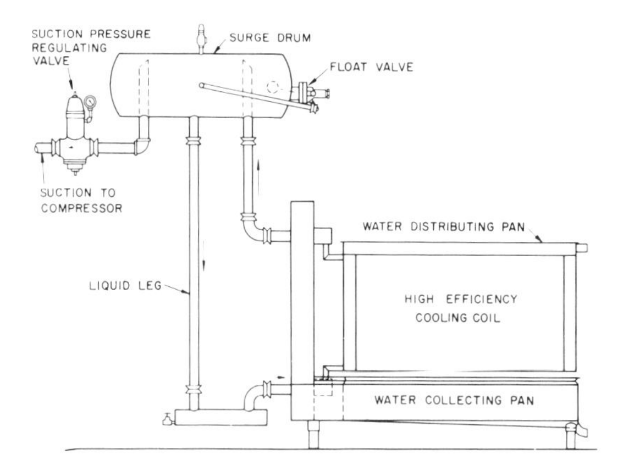

Figure 4–24 Baudelot cooler is used to cool liquid near its freezing point. This ammonia system uses a LPF type metering device and a flooded system. (By permission of *ASHRAE Handbook.*)

SECONDARY REFRIGERANTS

A secondary refrigerant is a liquid of some type that is cooled by the primary refrigerant and circulated by a pump to cool the product. Secondary refrigerants include water, brine and glycol solutions, methanol, and glycerin. One type of brine is made using calcium chloride or sodium chloride salts in a water solution. These salts are corrosive under certain conditions and are therefore limited in use. The trend in the use of secondary refrigerant for below-freezing temperatures (of water) is to use glycol. Ethylene and propylene glycols are most common and they are noncorrosive.

The lowest temperature that can be reached with a calcium chloride brine is $-67\,^\circ$F, with about 30% of the salt in the solution, by weight. Sodium chloride brine can only be used down to $-6\,^\circ$F, using about 23% salt by weight in solution.

Glycol solutions have the added advantage of not evaporating under normal operating conditions and are nonelectrolytic. Thus they can be used where dissimilar metals are present. Many skating rinks use an antifreeze glycol solution as the secondary refrigerant.

SELECTING A LIQUID COOLER

The following information is needed to select a liquid cooler to cool water:

1. Water inlet temperature
2. Water outlet temperature

3. Refrigerant suction temperature
4. Tonnage capacity or gpm rate of flow

The relationship of tonnage and gpm are as follows:

$$\text{tons} = \frac{\text{gpm} \times \text{range (°F)}}{24}$$

$$\text{gpm} = \frac{\text{tons} \times 24}{\text{range (°F)}}$$

For example, if it is desired to cool 40 gpm of water from 55 °F to 45 °F, the capacity is calculated as follows:

$$\text{capacity} = \frac{40 \text{ gpm} \times 10}{24} = 16.7 \text{ tons}$$

The required chiller can then be selected from the manufacturers' data.

An important item is the water pressure drop through the chiller. This is determined from manufacturers' data for the flow required.

Selecting a Cooler for a Liquid Other Than Water

In addition to the items needed for water, the following information is required:

5. Maximum allowable liquid pressure drop
6. Refrigerant type
7. Name of fluid to be cooled and the following characteristics:
   a. Specific heat
   b. Specific gravity
   c. Viscosity
   d. Thermal conductivity

The manufacturer's assistance is needed in the selection of special liquid chillers.

REVIEW EXERCISE

Select the letter representing your choice of the correct answer (there may be more than one).

4–1. A flooded evaporator is filled almost entirely with liquid refrigerant.

_______ a. True

_______ b. False

4–2. Which type of fin spacing is desirable for defrosting evaporators?

_______ a. 2 to 4 fins per inch

_______ b. 4 to 8 fins per inch

_______ c. 8 to 14 fins per inch

_______ d. 14 to 30 fins per inch

4–3. The capacity ratings in Btuh for evaporators is usually given by the manufacturer at a temperature difference (between suction temperature and box temperature) of:

_______ a. 5°F

_______ b. 10°F

_______ c. 15°F

_______ d. 20°F

4–4. What control stops the compressor on a system that uses pump-down?

_______ a. Room thermostat

_______ b. Automatic timer

_______ c. Low-pressure control (LPC)

_______ d. High-pressure control (HPC)

4–5. What unique feature is incorporated into the Thermobank defrost system?

_______ a. A heat storage bank

_______ b. A special thermostat

_______ c. Dual expansion valves

_______ d. A larger-capacity evaporator

4–6. In a reverse-cycle defrost system, how many thermal expansion valves are required on a single-evaporator system?

_______ a. None

_______ b. One

_______ c. Two

_______ d. Three

4–7. In a flooded liquid chiller, is the fluid being cooled in the shell or in the tubes?

_______ a. In the shell

_______ b. In the tubes

4–8. A Baudelot-type liquid cooler would be suitable for cooling a liquid where aeration is a factor, such as in cooling:

_______ a. Milk

_______ b. Wine

_______ c. Wort

_______ d. Water for carbonization

4-9. On an air-defrost system in a reach-in refrigerator, what control initiates the defrost cycle?

_________ a. High-pressure control

_________ b. Time clock

_________ c. Low-pressure control

_________ d. Thermostat

4-10. On a rise in suction temperature, does the capacity of the evaporator increase or decrease?

_________ a. Increase

_________ b. Decrease

5 Condensers and Water Economizers

OBJECTIVES

After studying this chapter, the reader will be able to:

- Identify the various types of condensers and water economizers.
- Select the type of condensers and water economizers that best fit the application.
- Apply low-ambient controls.

HEAT TRANSFER

A condenser is an essential element in the refrigeration cycle, having the special function of condensing the refrigerant delivered by the compressor. Some medium, usually air or water, at a temperature below the discharge temperature needs to be available to absorb the dissipated heat. The condenser, therefore, is a heat transfer device.

With reference to the refrigeration cycle, the condenser must remove not only the heat absorbed by the evaporator but also the heat added by the compressor, known as *heat of compression*. The heat of compression is greater per ton of refrigeration as the suction temperature is reduced, as shown in Figure 5-1. The total heat rejection capability of the condenser is reduced as the altitude increases and therefore must be taken into consideration at altitudes above 2000 ft.

SELECTION BASED ON TOTAL HEAT REJECTION

In selecting a condenser the total heat rejection must be determined. For example, a semihermetic compressor operating at 10 °F suction and 115 °F condensing temperature

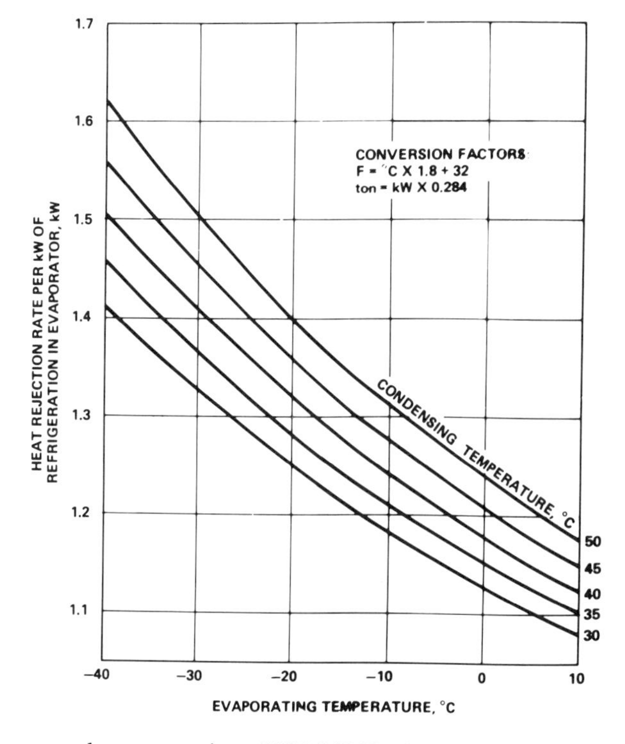

Figure 5-1 Heat removed in R-12 condenser. Curve shows the increase in condenser heat rejection per ton of refrigeration as the evaporating temperature is lowered. Data are given for various condensing temperatures. (By permission of 1983 ASHRAE Handbook.)

has a capacity of 280 MBH (thousands of Btu per hour) with 34.9 kW input. What is the total heat of rejection?

The heat of compression of a semihermetic compressor is 3075 Btu/kW and for an open compressor, 2295 Btu/kW. Since the compressor is semihermetic, the heat of compression is estimated to be

$$34.9 \text{ kW} \times 3075 \text{ Btu/kW} = 107{,}317 \text{ Btuh}$$

The total heat of rejection is, therefore

$$280{,}000 \text{ Btuh} + 107{,}317 \text{ Btuh} = 387 \text{ MBH}$$

There are three principal types of condensers:

1. Air-cooled
2. Water-cooled
3. Evaporative

AIR-COOLED CONDENSERS

The air-cooled condenser can be either natural draft or forced air. By far the most common is the fan or blower type. The air-cooled condenser can be part of the condensing unit package (Figure 5-2) or a separate unit (Figure 5-3). The air-cooled condenser is preferred on many installations due to its design simplicity. However, air-cooled condensing is usually less efficient than water-cooled condensing. It is common practice to size air-cooled con-

Figure 5–2 Remote air-cooled condensing unit. Package includes compressor, air-cooled condenser, liquid receiver, and control panel. (By permission of Dunham-Bush, Inc.)

(a) (b)

Figure 5–3 (a) Air-cooled condenser, commercial model. These condensers can be used with R-12, R-22, or R-502 refrigerants. The capacity range is from approximately 5 to 150 tons. (b) Air-cooled condensers, industrial models. The range of sizes is from approximately 10 to 300 tons. Units are available with subcooling coils and low-ambient control. (By permission of Dunham-Bush, Inc.)

densers to provide a condensing temperature 15 to 30 °F higher than design ambient. Thus, for a 95 °F outside temperature with a 20 °F TD, the condensing temperature would be 115 °F.

For most efficient operation a low TD is recommended to provide as low a condensing temperature as possible.

Air-cooled condensers can be supplied with multiple circuits when used for a number of individual compressor systems. Vertical models usually require wind deflectors.

Remote Air-Cooled Condensers

Remote air-cooled condensers are used for systems that are assembled on the job and usually placed outside on the ground or roof. The diagrams in Figure 5–4 show the recommended piping and accessories used for a single condenser with one or more circuits.

TYPICAL SYSTEM PIPING DETAILS

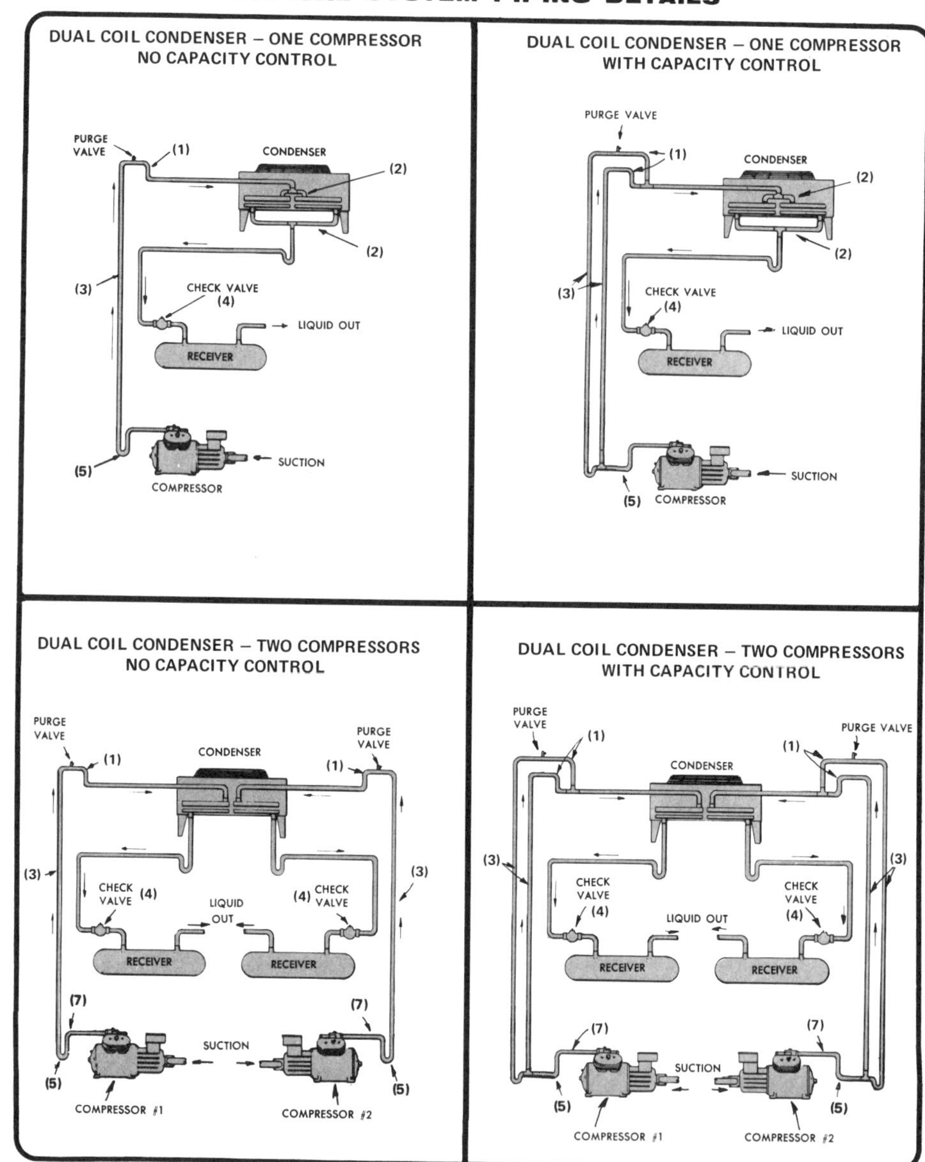

NOTES:

(1) "Over Traps" on top of risers must not be less than 6 inches.

(2) Refer to Page 24 for Optional Connection Manifold sizes (if applicable).

(3) When vertical lift exceeds 20 feet, insert close-coupled traps in riser at every 10 feet. See Page 23 for trap and other piping details.

(4) Receiver check valves are not necessary unless receiver is heated or in an ambient warmer than the condenser ambient.

(5) Trap at compressor should be a minimum of 18 inches.

(6) Head pressure control valves or manifold may be installed in equipment room and connected to hot gas and drain lines.

(7) If multiple compressors are utilized with a single evaporator, a discharge check valve must be utilized.

Figure 5–4 Typical piping diagrams for air-cooled condensers.

TYPICAL PIPING DETAILS

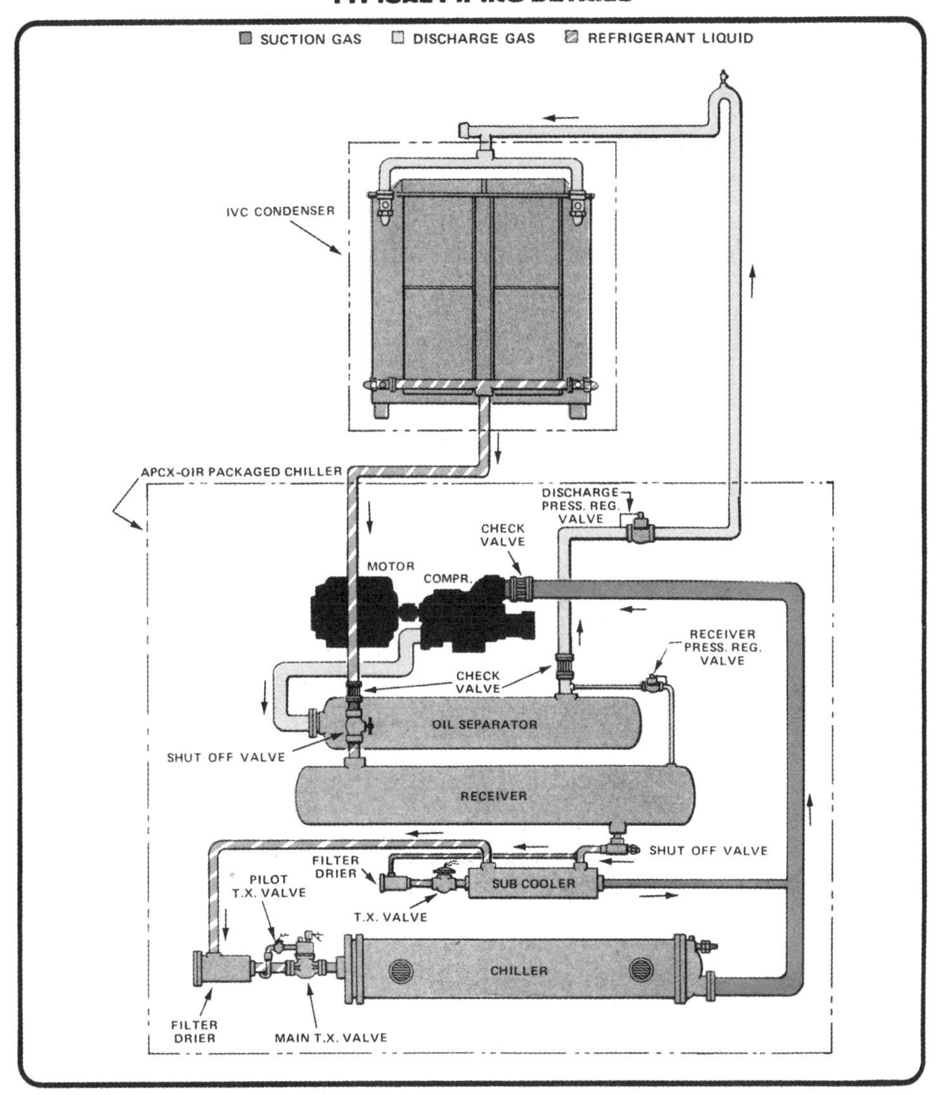

NOTES:

1. The DBX compressor and receiver units are shown as typical only. Reciprocating compressors and compressor units can be applied in an identical manner with the addition of an 18" (457 mm) discharge line trap at the compressor.
2. All piping and valves are provided by others.
3. Remote pumpdown receivers must be shielded from the sun or other heat sources.
4. Pipe sizing shall be done in accordance with the latest ASHRAE and/or ARI standards. Discharge risers must be sized so that gas velocities will be sufficient to properly entrain oil.
5. Diagrams are typical only. Consult D/B West Hartford Systems Application Department for further details.
6. Optional sealpot receiver can be mounted within the cabinet of the IVC condenser and factory piped between the condenser and subcooling coils. It is not to be used as a pumpdown receiver.
7. Consult D/B West Hartford Systems Application Department for applications requiring flooding type head pressure control piping arrangement.

Figure 5–5 Typical piping diagram for air-cooled condenser with chiller package.

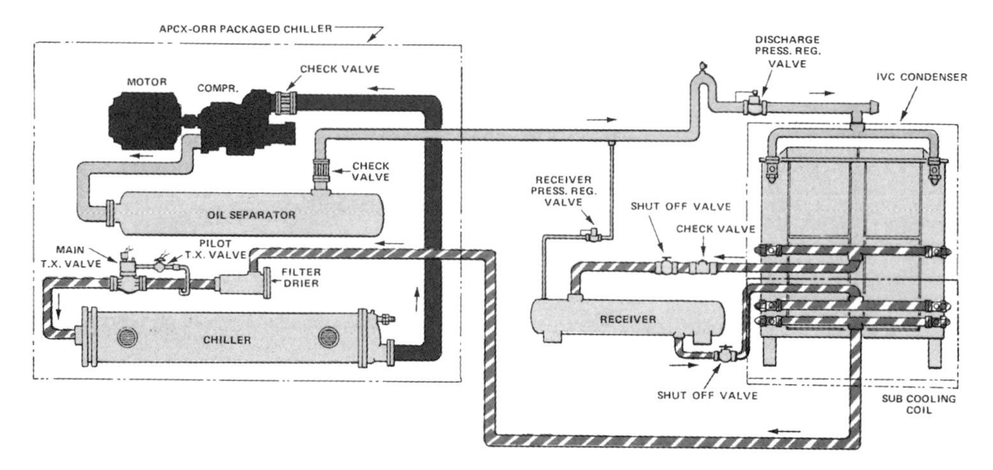

Figure 5-6 Typical piping diagram for air-cooled condenser with subcooler.

Another diagram of the piping for an air-cooled condenser is shown in Figure 5-5. This system uses an industrial-type condenser (Figure 5-3b) applied to a radial rotary compressor/chiller package. Note the use of an oil separator, discharge check valve, and separate subcooler. Figure 5-6 shows the subcooler as part of the air-cooled condenser.

Head Pressure Control

When air-cooled condensers are operated under low-ambient conditions, some provision needs to be made to control the head pressure. As the ambient temperature drops, the pressure difference between the high and low sides of the system drops, thus decreasing the flow through the expansion valve. To maintain a reasonable head pressure, the following types of control can be used:

1. Fan cycling
2. Use of dampers to control the airflow through the condenser
3. Fan speed control used on direct-drive fan motors
4. Condenser coil flooding arrangement

Both fan cycling and damper controls vary in effectiveness depending on the loading of the equipment and the temperature difference. Generally, they are only effective down to 25 or 30 °F ambient temperatures. Variable fan speeds have a greater effective range down to as low as 0 °F ambient. The coil flooding arrangement is suitable down to ambient temperatures of −20 °F.

Controls needed for the coil flooding system are shown in Figures 5-7 (adjustable) and 5-8 (nonadjustable). When low-ambient conditions make it difficult to start a system, the installation of an arrangement shown in Figure 5-9 may be suitable.

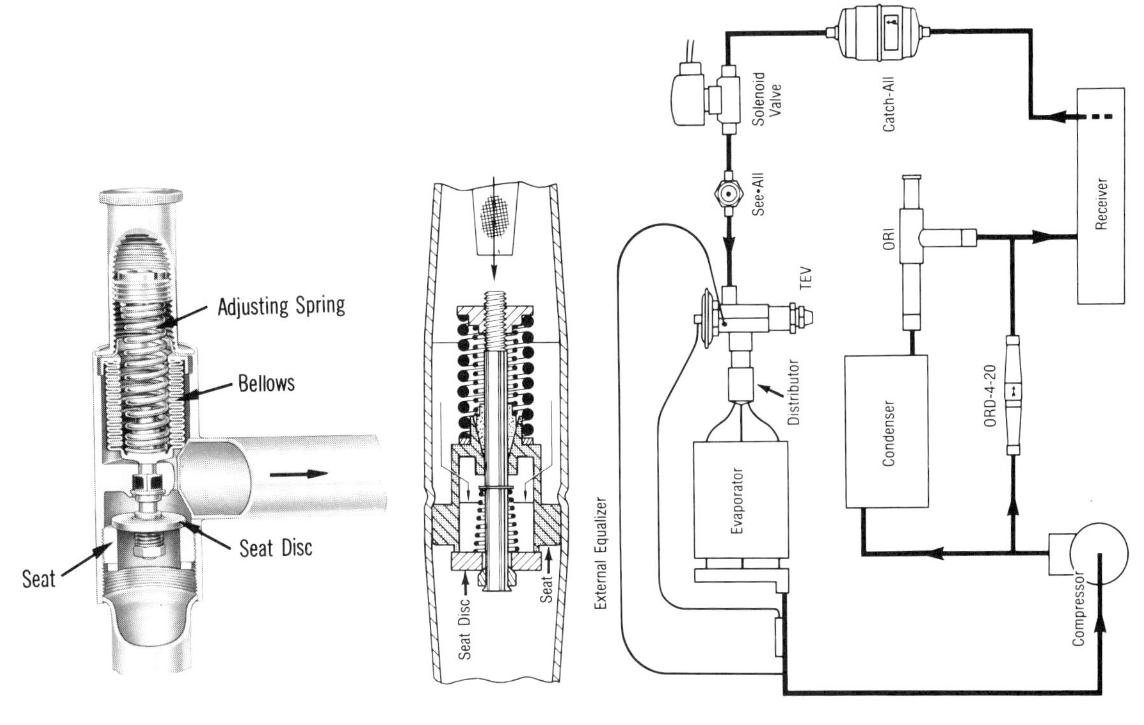

Figure 5–7 The two devices shown on the left view are arranged as indicated in the drawing on the right. The pressure regulating valve (ORI) is sensitive to inlet pressure and on a drop in pressure moves toward the closed position. This causes refrigerant liquid to build up in the condenser and inactivate parts of the condenser surface. The ORD valve is used to maintain receiver pressure. (By permission of Sporlan Valve Company.)

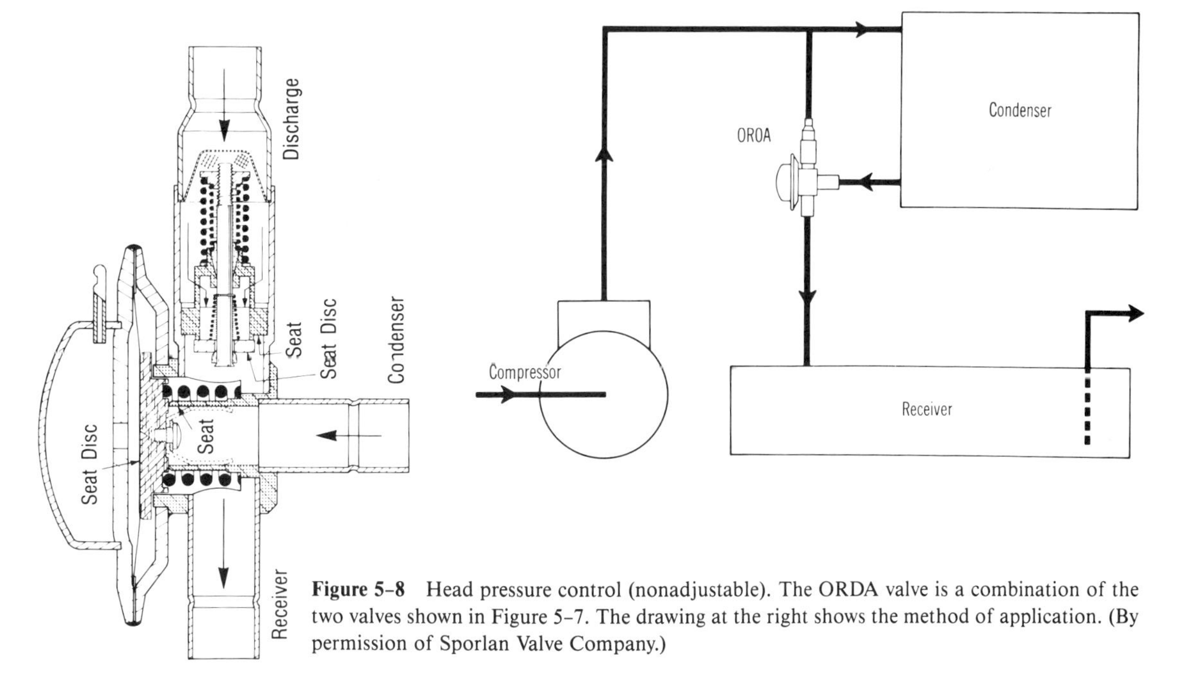

Figure 5–8 Head pressure control (nonadjustable). The ORDA valve is a combination of the two valves shown in Figure 5–7. The drawing at the right shows the method of application. (By permission of Sporlan Valve Company.)

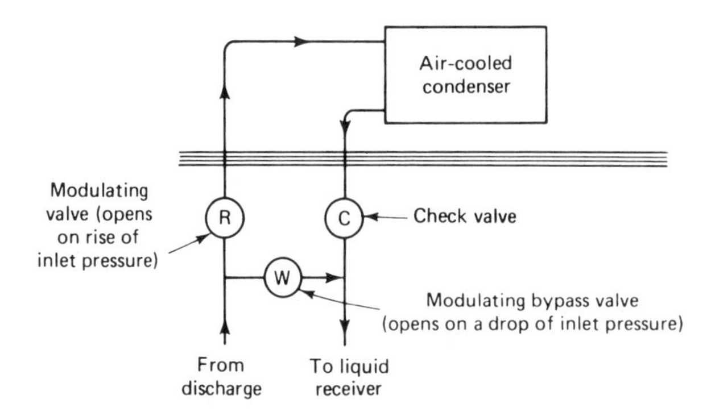

Figure 5-9 Arrangement for starting system under low-ambient conditions. When the compressor starts, valve W is open and valve R is closed, pressurizing the liquid refrigerant in the receiver. When normal pressures are reached, valve R opens and valve W modulates to closed position.

WATER-COOLED CONDENSERS

Water-cooled condensers are classified into the following groups:

1. Shell-and-tube
2. Shell-and-coil
3. Double tube
4. Atmospheric

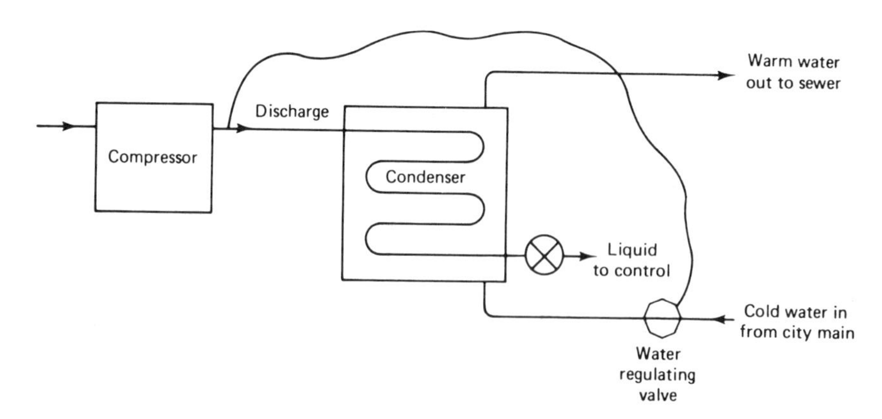

Figure 5-10 Wastewater system where refrigerant is in tubes and water in shell. Water regulating valve alters water flow to maintain condensing temperature.

The condenser may use water once, in which case the water is wasted (see Figure 5-10), or the water can be reused (see Figure 5-11). Reuse of the water is usually more economical. Some city codes require it. Two principal types of equipment are used for this purpose:

1. Cooling towers
2. Closed-circuit evaporative coolers

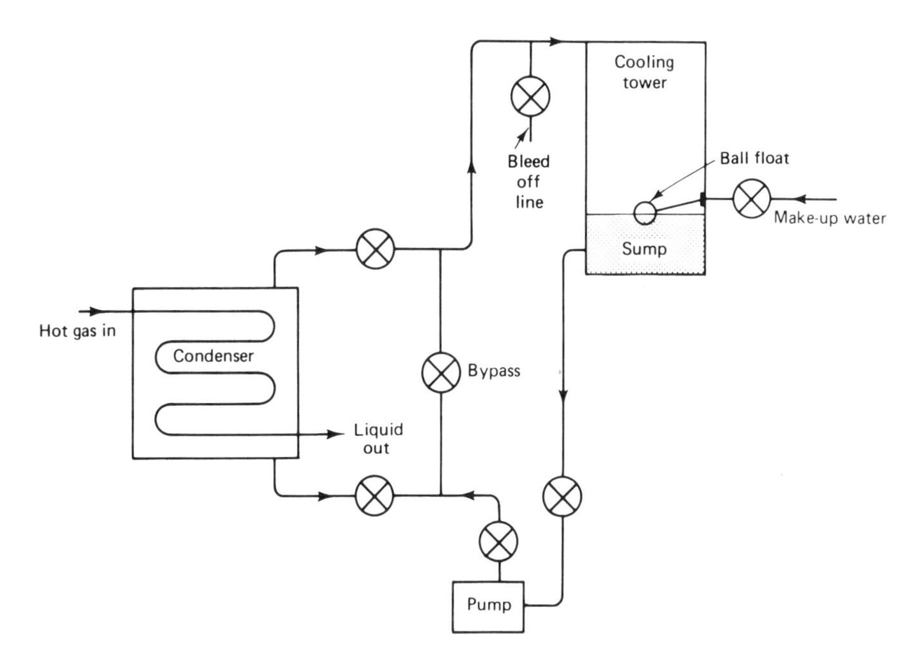

Figure 5-11 Recirculating water system uses cooling-tower water. Only water lost is through bleed-off and evaporation.

Evaporative condensers also conserve water and will be discussed in a separate section since they replace the need for a water-cooled condenser.

Series and Parallel Flow

Certain types of condensers, principally the shell-and-tube type, can be arranged for either series or parallel flow (see Figure 5–12). The headers can be adjusted to control the number of passes.

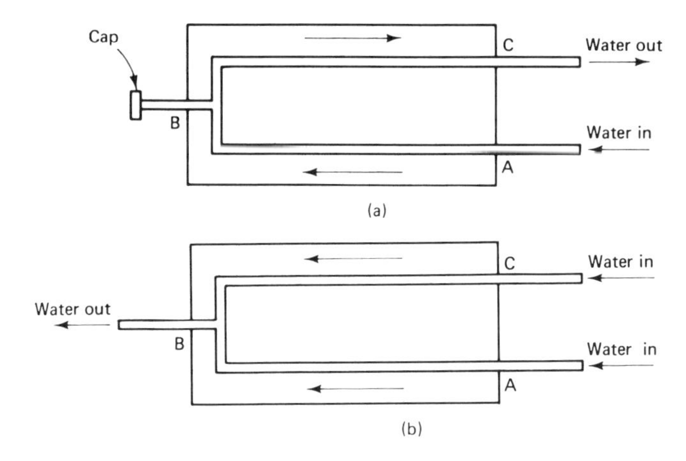

Figure 5–12 Condenser water flow arrangement for (a) series and (b) parallel flow.

For cooling-tower application, the number of water circuits is increased to permit about a 10 °F temperature rise through the water tubes. This may mean two or more water passes. If the condenser has 100 tubes and uses a two-pass circuit, the headers would be so arranged as to place 50 tubes in each pass: for a four-pass circuit there would be 25 tubes in each pass. The limitation on the number of passes is the pressure drop through the condenser.

Determining the Capacity

The capacity of a water-cooled condenser can easily be determined knowing the temperature rise of the water and the gpm flow. The relationship of gpm, temperature rise, and capacity are given in the formula

$$\text{Btuh} = \text{gpm} + 500 + \text{temperature rise}$$

For example, if 100 gpm flows through a condenser, raising the water temperature from 85 °F to 95 °F, the capacity of the condenser would be

$$\begin{aligned} \text{capacity} &= 100 \text{ gpm} \times 500 \times (95\,°\text{F} - 85\,°\text{F}) \\ &= 500{,}000 \text{ Btuh} \end{aligned}$$

The amount of heat rejection per ton increases as the suction temperature drops (see Figure 5–1). However, assuming 15,000 Btuh per ton of condenser capacity, the condenser would represent about 33.3 tons of refrigeration.

An important factor affecting the capacity of a water-cooled condenser is the fouling factor. As the condenser is used the surface develops a film and the tubes get dirty. The more debris there is on the tube, the less its effectiveness as a heat transfer unit. An increase in head pressure may be an indication that the condenser tubes need cleaning.

Effect of Noncondensables

Particularly when the condenser is first placed in operation, noncondensable gases collect in the condenser, reducing its capacity. This can occur even during operation. Noncondensables reduce the heat transfer rate of condensers. Even a small amount of air, for example 1%, can reduce the capacity of a condenser by 25 to 40%. Evidence of noncondensables can be detected by a condensing pressure above that corresponding to the temperature at which the refrigerant is actually condensing. Noncondensables can be eliminated to a large degree by purging.

Shell-and-Tube Condensers

Shell-and-tube condensers are constructed in either a horizontal (Figure 5–13a) or a vertical (Figure 5–13b) configuration. The horizontal type is most common. Two types of small horizontal shell-and-tube condensers are shown in Figure 5–14.

Shell-and-tube condensers are made in sizes ranging from 5 tons to those required for large installations. The refrigerant is in the shell and the water is circulated through the tubes in a single- or multipass circuit. The condenser must be located on the job in an area where there is sufficient room to clean or replace a tube. The units are constructed

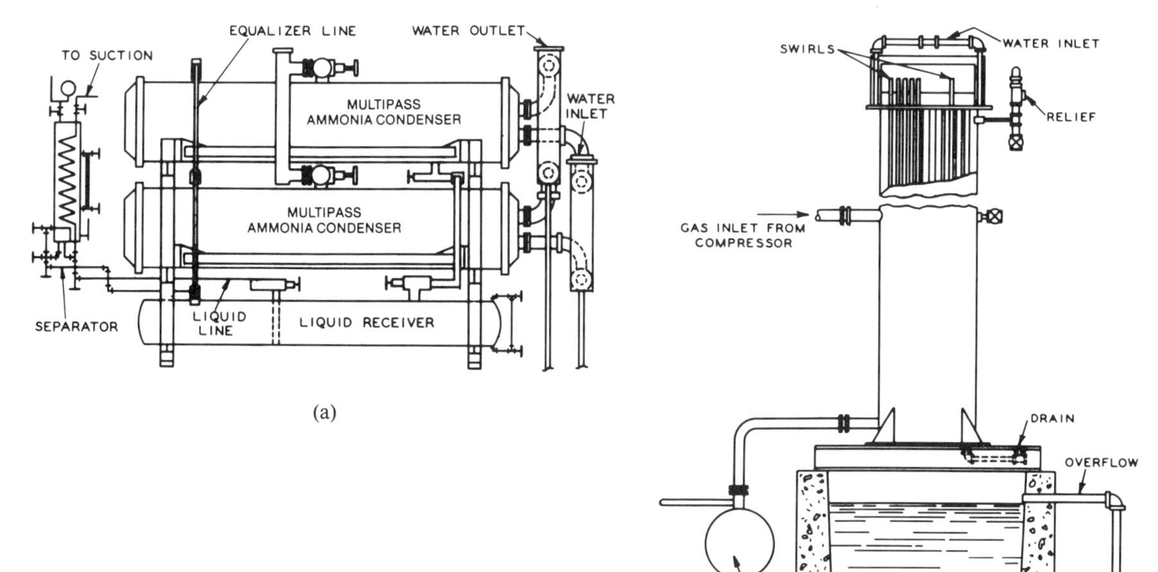

Figure 5–13 Ammonia refrigerant shell-and-tube condensers: (a) horizontal; (b) vertical. (By permission of *ASHRAE Handbook*.)

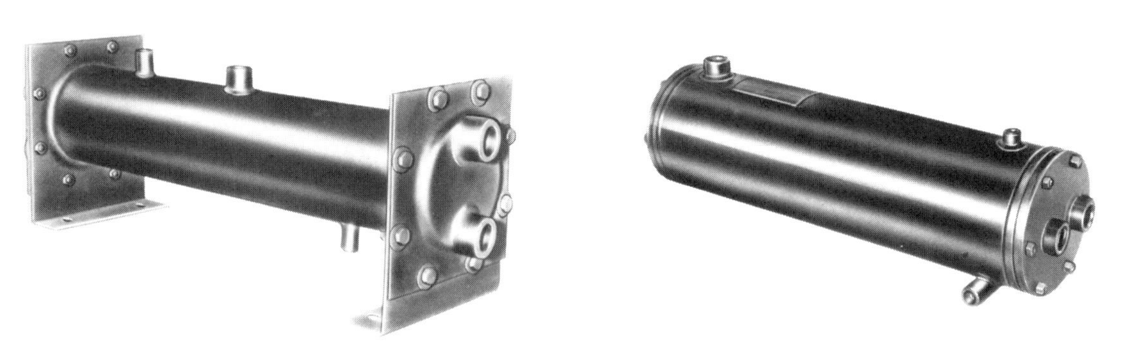

Figure 5–14 Two types of shell-and-tube condensers, one with mounting brackets. The advantage of this type of condenser is the ability to clean the tubes. Water is in the tubes. Refrigerant is in the shell. These are for halocarbon refrigerants. (By permission of Standard Refrigeration Company.)

with a fixed tube sheet and usually, straight tubes. A shell-and-U-tube condenser uses a water tube bent in a U configuration and attached to a single tube sheet.

Shell-and-Coil Condensers

The shell-and-coil condenser is relatively compact and its principal use is in packaged units of less than 10 tons. The refrigerant in the shell is condensed by the water in the tubes. Care should be taken not to overcharge the system since too high a liquid level will reduce the condensing surface. The shell acts as a receiver tank. The water tubes are cleaned by chemical treatment, usually an acid solution. The tubes must be thoroughly

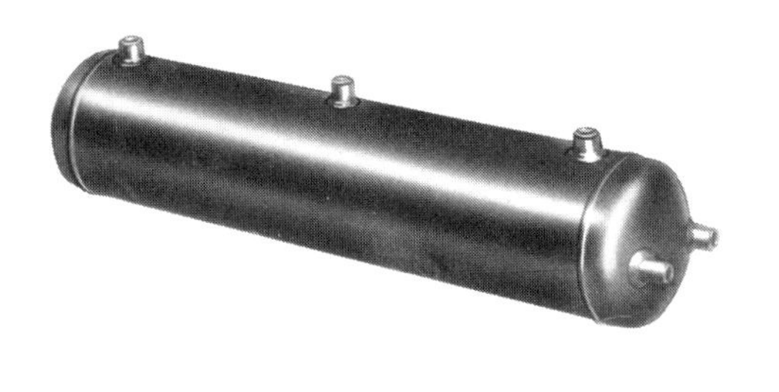

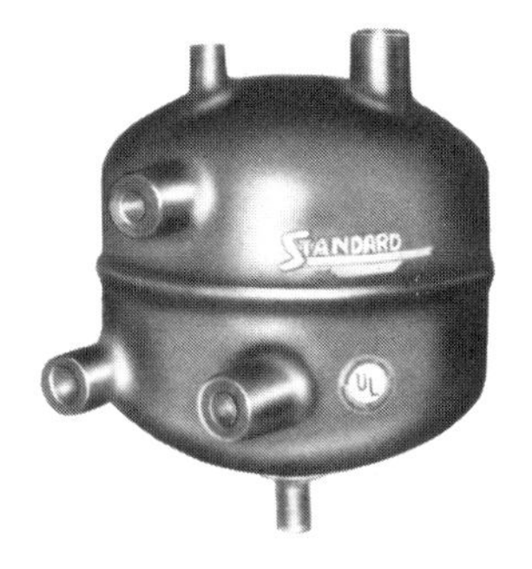

Figure 5–15 Two examples of shell-and-coil condensers. The advantage of this type of condenser is the compactness for use in packaged equipment. Acid needs to be used for cleaning (see instructions in Chapter 15). These condensers are for halocarbon refrigerants.

flushed out with clear water after cleaning to prevent corrosion and pitting of the tubes. The shell-and-coil condenser is illustrated in Figure 5–15.

Double-Tube Condensers

The tube-in-tube condenser, as the name implies, is made by placing the water inside the refrigerant tube and using the counterflow principle (Figure 5–16). Double-tube condensers can be constructed as shown in Figure 5–17, permitting mechanical cleaning of the water passages. Other types, such as those shown in Figure 5–18, require chemical cleaning. Tube-in-tube condensers are sometimes used as "booster" condensers for air-cooled condensers to handle peak-load requirements. Water regulating valves can be set to permit use of the water-cooled condenser only when the condensing pressure reaches a certain limit.

Figure 5–16 Double-tube condenser. A type of tube-in-tube condenser. Refrigerant flows in inner tube, water in outer tube, in counterflow directions. (By permission of Standard Refrigeration Company.)

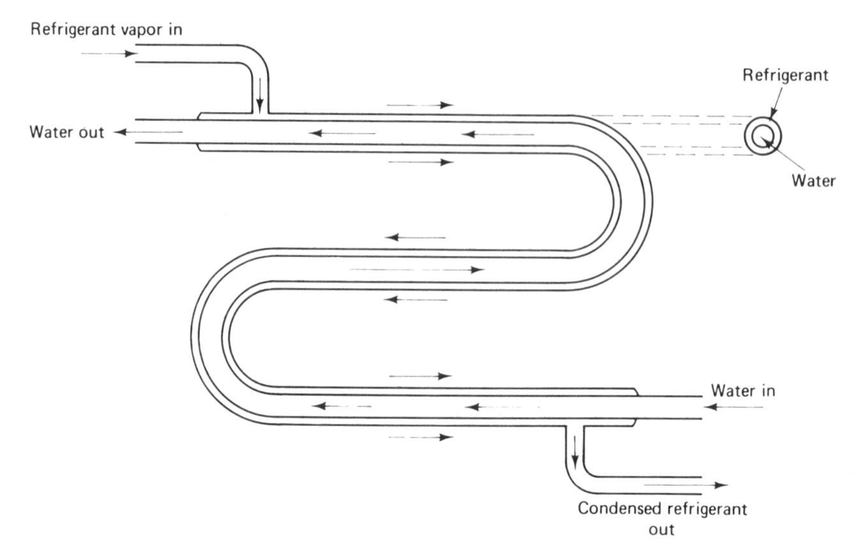

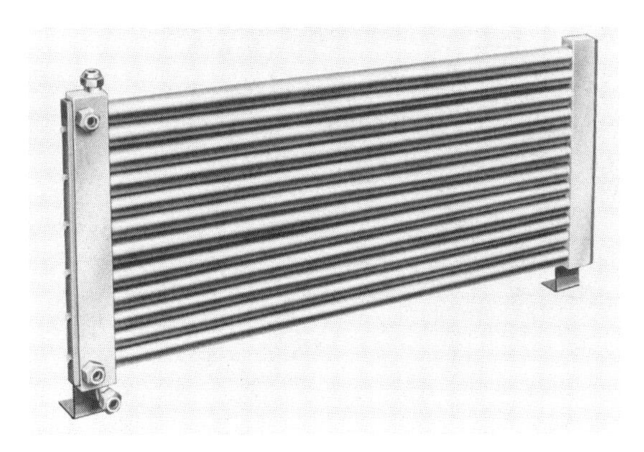

Figure 5-17 Cleanable-type water-cooled tube-in-tube condenser. Headers can be removed at both ends for cleaning. (By permission of Halstead Mitchell.)

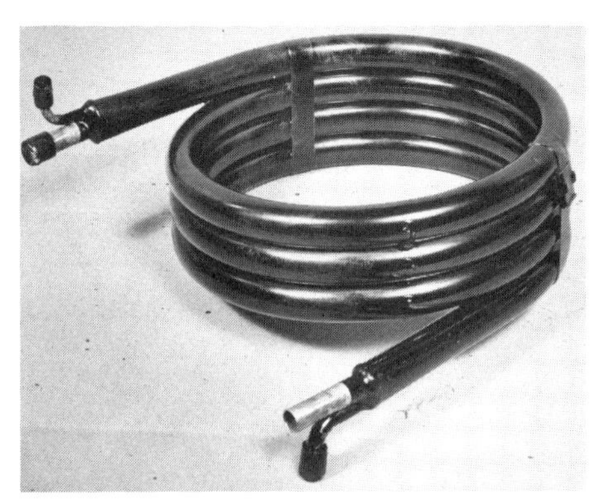

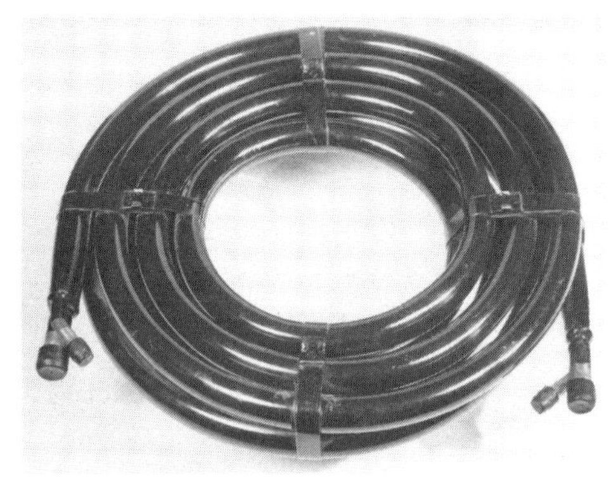

Figure 5-18 Coaxial-type tube-in-tube water-cooled condensers. These are specially designed to fit into a minimum space in packaged units. Similar configuration can be used for water chilling. (By permission of Edwards Engineering Corporation.)

Atmospheric Condensers

The condenser consists of a number of tubes, one above the other. Refrigerant is in the tubes and water is distributed over the top tube, flowing by gravity over the other tubes directly below. Sufficient water must be used to wet the tubes thoroughly.

EVAPORATIVE CONDENSERS

The evaporative condenser is a device for cooling the discharge gases from a compressor by means of evaporative cooling. Figure 5-19 illustrates the construction. Usually, a bare tube coil is used to facilitate cleaning. The water sprayed on the coil is reused to economize on water usage. Water treatment is desirable in most areas to prevent scale.

The amount of water usage by evaporation is about 2 gal/ton per hour. Some water is also lost by bleed-off and draft, depending on the design. The total water usage is usually between 3 and 4 gal/ton per hour.

An evaporative condenser is as effective as a water-cooled condenser in controlling head pressure. The design discharge temperature with 95 °F DB and 78 °F WB ambient

Figure 5-19 Evaporative condenser for use inside or outside. Options are centrifugal or axial fans. Unit has capacity controls with damper regulation or fan cycling. Coil section has all-prime surface, hot-dipped galvanized steel headers, and moisture eliminators. (By permission of Baltimore Aircoil.)

Figure 5-20 Typical evaporative condenser performance. Heat rejection capacities are in MBH (thousands of Btu per hour), based on 105 °F condensing temperature and 78 °F wet-bulb temperature, using halocarbon refrigerants (R-12, R-22, R-500, or R-502). (By permission of Baltimore Aircoil.)

Model No. VXC	Heat rejection (MBH)	Model No. VXC	Heat rejection (MBH)	Model No. VXC	Heat rejection (MBH)
10	147.0	185	2,719.5	590	8,673.0
15	220.5	N205	3,013.5	N600	8,820.0
20	294.0	N230	3,381.0	620	9,114.0
25	367.5	N250	3,675.0	650/N650	9,555.0
30	441.0	N275	4,042.5	680	9,996.0
38	558.6	N300	4,410.0	720/N720	10,584.0
46	676.2	320	4,704.0	760/N760	11,172.0
52	764.4	N325	4,777.5	N800	11,760.0
58	852.6	340	4,998.0	840	12,348.0
65	955.5	360/N360	5,292.0	900	13,230.0
72	1,058.4	380/N380	5,586.0	980	14,406.0
80	1,176.0	N400	5,880.0	1060	15,582.0
90	1,323.0	420	6,174.0	1100	16,170.0
100	1,470.0	450	6,615.0	1180	17,346.0
110	1,617.0	N460	6,762.0	1240	18,228.0
125	1,837.5	490	7,203.0	1300	19,110.0
135	1,984.5	N500	7,350.0	1360	20,000.0
150	2,205.0	530	7,791.0		
165	2,425.5	550/N550	8,085.0		

is 105 °F condensing temperature. The capacity of an evaporative condenser is based on the design wet-bulb temperature, due to its dependence on evaporation for cooling. (See the typical rating table shown in Figure 5–20.)

The advantage of an evaporative condenser is that it replaces both the water-cooled condenser and the cooling tower. The disadvantage is that a location must be selected to minimize the length of the refrigerant line connections.

De-superheating coils are often added to increase the capacity. These coils are placed on the inlet side of the standard coil. Hot gas passes through the de-superheating coils first, precooling the discharge gases. The receiver is often placed in the sump to subcool the condensed liquid refrigerant. A diagram of the operation of an evaporative condenser is shown in Figure 5–21.

Cooling Towers

A cooling tower is a type of evaporative cooler that reduces the temperature of water from the water-cooled condenser to permit reuse. A commercial-type cooling tower is illustrated in Figure 5–22. Typical selection information is given in Figure 5–23. A schematic

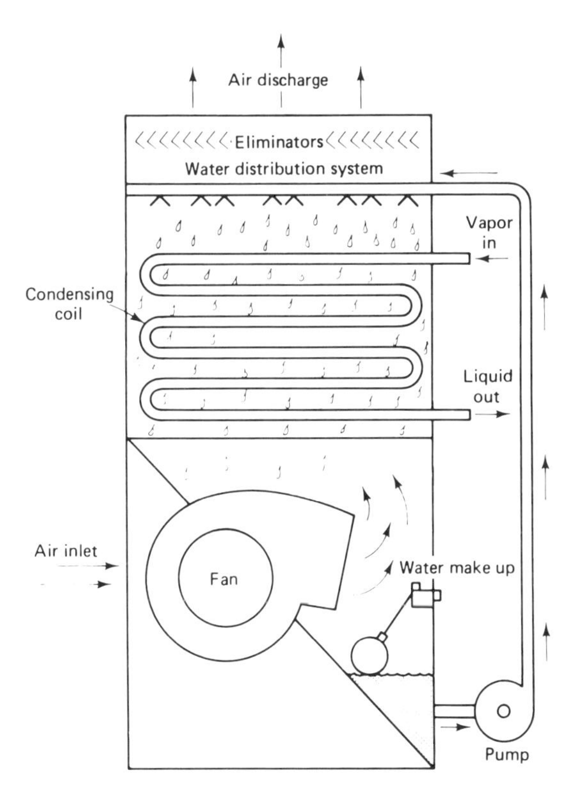

Figure 5–21 Schematic of evaporative condenser design showing principle of operation. The vapor to be condensed is circulated through a condensing coil which is continually wetted on the outside by a recirculating water system. Air is simultaneously blown upward over the coil, causing a small portion of the recirculated water to evaporate. This evaporation removes heat from the coil, cooling and condensing the vapor in the coil. (By permission of Baltimore Aircoil.)

Figure 5–22 Condenser water cooling tower. Capacities 10 to 4800 tons. This equipment can be arranged for winter or summer operation. Capacity control is by dampers or fan cycling. The wet deck surface is made of polyvinyl chloride (PVC). (By permission of Baltimore Aircoil.)

CHART 1 - Counterflow Cooling Tower Selection and Performance Chart

Entering Water Temp* ________ °F
Leaving Water Temp ________ °F
Range ________ °F

Leaving Water Temp ________ °F
Design Wet Bulb Temp ________ °F
Approach ________ °F

*Do not exceed 130°F with PVC wet deck surface.

Selection Example:

GIVEN: To cool 2,250 GPM of water from 103°F to 85°F at 76°F wet bulb.

1. Determine Range
   Range = Water on 103°F — Water off 85°F = 18°
2. Determine Approach
   Approach = Water off 85°F — Wet Bulb 76°F = 9°
3. Determine Selection Factor
   Enter the Wet Bulb Correction Section of Chart 1 on the 18° Range line as shown by the dotted line. From the intersection of the 18° Range line and the 76° Wet Bulb, curve project a straight line into the Approach Section to intersect the 9° Approach curve. From this point, extend a line horizontally into the Selection Factor Section intersecting the 18° Range line to obtain the selection factor. The factor is 1.18.
4. Unit Selection
   Enter Chart 2 reading across the Selection Factor Columns to find a factor EQUAL TO OR GREATER THAN the factor determined in Step 3. In this case enter the column headed by 1.2. Read downward until reaching a GPM EQUAL TO OR GREATER THAN the quantity to be cooled (2,250 gpm). Read the recommended unit selection from the unit column on the left. For the given design conditions, it is a Model VXT-945.

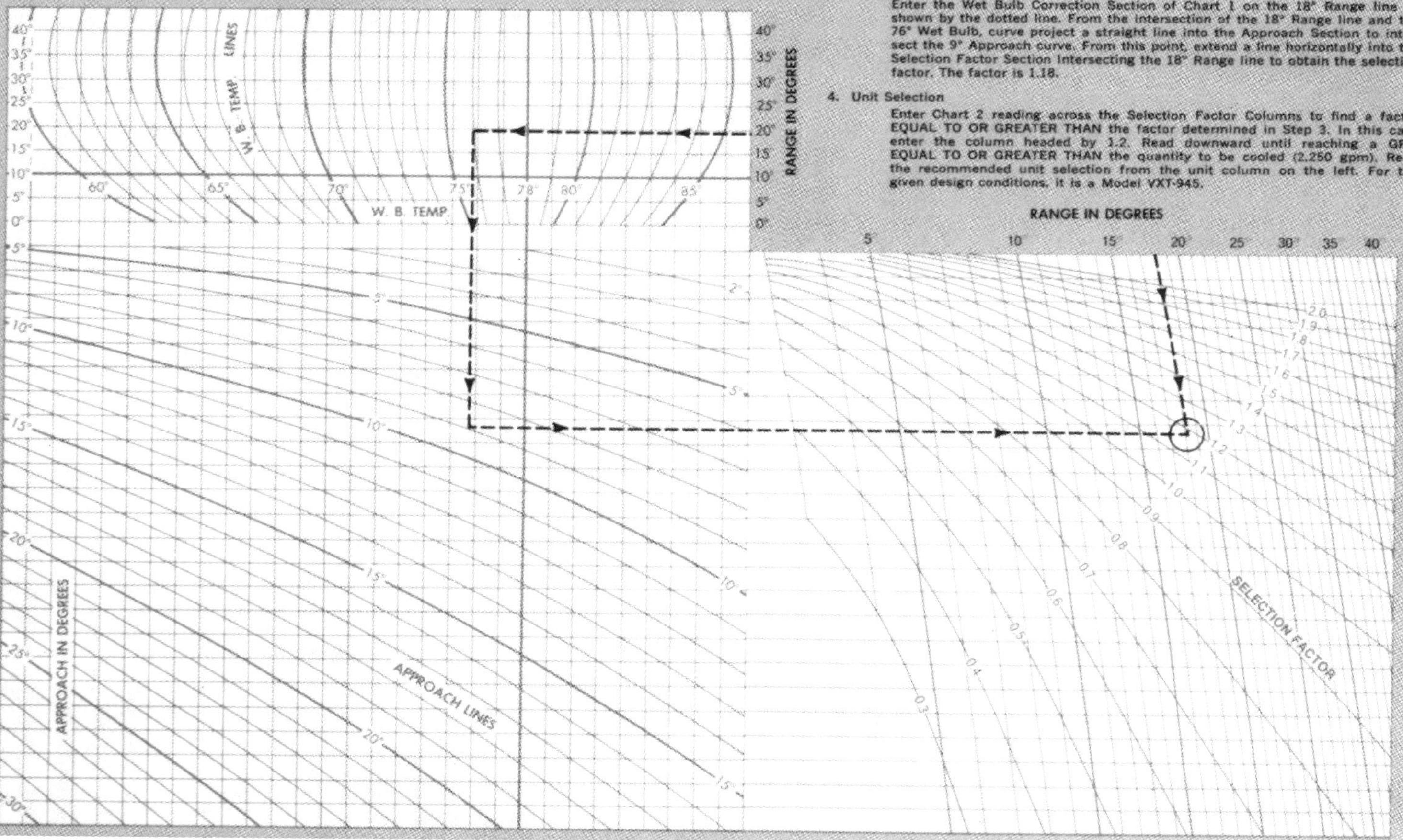

Figure 5–23 Selection example using cooling-tower performance chart and tables. (By permission of Baltimore Aircoil.)

CHART 2 - Recommended Selections in GPM / Selection Factors 0.40 to 0.85

Enter chart reading across the Selection Factor Columns to find a factor EQUAL TO OR GREATER THAN the design selection factor. Read downward until reaching a GPM EQUAL TO OR GREATER THAN design. Read the recommended unit selection from the unit column on the left. Interpolation is permitted between Selection Factors only.

UNIT	SELECTION FACTOR									
	0.40	0.45	0.50	0.55	0.60	0.65	0.70	0.75	0.80	0.85
VXT-10	105	93	84	75	67	61	55	51	45	41
VXT-15	130	115	105	94	86	78	72	67	61	57
VXT-20	135	135	128	116	106	97	90	84	78	73
VXT-25	135	135	135	135	127	116	108	101	94	89
VXT-30	260	235	211	192	174	158	147	136	124	116
VXT-40	280	280	255	233	214	196	183	170	157	148
VXT-45	280	280	280	254	233	215	202	187	174	164
VXT-55	280	280	280	280	278	255	238	223	208	197
VXT-65	430	430	410	373	343	315	292	272	251	236
VXT-70	430	430	430	396	367	337	314	291	269	252
VXT-75	430	430	430	420	387	356	332	310	287	270
VXT-85	430	430	430	430	430	396	370	345	322	304
VXT-95	575	575	575	540	495	455	425	395	370	346
VXT-105	575	575	575	575	540	495	465	431	404	380
VXT-120	575	575	575	575	575	560	520	485	456	429
VXT-135	575	575	575	575	575	575	575	540	505	479
VXT-150	810	810	810	810	765	710	660	615	575	540
VXT-165	810	810	810	810	810	770	720	665	625	590
VXT-185	810	810	810	810	810	810	795	740	695	655
VXT-N215	1140	1140	1140	1140	1085	1005	930	870	815	765
VXT-N240	1140	1140	1140	1140	1140	1115	1030	965	905	850
VXT-N265	1140	1140	1140	1140	1140	1140	1135	1060	990	935
VXT-N310	1740	1740	1740	1730	1590	1460	1350	1260	1185	1110
VXT-N345	1740	1740	1740	1740	1740	1610	1490	1390	1305	1225
VXT-N370	1740	1740	1740	1740	1740	1720	1590	1490	1395	1310
VXT-N395	1740	1740	1740	1740	1740	1740	1690	1580	1480	1395
VXT-N430	2280	2280	2280	2270	2170	2010	1860	1740	1630	1530
VXT-N480	2280	2280	2280	2280	2280	2230	2060	1930	1810	1700
VXT-N510	2280	2280	2280	2280	2280	2280	2180	2030	1910	1800
VXT-N535	2280	2280	2280	2280	2280	2280	2280	2140	2000	1880

Models VXT-N215 to VXT-N535 are furnished with 8 foot wide pan sections. Models VXT-315 to VXT-4800 are furnished with 10 foot wide pan sections. Refer to pages 15 to 17 for detailed dimensional information.

UNIT	SELECTION FACTOR									
	0.40	0.45	0.50	0.55	0.60	0.65	0.70	0.75	0.80	0.85
VXT-315	1750	1750	1750	1750	1600	1500	1380	1300	1205	1130
VXT-350	1750	1750	1750	1750	1750	1630	1515	1415	1330	1250
VXT-375	1750	1750	1750	1750	1750	1750	1615	1500	1415	1330
VXT-400	1750	1750	1750	1750	1750	1750	1715	1600	1500	1410
VXT-470	2600	2600	2600	2600	2400	2200	2070	1920	1800	1680
VXT-525	2600	2600	2600	2600	2600	2450	2280	2130	2000	1870
VXT-560	2600	2600	2600	2600	2600	2600	2410	2250	2110	2000
VXT-600	2600	2600	2600	2600	2600	2600	2570	2410	2260	2120
VXT-630	3500	3500	3500	3500	3200	3000	2760	2600	2410	2260
VXT-700	3500	3500	3500	3500	3500	3250	3030	2830	2660	2500
VXT-750	3500	3500	3500	3500	3500	3500	3230	3000	2830	2660
VXT-800	3500	3500	3500	3500	3500	3500	3430	3200	3000	2820
VXT-870	5250	5250	5250	4950	4500	4200	3880	3630	3370	3150
VXT-945	5250	5250	5250	5250	4800	4500	4150	3900	3620	3390
VXT-1050	5250	5250	5250	5250	5250	4900	4550	4250	3990	3750
VXT-1125	5250	5250	5250	5250	5250	5250	4850	4500	4250	4000
VXT-1200	5250	5250	5250	5250	5250	5250	5150	4800	4500	4240
VXT-1260	7000	7000	7000	7000	6400	6000	5520	5200	4820	4520
VXT-1400	7000	7000	7000	7000	7000	6500	6060	5660	5320	5000
VXT-1500	7000	7000	7000	7000	7000	7000	6460	6000	5660	5320
VXT-1600	7000	7000	7000	7000	7000	7000	6860	6400	6000	5640
VXT-1740	10500	10500	10500	9900	9000	8400	7760	7260	6740	6300
VXT-1890	10500	10500	10500	10500	9600	9000	8300	7800	7240	6780
VXT-2100	10500	10500	10500	10500	10500	9800	9100	8500	7980	7500
VXT-2250	10500	10500	10500	10500	10500	10500	9700	9000	8500	8000
VXT-2400	10500	10500	10500	10500	10500	10500	10300	9600	9000	8480
VXT-2520	14000	14000	14000	14000	12800	12000	11000	10400	9640	9040
VXT-2800	14000	14000	14000	14000	14000	13000	12100	11300	10600	10000
VXT-3000	14000	14000	14000	14000	14000	14000	12900	12000	11300	10600
VXT-3200	14000	14000	14000	14000	14000	14000	13700	12800	12000	11300
VXT-3480	21000	21000	21000	19800	18000	16800	15500	14500	13500	12600
VXT-3780	21000	21000	21000	21000	19200	18000	16600	15600	14500	13600
VXT-4200	21000	21000	21000	21000	21000	19600	18200	17000	16000	15000
VXT-4500	21000	21000	21000	21000	21000	21000	19400	18000	17000	16000
VXT-4800	21000	21000	21000	21000	21000	21000	20600	19200	18000	17000

Figure 5–23 (continued)

CHART 2 - Recommended Selections in GPM / Selection Factors 0.90 to 1.35

Enter chart reading across the Selection Factor Columns to find a factor EQUAL TO OR GREATER THAN the design selection factor. Read downward until reaching a GPM EQUAL TO OR GREATER THAN design. Read the recommended unit selection from the unit column on the left. Interpolation is permitted between Selection Factors only.

UNIT	SELECTION FACTOR									
	0.90	0.95	1.00	1.05	1.10	1.15	1.20	1.25	1.30	1.35
VXT-10	36	33	30	26	N.A.	N.A.	N.A.	N.A.	N.A.	N.A.
VXT-15	52	49	45	41	37	34	32	29	27	N.A.
VXT-20	67	64	60	56	52	49	46	43	40	37
VXT-25	83	80	75	71	67	64	60	57	54	51
VXT-30	105	98	90	83	75	69	63	58	N.A.	N.A.
VXT-40	136	128	120	112	105	99	92	86	80	75
VXT-45	153	144	135	127	120	112	106	100	94	89
VXT-55	185	176	165	157	149	141	133	127	120	114
VXT-65	221	208	195	183	172	161	151	142	127	124
VXT-70	236	223	210	198	187	175	165	155	145	137
VXT-75	252	239	225	213	202	190	180	169	159	152
VXT-85	284	268	255	239	228	217	207	196	185	175
VXT-95	324	306	285	268	252	238	226	213	201	189
VXT-105	358	338	315	300	284	267	252	238	226	214
VXT-120	405	384	360	343	326	310	294	278	263	250
VXT-135	454	432	405	388	370	354	337	321	305	291
VXT-150	509	482	450	430	407	386	367	346	326	310
VXT-165	558	527	495	471	448	426	405	385	364	347
VXT-185	621	591	555	530	507	482	460	436	415	395
VXT-N215	725	685	645	610	580	545	520	490	465	440
VXT-N240	800	760	720	685	650	615	585	555	530	505
VXT-N265	880	835	795	755	710	685	655	620	590	565
VXT-N310	1040	985	930	875	830	785	740	705	665	630
VXT-N345	1155	1090	1035	975	925	880	830	790	750	710
VXT-N370	1235	1170	1110	1050	1000	950	900	855	810	770
VXT-N395	1315	1245	1185	1120	1070	1020	970	925	880	835
VXT-N430	1450	1370	1290	1220	1160	1090	1040	980	930	880
VXT-N480	1600	1520	1440	1370	1300	1230	1170	1110	1060	1010
VXT-N510	1700	1615	1530	1455	1390	1320	1255	1200	1135	1080
VXT-N535	1780	1690	1605	1530	1455	1390	1320	1260	1205	1145

Models VXT-N215 to VXT-N535 are furnished with 8 foot wide pan sections. Models VXT-315 to VXT-4800 are furnished with 10 foot wide pan sections. Refer to pages 15 to 17 for detailed dimensional information.

UNIT	SELECTION FACTOR									
	0.90	0.95	1.00	1.05	1.10	1.15	1.20	1.25	1.30	1.35
VXT-315	1065	1000	945	900	850	800	765	725	685	645
VXT-350	1185	1115	1050	985	935	900	850	800	765	715
VXT-375	1265	1200	1125	1065	1015	965	915	865	835	785
VXT-400	1350	1265	1200	1135	1085	1035	985	935	900	850
VXT-470	1580	1490	1410	1325	1250	1190	1125	1070	1015	950
VXT-525	1760	1660	1575	1485	1410	1340	1270	1210	1145	1080
VXT-560	1880	1780	1680	1595	1510	1450	1370	1305	1240	1175
VXT-600	2010	1900	1800	1710	1620	1550	1480	1410	1345	1280
VXT-630	2130	2000	1890	1800	1700	1600	1530	1450	1370	1290
VXT-700	2370	2230	2100	1970	1870	1800	1700	1600	1530	1430
VXT-750	2530	2400	2250	2130	2030	1930	1830	1730	1670	1570
VXT-800	2700	2530	2400	2270	2170	2070	1970	1870	1800	1700
VXT-870	2950	2800	2610	2450	2300	2200	2070	1960	1845	1735
VXT-945	3200	3000	2835	2700	2550	2400	2300	2170	2050	1940
VXT-1050	3550	3350	3150	2950	2800	2700	2550	2400	2300	2150
VXT-1125	3800	3600	3375	3200	3050	2900	2750	2600	2500	2350
VXT-1200	4050	3800	3600	3400	3250	3100	2950	2800	2700	2550
VXT-1260	4260	4000	3780	3600	3400	3200	3060	2900	2740	2580
VXT-1400	4740	4460	4200	3940	3740	3600	3400	3200	3060	2860
VXT-1500	5060	4800	4500	4260	4060	3860	3660	3460	3340	3140
VXT-1600	5400	5060	4800	4540	4340	4140	3940	3740	3600	3400
VXT-1740	5900	5600	5220	4900	4600	4400	4140	3920	3690	3470
VXT-1890	6400	6000	5670	5400	5100	4800	4600	4340	4100	3880
VXT-2100	7100	6700	6300	5900	5600	5400	5100	4800	4600	4300
VXT-2250	7600	7200	6750	6400	6100	5800	5500	5200	5000	4700
VXT-2400	8100	7600	7200	6800	6500	6200	5900	5600	5400	5100
VXT-2520	8520	8000	7560	7200	6800	6400	6120	5800	5480	5160
VXT-2800	9480	8920	8400	7880	7480	7200	6800	6400	6120	5720
VXT-3000	10100	9600	9000	8520	8120	7720	7320	6920	6680	6280
VXT-3200	10800	10100	9600	9080	8680	8280	7880	7480	7200	6800
VXT-3480	11800	11200	10440	9800	9200	8800	8280	7840	7380	6940
VXT-3780	12800	12000	11340	10800	10200	9600	9200	8680	8200	7760
VXT-4200	14200	13400	12600	11800	11200	10800	10200	9600	9200	8600
VXT-4500	15200	14400	13500	12800	12200	11600	11000	10400	10000	9400
VXT-4800	16200	15200	14400	13600	13000	12400	11800	11200	10800	10200

Figure 5–23 (continued)

CHART 2 -Recommended Selections in GPM / Selection Factors 1.40 to 2.00

Enter chart reading across the Selection Factor Columns to find a factor EQUAL TO OR GREATER THAN the design selection factor. Read downward until reaching a GPM EQUAL TO OR GREATER THAN design. Read the recommended unit selection from the unit column on the left. Interpolation is permitted between Selection Factors only.

UNIT	SELECTION FACTOR									
	1.40	1.45	1.50	1.55	1.60	1.65	1.70	1.80	1.90	2.00
VXT-10	N.A.	N.A.	N.A.	N.A.	N.A.	N.A.	N.A.	N.A.	N.A.	N.A.
VXT-15	N.A.	N.A.	N.A.	N.A.	N.A.	N.A.	N.A.	N.A.	N.A.	N.A.
VXT-20	35	32	30	28	26	N.A.	N.A.	N.A.	N.A.	N.A.
VXT-25	48	45	43	40	38	36	34	30	26	N.A.
VXT-30	N.A.	N.A.	N.A.	N.A.	N.A.	N.A.	N.A.	N.A.	N.A.	N.A.
VXT-40	70	64	61	57	N.A.	N.A.	N.A.	N.A.	N.A.	N.A.
VXT-45	83	77	73	68	64	59	55	N.A.	N.A.	N.A.
VXT-55	108	102	98	92	87	82	78	69	61	54
VXT-65	113	108	102	95	89	83	N.A.	N.A.	N.A.	N.A.
VXT-70	128	121	114	107	101	94	89	N.A.	N.A.	N.A.
VXT-75	142	133	126	118	110	103	97	85	N.A.	N.A.
VXT-85	171	163	150	142	134	127	120	107	96	85
VXT-95	177	171	157	147	137	129	121	N.A.	N.A.	N.A.
VXT-105	202	190	180	169	163	150	140	123	109	N.A.
VXT-120	238	228	217	207	197	186	177	160	144	128
VXT-135	275	262	248	237	227	215	205	183	164	147
VXT-150	291	276	259	245	232	220	209	186	165	N.A.
VXT-165	328	312	295	280	262	250	237	213	190	168
VXT-185	374	358	340	321	305	287	270	240	216	193
VXT-N215	417	395	375	355	335	317	300	268	240	N.A.
VXT-N240	475	450	430	410	385	360	345	305	270	242
VXT-N265	535	510	486	462	440	423	395	355	316	282
VXT-N310	591	558	533	497	470	444	418	370	330	N.A.
VXT-N345	669	632	600	563	530	500	472	417	370	328
VXT-N370	730	690	650	615	580	545	515	457	405	358
VXT-N395	795	755	720	685	650	615	582	520	464	410
VXT-N430	835	790	750	710	670	635	600	536	480	N.A.
VXT-N480	950	900	860	820	770	725	690	610	540	484
VXT-N510	1020	970	920	875	825	780	740	660	585	520
VXT-N535	1090	1035	990	940	895	850	805	720	642	570

Models VXT-N215 to VXT-N535 are furnished with 8 foot wide pan sections. Models VXT-315 to VXT-4800 are furnished with 10 foot wide pan sections. Refer to pages 15 to 17 for detailed dimensional information.

UNIT	SELECTION FACTOR									
	1.40	1.45	1.50	1.55	1.60	1.65	1.70	1.80	1.90	2.00
VXT-315	605	570	540	510	480	450	425	380	335	N.A.
VXT-350	675	640	605	570	535	500	470	420	375	330
VXT-375	740	700	665	625	585	545	515	455	405	360
VXT-400	810	770	730	700	660	615	585	520	460	410
VXT-470	910	860	820	770	720	670	635	565	500	N.A.
VXT-525	1030	970	920	870	815	755	715	630	555	500
VXT-560	1120	1050	1000	940	880	820	775	680	600	535
VXT-600	1220	1160	1110	1055	1000	935	900	795	700	625
VXT-630	N.A.	N.A.	N.A.	N.A.	N.A.	N.A.	N.A.	N.A.	N.A.	N.A.
VXT-700	1350	1280	1210	1140	1070	1000	940	840	750	660
VXT-750	1480	1400	1330	1250	1170	1090	1030	910	810	720
VXT-800	1620	1540	1460	1400	1320	1230	1170	1040	920	820
VXT-870	N.A.	N.A.	N.A.	N.A.	N.A.	N.A.	N.A.	N.A.	N.A.	N.A.
VXT-945	1815	1710	1620	1530	1440	1350	1275	1140	1005	N.A.
VXT-1050	2025	1920	1815	1710	1605	1500	1410	1260	1125	990
VXT-1125	2220	2100	1995	1875	1755	1635	1545	1365	1215	1080
VXT-1200	2430	2310	2190	2100	1980	1845	1755	1560	1380	1230
VXT-1260	N.A.	N.A.	N.A.	N.A.	N.A.	N.A.	N.A.	N.A.	N.A.	N.A.
VXT-1400	2700	2560	2420	2280	2140	2000	1880	1680	1500	1320
VXT-1500	2960	2800	2660	2500	2340	2180	2060	1820	1620	1440
VXT-1600	3240	3080	2920	2800	2640	2460	2340	2080	1840	1640
VXT-1740	N.A.	N.A.	N.A.	N.A.	N.A.	N.A.	N.A.	N.A.	N.A.	N.A.
VXT-1890	3630	3420	3240	3060	2880	2700	2550	2280	2010	N.A.
VXT-2100	4050	3840	3630	3420	3210	3000	2820	2520	2250	1980
VXT-2250	4440	4200	3990	3750	3510	3270	3090	2730	2430	2160
VXT-2400	4860	4620	4380	4200	3960	3690	3510	3120	2760	2460
VXT-2520	N.A.	N.A.	N.A.	N.A.	N.A.	N.A.	N.A.	N.A.	N.A.	N.A.
VXT-2800	5400	5120	4840	4560	4280	4000	3760	3360	3000	2640
VXT-3000	5920	5600	5320	5000	4680	4360	4120	3640	3240	2880
VXT-3200	6480	6160	5840	5600	5280	4920	4680	4160	3680	3280
VXT-3480	N.A.	N.A.	N.A.	N.A.	N.A.	N.A.	N.A.	N.A.	N.A.	N.A.
VXT-3780	7260	6840	6480	6120	5760	5400	5100	4560	4020	N.A.
VXT-4200	8100	7680	7260	6840	6420	6000	5640	5040	4500	3960
VXT-4500	8880	8400	7980	7500	7020	6540	6180	5460	4860	4320
VXT-4800	9720	9240	8760	8400	7920	7380	7020	6240	5520	4920

Figure 5–23 (continued)

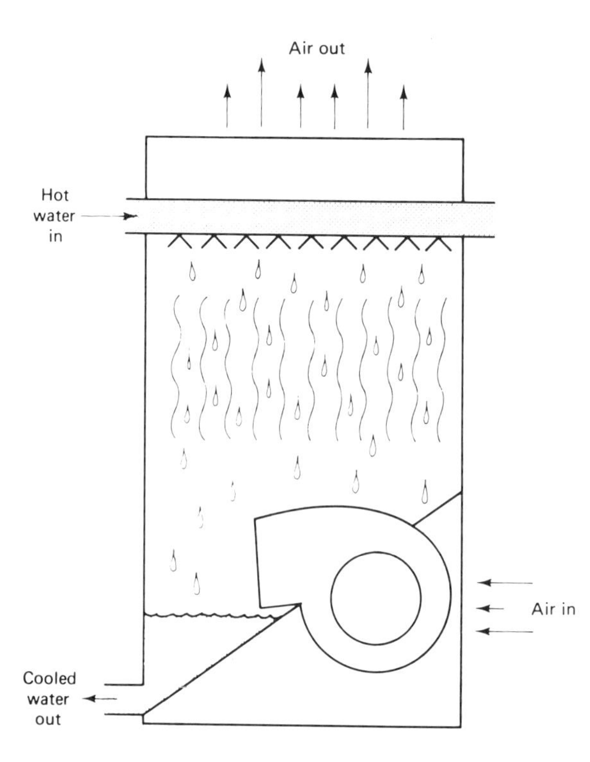

Figure 5-24 Schematic of cooling-tower design showing principle of operation. The water from the heat source is distributed over the wet deck surface by spray nozzles. Air is simultaneously blown upward over the wet deck surface, causing a small portion of the water to evaporate. This evaporation removes heat from the remaining water. The cooled water is collected in the tower sump and returned to the heat source. (By permission of Baltimore Aircoil.)

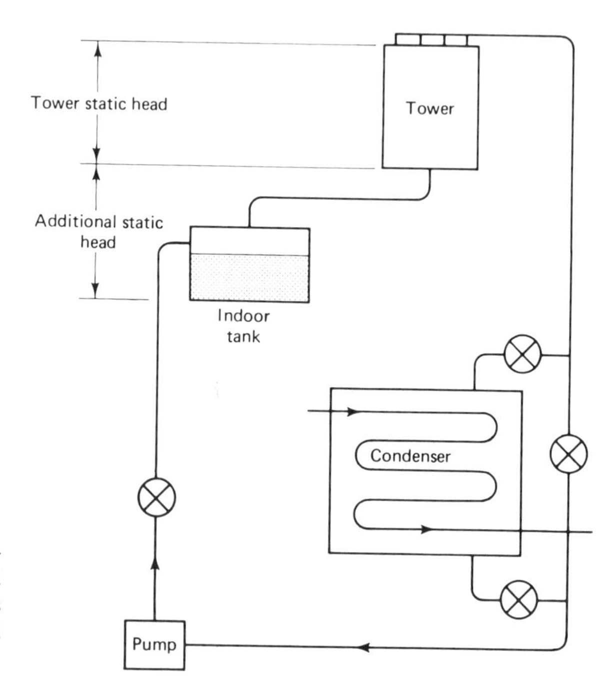

Figure 5-25 Freeze-protection arrangement using cooling tower. Water tank and pump are inside the building. Water flows only if the temperature can be maintained above freezing.

diagram is shown in Figure 5-24. Towers must, of course, reduce the water temperature in the same range as the temperature rise in the water-cooled condenser.

Like the evaporative condenser, its capacity is dependent on the ambient wet-bulb temperature. The leaving-water temperature can approach the wet-bult temperature within 7 to 10°F.

Cooling towers can be located inside the building using a suitable fan and ductwork, or they can be located outside on the ground or roof. Figure 5-11 shows a piping schematic and indicates the proper location of the bleed-off. In addition to determining the gpm, it is important to determine the pumping head for properly selecting the pump.

Items that must be considered in selecting the pump are as follows:

1. Pressure drop through the condenser
2. Pressure drop through the piping
3. Static pressure head lift from sump to water inlet area

Towers must be kept clean, and in most areas water treatment is required. A flow rate of 3 gpm per ton is design practice in most areas, although by means of a bypass, tower flow can be increased to increase capacity. An inside sump can be provided for winter use (Figure 5-25).

Closed-Circuit Evaporative Water Coolers

A closed-circuit evaporative cooler has the same configuration as the evaporative condenser shown in Figure 5-21, except that water flows through the condenser coil in place of refrigerant. Some modifications of the water coil are necessary to be able to circulate large quantities of water with a reasonable pressure drop.

If these units are used during the winter months, antifreeze must be used to prevent freeze-up. Water temperatures can be controlled to some extent by the use of dampers. Antifreeze solution decreases the transfer rate and increases the pumping head and therefore should be taken into consideration in the design of the system.

The advantage of this arrangement is that the condenser water is in a closed circuit and the water is kept clean. The full capacity of the water-cooled condenser is maintained.

In an installation such as a supermarket, water circuits from individual condensing units are manifolded to connect to the closed-circuit source of water for condensing.

This arrangement provides a uniform head pressure for all connected condensing units. Water regulating valves can be installed for each condenser if various condensing temperatures are to be maintained. The closed circuit must include an expansion tank.

WATER REGULATING VALVES

The water regulating valve is normally applied to wastewater systems to conserve the use of water. They are sometimes applied to recirculated water systems using cooling towers to control the head pressure (condensing temperature). A diagram of a typical wastewater

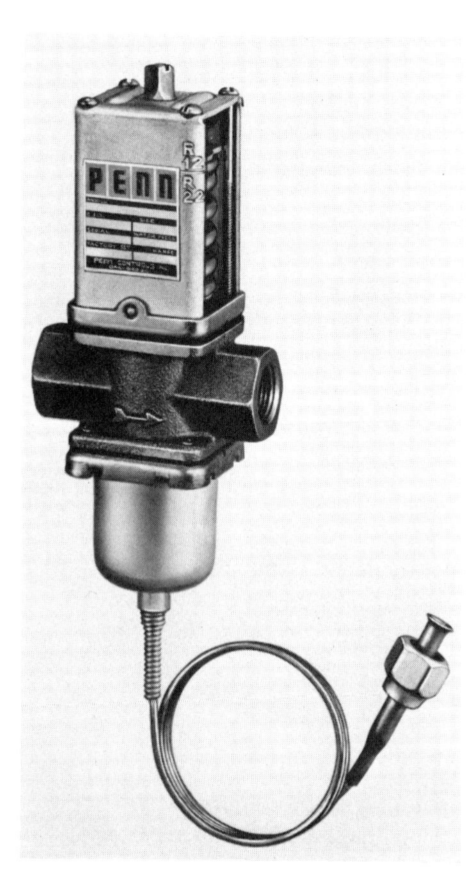

Figure 5–26 Water regulating valve, pressure actuated. (By permission of Johnson Controls, Inc.)

Figure 5–27 Selection chart for water regulating valves. (By permission of Johnson Controls, Inc.)

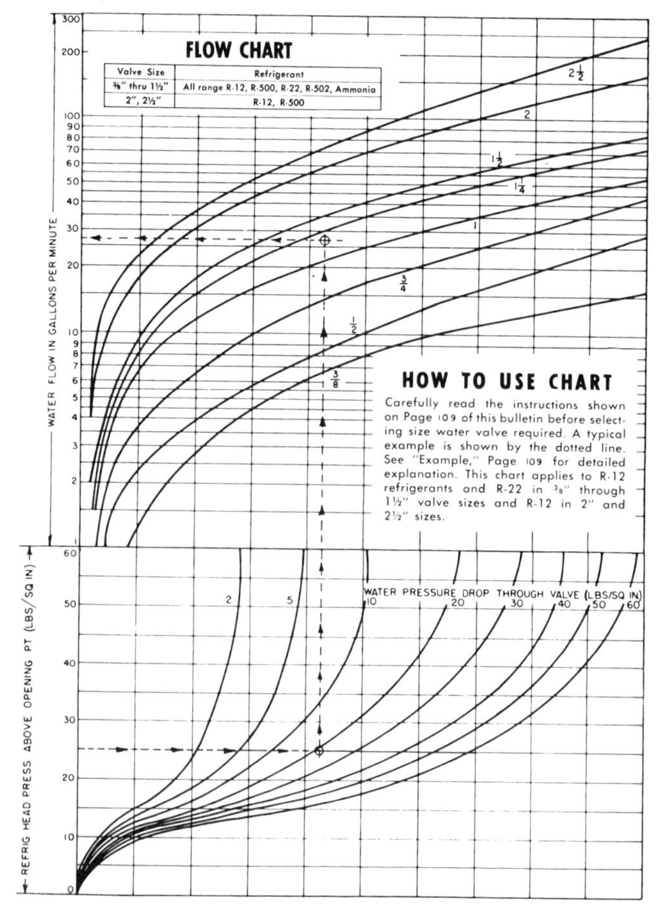

valve is shown in Figure 5–26. The bellows in the lower portion of the valve is connected into the refrigerant side of the condenser sensing head pressure. The valve is connected to the inlet side of the condenser. It modulates the water flow to maintain nearly constant head pressure. The pressure regulator is set to permit the valve to shut off the water when the compressor stops. Selection of the proper valve size is somewhat involved, but the following is an example (refer to the selection chart in Figure 5–27).

1. Determine the maximum water flow required. The manufacturer of the condensing unit can usually provide tables.
2. Draw a horizontal line across the upper half of the flowchart through the flow required.
3. Determine the refrigerant head pressure rise above the valve opening point.
    a. The valve closing point (to assure closure under all conditions) must be the refrigerant head pressure equivalent to the highest ambient air temperature expected at the time of the maximum load. Read this in pounds per square inch gauge from a "saturated vapor table" for the refrigerant selected.
    b. From the same table read the operating head pressure corresponding to the condensing temperature selected.
    c. The valve opening point will be about 7 psig above the closing point.
    d. Subtract the opening pressure from the operating pressure. This gives the head pressure rise.
4. Draw a horizontal line across the lower half of the flowchart through this value.
5. Determine the water pressure drop through the valve. This is the pressure actually *available* to force water through the valve.
    a. Determine the *minimum* water pressure available from city mains or other source.
    b. From condensing unit manufacturers' tables, read the pressure drop through the condenser corresponding to the flow required.
    c. Add to this the estimated or calculated drop through piping, and so on, between the water valve and the condenser, and from the condenser to the drain (or sump of a cooling tower).
    d. Subtract the total condenser and piping drop from the available water pressure. This is the *available* pressure drop through the valve.
6. On the lower half of the curve, mark the point on the horizontal head pressure line drawn in corresponding to the available water pressure drop through the valve. Interpolate between curves, or choose the curve for the nearest *lower* drop for which the curve is drawn (this gives an automatic safety factor).
7. From this point draw a line vertically upward until it intersects the water flow line in the upper half of the flowchart.
8. If the intersection falls on a valve size curve, this is the valve size.
9. If the intersection falls between two curves, the required valve size is the *larger* of the two.

Example

1. The required flow for an R-12 system is found to be 27 gpm. Condensing pressure is 125 psig and maximum ambient temperature is estimated at 86 °F. City water pressure is 40 psig and the manufacturers' table gives the drop through the condenser and accompanying piping and valves as 15 psi. The drop through installed piping is approximately 4 psi.
2. Draw a line through 27 pgm (see the dashed line in the upper half of the flowchart, Figure 5–27).
3. The closing point of the valve is the pressure of R-12 corresponding to 86 °F. Ambient = 93 psig.
4. The opening point of the valve is 93 + 7 = 100 psig.
5. The head pressure rise = 125 − 100 = 25 psig.
6. Draw a line through 25 psig (see the dashed line in the lower half of the flowchart Figure 5–27).
7. The available water pressure drop through the valve = 40 − 19 = 21 psi.
8. Interpolate just over the 20-psi curve (the circle on the lower half of the flowchart).
9. Draw a vertical line upward from this point to the flow line (the circle on the flowchart marks this intersection).
10. This intersection falls between curves for 1- and 1¼-in. valves. Therefore, a 1¼-in. *valve is required.*

Three-Way Water Regulating Valves

An illustration of the use of a water regulating valve to control head pressure on a multiple condenser application is shown in Figure 5–28. The manufacturer gives the following information concerning installation:

1. Install a three-way valve as shown in Figure 5–28. Port 1 is for connection to the condenser inlet and port 3 is the bypass connection.
2. With the tower pump operating and all compressors shut down, manually flush each valve by lifting the range spring follower with screwdrivers at two sides of the lower spring cap. This does not affect valve adjustment.
3. When used on a single-condenser system, the square head cock in the bypass should be adjusted with the compressor shut down and the tower pump operating. Adjust the cock so that the amount of water through the bypass is just sufficient to provide the minimum recommended nozzle pressure.
4. On a multiple-condenser system, the square head cocks in the bypass should be adjusted evenly with the compressors shut down and the tower pump operating. The total flow through all the bypasses should be just sufficient to provide the minimum recommended nozzle pressure.
5. The R-22 valves are factory set to start flow to the condenser at 165 psig and be full open at 215 psig. The R-12 valves are factory set to start flow to the con-

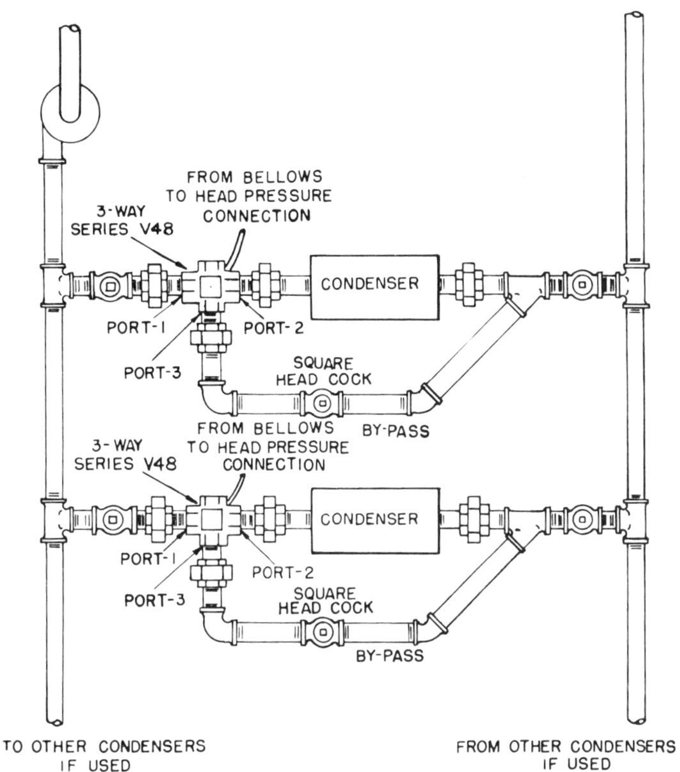

Figure 5–28 Three-way water regulating valve used to pipe multiple condenser. (By permission of Johnson Controls, Inc.)

denser at 95 psig and be fully open at 130 psig. The opening point may be increased or decreased by turning the adjustment screw counterclockwise or clockwise, respectively. Any increase or decrease in the opening point will result in a similar increase or decrease in the pressure at which the valve is fully open to the condenser.

CONDENSER CONTROLS

Some type of regulation needs to be provided to maintain the condensing temperature within reasonable limits. Controls are provided to serve the following functions:

1. To prevent too high a condensing temperature
2. To prevent too low a condensing temperature
3. To interlock the flow of the condensing medium to operate only when the compressor operates
4. To prevent freeze-up for all water-related equipment

To prevent too high a condensing temperature, a high-pressure control (HPC) is provided in the discharge side of the system, to stop the compressor if excessive head pressures are reached. This HPC control is common to most systems. Most of these controls are lockout devices and need to be reset before the compressor can be started. This provides an opportunity for the service person to determine the source of the problem before restarting the equipment.

The controls needed *to prevent too low a condensing temperature* vary somewhat with the type of system. For water-cooled condenser units using city water, a water regulating valve is used to control head pressure. Where the source of water for the condenser is from a cooling tower, the fan or fans can be cycled to control water temperature. When towers operate in the winter with an inside sump, the tower circuit pump can be cycled to control water temperature. For air-cooled condensers, dampers can be regulated, fans cycled, or liquid refrigerant used to flood part of the condenser coil. For evaporative condensers, the fan and/or pump can be cycled.

Electrical interlocks can be provided on the starting equipment to cause accessories such as a condenser fan and pumps to start and stop with the compressor.

There are various ways of *providing freeze-up protection* for water-related equipment. It is imporatnt to prevent the water from reaching freezing temperatures unless antifreeze is used. Antifreeze solutions in large quantities are expensive and are used only as a last resort. Towers that circulate water in the winter do so only when the water temperature is above freezing and quickly return the water to an inside sump. A similar arrangement is used for evaporative condensers. Water-cooled condensers must be located in a heated room in the winter.

CONDENSER MAINTENANCE

It is extremely important to keep condensers clean to maintain their capacity. The same is true of cooling towers. When water is reused, proper bleed-off needs to be provided. When the water hardness or the growth of algae is a problem, proper chemical treatment should be provided.

When rapid descaling is required, an inhibited solution of muriatic acid (18%) may be used. Refer to the illustration of descaling equipment in Figure 15-3. The condenser should be isolated from the rest of the system during descaling. After using acid the condenser must be thoroughly flushed with fresh water to remove all traces of acid.

REVIEW EXERCISE

Select the letter representing your choice of the correct answer (there may be more than one).

5-1. The condensing temperature of air-cooled condensers is usually how many degrees above design ambient?

_______ a. 10°F

_______ b. 20°F

_______ c. 30°F

_______ d. 40°F

5-2. Water-cooled condensers are usually selected on the basis of what condensing temperatures?

_______ a. 100°F

_______ b. 105°F

_______ c. 110°F

_______ d. 115°F

5-3. Which of the following is *not* a method of controlling head pressure on an air-cooled condenser?

_______ a. Fan cycling control

_______ b. Inlet air damper

_______ c. Pressure stabilizer

_______ d. Pump-down cycle

5-4. Does the Btu per minute removed by the refrigerant condenser per ton of refrigerating effect (in the evaporator) increase or decrease with a rise in suction temperature?

_______ a. Increases

_______ b. Decreases

5-5. The following is *not* a type of water-cooled condenser:

_______ a. Compound

_______ b. Shell-and-tube

_______ c. Shell-and coil

_______ d. Tube-in-tube

5-6. Assuming equal refrigeration capacity, which type of condenser uses the least amount of water?

_______ a. Wastewater condenser

_______ b. Evaporative condenser

5-7. If the water-cooled condenser capacity is 30,000 Btu and the flow through the condenser is 66 gpm, the temperature rise is:

_______ a. 5°F

_______ b. 10°F

_______ c. 15°F

_______ d. 20°F

5-8. Chemical cleaning is always required in the following type of water-cooled condenser:

________ a. Shell-and-tube

________ b. Tube-in-tube

________ c. Shell-and-coil

________ d. Evaporative

5-9. Which type of tube-in-tube condenser has the highest capacity?

________ a. Parallel flow

________ b. Counterflow

5-10. A special tool in an electric drill can be used to clean which type of water-cooled condenser?

________ a. Closed circuit

________ b. Shell-and-coil

________ c. Evaporative

________ d. Tube-in-tube

6 Refrigerant Metering Devices and Controls

OBJECTIVES

After studying this chapter, the reader will be able to:

- Identify the various types of metering devices and refrigerant flow controls.
- Describe the operation of the various types of metering devices.
- Select the type of metering device and refrigerant control that best fits the application.

TYPES OF METERING DEVICES

A metering device is the component in the refrigeration system used to control the flow of refrigerant from the high-pressure side to the low-pressure side of the system. These include:

1. Hand expansion valves
2. Automatic expansion valves (AEVs)
3. Thermostatic expansion valves (TEVs)
4. Capillary tubes
5. Low-side floats (LSFs)
6. High-side floats (HSFs)

There are some other refrigerant flow controls which are important accessories to many systems. These include:

7. *Solenoid valves:* used in the liquid line, suction line, or hot-gas line
8. *Pressure regulating valves* (PRVs): used to control evaporating pressure, condensing pressure, or for hot-gas bypass

HAND EXPANSION VALVES

The hand expansion valve was one of the earliest types of metering devices. It consists of a needle-nosed plunger that can be manually adjusted to regulate the flow through an orifice. The pressure drop across the valve is constant and the flow through the valve is dependent on the size of the orifice. Whenever the load changes, the valve must be readjusted, thus requiring the attention of an operator. When the system is shut down, the valve must be closed and opened again when the system is started (see Figure 6–1).

The hand valve is seldom used as an operating control on modern systems. It is occasionally installed in a bypass around an automatic valve as a standby facility in case the operating valve needs to be repaired.

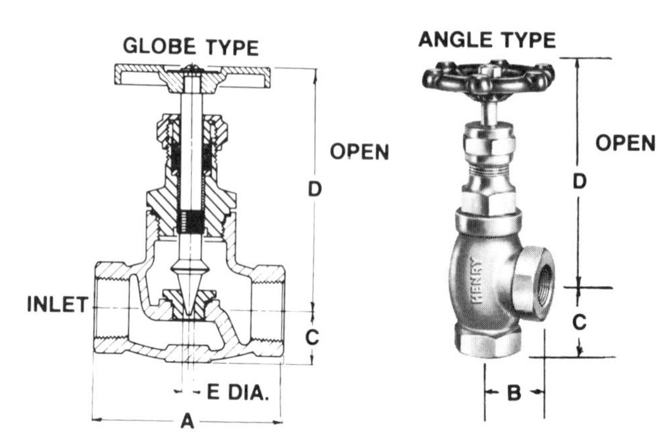

Figure 6–1 Hand expansion valve. Screw end design. Maximum working pressure, 400 psi. Maximum temperature rating is 300°F. This valve is suitable for ammonia. It can be backseated and repacked under pressure. (By permission of Henry Valve Company.)

AUTOMATIC EXPANSION VALVES

The automatic expansion valve (AEV) regulates the flow of refrigerant to maintain a constant evaporator pressure. A schematic diagram of the AEV is shown in Figure 6–2. The plunger position in the orifice opening is regulated by the position of the valve

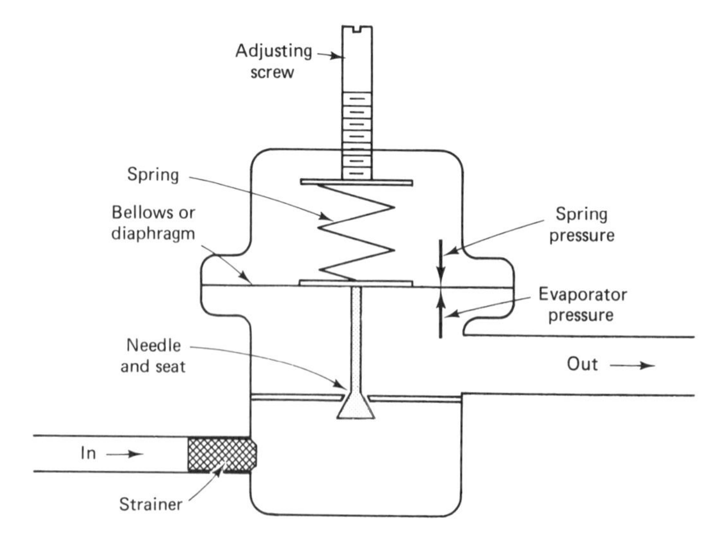

Figure 6–2 Automatic expansion valve, fully adjustable 0 to 80 psig. Internal or external equalizer. This valve is sometimes used for hot-gas bypass.

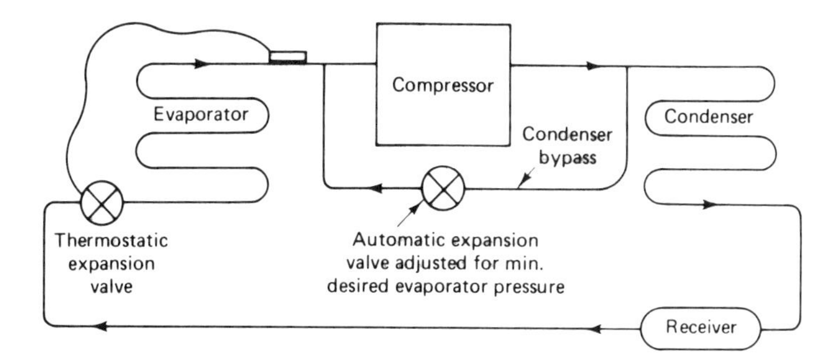

Figure 6–3 Condenser bypass using automatic expansion valve. The AEV supplies hot gas to the suction line to hold the suction pressure above a set minimum.

diaphragm. Two pressures regulate the flow: the suction pressure and the adjustable spring pressure. As indicated in the diagram, these pressures oppose each other. The best application of the AEV is a system where it is essential to maintain a constant evaporating temperature. One such use is in a drinking water cooler to hold the suction temperature above freezing.

The big disadvantage of the valve is its inability to increase flow with increased load. Actually, an increase in load will attempt to raise the suction pressure, causing the valve to move toward the closed position. A decrease in load will tend to open the valve. If constant suction pressure is desired, the valve will perform this function. If the refrigerant flow needs to be load oriented, the AEV should not be used. Actually, the AEV was an early attempt to control refrigerant flow automatically and has generally been superseded by other devices.

The automatic expansion valve can be used in a hot-gas bypass line as shown in Figure 6–3. This arrangement prevents the suction pressure from going below a set minimum.

THERMOSTATIC EXPANSION VALVES

The thermostatic expansion valve (TEV) is the most widely used type of metering device. There are a number of reasons for this:

1. *The TEV is load oriented.* Thus, when the load is increased, the valve opens to increase the refrigerant flow.
2. *The TEV protects the compressor* by vaporizing the liquid refrigerant before it leaves the evaporator.
3. *The TEV can be used to feed an evaporator* that has a higher-than-average pressure drop. This is accomplished by use of the external equalizer.

A cutaway view of a thermal expansion valve is shown in Figure 6–4. The valve operates as a result of the balance of three pressures: (1) suction pressure, (2) bulb pressure, and (3) spring pressure. Bulb pressure is exerted on the top of the diaphragm. It is opposed by the other two pressures: suction and spring pressures.

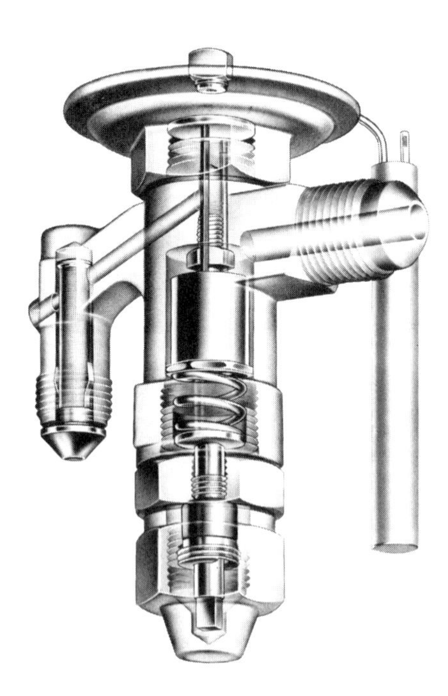

Figure 6-4 Thermostatic expansion valves regulate the flow of refrigerant to match the load. Superheat is adjustable. External equalizers can be supplied. Internal parts can be removed during installation. (By permission of Sporlan Valve Company.)

VALVE TYPES	NOMINAL CAPACITY	EVAPORATOR TEMPERATURE DEGREES F.																	
		40°						20°						0°					
		PRESSURE DROP ACROSS VALVE (Pounds Per Square Inch)																	
		40	60	80	100	120	140	60	80	100	120	140	160	60	80	100	120	140	160
G-NI-F	¼	0.20	**0.25**	0.29	0.32	0.35	0.38	0.25	**0.29**	0.32	0.35	0.38	0.41	0.25	0.29	0.32	0.35	0.38	0.41
G-S-NI-F	½	0.41	**0.50**	0.58	0.64	0.71	0.76	0.50	**0.58**	0.64	0.71	0.76	0.82	0.42	0.48	0.54	0.59	0.64	0.68
G-S-NI-F	1	0.82	**1.00**	1.15	1.29	1.41	1.53	1.00	**1.15**	1.29	1.41	1.53	1.63	0.80	0.92	1.03	1.13	1.22	1.31
G-S-H	1½	1.31	**1.60**	1.85	2.06	2.26	2.44	1.20	**1.38**	1.55	1.70	1.83	1.96	1.15	1.33	1.48	1.63	1.76	1.88
G(Ext.)-C-S	2	1.63	**2.00**	2.31	2.58	2.83	3.06	1.65	**1.91**	2.13	2.33	2.52	2.69	1.55	1.79	2.00	2.19	2.37	2.53
C-S	2½	2.04	**2.50**	2.89	3.23	3.54	3.82	2.10	**2.42**	2.71	2.97	3.21	3.43	2.00	2.31	2.58	2.83	3.05	3.26
C-S	3	2.45	**3.00**	3.46	3.87	4.24	4.58	2.60	**3.00**	3.36	3.68	3.97	4.24	2.30	2.66	2.97	3.25	3.51	3.76
H	4	3.27	**4.00**	4.62	5.16	5.66	6.11	3.10	**3.58**	4.00	4.38	4.73	5.06	2.65	3.06	3.42	3.75	4.04	4.33
C & S(Ext.)-H	5	4.08	**5.00**	5.77	6.45	7.07	7.64	3.60	**4.16**	4.65	5.09	5.50	5.88	3.00	3.46	3.87	4.24	4.58	4.90
S(Ext.)	6	4.90	**6.00**	6.93	7.74	8.48	9.16	4.32	**4.99**	5.58	6.11	6.60	7.05	3.60	4.16	4.65	5.09	5.50	5.88
P-H	8	6.12	**7.50**	8.66	9.68	10.6	11.5	6.50	**7.51**	8.39	9.19	9.93	10.6	5.20	6.00	6.71	7.35	7.94	8.49
P-H	12	9.55	**11.7**	13.5	15.1	16.5	17.9	10.0	**11.5**	12.9	14.1	15.3	16.3	8.60	9.93	11.1	12.2	13.1	14.0
O	6	4.90	**6.00**	6.90	7.80	8.50	9.20	4.60	**5.31**	5.94	6.51	7.03	7.51	3.80	4.39	4.91	5.37	5.80	6.21
O	9	7.35	**9.00**	10.4	11.6	12.7	13.8	7.50	**8.66**	9.69	10.6	11.5	12.3	6.50	7.51	8.39	9.19	9.93	10.6
O	12	9.55	**11.7**	13.5	15.1	16.5	17.9	10.0	**11.5**	12.9	14.1	15.3	16.3	8.60	9.93	11.1	12.2	13.1	14.0
O	16	12.6	**15.5**	17.9	20.0	21.9	23.7	14.0	**16.2**	18.1	19.8	21.4	22.9	12.0	13.8	15.5	17.0	18.9	19.6
O	23	18.8	**23.0**	26.6	29.7	32.5	35.1	22.0	**25.4**	28.4	31.1	33.6	35.9	20.0	23.1	25.8	28.3	30.6	32.7
O	40	32.7	**40.0**	46.2	51.6	56.6	61.1	35.0	**40.0**	45.2	49.5	53.5	57.2	30.0	34.6	38.7	42.4	45.8	49.0
M	15	12.7	**15.5**	17.9	20.0	21.9	23.7	14.2	**16.4**	18.3	20.1	21.7	23.2	11.5	13.3	14.8	16.3	17.6	18.8
M	20	16.3	**20.0**	23.1	25.8	28.3	30.6	16.0	**18.5**	20.7	22.6	24.4	26.1	14.0	16.2	18.0	19.8	21.4	22.9
M	25	20.4	**25.0**	28.9	32.3	35.4	38.2	19.0	**21.9**	24.5	26.9	29.0	31.0	16.0	18.5	20.7	22.6	24.4	26.1
V	35	28.6	**35.0**	40.4	45.2	49.5	53.5	32.0	**36.9**	41.3	45.3	48.9	52.3	24.0	27.7	31.0	33.9	36.7	39.2
V	45	36.7	**45.0**	52.0	58.1	63.6	68.7	40.5	**46.8**	52.3	57.3	61.9	66.1	30.5	35.2	39.4	43.1	46.6	49.8
V	55	44.9	**55.0**	63.5	71.0	77.8	84.0	47.0	**54.3**	60.7	66.5	71.8	76.7	36.0	41.6	46.5	50.9	55.0	58.8
W	80	69.4	**85.0**	91.8	110	120	130	72.6	**78.5**	94.1	103	111	119	55.6	60.1	72.0	78.5	85.1	90.9
W	110	93.0	**114**	132	147	161	174												

Figure 6-5 Selection chart for thermal expansion valve. This chart shows ton of refrigeration capacity using R-12 for various evaporating temperatures and pressure drops across the valve. Capacities shown are based on 100°F liquid temperatures. (By permission of Sporlan Valve Company.)

VALVE TYPES	NOMINAL CAPACITY	EVAPORATOR TEMPERATURE DEGREES F. −10°						−20°						−40°					
		PRESSURE DROP ACROSS VALVE (Pounds Per Square Inch)																	
		80	100	120	140	160	180	80	100	120	140	160	180	80	100	120	140	160	180
G-NI-F	¼	0.26	**0.30**	0.32	0.35	0.38	0.40	0.23	0.26	0.28	0.31	0.33	0.35	0.15	0.17	**0.18**	0.20	0.21	0.22
G-S-NI-F	½	0.42	**0.46**	0.50	0.55	0.59	0.62	0.35	0.39	0.42	0.46	0.49	0.52	0.23	0.26	**0.28**	0.31	0.33	0.35
G-S-NI-F	1	0.81	**0.90**	0.99	1.07	1.14	1.21	0.69	0.77	0.85	0.92	0.98	1.04	0.48	0.54	**0.59**	0.64	0.68	0.73
G-S-H	1½	1.00	**1.12**	1.23	1.33	1.42	1.51	0.89	0.99	1.09	1.18	1.26	1.33	0.58	0.64	**0.71**	0.76	0.82	0.87
G(Ext.)-C-S	2	1.30	**1.46**	1.60	1.73	1.84	1.96	1.15	1.29	1.41	1.53	1.63	1.73	0.77	0.86	**0.95**	1.02	1.09	1.16
C-S	2½	1.62	**1.81**	1.98	2.14	2.29	2.42	1.38	1.55	1.70	1.83	1.96	2.08	0.92	1.03	**1.13**	1.22	1.31	1.38
C-S	3	2.08	**2.32**	2.54	2.75	2.94	3.12	1.85	2.06	2.26	2.44	2.61	2.77	1.15	1.29	**1.41**	1.53	1.63	1.73
H	4	2.48	**2.77**	3.04	3.28	3.51	3.72	2.08	2.32	2.55	2.75	2.94	3.11	1.27	1.42	**1.55**	1.68	1.79	1.90
C & S(Ext.)-H	5	2.89	**3.23**	3.54	3.82	4.08	4.33	2.31	2.58	2.83	3.06	3.26	3.46	1.38	1.55	**1.69**	1.83	1.96	2.08
S(Ext.)	6	3.47	**4.48**	4.91	5.30	5.67	6.01	2.77	3.58	3.92	4.23	4.52	4.80						
P-H	8	4.27	**4.78**	5.23	5.65	6.04	6.41	3.00	3.36	3.68	3.97	4.24	4.50						
P-H	12	7.27	**8.13**	8.90	9.62	10.3	10.9	5.43	6.07	6.65	7.18	7.67	8.15						
O	6	3.40	**3.80**	4.16	4.50	4.81	5.10	1.80	2.01	2.20	2.36	2.55	2.70						
O	9	5.80	**6.48**	7.10	7.67	8.20	8.70	4.60	5.14	5.63	6.09	6.51	6.90						
O	12	8.00	**8.94**	9.80	10.6	11.3	12.0	7.00	7.83	8.57	9.26	9.90	10.5						
O	16	10.0	**11.2**	12.2	13.2	14.1	15.0	9.1	10.2	11.1	12.0	12.9	13.7						
O	23	15.0	**16.8**	18.4	19.8	21.2	22.5	12.0	13.4	14.7	15.9	17.0	18.0						
O	40	20.0	**22.4**	24.5	26.5	28.3	30.0	15.0	16.8	18.4	19.8	21.2	22.5						
M	15	12.1	**13.6**	14.6	16.0	17.2	18.2	10.9	12.3	13.4	14.5	15.5	16.5						
M	20	14.7	**16.4**	18.0	19.4	20.7	22.0	12.1	13.6	14.8	16.0	17.1	18.2						
M	25	16.2	**18.1**	19.8	21.4	22.9	24.3	13.7	15.4	16.8	18.2	19.4	20.6						
V	35	22.1	**24.7**	27.0	29.2	31.2	33.1	20.8	23.2	25.5	27.5	29.4	31.2						
V	45	29.1	**32.5**	35.6	38.5	41.1	43.6	24.8	27.8	30.4	32.8	35.1	37.2						
V	55	37.6	**42.1**	46.1	49.8	53.2	56.5	29.2	32.7	35.8	38.6	41.3	43.8						
W	80	54.4	**65.2**	71.1	77.1	82.4	87.4	42.2	50.7	55.2	59.7	63.8	67.7						

REFRIGERANT LIQUID TEMPERATURE CORRECTION FACTORS

Refrigerant Liquid Temperature °F.	40°	50°	60°	70°	80°	90°	100°	110°	120°	130°	140°
Correction Factor	1.36	1.30	1.24	1.18	1.12	1.06	1.00	0.94	0.88	0.82	0.75

EXAMPLE: Actual capacity of nominal 3 ton valve at −20°F. evaporator, 140 psi pressure drop and 60°F. liquid temperature = 2.44 tons × 1.24 = 3.03 tons.

These factors include corrections for liquid refrigerant density and net refrigerating effect and are based on an average evaporator temperature of 0°F. **However they may be used for any evaporator temperature from −40°F. to 40°F. since the variation in the actual factors across this range is insignificant.**

Figure 6–5 (continued)

The actual function of the valve is to control the amount of superheat in the evaporator. Usually, factory setting of the spring adjustment is to maintain 10°F of superheat. Thus, if the evaporator pressure is 21.05 psig (20°F for R-12), the bulb pressure is 28.46 psig (30°F for R-12), and the spring pressure is 7.41 psig, the valve would control 10°F of superheat (30°F − 20°F = 10°F). The spring pressure (7.41 psig) + the evaporator pressure (21.05 psig) = the bulb pressure (28.46 psig) when the valve is in equilibrum. The capacity of the valve is dependent on the size of the orifice and the pressure difference across it. A typical selection chart is shown in Figure 6–5.

For larger systems, where a single valve will not offer proper control, a pilot-operated valve is used (Figure 6–6). A small solenoid valve in the pilot circuit converts the expansion valve into a stop valve when the solenoid is closed.

Another variation of the expansion valve is the electric-operated valve (Figure 6–7). This valve also controls the flow of refrigerant in response to the load conditions.

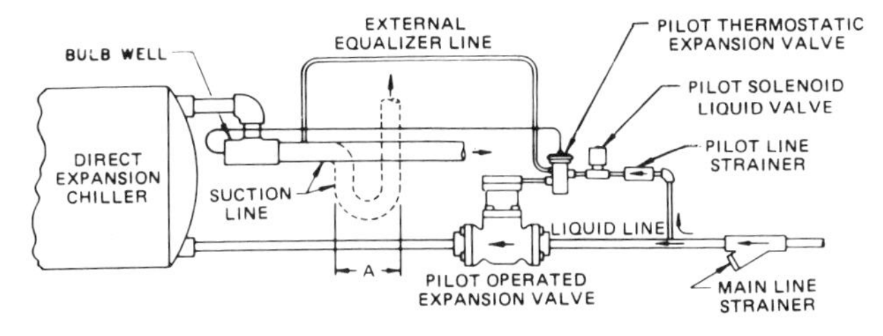

Figure 6–6 The pilot-operated thermostatic expansion valve is used on large systems where the capacity required is beyond the range of a direct operated valve. (By permission of *AHSRAE Handbook*.)

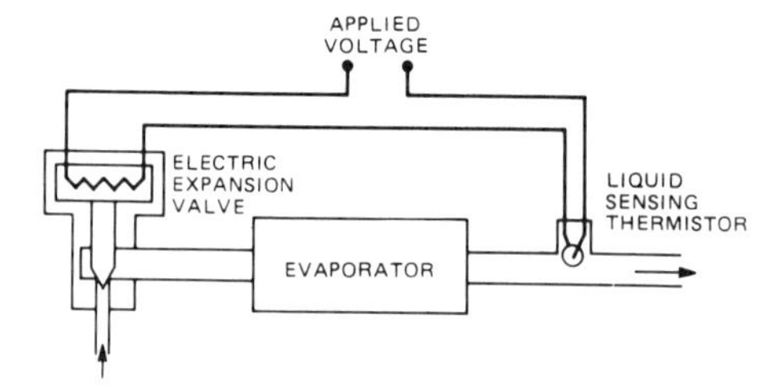

Figure 6–7 The electric expansion valve has the same function as a thermostatic expansion valve. The electric heat motor is operated by signals received from a liquid-sensing thermister. The valve is only sensitive to voltage and, therefore, can be used without an external equalizer. (By permission of *ASHRAE Handbook*.)

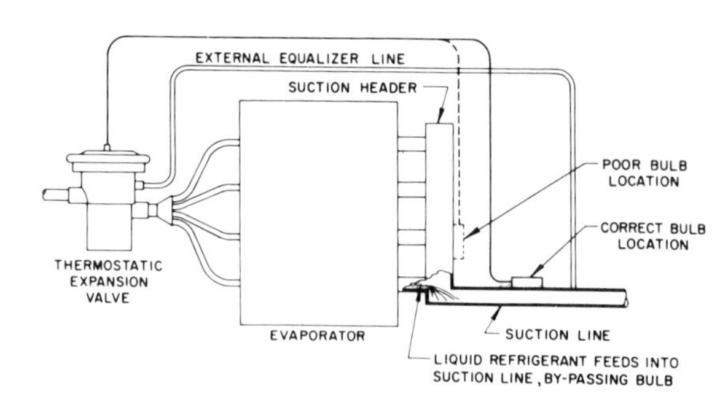

Figure 6–8 Thermal expansion valve with external equalizer. Bulb goes on suction line on the upstream side of the equalizer connection. (By permission of *ASHRAE Handbook*.)

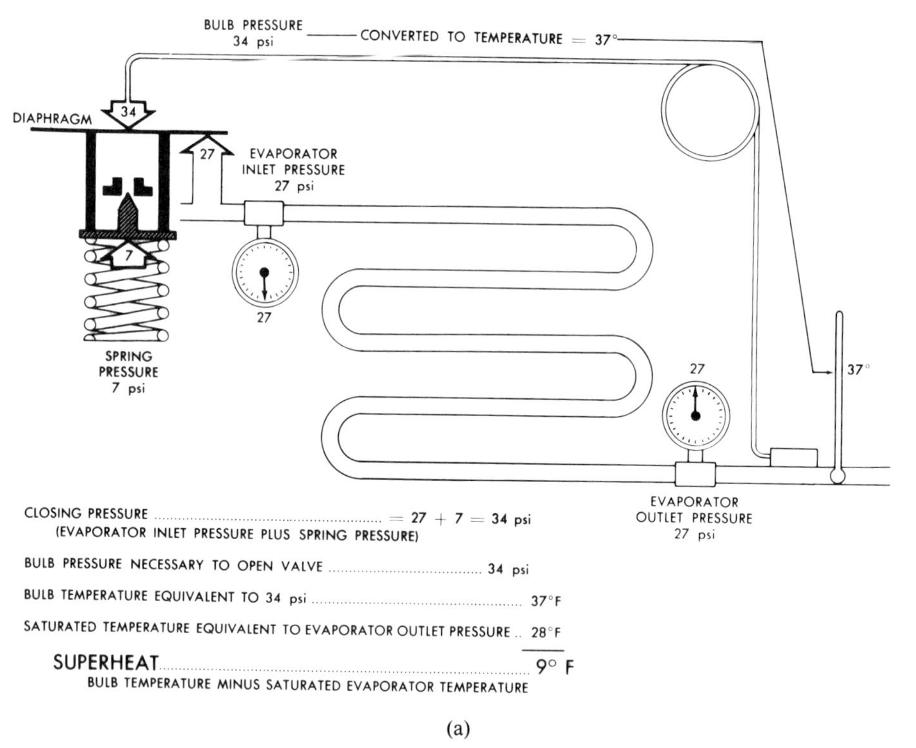

(a)

Figure 6–9 (a) System pressures using a standard evaporator with low-pressure drop; (b) system pressures using an expansion valve with an external equalizer to compensate for an evaporator with a high-pressure drop. (By permission of Sporlan Valve Company.)

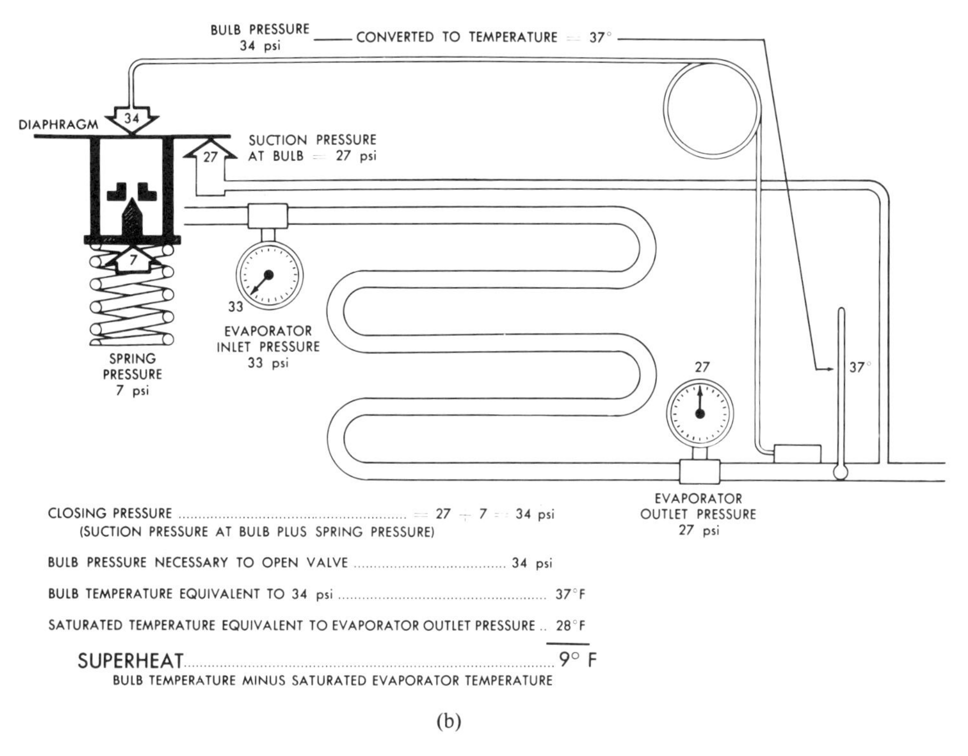

(b)

Figure 6–9 (continued)

When the pressure drop through the evaporator is greater than the equivalent of 2 °F, an external equalizer should be used. A diagrammatic view of the external equalized thermal expansion valve application is shown in Figure 6–8. The external equalizer is connected to the suction line on the compressor side of the bulb location. It senses the suction pressure near the bulb location, permitting an accurate control of superheat. Figure 6–9 shows the system pressures with and without an external equalizer.

Setting the Superheat

The procedure for setting the superheat on an expansion valve is as follows:

1. Read the temperature of the suction line at the point where the bulb is clamped.
2. Obtain the suction pressure at the bulb location. If an external equalizer is used, place a gauge in the equalizer line. If the valve is not externally equalized, read the suction pressure at the compressor and add the estimated pressure drop through the suction line from the bulb to the compressor. Convert the pressure at the bulb to a temperature by using the saturated pressure/temperature chart.
3. Determine the difference in temperature between the readings in steps 1 and 2.
4. Adjust the spring tension on the bulb to give the proper superheat setting.

Figure 6–9 illustrates the use of this procedure to determine the superheat.

Distributors

A distributor is required to feed the refrigerant properly for most designs of multicircuit evaporators. Typical distributors are shown in Figure 6–10. An external equalized expansion valve is required for these installations, due to the pressure drop through the distributor.

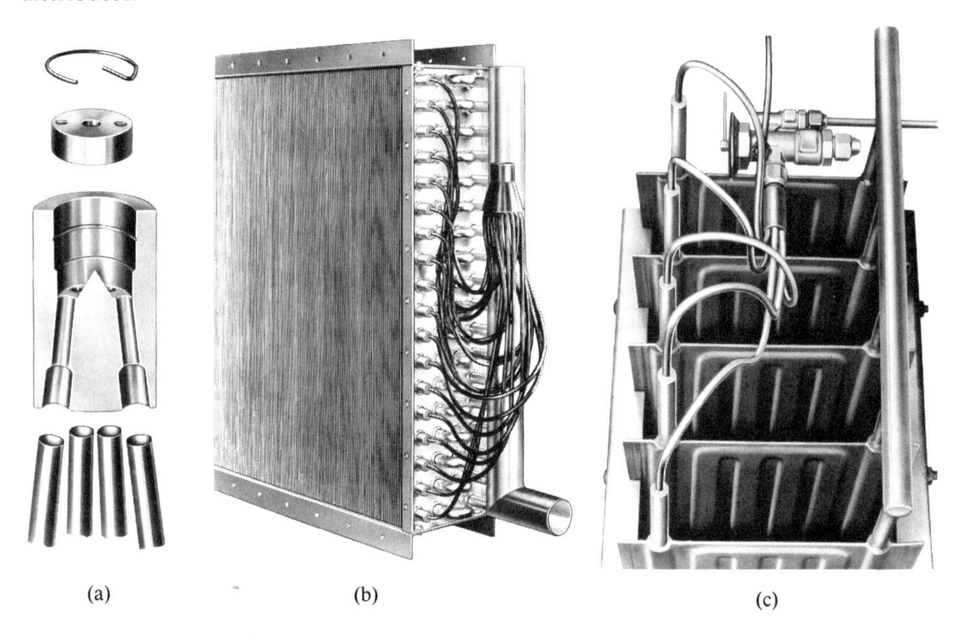

Figure 6–10 Refrigerant distributors: (a) exploded view; (b) distributor used on finned evaporator; (c) distributor used on plate evaporator. A portion of liquid flashes into vapor. Without a distributor some tubes of the coil can receive all vapor. (By permission of Sporlan Valve Company.)

Pressure-Limiting Valves

A pressure-limiting expansion valve is used to control the load on the compressor during pull-down. Usually, these valves have a bulb charged with sufficient refrigerant to permit all the liquid in the bulb to be vaporized at a certain pressure (the limiting pressure). Thus, these valves are sometimes termed *gas-charged expansion valves.* When the refrigerant in the bulb is completely vaporized, the pressure on the bellows will practically remain constant with an increase in temperature. Another type of pressure-limiting valve uses a spring to oppose the movement of the diaphragm at the limiting pressure.

Thermostatic Expansion Valve Charges

The thermostatic expansion valve controls the superheat in the evaporator by sensing the temperature of the refrigerant leaving the coil. Various types of charges have been used to control bulb pressure on the diaphram. One of these was described in the preceding section on pressure-limiting valves. Each type of charge has a specific use. A list of some of the more common charges is as follows:

1. Liquid charge
2. Liquid cross-charges
3. Gas cross-charges
4. Adsorption charges

Using the *liquid charge,* the power element contains the same refrigerant as the system. Thus the pressure–temperature curve for the charge is similar in shape and parallel to the system refrigerant charge. Liquid charges have both advantages and disadvantages. They are not subject to cross-ambient control losses. They have little or no superheat at startup. The superheat increases at lower evaporator pressures. However, they have a slow suction pressure pull-down after startup.

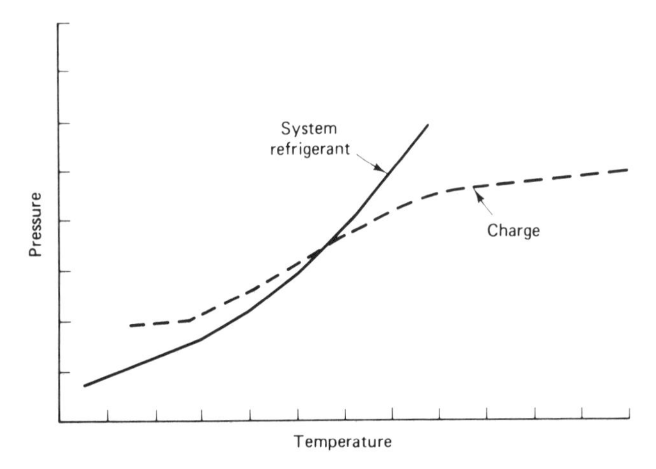

Figure 6–11 Typical liquid cross-charge pressure–temperature relationship. (By permission of Alco Controls.)

The *liquid cross-charge* contains liquid refrigerant that is differnt from the system refrigerant. The pressure/temperature curve of the charge crosses the curve of the system's refrigerant (see Figure 6–11). The spread of the two curves shown in Figure 6–11 during pull-down permits a large temperature change with a small amount of change in diaphragm pressure, improving the pull-down features of the valve and providing some limitation in the overload characteristics of the valve. The advantages include:

1. Moderately slow pull-down
2. Insensitivity to cross-ambient control
3. Dampened response to suction-line temperature changes, minimizing "hunting"
4. Superheat characteristics that can be tailored for specific applications

The *gas charge* has a different characteristic since it compresses. At some predetermined temperatures, the gas becomes superheated, limiting the force it exerts. This temperature is indicated as maximum operating pressure (MOP). However, gas charges are susceptible to cross ambient control when the power element is colder than the remote bulb. To control the MOP, some gas charges use a different refrigerant, producing a cross-charge effect.

The *adsorption charge* uses some type of solid, such as charcoal, silica gel, or activated aluminum, to hold large quantities of gas. The quantity of vapor adsorbed varies with the temperature and the system. Thus it can be used to exert operating pressure as a function of temperature. The great advantage of the adsorption charge is its ability to function between temperatures of +50 and −50 °F, making it possible to use the same valve for applications within a wide temperature range. The curve of the adsorption charge is similar to the cross-charge shown in Figure 6–11.

Expansion Valve and Bulb Location

The expansion valve should be located as close to the evaporator as possible. When a liquid-filled bulb is used, the expansion valve can be located in any convenient location: right side up, upside down, and so on. With a gas-filled bulb the valve should always be in a warmer location than the bulb, with the power side up.

The bulb should be firmly clamped to a horizontal section of the suction line, preferably in a 10 o'clock or 2 o'clock position. The bulb can best be located within the refrigerated space. If the bulb is located outside the refrigerated space, both the valve and the suction line should be well insulated.

When there is a possibility of the valve opening when the compressor is off, a solenoid valve should be installed in the liquid line and the system operated on a "pump-down" cycle.

The bulb should never be located in a liquid trap. The proper application of the bulb is shown in Figure 6–12.

TEV Hunting

Hunting is the condition where the valve cycles toward an open position and then toward a closed position without reaching a stabilized condition. This can be caused by the use of a larger valve than necessary, but not always. Where there are changes in the load the valve adjusts the flow to maintain superheat. Since there is a time lag between the time the change is made in the orifice size and the time the change is sensed at the bulb location, there is bound to be some overfeeding and starving of the refrigerant feed to the evaporator.

CAPILLARY TUBES

The capillary tube is a type of refrigerant metering device constructed of tubing with a small orifice. It is used primarily on hermetically sealed systems where the tubing containing the metering orifice can be brazed into the system. These systems require a critical charge of refrigerant, usually measured and indicated on the system by the manufacturer. The use of a capillary tube makes possible unloading the compressor on shutdown, permitting the use of low-torque-compressor motors.

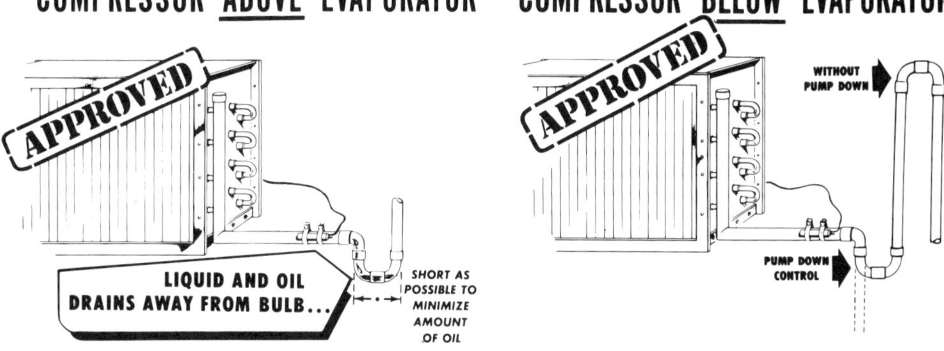

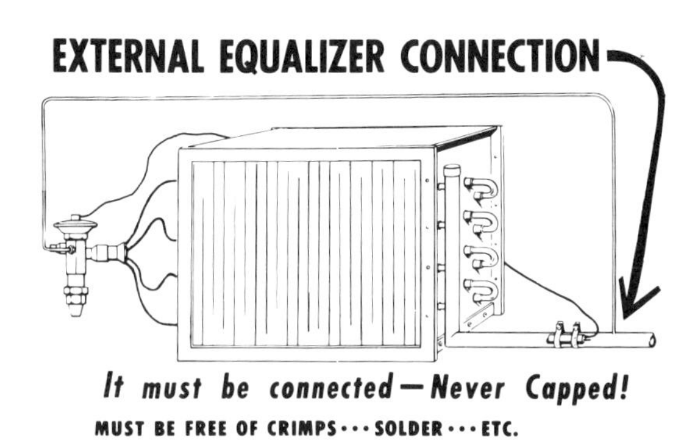

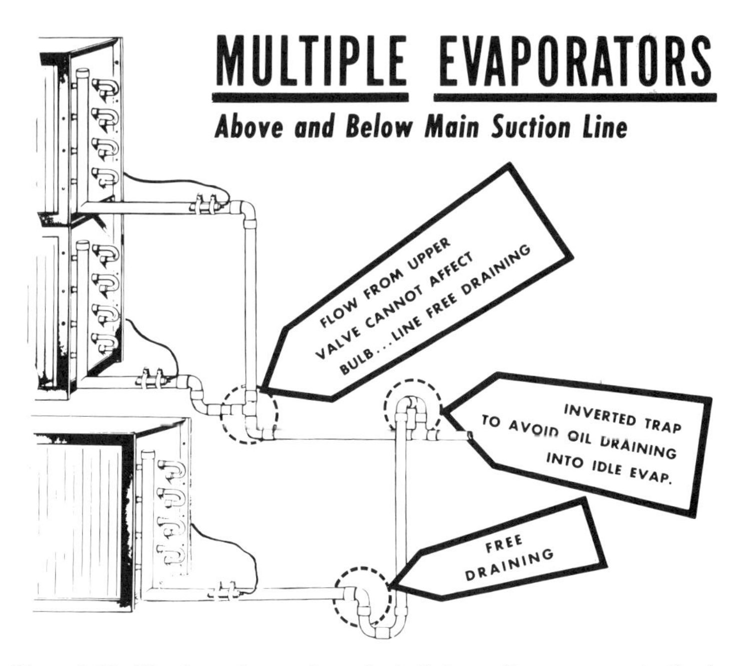

Figure 6–12 The thermal expansion valve bulb is usually most conveniently placed on a horizontal portion of the suction line leaving the evaporator as close to the evaporator as possible. On suction lines ⅞ in. O.D. and larger, the best location is at a 4-o'clock position. The bulb should not be located in a trap. (By permission of Sporlan Valve Company.)

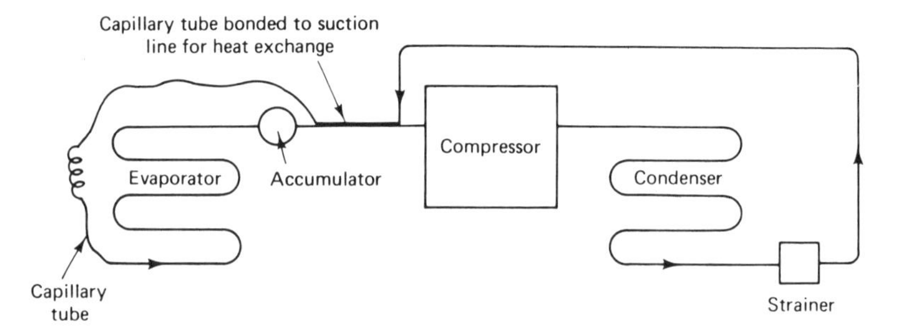

Figure 6–13 Capillary tube system. Capillary tubes act as a restrictor or orifice between condenser and evaporator. A critical refrigerant charge is required. Heat exchanger improves performance of the system.

The amount of refrigerant fed through the capillary tube is dependent on the pressure difference between the two ends of the tube and the size and length of the orifice. The system operates best at design conditions. However, it is load oriented since the pressure drop across the orifice varies directly as the load.

It is general practice to bond the liquid line to the suction line for a considerable length (Figure 6–13). This provides a heat exchanger, which improves the performance of the system.

The capillary tube is generally not subject to service problems, provided that the system of which it is a part is used properly. If the condenser becomes blocked, the discharge temperature can reach a high temperature, causing a breakdown of the oil in the system, clogging the inlet to the tube. In these instances the hermetic systems must be entered. Reaming devices under hydraulic pressure are available for opening up a clogged capillary tube. It is a major operation to open the system and clean the capillary tube.

When the system is recharged, it is important to evacuate the system properly and charge it with the correct amount of refrigerant. Manufacturers are required to state the refrigerant and charge on a suitable plate attached to a critically charged unit.

LOW-SIDE FLOATS

The low-side float is used primarily on ammonia systems with flooded evaporators (see Figure 6–14). Using a low-side float and regulating the liquid level in an accumulator header permits operating a number of evaporators on the same metering device. This type of system operates with practically zero superheat and, with a flooded evaporator, provides excellent efficiency. The float will close when the compressor stops, shutting down the flow of refrigerant. The float system is load oriented since the flow is directly proportionate to the rate of evaporation.

The low-side float maintains a constant level of liquid in the evaporator and feeds liquid to match the rate of evaporation. Some floats are installed in a separate connected circuit to the accumulator (called a quiet area). This arrangement prevents improper opening of the valve due to turbulence in the accumulator. Stop valves are usually installed in all connecting lines to the float chamber to facilitate removal in case service is required.

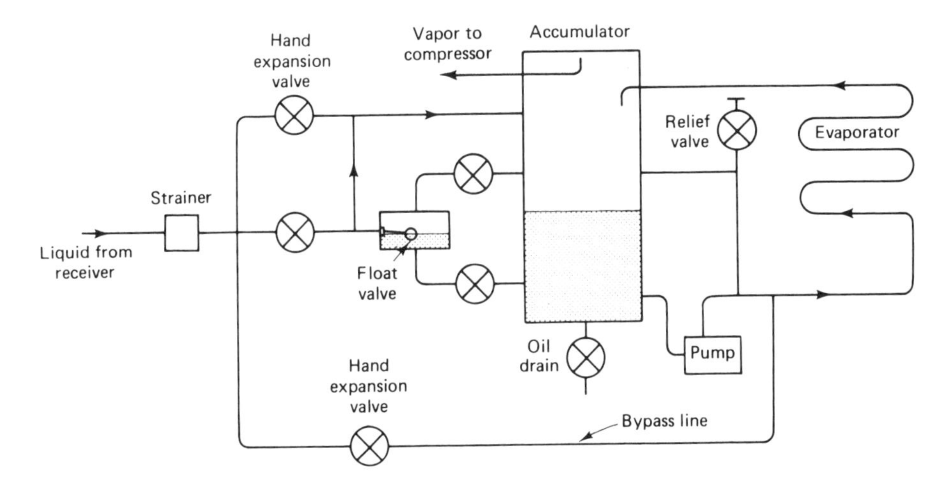

Figure 6-14 Low-side float valve, regulates flow of liquid from the receiver. In the illustration the float senses the level in the accumulator and maintains a constant level. The pump circulates liquid through the evaporator.

A bypass with a hand valve is often installed to provide system operation while the float valve is being repaired.

HIGH-SIDE FLOATS

A diagram showing the use of the high-side float is shown in Figure 6-15. The high-side float is located at the outlet of the condenser and the flow rate is regulated by the amount of refrigerant condensed. When the float is located a considerable distance from the evaporator it is usually necessary to install a weight valve near the evaporator to prevent premature evaporation of the liquid.

Figure 6-15 High-side float valve. Float valve regulates flow into the evaporator at the rate refrigerant is condensed. Weight valve prevents flash gas in the liquid line.

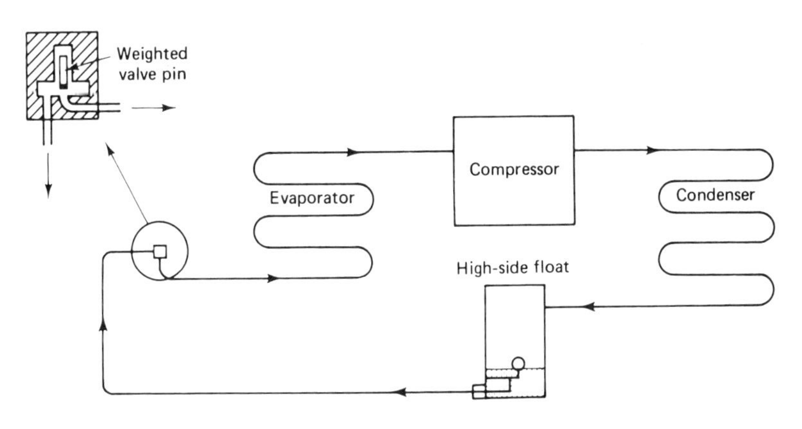

The refrigerant charge on a high-side-float system is critical since most of the refrigerant charge is in the evaporator. It can be used either on a dry expansion or flooded system. When the compressor stops, the high-side float valve closes, thus stopping the flow of refrigerant to the evaporator. The high-side float valve automatically and indirectly maintains a constant level of refrigerant in the evaporator.

SOLENOID VALVES

A solenoid valve is an electrically operated stop valve. It may be used in the liquid line, suction line, or hot-gas line. A liquid-line solenoid valve is shown in Figure 6–16 and a cutaway view in Figure 6–17. The main use of a liquid-line solenoid is to stop the flow of refrigerant during the off cycle of the compressor. The best arrangement is to use a pump-down cycle. The space thermostat operates the liquid lines solenoid valve. The compressor is turned on and off by a low-pressure control (LPC).

When solenoid valves are used in the suction line they must be selected with a minimum pressure drop, usually not to exceed 1 psi.

A common use of solenoid valves is in hot-gas defrost systems. Some types of valves are limited to 180 °F temperatures. However, high-temperature coils that operate up to 300 °F can be secured.

It is important to precede the solenoid valve with a filter-drier to be certain that the valve seat is kept clean. Foreign material may lodge in the seat and keep the valve open.

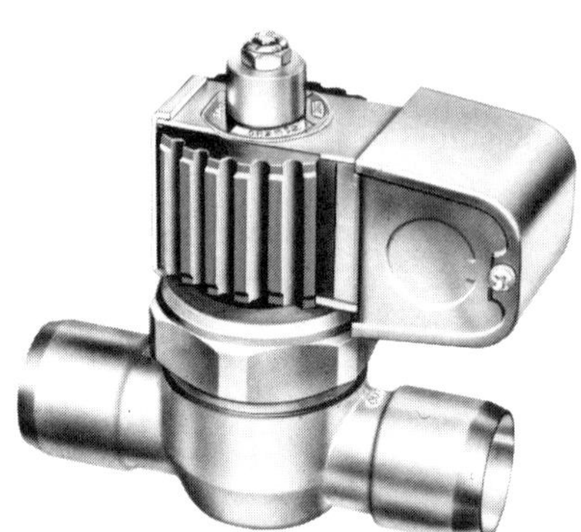

Figure 6–16 Electrically operated solenoid valve. The same valve can be used on R-12, R-22, and R-502. It is used to stop the flow of refrigerant in response to electrical current input. It can be obtained for normally open and normally closed application. (By permission of Sporlan Valve Company.)

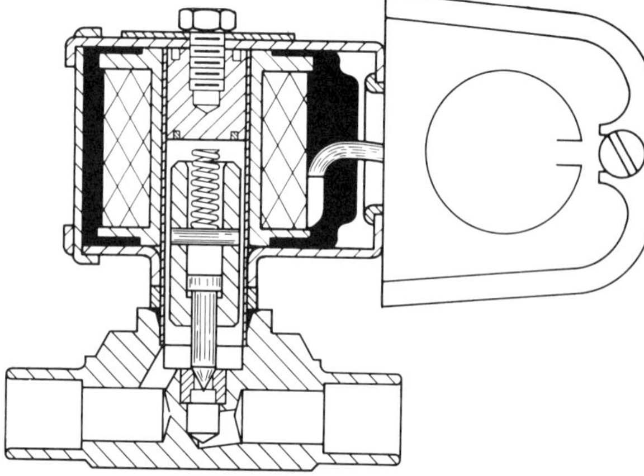

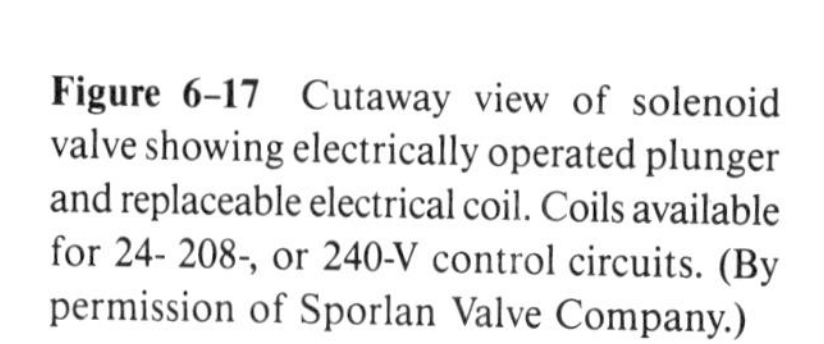

Figure 6–17 Cutaway view of solenoid valve showing electrically operated plunger and replaceable electrical coil. Coils available for 24- 208-, or 240-V control circuits. (By permission of Sporlan Valve Company.)

Valves are available with a variety of coil voltages. In replacing a coil, the new coil must have the same voltage as the one being replaced. If a transformer is used to supply control voltage to the solenoid, it must have a sufficient volt-ampere output to power the valve.

PRESSURE REGULATING VALVES

Pressuré regulators are modulating-type or snap-action devices used to control refrigerant pressure for a number of important functions:

1. *Evaporator pressure regulating valves* are used to prevent the evaporating pressure from falling below a predetermined value at which the valves have been set. A standard valve is shown in Figure 6–18 and the use of the valve in controlling evaporator pressures in a multiple-evaporator system is shown in Figure 6–19.

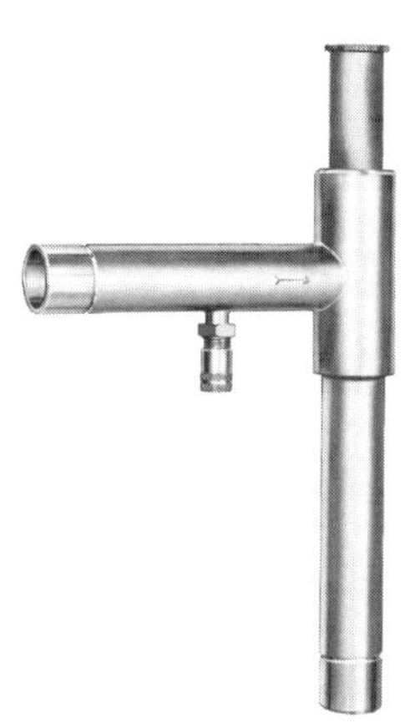

Figure 6–18 Evaporator pressure regulating valve (EPRV) is used to control suction pressure in an evaporator.

Figure 6–19 The suction pressure regulating valve is used in multievaporator systems to prevent the suction pressure in a single coil from dropping below a minimum setting. To select the proper valve, the following information is required: operating evaporator temperature, type of refrigerant, and pressure drop across the valve. The illustration shows two evaporators, each controlled by an evaporator pressure regulating valve (EPRV). (By permission of Sporlan Valve Company.)

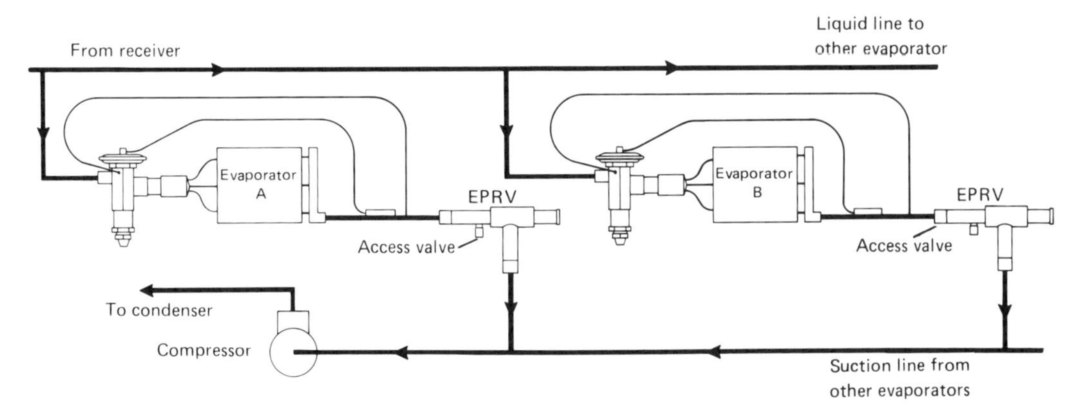

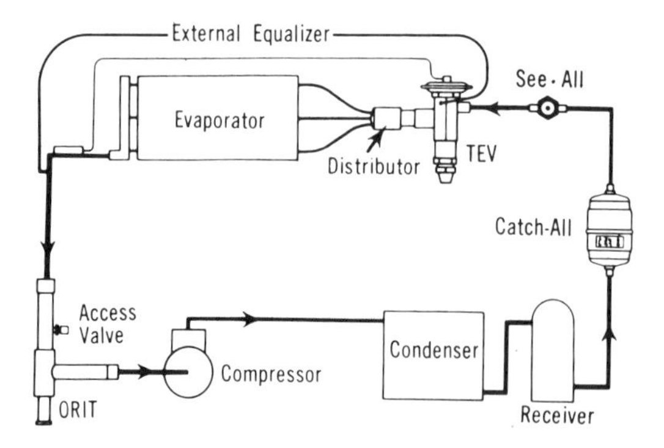

Figure 6–20 The crankcase pressure regulator (CPRV) is designed to prevent overloading the compressor motor by limiting the crankcase (suction) pressure during and after a defrost cycle or during a pull-down period. The normal adjustment is from 0 to 60 psig. It is necessary to throttle the vapor flow from the evaporator on startup to keep the compressor on the line. (By permission of Sporlan Valve Company.)

2. *Condenser pressure regulators* are used to flood the condenser tubes during low-ambient conditions to maintain proper head pressure.
3. *Crankcase pressure regulators* are used to prevent overloading the compressor motor by limiting the crankcase pressure during and after a defrost cycle or after a normal shutdown period. A typical application is shown in Figure 6–20.
4. *Hot-gas bypass control* uses a pressure regulator to bypass hot gas into the suction line during low-load conditions.

Evaporator Pressure Regulating Valves

Evaporator pressure regulators offer an efficient means of balancing the system capacity and the load requirements and maintaining different evaporator conditions on multiple temperature systems. These valves automatically throttle the vapor flow from the evaporator in order to maintain the desired minimum evaporator pressure. Evaporator pressure regulators respond only to variations in their inlet pressure (evaporator pressure). When the evaporator load changes, the valve opens and closes in response to the change in evaporator pressure. On a single evaporator system such as a water chiller, the valve is used to prevent freeze-up at light loads. On a multi-temperature refrigeration system with evaporators operating at different temperatures, one or more evaporators may be required to maintain pressures higher than the common suction line.

Two nominal adjustment ranges are available , 0 to 50 psi and 30 to 100 psi. Inlet strainers are supplied on most models. The selection is based on five basic system conditions:

1. Refrigerant
2. Evaporator design capacity in tons
3. Evaporator design temperature or pressure
4. Minimum evaporator pressure or temperature
5. Available pressure drop across the valve at design capacity

Head Pressure Controls

These valves are used to control the head pressure on an air-cooled condenser under low-ambient conditions. Two types are available: adjustable and nonadjustable. The valve serves to flood the condenser with liquid refrigerant at low-ambient conditions to reduce the condensing surface and hold up the head pressure. A complete description of these valves was given in Chapter 4.

Crankcase Pressure Regulating Valves

These valves are sensitive only to outlet pressure (crankcase pressure or suction pressure). They close on a rise of outlet pressure. As long as the valve outlet pressure is greater than the valve setting, the valve will remain closed. As the outlet pressure is reduced the valve will open and pass refrigerant to the compressor. The valve is placed at the evaporator outlet and connected to the compressor suction. An access valve is available for measuring the inlet pressure. The valve serves as a pressure-limiting device throttling the compressor suction until the machine can handle the load.

Valves are available with nominal operating pressures of 0 to 60 psi. Three factors are needed for proper selection of a valve:

1. Design suction pressure after pull-down
2. Maximum allowable suction pressure
3. Pressure drop across the valve at design load

Hot-Gas Bypass Valves

On many systems it is desirable to limit the minimum evaporating pressure during periods of low load. Thus a compressor can be kept on the line supplying the required refrigeration at the desired temperature even though the load is greatly reduced. This is done by bypassing a portion of the hot discharge gas directly into the low side of the system. The valve responds to changes in suction pressure. The valve opens on a drop in suction pressure.

Some evaporator distributors have a connection suitable for the outlet of hot gas from the discharge bypass valve. This allows a convenient connection to the low side of the system. When low evaporator loads are required, it is advisable to use a desuperheater to prevent overheating the suction gas going to the compressor.

Valves are available in two ranges, 0–30 psig and 0–80 psig. The adjustment ranges are 10 psi for R-12 and 15 psi for R-22. For some compressors, where unloaders for capacity reduction are not available, the discharge bypass valve offers a means of automatic capacity reduction.

Note that an auxiliary solenoid valve is used in the hot-gas bypass piping on the upstream side of the discharge bypass valve. This is necessary when a pump-down cycle is used, and the auxiliary solenoid valve should be wired to operate at the same time as the liquid-line solenoid valve.

AMMONIA VALVES

Ammonia was one of the early refrigerants, with use dating back to the 1880s. The latent heat available per pound of refrigerant is many times the amount for the more modern halocarbons. To review some of its characteristics:

1. It boils at $-28\,^{\circ}F$ at atmospheric pressure.
2. The liquid is lighter than water and the vapor is lighter than air.
3. Commercial-grade ammonia is approximately 99% pure NH_3.
4. Although it has no effect on metals when dry, it will completely destroy copper when wet.
5. Its irritating odor provides a means for early detection of a leak.
6. It can be used only where group 2 refrigerants are permitted.

Some of the more popular applications are:

1. Ice-making plants
2. Cold storage warehouses
3. Ice cream factories
4. Dairies
5. Breweries
6. Soda bottling plants
7. Canneries
8. Ice-skating rinks
9. Fisheries
10. Meat-packing plants
11. Frozen-food plants

Some of the characteristics of ammonia piping that differ from the halocarbons and need to be considered in selecting valves are:

1. All control valves connected to the piping should be flanged for easy removal.
2. A strainer should be used ahead of all valves to protect the valve.
3. It is advisable to use stop valves before and after operating controls, as well as equipment to provide a means of easy removal for service.
4. Valve stems should be vertical whenever possible since there is less chance for dirt or scale to be trapped in the seat or disk.

In most ammonia systems there is more refrigerant in the system than the receiver will hold, making it impractical to pump down the system. Therefore, the use of many valves in the piping as outlined in characteristic 3 is extremely important.

A typical strainer is shown in Figure 6–21 and a typical hand valve is shown in Figure 6–22.

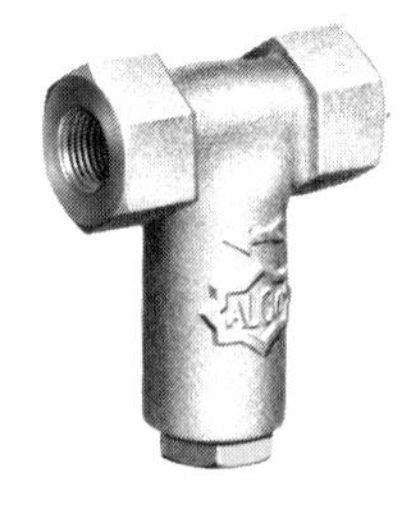

Figure 6–21 Ammonia system strainers. It is good practice on an ammonia system to install a strainer ahead of every automatic valve. These strainers can be opened for cleaning without breaking the refrigerant line. (By permission of Alco Valve Company.)

Figure 6–22 Shutoff valves: handwheel type, screw end design, and globe type. These can be backseated and repacked under pressure. Maximum cold working pressure is 400 psi for sizes ¼ through 4 in. (By permission of Henry Valve Company.)

Ammonia Expansion Valves

The ammonia TEV differs from the halocarbon type in that the ammonia valve uses a double pressure drop. This reduces the pressure at the orifice to prevent the effect of "wire drawing." A typical ammonia expansion valve is shown in Figure 6–23. The standard ratings are based on a 5 °F evaporator temperature, 140 psi pressure drop across the valve, and 86 °F vapor-free liquid entering the valve.

Float Valves

A high-side float valve is shown in Figure 6–15, and an ammonia low-side float valve in Figure 6–24. These valves function in a manner similar to the descriptions given previously.

The electric float switch has gained popularity due to its ease of installation, adjusting, and service (see Figure 6–25). The float consists of a pressure chamber that houses a float ball. As the ball rises and falls it makes and breaks an electrical circuit which operates a solenoid valve in the liquid line to control the flow of liquid to the evaporator. A hand valve is placed in series with the electric float switch to regulate the flow through it.

Another type of flooded chamber automatic control is called *bulb level control* (see Figure 6–26). This consists of a thermal expansion valve placed at the inlet side of a suction trap. The thermal bulb with built-in small heaters is inserted in a well to sense the level of the liquid in the float chamber. As the level drops the heaters warm the bulb sufficiently to cause the TEV to open. When the liquid rises to surround the bulb, the chilling effect closes the TEV.

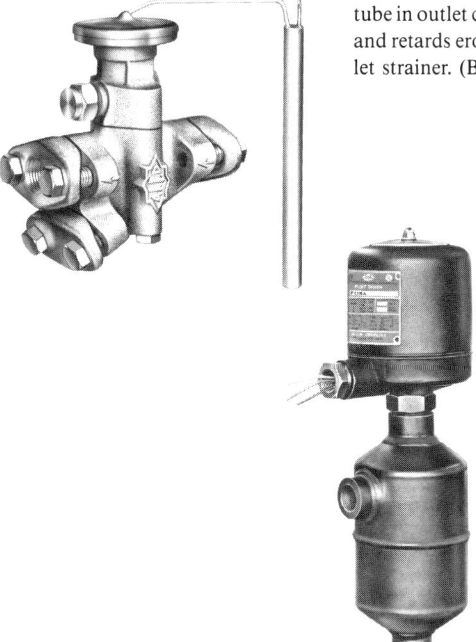

Figure 6–23 Ammonia thermal expansion valve. Discharge tube in outlet controls capacity, prevents frosting of the valve, and retards erosion of pin and seat. It is equipped with an inlet strainer. (By permission of Alco Valve Company.)

Figure 6–24 Low-side float and chamber for flooded ammonia systems. (By permission of H. A. Phillips and Company.)

Figure 6–25 Electric float switch provides electrical switching action in response to changes in level of liquid. (By permission of Alco Valve Company.)

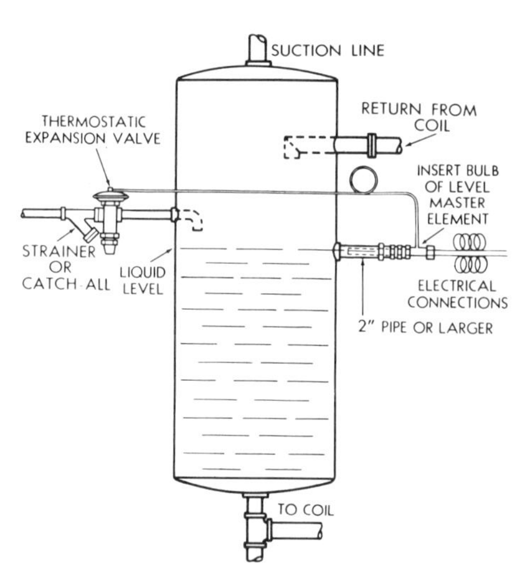

Figure 6–26 Bulb-level control provides modulated flow and maintains the liquid level in the low side. It is a standard thermostatic expansion valve with a level element. Insert bulb includes low-wattage heater. As the level drops, electrical heat is added, increasing the bulb pressure and opening the valve. (By permission of Sporlan Valve Company.)

Ammonia Solenoid Valves

For most uses, ammonia solenoid valves have the same function to perform as those of the halocarbon types. However, there are a few uses that are unique to ammonia systems. For example, on a flooded recirculation system, a liquid pump is used to circulate the liquid refrigerant through the evaporator. The pressure of this pump does not usually exceed 35 psi. Therefore, a liquid-line solenoid placed in the circuit should have a relatively low pressure drop compared to the normal pressure drop associated with the halocarbon system liquid solenoid.

A very interesting ammonia stop valve is sometimes used in the suction line of a hot-gas defrost system. It is called a *condenser gas-powered positive stop valve.* Hot-gas pressure opens the valve; suction pressure, with the help of a spring, closes the valve. Both the hot-gas connection and the suction connection to the valve have solenoid control. Due to the unique construction of this valve, it has a very low pressure drop in the suction line when it is open.

Ammonia Pressure Regulators

Due to the fact that ammonia systems often have many different temperature areas operating on a simple system, many pressure regulating valves are used. The most common is the evaporator pressure regulator (EPR) used to control evaporator temperatures in a multitemperature system (Figure 6–27).

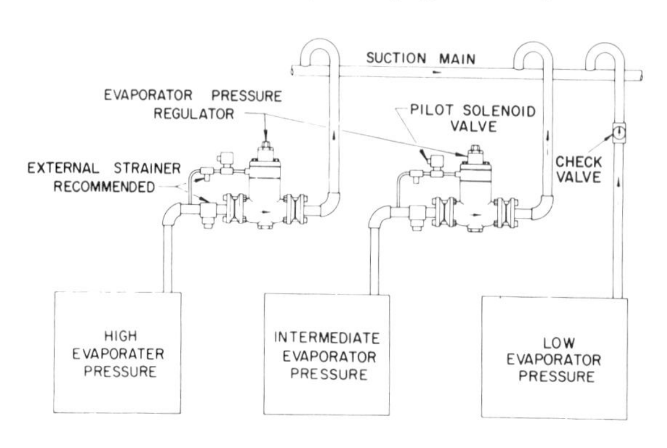

Figure 6–27 The use of evaporator pressure regulators for controlling multiple evaporators operating at different evaporator temperatures on a single compressor. A check valve is placed in the suction line to the lowest-pressure evaporator to prevent migration of refrigeration to this coil during shutdown. (By permission of *ASHRAE Handbook.*)

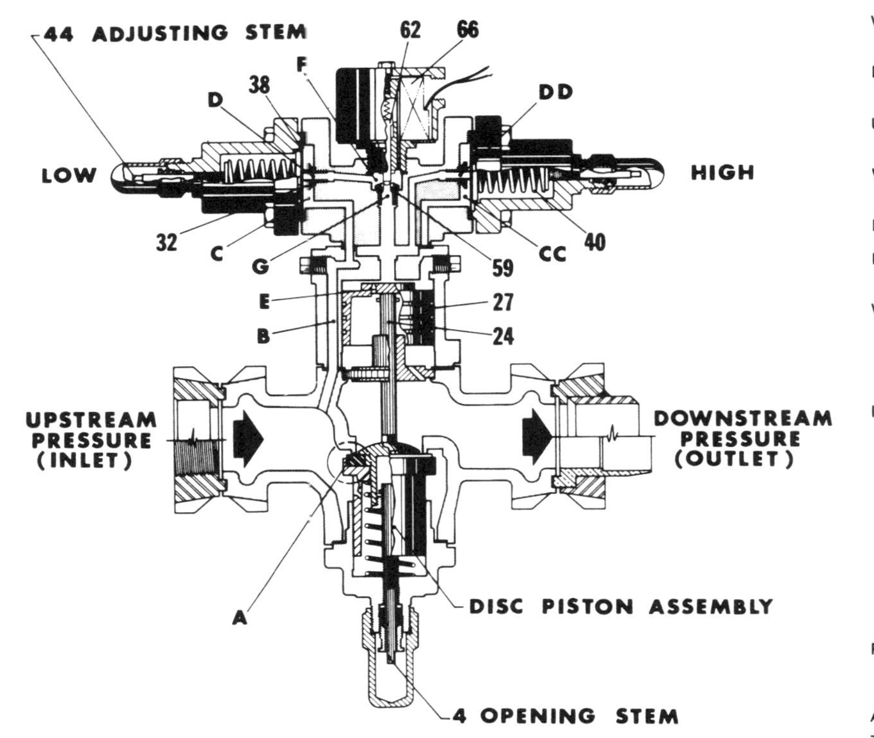

Upstream pressure is held back at point "A" by SEAT DISC contact with MAIN BODY SEAT.

Upstream pressure is transmitted to SENSING CHAMBER "C" through INTERNAL EQUALIZER line "B".

Spring loaded DIAPHRAGMS (38) contact polished radius of PILOT SEAT BEAD (32) creating positive SEAL.

When pressure in SENSING CHAMBER "C" exceeds SPRING LOAD (40) DIAPHRAGMS (38) are flexed permitting flow of upstream pressure through port "D" to pilot solenoid CHAMBER "F".

LOW – When COIL (66) is energized, DRAWBAR ASSEMBLY (62) is picked up electrically. Upstream pressure from CHAMBER "F" flows through port "G" to chamber above POWER PISTON (27).

Upstream pressure on top of POWER PISTON (27) causes PUSH ROD (24) to force DISC PISTON ASSEMBLY down permitting flow through main body orifice at point "A".

When upstream pressure in SENSING CHAMBER "C" is lower than SPRING LOAD (40), DIAPHRAGMS (38) reseat on PILOT SEAT BEAD (32) closing flow through port "D".

Pressure in chamber above POWER PISTON (27) bleeds through port "E" permitting DISC PISTON ASSEMBLY to reseal at point "A".

HIGH – When COIL (66) is de-energized, DRAWBAR ASSEMBLY (62) seats on teflon disc of SEAT BEAD (59) closing flow through port "G". Upstream pressure diverts internally to SENSING CHAMBER "CC"

When pressure in SENSING CHAMBER "CC" exceeds SPRING LOAD (40) DIAPHRAGMS (38) are flexed permitting flow of upstream pressure through port "DD" to chamber above POWER PISTON (27). Regulator then functions the same on High as above on Low.

PRESSURE (TEMPERATURE) ADJUSTMENT

Install GAUGE and GAUGE VALVE in GAUGE PORT of Regulator, STANDARD DUAL PRESSURE REGULATORS (Range S) are factory pre-set to maintain an upstream pressure of approximately 40 P.S.I.G. on LOW head and 70 P.S.I.G. on HIGH HEAD. De-energize COIL (66) to adjust upstream pressure on High head. Energize COIL (66) to adjust upstream pressure on Low head.

To RAISE upstream pressure setting, turn adjusting stem (44) IN (CLOCKWISE).

To LOWER upstream pressure setting, turn adjusting stem (44) OUT (COUNTER-CLOCKWISE).

Each FULL TURN changes upstream pressure setting by approximately 9 P.S.I.G.

DO NOT TURN IN (CLOCKWISE) BEYOND MILLED FLATS ON ADJUSTING STEM.

Fine adjustment is made after system has been permitted to stabilize. LOWER UPSTREAM PRESSURE GRADUALLY TO PREVENT FREEZE UP.

MANUAL OPERATION

ALL REGULATORS are arranged at the factory for AUTOMATIC OPERATION.

To manually open the regulator turn opening stem OUT (COUNTER-CLOCKWISE) to stop.

For automatic operation turn opening stem IN (CLOCKWISE) to the point where milled flats on opening stem begin to enter packing nut.

Figure 6–28 Dual-pressure regulator valve will maintain either of two predetermined pressure settings by opening or closing the integrated pilot solenoid valve. (By permission of Hubbell Corporation.)

More than one pressure can be used to control an EPR valve using the dual EPR shown in Figure 6–28. This permits one valve to have two functions. With the pilot solenoid energized, the low-pressure head controls upstream pressure. With the pilot solenoid deenergized, the high-pressure head controls the upstream pressure. An electric control device determines the switching of the pilot solenoid valve.

Another dual-function pressure regulating valve is the EPR with stop duty (Figure 6–29). By energizing or de-energizing a pilot solenoid, EPR may be made to do suction stop service or regular EPR duty.

A pilot-operated hot-gas bypass valve is shown in Figure 6–30. This is a type of pressure regulating valve used to control the low-side pressure bypassing hot gas into the suction side of the system.

Figure 6–29 Evaporator pressure regulator with stop features, uses high-side pressure for pilot operation. The adjustable range is from 0 to 75 psig. (By permission of Alco Valve Company.)

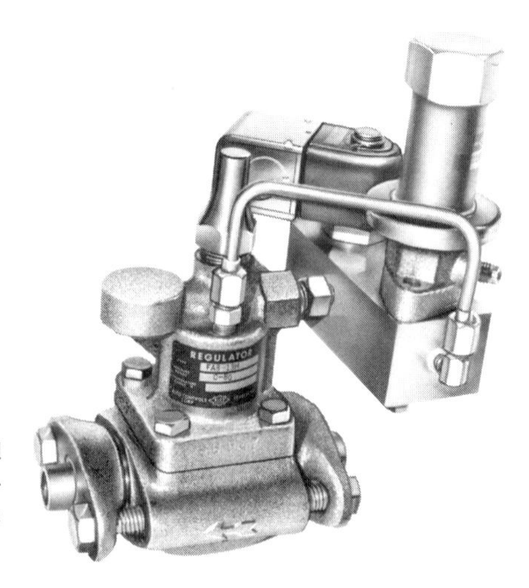

Figure 6–30 Hot-gas bypass valve, designed for automatic system capacity control. Adjustable range is 0 to 80 psig. (By permission of Alco Valve Company.)

REVIEW EXERCISE 1

Select the letter representing your choice of the correct answer (there may be more than one).

6-1. Which of the following is *not* a type of metering device:

________ a. Solenoid valve

________ b. TEV

________ c. AEV

________ d. Capillary tube

6-2. "AEV" stands for:

________ a. Automatic evaporator valve

________ b. After evaporator valve

________ c. Automatic expansion valve

________ d. Additional energy velocity

6-3. The number of pressures that control the AEV are:

________ a. One

________ b. Two

________ c. Three

________ d. Four

6-4. The principal advantage of the TEV is:

________ a. Simplicity of design

________ b. Controlled by one pressure

________ c. More readily available

________ d. Ability to react to load changes

6-5. Which type of metering device can be used in a hermetically sealed system?

________ a. High-side float

________ b. Capillary tube

________ c. TEV

________ d. Low-side float

6-6. Using a TEV with a superheat setting of 10 °F, the pressure exerted by the spring at a 20 °F evaporating temperature using R-12 is:

________ a. 7.41 psig

________ b. 21.05 psig

________ c. 28.46 psig

________ d. 0.0 psig

6-7. A TEV with an external equalizer should be used when the coil pressure drop exceeds the psig equivalent of how many degrees F?

________ a. 2 °F

________ b. 4 °F

________ c. 6 °F

________ d. 8 °F

6-8. A pressure-limiting TEV is used to:

________ a. Prevent using too large a valve

________ b. Control the load on the compressor during pull-down

________ c. Control condensing pressure

________ d. Provide closer control of evaporating temperature

6-9. The low-side float is used with which type of evaporator?

________ a. Flooded

________ b. Dry expansion

6-10. A crankcase pressure regulator is a type of:

________ a. Solenoid valve

________ b. Metering device

________ c. Check valve

________ d. Refrigerant control

REVIEW EXERCISE 2

Shown in Figure 6-31 is a demonstrator that can be operated for a number of types of systems. These include the (1) capillary tube system, (2) direct expansion system, (3) low-side float system, and (4) reverse-cycle system. The illustration shows the valves that must be operated for changing from one system to another.

Fill in the following table for the valve positions for each type of system. Indicate whether the valve should be "open" or "closed":

	Capillary tube	Direct expansion	Low-side float	Reverse cycle
Valve 1				
Valve 2				
Valve 3				
Valve 4				
Valve 5				
Valve 6				
To float valve				
To evaporator				

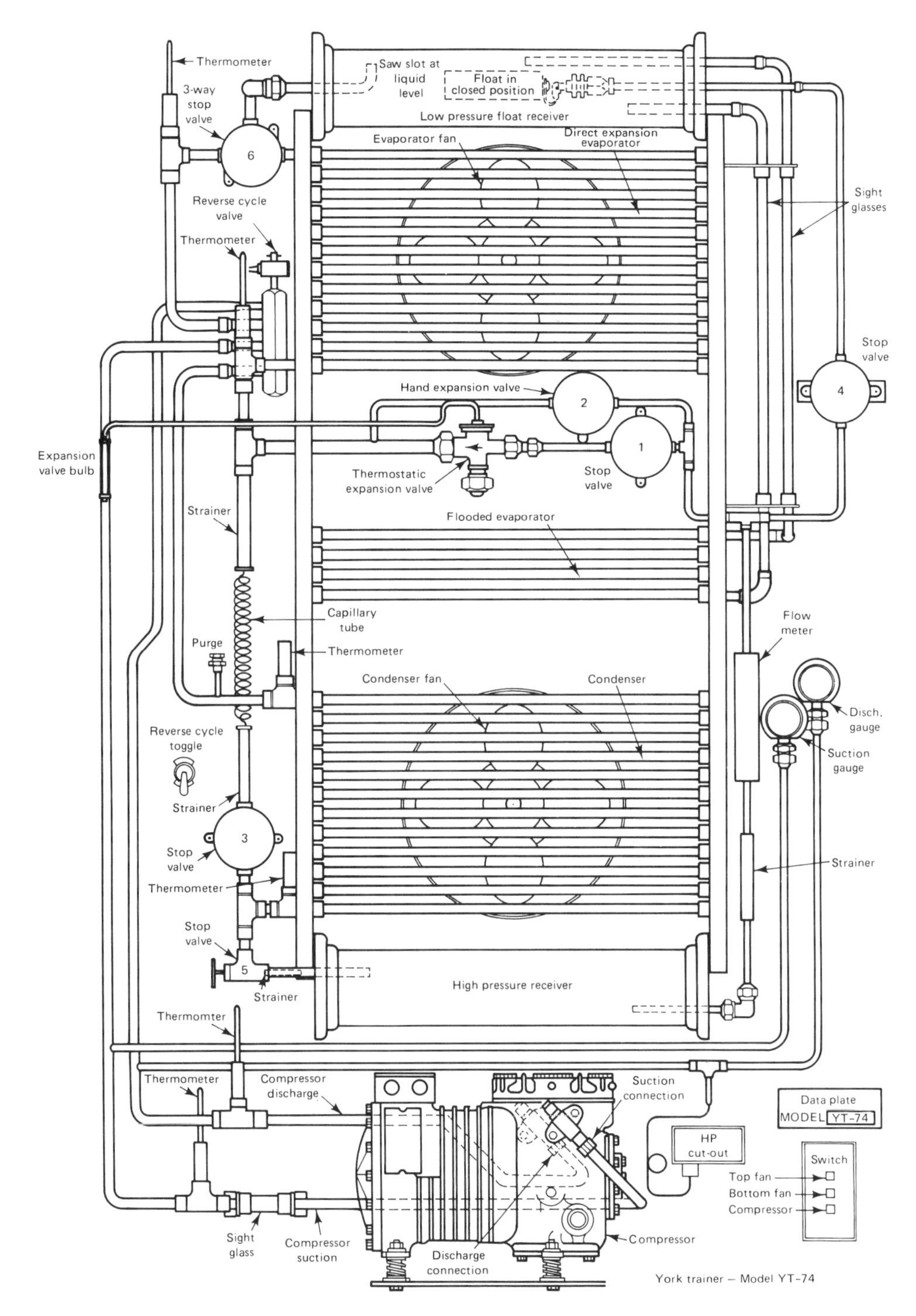

Figure 6–31 Diagram of refrigeration system demonstration unit. Hand valves are used to change the type of metering device and piping circuit used.

7 Refrigerant Piping and Accessories

OBJECTIVES

After studying this chapter, the reader will be able to:

- Make a layout of a refrigerant piping system.
- Troubleshoot an existing piping system to determine corrective action.
- Select the materials for installing a refrigerant piping system.

PIPING FUNCTIONS

All refrigerant piping should be installed to provide the following principal functions:

1. A leakproof path for the refrigerant to follow
2. Proper oil return to the compressor
3. Properly sized piping to permit full capacity and efficiency of system
4. Protection for the compressor to prevent liquid slugging

MATERIALS

The minimum requirements for refrigeration piping insofar as materials and joints are concerned are given in the Safety Code for Mechanical Refrigeration, ANSI/ASHRAE 15-74. All piping must also comply with local codes and regulations. Piping materials can be copper, brass, black steel, or wrought iron. The materials selected must conform with the refrigerant used, the size of the job, and the nature of the application.

Copper is the choice for halogenated hydrocarbon refrigerants. Copper is not suitable for ammonia since, in the presence of water, ammonia attacks nonferrous materials.

Copper is either hard drawn or soft temper. Refrigerant-grade copper tubing is sealed to keep it clean and dry. It is measured in outside dimensions (O.D.) such as ½-in. O.D., 1⅝-in. O.D., and 4⅛-in. O.D. Small copper tubing is available in soft temper and is usually used for final connections when much bending is required. Only types K and L copper are suitable for refrigeration work.

JOINTS

Most codes require brazed copper joints using brazing materials with melting points over 1000 °F. These temperatures usually require an oxyacetylene torch and a silver-alloy solder. The silver content is generally between 35 and 40%. The material melts at 1120 °F and flows at 1145 °F. Soft copper can use flared compression fittings.

Welding is the most common way of joining steel or iron piping. Swivel joints are limited to sizes 3 in. and smaller for normal refrigeration pressures.

SUPPORTS

All piping must be properly supported with hangers or brackets. When piping passes through a wall or floor, a suitable sleeve must be provided.

Solid connections to the moving compressor may cause noise and vibration. To prevent this, vibration eliminators can be provided, as in Figure 7–1, or tubing loops, as shown in Figure 7–2. On long runs of tubing, allowance must be made in the piping for expansion and contraction.

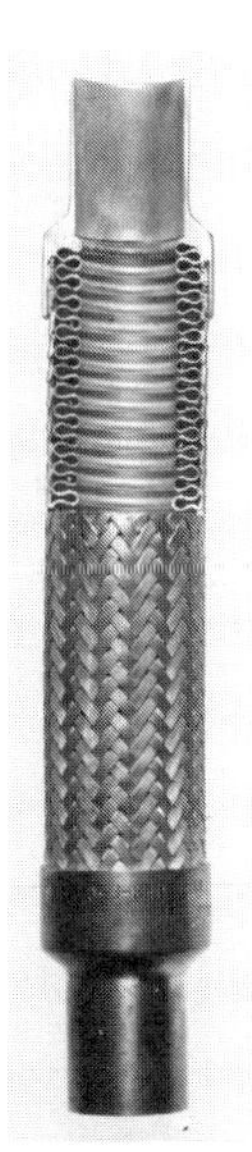

Figure 7–1 Flexible corrugated bronze hose with bronze wire overbraid is available in sizes O.D. copper, ¼ to 4⅛ in. It is used for straight-line installations. (By permission of Johnson Metal Hose, Inc.)

Figure 7–2 All piping to the compressor must allow for movement and vibration. Expansion loops are useful for this purpose. These loops should not constitute oil traps.

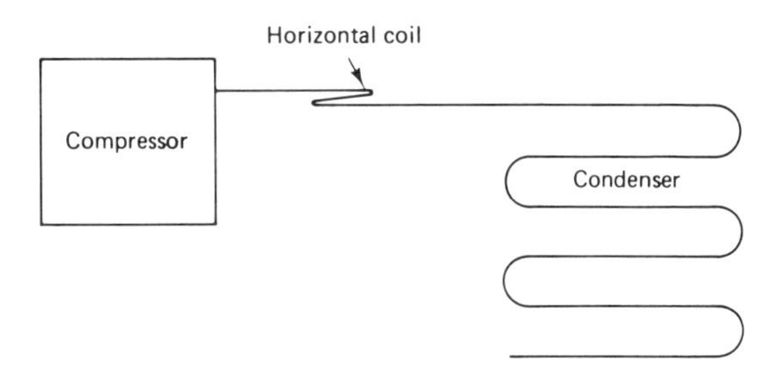

Occasionally, discharge tubing will vibrate due to the pulsations from the compressor. The movement can break hangers on a discharge line to an air-cooled or evaporative condenser. This can usually be corrected by the use of a muffler. This condition is caused by the pulsations of the compressor setting up a resonance in the piping. Anything to change the frequency of either the compressor or the piping will help.

PIPE SIZING

Information for pipe sizing can be obtained from Figures 7-3 through 7-10.

Line Size Type L Copper, OD	Suction Lines ΔT = 2F, Saturated Suction Temp F: −40, ΔP = 0.49	−20, ΔP = 0.72	0, ΔP = 1.01	20, ΔP = 1.38	40, ΔP = 1.82	Discharge Lines ΔT = 1.0F, ΔP = 1.83, Saturated Suction Temp F: −40	0	40	Liquid Lines*: Line Size Type L Copper OD	Velocity = 100 fpm	ΔT = 1 F, ΔP = 1.9 psi
1/2	—	—	—	0.20	0.30	0.38	0.43	0.46	1 1/2	1.9	2.2
5/8	—	0.16	0.25	0.38	0.56	0.72	0.80	0.87	5/8	3.0	4.1
7/8	0.25	0.42	0.67	1.0	1.5	1.9	2.10	2.30	7/8	6.2	10.8
1 1/8	0.51	0.86	1.4	2.1	3.0	3.8	4.3	4.6	1 1/8	10.5	21.9
1 3/8	0.90	1.5	2.4	3.6	5.3	6.7	7.4	8.1	1 3/8	16.0	38.3
1 5/8	1.4	2.4	3.8	5.7	8.3	10.6	11.7	12.8	1 5/8	22.7	60.7
2 1/8	3.0	5.0	7.8	11.8	17.3	21.9	24.2	26.4	2 1/8	39.5	126.4
2 5/8	5.3	8.8	13.9	21.0	30.5	38.6	42.8	46.8	2 5/8	60.9	223.8
3 1/8	8.4	14.1	22.1	33.4	48.6	61.6	68.2	74.5	3 1/8	86.9	357.7
3 5/8	12.6	20.9	32.9	49.7	72.2	91.4	101.2	110.6	3 5/8	117.6	532.3
4 1/8	17.8	29.5	46.5	70.1	101.9	128.8	142.6	155.8	4 1/8	152.9	751.7
5 1/8	31.9	53.0	83.3	125.3	182.0	229.8	254.4	278.0			
6 1/8	51.4	85.2	133.8	201.6	292.3	368.5	408.1	445.9			

Steel IPS	SCH	Suction −40	−20	0	20	40	Discharge −40	0	40	Steel PS	SCH	Velocity = 100 fpm	ΔT = 1 F
1/2	40	—	—	0.30	0.44	0.64	0.8	0.9	1.0	1/2	80	3.0	3.4
3/4	40	0.25	0.40	0.63	0.93	1.3	1.7	1.9	2.0	3/4	80	5.5	7.7
1	40	0.47	0.76	1.2	1.8	2.5	3.2	3.5	3.8	1	80	9.2	15.1
1 1/4	40	0.97	1.6	2.4	3.6	5.2	6.5	7.2	7.9	1 1/4	80	16.4	32.5
1 1/2	40	1.5	2.4	3.7	5.5	7.8	9.8	10.9	11.9	1 1/2	80	22.6	49.5
2	40	2.8	4.6	7.1	10.6	15.1	18.9	20.9	22.9	2	40	42.8	115.0
2 1/2	40	4.5	7.3	11.3	16.8	24.1	30.1	33.4	36.5	2 1/2	40	61.1	183.6
3	40	8.0	13.0	20.0	29.7	42.6	53.2	59.0	64.4	3	40	94.3	324.3
4	40	16.3	26.4	40.8	60.6	86.8	108.3	119.9	131.0	4	40	162.5	662.4
5	40	29.4	47.7	73.7	109.3	156.3	195.5	216.5	236.6				
6	40	47.7	77.3	119.2	176.8	252.8	316.2	350.1	382.6				
8	40	97.9	158.1	243.8	361.7	517.1	646.9	716.3	782.7				
10	40	177.3	286.2	441.5	654.1	936.8	1168.7	1294.0	1414.0				
12	ID	283.5	457.9	705.3	1045.2	1496.6	1867.0	2067.0	2259.0				

NOTES:

For other ΔT's and Equivalent Lengths, L_e

(1) Line Capacity (Tons)

$$= \text{Table Tons} \times \left(\frac{\text{Table } L_e}{\text{Actual } L_e} \times \frac{\text{Actual } \Delta T \text{ Loss Desired}}{\text{Table } \Delta T \text{ Loss}} \right)^{0.55}$$

(2) For other Tons and Equivalent Lengths for a given pipe size

$$\Delta T = \text{Table } \Delta T \times \frac{\text{Actual } L_e}{\text{Table } L_e} \times \left(\frac{\text{Actual Tons}}{\text{Table Tons}} \right)^{1.8}$$

(3) Values based on 105 F Condensing Temp. For capacities at other condensing temp. multiply table value by line capacity multiplier below:

Line	Condensing Temp, F: 80	90	100	110	120	130	140	150
Suction Lines	1.11	1.06	1.02	.98	.94	.87	.82	.76
Discharge Lines	.81	.89	.97	1.05	1.13	1.20	1.26	1.33

Figure 7-3 Refrigeration line capacities in tons, for R-12 (single- and high-stage applications). By permission of *ASHRAE Handbook.*)

Line Size Type L Copper, OD	Suction Lines ΔT = 2F — Saturated Suction Temp, F: −40, ΔP = 0.79	−20, ΔP = 1.15	0, ΔP = 1.6	20, ΔP = 2.22	40, ΔP = 2.91	Discharge Lines ΔT = 1.0 F, ΔP = 3.05 — Saturated Suction Temp: −40	40	Liquid Lines — Line Size Type L Copper, OD	Velocity = 100 fpm	ΔT = 1 F ΔP = 3.05
1/2				0.40	0.6	0.75	0.85	1/2	2.3	3.6
5/8		0.32	0.51	0.76	1.1	1.4	1.6	5/8	3.7	6.7
7/8	0.52	0.86	1.3	2.0	2.9	3.7	4.2	7/8	7.8	18.2
1 1/8	1.1	1.7	2.7	4.0	5.8	7.5	8.5	1 1/8	13.2	37.0
1 3/8	1.9	3.1	4.7	7.0	10.1	13.1	14.8	1 3/8	20.2	64.7
1 5/8	3.0	4.8	7.5	11.1	16.0	20.7	23.4	1 5/8	28.5	102.5
2 1/8	6.2	10.0	15.6	23.1	33.1	42.8	48.5	2 1/8	49.6	213.0
2 5/8	10.9	17.8	27.5	40.8	58.3	75.4	85.4	2 5/8	76.5	376.9
3 1/8	17.5	28.4	44.0	65.0	92.9	120.2	136.2	3 1/8	109.2	601.5
3 5/8	26.0	42.3	65.4	96.6	137.8	178.4	202.1	3 5/8	147.8	895.7
4 1/8	36.8	59.6	92.2	136.3	194.3	251.1	284.4	4 1/8	192.1	1263.2
5 1/8	66.0	106.9	164.5	243.5	346.6	448.2	507.6			
Steel IPS SCH								**Steel IPS SCH**		
1/2 40		0.38	0.58	0.85	1.2	1.5	1.7	1/2 80	3.8	5.7
3/4 40	0.50	0.8	1.2	1.8	2.5	3.3	3.7	3/4 80	6.9	12.8
1 40	0.95	1.5	2.3	3.4	4.8	6.1	6.9	1 80	11.5	25.2
1 1/4 40	2.0	3.2	4.8	7.0	9.9	12.6	14.3	1 1/4 80	20.6	54.1
1 1/2 40	3.0	4.7	7.2	10.5	14.8	19.0	21.5	1 1/2 80	28.3	82.6
2 40	5.7	9.1	13.9	20.2	28.5	36.6	41.4	2 40	53.8	192.0
2 1/2 40	9.2	14.6	22.1	32.2	45.4	58.1	65.9	2 1/2 40	76.7	305.8
3 40	16.2	25.7	39.0	56.8	80.1	102.8	116.4	3 40	118.5	540.3
4 40	33.1	52.5	79.5	115.9	163.2	209.5	237.3	4 40	204.2	1101.2
5 40	59.7	94.7	143.6	208.7	294.3	377.4	427.5			
6 40	96.5	153.3	232.3	337.6	476.0	610.4	691.4			
8 40	197.9	313.6	475.1	691.4	973.8	1249.0	1414.5			
10 40	358.4	568.0	859.3	1250.8	1759.2	2256.0	2555.0			
12 ID	572.5	907.4	1374.7	1998.2	2810.5	3609.0	4088.0			

NOTES:

(1) For Other ΔT's and Equivalent Lengths, L_e
Line Capacity (Tons)

$$= \text{Table Tons} \times \left(\frac{\text{Table } L_e}{\text{Actual } L_e} \times \frac{\text{Actual } \Delta T \text{ Loss Desired}}{\text{Table } \Delta T \text{ Loss}} \right)^{0.55}$$

(2) For other Tons and Equivalent Lengths in a given pipe size

$$\Delta T = \text{Table } \Delta T \times \frac{\text{Actual } L_e}{\text{Table } L_e} \times \left(\frac{\text{Actual Tons}}{\text{Table Tons}} \right)^{1.8}$$

(3) Values are based on 105 F condensing temperature. For other condensing temperatures, multiply table tons by the following factors:

Condensing Temp F	Suction Lines	Hot Gas Lines
80	1.11	0.79
90	1.07	0.88
100	1.03	0.95
110	0.97	1.04
120	0.90	1.10
130	0.86	1.18
140	0.80	1.26

Figure 7–4 Refrigerant line capacities for R-22 (single- and high-stage applications). (By permission of *ASHRAE Handbook.*)

Line Size Type L Copper, OD	Suction Lines $\Delta T = 2$ F Saturated Suction Temp, F −60 $\Delta P = 0.61$	−40 $\Delta P = 0.92$	−20 $\Delta P = 1.33$	0 $\Delta P = 1.84$	20 $\Delta P = 2.45$	40 $\Delta P = 3.18$	Discharge Lines $\Delta T = 1.0$ F, $\Delta P = 3.15$ Saturated Suction Temp. −40	0	40	Liquid Lines[a] Line Size Type L Copper, OD	Velocity = 100 fpm	$\Delta T = 1$ F $\Delta P = 3.15$
1/2	0.05	0.08	0.14	0.22	0.33	0.49	0.56	0.63	0.70	1/2	1.5	2.4
5/8	0.09	0.16	0.27	0.42	0.63	0.91	1.0	1.2	1.3	5/8	2.3	4.5
7/8	0.24	0.43	0.70	1.1	1.7	2.4	2.7	3.1	3.4	7/8	4.9	11.8
1 1/8	0.50	0.87	1.4	2.2	3.4	4.8	5.5	6.3	7.0	1 1/8	8.3	24.1
1 3/8	0.87	1.5	2.5	3.9	5.8	8.4	9.6	10.9	12.1	1 3/8	12.6	42.0
1 5/8	1.4	2.4	4.0	6.2	9.2	13.3	15.2	17.2	19.1	1 5/8	17.9	66.4
2 1/8	2.9	5.0	8.2	12.8	19.1	27.5	31.4	35.6	39.5	2 1/8	31.1	138.0
2 5/8	5.1	8.8	14.5	22.6	33.7	48.4	55.3	62.8	69.5	2 5/8	48.0	243.7
3 1/8	8.2	14.1	23.2	36.0	53.7	77.0	87.9	99.8	110.5	3 1/8	68.4	389.3
3 5/8	12.1	21.0	34.4	53.5	79.7	114.3	130.5	148.1	164.0	3 5/8	92.6	579.0
4 1/8	17.1	29.7	48.5	75.4	112.3	161.0	183.7	208.4	230.9	4 1/8	120.3	816.9
5 1/8	30.7	53.2	86.7	134.6	200.3	287.1	327.3	371.3	411.3	—	—	—
6 1/8	49.5	85.6	139.5	216.2	321.3	460.6	525.2	595.9	660.1	—	—	—

NOTES:

(1) For other ΔT's and Equivalent Lengths, L_e
Line Capacity (Tons)

$$= \text{Table Tons} \times \left(\frac{\text{Table } L_e}{\text{Actual } L_e} \times \frac{\text{Actual } \Delta T \text{ Loss Desired}}{\text{Table } \Delta T \text{ Loss}} \right)^{0.55}$$

(2) For other Tons and Equivalent Lengths in a given pipe size

$$\Delta T = \text{Table } \Delta T \times \frac{\text{Actual } L_e}{\text{Table } L_e} \times \left(\frac{\text{Actual Tons}}{\text{Table Tons}} \right)^{1.8}$$

(3) Values are based on 105 F condensing temperature. For other condensing temperatures, multiply table tons by the following factors:

Condensing Temp F	Suction Lines	Hot Gas Lines
80	1.20	.83
90	1.12	.91
100	1.04	.97
110	.96	1.02
120	.88	1.08
130	.80	1.16

Figure 7–5 Refrigerant line capacities for R-502 (single- and high-stage applications.) (By permission of *ASHRAE Handbook*.)

Line Size IPS	SCH	Suction Lines $\Delta T = 1$ F Saturated Suction Temperature F −40 $\Delta P = 0.31$	−20 $\Delta P = 0.49$	0 $\Delta P = 0.73$	20 $\Delta P = 1.06$	40 $\Delta P = 1.46$	$\Delta T = 1$F Discharge Lines $\Delta P = 2.95$	Liquid Lines[a] Line Size IPS	SCH	Velocity = 100 fpm	$\Delta P = 2.0$ $\Delta T = 0.7$
3/8	80							3/8		8.6	12.1
1/2	80						3.1	1/2	80	14.2	24.0
3/4	80				2.6	3.8	7.1	3/4	80	26.3	54.2
1	80		2.1	3.4	5.2	7.6	13.9	1	80	43.8	106.4
1 1/4	40	3.2	5.6	8.9	13.6	19.9	36.5	1 1/4	80	78.1	228.6
1 1/2	40	4.9	8.4	13.4	20.5	29.9	54.8	1 1/2	80	107.5	349.2
2	40	9.5	16.2	26.0	39.6	57.8	105.7	2	40	204.2	811.4
2 1/2	40	15.3	25.9	41.5	63.2	92.1	168.5	2 1/2	40	291.1	1292.6
3	40	27.1	46.1	73.5	111.9	163.0	297.6	3	40	449.6	2287.8
4	40	55.7	94.2	150.1	228.7	333.0	606.2	4	40	774.7	4662.1
5	40	101.1	170.4	271.1	412.4	600.9	1095.2	5	40		
6	40	164.0	276.4	439.2	667.5	971.6	1771.2	6	40		
8	40	337.2	566.8	901.1	1366.6	1989.4	3623.0	8	40		
10	40	611.6	1027.2	1634.3	2474.5	3598.0		10	40		
12	ID	981.6	1644.5	2612.4	3963.5	5764.6		12	ID		

NOTES:

(1) Basis of Table: 90 F Condensing Temperature, 1F ΔT per 100 ft Equivalent Length. Discharge and Liquid Lines based on 20 F suction.

(2) For other ΔT's and Equivalent Lengths, L_e
Line Capacity (Tons)

$$= \text{Table Tons} \times \left(\frac{\text{Table } L_e}{\text{Actual } L_e} \times \frac{\text{Actual } \Delta T \text{ Loss Desired}}{\text{Table } \Delta T \text{ Loss}} \right)^{0.55}$$

(3) For other Tons and Equivalent Lengths in a given pipe size

$$\Delta T = \text{Table } \Delta T \times \frac{\text{Actual } L_e}{\text{Table } L_e} \times \left(\frac{\text{Actual Tons}}{\text{Table Tons}} \right)^{1.8}$$

(4) Values based on 90 F condensing temp. For capacities at other condensing temp, multiply table value by line capacity multiplier:

Line	Condensing Temperature, F 70	80	90	100
Suction Lines	1.05	1.02	1.00	0.98
Discharge Lines	0.78	0.89	1.0	1.11

(5) Based on 0 F evaporator temperature. The capacity is effected less than 3% when applied from −40 F to 40 F extremes

Figure 7–6 Refrigerant line capacities for R-717 (single- and high-stage applications). (By permission of *ASHRAE Handbook*.)

Refrigerant and ΔT Equivalent of Friction Drop*	Line Size Type L Copper OD	Suction Lines* Saturated Suction Temp F −90	−80	−70	−60	−50	−40	−30	Discharge Lines* −50	Line Size Type L Copper OD	Liquid Lines
Refrigerant 12	1/2									1/2	
	5/8									5/8	
	7/8				0.23	0.30	0.40	0.51	0.9	1 7/8	
	1 1/8	0.17	0.25	0.34	0.46	0.62	0.81	1.1	1.9	1 1/8	
	1 3/8	0.31	0.44	0.60	0.81	1.1	1.4	1.8	3.3	1 3/8	
	1 5/8	0.49	0.69	0.95	1.3	1.7	2.3	2.9	5.3	1 5/8	
	2 1/8	1.0	1.4	2.0	2.7	3.6	4.7	6.1	11.0	2 1/8	
2 F ΔT per 100 ft	2 5/8	1.8	2.6	3.6	4.8	6.4	8.4	10.8	19.5	2 5/8	
Equiv. Length	3 1/8	3.0	4.2	5.7	7.7	10.3	13.4	17.2	31.1	3 1/8	See
	3 5/8	4.4	6.2	8.5	11.6	15.3	20.0	25.6	46.2	3 5/8	Fig. 7-3
	4 1/8	6.2	8.8	12.1	16.3	21.7	28.2	36.2	65.2	4 1/8	
	5 1/8	11.3	15.8	21.8	29.4	38.9	50.7	65.0	116.7	5 1/8	
	6 1/8	18.2	25.6	35.2	47.4	62.8	81.6	104.8	188.0	6 1/8	
	1/2									1/2	
	5/8								0.7	5/8	
Refrigerant 22	7/8	0.18	0.25	0.34	0.46	0.61	0.79	1.0	1.9	7/8	
	1 1/8	0.36	0.51	0.70	0.94	1.2	1.6	2.1	3.8	1 1/8	
	1 3/8	0.6	0.9	1.2	1.6	2.2	2.8	3.6	6.6	1 3/8	
	1 5/8	1.0	1.4	1.9	2.6	3.4	4.5	5.7	10.5	1 5/8	
	2 1/8	2.1	3.0	4.1	5.5	7.2	9.3	11.9	21.7	2 1/8	
2 F ΔT per 100 ft	2 5/8	3.8	5.3	7.2	9.7	12.7	16.5	21.1	38.4	2 5/8	
Equiv. Length	3 1/8	6.1	8.5	11.6	15.5	20.4	26.4	33.8	61.4	3 1/8	See
	3 5/8	9.1	12.7	17.3	23.1	30.4	39.4	50.2	91.2	3 5/8	Fig. 7-4
	4 1/8	12.9	18.0	24.5	32.7	43.0	55.6	70.9	128.6	4 1/8	
	5 1/8	23.2	32.3	43.9	58.7	77.1	99.8	126.9	229.5	5 1/8	
	6 1/8	37.5	52.1	71.0	94.6	124.2	160.5	204.2	369.4	6 1/8	
	Steel IPS SCH				−60	−50	−40	−30	−50	Steel IPS SCH	
										3/8 80	
	1/2 40				0.25	0.34	0.46	0.61	1.3	1/2 80	
	3/4 40				0.54	0.74	0.99	1.3	2.7	3/4 80	
	1 40				1.0	1.4	1.9	2.5	5.0	1 80	
Refrigerant 717	1 1/4 40				2.2	3.0	4.0	5.2	10.5	1 1/4 80	
(Ammonia)	1 1/2 40				3.3	4.5	6.0	7.8	15.8	1 1/2 80	
	2 40				6.4	8.7	11.6	15.2	30.5	2 40	See
	2 1/2 40				10.2	13.9	18.7	24.4	48.7	2 1/2 40	Fig. 7-6
1 F ΔT Per 100 ft	3 40				18.3	24.8	33.1	43.3	86.5	3 40	
Equiv. Length	4 40				37.6	51.0	68.1	88.8	176.6	4 40	
	5 40				68.2	92.4	123.4	160.7	318.8	5 40	
	6 40				111.2	150.2	200.2	260.8	516.5	6 40	
	8 40				229.3	309.6	411.9	536.1	1060.0	8 40	

NOTES:

*(1) Values in this table are tons of refrigeration resulting in a line friction drop per 100 ft of equivalent pipe length corresponding to the change in saturation temp indicated under the refrigerant designation.

(2) Values based on 0 F condensing temp. For capacities at other condensing temp. multiply table value by proper line capacity multiplier from Table 26.

(3) For other ΔT's and Equivalent Lengths, L_e
Line Capacity (Tons)

$$= \text{Table Tons} \times \left(\frac{\text{Table } L_e}{\text{Actual } L_e} \times \frac{\text{Actual } \Delta T \text{ Loss Desired}}{\text{Table } \Delta T \text{ Loss}} \right)^{0.55}$$

(4) For other Tons and Equivalent Lengths in a given pipe size,

$$\Delta T = \text{Table } \Delta T \times \frac{\text{Actual } L_e}{\text{Table } L_e} \times \left(\frac{\text{Actual Tons}}{\text{Table Tons}} \right)^{1.8}$$

(5) For pressure drop (psi) corresponding to ΔT, refer to Refrigerant property tables.

Figure 7–7 Refrigerant line capacities for intermediate and low stage duty (Tons) for Refrigerants-12, -22 and ammonia. (By permission of *ASHRAE Handbook*.)

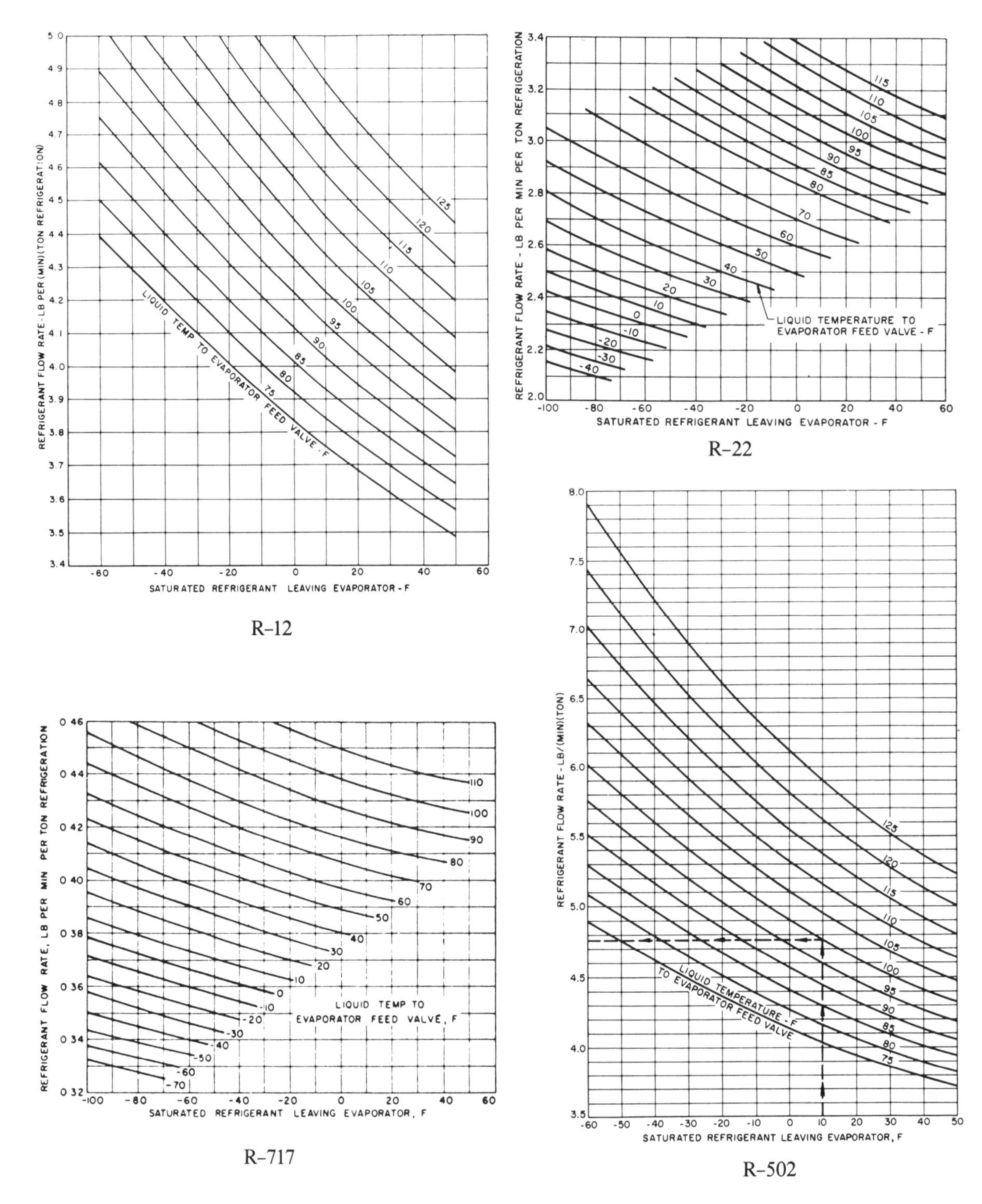

Figure 7–8 Refrigerant flow rates in pounds per minute per ton of refrigeration, as a function of liquid temperature entering the coil and evaporating temperature. (By permission of *ASHRAE Handbook*.)

Nominal Pipe or Tube Size (in.)	Globe[b]	60°-Y	45°-Y	Angle[b]	Gate[c]	Swing Check[d]	Lift Check
3/8	17	8	6	6	0.6	5	
1/2	18	9	7	7	0.7	6	Globe &
3/4	22	11	9	9	0.9	8	Vertical Lift
1	29	15	12	12	1.0	10	Same as
1 1/4	38	20	15	15	1.5	14	Globe
1 1/2	43	24	18	18	1.8	16	Valve[e]
2	55	30	24	24	2.3	20	
2 1/2	69	35	29	29	2.8	25	
3	84	43	35	35	3.2	30	
3 1/2	100	50	41	41	4.0	35	
4	120	58	47	47	4.5	40	
5	140	71	58	58	6	50	
6	170	88	70	70	7	60	
8	220	115	85	85	9	80	
10	280	145	105	105	12	100	Angle Lift
12	320	165	130	130	13	120	Same as Angle
14	360	185	155	155	15	135	Valve
16	410	210	180	180	17	150	
18	460	240	200	200	19	165	
20	520	275	235	235	22	200	
24	610	320	265	265	25	240	

[a] Losses are for all valves in fully open position.
[b] These losses do not apply to valves with needle point type seats.
[c] Regular and short pattern plug cock valves, when fully open, have same loss as gate valve. For valve losses of short pattern plug cocks above 6 in, check manufacturer.
[d] Losses also apply to the in-line, ball type check valve.
[e] For *Y* pattern globe lift check valve with seat approximately equal to the nominal pipe diameter, use values of 60° - *Y* valve for loss.

Figure 7-9 Valve losses in equivalent feet of pipe. (By permission of *ASHRAE Handbook*.)

Nominal Pipe or Tube Size (in.)	Smooth Bend Elbows						Smooth Bend Tees			
	90° Std*	90° Long Rad.†	90° Street*	45° Std*	45° Street*	180° Std*	Flow Through Branch	Straight-Through Flow: No Reduction	Straight-Through Flow: Reduced 1/4	Straight-Through Flow: Reduced 1/2
3/8	1.4	0.9	2.3	0.7	1.1	2.3	2.7	0.9	1.2	1.4
1/2	1.6	1.0	2.5	0.8	1.3	2.5	3.0	1.0	1.4	1.6
3/4	2.0	1.4	3.2	0.9	1.6	3.2	4.0	1.4	1.9	2.0
1	2.6	1.7	4.1	1.3	2.1	4.1	5.0	1.7	2.2	2.6
1 1/4	3.3	2.3	5.6	1.7	3.0	5.6	7.0	2.3	3.1	3.3
1 1/2	4.0	2.6	6.3	2.1	3.4	6.3	8.0	2.6	3.7	4.0
2	5.0	3.3	8.2	2.6	4.5	8.2	10	3.3	4.7	5.0
2 1/2	6.0	4.1	10	3.2	5.2	10	12	4.1	5.6	6.0
3	7.5	5.0	12	4.0	6.4	12	15	5.0	7.0	7.5
3 1/2	9.0	5.9	15	4.7	7.3	15	18	5.9	8.0	9.0
4	10	6.7	17	5.2	8.5	17	21	6.7	9.0	10
5	13	8.2	21	6.5	11	21	25	8.2	12	13
6	16	10	25	7.9	13	25	30	10	14	16
8	10	13	—	10	—	33	40	13	18	20
10	25	16	—	13	—	42	50	16	23	25
12	30	19	—	16	—	50	60	19	26	30
14	34	23	—	18	—	55	68	23	30	34
16	38	26	—	20	—	62	78	26	35	38
18	42	29	—	23	—	70	85	29	40	42
20	50	33	—	26	—	81	100	33	44	50
24	60	40	—	30	—	94	115	40	50	60

*R/D approximately equal to 1.
†R/D approximately equal to 1.5.

Figure 7-10 Fitting losses in equivalent feet of pipe. (By permission of *ASHRAE Handbook*.)

Nom. Pipe or Tube Size (in.)	Sudden Enlargement* d/D			Sudden Contraction* d/D			Sharp Edge*		Pipe Projection*	
	1/4	1/2	3/4	1/4	1/2	3/4	Entrance	Exit	Entrance	Exit
3/8	1.4	0.8	0.3	0.7	0.5	0.3	1.5	0.8	1.5	1.1
1/2	1.8	1.1	0.4	0.9	0.7	0.4	1.8	1.0	1.8	1.5
3/4	2.5	1.5	0.5	1.2	1.0	0.5	2.8	1.4	2.8	2.2
1	3.2	2.0	0.7	1.6	1.2	0.7	3.7	1.8	3.7	2.7
1 1/4	4.7	3.0	1.0	2.3	1.8	1.0	5.3	2.6	5.3	4.2
1 1/2	5.8	3.6	1.2	2.9	2.2	1.2	6.6	3.3	6.6	5.0
2	8.0	4.8	1.6	4.0	3.0	1.6	9.0	4.4	9.0	6.8
2 1/2	10	6.1	2.0	5.0	3.8	2.0	12	5.6	12	8.7
3	13	8.0	2.6	6.5	4.9	2.6	14	7.2	14	11
3 1/2	15	9.2	3.0	7.7	6.0	3.0	17	8.5	17	13
4	17	11	3.8	9.0	6.8	3.8	20	10	20	16
5	24	15	5.0	12	9.0	5.0	27	14	27	20
6	29	22	6.0	15	11	6.0	33	19	33	25
8	—	25	8.5	—	15	8.5	47	24	47	35
10	—	32	11	—	20	11	60	29	60	46
12	—	41	13	—	25	13	73	37	73	57
14	—	—	16	—	—	16	86	45	86	66
16	—	—	18	—	—	18	96	50	96	77
18	—	—	20	—	—	20	115	58	115	90
20	—	—	—	—	—	—	142	70	142	108
24	—	—	—	—	—	—	163	83	163	130

*Enter table for losses at smallest diameter *d*.

Figure 7-10 (continued)

SYSTEM PRACTICES FOR HALOCARBON REFRIGERANTS

The information given in the next group of pages is designed primarily for halocarbon refrigerants. System practices for ammonia refrigeration are described later in a separate section.

To provide a properly designed piping system, not only are the sizes important but also the layout of the piping to provide a suitable arrangement to return the entrapped oil to the compressor. The following descriptions and drawings therefore relate to piping design. Four different areas are discussed: (1) suction lines, (2) hot-gas lines, (3) condenser-to-receiver lines, and (4) liquid lines.

Suction Lines

The suction-line design is the most critical since a pressure drop in the suction line means a greater loss of system capacity. The proper piping for a suction riser from a single evaporator is shown in Figure 7-11. The sump or loop at the outlet of the coil permits good coil drainage and, with proper riser gas velocities, good return of oil to the compressor. The riser velocity should be not less than 1000 fpm or not more than 4000 fpm.

The piping selection tables are based on a 2 °F temperature drop for 100-ft equivalent length for halocarbon refrigerants and a 1 °F temperature drop per 100-ft equivalent length for ammonia.

When the system operates over a wide capacity range, a double suction riser should be used to ensure proper oil return (Figure 7-12). At minimum capacity the small pipe

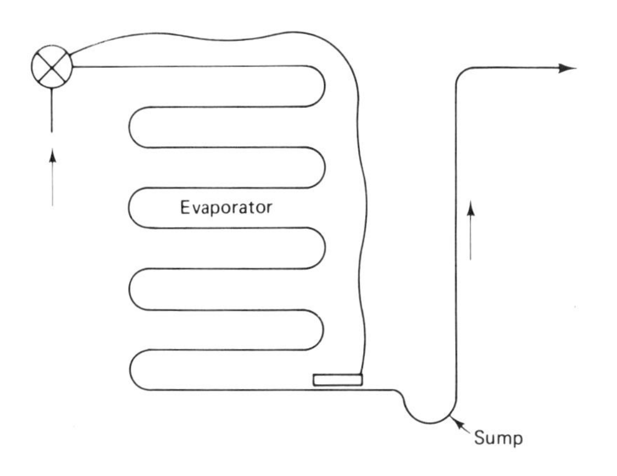

Figure 7-11 Suction riser pipe can be reduced in size to ensure oil return, particularly where compressor capacity control is used.

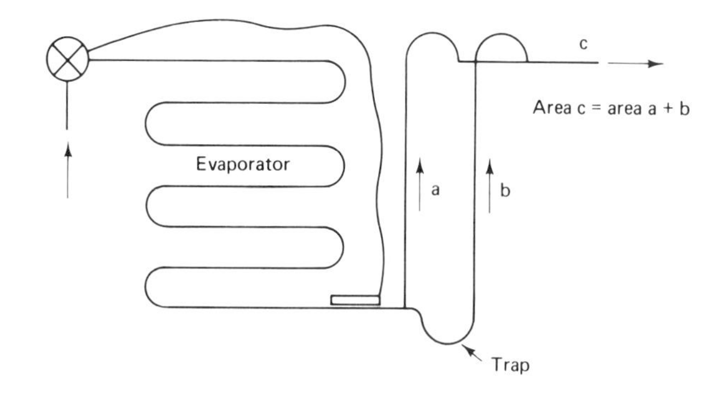

Figure 7-12 This double suction riser arrangement is used where system capacity varies over a wide range. On low capacity, oil fills trap, directing suction gas into single riser. At full capacity, both risers are used.

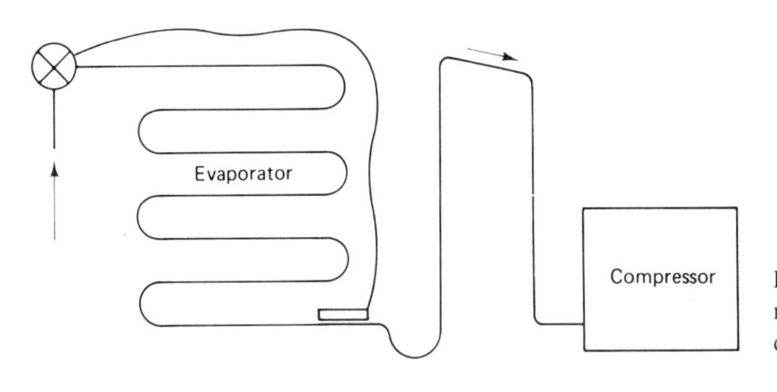

Figure 7-13 Suction loop prevents refrigerant and oil draining to compressor during "off" cycle.

handles the load. At full capacity both risers are put into use. The size of the horizontal suction lines is equivalent to the sum of the areas of the two risers. The small riser is sufficient to handle the minimum load. At minimum load the trap, located between the two risers, fills up with oil, sealing off the large riser. Note that both risers form a loop as they enter the main suction line at the top. This prevents drainage into the idle riser when the system is operating at partial capacity.

If the compressor is located below the evaporator or on the same level, a suction loop needs to be used to prevent liquid drainage from the evaporator to compressor during shutdown (see Figure 7-13). The loop must extend as high as the top of the evaporator coil. If the system is provided with automatic pump-down, so that the refrigerant is pumped out of the evaporator before shutdown, the suction loop can be eliminated.

Hot-Gas Lines

There are two conditions that require separate treatment. If the compressor is above the condenser a simple piping arrangement pitches down toward the condenser (see Figure 7–14). If the compressor is below the condenser it is important to provide a trap at the outlet of the compressor the full height of the compressor and to size the riser to maintain adequate velocity to carry the oil to the condenser. A double riser can be used when there is wide variation of load (Figure 7–12).

When the compressor is located in an area that can become colder than the condenser, it is advisable to use a check valve at the inlet to the condenser. This prevents migration of refrigerant during the "off" cycle (see Figure 7–15).

On halocarbon systems, if an oil separator is used it should be located between the compressor and the hot-gas loop (Figure 7–16). This keeps any condensed liquid refrigerant out of the oil separator.

The best location for the hot-gas muffler is in the down side of the hot-gas loop (see Figure 7–17). This prevents liquid trapping in the muffler.

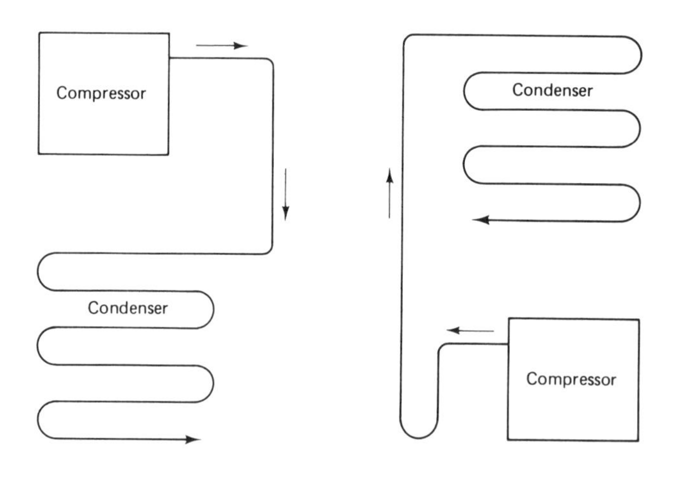

Figure 7–14 Hot-gas line piping from compressor to condenser. Piping must pitch downward from the compressor. Note that if the condenser is above the compressor a protection loop is provided.

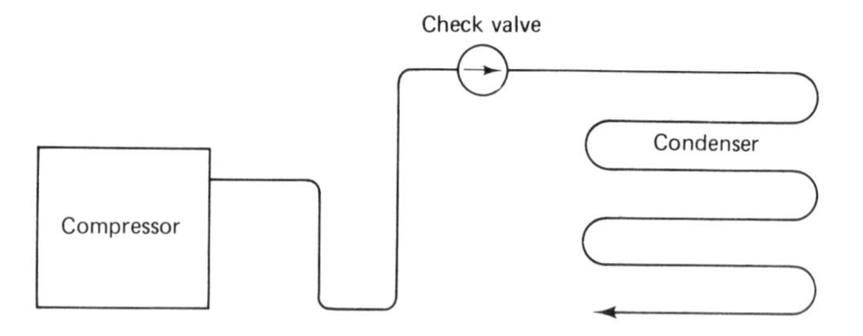

Figure 7–15 On discharge-line piping, both check valve and loop are used to prevent migration of refrigerant liquid to head of compressor during "off" cycle. This is particularly important where the compressor is in a cooler location than is the condenser.

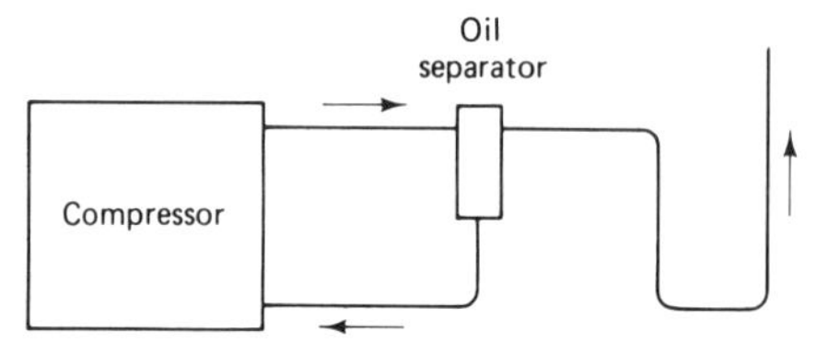

Figure 7–16 Where oil separators are used with halocarbon refrigerants, they should be placed close to the compressor and before the protective hot-gas loop. Insulating the oil separator is good practice to prevent condensation of refrigerant.

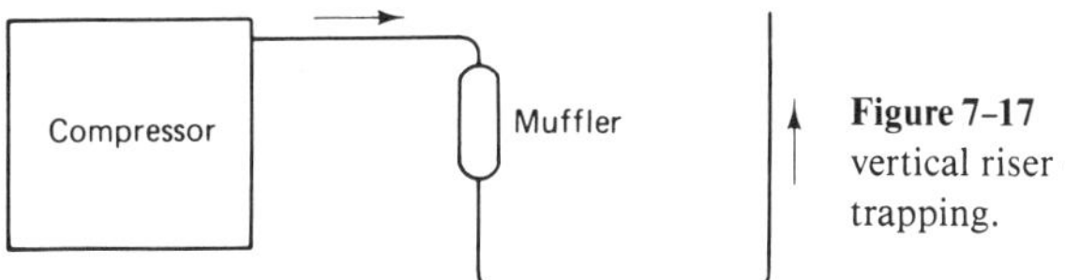

Figure 7–17 Muffler should be installed in vertical riser of hot-gas loop to prevent oil trapping.

Condenser-to-Receiver Lines

The refrigerant piping between the condenser and receiver should be free-draining. In a close-coupled arrangement a single pipe can carry the liquid to the receiver and at the same time permit the flow of replaced vapor in the opposite direction back to the condenser. The pipes should be of ample size to allow this double-flow condition (see Figure 7–18).

For large condensers and where the horizontal distance to the receiver is more than 6 ft, an equalizer line between the inlet to the condenser and the top of the receiver should be installed (see Figure 7–19). The vertical distance between the condenser and receiver must be sufficient to overcome the piping friction.

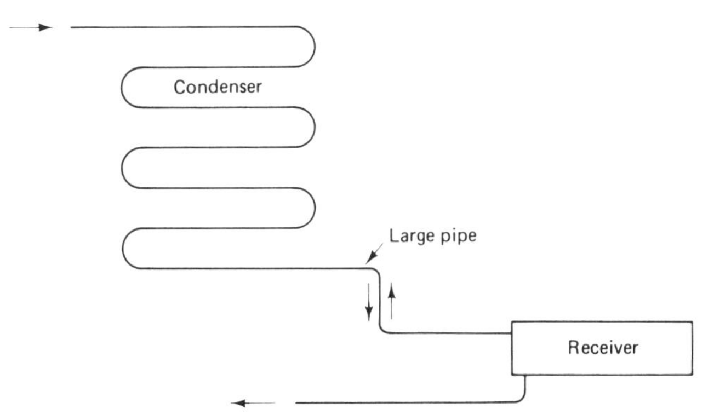

Figure 7–18 Condenser to receiver piping should be oversized to allow for counterflow of liquid and vapor refrigerant. Piping should be free draining.

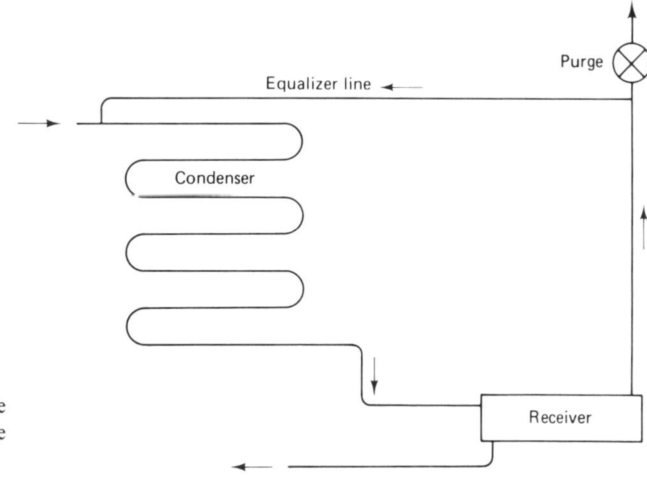

Figure 7–19 An equalizer line should be used where the condenser and receiver are horizontally more than 6 ft apart.

Liquid Lines

Fortunately, the return of oil in the liquid line is accomplished without special attention. Even traps in the liquid line do not interfere with the return of oil. Two items are important, however:

1. Excessive pressure drop will reduce the capacity unless special allowance is made. The flow through the metering device is directly dependent on the pressure drop across it.
2. A pressure drop in the liquid line can cause the formation of flash gas unless the liquid has adequate subcooling.

The pressure drop in the liquid line is the total of loss through piping, fittings, accessories, and vertical lift. One psi is lost for each 1.8 ft of vertical lift for halocarbon refrigerants. Thus, for an 18-ft lift the pressure drop is 10 psi. When the refrigerant lift is considerable it is advisable to determine the amount of subcooling required. For example, with an 18-foot lift (10 psi), a pipe friction of 3 psi and a 7-psi drop through the accessories, the total pressure drop would be 20 psi. If the discharge pressure was 190.86 psi (128 °F for R-12), subtracting 20 psi gives 170.86 psi (119 ° for R-12) available at the expansion valve. The amount of subcooling required would be 9 °F (128 °–119 °F).

Installation

It is usually good practice to insulate suction lines. For one reason, condensate dripping from the pipe is usually undesirable. Unless heat is needed to evaporate raw liquid in the suction line, heat added outside the refrigerated space lowers the efficiency of the system. Hot-gas lines are usually at a higher-than-ambient temperature and do not need to be insulated. Liquid lines are seldom insulated unless they are exposed to the sun or pass through a heated space.

Adequate piping support is important. When insulated suction lines are supported a metal hanger should not touch the pipe or compress the insulation.

The location of the sight glass in the piping should be after the drier and before the metering device. This location will detect a stopped-up drier as well as indicating the quantity of the liquid refrigerant. Occasionally, driers are also placed at the outlet of the receiver to detect an adequate supply of liquid from the receiver.

MULTIPLE SYSTEM PRACTICES FOR HALOCARBON REFRIGERANTS

This section pertains to refrigerant piping for multiple compressors, evaporators, and condensers. There are three considerations that must be included in the piping system:

1. Oil return to the compressor
2. Protection to the compressor against "slugging" with liquid
3. Protection to the compressor during the "off" cycle

Hot-Gas Lines

The best piping arrangement for connecting hot-gas lines on multiple compressors is shown in Figure 7–20. This arrangement prevents the oil or liquid refrigerant from draining from a running compressor into the idle compressor. Attention must be given to allowance for vibration. In most cases it is advisable to use vibration absorbers at the outlet of each compressor. The use of a bullhead tee is never advisable (Figure 7–21).

When connecting multiple water-cooled condensing units, a hot-gas equalizer line is required to equalize the pressures to the condensers. Liquid-line outlets should be at the same level. When the liquid lines connect to a common header, at least a 12-in. drop is required between the condenser outlets and the common liquid line.

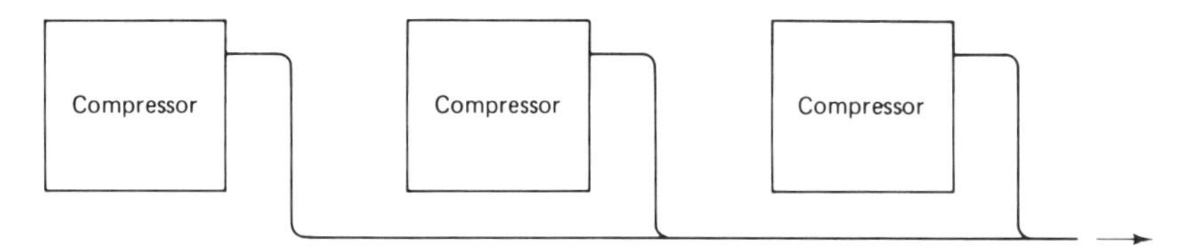

Figure 7–20 Method of connecting the discharge lines of multiple compressors connected to a common condenser. A convenient location for the header is at floor level.

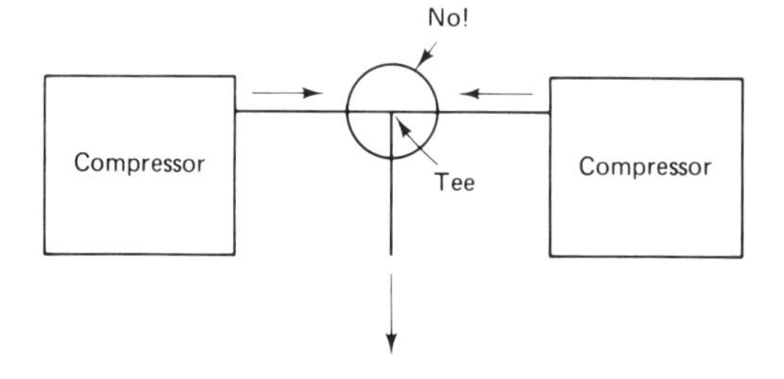

Figure 7–21 Bullhead tees should not be used to connect the discharge lines from two compressors.

Suction Lines

When suction lines are manifolded above the compressor, a horizontal takeoff should be used between the compressors (Figure 7–22). This prevents oil from draining to the idle compressor by gravity. When the suction header is below the compressors the takeoff should be made to equalize the return of oil from each compressor. (Figure 7–23).

Figure 7–22 Suction-line manifold should be arranged to prevent oil draining to the idle compressor. A crankcase oil-levelizer mechanism is advisable.

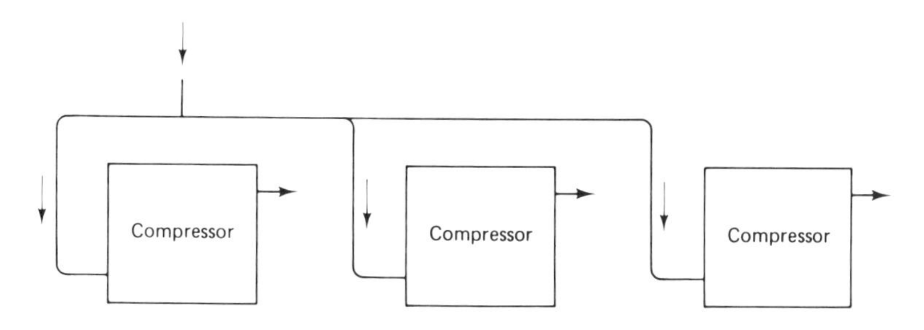

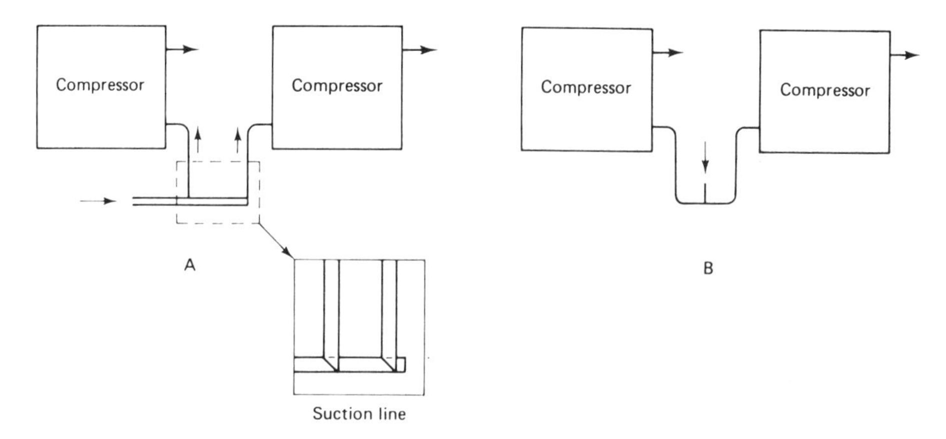

Figure 7–23 Method of connecting the suction header to multiple elevated compressors to provide equal oil return to each compressor.

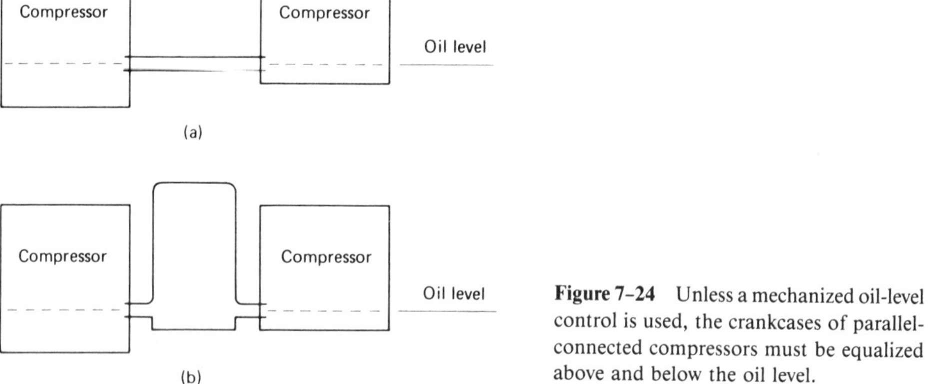

Figure 7–24 Unless a mechanized oil-level control is used, the crankcases of parallel-connected compressors must be equalized above and below the oil level.

When compressors are connected in parallel, the crankcases of the compressors must be equalized above and below the oil level (see Figure 7–24). An alternative arrangement is the use of oil levelizers. This is a mechanical arrangement for maintaining the proper oil level in each compressor.

Evaporator Piping

The proper piping of multiple evaporators is shown for a number of common variations:

1. Evaporators above the compressor, on the same level
2. Evaporators above the compressor, on different levels
3. Evaporators below the compressor, on the same level
4. Evaporators below the compressor, on different levels

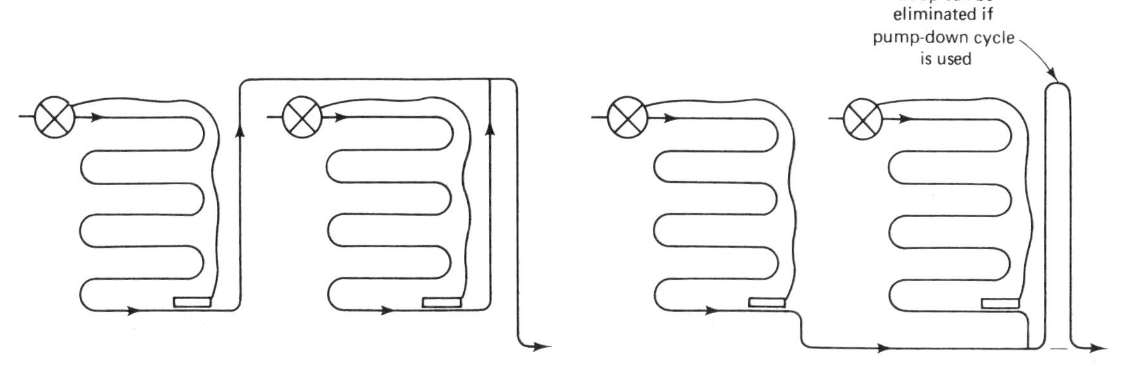

Figure 7–25 Method of connecting suction lines to evaporators on the same level above compressors.

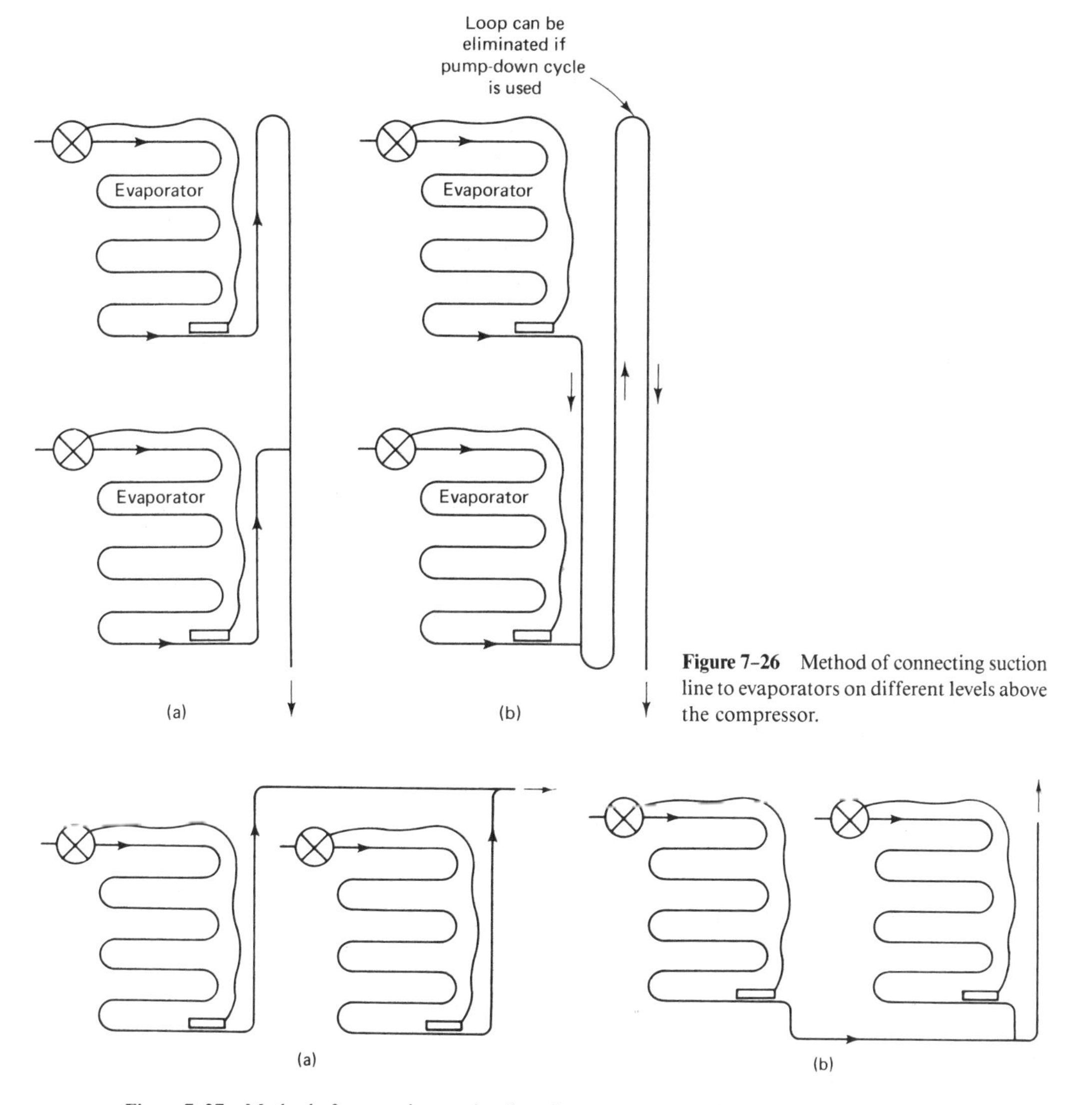

Figure 7–26 Method of connecting suction line to evaporators on different levels above the compressor.

Figure 7–27 Method of connecting suction lines for evaporators on the same level below the compressor.

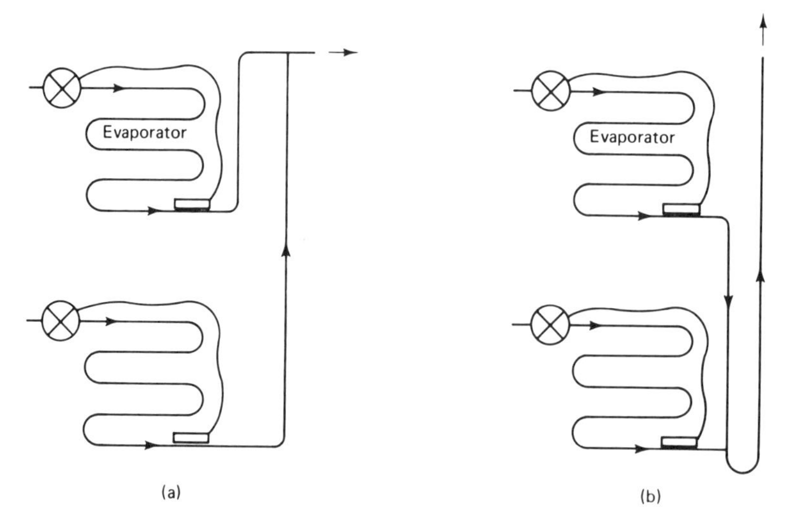

Figure 7–28 Method of connecting suction line to evaporators on different levels below the compressor.

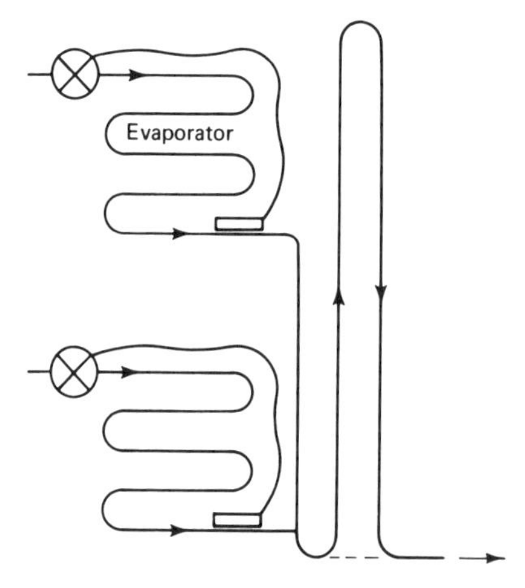

Figure 7–29 The dashed line shows suction connection to evaporators at different levels where pump-down cycle is used.

Figure 7–25 meets the requirements of arrangement 1. Figure 7–26 meets the requirements of arrangement 2. Figure 7–27 meets the requirements of arrangement 3. Figure 7–28 meets the requirements of arrangement 4. The suction loop in Figure 7–29 can be omitted when a pump-down cycle is used.

Condenser Piping

The preferred arrangement for piping multiple condensers is shown in Figure 7–30. An equalizer line should be used for a common receiver as shown in Figure 7–31.

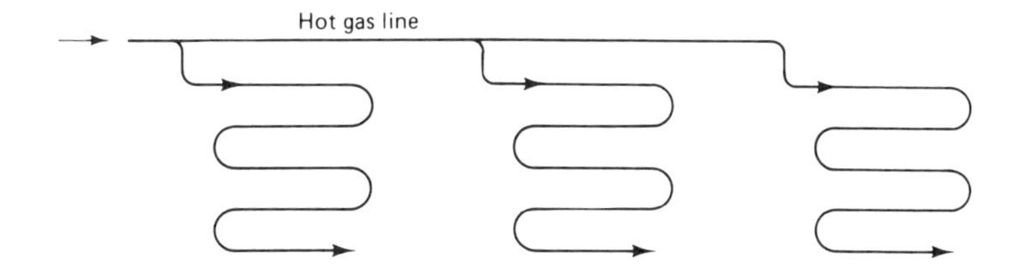

Figure 7-30 Method of connecting discharge lines to multiple air-cooled condensers.

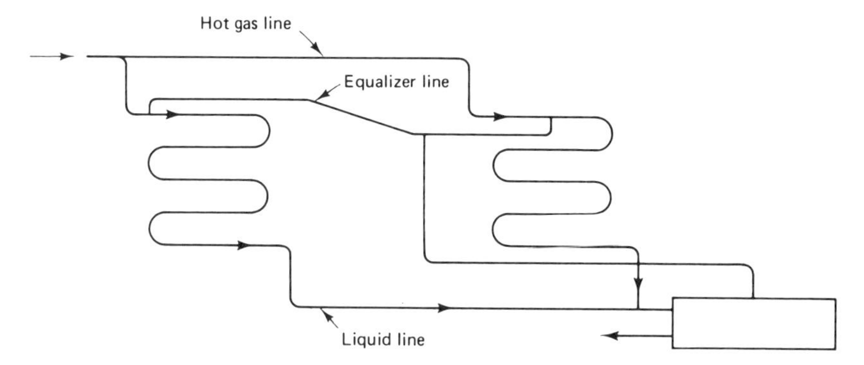

Figure 7-31 Method of connecting piping to multiple evaporators with a common receiver.

ACCESSORIES

There are a number of imporatant refrigeration system accessories which have specialized uses and are applied for safety reasons, to enhance the operating characteristics of the system, or to facilitate service. Some of the more useful accessories are: (1) vibration eliminators, (2) liquid indicators, (3) pressure relief valves, (4) oil separators, (5) heat exchangers, (6) receiver tank valves, (7) compressor service valves, (8) manual valves, and (9) dehydrators.

Vibration Eliminators

A typical vibration eliminator is shown in Figure 7-1. These units are made of accordion-shaped seamless flexible tubing covered with high-tensile braided wire. They are usually applied at the compressor for discharge or suction-line connections. The end of the vibration eliminator not connected to the compressor should be held rigidly. The end connected to the compressor should move in a plane perpendicular to the vibration eliminator. This accessory should not be used in compression, but only with lateral motion.

Liquid Refrigerant Indicators

Liquid indicators are useful in determining the refrigerant charge in a system. These consist of a sight glass that reveals the nature of the fluid flowing through the liquid line. A solid stream of liquid is desirable. Bubbles indicate flash gas or low refrigerant charge.

Liquid indicators are also available that show the moisture content of the liquid refrigerant. If the indicator shows a *wet* refrigerant, the system must be dried out.

Typical liquid indicators are shown in Figure 7-32. The key to the indicator's information is usually shown on the cap or at the side of the sight glass.

Figure 7-32 Moisture-indicating sight glass. The center dot changes color to indicate the presence of moisture in refrigerant. (By permission of Sporlan Valve Company.)

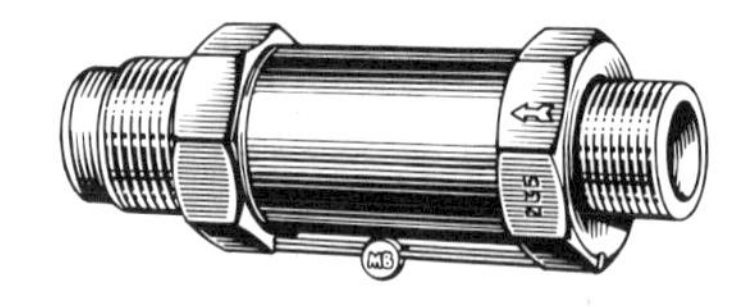

Figure 7-33 Spring-loaded pressure relief valve. It must be selected to unload the refrigerant at the required pressure, depending on refrigerant used and the type of system. (By permission of Mueller Brass Company.)

Pressure Relief Valves

Pressure relief valves are of two types: spring loaded (see Figure 7-33) and fusible plug. The location and number of pressure relief valves are usually specified by the refrigeration code. The advantage of the spring relief valve is that *not all* of the refrigerant is lost if the valve opens. The fusible plug operates on temperature. When it opens, the entire charge is lost. A vent is usually placed to release the charge to the outdoors.

Oil Separators

Oil separators should be used for the following applications:

1. On all systems using nonmiscible refrigerants
2. Systems operated at low temperatures
3. Systems using flooded evaporators
4. Other systems where there is high oil entrainment

Oil separators do not remove all the oil carried with the refrigerant, but they can remove 90 or 95% of it. Provisions must be made for properly handling the oil that is not removed by the oil separator. A typical oil separator is shown in Figure 7-16. The oil drops out due to a change in velocity and collision with an impingement screen. The level of oil is sensed by a float valve. When the sump fills up, oil is returned to the compressor or delivered to a sump.

There is some danger of liquid refrigerant condensing in the oil separator during the off cycle and being dumped by the oil separator in the crankcase of the compressor. This can be eliminated by returning the oil from the separator to the suction line of the compressor rather than to the crankcase. This arrangement must be designed carefully to gradually feed the oil to the suction line. A hand valve and a sight glass in the oil return line are useful in regulating the oil flow.

Oil separators for ammonia systems. Special provision needs to be made in an ammonia system for the handling of oil that is carried with the refrigerant in the discharge from the compressor. Since ammonia is a nonmiscible refrigerant, oil does not mix with the refrigerant. It is important to use an oil separator in the discharge line to remove the entrained oil. This oil separator is often placed a considerable distance from the compressor since the oil separates best when the discharge gas is cooler. Oil is not usually returned to the compressor because it often collects other contaminants.

The oil that is not removed by the oil separator collects in the low places in the system since oil is heavier than liquid ammonia. Sumps with drain valves are provided in various components of the system, such as receivers, evaporators, accumulators, and condensers. Oil is periodically drained manually from these sumps. On some systems a "blow-down" arrangement is provided, using hot gas in the piping to force oil into an oil receiver. Periodically, oil needs to be added to the compressor to replace that which is lost through entrainment.

Heat Exchangers

There are many types of heat exchangers used in refrigeration work. An evaporator, a condenser, or an inner stage cooler are all types of heat exchangers.

The liquid-to-suction line heat exchanger (Figure 7-34) has the special function of superheating the suction gas and subcooling the liquid. This protects the compressor and gives additional subcooling to the liquid to prevent flash gas.

Figure 7-34 Liquid-to-suction heat exchanger is used to superheat the suction vapor and subcool the liquid. It therefore serves to protect the compressor from liquid slugs as well as to prevent flash gas in the liquid line. It is useful in subcooling liquid which has a vertical rise to the expansion valve.

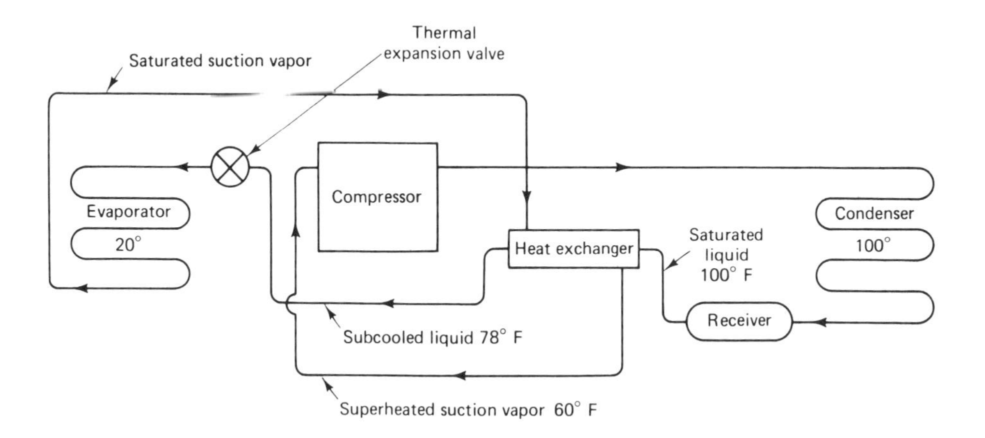

Receiver Tank Valves

Referring to Figure 7–35, these valves make it possible to pump down the charge of refrigerant into the receiver. A dip tube is placed on the outlet valve so that liquid refrigerant can be drawn from the bottom of the receiver.

Compressor Service Valves

Referring to Figure 7–36, these valves have three positions. The back-seat position is for normal operation or for connecting gauges. The midposition permits reading the pressures in the system during operation. The front-seat position shuts off the piping connection to the compressor. The front seat is often opened slowly in a suction valve during pull-down. The front-seat position on a discharge valve is seldom used due to the danger of building up a damaging head pressure.

Manual Valves

Figure 7–37 shows valves that are globe design. They may be a packed or a packless type. If a packing gland is used, the packing nut must be loosened when the valve is used and retightened after the system is put back into service.

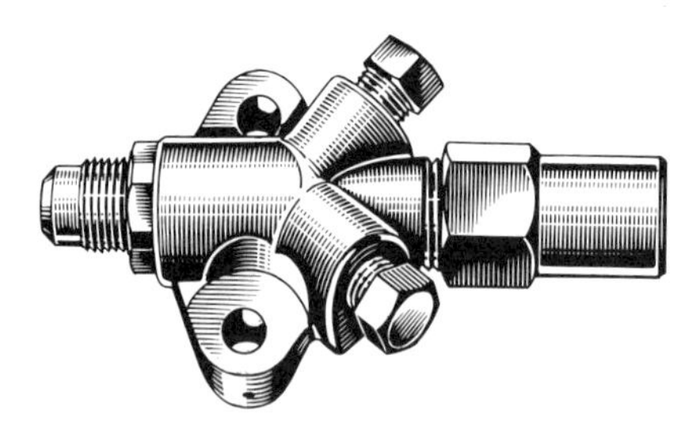

Figure 7–35 Refrigerant receiver tank shutoff valve. Opened or closed by valve wrench. Light outlet, called King valve, closed when system is manually pumped down. (By permission of Mueller Brass Company.)

Figure 7–36 Compressor service valve. Three positions: front seat, back seat, and midposition. (By permission of Mueller Brass Company.)

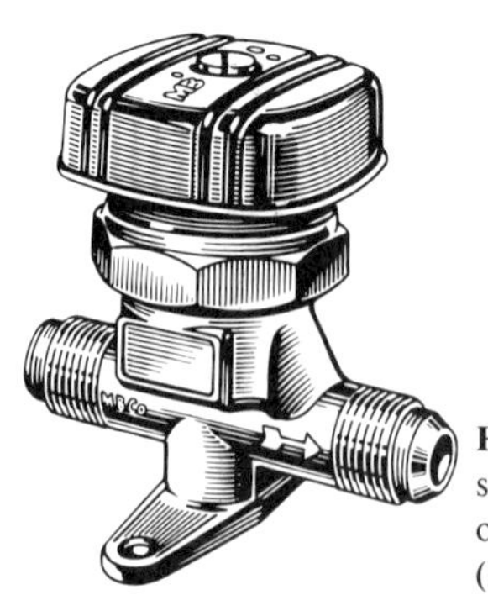

Figure 7–37 Diaphragm-type refrigerant shutoff valve. Diaphragm prevents leakage of refrigerant from valve, in any position. (By permission of Mueller Brass Company.)

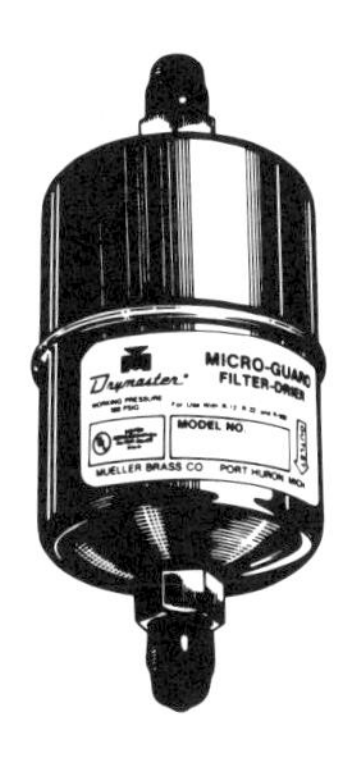

Figure 7-38 Dehydrator for removing moisture from halocarbon refrigerants. See manufacturers' information for proper selections. (By permission of Mueller Brass Company.)

Dehydrators

Dehydrators or driers are used to remove moisture from the system (see Figure 7-38). When saturated they must be replaced. Larger sizes have replaceable inserts. Recommended sizes are usually specified by the manufacturer.

SYSTEM PRACTICES FOR AMMONIA REFRIGERATION

From a refrigeration standpoint, ammonia has excellent properties. The heat absorbed per pound of refrigerant is far greater than any other one of the common refrigerants. Another advantage of ammonia is the ability it provides to detect leaks. Caution must be used in its application due to its toxicity in case of a leak. With certain mixtures of air it is explosive.

It is extremely important that ammonia pipe be cleaned internally when it is installed. Ammonia is an excellent solvent and will pick up dirt and other extraneous matter to clog valve ports and will harm compressor cylinders unless they are thoroughly cleaned. Pipes should be power rotary brushed and blown out with compressed air before use.

Iron and steel are suitable for piping following the guidelines of ANSI B31.5 code for pressure piping. Fittings should be forged steel. Stop valves should be used on both the inlet and outlet of condensers, receivers, and evaporators to permit isolation in case of leaks or service requirements. All control valves should be preceded by a strainer. Valve stems should be horizontal wherever possible, allowing less chance of scale deposits on the valve seat. Solenoid valves should be located upright.

Oil is fusible with ammonia in very small quantities; however, most of the oil in the system separates out. Since oil is heavier than ammonia, it collects in the bottom of pressure vessels. Oil must be removed from the system periodically and replaced in the compressor with new oil.

Compressor Piping

Figure 7-39 shows two compressors operating from the same suction main. Service valves are supplied in both suction and discharge lines to the compressor, which is useful in case the standard compressor valves do not close tightly.

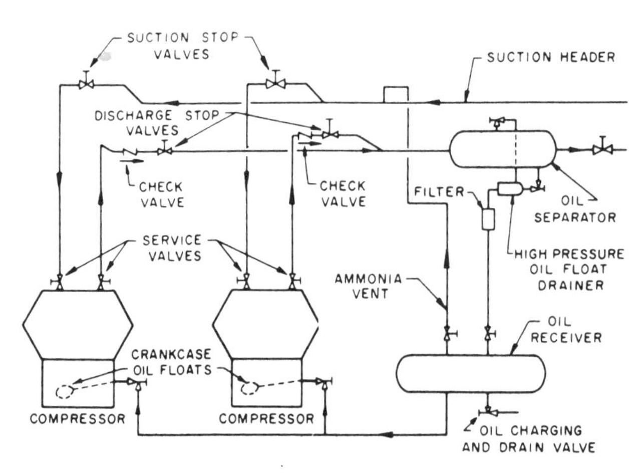

Figure 7–39 Parallel compressors with common oil separator. (By permission of *ASHRAE Handbook*.)

An oil separator should be located as far from the compressors as possible since the oil separates more easily at lower temperatures. Figure 7–39 shows a common oil separator for the two compressors and a single oil receiver.

Check Valves

Note in Figure 7–39 that check valves are installed in the discharge line to each compressor. Where an individual oil separator is provided for each compressor, the check valve can be installed on the downstream side of the oil separator. This arrangement minimizes the flow of hot gas back to the compressor during shutdown.

Condenser and Receiver Piping

Figure 7–40 shows a horizontal water-cooled condenser with through-type receiver. Head pressure controls are not usually necessary. Air vents should be provided for manual purging. Any ammonia vessel that can be valved off must be installed with relief valves. A noncondensable gas automatic purge unit is useful on most large plants, particularly where the suction pressure is below atmospheric pressure.

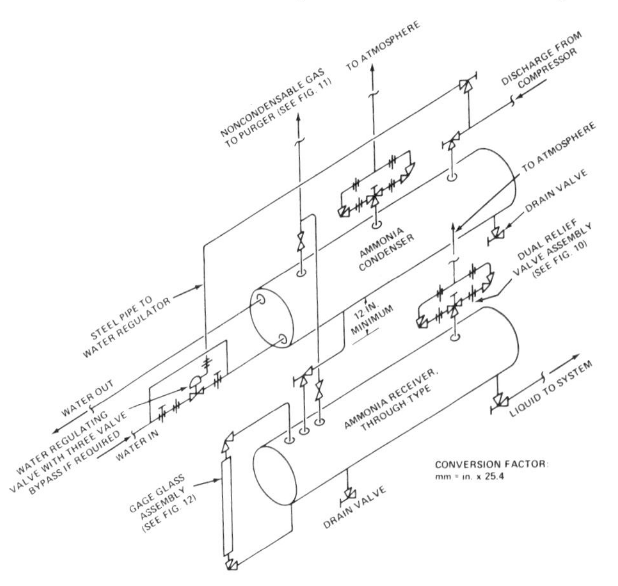

Figure 7–40 Horizontal condenser with through-receiver piping. (By permission of *ASHRAE Handbook*.)

Parallel Condensers

The piping for two condensers in parallel is shown in Figure 7–41. Note the equalizer line connecting the top of both condensers. Also note the dual relief valves on the receiver. This arrangement permits testing the valves. One valve is removed while the other is kept in service.

Evaporative Condensers

It is common practice to use evaporative condensers with ammonia systems. The ammonia receiver must always be at a lower pressure than the condensing pressure. When evaporative condensers are used in the winter, an arrangement similar to that shown in Figure 7–42 is good practice. Water is only circulated until it reaches near freezing temperatures. Piping for parallel evaporative condensers is shown in Figure 7–43. The receiver is normally placed at least 3 ft below the outlet of the condensers.

Air Blowers

Figure 7–44 shows the piping for an *air blower using thermal expansion valve with automatic hot-gas defrost.* The drain pan is usually kept free of frost by placing a hot-gas coil in the drain pan. Hot gas enters the drain pan coil before it enters the main coil for defrost.

To properly utilize the hot-gas defrost arrangement the system should contain multiple evaporators so that the compressor will be running when the evaporator to be defrosted is shut down. The hot-gas header must be kept in a space where the ammonia will not condense in the pipe. Otherwise, the coil will receive liquid ammonia at the start of defrosting and will not be able to take full advantage of the latent heat of hot-gas condensation coming into the coil.

The liquid and suction solenoids are open during the normal cycle, and during the defrost cycle these valves are closed. When the defrost cycle is started the defrost solenoid valve opens. The pressure regulator is used to maintain about 70 psi in the coil. A hand valve may also be used to regulate pressure in the main.

A *flooded evaporator* (shown in Figure 7–45) uses an air blower and water defrost. The lower float switch controls the liquid solenoid, regulating the level of liquid in the suction trap. The higher float switch is wired to an alarm circuit. A combination evaporator pressure regulator and stop valve is used in the suction line to the suction trap to regulate the pressure. The stop mechanism is closed during defrost. A spring-loaded relief valve is used around the suction pressure regulator. The outlet damper of the air blower is closed during defrost. The water system is completely drained after defrost.

Suction Traps

Suction traps are recommended on all systems for separating out any liquid that returns through the suction line. A properly designed suction trap assures that only vapor refrigerant will enter the compressor.

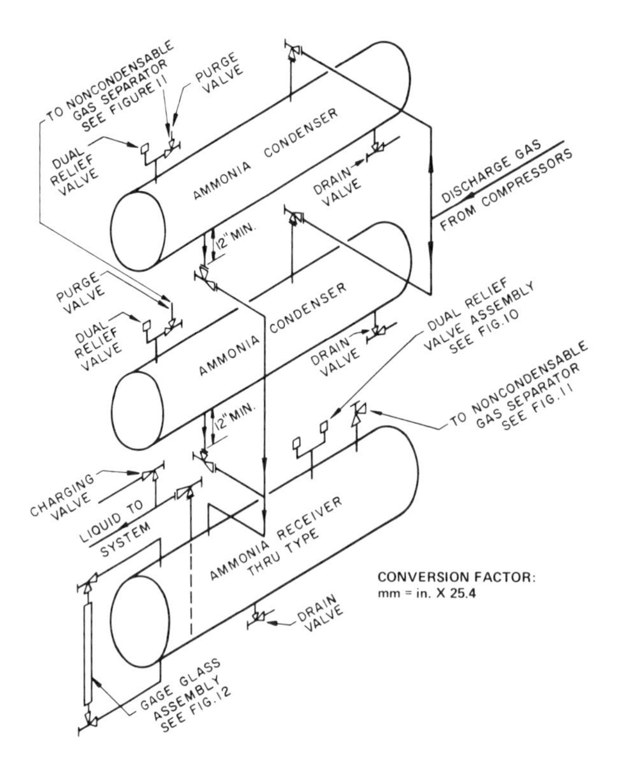

Figure 7-41 Two condensers in parallel. (By permission of *ASHRAE Handbook*.)

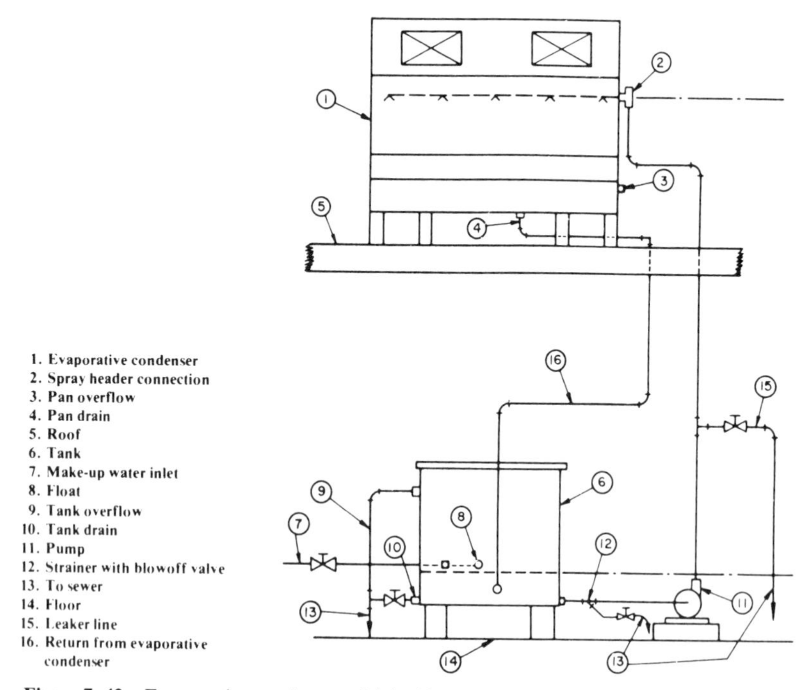

Figure 7-42 Evaporative condenser with inside water tank. (By permission of *ASHRAE Handbook*.)

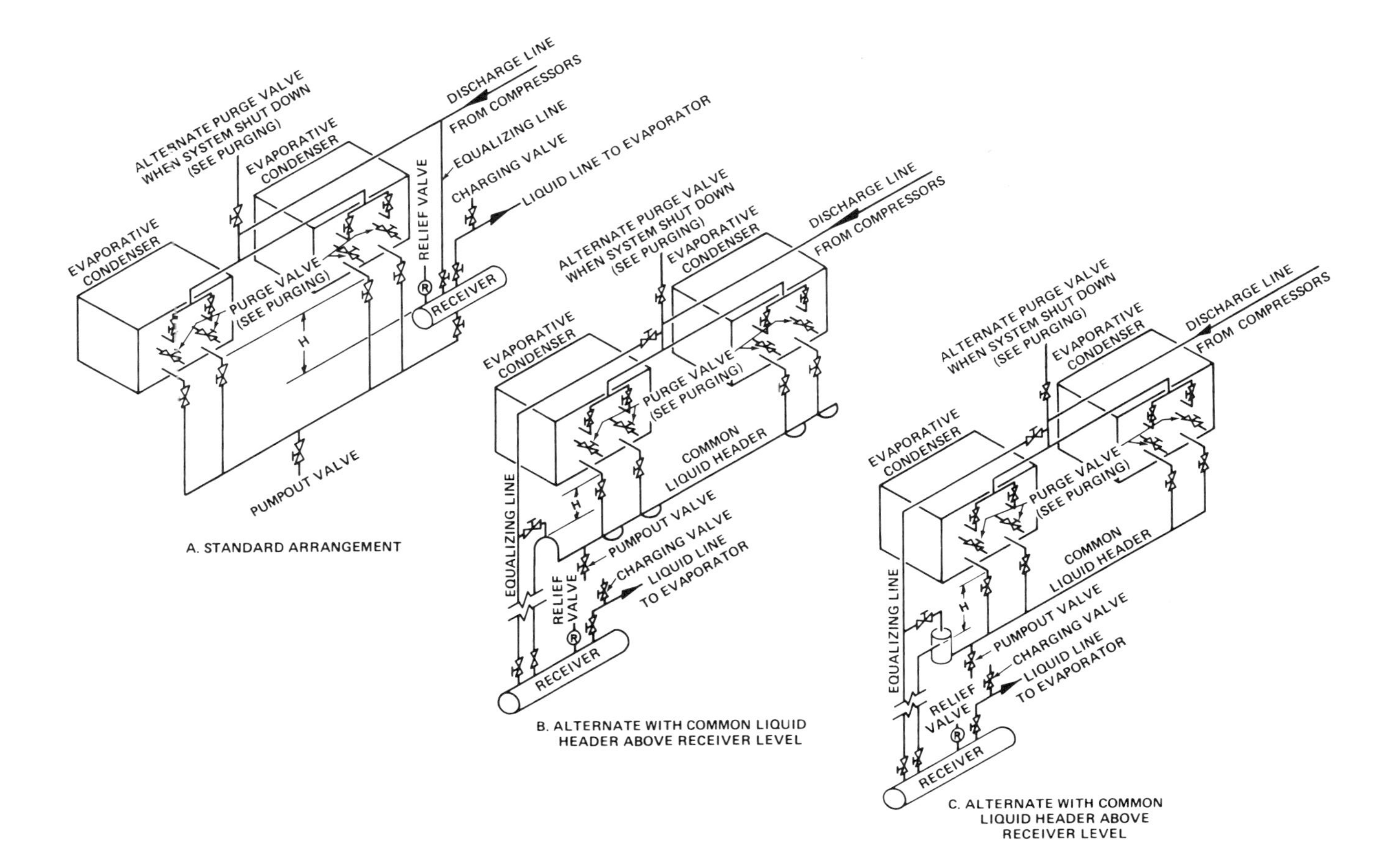

Figure 7–43 Paralleling evaporative condensers. (By permission of *ASHRAE Handbook*.)

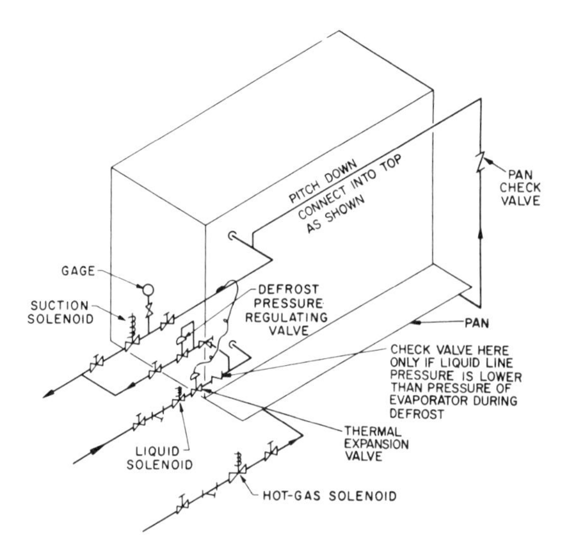

Figure 7-44 Piping for air blower with thermal expansion valve and hot-gas defrost. (By permission of *ASHRAE Handbook*.)

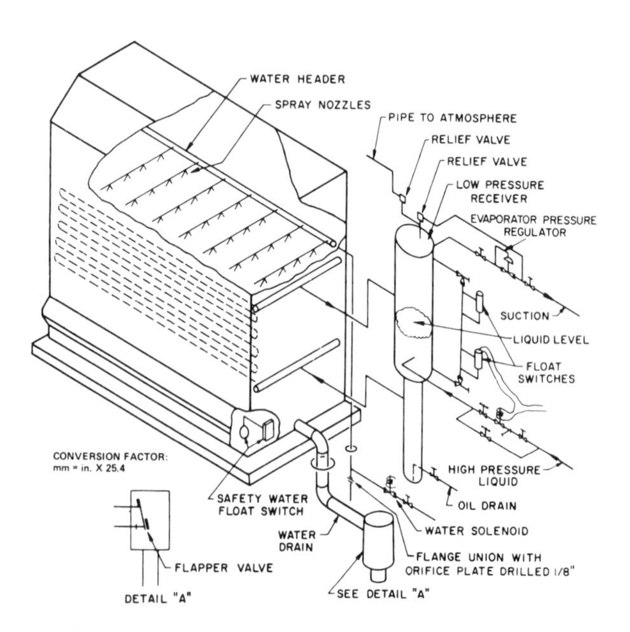

Figure 7-45 Piping for air blower with flooded coil and water defrost. (By permission of *ASHRAE Handbook*.)

Figure 7–46 shows a vertical suction trap with a high-pressure liquid coil. The high-pressure, high-temperature liquid line serves to evaporate the returned liquid refrigerant.

Figure 7–47 shows a vertical suction trap with three float switches piped into the equalizer header. Two float switches are used to start and stop the pump. The third float switch operates the alarm circuit. Note the use of two check valves in series with the pump outlet to prevent reverse flow when the pump is stopped. Sudden changes in the load should be avoided to keep agitation to a minimum.

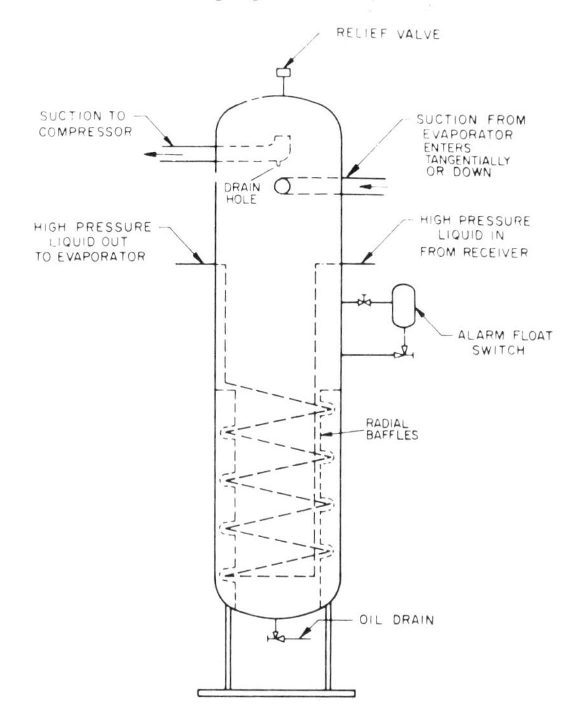

Figure 7–46 Vertical suction trap with high-pressure liquid coil. (By permission of *ASHRAE Handbook.*)

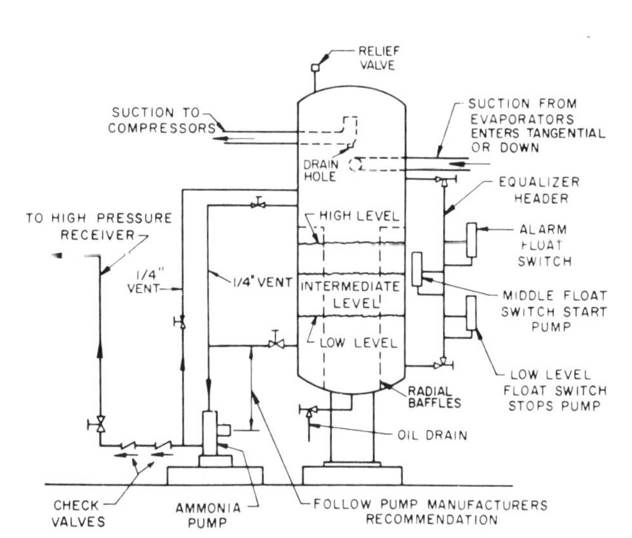

Figure 7–47 Piping for vertical suction trap with high head pump. (By permission of *ASHRAE Handbook.*)

REVIEW EXERCISE

Select the letter representing your choice of the correct answer (there may be more than one).

7-1. The following is *not* a proper function of refrigerant piping:

________ a. A path for the refrigerant to follow
________ b. Permits proper oil return to the compressor
________ c. Protects compressor against liquid slugging
________ d. Limits the capacity of the compressor

7-2. Which materials are *not* suitable for ammonia piping?

________ a. Copper
________ b. Iron
________ c. Steel
________ d. Brass

7-3. Brazed joints use alloys with a melting point above:

________ a. 500°F
________ b. 750°F
________ c. 1000°F
________ d. 1500°F

7-4. Which type of valve has the greatest pressure drop?

________ a. Gate valve
________ b. Globe valve

7-5. Which type of piping is most critical as far as pressure drop is concerned?

________ a. Condenser to receiver piping
________ b. Hot-gas line
________ c. Liquid line
________ d. Suction line

7-6. What is the minimum riser velocity for a suction line?

________ a. 500 fpm
________ b. 1000 fpm
________ c. 1500 fpm
________ d. 2000 fpm

7-7. What is the design temperature drop used in sizing suction line refrigerant piping for an R-12 system?

________ a. 2°F
________ b. 3°F
________ c. 4°F
________ d. 5°F

7-8. What types of loads require a double suction riser?

_______ a. All commercial refrigeration loads

_______ b. All low-temperature loads

_______ c. A constant-capacity load

_______ d. One that operates over a wide capacity range

7-9. Which refrigerant pipe has counterflow (flow in both directions)?

_______ a. Suction line

_______ b. Liquid line

_______ c. Pipe between condenser and receiver

_______ d. Entire hot-gas line

7-10. The most important item in sizing liquid lines is:

_______ a. To avoid oil traps

_______ b. To allow for refrigerant evaporation

_______ c. To prevent excessive pressure drop

_______ d. To avoid piping that is too large

8 Refrigeration Systems

OBJECTIVES

After studying this chapter, the reader will be able to:

- Identify the various common types of refrigeration systems.
- Name the essential components included in each type of system.
- Discuss the advantages and disadvantages of each type of system.

CLASSIFICATION OF SYSTEMS

The basic components of a refrigeration system are presented in Chapter 1. The essential parts consist of:

1. A compressor
2. A condenser
3. A metering device
4. An evaporator

Together with the connecting piping, these four components make up a simple or elementary refrigeration system. As indicated in Chapter 7, many other devices can be added to enhance the operation of the system or to make it usable for a given application.

Some of the more useful additional parts are: (1) a receiver, (2) an accumulator, (3) a drier, (4) a sight glass, (5) an evaporator fan, and (6) a liquid solenoid valve. Whatever the final selection of components includes, the parts work together to form a refrigeration system, designed to perform a certain function.

Because refrigeration is a competitive business, many manufacturers devise innovative arrangements to provide unique and useful variations of the standard refrigeration cycle. Chapter 12 describes a number of interesting applications of system components to commercial products. Some of them are available as a package, while others are incorporated into a system constructed on the job. Some of the principal types of systems are:

1. *Basic (self-contained) refrigeration systems.* Examples are water coolers, reach-in refrigerators, and island-type supermarket merchandisers.
2. *Automatic defrost systems.* A typical system of this type is used in refrigeration storage rooms, commonly found in sizable food stores or supermarkets. One of the most innovative systems is the Thermobank system using hot-gas defrost.
3. *Automatic-ice-making systems.* Making ice is a big business and there are many types of these systems. Some are self-contained, other are built-up to produce large quantities of ice. Some ice makers store cooling capacity in an ice bank for later use.
4. *Supermarket refrigeration systems.* There is a large market for this type of refrigeration equipment. The most common state-of-the-art equipment now uses a bank of compressors connected to a group of display cases with a number of innovative features to conserve energy. These systems involve the use of multiple evaporators with variations in evaporative temperatures.
5. *Compound compressor systems.* These systems use one or more compressors connected in series to produce low temperatures. They are used by dairy product plants for producing the low temperatures required for ice cream. They commonly use ammonia as a refrigerant.
6. *Cascade compressor systems.* Cryogenic, or very low, temperatures are produced by cascade systems. These systems are used to produce temperatures that go below those produced by compound systems. They often use a different refrigerant in each stage. These systems have a unique arrangement for connecting one refrigeration circuit to another without interchanging refrigerants.
7. *Heat pump systems.* For this application, compressors are used to produce both heating and cooling. The piping arrangement makes possible switching from heating to cooling by activating a reversing valve. A common use is to maintain the temperature in a refrigerated truck body both winter and summer.
8. *Secondary refrigerant systems.* This is a process whereby the primary refrigerating system cools a second fluid, called a "brine," which carries cooling to the final product. An example of this arrangement is the refrigeration system for an ice rink. Here the basic system cools a calcium chloride or glycol brine, which is circulated through the piping to freeze the water.

BASIC SYSTEMS

A good example of the basic refrigeration system is the circuit used in the drinking water cooler shown in Figure 8-1. This unit is hermetically sealed and uses a capillary tube as a refrigerant metering device. The condenser is air cooled, using forced-air circula-

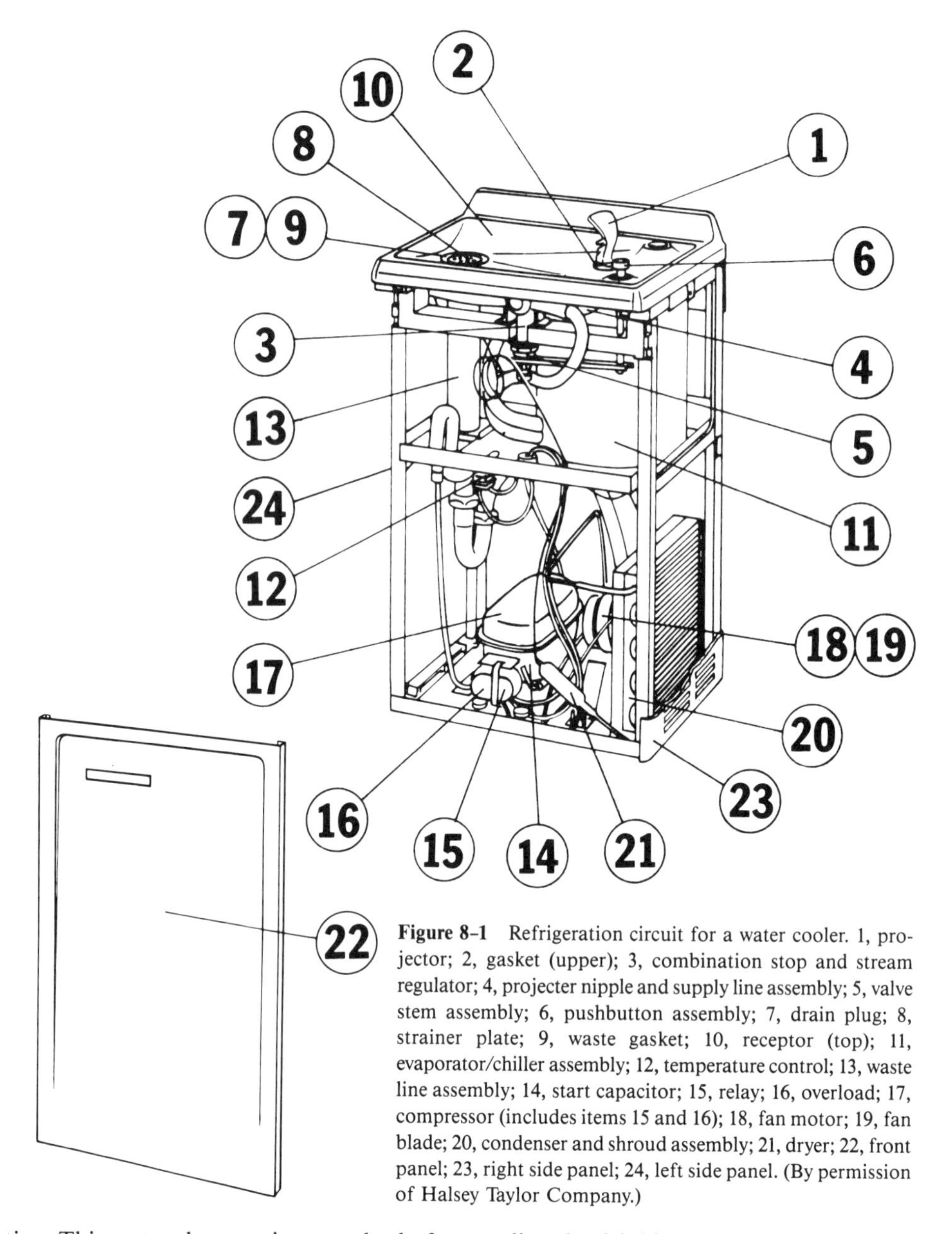

Figure 8–1 Refrigeration circuit for a water cooler. 1, projector; 2, gasket (upper); 3, combination stop and stream regulator; 4, projecter nipple and supply line assembly; 5, valve stem assembly; 6, pushbutton assembly; 7, drain plug; 8, strainer plate; 9, waste gasket; 10, receptor (top); 11, evaporator/chiller assembly; 12, temperature control; 13, waste line assembly; 14, start capacitor; 15, relay; 16, overload; 17, compressor (includes items 15 and 16); 18, fan motor; 19, fan blade; 20, condenser and shroud assembly; 21, dryer; 22, front panel; 23, right side panel; 24, left side panel. (By permission of Halsey Taylor Company.)

tion. This system has a unique method of precooling the drinking water by using a heat exchanger on the drain tube. The wastewater from the fountain precools the incoming water. The evaporator tubing is bonded to the drinking water storage tank. The compressor is operated by a water temperature control.

There are many variations of the basic system. For example, other types of metering devices are used and not all systems are hermetically sealed. Another good example is the reach-in refrigerator. A semihermetic condensing unit is located in the base of the cabinet. The liquid line is connected to the fan-driven evaporator unit through an expansion valve and the suction line returns to the compressor. The compressor is operated by a low-pressure control (LPC) connected in the suction line. The compressor is oversized to permit about

16 hours of running time and 8 hours of air defrost time. Air within the cabinet can be used for defrost since the box is maintained at about 40°F. The low-pressure control (LPC) maintains satisfactory temperatures in the box. The cut-in point, cutout point, and differential of the LPC must be properly set to comply with the box usage.

AUTOMATIC DEFROST SYSTEMS

There are many types of automatic defrost systems. A detailed description of representative systems is given in Chapter 4. Automatic defrost is used to clear the evaporator coil of ice accumulation where the evaporator must cool moist air while maintaining a below-freezing evaporating temperature.

One unique system using hot gas for defrost is the Thermobank system. A diagram of the refrigeration piping for this system is shown in Figure 8–2. This system has three phases: (1) the normal cycle, (2) the defrost cycle, and (3) the post-defrost cycle.

During the normal refrigeration cycle, the discharge piping first enters the "bank," transferring heat to the stored water. The vapor then goes to the condenser, where

Figure 8–2 Thermobank: refrigeration system for hot-gas defrost. A, bank; B, condenser; C, receiver; D, heat exchanger; E, TXV; F, evaporator; G, solenoid valve; H, evaporator fan; J, H.G. solenoid; K, liquid solenoid; L, suction solenoid; M, check valve; P, holdback valve; R, compressor; T, box stat. (By permission of Kramer Trenton Company.)

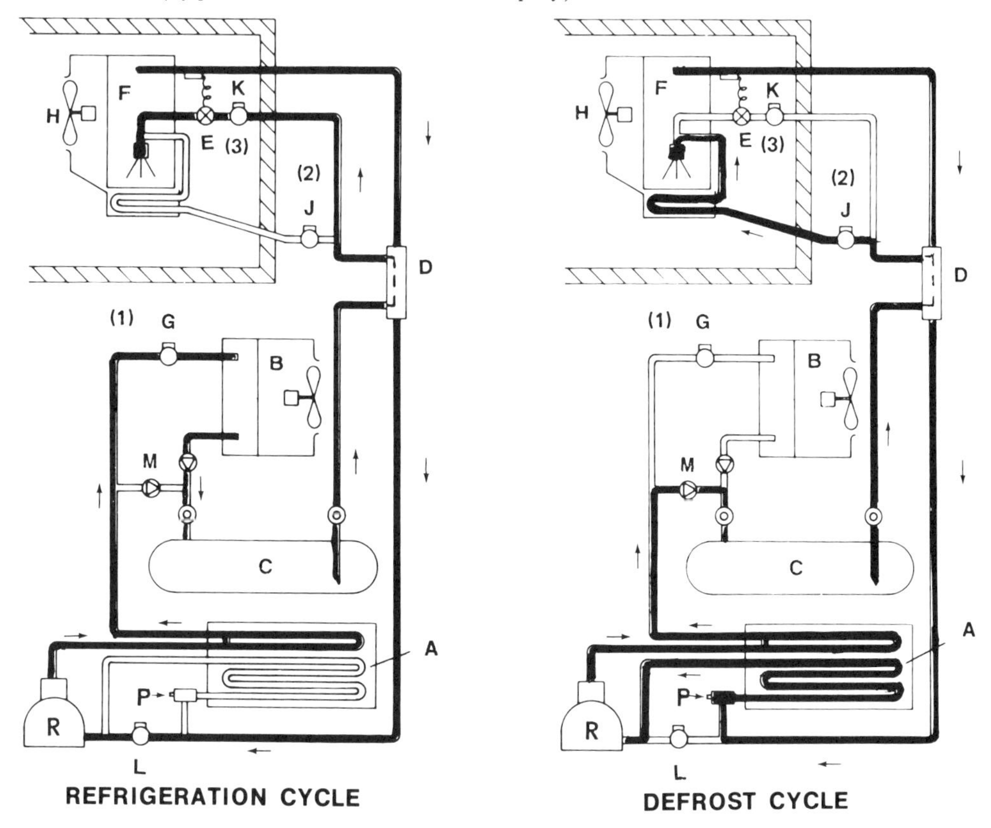

refrigerant condensation takes place. The liquid refrigerant proceeds through the receiver and through the expansion valve to the evaporator. Suction gas returns to the compressor.

The time clock automatically places the system in the defrost cycle. A number of actions take place:

1. The evaporator fans are stopped.
2. The discharge gas solenoid valve (1) to the condenser is closed.
3. The hot-gas solenoid valve (2) to the evaporator is opened.
4. The liquid line solenoid (3) at the evaporator is closed.
5. The suction line solenoid (4) near the compressor is closed.

First, the warm liquid from the receiver enters the evaporator, quickly starting the defrost process. This is followed by the hot gas from the operating compressor. During defrost the evaporator acts as a condenser, returning liquid refrigerant through the suction line. Due to the closing of the suction solenoid, the returning refrigerant is directed through the hold-back valve and picks up heat from the "bank" before entering the compressor. The bank acts as an evaporator, heating and vaporizing any returning liquid before it enters the compressor. In this manner the heat from the bank is added to the compressor heat during defrost to increase the amount of heat available to melt the ice. The hold-back valve keeps the hot-gas pressure and corresponding vapor temperature high enough to do an effective job of defrosting.

The defrost cycle is terminated by a pressure switch and starts the post-defrost period as follows:

1. The discharge gas solenoid valve (1) to the condenser is opened.
2. The hot gas solenoid valve (2) to the evaporator is closed.

The post defrost period permits the exterior surface of the evaporator to be cleared of excess moisture before the fan is turned on. At the end of the pressure-terminated post-defrost period:

1. The liquid-line solenoid (3) to the evaporator is opened.
2. The suction-line solenoid (4) to the compressor is opened.
3. The evaporator fans are started.

The system then operates on the normal refrigeration cycle.

AUTOMATIC-ICE-MAKING SYSTEMS

There are many types of automatic ice makers. The one described here is a packaged, self-contained unit that produces flake ice. Flake ice is produced in thin sheets and is commercially used for cooling chicken or fish. It is usually cloudy (not clear) ice since air is entrained during freezing.

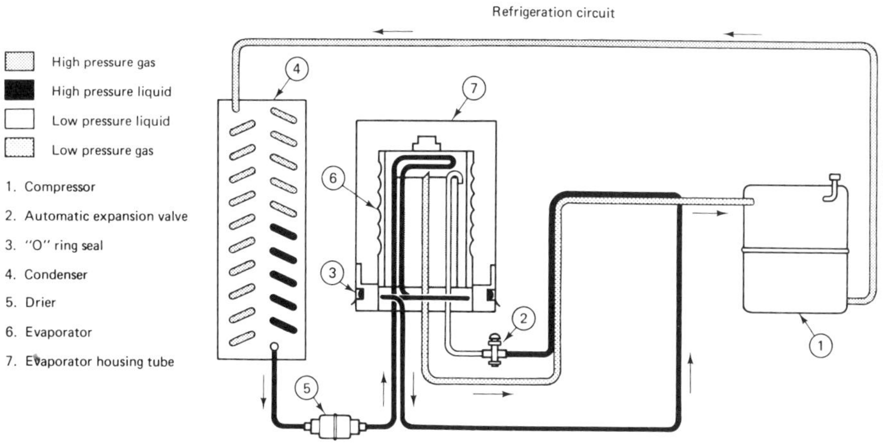

Refrigeration circuit
High pressure gas
High pressure liquid
Low pressure liquid
Low pressure gas
1. Compressor
2. Automatic expansion valve
3. "O" ring seal
4. Condenser
5. Drier
6. Evaporator
7. Evaporator housing tube
1
2
3
4
5
6
7

Figure 8–3 Refrigeration circuit for flake ice machine. (By permission of Crystal Tip Ice Products, McQuay Perfex, Incorporated.)

Gear motor
Hex nut
Gear motor adapter plate
Felt washer
Oil slinger
Gearmotor mounting
Grommet (4)
Spacer (4)
Washer (4)
Lockwasher (4)
Capscrew (4)
Housing cap
Housing tube
Top thrust bearing
Tie bolt
Cage
Top cage bearing
Bottom bearing
Evaporator and base assembly
O-ring seal

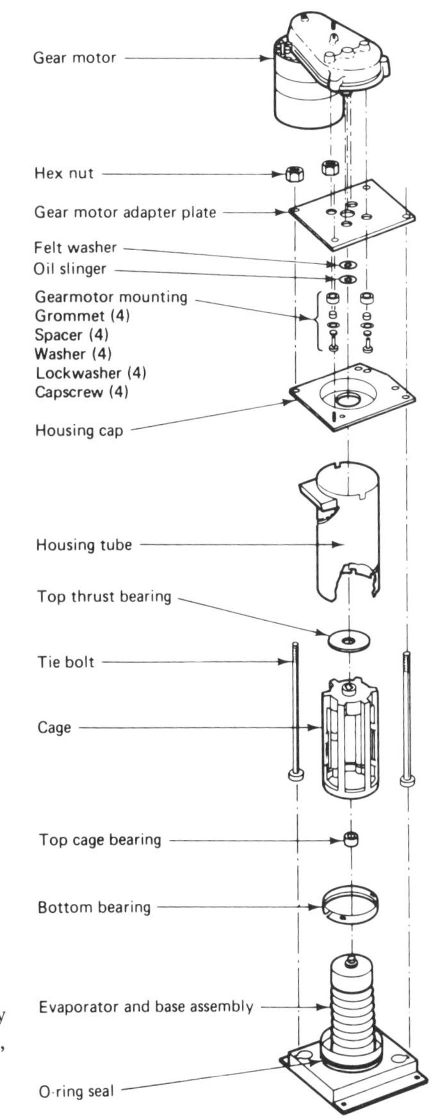

Figure 8–4 Evaporator assembly for flake ice machine. (By permission of Crystal Tip Ice Products, McQuay Perfex, Incorporated.)

Figure 8–3 shows a diagram of a typical refrigeration cycle. The evaporator surface is a vertical cylindrical wall with horizontal corrugations. The evaporation of the refrigerant takes place inside the cylinder (6), keeping the surface below freezing temperatures. A second cylinder surrounds the evaporator (7). This enclosure has a watertight seal made by an O-ring (8) located at the bottom.

As the ice is formed on the surface of the evaporator it is removed by a revolving cage assembly with cutter bars. The cage assembly also has a spiral groove which propels the ice upward to the chute opening, where it drops into the bin. The gear motor at the top of the assembly (Figure 8–4) turns the cage assembly. As the bin fills, the stored ice contacts the thermostat bulb located near the top of the bin, shutting off the unit. Note that this system uses an automatic expansion valve (2) to maintain a constant evaporating temperature.

Figure 8–5 Typical refrigeration piping for a compressor on a supermarket system. 1, Compressor; 2, oil-level regulator; 3, head cooling fan (where required); 4, oil separator; 5, differential pressure regulator; 6, condenser (may be remote air or water); 7, hot-gas defrost manifold; 8, hot-gas defrost hand valve; 9, reversing valve solenoid (optional); 10, suction stop solenoid (optional); 11, evaporator pressure regulator (optional); 12, suction manifold; 13, suction-line filter; 14, suction accumulator; 15, oil reservoir; 16, oil return filter/drier; 17, liquid receiver; 18, inlet valve; 19, outlet "king" valve; 20, liquid-line drier; 21, liquid-line moisture indicator; 22, master liquid solenoid with bypass check valve; 23, liquid-line solenoid (optional).

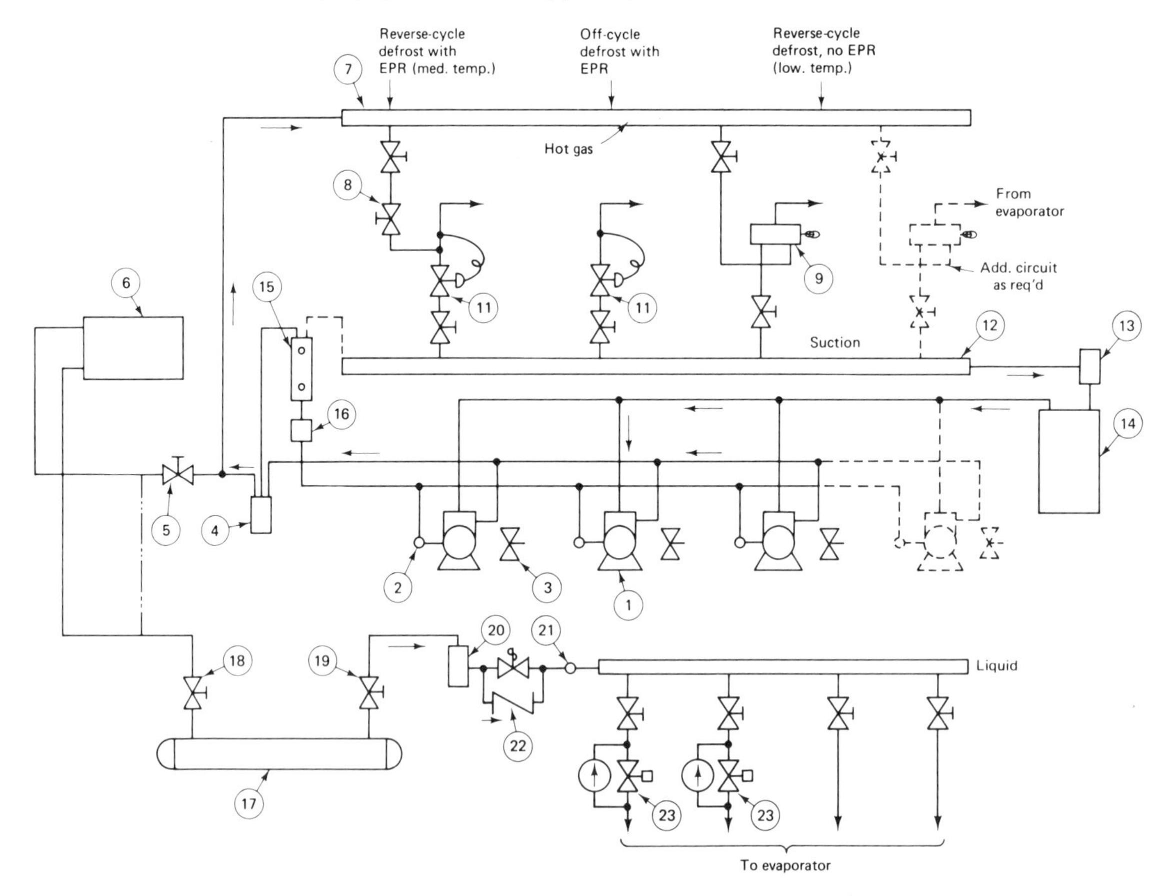

SUPERMARKET SYSTEMS

The modern refrigeration system for a supermarket consists of a number of banks of parallel compressors connected by refrigerant piping to the evaporators located in refrigerated rooms and cases throughout the store. Each bank of compressors consists of two or more compressors connected in parallel, a control panel, and a receiver mounted on a common base. This is illustrated in Figure 12–54. The bank of compressors is connected to a common air-cooled or water-cooled condenser. A heat recovery arrangement may also be incorporated into the system.

A refrigeration piping schematic is shown in Figure 8–5. The diagram shows four compressors (1) connected in parallel. An oil separator (4) is placed in the discharge piping. One branch goes to the condenser and the other branch goes to the hot-gas manifold. During the normal cycle the liquid refrigerant leaves the condenser, enters and leaves the receiver, and feeds to the evaporator through the liquid manifold. The suction gas returns to the compressor through the suction manifold. An accumulator (14) is placed in the suction line piping to the compressors to assure dry vapor entering the compressors.

When the timer initiates a defrost cycle, the reversing valve solenoid (9) is positioned to supply hot gas to the evaporators through the suction line. The condensed liquid returns to the liquid manifold through a bypass around the liquid-line solenoid (23), supplying refrigerant to the operating evaporators. Note that a maximum of 25 percent of the evaporators are defrosted at one time.

Figure 8–6 Hot-gas diverting method with EPR valve control.

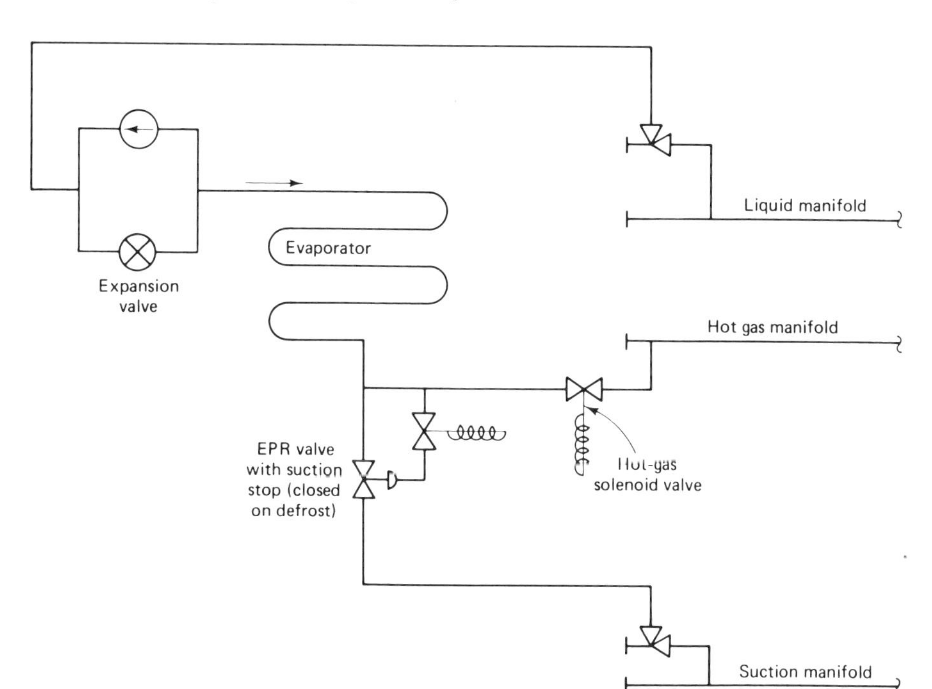

To assure liquid returning to the liquid manifold during defrost, a pressure regulating valve (5) is placed in the discharge line to provide a differential pressure of 20 psi or more between the hot-gas supply and the liquid-refrigerant manifold. The pressure regulator (5) is active only during defrost, and the valve is held wide open during the normal cycle.

The control of refrigerant through the evaporator is shown in Figure 8–6. During the normal cycle the liquid is supplied to the evaporator through the expansion valve. The evaporating pressure is controlled by the EPR valve in the suction line. During defrost the hot gas melts the ice on the coil and the condensed liquid bypasses the expansion valve returning to the liquid manifold to feed the active evaporators.

On medium-temperature evaporators the normal refrigeration resumes immediately after the defrost cycle. On low-temperature evaporators a time delay occurs to drain off the defrost water. Required defrost times are determined by actual operation when the system is placed in service.

ULTRALOW (CRYOGENIC) TEMPERATURE SYSTEMS

Ultralow-temperature systems are those operating at evaporating temperatures in the range −40 to −150 °F. Commercially available equipment can be used for these applications. There are, however, certain aspects of these systems that are critical and a service person must thoroughly understand the equipment that can be used and the application requirements. No guesswork can be tolerated. All losses must be kept to a minimum. Some of the critical factors are listed below.

Heat leakage. The heat leakage for a standard cabinet would be enormous. Maintaining a box temperature of −100 °F in an 80 °F room amounts to a temperature differential of 180 °F. To prevent excessive heat leakage requires four and one-half times the standard 2-inch insulation thickness for a 40 °F box, or nine inches of insulation.

Loads. The *product load* can be a significant factor, too. A product cooled from 80 °F to −100 °F requires four and one-half times as much cooling as the same product cooled to 40 °F.

The *service load* can be a major factor. The frequency of cabinet openings to add or remove products can temporarily raise the box temperature to near ambient conditions. The service load can be the highest single load for the system. A generous amount of cooling needs to be provided for this loss.

Any *internal loads,* such as fans or light, however small, can materially increase the load.

Pull-down time. The pull-down time is an important factor. How long must the equipment run when the box is originally loaded in order to reach the design box temperature? A pull-down time of 1 hour rather than 2 hours may double the refrigeration load.

Vapor seal. The *vapor seal* on the cabinet is important. It must be practically perfect. Any air or moisture leakage can critically affect the load. Moisture entering the cabinet will cause ice to form on the evaporator and lower the transfer rate.

Many ultralow-temperature systems use plate-type evaporators. In fact, every available surface inside the cabinet may be lined with plates in order to provide enough evaporative surfaces. Small fans may be used for circulation to increase the effectiveness of the surfaces. These evaporators must be kept clean to maintain the required temperatures.

Closely controlled brine coolers are used for some applications where cooling must be transferred to a special product cooler.

Compression ratio. Like other types of refrigeration, the greatest efficiencies are maintained by keeping the discharge pressure low and the evaporating temperatures high. This is extremely important in low-temperature work. Another way of stating the same fact is: keep the compression ratio low and the volumetric efficiency high.

The compression ratio is the number obtained from dividing the discharge pressure by the suction pressure after converting both to pressure per square inch, absolute (psia). The volumetric efficiency of a compressor is the ratio (expressed in percent) of the volume of vapor actually pumped by the compressor, divided by the theoretical volume obtained by calculating the volume from the bore, stroke, and number of cylinders. The clearance volume as well as discharge pressure and suction pressure affect the volumetric efficiency of the compressor. Increasing discharge pressure and lowering suction pressure lowers the volumetric efficiency. Thus the compression ratio can be so high that the compressor "quits pumping" since the compressor cannot lower suction pressure.

In order to use commercially available compressors for low-temperature work, the compression ratio should not exceed 10 as a maximum, with 7 or 8 as the preferred limit. For example, if the requirement is for $-75\,^{\circ}\mathrm{F}$ evaporating temperature, the equivalent absolute pressure is 5.6 psia using R-22 refrigerant. Based on a head pressure of 145 psig or 159.7 psia (145 + 14.7), the compression ratio is

$$\mathrm{CR} = \frac{159.7}{5.6} = 28.5$$

Figure 8-7 Typical two-stage compound compression system. (By permission of *ASHRAE Handbook*.)

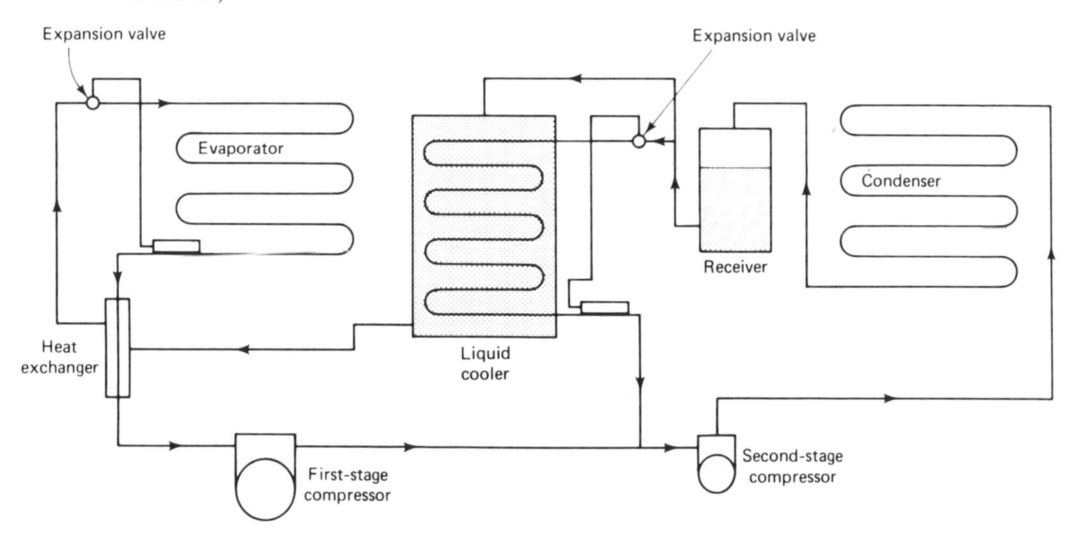

which is excessive for any commercially available compressor. Thus, to obtain the desired performance, the compression is separated into a two-step process using either a compound system or a cascade system.

Compound Systems

Compound systems use two or more compressors in series to achieve ultralow evaporating temperatures. A simplified drawing of such a system is shown in Figure 8–7. Note that in this system the discharge of the low-stage compressor enters the suction line of the high-stage compressor. Thus the two compressors are said to be connected in series. There are a number of additional components that are needed in these systems to permit them to operate efficiently.

Using Figure 8–7, the following example shows the operating characteristics of a two-stage system. The overall arrangement is designed to produce an evaporating temperature of $-75\,^\circ$F and a condensing temperature of $80\,^\circ$F. Doing this with a single-stage compressor would be prohibitive, as shown previously, because of the 28.5 compression ratio. However, this performance is easily achieved using a two-stage design. The refrigerant selected is R-22.

The low-stage compressor has a suction pressure of 18.5 inches of mercury. This is equivalent to 5.6 psia. It is calculated as follows:

$$\frac{29.9 \text{ in.} - 18.5 \text{ in.}}{2.036} = 5.6 \text{ psia}$$

An intermediate discharge pressure of 13.3 psig (13.5 psig + 14.7 psi = 28.0 psia) is selected for the low-stage compressor. The compression ratio of the first-stage machine is, therefore, 5 (28.0 psia/5.6 psia), which is satisfactory.

The high-stage compressor has a suction pressure of 13.3 psia (28.0 psia) and a discharge pressure of 145 psig (145 psig + 14.7 psi = 159.7 psia). The compression ratio of the high-stage machine is, therefore, 5.7 (159.7 psia/28.0 psia), which is also satisfactory.

The *interstage* is the area between the discharge of the low-stage compressor and the suction of the high-stage compressor. When the vapor leaves the low-stage compressor at a pressure of 13.3 psi, the vapor is superheated to approximately $63\,^\circ$F. Since the saturated vapor temperature equivalent to 13.3 psi is $-15\,^\circ$F, this amounts to a superheat of $78\,^\circ$F (from $-15\,^\circ$F to $65\,^\circ$F). To improve the performance of the high-stage machine, this superheat must be reduced to approximately $0\,^\circ$F before the discharge vapor of the first machine enters the second. One of the best ways to accomplish this is by means of an intercooler (Figure 8–7).

The intercooler has two functions: (1) to precool the liquid R-22 before it goes to the $-75\,^\circ$ evaporator, and (2) to cool the discharge vapor from the first compressor before it enters the suction of the second compressor. The expansion valve on the intercooler can be adjusted to cool the vapor entering the high-stage compressor to $0\,^\circ$F, still permitting $15\,^\circ$F of superheat ($-15\,^\circ$F to $0\,^\circ$F = $15\,^\circ$F).

The second compressor takes the 13.3 psig $0\,^\circ$F vapor and compresses it to 148 psig (saturated temperature $80\,^\circ$F) at a discharge temperature of $153\,^\circ$F or greater. The condenser then liquefies the R-22 at a condensing temperature of $80\,^\circ$F.

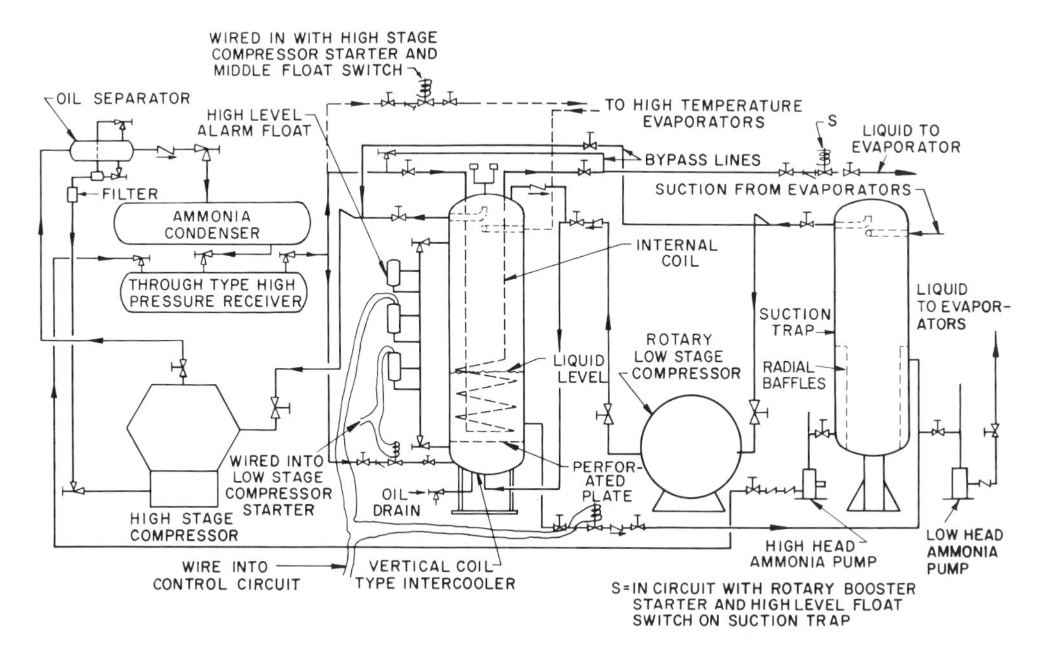

Figure 8–8 Two stage compressor system with low and high temperature evaporators.

Some manufacturers provide a two-stage compressor in a single machine using possibly three cylinders for the low stage and one cylinder for the high stage. Provision is made for an intercooler between stages. There is also a wide selection of refrigerants, depending on the application.

A two-stage compound system using ammonia refrigerant is shown in Figure 8–8. Note that both low-temperature and high-temperature evaporators are connected into this system.

Cascade Systems

Cascade systems provide another method for producing ultralow temperatures. A simplified drawing of a two-stage cascade arrangement is shown in Figure 8–9. The distinguishing characteristic of this system is the interstage arrangement whereby the condenser of the high-stage compressor is used as the evaporator for the low-stage compressor. Referring to Figure 8–9, assume that the low-stage system uses R-22 and that the evaporating temperature is $-75\,°$F. The suction pressure is 18.5 inches of mercury or 5.6 psia. Assume a discharge pressure for the low-stage compressor of 24.1 psig or 38.8 psia. The compression ratio is 6.9 (38.8 psia/5.6 psia). This is equivalent to a condensing temperature of 0 °F. Thus, in the cascade system the high-stage evaporator is used to cool the low-stage condenser down to 0 °F.

Assuming that the high-stage compressor also uses R-22 and that a 10 ° differential is required, the evaporating temperature of the second stage is $-10\,°$F with a saturated evaporating pressure of 16.6 psig or 31.3 psia. With a condensing pressure of 145 psig (159.7 psia) the compression ratio of the high-stage compressor is 5.1.

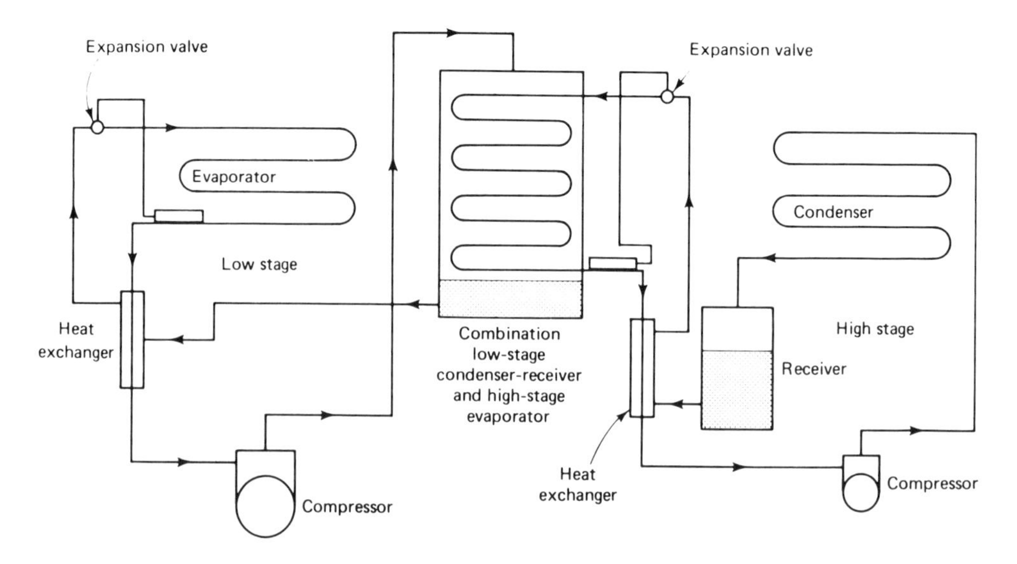

Figure 8–9 Simplified two-stage cascase system. (By permission of *ASHRAE Handbook*.)

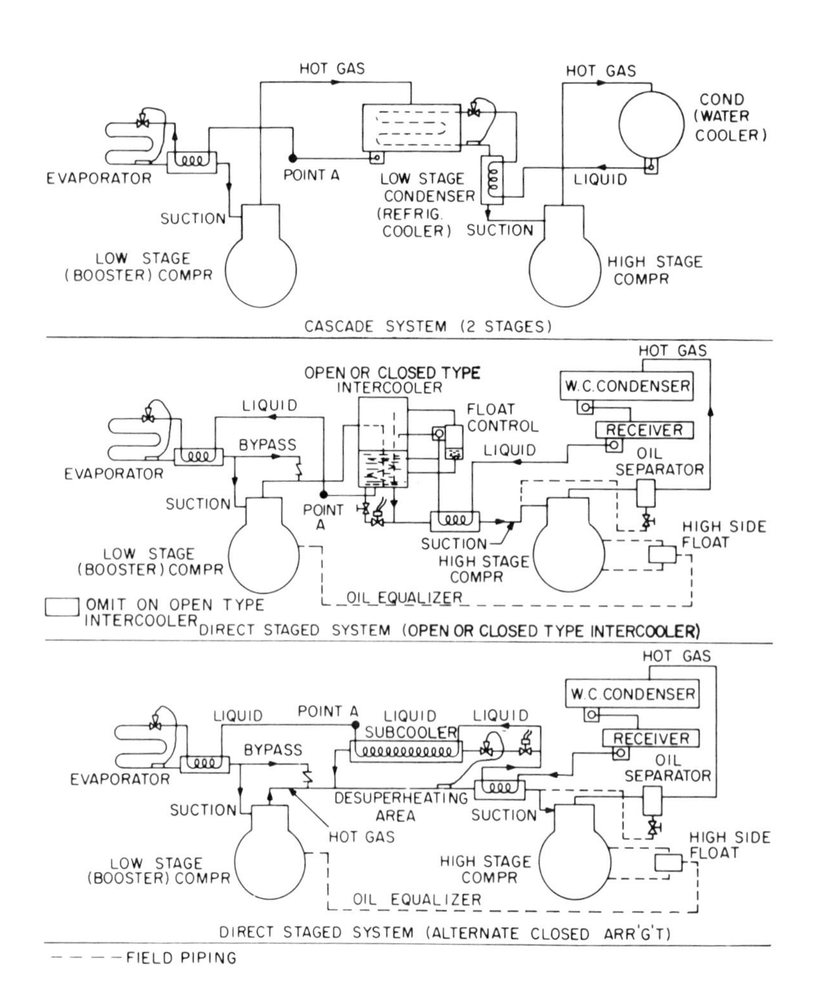

Figure 8–10 Types of intercoolers. (By permission of Carrier Corporation.)

Comparing the interstage pressures of the cascade system with those of the compound system, variations are required in the cascade system to permit maintaining the 10 °F difference between the low-stage condensing temperature and the high-stage evaporating temperature. Three types of intercoolers are shown in Figure 8–10.

There are a number of other differences between the two ultralow-temperature systems:

Startup. When the compound system is started, both compressors are started at once. The entire system gradually pulls down to design temperatures. On the cascade system the high-state compressor is started first and pulls down to the required evaporating temperature before the low-stage compressor can be started.

Refrigerant. The compound system uses one refrigerant. The cascade system can use a separate refrigerant in each stage. The best refrigerant can be selected for each operation. For example, assume that the low stage has a −75 °F evaporating temperature and 0 °F condensing temperature; the high stage has a −10 °F evaporating temperature and an 80 °F condensing temperature. The following table gives the pressures and compression ratios for the various refrigerants that could be selected:

Refrigerant	Pressure (psig)		Comp. ratio
	−75 °F Evap.	0 °F Cond.	
Low stage			
Methyl chloride	25″	4.2	5.8
Refrigerant-12	23″	9.2	7.0
Refrigerant-22	18.5″	24.19	7.4
Propane	17″	23.5	6.0
Kulene 131	1.4	56.4	5.1
Refrigerant-13	28	165.0	4.2
High stage			
Methyl chloride	.27	71.6	5.8
Refrigerant-12	4.5	84.1	5.2
Refrigerant-22	16.6	145.0	5.1
Propane	16.7	128.1	4.5
Kulene 131	44.6	227.4	4.1
Refrigerant-13	140.0	550.0	3.6

This shows R-13 to be the best refrigerant for the low stage and R-22 to be the best for the high stage (see Figure 8–11).

Shutdown. The compound system can be shut down in the normal way. However, with the cascade system the gas volume on the R-13 should be increased by adding an expansion tank to the first stage as shown in Figure 8–11, so that at room temperature the pressures would not become excessive. A relief valve in the high side of the refrigerant system discharges into the expansion tank. Otherwise, it would be necessary to transfer refrigerant to storage cylinders during shutdown.

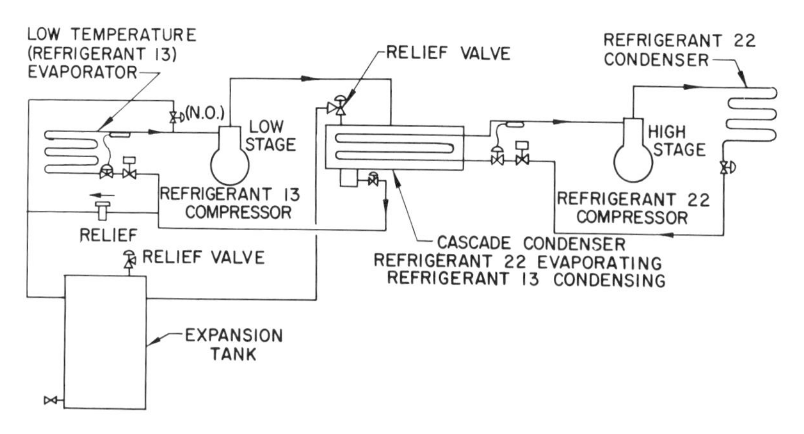

Figure 8-11 Two stage cascade system, showing use of two refrigerants. (By permission of *ASHRAE Handbook*.)

HEAT PUMPS

The heat pump consists of an arrangement of the refrigeration piping to permit using the cycle for both heating and cooling. A good example of a heat pump used in commercial refrigeration is the application for truck refrigeration.

A schematic diagram of a heat pump is shown in Figure 8-12. Note that one important element of the equipment is the use of the reversing valve. This valve permits the discharge gas from the compressor to go to either the outside coil or the inside coil. In the normal cycle the discharge gas goes to the outside coil and is condensed to liquid refrigerant. The liquid then passes through the metering device, expanding and providing cooling in the refrigerated space. When heating is required the reversing valve changes position, delivering the hot gas to the inside coil. The inside coil then acts as a condenser, furnishing liquid refrigerant through a metering device to the outside coil. Thus, under these conditions, the outside coil acts as an evaporator, picking up heat from the outdoor (ambient) air.

The air-to-air heat pump has these distinct modes: (1) the cooling cycle, (2) the heating cycle, and (3) the defrosting cycle.

Defrosting of the outside coil is required at certain times when the outside coil is acting as an evaporator and collecting ice. This is due to the suction temperature (below freezing) and the condensation of moisture (from the outside air) on the surface of the coil.

Figure 12-78 shows the truck refrigeration system operating in the normal cycle. Figure 12-79 shows the truck refrigeration unit operating in the defrost or heating cycle. The legend on the drawing indicates the names of the various parts involved.

Follow the arrows through the normal cycle (Figure 12-78). The three-way valve (5) directs the hot gas to the condenser (6). The liquid refrigerant travels from the receiver (8) through the heat exchanger (13) to the expansion valve (14) and into the evaporator (18). The suction gas goes through the heat exchanger (13) and back to the compressor (1).

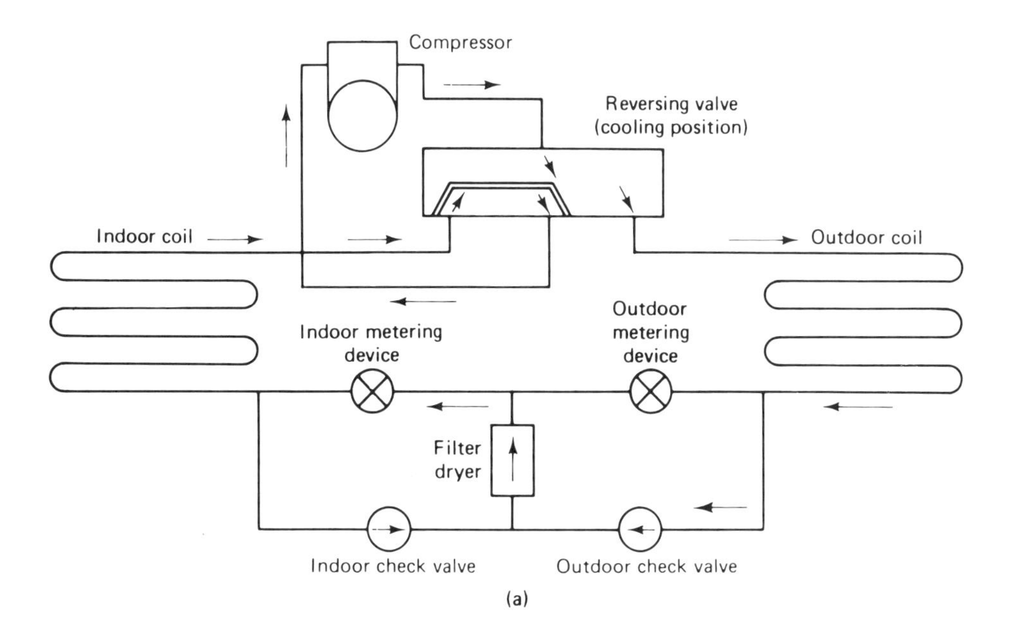

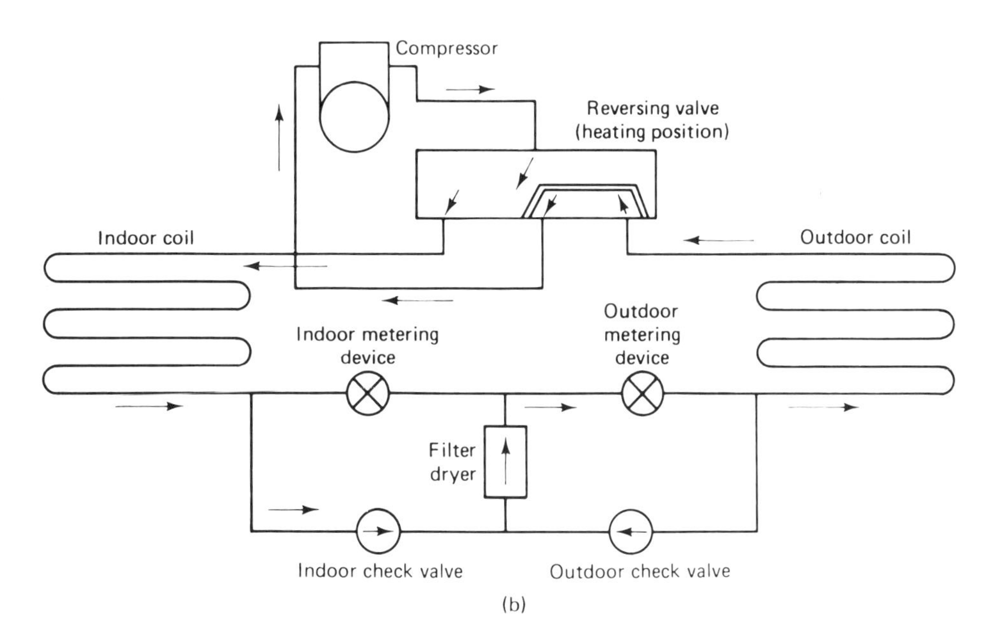

Figure 8–12 Air-to-air heat pump schematic: (a) cooling; (b) heating. (By permission of *ASHRAE Handbook*.)

During the heating cycle (Figure 12–79), the three-way valve (5) directs the discharge gas to the inside coils (18) by passing the expansion valve (14). The inside coil (18) acts as a condenser. Refrigerant then goes through the heat exchanger (13) and the accumulator (2), back to the compressor (1). Note the "T" in the discharge line after it leaves the three-way valve. During defrost some of the hot gas is bypassed through the receiver (8) and to the heat exchanger (13) to vaporize liquid returning from the inside coil (18). The accumulator (20) serves to protect the compressor against liquid "slugging."

SECONDARY REFRIGERATION (BRINE) SYSTEMS

Liquids, other than water, cooled by a primary refrigeration system and used to transfer heat (or refrigeration) without a change in state are called *secondary refrigerants* or *brine*. A good example is the application to ice-skating rinks.

The advantages for this application are many:

1. The properly insulated brine lines can be run longer distances than refrigerant lines.
2. Brine temperatures can be controlled more closely than can the refrigerant evaporating temperature.
3. Water leaks are easier to locate and repair than are refrigerant leaks.
4. Sharp peak loads are easier to absorb, particularly if a brine storage system is used.

The disadvantages are:

1. Some brines, such as calcium chloride, are corrosive and provision needs to be made to avoid the possibility of corrosion.
2. Extra equipment is required, such as pumps, motors, valves, and control equipment.
3. There is a possibility of damage due to freezing.
4. Power costs may be higher.

A typical brine system is shown in Figure 8–13. This is a two-pipe system with a revised return. Note that an expansion tank needs to be used to allow for the expansion and contraction of the liquid due to temperature changes.

Figure 8–14 shows a number of typical brines and the range of temperatures that can be used. Figure 8–13 shows the use of a brine strength unit for salt brines incorporated into the system piping. Salt brines must be maintained at the required strength to retain their freezing-point temperatures. The freezing point should be 5 to 15 °F below operating temperatures. To keep a vacuum from developing, certain types of brines require pressure-regulated dry nitrogen supplied to the expansion tank (Figure 8–13).

Corrosion characteristics of the brine solution are prevented by choice of the proper inhibitors, regulated by a routine test of the pH (acid condition value). Steel, iron, copper,

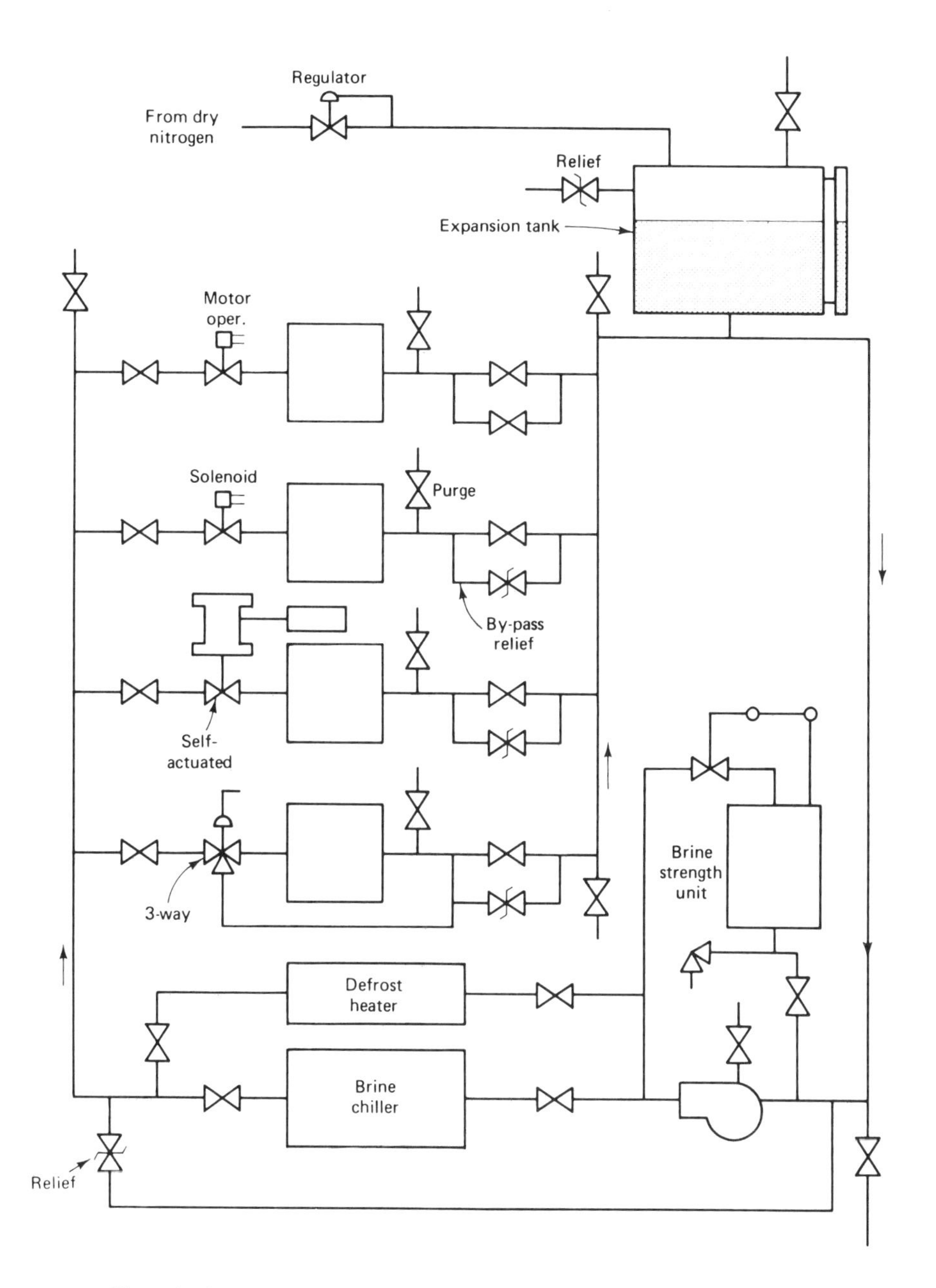

Figure 8-13 Typical closed brine system. (By permission of *ASHRAE Handbook.*)

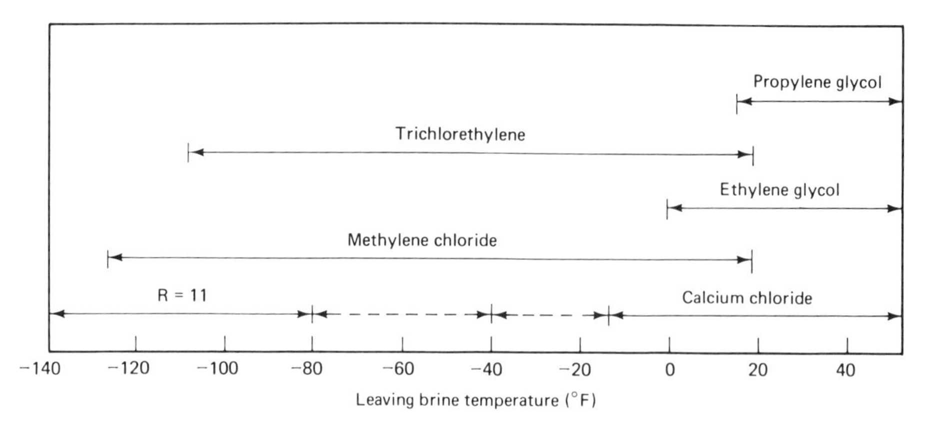

Figure 8-14 Temperature range for brine systems. (By permission of *ASHRAE Handbook.*)

or PVC piping can be used for most brines; however, it is safer to use steel or iron for salt brines in case the pH is not accurately controlled.

It is customary to use multistage compressors for low-temperature brine systems. One very common application is the making of some ice cream products where ammonia is used as the refrigerant.

REVIEW EXERCISE

Select the letter representing your choice of the correct answer (there may be more than one).

8-1. Which one of the following components is *not* part of the basic refrigeration system?

________ a. Compressor

________ b. Condenser

________ c. Evaporator

________ d. Control relay

8-2. Which of the following is *not* an accessory?

________ a. Evaporator

________ b. Sight glass

________ c. Drier

________ d. Solenoid valve

8-3. A hermetically sealed refrigeration system is a good example of what type of system?

________ a. A cascade system

________ b. A basic system

________ c. A compound system

________ d. A secondary refrigerant system

8-4. Which of the following is a good example of an air defrost system?

________ a. Ice maker

________ b. Water chiller

________ c. Reach-in refrigerator

________ d. Truck refrigeration unit

8-5. What type of defrost system does a Thermobank use?

________ a. Air

________ b. Hot gas

________ c. Water

________ d. Glycol

8-6. How are the compressors connected on a modern supermarket refrigeration system?

________ a. Parallel

________ b. Series

________ c. One compressor for two evaporators

________ d. Two compressors for one evaporator

8-7. What is the maximum number of evaporators that should be defrosted at one time in a supermarket system?

________ a. 5%

________ b. 15%

________ c. 25%

________ d. 50%

8-8. Ultralow-temperature systems operate in the following temperature range:

_______ a. 0 to −10 °F

_______ b. −10 to −250 °F

_______ c. −20 to −30 °F

_______ d. −40 to −150 °F

8-9. If a 40 °F cabinet in an 80 °F room requires 2 inches of insulation, what thickness of the same insulation would you expect a −100 °F cabinet to require in a similar temperature room?

_______ a. 5 in.

_______ b. 7 in.

_______ c. 9 in.

_______ d. 11 in.

8-10. What type of system has the discharge of one compressor feeding into the suction of a second compressor?

_______ a. Compound

_______ b. Cascade

_______ c. Basic

_______ d. Supermarket

9 Electrical Theory

OBJECTIVES

After studying this chapter, the reader will be able to:

- Define the commonly used electrical terms.
- Identify the various types of electrical circuits and power supplies.
- Use the common types of electrical meters.

VOLTS-AMPERES-OHMS

The transfer of energy by the movement of electrons is called electric current. When a material such as copper is used, a free movement of electrons occurs. Thus, copper is a good *conductor* of electricity. In materials such as glass or porcelain, a high resistance is offered to the movement of electrons. These types of materials are called *insulators.*

The force that causes the flow of electric current is called a potential difference or an electromotive force (EMF) and is measured in *volts.* The amount of electric current flowing through a conductor is measured in *amperes.* The resistance offered by the conductor to the flow of current is measured in *ohms.* The resistance in any circuit is the component that uses power. Thus, most loads in an electrical circuit can be represented by a resistor.

SIMPLE CIRCUITS

All circuits are made up of four elements:

1. *Load:* a device that has resistance to the flow of current
2. *Source of power:* such as a battery or generator

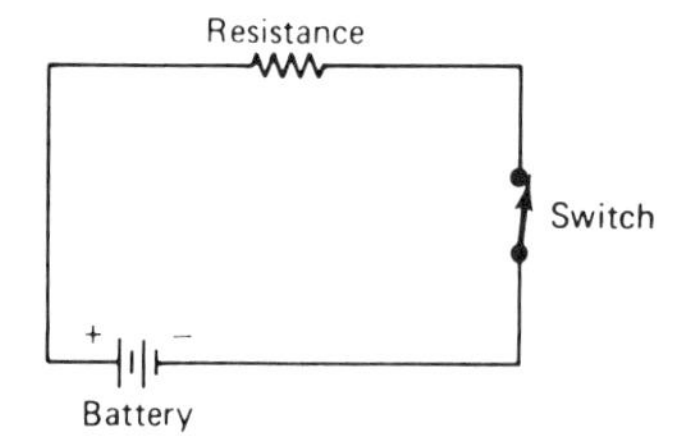

Figure 9–1 Simple circuit. A simple circuit consists of a load, a path for the electricity to follow, and a course of power. The switch shown in the illustration is an accessory.

3. *Path:* a wire conductor to carry the current through the circuit
4. *Control:* usually, a switch to regulate the flow of electricity in the circuit

A diagram of a simple circuit is shown in Figure 9–1.

OHM'S LAW

Ohm's law is a mathematical equation that gives the relationship between E [electromotive force in volts (V)], R [resistance in ohms (Ω)], and I [intensity or current flow in amperes (A)]. The formula can be stated in three ways:

$$E = I \times R \tag{1a}$$

$$I = \frac{E}{R} \tag{1b}$$

$$R = \frac{E}{I} \tag{1c}$$

The formula used depends on which quantity is unknown. Having given any two of the quantities, the other can be calculated. For example, in a simple circuit, if

$$R = 10\ \Omega \qquad \text{and} \qquad I = 5\ \text{A}$$

then

$$E = 5 \times 10 = 50\ \text{V}$$

Occasionally, it is necessary to divide an ampere into smaller parts. Since the ampere is a term that is common to both the English and the International System (SI), it is customary to use the metric system for smaller units:

$$1\ \text{ampere} = 1000\ \text{milliamperes (mA)}$$

The "milli" prefix is also used in describing the smaller portions of a volt. For example, a flashlight battery has a voltage of 1500 mA or 1½ V.

POWER

The formula for power is also useful:

$$P = E \times I$$

where P = power, watts

E = electromotive force, volts

I = intensity or current, amperes

This formula can also be stated in three ways:

$$P = E \times I \tag{2a}$$

$$I = \frac{P}{E} \tag{2b}$$

$$E = \frac{P}{I} \tag{2c}$$

As an example of the use of the power formula, if a resistance has an applied voltage of 120 V and a current of 5 A flowing through it, the power being used is

$$P = 120 \times 5 = 600 \text{ W}$$

POWER FACTOR

Normally formulas (2) are applied only to resistance loads. When an alternating-current (ac) motor is used in the circuit, a new type of resistance to the flow enters into the formula, known as *inductance.* Inductance is the resistance caused by the counter EMF generated by a current-carrying coil, which opposes the flow of current. Another restriction occurs in an ac circuit when a *capacitor* (to be described later) is used. This resistance to flow is known as *capacitance.* The combination of capacitance and inductance is called *impedance.* This type of resistance does an interesting thing: in an ac circuit, it prevents the voltage and current from peaking at the same time. Thus, the instantaneous product of volts × amperes is less than the product of the peaks being generated by the power company.

This reduction in actual power, compared to the product of "peak" volts × amperes, is known as the *power factor.* Since meters read peak volts and peak amperes, the actual watts are determined by the formula

$$\text{actual watts} = \text{volts} \times \text{amperes} \times \text{power factor}$$

The power factor can be determined by the use of three meters:

$$\text{power factor} = \frac{\text{watts}}{\text{volts} \times \text{amperes}}$$

A wattmeter takes the power factor into consideration. For example, if a reach-in refrigerator on 120 V uses 10 A of current at a power factor of 0.90, the watts used are:

$$\text{power} = 120 \times 10 \times 0.90 = 1080 \text{ W}$$

HORSEPOWER

Horsepower is a useful term to indicate the rate of doing work. Its value is established by the formula

$$\frac{\text{force} \times \text{distance}}{\text{time}}$$

One horsepower is equivalent to the work done at the rate of 33,000 foot-pounds per minute. A useful relationship is the conversion factor to watts.

1 horsepower = 746 watts of electrical power

The larger unit of watts is called *kilowatts.* "Kilo" is the SI prefix for 1000. Thus

1 horsepower = 0.746 kilowatt (kW)

DIRECT AND ALTERNATING CURRENT

Direct current (dc) flows in only one direction, from negative to positive. A battery supplies dc current. Alternating current reverses itself periodically: 60-cycle (hertz) current reverses itself every $\frac{1}{120}$ of a second. Most of the current used commercially is ac current since it is easier to produce and transmit.

ELECTROMAGNETISM

When electricity passes through a coil of wire a magnetic force is created perpendicular to the plane of the coil. This is known as an *electromagnetic force.* This characteristic of electrical energy is useful in constructing solenoid valves, transformers, motors, and electrical relays.

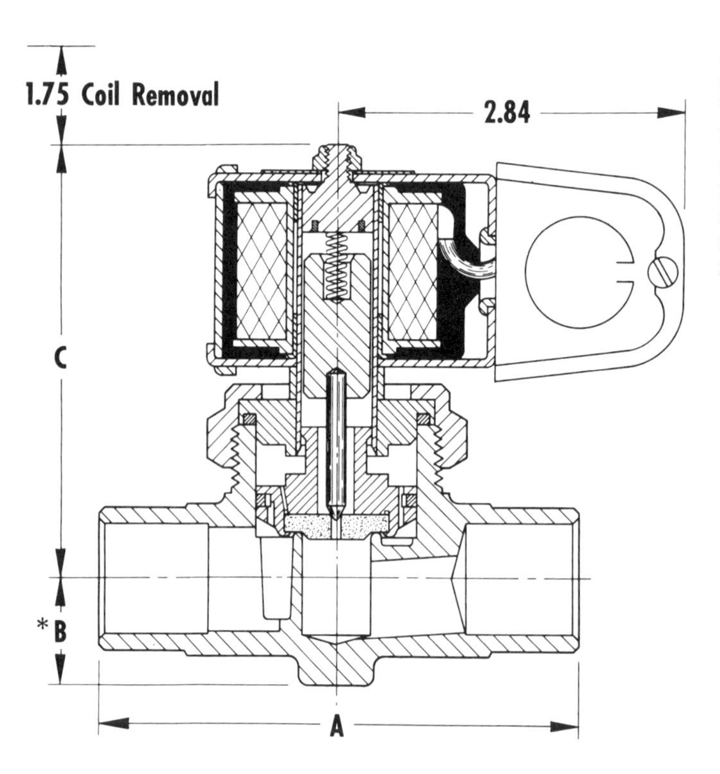

Figure 9-2 Solenoid valves are commonly used to stop/start the flow of refrigerant. They are made in sizes ranging from ¼ to 2 in. piping connections. They can be normally open or normally closed. It is good practice to use a filter-drier ahead of a solenoid valve. Coil electrical ratings cover most standard voltages. (By permission of Sporlan Valve Company.)

For example, in the construction of a solenoid valve, a coil of wire is placed around the plunger. The plunger can be connected to a valve. When current flows through the coil, the plunger raises and the valve opens (see Figure 9–2).

CONDUCTANCE

There are three types of materials used for electrical work: conductors, semiconductors, and insulators. Each relates to the ability of the material to accept the flow of electricity. Copper wire is an excellent conductor. Rubber is an excellent insulator. The semiconductors are between these two extremes and have found common use in electronic work. Silicon is a good example of a semiconductor.

TYPES OF CIRCUITS

Circuits fall into three categories: *series, parallel,* and *a combination of the two.* In a series circuit the current can follow only in one path through a series of loads (see Figure 9–3). Using resistance loads, a *series circuit* can be analyzed, using the following formulas:

$$E_T = E_1 + E_2 + E_3 + \ldots \quad (3)$$

$$I_T = I_1 = I_2 = I_3 = \ldots \quad (4)$$

$$R_T = R_1 + R_2 + R_3 + \ldots \quad (5)$$

These formulas can be stated as follows:

1. The sum of the voltage drops across each resistance is equal to the supply voltage.
2. The current flow is the same through each resistance.
3. The total resistance is the sum of the individual resistances.

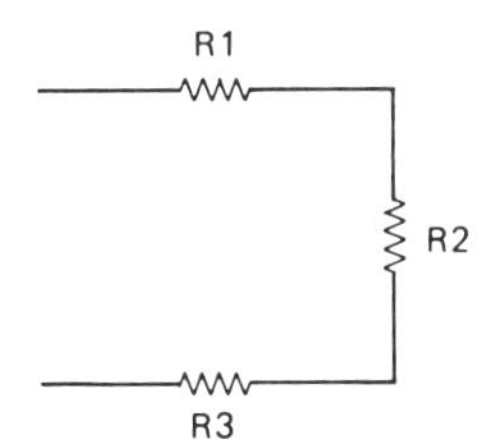

Figure 9–3 Series loads are connected to each other like the links in a chain. If the individual resistances are known, together with the voltage of the power source, the voltage drops across each resistance can be calculated.

Using Figure 9–3, assume that $R_1 = 5\ \Omega$, $R_2 = 10\ \Omega$, and $R_3 = 20\ \Omega$. The circuit voltage is 140 V. The total resistance in the circuit is found by:

$$R_T = 5 + 10 + 20 = 35\ \Omega \quad (5)$$

The total current flowing is found using the formula

$$I = \frac{E}{R} = \frac{140}{35} = 4\text{A}$$

Since all resistances have the total current flowing through them,

$$I = I_1 = I_2 = I_3 = 4 \text{ A} \tag{4}$$

The voltage drop across each resistance is found by using Ohm's law:

$$E_{R1} = 4 \times 5 = 20 \text{ V}$$
$$E_{R2} = 4 \times 10 = 40 \text{ V}$$
$$E_{R3} = 4 \times 20 = 80 \text{ V}$$

The total voltage is found by formula (3) and should check with the circuit voltage:

$$E_T = 20 + 40 + 80 = 140 \text{ V} \tag{3}$$

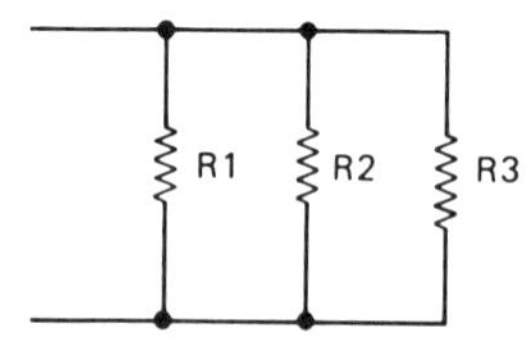

Figure 9-4 Parallel loads. Each of the parallel loads is connected to the same source of power. If the resistances are known and the voltage of the source of power is also known, the amperes of current through each resistance can be calculated.

A *parallel circuit* is one through which the current has more than one path to follow (see Figure 9-4). The formulas for analyzing *parallel circuits* are as follows:

$$E_T = E_1 = E_2 = E_3 = \ldots \tag{6}$$

$$I_T = I_1 + I_2 + I_3 + \ldots \tag{7}$$

$$R_T = \frac{R_1 \times R_2}{R_1 + R_2} \text{ (for 2 resistances only)} \tag{8a}$$

or

$$\frac{1}{R_T} = \frac{1}{R_1} + \frac{1}{R_2} + \frac{1}{R_3} + \ldots \tag{8b}$$

The formulas can be stated as follows:

1. The supply voltage is the same for all circuits.
2. The total current is the sum of the currents drawn by each load.
3. The total resistance is less than any one individual resistance and can be obtained by use of the formulas given.

Using Figure 9-4, assume that $R_1 = 5\ \Omega$, $R_2 = 10\ \Omega$, and $R_3 = 20\ \Omega$; assume that the circuit voltage is 100 V. The voltage for each resistance is the same as the circuit voltage:

$$E_T = E_1 = E_2 = E_3 = 100 \text{ V} \tag{6}$$

The current through each resistance is found by using Ohm's Law:

$$I_{R1} = \frac{100}{5} = 20\text{A}$$

$$I_{R2} = \frac{100}{10} = 10\text{A}$$

$$I_{R3} = \frac{100}{20} = 5\text{A}$$

The total amperage is found by:

$$I_T = 20 + 10 + 5 = 35 \text{ A} \tag{7}$$

The total resistance can easily be found by using Ohm's law:

$$R_T = \frac{E_T}{I_T} = \frac{100}{35} = 2.86 \ \Omega$$

or formula (8b) can be used:

$$\frac{1}{R_T} = \frac{1}{5} + \frac{1}{10} + \frac{1}{20} \tag{8b}$$

$$R_T = 2.86$$

which checks with the calculation determined by Ohm's law.

VOLTAGES

The type of voltages available originates from the type of transformer supplied by the power company. Basically there are two types of systems, Y (wye) and Δ (delta). Figure 9–5 shows the voltages available from the two types of service.

Figure 9–5 (a) Wye and (b) delta systems. Both the wye and the delta power systems use four wires. The three "hot" lines constitute three-phase power. Any two "hot" lines constitute single-phase power at the three-phase voltage. A "hot" wire and the neutral wire provide single-phase power at reduced voltage.

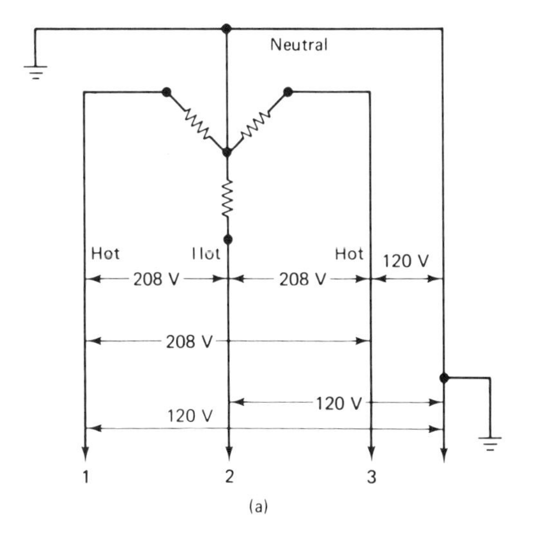

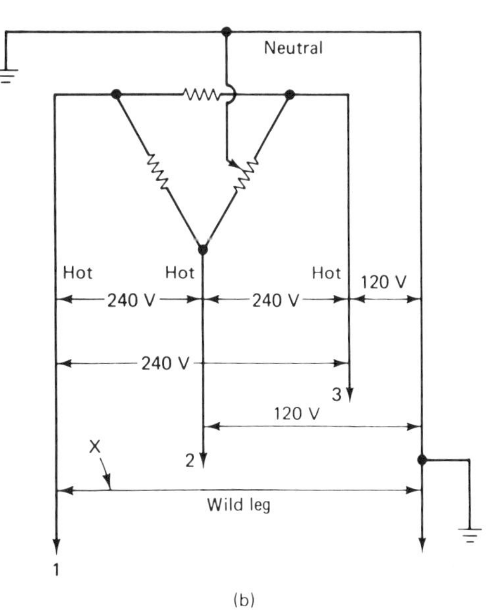

Note in the delta circuit that the neutral is obtained at a midpoint of one of the field coils. The power available is 240-V three-phase, 240-V single-phase, and 120 V single-phase. In the wye circuit the neutral is available at the junction point of the field windings. The power available is 208-V three-phase, 208-V single-phase, and 120-V single-phase.

PHASE

All power is generated as three phase: this means that there are three hot wires. Single phase is made by using any two of these hot wires. Alternating-current power is generated in the waveform shown in Figure 9–6. For 60-hertz power, the cycle is repeated every 1/60 of a second. For three-phase power, three separate waves are generated, all out of phase with each other by 120 °F. This out-of-phase relationship is advantageous in its application to three-phase motors.

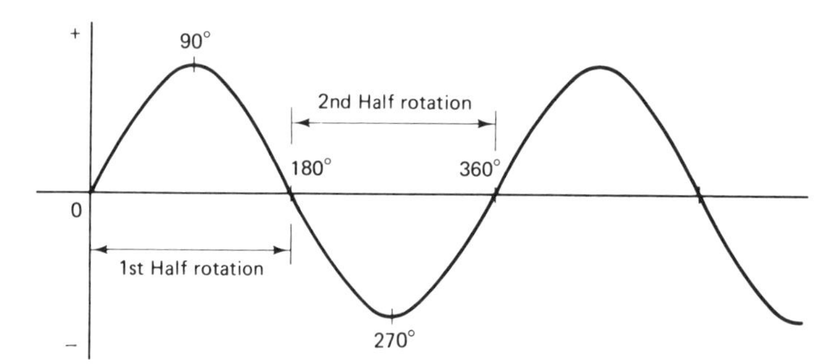

Figure 9–6 Sine wave. The sine wave represents the flow of power from a single conductor when alternating current is generated. A full wave is 1 Hz.

CIRCUIT CONDITIONS

The *open* circuit is experienced when the circuit path is not completed from one side of the power supply, through the load, to the other side of the power supply. Current is not flowing where it should be.

The *short* circuit is when the path from one side of the power supply to the other does not pass through a load. This usually blows a fuse or opens a circuit breaker.

A *grounded* circuit is one where some or all of the current is bypassing the required path and going to ground. For example, on a motor where the winding insulation is worn and the current passes to the grounded motor housing, the motor will not run. It is common practice to ground enclosures of electrical equipment for safety reasons.

MOTOR NAMEPLATES

Much information can be derived from examining a typical motor nameplate (Figure 9–7). Particular attention should be given to:

Mod. 44-AL3 Ser. #68C1571 Type. YX-26 Hp. 5 Ph. 3 Cy. 60
Volts: 220/440 F.L.A. 14.8/7.4 L.R.A. 57.2/40.0 S.F. 1.25
R.P.M. 1745 Frame: 69-T Temperature Rise: 40 Deg. Centigrade

Figure 9-7 Motor nameplate information supplies useful data relative to the characteristics of the motor.

1. Full-load amperes (FLA)
2. Locked-rotor amperes (LRA)
3. Service factor (SF)
4. Temperature rise
5. Horse power
6. Voltage/phase/cycles

The FLA is the amperage of the motor at full load and at rated voltage. This is the amperage that indicates the size of fuse for the circuit. Generally speaking, the fuse should be 1.25 to 1.50 times the FLA, depending on local electrical code regulations. A time-delay fuse should be used to protect a motor.

The LRA indicates the current draw for a duration of less than 1 second at startup. If a transformer feeds the circuit, an allowance must be made for this amperage.

The SF indicates the maximum horsepower that a motor can supply. For example, a 10-horsepower motor with a 1.25 SF could handle a 12.5-brake horsepower (bhp) load.

The voltage rating is important since it must be matched with the supply voltage. If the motor is rated 240/480 V, the motor can be connected to either source of power following the instruction on the motor. Voltage can vary 10% plus or minus from either of these voltages and still be satisfactory. This particular motor would not be suitable for 208 V.

The temperature rise is usually given in Celsius degrees and can be converted to Fahrenheit degrees. This is the rise in temperature over ambient that the motor will tolerate.

MANUFACTURERS' INFORMATION

Most manufacturers supply complete electrical data to assist in properly applying their equipment. In all cases it is important to obtain and review this information. Below is a description of a typical manufacturer's electrical data for an air cooled packaged chiller:

Model	Volts (60 Hz)			MCA	MOPA	ICF
	Nameplate	Limits				
		Min.	Max.			
500	203-230	187	254	397	500	766

MCA = Minimum circuit amperes (for wire sizing)
MOPA = Maximum over current protective device amperes
ICF = Maximum instantaneous current flow at starting

ELECTRICAL METERS

The most commonly used electrical meters are: (1) Ohmmeters, (2) clamp-on ammeters, (3) voltmeters, (4) wattmeters, and (5) capacitor checker meters. Often, a number of these meters is combined into a single meter, which enhances usefulness.

Ohmmeters

The ohmmeter is used to measure the resistance of a load when no external current is connected. It must be used with the power supply off. This meter has its own battery, which must be in good condition for the meter to function. One type of ohmmeter is shown in Figure 9-8.

In using the ohmmeter, the following procedure is recommended:

1. Turn off the power to the device being tested.
2. Turn the selector switch to the proper scale. The meter usually has a number of scale positions often indicated by R×1, R×100, R×50,000, and so on. The number indicates the multiplier to be used. If the reading is 10 and the R×100 position is used, the actual reading is 1000 Ω (10 × 100).
3. Test the meter by placing the two leads together. The meter should read "0." If not, adjust it to "0."

Figure 9-8 Standard and digital VOM meters. (By permission of Simpson Electric Company.)

4. Isolate the load to be tested. Disconnect at least one side of the power supply.
5. Apply the two leads to the two terminals of the load and read the resistance. A zero reading means a short circuit, no resistance. A reading of infinity (∞) indicates an open circuit. If the reading is a resistance value, a measurable resistance is obtained and the load is usually satisfactory.

Voltmeters

The voltmeter is used to measure the EMF of a power supply (see Figure 9–8). Always enlist the highest scale first and drop down to lower scales if required to read the value accurately. The most accurate readings are in the center portion of the scales. A zero reading indicates no voltage drop which is typical when reading the voltage across a closed switch with the power on.

Clamp-On Ammeters

The clamp-on ammeter is used to measure current flow to an electrical load while the equipment is running (see Figure 9–9). The correct scale must be selected, using the highest first and dropping down if necessary. The jaws of the meter are opened to surround *one* wire of the power supply.

A multiplier coil may be used to read small values as shown in Figure 9–10. The actual reading, divided by the number of loops around the clamp, gives the true value. For example, if the multiplier has 10 loops and the reading is 5, the true value is 0.5 A ($\frac{5}{10}$).

Figure 9–9 Combination volt-ohm-ampere meter. (By permission of Amprobe Instruments, Inc.)

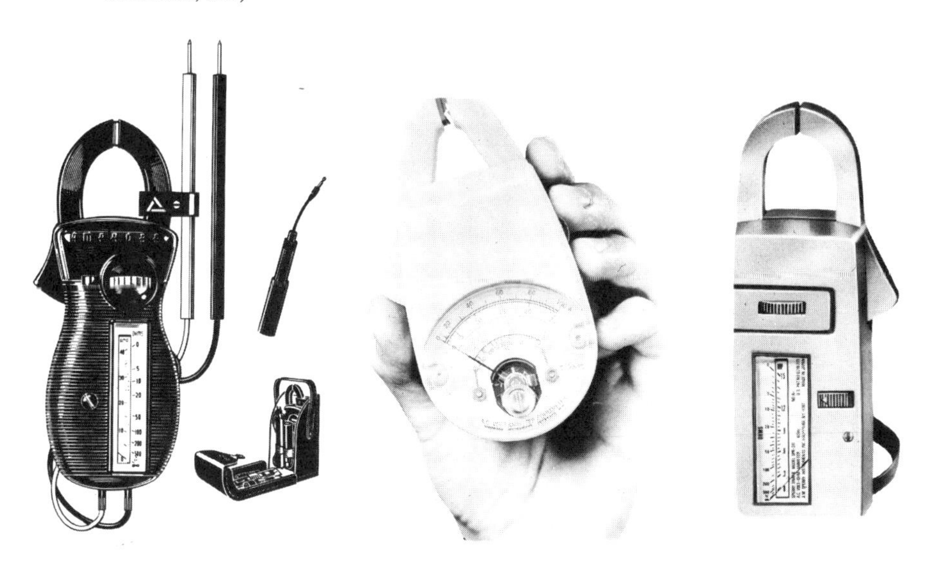

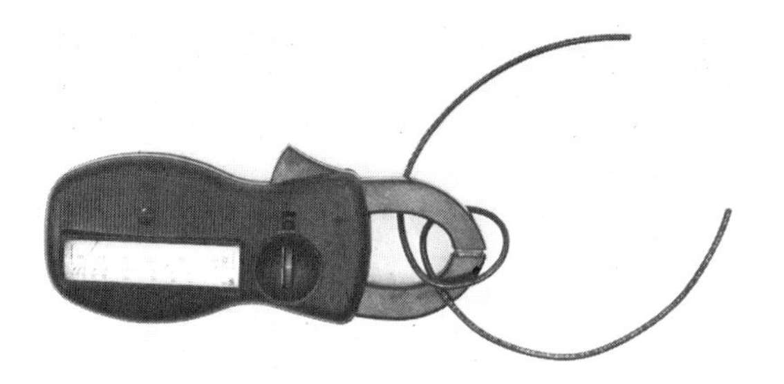

Figure 9–10 Multiplier coil on clamp-on ammeter. (By permission of Amprobe Instruments, Inc.)

Wattmeters

The wattmeter indicates the power drawn by electric devices. The wattmeter automatically measures the instantaneous volts × amperes.

Capacitor-Checker Meters

The capacitor-checker meter measures capacitance. It is useful in troubleshooting. The capacitor checker reads in microfarads. As a safety precaution, a discharge device should be used to discharge the capacitor before and after using the capacitor checker. A good discharge device is a 20,000 Ω, 2-W resistor with suitable insulated leads.

REVIEW EXERCISE

Select the letter representing your choice of the correct answer (there may be more than one):

9-1. What is the power factor for an ac current, with the following meter readings: 120 V, 10 A, 1080 W?

________ a. 0.9

________ b. 0.8

________ c. 0.7

________ d. 0.6

9-2. Using Ohm's law, what is the resistance with the following meter readings: 120 V, 10 A?

________ a. 8 Ω

________ b. 10 Ω

________ c. 12 Ω

________ d. 14 Ω

9-3. The following is *not* a requirement for an electrical circuit:

________ a. Path for current to follow

________ b. Source of power

________ c. A load (resistance)

________ d. A switch

________ e. A thermal expansion valve

9-4. In a series circuit connecting a number of resistance loads, the total current is equal to:

________ a. The sum of the currents through each resistance

________ b. The current through any one of the resistances

________ c. Amount determined by the formula $I_T = I_1 + I_2 + I_3 + \dots$

________ d. The total resistance divided by the total voltage

9-5. In a parallel circuit, $I_T = I_1 = I_2$.

________ a. True

________ b. False

9-6. In a three-phase, 4-wire, Wye system rated at 240 V, the single-phase voltage between the neutral and one hot leg is:

________ a. 120 V

________ b. 277 V

________ c. 207.8 V

________ d. 440 V

9-7. Most single-phase motors will operate with a voltage within what percent of the nameplate voltage?

________ a. 12%

________ b. 10%

________ c. 8%

________ d. 6%

9-8. "LRA" stands for:

________ a. Large resistance armature

________ b. Little resistance armature

________ c. Laminated rotor amperes

________ d. Locked-rotor amperes

9-9. One ampere is equal to how many milliamperes?

________ a. 10

________ b. 100

________ c. 1000

________ d. 10,000

9-10. If 120 V is supplied to an appliance that uses 5 A, how much power is required? (Assume 100% power factor.)

________ a. 5 W

________ b. 100 W

________ c. 600 W

________ d. 750 W

10 Electrical Components and Wiring Symbols

OBJECTIVES

After studying this chapter, the reader will be able to:

- Identify the electrical characteristics of loads and switches.
- Be able to draw or recognize the symbols of electrical components used in wiring diagrams.
- Describe the operation and function of various electrical components used in refrigeration systems.

ELECTRICAL EQUIPMENT

Alternating-current systems are made up primarily of four types of electrical devices:*

1. Loads
2. Switches
3. Combination of a load and a switch
4. Capacitors

All *loads* offer a resistance to the electric current. They produce heat, create motion, or produce a light. All loads require electrical energy to operate. A group of symbols for load devices is shown in Figure 10–1.

Switches are used to turn loads on and off. They do not consume but direct power to the load or loads they operate. There are many types of switches, depending on the

*This list of electrical devices does not include some of the solid-state units made with semiconductor material such as diodes, transistors, and thermisters that are used in dc electronic circuitry.

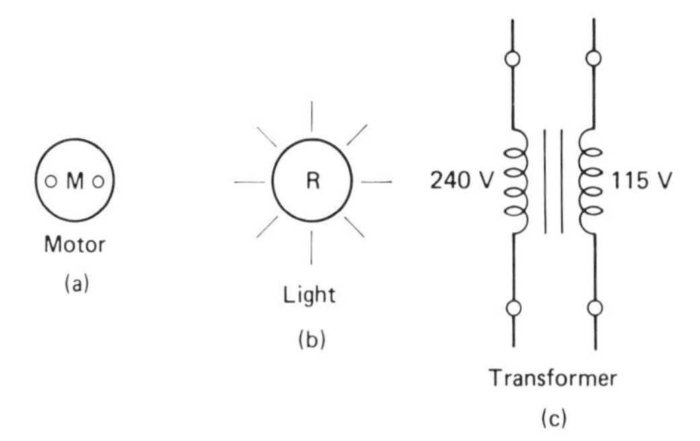

Figure 10–1 Examples of electrical loads: (a) motor; (b) light; (c) transformer.

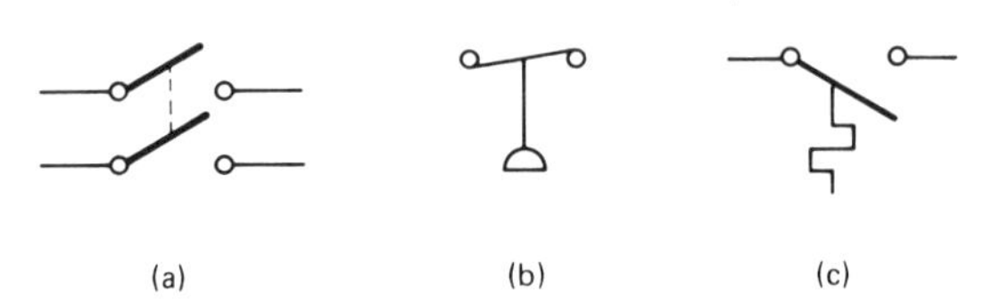

Figure 10–2 Electrical switches: (a) motor starter; (b) pressure switch; (c) thermostat. Switches have the function of turning loads on and off.

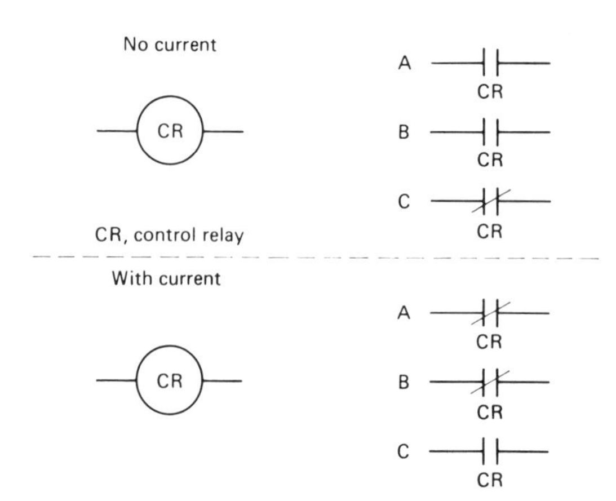

Figure 10–3 Combination load and switch.

force that is used to operate them: manual switches, pressure switches, temperature switches or thermostats, light-operated switches, and many others (see Figure 10–2).

A *combination* device uses a load to operate a switch. A good example of this is a relay, a contactor, or a starter (see Figure 10–3).

A *capacitor* is a unique accessory used primarily to assist in starting single-phase motors. Certain types of capacitors are used also to increase the efficiency of single-phase motors.

Electrical diagrams are used to show the wiring connections between electrical devices. On these diagrams certain symbols represent the electrical devices. In the following information the various types of common *loads* will be discussed together with an appropriate symbol to represent them in the wiring diagram.

LOAD DEVICES

All load devices use power and must be operated by properly located switches. Common load devices are: (1) motors, (2) solenoids, (3) transformers, (4) lights, and (5) heaters.

Motors

Probably the most important load device is the electric motor. A number of symbols can be used for a motor. The most common symbol is a circle, as shown in Figure 10–4. A wiring representation can also be used. Regardless of the type of symbol used, it must be identified, usually by an abbreviation, in the legend of the wiring diagram. The legend in Figure 10–4 shows that "Comp." stands for compressor and "OFM" stands for outside fan motor.

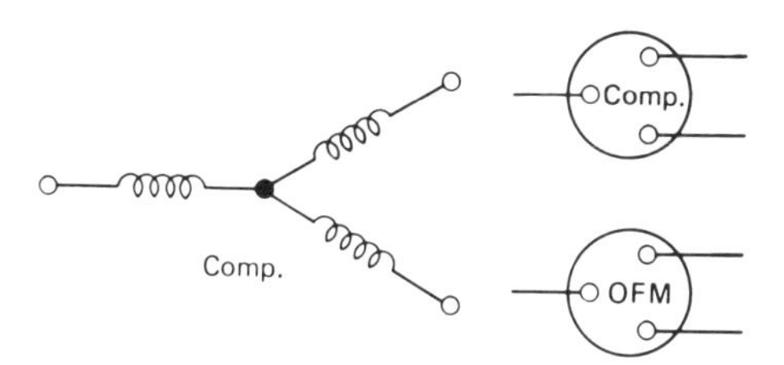

Figure 10–4 Motor symbols. Various symbols are used for motors. The most popular one is the circle with connecting terminals and letter to indicate the use of the motor. Comp., compressor; OFM, outdoor fan motor.

Much of the information on the use of motors in refrigeration work is given in Chapter 11. However, in this section, the reader should pay particular attention to the motor characteristics, which should be clearly evident in the wiring diagram. There must be a match between the motor actually used and the electrical characteristics shown in the wiring diagram.

Motors used for refrigeration work are normally applied to run compressors, fans, or pumps. Information that should be known about the motor before it can be properly applied includes:

1. The duty
2. The service requirements
3. The power requirements
4. The starting device
5. The motor protection
6. The motor accessories
7. The motor control
8. The speed regulation (if required)

Look in any wholesaler's catalog and note the wide variety of motors that are available. Motors are constructed for specific applications and special care must be exercised in selecting and applying them.

Figure 10–5 Illustration for typical fractional-horsepower motor, permanent split capacitor type. (By permission of Component Products.)

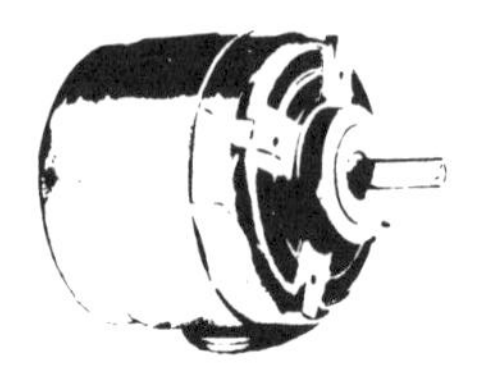

Figure 10–6 Typical illustration for a frame-type fan-duty motor, shaded-pole type. (By permission of Component Products.)

Figures 10–5 and 10–6 show two typical illustrations for fractional-horsepower motors. The motor shown in Figure 10–5 is a fan-duty motor. It has clockwise (CW) rotation viewing the motor from the drive end. The power requirement is 115 V, single phase, 60 Hz. Due to its low amperage of 1.0 for the $\frac{1}{15}$-hp size, it can be started directly across the line. The motor has inherent overload protection. It requires a 5-μF (microfarad) capacitor rated at 370 V. It has a top speed of 1600 rpm with two other speeds available.

Figure 10–6 also shows a fan-duty motor. It is available in either clockwise (CW) or counterclockwise (CCW) rotation. The motor is designed for either single voltage, 115 or 230 V; or dual voltage, 115/230 V. The higher voltage can be applied to either 208- or 230-V power sources. Due to its low horsepower it can be started directly across the line. No special accessories such as capacitors are required. It can be operated continuously and has overload protection. It can be placed at any angle. Base mountings are available. The motor is single speed but is available for 1050 rpm (six pole) or 1550 rpm (four pole). The shaft is $\frac{3}{8}$ in.

Solenoids

A solenoid is a coil of wire that is used to produce magnetism. This force is used to operate valves and switches. The symbol for the solenoid is shown in Figure 10–7. The identifying letter, explained in the legend, tells how the solenoid is being used. In the example shown, "RVS" refers to a reversing-value solenoid and "S" refers to a solenoid used on a stop valve.

There are two types of solenoids:

1. Movable armature, such as that used to operate a flow valve
2. Fixed armature, such as that used to operate a relay

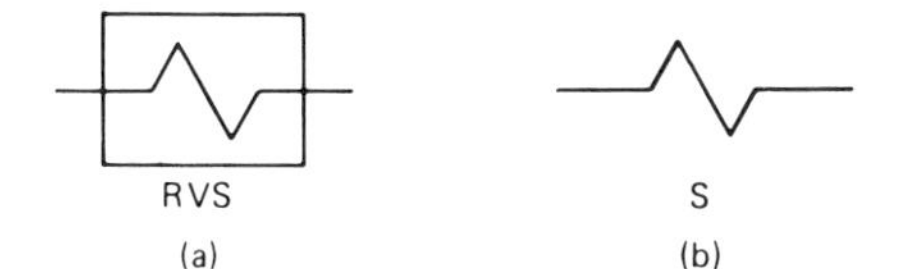

Figure 10-7 Solenoid symbols: (a) reversing valve solenoid; (b) stop valve solenoid. The most common symbol for the solenoid is shown in part (b) ("S" refers to stop valve). The reversing valve solenoid is marked RVS.

Solenoid valves are widely used in refrigeration work, principally as stop valves. The most common use is in the liquid line to stop or start the flow of refrigerant to the evaporator. Generally, the solenoid valve is activated by a temperature sensor (thermostat). Solenoid valves can be selected for normally closed or normally open operations. The "normal position" of the valve is the condition when no current is flowing through the solenoid coil. They are usually selected "normally closed."

A good example of the use of a liquid-line solenoid valve is in the pump-down cycle (see Figure 10-8). When the thermostat is satisfied, the solenoid valve is closed and the refrigerant stops flowing through the expansion valve. The compressor continues to run, pumping down the coil, and the compressor is stopped by the low-pressure control. When cooling is needed the thermostat opens the solenoid valve, starting the flow of refrigerant. When the pressure in the suction line reaches the cut-in point of the low-pressure control, the compressor is started. This control arrangement protects the compressor against "slugging."

The same type of valves can be used as suction or discharge stop valves. A typical liquid-capacity selection table is shown in Figure 10-9 and a typical hot-gas and suction-line-capacity selection table is shown in Figure 10-10.

Figure 10-8 Pump-down cycle. When thermostat is satisfied, the solenoid valve closes, shutting off liquid refrigerant from the receiver. Compressor continues to operate until stopped by the low-pressure control.

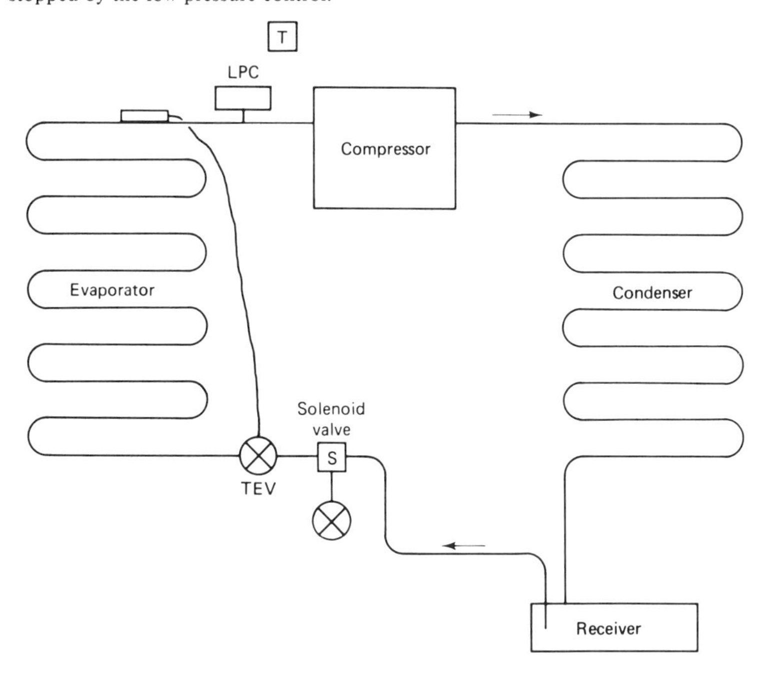

TYPE NUMBER						CONNECTIONS Inches	PORT SIZE Inches	TONS OF REFRIGERATION — PRESSURE DROP — psi																			
"A" and "B" Series Valves			"E" Series Extended Connections					1				2				3				4				5			
Without Manual Lift Stem		With Manual Lift Stem	Without Manual Lift Stem		With Manual Lift Stem																						
Normally Closed	Normally Open	Normally Closed	Normally Closed	Normally Open	Normally Closed			12	22	500	502	12	22	500	502	12	22	500	502	12	22	500	502	12	22	500	502
A3P1	—	—	—	—	—	3/8 NPT Female	.101	.7	.9	.8	.6	1.0	1.3	1.1	.8	1.2	1.6	1.4	1.0	1.4	1.9	1.6	1.2	1.6	2.1	1.9	1.4
A3F1	—	—	—	—	—	1/4 SAE Flare																					
A3S1	—	—	—	—	—	1/4 or 3/8 ODF Solder																					
—	—	—	E5S120	—	—	1/4 ODF Solder	.150	1.2	1.6	1.5	1.1	1.8	2.3	2.0	1.5	2.2	2.8	2.6	1.9	2.5	3.3	2.9	2.1	2.8	3.6	3.3	2.4
—	—	—	E5S130	—	—	3/8 ODF Solder																					
B6P1	—	MB6P1	—	—	—	3/8 NPT Female	3/16	2.2	2.9	2.6	1.9	3.1	4.0	3.5	2.6	3.8	4.9	4.4	3.2	4.4	5.7	5.0	3.7	4.9	6.4	5.6	4.1
B6F1	—	MB6F1	—	—	—	3/8 SAE Flare																					
B6S1	—	MB6S1	E6S130	—	ME6S130	3/8 ODF Solder																					
B6S1	—	MB6S1	E6S140	—	ME6S140	1/2 ODF Solder																					
B9P2	OB9P2	MB9P2	—	—	—	3/8 NPT Female	9/32	3.6	4.7	4.1	3.0	5.1	6.6	5.9	4.3	6.2	8.0	7.1	5.2	7.2	9.4	8.3	6.1	8.0	10.4	9.1	6.7
B9F2	OB9F2	MB9F2	—	—	—	3/8 SAE Flare																					
—	—	—	E9S230	OE9S230	ME9S230	3/8 ODF Solder																					
B9S2	OB9S2	MB9S2	E9S240	OE9S240	ME9S240	1/2 ODF Solder																					
B10F2	OB10F2	MB10F2	—	—	—	1/2 SAE Flare	5/16	4.9	6.4	5.6	4.1	7.0	9.1	8.0	5.9	8.5	11.1	9.8	7.2	9.8	12.8	11.2	8.2	11.0	14.3	12.7	9.3
—	—	—	E10S240	OE10S240	ME10S240	1/2 ODF Solder																					
B10S2	OB10S2	MB10S2	E10S250	OE10S250	ME10S250	5/8 ODF Solder																					
B14P2	OB14P2	MB14P2	—	—	—	1/2 NPT Female	7/16	7.0	9.1	8.0	5.9	9.9	12.9	11.3	8.3	12.1	15.8	13.9	10.2	13.9	18.1	16.0	11.7	15.6	20.3	17.9	13.1
B14S2	OB14S2	MB14S2	E14S250	OE14S250	ME14S250	5/8 ODF Solder																					
B19P2	OB19P2	MB19P2	—	—	—	3/4 NPT Female	19/32	10.7	13.9	12.3	9.0	15.2	19.8	17.5	12.8	18.6	24.2	21.3	15.6	21.5	28.0	24.7	18.1	24.0	31.3	27.6	20.3
B19S2	OB19S2	MB19S2	E19S250	OE19S250	ME19S250	5/8 ODF Solder																					
B19S2	OB19S2	MB19S2	E19S270	OE19S270	ME19S270	7/8 ODF Solder																					
B25P2	OB25P2	MB25P2	—	—	—	1 NPT Female	25/32	18.3	23.8	21.0	15.4	25.9	33.7	29.7	21.8	31.8	41.4	36.6	26.8	36.7	47.8	42.1	30.9	41.0	53.4	47.1	34.5
B25S2	OB25S2	MB25S2	E25S270	OE25S270	ME25S270	7/8 ODF Solder																					
B25S2	OB25S2	MB25S2	E25S290	OE25S290	ME25S290	1 1/8 ODF Solder																					
B33S2	OB33S2	MB33S2	E33S290	OE33S290	ME33S290	1 1/8 ODF Solder	1	25.5	33.2	29.2	21.4	36.0	46.9	41.3	30.3	44.2	57.6	50.7	37.2	51.0	66.5	58.5	42.9	57.0	74.3	65.3	47.9
B33S2	OB33S2	MB33S2	E33S2110	OE33S2110	ME33S2110	1 3/8 ODF Solder																					
B33S2	OB33S2	MB33S2	—	—	—	1 5/8 ODF Solder																					
—	—	—	E42S2130	OE42S2130	ME42S2130	1 5/8 ODF Solder	1 15/16	56.3	73.4	64.9	47.6	79.7	104	91.4	67.0	97.6	127	112	82.1	113	147	130	95.1	126	164	145	106
—	—	—	E42S2170	OE42S2170	ME42S2170	2 1/8 ODF Solder																					
—	—	MA42P3	—	—	—	1 1/2 NPT Female	1 15/16	47	59	53	39	68	85	76	56	80	102	91	67	88	111	100	73	93	118	105	77
—	—	MA42S3	—	—	—	1 5/8 or 2 1/8 ODF Solder																					
—	—	MA50P3	—	—	—	2 NPT Female	1 9/16	92	106	104	77	120	152	136	100	141	178	160	117	156	198	176	129	169	214	191	140
—	—	MA50S3	—	—	—	2 1/8 or 2 5/8 ODF Solder																					

REFRIGERANT 717

Type						Connections	Port	1	2	3	4	5
MA5A3	—	—	—	—	—	1/4, 3/8, 1/2 NPT Female	.140	8.0	11.3	13.7	16	17.8
MA17A3	—	—	—	—	—	1/2, 3/4 or 1 NPT Female	17/32	73	95	120	143	160
MA32P3	—	—	—	—	—	1 or 1 1/4 NPT Female	1	125	176	225	250	280
MA42P3	—	—	—	—	—	1 1/2 NPT Female	15/16	275	390	500	550	625
MA50P3	—	—	—	—	—	2 NPT Female	19/16	500	725	875	1000	1110

Liquid capacities for Refrigerants 12, 22, 500, and 502 shown in the above table are based on 40°F. evaporating and 100°F. liquid case. Refrigerant 717 capacities are based on 5°F. evaporating and 86°F. liquid.

REFRIGERANT LIQUID TEMPERATURE CORRECTION FACTORS

Refrigerant Liquid Temperature °F.		40°	50°	60°	70°	80°	90°	100°	110°	120°	130°	140°
R-12	Correction	1.36	1.30	1.24	1.18	1.12	1.06	1.00	0.94	0.88	0.82	0.75
R-22	Factor	1.34	1.29	1.23	1.17	1.12	1.06	1.00	0.94	0.88	0.82	0.76

These factors include corrections for liquid refrigerant density and net refrigerating effect and are based on an average evaporator temperature of 40°F. For each 10°F. reduction in evaporating temperature, capacities are reduced by approximately 1 1/2%.

Refrigerant Liquid Temperature °F.		40°	50°	60°	70°	80°	90°	100°	110°	120°	130°	140°
R-500	Correction	1.24	1.20	1.16	1.13	1.09	1.04	1.00	0.95	0.91	0.86	0.81
R-502	Factor	1.52	1.44	1.35	1.26	1.18	1.09	1.00	0.91	0.82	0.73	0.64

Refrigerant Liquid Temperature °F.		0°	10°	20°	30°	40°	50°	60°	70°	80°	86°	90°	100°
R-717	Correction Factor	1.27	1.24	1.20	1.17	1.14	1.11	1.08	1.05	1.02	1.00	0.99	0.96

These factors include corrections for liquid refrigerant density and net refrigerating effect and are based on an average evaporator temperature of 5°F. For each 10°F. reduction in evaporating temperature, the above capacities are reduced by approximately 3/4%.

Figure 10–9 Selection chart for liquid-line solenoid valves based on pressure drop and tons of refrigeration capacity. (By permission of Sporlan Valve Company.)

VALVE SERIES	DISCHARGE GAS CAPACITIES — Tons Pressure Drop Across Valve psi					
	2	5	10	25	50	100
REFRIGERANT 12						
A3	.16	.26	.35	.51	.61	.61
E5	.39	.61	.84	1.2	1.4	1.4
B6 & E6	.68	1.01	1.43	2.0	2.3	2.3
B9 & E9	.97	1.5	2.1	3.1	4.0	4.4
B10 & E10	1.5	2.3	3.2	4.7	5.7	5.8
B14 & E14	2.1	3.2	4.4	6.5	8.1	8.5
B19 & E19	3.0	4.7	6.5	9.7	12.3	13.3
B25 & E25	4.8	7.5	10.5	15.5	19.5	21.1
B33, E33, and MA32	6.3	10.0	14.0	21.7	29.8	39.4
E42	14.1	22.0	30.4	45.2	56.9	61.3
MA42	12.5	19.8	28.0	44.3	54.9	54.9
MA50	21.0	32.2	43.2	56.9	58.4	58.4
REFRIGERANT 22						
A3	.24	.37	.51	.77	.99	1.14
E5	.56	.88	1.22	1.83	2.36	2.70
B6 & E6	.97	1.51	2.10	3.10	3.88	4.17
B9 & E9	1.38	2.17	3.04	4.64	6.19	7.69
B10 & E10	2.15	3.37	4.69	7.06	9.16	10.6
B14 & E14	2.92	4.58	6.40	9.71	12.8	15.4
B19 & E19	4.31	6.77	9.46	14.4	19.1	23.5
B25 & E25	6.89	10.8	15.1	23.0	30.5	37.4
B33, E33, and MA32	9.0	14.2	20.0	31.3	43.4	59.1
E42	20.1	31.5	44.0	67.1	88.9	108.9
MA42	17.6	27.8	39.3	62.2	77.8	94.5
MA50	30.1	46.8	64.1	91.7	107	107
REFRIGERANT 502						
A3	.19	.30	.42	.64	.83	.97
E5	.46	.72	1.0	1.5	2.0	2.3
B6 & E6	.80	1.3	1.7	2.6	3.2	3.5
B9 & E9	1.1	1.8	2.5	3.8	5.1	6.4
B10 & E10	1.8	2.8	3.8	5.9	7.6	9.0
B14 & E14	2.4	3.8	5.3	8.0	10.6	13.0
B19 & E19	3.6	5.7	8.0	12.2	16.3	20.2
B25 & E25	5.7	8.9	12.5	19.0	25.4	31.4
B33, E33, and MA32	9.4	14.6	20.4	31.0	40.7	50.1
E42	16.5	26.0	36.3	55.5	73.9	91.5
MA42	14.0	22.1	30.9	47.2	62.1	77.8
MA50	24.9	38.7	53.1	76.5	00.5	91.5

VALVE SERIES	SUCTION CAPACITY — Tons at 1 PSI Pressure Drop and Evaporating Temperatures of:		
	40°F.	20°F.	0°F.
REFRIGERANT 12			
A3	.08	.07	.05
E5	.19	.16	.13
B6 & E6	.32	.27	.22
B9 & E9	.47	.39	.31
B10 & E10	.72	.60	.48
B14 & E14	.98	.81	.66
B19 & E19	1.45	1.20	.98
B25 & E25	2.32	1.92	1.56
B33, E33, and MA32	3.04	2.53	2.06
E42	6.8	5.6	4.5
MA42	6.2	5.1	4.1
MA50	10.1	8.3	6.6
REFRIGERANT 22			
A3	.12	.10	.08
E5	.28	.23	.19
B6 & E6	.48	.40	.32
B9 & E9	.68	.57	.46
B10 & E10	1.06	.88	.72
B14 & E14	1.45	1.20	1.00
B19 & E19	2.13	1.77	1.45
B25 & E25	3.41	2.83	2.31
B33, E33, and MA32	4.45	3.71	3.03
E42	9.9	8.3	6.7
MA42	8.7	7.3	5.9
MA50	14.9	12.3	10.0
REFRIGERANT 502			
A3	.10	.08	.07
E5	.23	.19	.16
B6 & E6	.39	.33	.27
B9 & E9	.56	.47	.39
B10 & E10	.87	.73	.60
B14 & E14	1.85	1.00	.82
B19 & E19	1.75	1.47	1.21
B25 & E25	2.80	2.35	1.94
B33, E33, and MA32	4.3	3.3	2.6
E42	8.2	6.9	5.7
MA42	7.1	5.9	4.9
MA50	12.2	10.2	8.4

EVAPORATOR TEMPERATURE CORRECTION FACTORS

Evap. Temp. °F.	40°	30°	20°	10°	0°	−10°	−20°	−30°	−40°
Multiplier	1.00	.96	.93	.90	.87	.84	.81	.78	.75

Capacities based on 100°F. Condensing Temperature, Isentropic Compression plus 50°F., 40°F. Evaporator and 65°F. Suction Gas. For capacities at other conditions use the multipliers in above table.

Figure 10–10 Selection chart for hot-gas and suction-line solenoid valves based on pressure drop and tons of refrigeration capacity. (By permission of Sporlan Valve Company.)

It is usually considered good practice to select liquid-line solenoids on the basis of a 2-psi pressure drop across the valve for R-12 and a 3-psi drop for R-22 and R-502. The maximum operating pressure differential (MOPD) for most valves is 300 psi and the safe working pressure (SWP) is 500 psi. Valves may be supplied with a manual operating stem if requested. A cutaway view of a typical valve is shown in Figure 10–11.

Figure 10–11 Cutaway view of typical solenoid valve showing directions of flow through valve. (By permission of Sporlan Valve Company.)

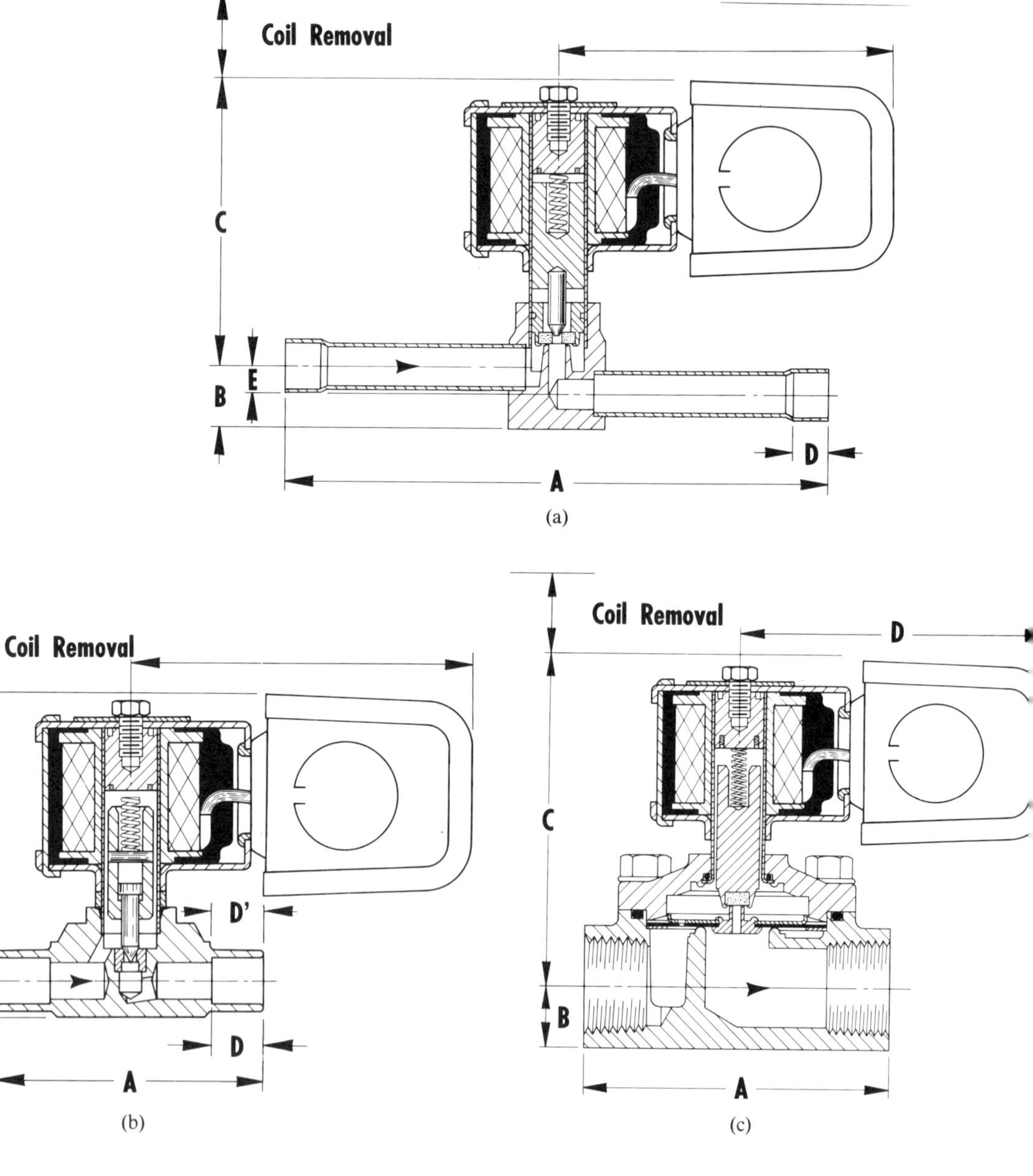

There are a number of special-purpose solenoid valves which are extremely useful in certain applications:

1. The solenoid pilot control valve
2. The three-way hot-gas defrost valve or heat reclaim valve
3. The industrial solenoid valve
4. The reversing valve

A typical application of a *pilot valve* is shown in Figure 10-12. This valve has two parts: one high-pressure and one low-pressure. When the solenoid is energized, the plunger moves upward closing off the high-pressure port. Under these conditions, the true suction pressure acts on the expansion valve diaphragm. When the solenoid is deenergized, the low-pressure port is closed and the high-pressure port is opened. The high-side pressure overcomes the bulb pressure and assists the spring pressure, closing the expansion valve.

The *three-way hot gas valve* is widely used in applications of the type shown in Figure 10-13. During the normal refrigeration cycle the suction gas returns to the compressor through the valve in a deenergized position. During the defrost, the flow of the

Figure 10-12 Diagram of piping using pilot-operated solenoid valve. (By permission of Sporlan Valve Company.)

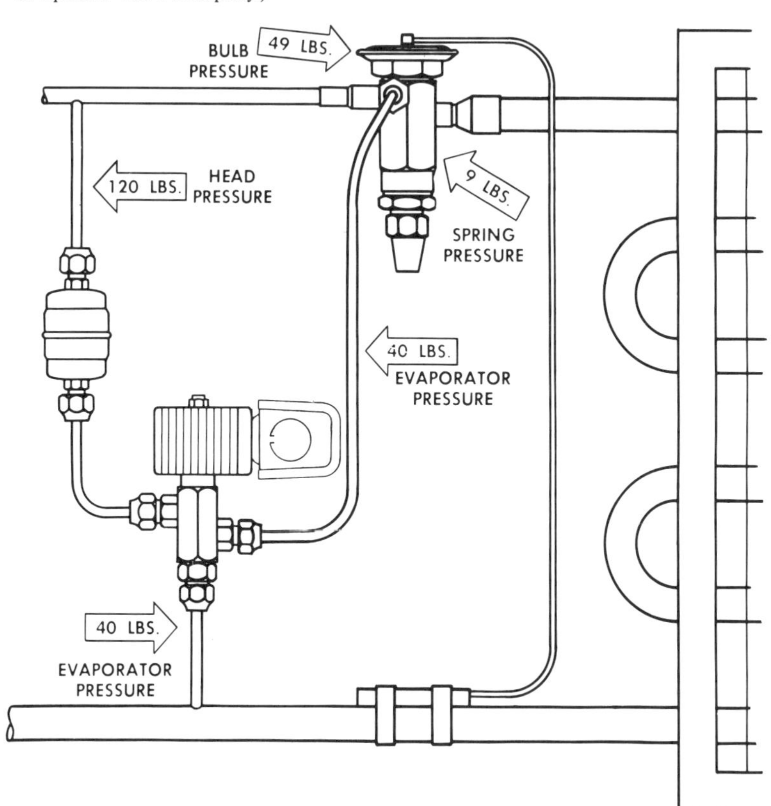

Figure 10-13 Diagram of piping using three-way solenoid valve to activate defrost cycle. (By permission of Sporlan Valve Company.)

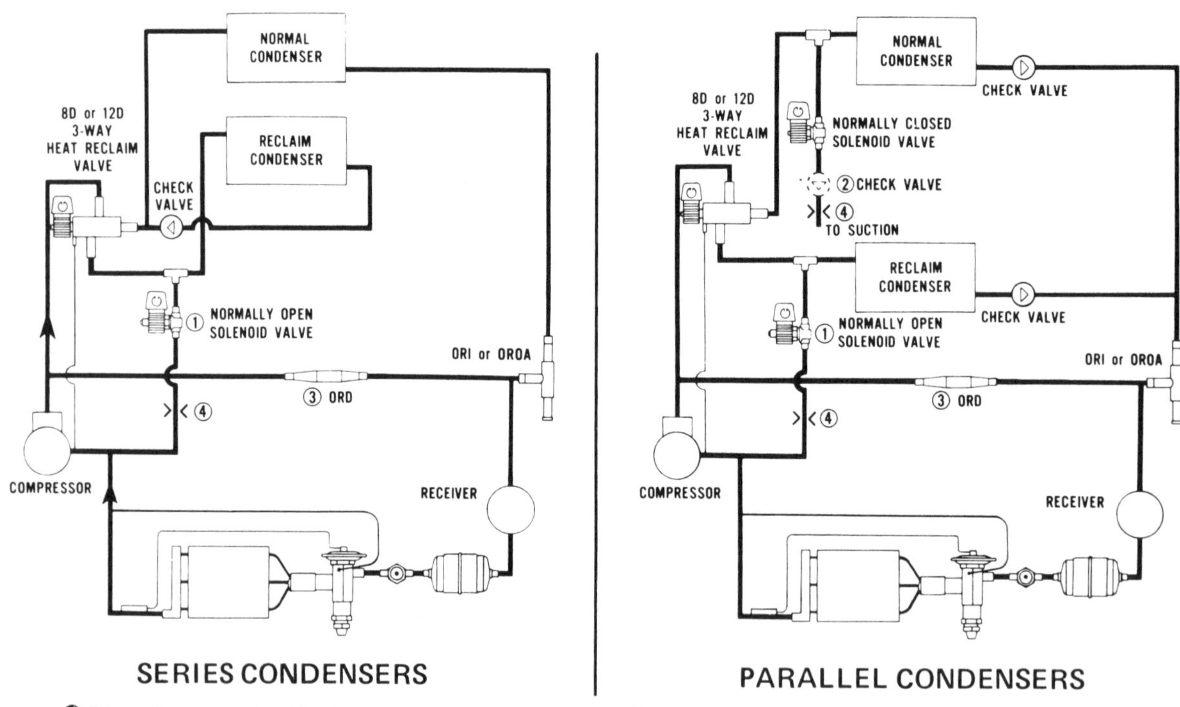

① Normally open solenoid valve used with "C" type only, see ④. Omit on systems using "B" type heat reclaim valves.
② This check valve required if lowest operating ambient temperature is lower than evaporator temperature.
③ Not used with OROA (B, C or D).
④ Restrictor may be required to control pump out rate on inactive condenser.

Figure 10-14 Diagram of piping of head pressure control, for heat-reclaim system. (By permission of Sporlan Valve Company.)

refrigerant is reversed and the valve is energized. The hot gas flows through the valve to the evaporator to defrost the coil.

Another use of the same valve is in reclaiming heat from the hot gas. See the typical piping schematics in Figure 10–14 for both series and parallel condensers.

When using the heat-reclaim arrangement, head pressure controls should be included not only to maintain liquid pressure at the expansion valve, but also to assure the availability of hot gas at the reclaim heat exchanger.

Industrial solenoid valves are used to control air, water, steam, or light oil. They can be used to control steam up to 10 psig (240 °F). Stainless steel seats are used in place of the standard brass seats in areas where water is extremely hard. Typical capacities are shown in Figure 10–15.

Figure 10–15 Typical capacities for industrial solenoid valves. (By permission of Sporlan Valve Company.)

WATER and AIR CAPACITIES

TYPE	Water – GPM					*Air – CFMFA		Coil
	1 psi △ P	3 psi △ P	5 psi △ P	10 psi △ P	20 psi △ P	5 psi △ P	10 psi △ P	
R183P1 BR183P1 KR183P1 KBR183P1	3.6	6.2	8.1	11.4	16.1	32.5	47.4	MKC-1
R184P1 BR184P1 KR184P1 KBR184P1	4.1	7.1	9.2	13.0	18.3	37	54	MKC-1
R246P1 BR246P1 KR246P1 KBR246P1	5.8	10.0	12.9	18.3	25.8	52.3	76.4	MKC-1

STEAM CAPACITIES

TYPE	*Pounds Per Hour		Coil
	5 psi △ P	10 psi △ P	
R183P1 KR183P1	105	152	HMKC-1
R184P1 KR184P1	119	173	HMKC-1
R246P1 KR246P1	168	245	HMKC-1

For steam applications use only Teflon diaphragm types.
*Exhaust to atmosphere.

225°F. Maximum Fluid Temperature, Buna-N Diaphragm
240°F. Maximum Fluid Temperature, Teflon Diaphragm

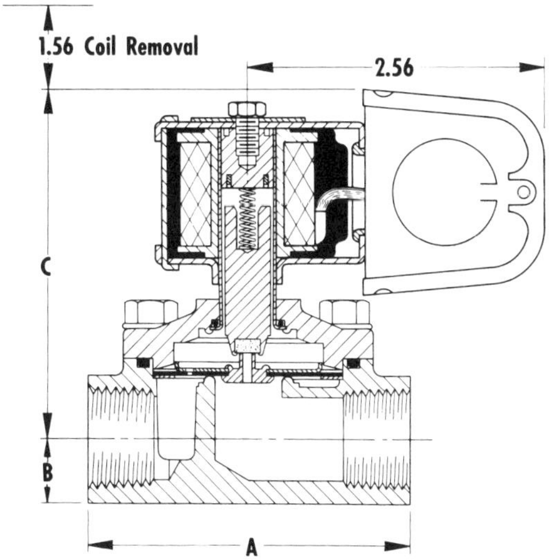

1.56 Coil Removal
2.56
C
B
A

DIMENSIONS — Inches

TYPE	CONNEC-TIONS (Pipe)	A	B	C
R183P1 BR183P1 KR183P1 KBR183P1	3/8	2.75	.530	2.97
R184P1 BR184P1 KR184P1 KBR184P1	1/2	2.75	.530	2.97
R246P1 BR246P1 KR246P1 KBR246P1	3/4	3.06	.570	3.09

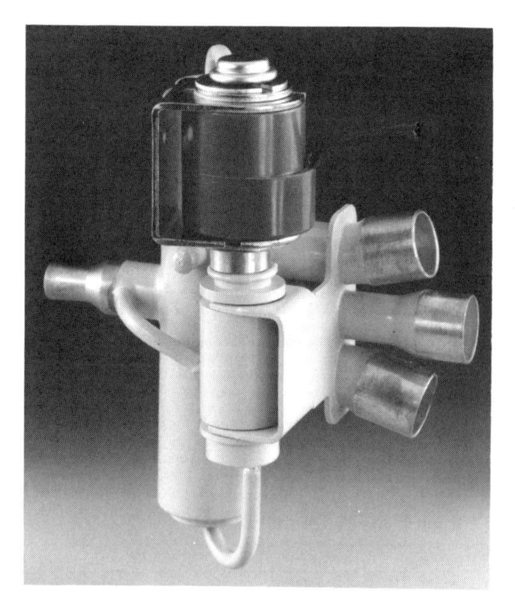

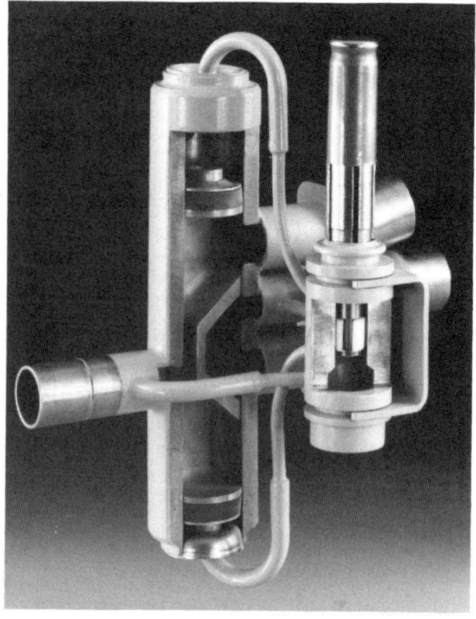

Figure 10–16 Reversing solenoid valve for heat pumps. (By permission of Alco Valve Division of Emerson Electric Co.)

Solenoid reversing valves are used for heat pumps. This type of valve has four piping connections. Two connections go to the compressor and the other two connect to the inlets to the evaporator and condenser, respectively. Usually, in the normal deenergizing position, the compressor discharge goes to the outdoor coil and the indoor coil outlet goes to the suction side of the compressor. When the valve is energized, the connections are reversed: the compressor discharge goes to the indoor coil and the outlet of the outdoor coil goes to the suction side of the compressor. This is illustrated in Figure 10–16.

The available *solenoid coil ratings* are usually 24 V/50–60 Hz, 120 V/50–60 Hz, and 208–240 V/50–60 Hz. Usually, the coils can be selected separately from the valve to match the power requirement.

Special care must be exercised in installing sweat-type solenoid valves to prevent overheating. The electrical coil and moving parts should be removed during the brazing operation. Direct flame should not touch the valve body.

Transformers

The transformer is another power-consuming device. The primary is the input side. The secondary is the output side. The symbol for the transformer is shown in Figure 10–17. The symbol also has a notation to show the voltage for both the primary and the secondary connections. Transformers may be used to reduce (step down) the voltage or increase (step up) the voltage.

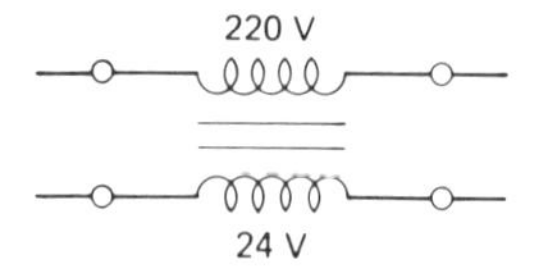

Figure 10–17 Transformer symbol. The transformer symbol usually shows both the primary and secondary voltages.

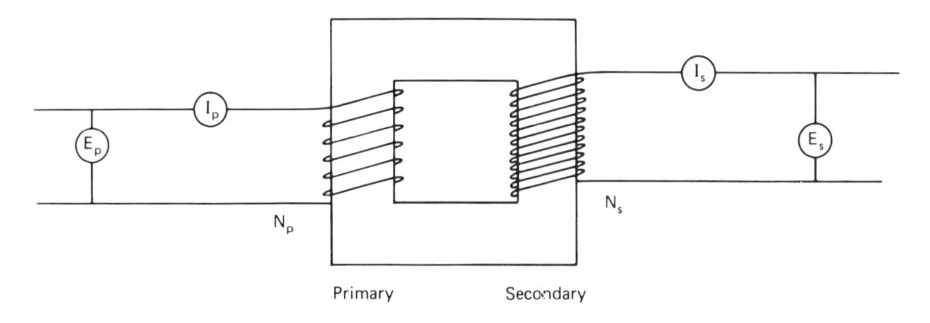

Figure 10–18 Schematic diagram of a transformer.

A transformer is a type of load because it uses current, or rather, consumes power. The power loss is evidenced by heat. A simple diagram is shown in Figure 10–18.

The primary of a transformer is a coil of wire, usually insulated copper wire. The coil of wire offers resistance to the flow of electricity. When the current flows through the coil, it produces a magnetic field. The secondary coil is placed in close proximity to the primary coil. The field of the two coils are connected by a *core.* The core, usually iron, forms a coupling between the two fields and permits power to be transferred from the primary coil to the secondary coil. What is transferred is actually watts of power.

These interesting relationships exist:

$$\frac{E_{in}}{E_{out}} = \frac{N_{primary}}{N_{secondary}} \tag{1}$$

$$\frac{I_{out}}{I_{in}} = \frac{N_{primary}}{N_{secondary}} \tag{2}$$

$$\frac{I_{out}}{I_{in}} = \frac{E_{in}}{E_{out}} \tag{3}$$

where:

E_{in} = voltage input
E_{out} = voltage output
I_{in} = amperes input
I_{out} = amperes output
$N_{primary}$ = number turns of wire on the primary
$N_{secondary}$ = number turns of wire on the secondary

Referring to equation (3) and cross-multiplying terms gives

$$EI_{in} = EI_{out} \tag{3a}$$

or

$$\text{volt-amperes input} = \text{volt-amperes output}$$

Transformers are rated in volt-amperes and must be of sufficient size to match the load.

Figure 10–19 illustrates a general-purpose control transformer. The ratings of various sizes are shown in Figure 10–20. The load on the secondary is the controlling factor in the selection and must not exceed the volt-amperes marked on the transformer.

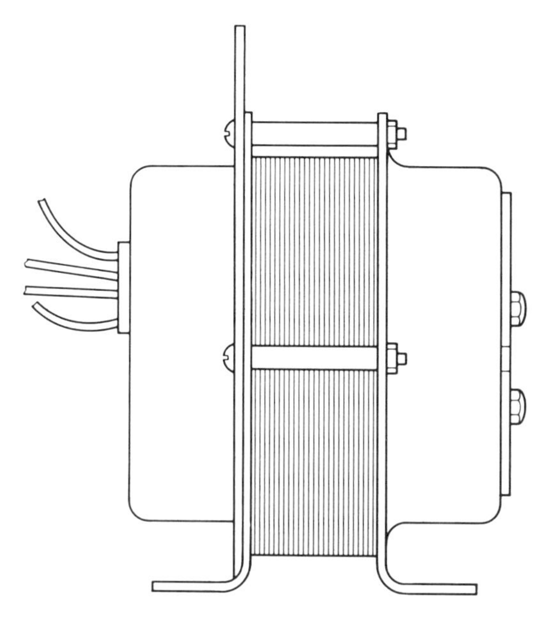

Figure 10–19 Typical control transformer.

Order number	Primary Voltage (50/60 Hz)	Primary Connections	Secondary Voltage (open circuit)	Secondary Connections	Output (100 % power factor)	Mounting	Overload protection
AT87A1007	120	9 in. (228.6 mm) leads	26.5	2 quick-connect terminals.	48 VA	4×4 in. plate	Fusible link in primary
AT87A1056	120/208/240	13 in. (330.2 mm) leads	28.0	13 in. (330.2 mm) leads.	48 VA	Foot-mounted	3.2Amp fuse in secondary
AT87A1106[a]	120/208/240		28.0		50 VA	Foot-mounted or 4×4 in. plate	Built-in protection
AT87A1155	480	12 in. (304.8 mm) leads	26.5	12 in. (304.8 mm) leads.	48 VA	Foot-mounted	Fusible link in primary
AT87A1189	277		26.5		48 VA		
AT88A1005	120		27.0		75 VA		3.2-Amp fuse in secondary
AT88A1021	208/240[b]		27.0		75 VA		Fusible link in primary
AT88A1047	480		27.0		75 VA		3.2-Amp fuse in secondary

[a]Super Tradeline model.

[b]60 Hz.

Figure 10–20 Selection chart for control circuit transformers. (By permission of Honeywell, Inc.)

Lights

The symbol for light is shown in Figure 10–21. The notation in the center of the radiating circle represents the color of the light. Lights are often used in wiring to signal when some power circuit is operating where the actual device being run cannot be directly observed.

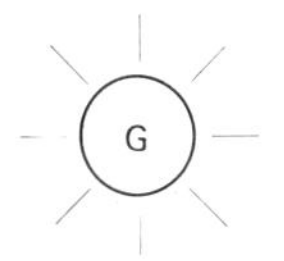

Figure 10–21 Light symbol. Signal lights are used to indicate visually the condition of a circuit. The letter in the symbol shows the color of the light: G, green; R, red; W, white.

Heaters

Another load that often appears in wiring diagrams is the heater. Heaters are used for a number of purposes, such as for a compressor crankcase, as a defrost for evaporators, or for auxiliary heat for a heat pump. The symbol for a heater is a zigzag line with terminals at both ends to indicate the connection to power.

SWITCHES

Switches are used to activate and deactivate loads. Common switch devices are:

1. Pressure controls
2. Thermostats or temperature controls
3. Time clocks
4. Oil lubrication switches
5. Fuses and circuit breakers
6. Overloads

Switching Actions

There are a number of common types of switch actions, as follows:

1. Single pole, single throw (SPST)
2. Single pole, double throw (SPDT)
3. Double pole, single throw (DPST)
4. Double pole, double throw (DPDT)

Examples of these switches and their symbols are shown in Figure 10–22. The number of poles indicates the number of circuits through the switch. The number of throws indicates the places the current can go.

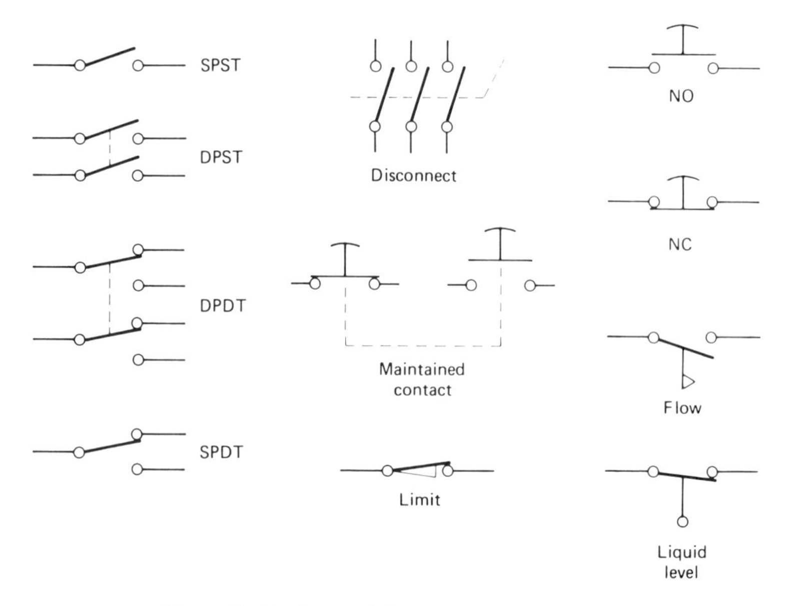

Figure 10–22 Some of the many types of switches.

There are many other types of switches, each made for a particular type of application. In the wiring diagram it may be necessary to separate multiple switches to indicate clearly how they function in the electrical system.

Pressurestats and Thermostats

The pressurestat and the thermostat are often called primary controls since, in many cases, they initiate the action of the system. They are also used as limit controls.

The symbol of the *pressurestat* serves to indicate the force that operates it. The lower part of the symbol represents a bellows or diaphragm. Referring to Figure 10–23,

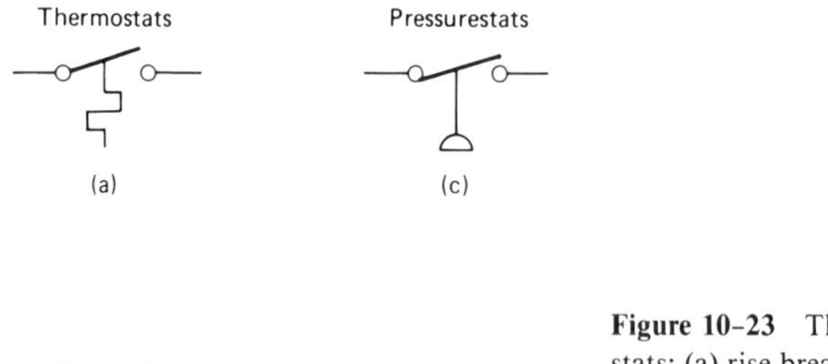

Figure 10–23 Thermostats and pressurestats: (a) rise breaks circuit; (b) rise makes circuit. These devices are primary controls. The symbols show the position of the switch. Room thermostats are usually shown normally open (NO).

the upper drawing shows a normally open pressure switch that opens on a rise in pressure. The lower drawing shows a normally closed pressurestat that opens on a drop in pressure. The exact use of these controls is shown in a wiring diagram. The normal position is usually the condition of the switch when the system is operating properly. The thermostat, however, is usually shown in an open position.

The symbol for the thermostat shows a switch connected to a zigzag line that represents a bimetallic sensor. Bimetal reacts to temperature change by a movement that operates a switch. Figure 10–23a shows an open switch that closes on a drop in temperature. Such a control could be used as a heating thermostat. Figure 10–23b shows a closed switch that opens on a drop in temperature. Such a control could be used as a cooling thermostat.

Refrigeration pressure controls. The symbol for a pressure control shows a switch connected to a bellows, representing a pressure chamber. On a rise in pressure the pressure switch shown in Figure 10–23(c) opens. This action is used on a limit control. In Figure 10–23 (d) the switch opens on a drop in pressure. This action is used on a low pressure control (LPC).

Refrigeration pressure controls may be used as either operating controls or safety controls. As operating controls they start and stop the compressor to maintain an operating pressure. A low-pressure control is often used for this purpose. If the refrigerant evaporating pressure is above the operating pressure, the compressor runs; if below the operating pressure, the compressor stops. The safety control protects the compressor against operating at excessive pressures. When the safety pressure setting is reached, the compressor is stopped.

All pressure controls are designed to function using a differential pressure. Thus, there is always a pressure difference between the cut-in and the cutout setting, or vice versa. For example, a safety control in the R-12 system could be set to cut out at 200 psig and cut in at 175 psig. The differential in this case is 25 psi. A low-pressure control could be set to cut out at 32 psig and cut in at 37 psig with a differential of 5 psi.

Some boxes, such as reach-in refrigerators, use a low-pressure control not only to control the temperature in the box but also to control the defrost cycle. These boxes usually maintain about 40 °F box temperature and have an air defrost cycle every time the compressor shuts off. This is possible since the air in the cabinet is above freezing although the coil may ice-up due to the need for an evaporating temperature below freezing. For example, a reach-in refrigerator may have an average evaporating temperature of 25 °F and an average box temperature of 40 °F.

When the low-pressure control is used as a operating control, both the differential setting and the range setting are important. The differential is the difference between the cut-in and the cutout points. The range is the actual temperature settings of the cut-in and cutout points in relation to the box temperature. For example, to maintain a 40 °F average box temperature the cut-in point could be 33 psig and the cutout point 29 psig. If the load were light, it might be advisable to change the cut-in point to 37 psi and the cutout point to 33 psig. In both cases the differential is 4 psi. If the box had a tendency to short-cycle it would be advisable to increase the range to a cutout of 30 psig and a cut-in of 37 psig, a range of 7 psi. This would provide a longer-running cycle and a fewer number of cycles per hour. Many of these control settings are experimentally determined when the cabinet is in use. Adequate time should be given between changes for the temperature to stabilize before making additional adjustments.

A typical low pressure control is shown in Figure 10–24. Note the adjustments for cutout points and differential. The cut-in point is the cutout point minus the differential. To change the temperature of the box, adjust the range. To change the length of the running cycle, adjust the differential.

The high–low pressure control is a dual-function unit. It incorporates both a high-limit and a low-limit control. The high limit opens on a rise in temperature and the low limit opens on a drop in temperature. Two piping connections need to be provided, one to the high-pressure side of the system and one to the low-pressure side of the system. A typical high–low pressure control is shown in Figure 10–25.

For small units up to about 3 hp a direct electrical connection to the motor circuit can be made. For large units the electrical connection is in the motor control circuit. This is illustrated in Chapter 11.

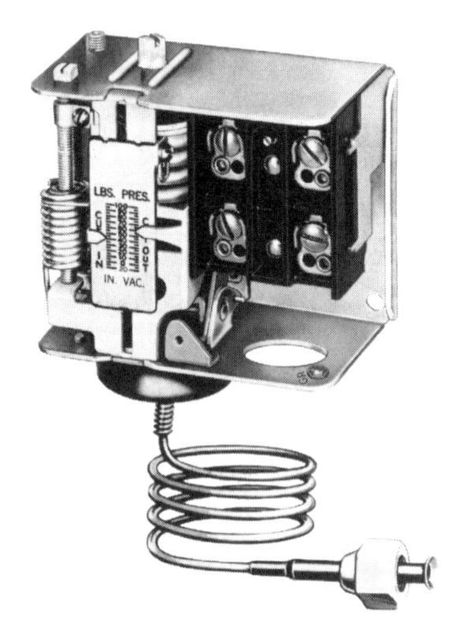

Figure 10–24 Interior view of low-pressure control (LPC). (By permission of Johnson Controls. Inc.)

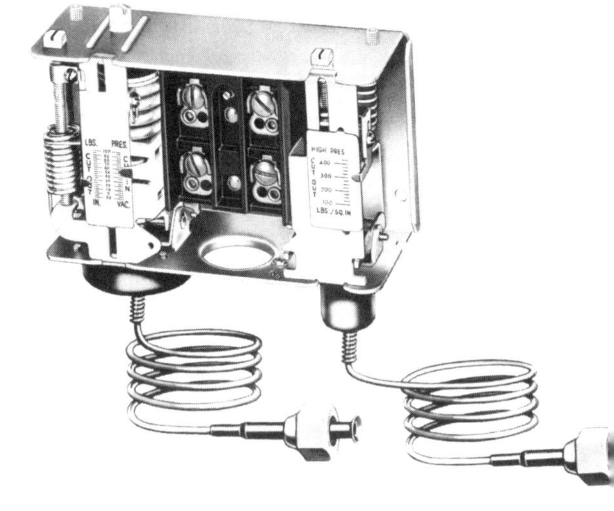

Figure 10–25 Interior view of two-pole dual pressure control. (By permission of Johnson Controls, Inc.)

Refrigeration temperature controls. Temperature controls are used in refrigeration work for a variety of purposes. They can be used to control liquid-line solenoid valves, cooling tower fans, to start and stop compressors, or for many other functions. They provide a switching action to start and stop some form of load device. They sense the temperature by means of a bimetallic element or the expansion and contraction of a fluid which initiates opening or closing a switch. The switch is usually either SPST or SPDT. They are placed in series with some type of load device that performs some useful action. For example, a room-temperature control is placed in a freezer room to maintain a temperature of 0°F. The differential is usually set at about 3°F. When the room temperature reaches 0 °F, the solenoid valve in the liquid line opens to produce refrigeration in the evaporator coil. When the room temperature reaches −3 °F, the liquid-line solenoid is closed. Generally the solenoid used is normally closed (NC); thus, when the thermostat contacts close, requiring cooling, the solenoid is energized and opens.

It is important in selecting a thermostat to consider a number of operating characteristics:

1. The voltage and the current requirements of the operating load
2. The type of switching action required: SPST or SPDT
3. The temperature range and differential characteristics required
4. The conditions of installation, such as the length of the thermostat control bulb needed, the type of mounting, the method of adjusting, and so on
5. The function of the control, such as operating a compressor, acting as a limit control, and so on

Knowing these requirements, the control can be properly selected, installed, and operated. Manufacturers' information sheets shipped with the controls are extremely helpful in properly utilizing the control devices.

Some of the common refrigeration temperature controls are:

1. The general-purpose remote bulb control
2. The defrost duration and fan delay control
3. The cooling tower or air-cooled condenser fan control
4. Freeze protection control

The *general-purpose remote bulb control* is available with either a SPST or SPDT switching action, and an automatic or manual reset. Controls are supplied with an adjustable range and an adjustable or nonadjustable differential. The differential adjustment is shown in Figure 10–26. The control is mounted to a surface through holes in the back. These controls can be used for closed-tank applications or supplied with a bulb well for installation in piping. Low- or high-limit stops are supplied on some models. They can be used as space thermostats with a range adjustment knob and an integral air bulb.

The *defrost duration and fan-delay control* is used in conjunction with a defrost initiation timer or clock on both electric and hot-gas defrost systems. This is illustrated in Figure 10–27. It allows adjustment of the defrost termination temperature without affecting the set temperature at which the circulating fan is allowed to start after the end of the defrost period. The defrost termination temperature can be adjusted within a range of 40 °F. The temperature to which the coil must drop before the circulating fan is allowed to start has a limited adjustment of 5 °F. The purpose of this control is to provide the fastest possible pull-down to operating temperatures by preventing the circulation of warm, moist air existing immediately after the end of the defrost period.

The *cooling-tower or air-cooled condenser fan control* is designed to maintain optimum head pressure by controlling the operation of the cooling-tower or condenser fans. The two-stage control is illustrated in Figure 10–28. For the cooling-tower model the control has a neoprene-coated bulb and capillary for sump water temperature con-

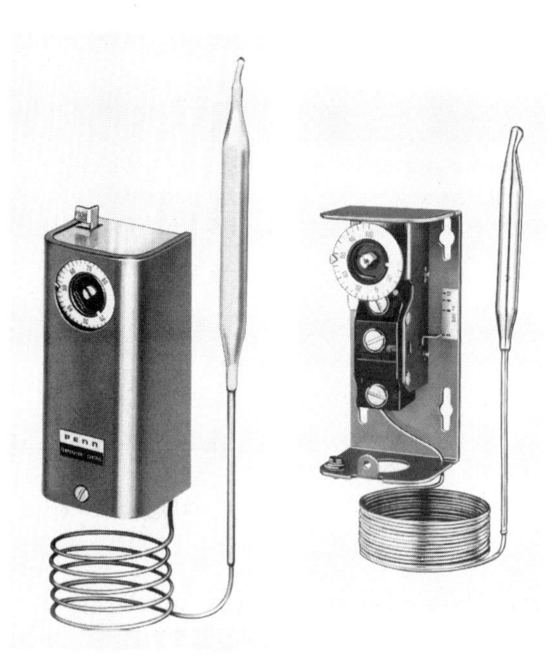

Figure 10–26 General-purpose temperature controls (SPST). (By permission of Johnon Controls, Inc.)

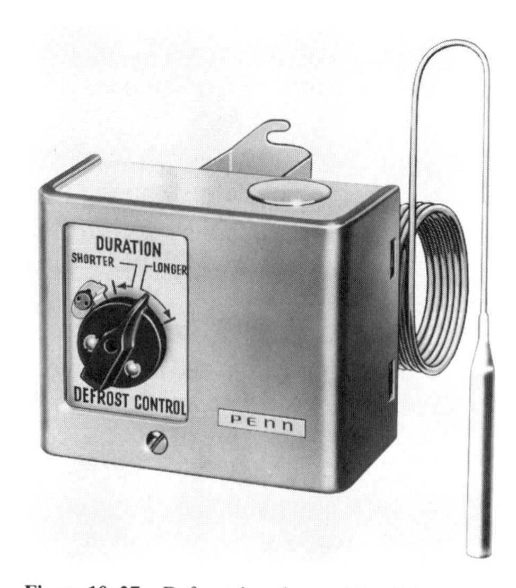

Figure 10–27 Defrost duration and fan-delay control. (By permission of Johnson Controls, Inc.)

Figure 10–28 Two-stage tower fan or condenser fan control. (By permission of Johnson Controls, Inc.)

trol. For the air-cooled condenser fan control, the bulb and capillary are tin plated for clamping to the condenser or liquid line. The two-stage models can be used to operate either a two-speed fan motor or dual fans.

The *freeze protection controls* are supplied with either SPST or SPDT switching action. Controls with lockout require manual reset. They can be secured with adjustable range and adjustable or fixed differential. The bulb can be installed directly in a closed tank or in a bulb well. Ranges of 20 to 80 °F and −30 to + 50 °F are available.

Time Clocks

Time clocks are commonly used on refrigeration systems, particularly where defrost cycles are required. One type is shown in Figure 10–29. This device uses a 24-hour clock. The timed defrost initiation is adjustable from one to eight cycles per day. A minimum of 3 hours is required between cycles. Pump-down drain or fan-delay cycles are adjustable from 0 to 30 minutes. The clock has two normally open (NO) switches and two normally closed (NC) switches.

Clocks are available that will permit time initiated–time terminated, time initiated–temperature terminated, or time initiated–pressure terminated arrangements.

When the system uses electric defrost with three-phase current, timers are available to handle 55 A heaters and one compressor contact (see Figure 10–30). These clocks usually have three NO contacts and one NC contact.

Figure 10–29 Time clock used to activate refrigeration defrost controls. (By permission of Paragon Electric Co., Division of AMF, Inc.)

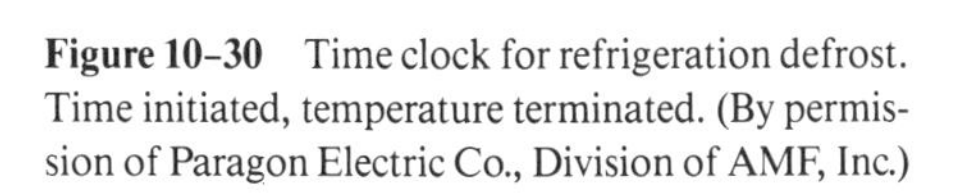

Figure 10–30 Time clock for refrigeration defrost. Time initiated, temperature terminated. (By permission of Paragon Electric Co., Division of AMF, Inc.)

Figure 10–31 Refrigeration defrost time control for multiple systems. (By permission of Paragon Electric Co., Division of AMF, Inc.)

For supermarket installations a multicircuit timer is available (see Figure 10–31). These are designed for time-termination and solenoid-initiated controls. Motors drive up to 24 program modules. They are adjustable from 1 to 12 defrost operations per module per day, with a minimum of 2 hours between cycles.

Where timers are shown in the wiring diagram, each switch is placed in the circuit it operates. The legend shows the symbol used to identify all the switches related to a specific time clock.

Oil Lubrication Safety Switches

The lubrication oil safety switch is used to protect compressors against low oil pressure delivered by the lubrication oil pump, illustrated in Figure 10–32. This is a differential control and has two pressure bellows: one is connected to the oil-pressure side of the pump and the other to crankcase pressure. The effective pressure of the oil pump is the difference between the oil pressure and the crankcase pressure. Manufacturers indicate

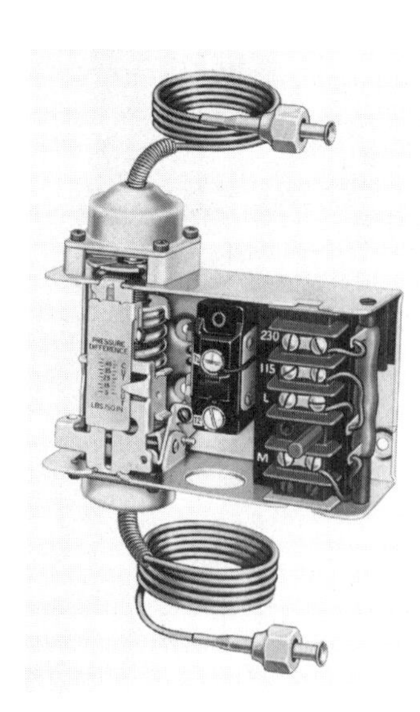

Figure 10–32 Lubrication and protection control (OPC). (By permission of Johnson Controls, Inc.)

the net pressure required to properly lubricate the compressors they manufacture. The pressure differential switch is set to turn off the compressor after initial startup if the required pressure is not reached.

A diagram of the internal wiring of the control is shown in Figure 10–33 and a diagram of the control connected to a compressor circuit is shown in Figure 10–34. The two power leads are connected to terminal 2 and terminal 240 or 120, whichever voltage is used. The compressor motor starting circuit is connected to L and M. The control switch SC_1 (NC) is in series with the holding coil in the starter. H is a time-delay heater. If H is energized for the full time-delay period, switch SC_1 opens, stopping the compressor. If the oil pressure builds up to the required amount before the end of the time-delay period, the pressure switch PC_1 will open, disconnecting the time-delay heater. Con-

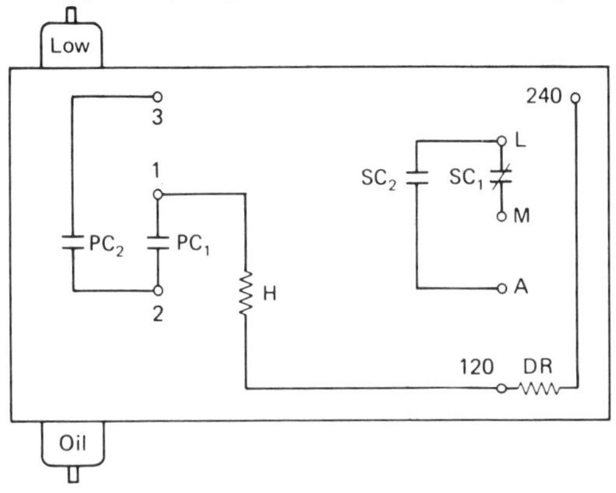

Figure 10–33 Internal wiring diagram for oil failure cutout device. (By permission of Johnson Controls, Inc.)

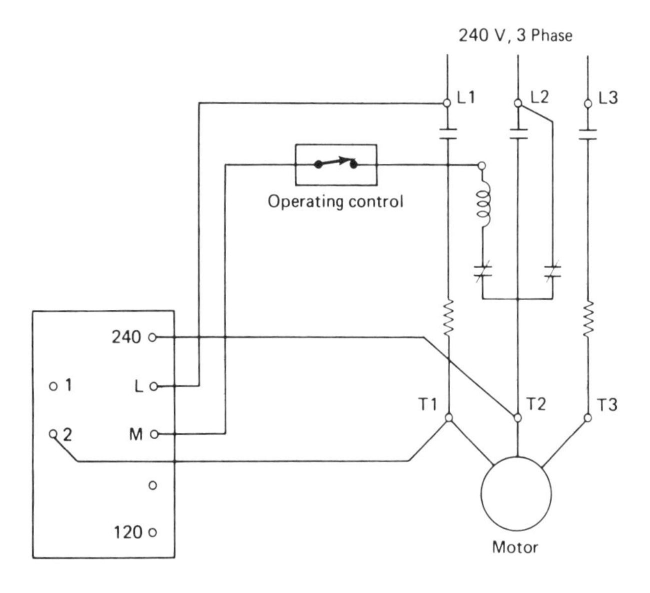

Figure 10–34 External wiring diagram for oil failure cutout. (By permission of Johnson Controls, Inc.)

trols are available with 30-, 45-, 60-, 90-, or 120-second time delays. With the time-delay heater deenergized, the switch SC_1 remains closed and the compressor continues to run.

The factory settings of the control shown are cutout 7 psig, cut-in 14 psig. These settings can be changed in the field. By turning the adjusting wheel clockwise the cutout is raised. The cutout point can be set 6 to 8 psi below the running net oil pressure.

To test the control for shutdown a jumper is placed across terminals 1 and 2, shown in Figure 10–33. The line switch on the compressor is closed. The time-delay switch will stop the compressor after the time-delay interval.

Fuses and Circuit Breakers

Both fuses and circuit breakers are used to protect electrical circuits against overloading and short circuits. The symbols used for fuses are shown in Figure 10–35. Screw-in fuses are used for circuits with a maximum ampere rating of 30 A. Above that amount, cartridge-type fuses are required. When circuits using motors are protected by fuses, a time-delay type is needed. They permit the inrush of power to start the motor (LRA), which usually occurs for a fraction of a second, but will "blow" if the running current (FLA) is exceeded for a few seconds.

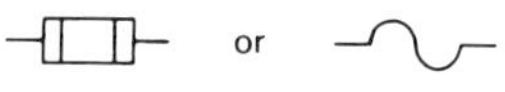

or

Figure 10–35 Fuses. The fuse is a type of overload protection. A fuse is always normally closed (NC).

An electrical fuse can easily be checked with an ohmmeter. The fuse is first removed. The leads of the ohmmeter are touched to the two metal ends of the fuse (scale R×1 is satisfactory for the ohmmeter). If the ohmmeter reads "0", the fuse is good.

All commercial installations of electrical equipment require a fused disconnect or circuit breaker to isolate the power supply to major load devices. These disconnects must be near the equipment they serve and do not replace the overloads which are supplied with motor starter equipment.

Figure 10–36 Thermal overloads. These require an increase in temperature to open the switch. In some cases the switch and the source of heat are in two separate circuits.

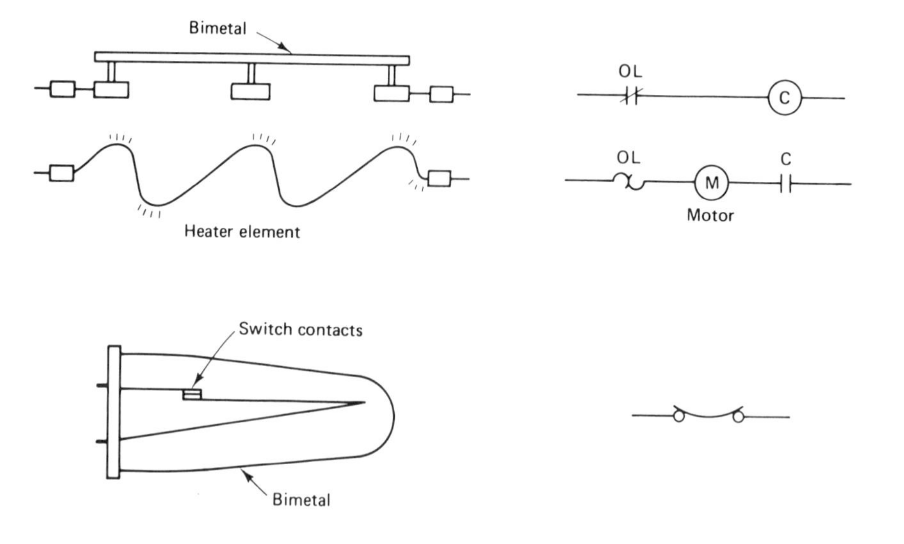

Thermal Overloads

Certain types of overloads are purely thermal switches. The increases in temperature causes a bimetallic element to warp, opening a switch. An example is shown in Figure 10-36 together with the appropriate symbol.

COMBINATION DEVICES

This equipment combines a load and a switch. Common types include: (1) switching relays, (2) contactors, (3) starters, and (4) dual circuit overloads.

When relays are shown in a wiring diagram, the coil is usually separated from the switch or switches. The legend used to identify the switches is the same as that used to identify the coil that activates them. A typical illustration of relay symbols is shown in Figure 10-37. Switches are shown normally open or normally closed, depending on the position of the switch when the coil is deenergized.

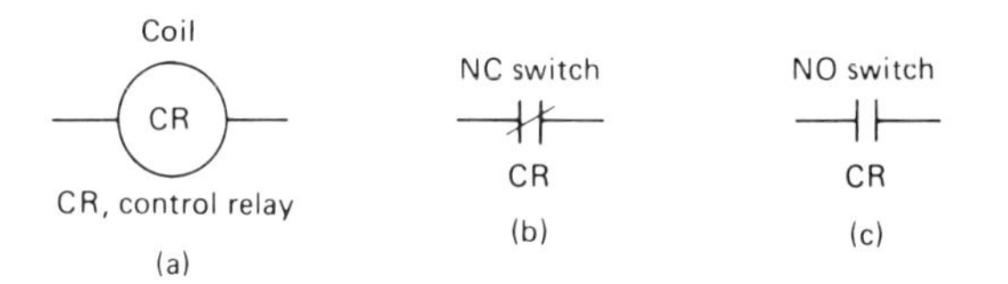

Figure 10-37 Control relay (CR): (a) coil; (b) NC switch; (c) NO switch. The control relay usually has one coil and a number of switches. In the illustration there is one NO switch and one NC switch.

Relays

A relay is a solenoid with a fixed armature and, as such, constitutes a load. Relay coils draw a relatively small amount of current and are used to perform some type of switching action. When used as contactors or starters, a small current through the control circuit can be used to connect (through suitable switches) larger loads or currents required for motors that drive compressors, pumps, or fans.

Relays can be classified in a number of ways. For purposes of discussion, they can be separated by the function they perform, as follows:

1. Switching relays
2. Contactor (relays)
3. Starter (relays)

Figure 10-38 shows a typical *switching relay.* Figure 10-39 shows some of the switching configurations available. The coil ratings are shown in Figure 10-40 and the contact

Figure 10-38 General-purpose switching relays. (By permission of Honeywell, Inc.)

SWITCHING CONFIGURATION	TERMINALS	R4222 OR R8222 MODEL SUFFIX	
		POWER RATED	PILOT DUTY RATED
Spst, N.O.	1–3	A	K
Spdt	1–3, 1–2	B	
Dpst, N.O.	1–3, 4–6	C	M
Dpdt	1–3, 1–2, 4–6, 4–5	D	N
Spst, N.C.	1–2		P
Dpst, 1-N.O. and 1-N.C.	1–3, 4–5	F	
Dpst N.O. (one power and one pilot duty)	1–3, 4–6	U[a]	
Dpdt (one power and one pilot duty)	1–3, 1–2, 4–6, 4–5	V[a]	

[a]Power rated contacts on silver-colored terminals and pilot duty rated contacts on brass-colored terminals.

Figure 10–39 Switching configuration for general-purpose relays. (By permission of Honeywell, Inc.)

COIL RATINGS:

RATED COIL VOLTAGE	MAX. PICKUP VOLTAGE	MAX. SEALED VA AT RATED VOLTAGE	MAX. INRUSH VA AT RATED VOLTAGE
24	18V	10	20
120	96V		
208/240	176V		
277	220V		
480	384V		

Figure 10–40 Coil ratings for general-purpose relays. (By permission of Honeywell, Inc.)

CONTACT RATINGS:

Power Poles (amperes per pole)—

R4222, R8222	120V ac	208/240/277V ac	480V ac
Full Load	12	6	3
Locked Rotor	60	35	18
Resistive—			
A and C models[a]	20.8	20.8	10
All others	15	15	10
Horsepower	3/4 hp	3/4 hp	3/4 hp

[a]Also rated 5 amp resistive at 600 volts.

R4228A,B; R8228A,B	120V ac	208/240V ac	277V ac	480V ac
Full Load	16 / 18	18	12	5
Locked Rotor	96 / 72	72	72	30
Resistive[a]	25	25	25	15
Horsepower	1 hp	2 hp	2 hp	1.5 hp

[a]Also rated 10 amp resistive at 600 volts.

R4228D; R8228C,D	120V ac	208/240/277V ac	480V ac
Full Load	5.5	5.5	3.0
Locked Rotor	15.0	15.0	8.0
Resistive[a]	25.0	25.0	12.5

[a]Also rated 10 amp resistive at 600 volts.

Figure 10–41 Contact ratings for general-purpose relays. (By permission of Honeywell, Inc.)

ratings are shown in Figure 10–41. These relays are used in control circuits to perform the switching action required and must be selected to operate with the power characteristics of the circuit in which they are placed. The switching configuration is usually mounted on the face of the relay where the terminals are located.

Although *contactors* are a type of relay, they are usually spoken of as "contactors" rather than as "contactor relays." Contactors are supplied with two poles, three poles, and four poles. A typical group of contactors is shown in Figure 10–42. The coil ratings for these contactors is shown in Figure 10–43. A typical wiring diagram for a two-pole contactor is shown in Figure 10–44. The contact ratings are shown in Figure 10–45. The complete model number is given in Figure 10–46.

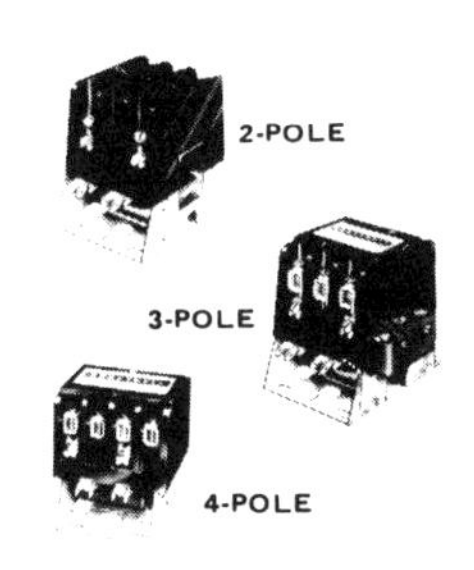

Figure 10–42 Two-, three-, and four-pole contactors. (By permission of Honeywell, Inc.)

RATED COIL VOLTAGE AT 50/60 Hz	COLOR CODE	25, 30, 40 AMP CONTACTORS			50 AMP CONTACTORS		
		SEALED VA AT RATED VOLTAGE	INRUSH VA AT RATED VOLTAGE	MAXIMUM PICKUP VOLTAGE	SEALED VA AT RATED VOLTAGE	INRUSH VA AT RATED VOLTAGE	MAXIMUM PICKUP VOLTAGE
24V	Black	9.0	90	15.6	11.5	115	18.0
120V	Red	9.6	96	90.0	11.0	110	96.0
208/240V	Green	9.6	96	156.0	12.0	120	177.0
277V	Blue	10.0	100	208.0	11.0	110	222.0
480V	Nat. Gray	10.0	100	360.0	11.0	110	384.0
600V	Brown	10.0	100	412.0	11.0	110	465.0

Figure 10–43 Coil ratings for contactors. (By permission of Honeywell, Inc.)

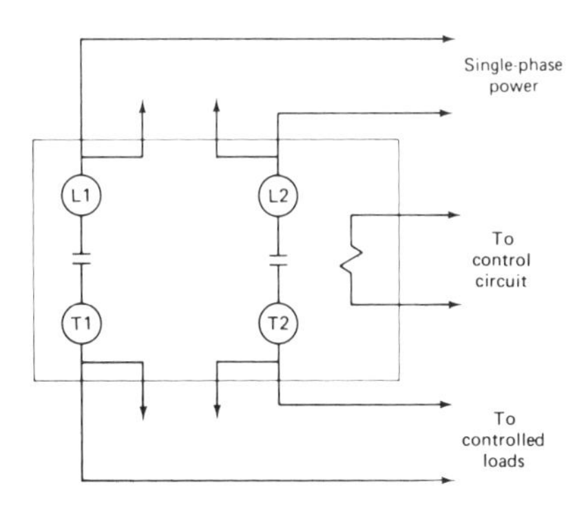

Figure 10–44 Typical wiring for two-pole contactors. (By permission of Honeywell, Inc.)

Figure 10–45 Contact ratings for two-pole contactors. (By permission of Honeywell, Inc.)

CONTACT RATINGS (amperes):
Main Poles—

NOMINAL RATING (AMPERES)	LINE VOLTS	MOTOR RATING[b] 3Ø 3P OR 1Ø 2P			RESISTIVE RATING (PER POLE)	PART WINDING START	
		FULL LOAD	LOCKED ROTOR	RECYCLE		FULL LOAD	LOCKED ROTOR
25 amp (R4210, R8210)	240	25	150	150	35[a]	25	120
	480	25	125	125	35	25	100
	600	25	100	100	35	25	80
30 amp (R4212, R8212)	240	30	180	180	40[a]	30	144
	480	30	150	150	40	30	120
	600	30	120	120	40	30	96
40 amp (R4214, R4215, R8214)	240	40	240	240	50[a]	40	190
	480	40	200	165	50	40	160
	600	40	160	160	50	40	130
50 amp (R4220, R8220)	277	50	300	300	60	50	240
	480	50	250	250	60	50	200
	600	50	200	200	60	50	160

[a] 240 resistive ratings also apply at 277V.

[b] 25 and 30 amp model ratings are per pole. Two-pole, 40 amp models have a per pole rating if used with loads of 240V or less and locked rotor rating of 180 amp or less. All other 40 amp models have 3Ø 3P or 1Ø 2P ratings.

ORDER NUMBER	POLES	COIL VOLTAGE (50/60 Hz)	NOMINAL CONTACT RATING (AMP)	TERMINAL ASSEMBLY	INCLUDES
R4210G1002	3	208/240	25	No. 10 binding head screw: with screw access. on poles 1,2, and 3; with double quick-connect access. on poles 1 and 3.	
R4210G1036	3	120	25	No. 10 binding head screw with double quick-connect access. on poles 1 and 3.	
R4210G1325	3	120	25	No. 10 binding head screws.	
R4210P1085	4	208/240	25	No. 10 binding head screw with double quick-connect access. on poles 1 and 3.*	8 terminal screws (packed separately); 2 screws for coil terminal.
R4212A1045	3	120	30	No. 10 binding head screw with screw access.	Two 1/4 in. [6.4 mm] double quick-connects for contact terminals.
R4212G1026	3	120	30	No. 10 binding head screw with screw access. on poles 1 and 3.	Two 1/4 in. [6.4 mm] double quick-connects for contact terminals; 2 screws for coil terminal.
R4212G1323	3	120	30	Pressure lug with screw access. on poles 1 and 3.	Mounting screws, four 1/4 in. [6.4 mm] double quick-connects for contact terminals; 2 screws for coil terminals.
R4212G1331	3	208/240	30		
R4214G1024	3	208/240	40	Pressure lug with double quick-connect access. on poles 1 and 3.	
R4214G1032	3	120	40		6 terminal screws (packed separately).
R4214G1347	3	120	40	Pressure lug with screw access. on poles 1 and 3.	Mounting screws, four 1/4 in. [6.4 mm] double quick-connects for contact terminals; 2 screws for coil terminals.
R4214G1354	3	208/240	40		
R4214K1017	4	208/240	40	Pressure lug with double quick-connect access. on poles 1 and 3.	
R4214P1008	4	120	40	Pressure lug with screw access. on poles 1 and 3.	Mounting screws, four double 1/4 in. [6.4 mm] quick-connects for contact terminals, two 1/4 in. [6.4 mm] quick-connects for coil terminals; hardware to convert fourth pole from a N.O. fan pole to a N.C. auxiliary pole.
R4214P1016	4	208/240	40		
R4215G1023[a]	3	120	40	Pressure lug with quick-connect access. on poles 1 and 3.	
R4220A1003	2	120	50	Pressure lug with double quick-connect access.	
R4220A1011	2	208/240	50		
R4220B1092	3	120	50	Pressure lug with double quick-connect access. on poles 1 and 3.	Mounting screws, four 1/4 in. [6.4 mm] double quick-connects for contact terminals; 2 screws for coil terminals.
R4220B1100	3	208/240	50		
R8210A1063	2	24	25	Pressure lug with double quick-connect access. on poles 1 and 3.	
R8210G1003	3	24	25	No. 10 binding head screw with double quick-connect access. on poles 1 and 3.	
R8210G1011	3	24	25	Pressure lug with quick-connect access.	

Figure 10–46 Complete model numbers for contactors. (By permission of Honeywell, Inc.)

For example, if a contactor is selected to operate a 25-A full-load motor using a 24-V 60-Hz coil, an R8210A1063 or an R8210G1003 contactor could be selected, depending on whether two poles or three poles are desired. If single-phase two-wire supply power is available, a three-pole contactor would have the advantage of offering an extra pole for turning on an auxiliary load at the same time as the primary motor.

Starters are large relays used to start motors incorporating overload protection. Many types of starters are available and must be selected to comply not only with the load being connected but also with the local power requirements. Most starters are used with three-phase motors. A complete discussion of the types of three-phase starters is supplied in Chapter 11.

Dual-Circuit Overloads

Two types are shown: thermal (Figure 10–36) and magnetic (Figure 10–47). Appropriate symbols are illustrated for each type. The circuit controlling the motor is opened when the overload condition in the primary circuit heats the thermal switch in the control circuit.

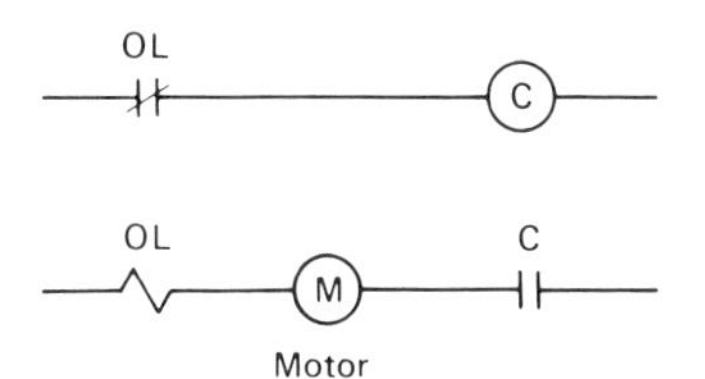

Figure 10–47 Magnetic overloads. This overload requires two circuits. If the current is excessive through the motor, a relay coil is energized and the switch opens, removing the source of power to the motor control circuit.

For the magnetic overload (Figure 10–47) the excess current in the primary circuit creates a magnetic effect that opens a switch in the control circuit.

CAPACITORS

A capacitor is an electrical device made up by alternating conductors and insulators, which permits the storage of electricity.

The unit of capacitance is the farad (F). A farad is a rather large unit and, therefore, most capacitors are rated in microfarads (μF).

The electrical symbol used to represent a capacitor is shown in Figure 10–48. The symbols "sc" denote start capacitor. The "run" capacitor comes in an oval enclosure and the "start" capacitor is in a cylindrical enclosure.

SC

Figure 10–48 Electrical symbol for capacitors used in wiring diagrams. (By permission of Component Products.)

The run capacitor has a rather low microfarad rating and can be left in the circuit continuously. The start capacitor has a comparatively high microfarad rating, generating considerable heat, and must be disconnected after the starting operation is completed. Both types of capacitors are used to increase the efficiency of single-phase ac motors.

Together with the capacitance rating, capacitors also have a voltage rating. The voltage rating is the highest electrical potential that can be in the circuit in which the capacitor is applied.

The total capacitance for two capacitors connected in series is found using the formula

$$C_T = \frac{C_1 \times C_2}{C_1 + C_2}$$

where C_T = total capacitance, μF
C_1 = rating of first capacitor, μF
C_2 = rating of second capacitor, μF

The series arrangement is occasionally useful to a service person to provide a replacement from a limited stock.

The total capacitance for two capacitors connected in parallel is found using the formula

$$C_T = C_1 + C_2$$

Thus, when space is at a premium, some manufacturers use two capacitors connected in parallel rather than a single unit.

REVIEW EXERCISE

Select the letter representing your choice of the correct answer (there may be more than one).

10-1. What notation is used on a motor nameplate to indicate clockwise rotation?

_______ a. CKW
_______ b. CR
_______ c. CW
_______ d. CKR

10-2. What is the normal speed of a six-pole single-phase (shaded-pole) motor?

_______ a. 1550 rpm
_______ b. 1050 rpm
_______ c. 1200 rpm
_______ d. 1800 rpm

10-3. What does the legend "LLS" mean, describing a solenoid valve?

_______ a. Symbol for voltage
_______ b. Name of manufacturer
_______ c. Electrical current characteristic
_______ d. Liquid-line solenoid

10-4. When selecting a liquid-line solenoid for an R-12 system, what pressure drop should be used as a basis for selection?

_______ a. 1 psi
_______ b. 2 psi
_______ c. 3 psi
_______ d. 4 psi

10-5. If a transformer has an input of 120 V and 1 A, what would the output voltage be if the output is 5 A?

_______ a. 24 V
_______ b. 120 V
_______ c. 1 V
_______ d. 5 V

10-6. If a high unit pressure control operates with a differential of 25 psi and the cutout point is 200 psi, what is the cut-in presure?

_______ a. 225 psi
_______ b. 200 psi
_______ c. 175 psi
_______ d. 150 psi

10-7. If a reach-in refrigerator operates at too high an interior temperature, which is adjusted on the LPC, the range or the differential?

_______ a. Range
_______ b. Differential

10–8. What type of switching action does a refrigeration temperature control use?

________ a. DPDT

________ b. DPST

________ c. SPST or SPDT

10–9. Liquid-line solenoid valves used in a pump-down cycle are usually selected with the following action:

________ a. NC

________ b. NO

10–10. On a time clock used for refrigeration defrost, the fan-delay cycle can be adjusted up to a maximum time of how many minutes?

________ a. 5 minutes

________ b. 10 minutes

________ c. 15 minutes

________ d. 30 minutes

11 Electric Motors and Starters

OBJECTIVES

After studying this chapter, the reader will be able to:

- Identify the various types of electrical motors and motor starters.
- Determine the operating characteristics of various types of power circuits from power wiring diagrams.
- Select the most suitable starter for an electric motor.

ALTERNATING CURRENT MOTORS

Alternating-current motors are available for either single-phase or three-phase power service and for various voltages: 120 V, 208 V, 240 V, 480 V, and so on. Motors used in the United States usually operate at 60 Hz (cycle). Motors for foreign service often operate at 50 Hz with appropriate voltage. Alternating-current motors for operation on single-phase current are available from fractional-horsepower sizes up to 10 hp. Motors for three-phase current start at ½ hp and go up in size to hundreds of horsepower or higher.

WYE AND DELTA POWER SYSTEMS

All ac electrical power in the United States is produced by three-phase generator. The availability of different voltages and phases is dependent on the type of transformer supplied by the power company at the point of use. Diagrams of two typical transformers are shown in Figure 9–5. In areas where the use is primary single-phase, the Y-type transformer is preferred since the single-phase power can easily be distributed to each of the phases as shown. The center connection of the Y is in the neutral position. 120-V single-phase current is available from the neutral and any one of the phases. 208-V single-phase power is available using any two of the terminal connections.

When the power used is largely three phase, the delta (Δ) transformer is generally preferred. Any two terminals produce 240-V single-phase current. The neutral for 120-V single-phase current must come from a midpoint tapping, as shown in Figure 9-5.

SINGLE-PHASE MOTORS

Single-phase motors cannot take advantage of the rotating effect of three-phase current (see Figure 11-1). The rotor of a single-phase motor will stop on dead center when the opposing poles are opposite each other. Therefore, some special means needs to be used to get them started. After they are started, the induction effect of the stator cutting the magnetic line of force induced in the rotor will keep them running. Single-phase motors are classified as follows: (1) split phase, (2) capacitor start (CS), (3) capacitor start, capacitor run (CSR), (4) permanent split capacitor (PSC), (5) shaded pole, and (6) wound rotor. They differ in the amount of starting torque and the method of obtaining it.

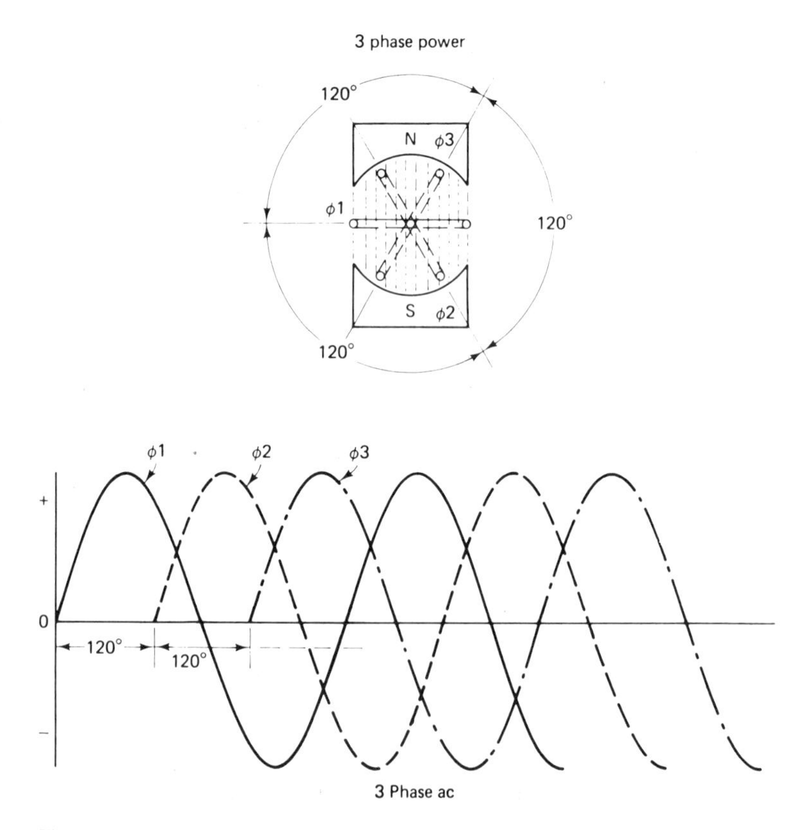

Figure 11-1 Three-phase power is produced by a generator with three sets of conductors physically spaced 120° apart.

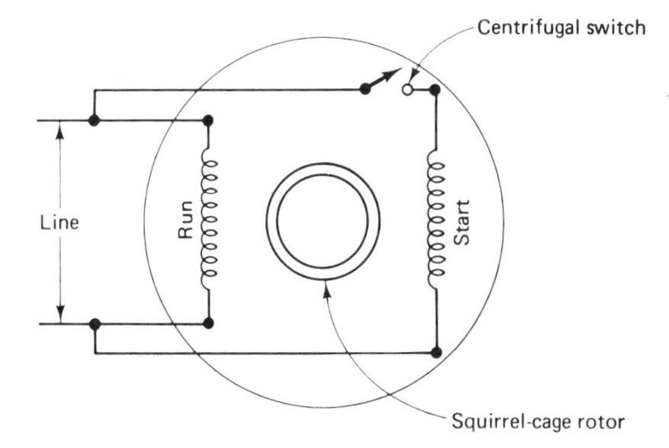

Figure 11–2 Split-phase motor. To produce a starting torque, two windings are used in this motor, connected in parallel. The running winding is heavy wire with relatively low resistance. The starting winding is smaller wire with a higher resistance. The difference in resistance causes the starting winding current to be about 30° out of phase with the running winding. The torque created starts the motor. After the initial start the starting winding is deenergized.

Split-Phase Motors

The split-phase motor has an extra set of windings called starting windings, placed midway between the poles (see Figure 11–2). This winding uses many turns of small wire compared to the field windings of larger wire and fewer turns. Due to the high resistance and low inductance of the starting winding, the current in the running winding lags behind the current going into the starting winding. This lag produces an out-of-phase relationship of about 30 electrical degrees. The starting windings are energized at the same time as the running windings and, due to the phase difference, the motor starts. At about 70 percent of full speed a centrifugal switch or a relay cuts out the start winding, and the motor runs at full speed on the run windings.

The amount of starting torque of a split-phase motor is relatively low. These motors have application for small fans, pumps, and blowers. The maximum horsepower is about ⅓ hp.

Capacitor-Start Motors

The capacitor-start motor is a split-phase motor with a start capacitor placed in series with the start winding (see Figure 11–3). The use of the capacitor causes the current to

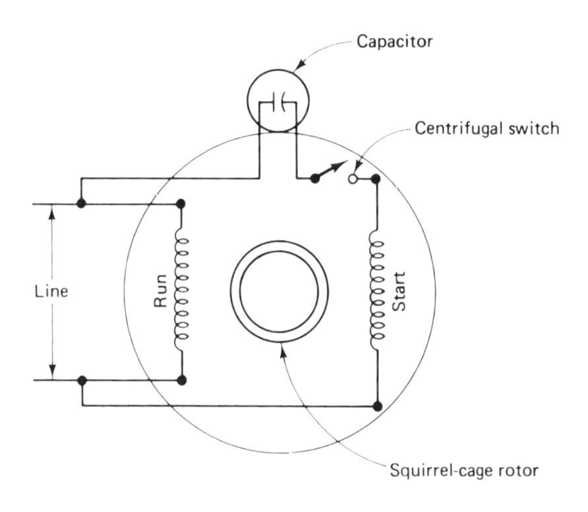

Figure 11–3 Capacitor-start motor. This motor is similar in construction to the split-phase motor except that a start capacitor is wired in series with the start winding. This causes a phase shift of about 60° and gives this motor higher starting torque than the split-phase motor.

lead the voltage in the start winding by as much as 90 electrical degrees as compared to the field pole windings. This is, in effect, a two-phase action and produces a high starting torque. These motors are often used for refrigeration compressors. The size range is usually ⅙ to about ¾ hp.

Capacitor-Start, Capacitor-Run Motors

The capacitor-start, capacitor-run (CSR) motor is similar to the capacitor-start (CS) motor except that a running capacitor has been placed parallel to the start capacitor (see Figure 11–4). This motor has excellent starting characteristics since it has two capacitors to assist the motor in starting. The running capacitor is not taken out of the circuit when the motor reaches full speed. Thus, both windings of the motor are used at full speed. The purpose of the running capacity is to correct the power factor. It has the effect of reducing the current through the start winding to operable levels. The CSR motor has high starting torque and good running efficiency. It is used for driving refrigeration compressors in sizes ranging from ½ to about 10 hp.

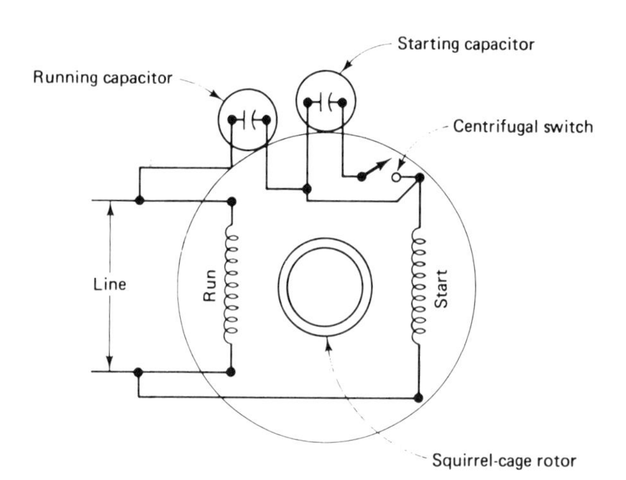

Figure 11–4 Capacitor-start, capacitor-run motor. This motor is similar to the capacitor-start motor except that a running capacitor is connected in parallel with the start capacitor and stays in the circuit after the start capacitor is removed. The running capacitor corrects the power factor. This motor has excellent starting and running torque and is used for compressor drives up to 10 hp.

Permanent Split Capacitor Motors

This motor uses a running capacitor connected between the run and start terminals of the compressor for both starting and running conditions. It does not use a start capacitor or a relay. The start winding remains in the circuit continuously since its current supply is restricted by the run capacitor (see Figure 11–5). Running capacitors have lower microfarad ratings than those of start capacitors; therefore, the starting torque of the PSC is low. It is used for fan pumps and for compressors that can be started unloaded. It has the advantage of not requiring a switch.

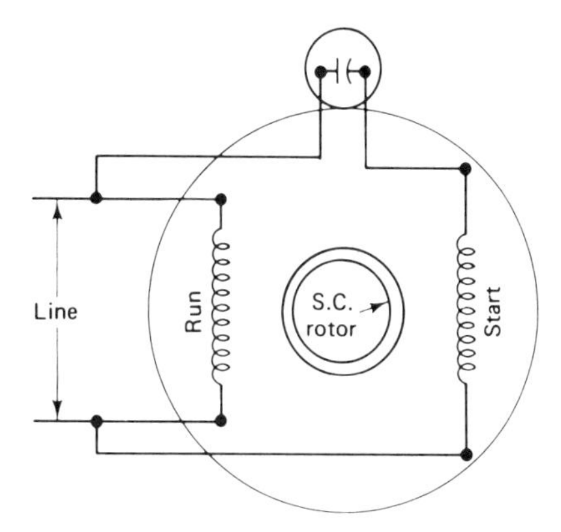

Figure 11–5 Permanent split capacitor motor. The running capacitor is connected across the two terminals of the motor similar to the CSR motor. It is left in the circuit to correct the power factor. No relay is required. This motor has excellent running characteristics but not as good starting torque as the CSR motor.

Shaded-Pole Motors

Shaded-pole motors have stator windings only on the poles. No start winding is necessary. They obtain their starting torque by the use of a copper ring which surrounds a portion of the stator pole (see Figure 11–6). This shading coil produces sufficient torque to get the motor started. The starting torque is very low. Thus, these motors are used primarily for small fans usually requiring 1/20 hp or less. The advantage of the shaded-pole motor is its simple design, providing good service at a low cost.

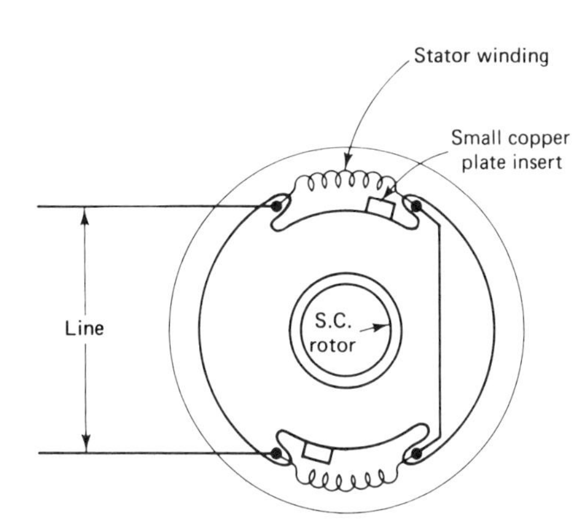

Figure 11–6 Shaded-pole motor. This motor is used for small fans and loads that require a low starting torque. It has a single winding (unlike the split-phase motor). A loop of large wire is used around a corner of each of the motor poles, which distorts the magnetic field and produces a small starting torque.

Wound-Rotor Motors

The wound rotor motor is a repulsion-start induction-run motor. The rotor windings are connected to insulated bars on the commutator. The rotor current is fed by brushes in contact with the rotating commutator. When the motor starts, the repulsion effect of the rotor magnetic field produces an exceptionally high starting torque. On some motors, at 70 or 75 percent of full speed the commutator current is disconnected and the motor runs as an induction motor. The direction of rotation can be changed by altering the position of the brushes.

This motor is expensive and subject to service requirements. It has largely been replaced with CSR motors for compressor operation.

THREE-PHASE MOTORS

Three-phase motors are of three types: (1) squirrel cage, (2) wound rotor, and (3) synchronous. Due to the nature of the power supply, three-phase motors do not require a starting winding or accessories such as relays and capacitors. They are self-starting.

Squirrel-Cage Motors

The squirrel-cage rotor is made up of a series of laminations. The magnetic field in this rotor is induced by the field windings.

When three-phase current flows through the stator windings, a rotating magnetic field is produced since each phase peaks at different times in sequence as shown in Figure 11–1. The magnetic poles of the rotor are attracted by the stator poles and attempt to follow the shift in current and polarity.

Squirrel-cage three-phase motors with laminated rotors have a starting torque between 125 and 275% of full-load torque, depending on the design. They are used for most refrigeration applications.

Wound-Rotor Motors

The wound-rotor motor has a series of coils in the rotor, the same number as the stator poles. Each of these rotor coils is connected to a slip ring. Current is delivered to the rotor windings through brushes contacting the slip rings. These windings produce a magnetic field in the rotor that reacts with the field produced in the stator, causing the motor to turn.

On a wound-rotor motor the rotor windings are short-circuited at full speed. The slip-ring design is used only to produce high starting torque.

Synchronous Motors

At full speed a synchronous motor operates without slippage. The motors have a squirrel cage winding called a *damper* which is used to start the motor. When it gets up to speed, the field windings are supplied with direct current from an *exciter.* Alternate poles are north and south. This locks the rotor into a synchronous speed.

The synchronous motor operates with a leading power factor, and this serves to compensate for the lagging power factor of induction motors. When power-factor correction is needed to improve the utility rate, the synchronous motor is a valuable asset.

MOTOR SPEED

A single-phase motor with a minimum number of poles (two) would rotate at a nearly synchronous speed of 3600 rpm (3600 rpm = 60 cycles × 60 seconds in a minute). A four-pole motor would operate at a nearly synchronous speed of 1800 rpm. A six-pole motor would operate at a nearly synchronous speed of 1200 rpm. To determine the synchronous speed of a single-phase motor, divide 3600 by half the number of poles.

The field windings in a three-phase motor are in multiples of three: one for each phase, placed in the stator 120 electrical degrees apart.

A three-phase motor with a minimum number of poles (three) would rotate at nearly synchronous speed of 3600 rpm (3600 rpm = 60 cycles × 60 seconds in a minute). A six-pole three-phase motor would operate near a synchronous speed of 1800 rpm. A nine-pole three-phase motor would operate at a nearly synchronous speed of 1200 rpm. To determine the synchronous speed, divide 3600 by one-third the number of three-phase poles.

In any ac inducation motor there is some *slippage.* The rotor does not exactly follow the rotating field. This is necessary due to the need for the rotor field to cut the lines of force of the stator field. The greater the load, the greater the slippage.

The slippage is on the order of 2.5 to 3.5%. Thus a four-pole single-phase motor is rated at a speed of 1725 rpm instead of the synchronous speed of 1800 rpm and so on.

HERMETIC COMPRESSOR MOTORS (SINGLE PHASE)

On hermetically sealed single-phase compressors it is *not* practical to place starting equipment, switches, and capacitors inside the hermetic shell. A centrifugal switch, used on an open motor, is replaced by a starting relay. There are three types of starting relays:

1. Hot-wire relay
2. Current relay
3. Potential relay

The *hot-wire relay* has two switches. One is in series with the start winding and the other is in series with the run winding. The hot wire depends on the heating effect of the starting current to warp the bimetallic element that breaks the circuit to the start winding. The normal running current keeps the switch open. If an overload occurs, the additional heat generated by the hot wire will warp the bimetallic overload and open the switch to the run winding. This relay is sized to fit the requirements of the motor it serves.

The *current relay* is shown in Figure 11–7. The coil of the relay is in series with the run winding, and the NO switch is in series with the start winding. The starting current closes the relay switch to permit the use of the start winding (and the start capacitor) to start the motor. The running current of the motor is not sufficient to hold the relay in; thus the starting winding (and the start capacitor) are relased during the run cycle.

This relay is limited in size to relatively small compressors since the full running current must pass through the relay coil.

The *potential relay* is used for the larger single-phase compressors and operates by an entirely different principle. The switch on a potential relay is normally closed. The coil is connected between the common and start terminals of the compressor. The switch is in series with the start winding and start capacitor (see Figure 11–8). When the motor turns at nearly full speed, a voltage (potential) is induced in the starting winding as high as 150% of connected voltage. This induced voltage is sufficient to energize the relay coil and open the switch. Thus the start winding and start capacitor are taken out of the circuit during full-speed operation of the compressor.

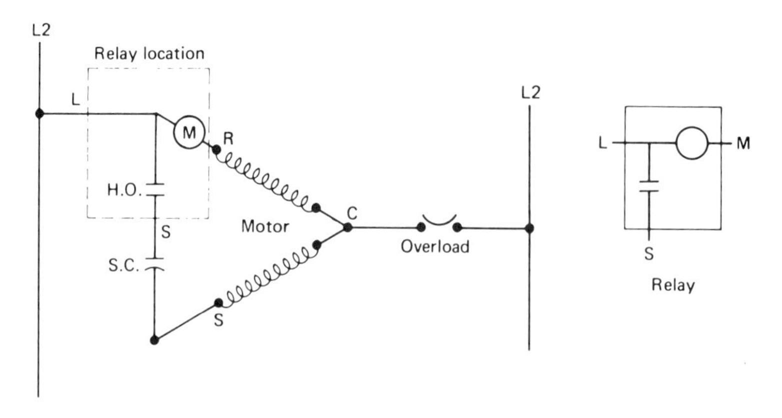

Figure 11-7 Current relay. This relay is used with relatively small motors to disconnect the start winding and, if used, the start capacitor. It is used with hermetic compressors where all switching must be outside the hermetic shell. The relay coil is in series with the run winding and the switch is normally open.

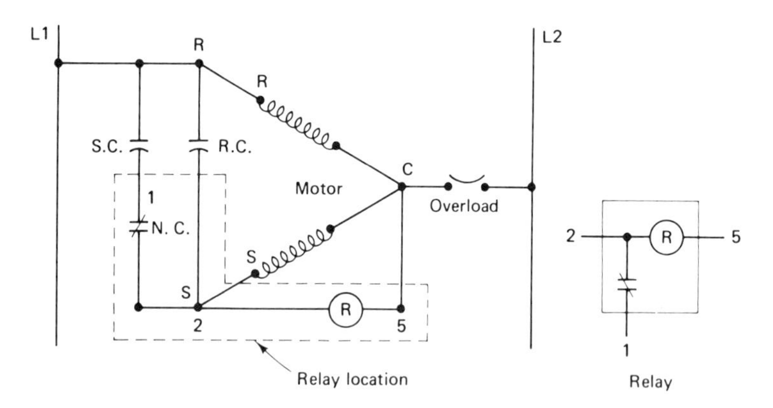

Figure 11-8 Potential relay. Used with hermetic motors, it is capable of operating with the full size range up to 10 hp, single phase. The relay coil is parallel to the start winding. Induced current energizes the coil when the motor is near full speed, opening the normally closed switch to disconnect the start capacitor. This relay is primarily used with CSR motors.

Thermal Overload Protection for Hermetic Motors

It is common practice to use a thermal overload between the common compressor connection and the source of power, as shown in Figures 11-7 and 11-8.

The overload metal enclosure usually makes contact with the compressor shell to sense excessive temperature and often is supplied with a heater wire to sense excessive current.

Other types of overloads, such as an internal thermal overload, are used by some compressor manufacturers. Although these internal overloads are extremely effective, they do require considerable time to cool to permit normal operation after the overload condition is removed.

MOTOR STARTERS

Although some small motors may be started manually by closing a power switch in series with the motor, it is general practice for medium-sized and larger motors to use a magnetic starter or contactor to perform this function. A typical magnetic across-the-line motor starter is shown in Figure 11–9. The advantage of the starter is that it permits the use of a control circuit of selected voltage and low current to start the motor. Figure 11–9 shows both a schematic and pictorial diagram of the motor starter arrangement.

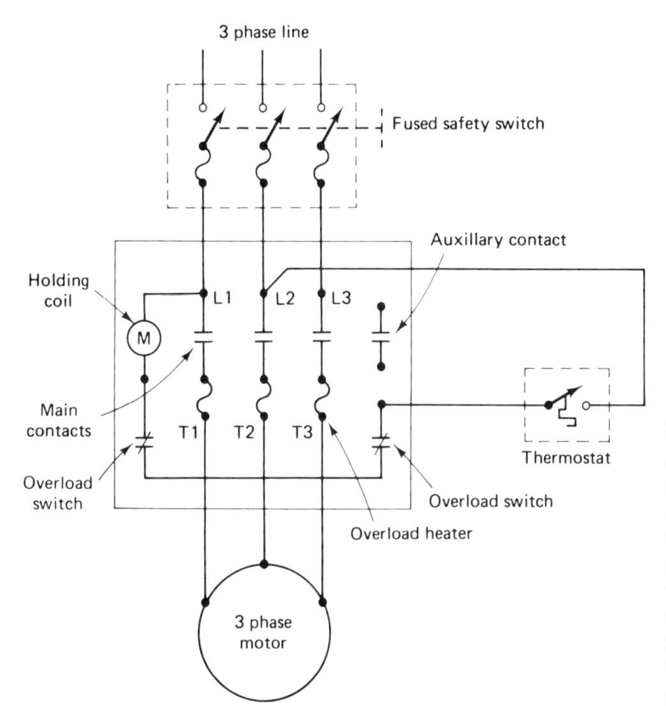

Figure 11–9 Three-phase magnetic starter. Across-the-line starters are used to connect sizable electric motors directly to the power supply by activating a magnetic relay operated by a separate control circuit. The magnetic coil can be selected to use line voltage or low voltage. The coil is placed in series with a thermostat, pressure switch, or manual switch. Thus a relatively low control circuit current can be used to start a large motor.

Where the power company will not permit across-the-line starting, reduced voltage starters must be used. There are a number of different types of three-phase reduced-voltage starters: (1) wye–delta, (2) part-winding, (3) primary resistance, and (4) autotransformer magnetic.

Wye–Delta Starters

The wye–delta starter for three-phase motors is a means of reducing the starting voltage and current to meet power source requirements. A schematic diagram of the arrangement is shown in Figure 11–10. The wye (Y) connections to the motor windings are used for starting. The electrical connections provided for starting place two windings in series across each set of terminals. This configuration reduces the starting current to 58 percent of the across-the-line value. The second step uses the delta (Δ) connections and places each winding across a set of terminals.

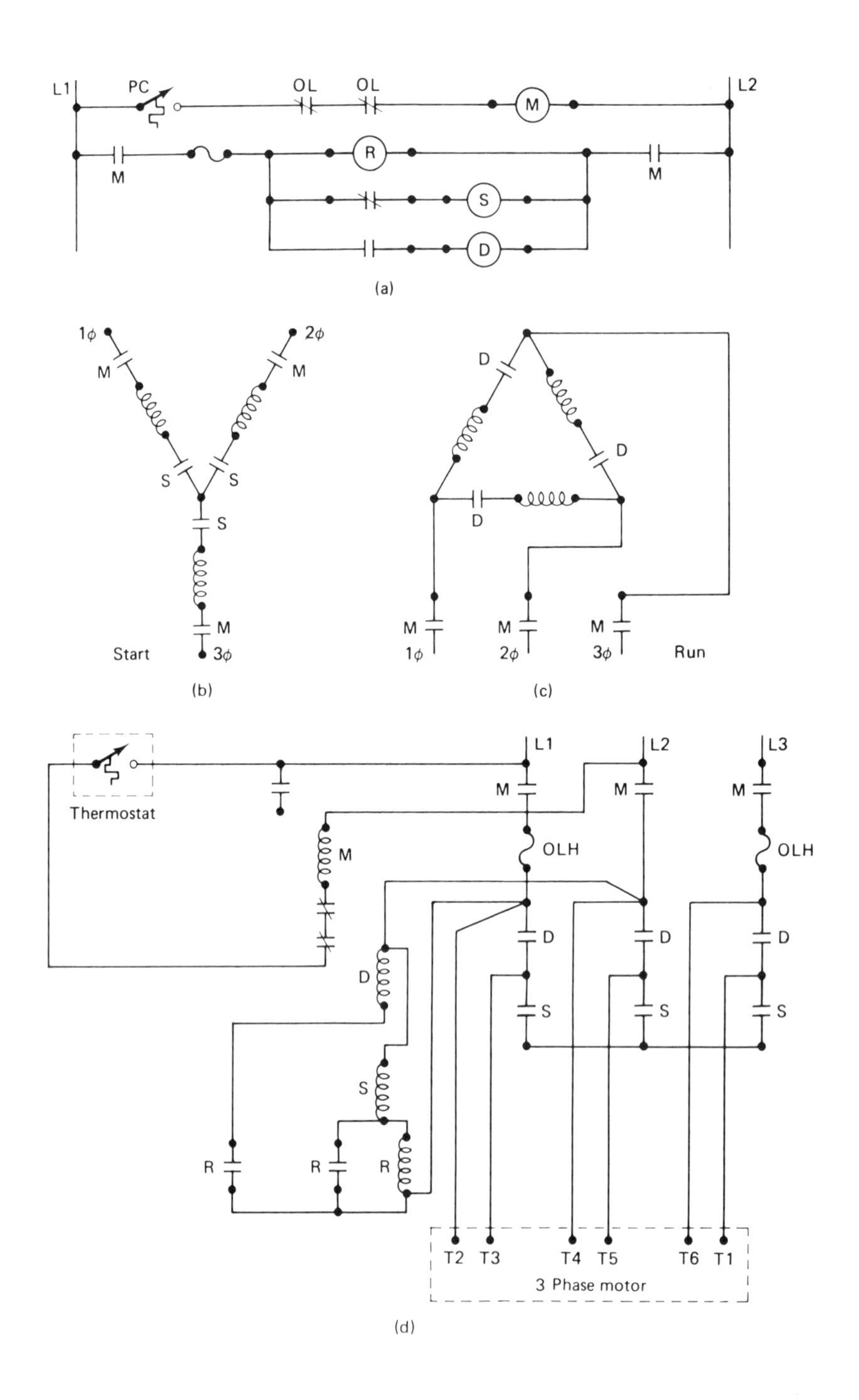

Figure 11–10 Wye–delta starter. This is a type of reduced voltage starter using a control circuit to initiate the action. The first step connects the motor windings in Y configuration. The second step places the windings on full line voltage, connecting the windings in a △ configuration. (a) Schematic control diagram, (b)starting arrangement, (c) running arrangement, (d) connection wiring diagram.

The control arrangement is shown in the upper diagram in Figure 11–10a. When the pressure control (PC) is made, the starter relay coil M is energized, closing the two M contacts (Figure 11–10a). This causes power to flow to the time-delay relay coil R and the start relay coil S. The S relay switches, together with the M relay switches, form the wye (Y) connections for starting, as shown in Figure 11–10b.

When the required time has elapsed (usually 2 or 3 seconds), the time-delay relay coil R is energized and the R switches change position. This deenergizes the S relay and energizes the D relay connecting the windings in a delta (Δ) position, across-the-line for running (Figure 11–10c). The connection diagram is shown in Figure 11–10d.

Part-Winding Starters

Of all the arrangements for reduced current starting, the part-winding type has the lowest cost. However, if this arrangement is desired, the compressor must be modified at the factory and the supplier notified of this requirement at the time the order is placed. It is not practical to add this modification in the field.

Figure 11–11a shows the connections that are provided in the motor windings. Note that 1 to 4 and 7 to 10 make up the windings on one pole; 2 to 5 and 8 to 11 make up the windings on the second pole; 3 to 6 and 9 to 12 make up the windings on the third pole. For part-winding starting 4, 5, and 6 are connected together, also, 10, 11, and 12 are connected together. Power is supplied to terminals 1, 2, and 3 by the first starter. The current required for the part winding-start is 60 to 70% of the locked rotor current. After the motor is started, power is supplied to terminals 7, 8, and 9 by the second starter. The second starter places the motor at full speed.

The control schematic is shown in Figure 11–11a and the connection diagram in Figure 11–11(b). When the pressure control (PC) closes, the holding coil of the first starter (M1) is energized. At the same time, the time-delay relay, TR, is energized. After a set lapse of time the TR relay energizes the holding coil of the second starter (M2) and the motor runs at full speed.

Primary-Resistance Starters

Figure 11–12 is a diagram of a two-step resistance starter. The resistance-type reduced-voltage starter uses resistances in series with each terminal for starting. These resistances are shunted out for full-load running.

In Figure 11–12b holding coil M closes contacts for reduced-voltage starting. After the set time delay, using relay TR, the D holding coil is energized, closing the D contacts, shunting the resistors. The amount of current reduction for starting depends on the number of resistance steps used.

Autotransformer Magnetic Starters

The use of an adjustable autotransformer in series with a motor winding makes possible regulation of the starting current on the job to comply with requirements. The autotransformer is used in place of the resistance in the resistance starter to lower the current in winding for starting.

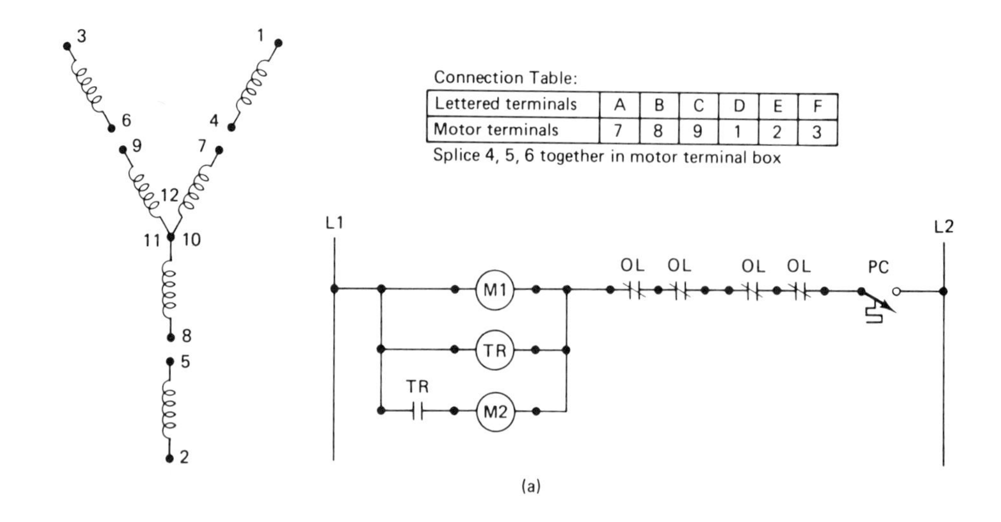

Lettered terminals	A	B	C	D	E	F
Motor terminals	7	8	9	1	2	3

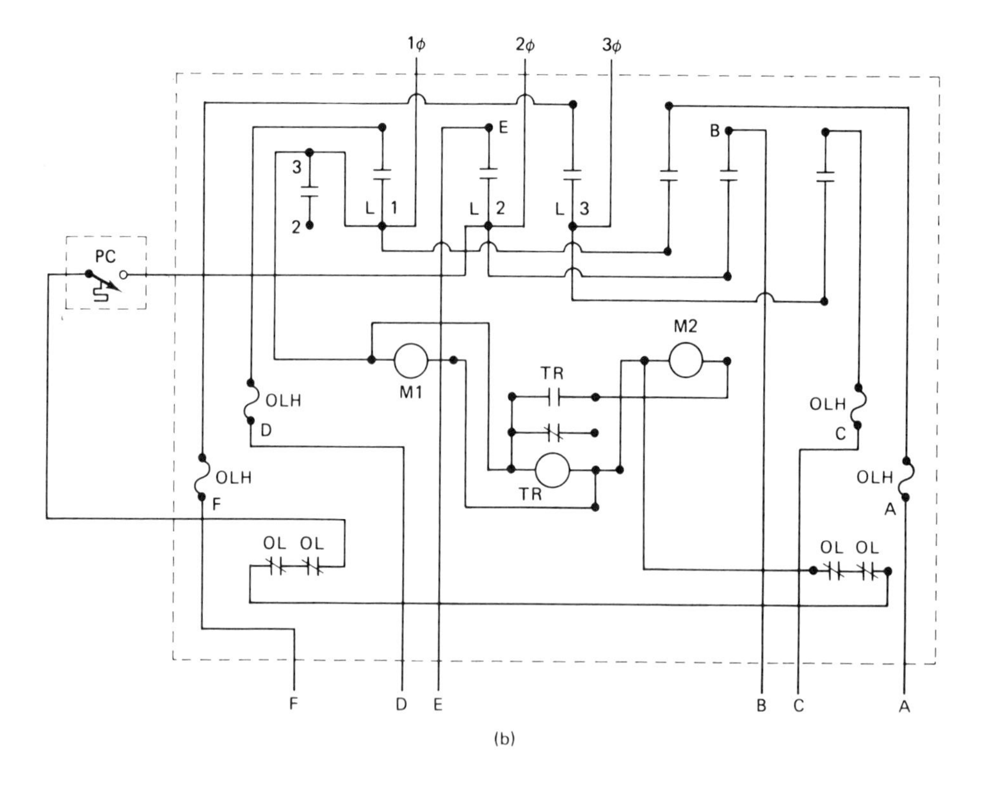

Figure 11–11 Part-winding starter. This starter is probably the least expensive of the reduced voltage starters. Half the windings are energized on the first step and all windings are energized on the second step. (a) Schematic wiring diagram, (b) connection wiring diagram.

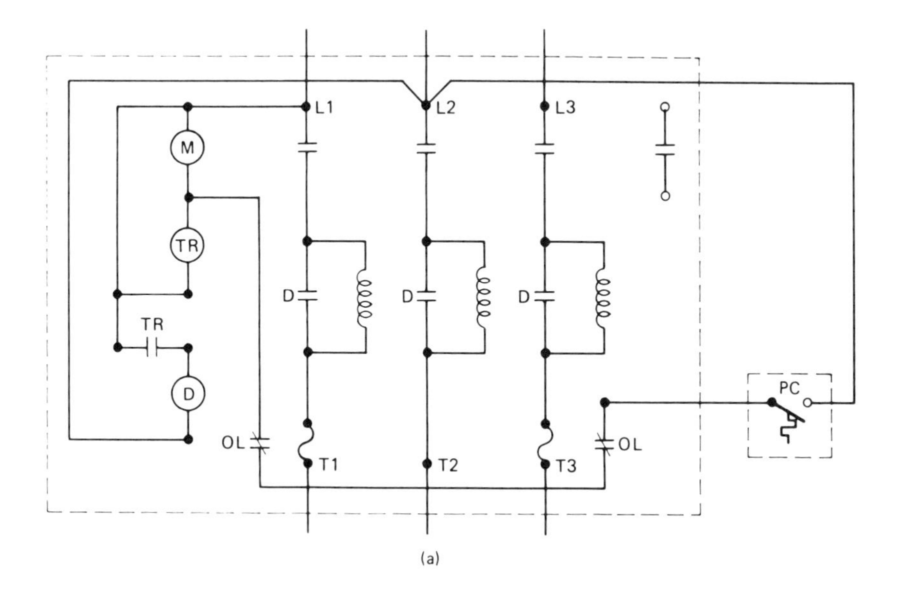

(a)

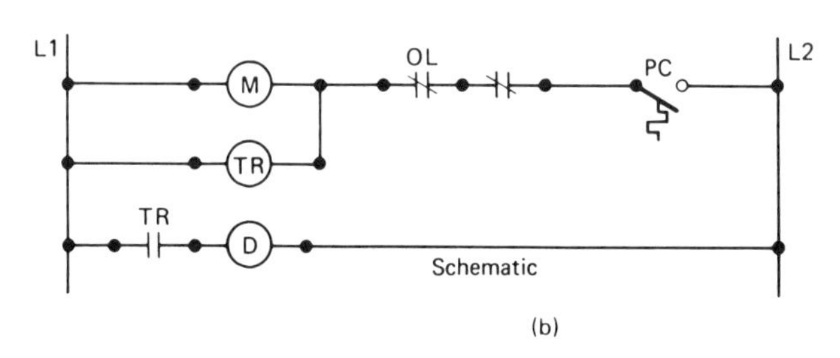

(b)

Figure 11–12 Primary resistance-type starter. This starter can use two or more steps. A resistance is placed in series with each winding to reduce the voltage in starting. Where a number of steps are used, the resistances are selected to progressively increase the voltage on the motor until in the final step the motor is placed directly across the line. (a) Connection diagram, (b) schematic control diagram.

Figure 11–13a shows a schematic diagram of the connection arrangement. Figure 11–13b shows the control diagram.

When the start button is pressed, the holding circuit to the M relay is made by closing contact M. With the pilot motor (PM) in its start position, holding coil S is energized. This applies voltage through the autotransformer and the motor starts. Power is supplied to the pilot motor (PM) and after a proper lapse of time, timer contacts TR1, TR2, and TR3 open and timer contacts TR4 and TR5 close. This deenergizes relay S and energizes relay R, putting the motor across the line for full speed. The pilot motor (PM) stops during the running cycle.

When the stop button is pressed, relays M and R are deenergized, stopping the motor. When relay R is deenergized, the contacts R in the control circuit close, energizing PM, which runs to reset all the TR contacts for restarting.

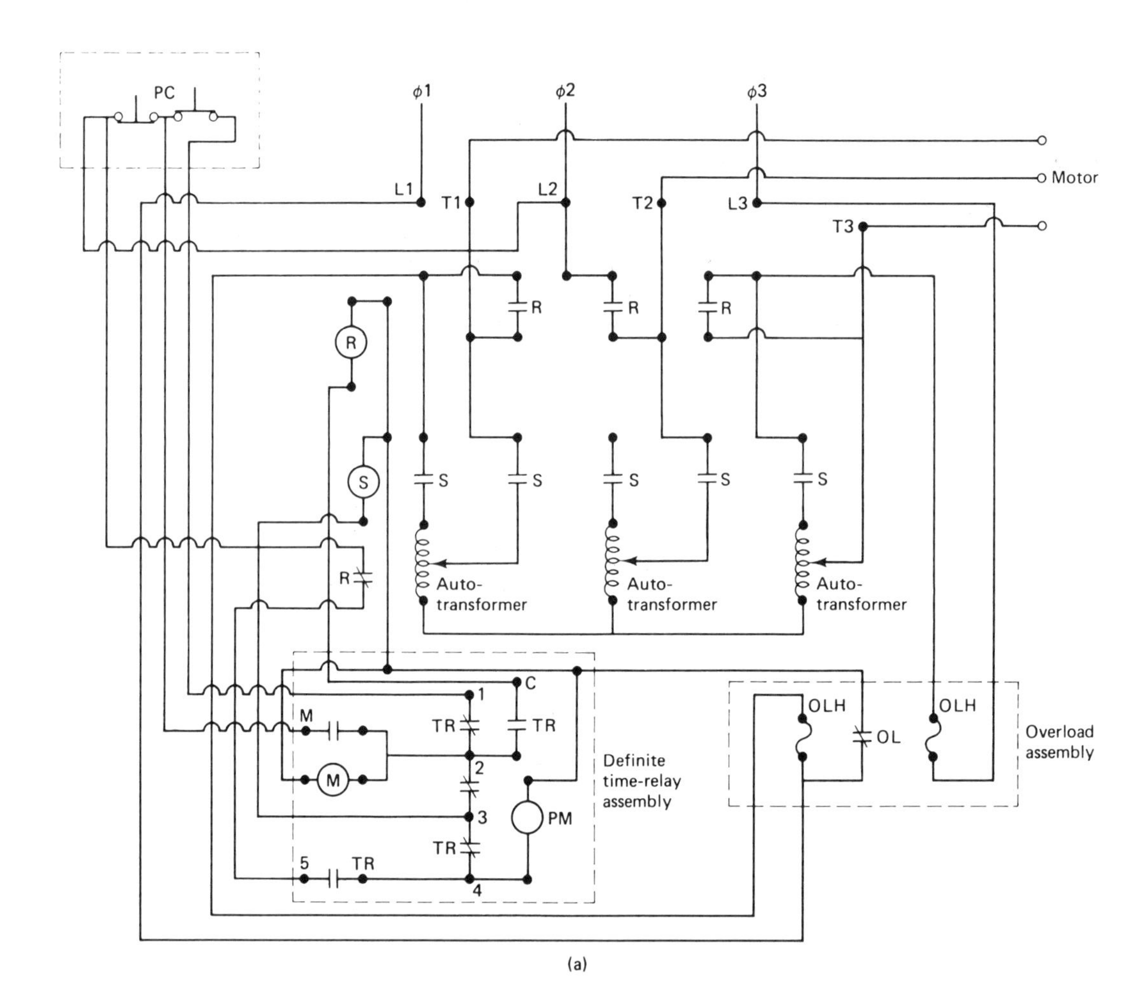

(a)

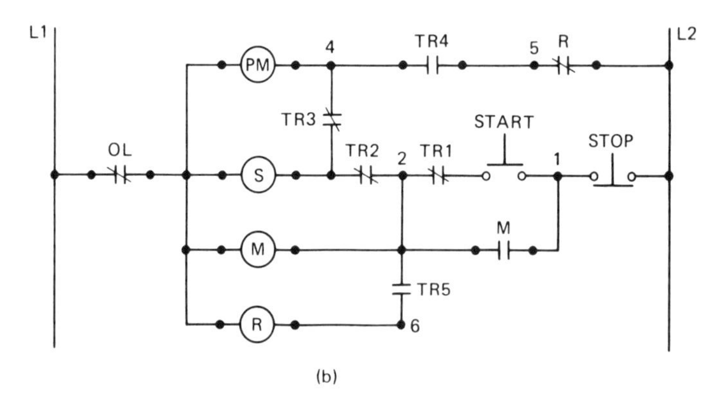

(b)

Figure 11–13 (a) Autotransformer magnetic motor starter. It is similar to the resistance-type starter except that autotransformers are used to reduce the motor voltage for starting. The control circuit is somewhat more complex, as shown in part (b). (a) Connection diagram, (b) schematic control diagram.

OIL LUBRICATION SAFETY CONTROLS

An oil lubrication safety control is an important requirement for most large compressors. They do not permit the compressor to run when there is insufficient oil pressure. The safety switch is placed in series with the holding coil on the starter. Refer to Chapter 10 for details on the operation of this control.

INTERLOCKING CIRCUITS

It is generally considered good practice to condition the compressor starting on having the evaporator fan, condenser fan, and pumps (if used) operating. This does not preclude the possibility of cycling a condenser fan to maintain head pressure. The method of being sure that these related devices are in operation is provided by an electrical interlock.

One of the best methods of providing this interlock is to use the auxiliary contacts on the starters (Figure 11–14). Assuming that all disconnect switches are closed, the system is started by closing the manual switch starting the evaporator fan. The thermostat or cycling control is then in position, on a call for cooling, to start both the compressor and the condenser fan at the same time.

Figure 11–14 Interlocking accessories to start when the main motor starts. This arrangement is often used to start cooling towers and pumps at the same time as the compressor.

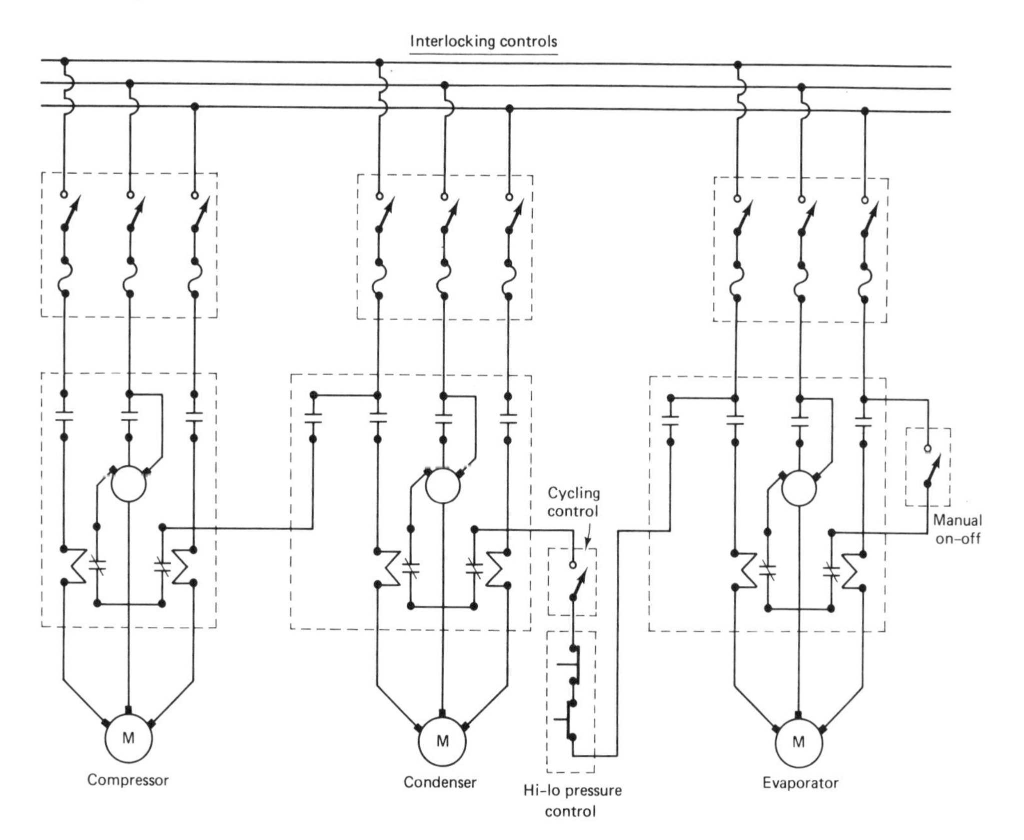

REVIEW EXERCISE

Select the letter representing your choice of the correct answer (there may be more than one).

11-1. The upper limit for the use of single-phase motors is:

________ a. 4 hp
________ b. 6 hp
________ c. 8 hp
________ d. 10 hp

11-2. The speed of a six-pole motor running on 60 Hz ac current, not including slippage, is:

________ a. 800 rpm
________ b. 1200 rpm
________ c. 1600 rpm
________ d. 2000 rpm

11-3. The maximum horsepower of a split-phase motor is about:

________ a. 1/16 hp
________ b. 1/3 hp
________ c. 1/2 hp
________ d. 1 hp

11-4. Capacitor-start motors have a range of sizes usually from:

________ a. 1/6 to 3/4 hp
________ b. 1/2 to 1 1/2 hp
________ c. 1 to 5 hp
________ d. 5 to 100 hp

11-5. A capacitor start, capacitor run motor has how many capacitors?

________ a. 0
________ b. 1
________ c. 2
________ d. 3

11-6. Which of the following motors do *not* have a start winding?

________ a. Permanent split capacitor
________ b. Split phase
________ c. Capacitor start
________ d. Shaded pole

11-7. A current relay is used to start open motors.

________ a. True
________ b. False

11-8. A potential relay has a:

________ a. Normally open switch
________ b. Normally closed switch

11-9. Which type of relay provides overload protection?

________ a. Current relay

________ b. Potential relay

________ c. Hot-wire relay

________ d. Pneumatic relay

11-10. Which type of reduced-voltage starting equipment requires modification of the basic compressor-motor electrical systems at the factory before the equipment is shipped?

________ a. Part winding

________ b. Wye–Delta

________ c. Primary resistance

________ d. Autotransformer

12 Refrigeration Equipment

OBJECTIVES

After studying this chapter, the reader will be able to:

- Identify and describe the various types of commercial, industrial, and institutional refrigeration equipment.
- Describe the type and location of the various refrigeration components used in assembled equipment.
- Determine the refrigerating capacity of various types of assembled equipment.

CLASSIFICATION OF EQUIPMENT

There are many ways to classify refrigeration equipment. One of the simpler ways is to separate self-contained units from the field-assembled units. Another division is to separate food storage equipment from strictly refrigeration processing units or industrial units. Water cooling units and ice-making equipment are a large group that can either be related to food or to some industrial process that requires cold water or ice. There are many other groupings, such as high temperature and low temperature, small equipment and large equipment, frequently used or seldom used, and so on. Generally speaking, it is the purpose of this text to discuss the more popular or more common types of units, possibly starting with the small units and going to the large ones. Where each fits in the broad industry is best described in terms of applications (see Chapter 17).

The types of equipment that will be described in this chapter include:

1. Drinking water coolers
2. Ice cube makers

3. Flake ice makers
4. Large ice-making machines
5. Tube ice makers
6. Ice builders
7. Refrigerated buildings
8. Walk-in coolers and freezers
9. Dairy product and delicatessen display cases
10. Frozen-food and ice cream display cases
11. Meat display cases
12. Produce display cases
13. Supermarket compressor systems
14. Reach-in refrigerators
15. Packaged liquid coolers
16. Large industrial liquid coolers
17. Transport refrigeration units

The first unit that will be described has universal application and is essential to health and welfare: the drinking water cooler.

DRINKING WATER COOLERS

For purposes of discussion, drinking water coolers can be divided into two groups: self-contained and remote. The self-contained units are shown in Figures 12–1 through 12–5. The two projector water stream is shown in Figure 12–6. The two parts of the remote cooler are shown in Figures 12–7 and 12–8.

Deluxe Floor Standing

Models: SCWT-4, SCWT-8, SCWT-14, SCWT-14W

Can be used freestanding or wall tight to avoid unsightly connections.

Exclusive Projector Design: 2-stream mound-building projector and automatic stream regulator.

Unit Size: 16″ W x 13¼″ D x 40″ H

Plumbing Connections: Enter directly into cabinet. Waste 1¼″ IPS. Supply ⅜″ IPS. Back panel has access window for variable pipe positioning.

Model No.	Rated GPH Cooled to 50° F	Watt Usage Per Hour	Amps	Compressor Size	Condenser Cooling*	Shpg Wt
SCWT-4	4.0	190	2.4	1/6	A	72
SCWT-8	8.0	240	3.0	1/6	A	74
SCWT-14	14.0	400	4.8	1/5	A	79
SCWT-14W	16.0	360	5.0	1/5	W	73

*A-Air; S-Static; W-Water

Pushbutton glass filler available factory-installed as accessory Model 8560. Coffee Server accessory available on Models SCWT-4, 8, and 14. Order as Models SCWT-4, 8, and 14-CB.

Figure 12–1 Floor-mounted water cooler. Can be used free standing or wall tight to avoid unsightly connections. Capacity range 4 to 16 gph with water cooled to 50°F. Wattage usage per hour ranges from 190 to 360 W. (By permission of Halsey-Taylor Division, King-Seeley Thermos Company.)

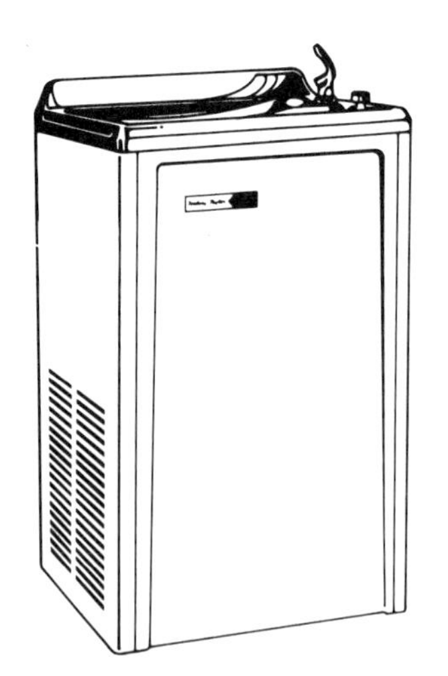

Deluxe Wall Mounted

Models:
WM-8A-1, WM-14A-1, WM-16A-1, WM-16W

Beautiful tapered cabinet is topped with stainless steel receptor with no crevices to catch dirt or cause splashing. Sturdy hanger bar mounts on any wall with all plumbing connections out of sight.

Exclusive Projector Design: 2-stream mound-building projector and automatic stream regulator.

Cabinet Apron: Heavy-gauge steel panels removable with frame still on wall.

Unit Size: 18⅛" W x 14¼" D x 28" H

Plumbing Connections: Enter directly into cabinet. Waste 1¼" OD slip connection. Supply ⅜" IPS.

Model No.	Rated GPH Cooled to 50° F	Watt Usage Per Hour	Amps	Compressor Size	Condenser Cooling*	Shpg Wt
WM-8A-1	8.0	260	3.0	1/5	A	70
WM-14A-1	14.0	400	4.8	1/5	A	75
WM-16A-1	16.0	450	5.7	1/4	A	79
WM-16W	17.0	400	4.7	1/5	W	97

*A-Air. S-Static. W-Water

Pushbutton glass filler available factory-installed as accessory Model 8560

Coffee Server available factory-installed as accessory Model WM-8A-1CB or WM-14A-1CB

Figure 12-2 Wall-mounted water cooler. Hanger bar mounts on the wall with all plumbing connections out of sight. Capacity ranges from 8 to 17 gph with water cooled to 50 °F. Wattage usage ranges from 260 to 450 W. (By permission of Halsey-Taylor Division, King-Seeley Thermos Company.)

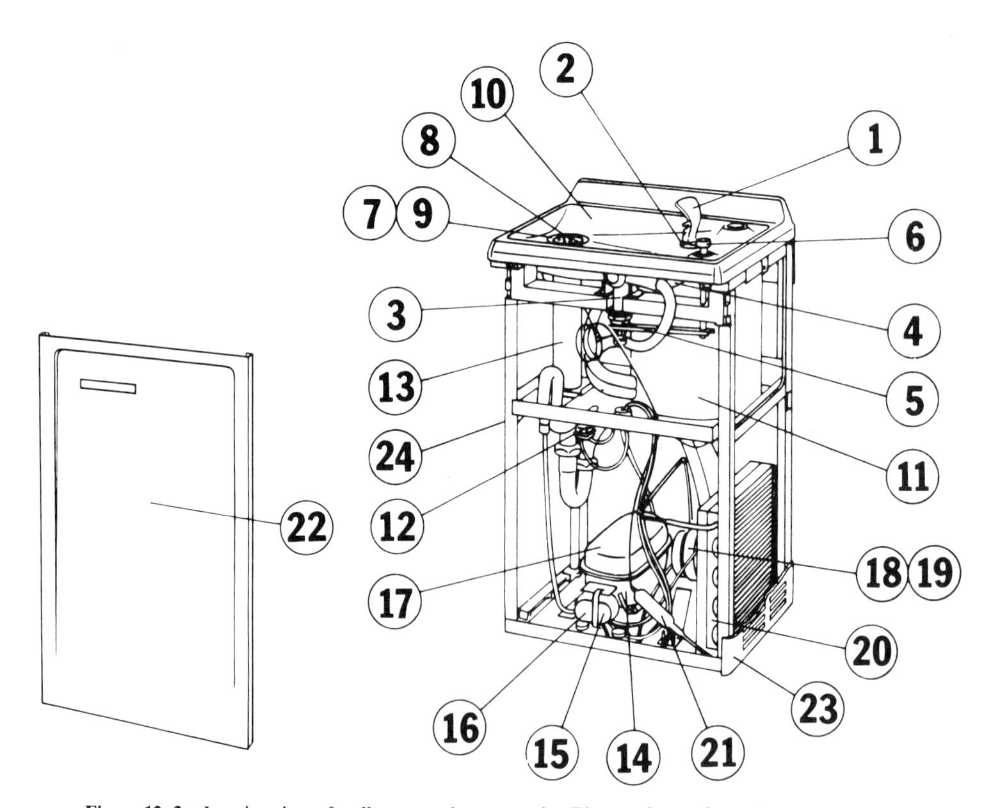

Figure 12-3 Interior view of wall-mounted water cooler. The numbers refer to the parts list description in Figure 5-4. (By permission of Halsey-Taylor Division, King-Seeley Thermos Company.)

Item no.	Description
1	Projector
2	Gasket (upper and lower)
3	Combination stop and stream regulator
4	Projector nipple and supply line assembly
5	Valve stem assembly
6	Push button assembly
7	Drain plug
8	Strainer plate
9	Waste gasket
10	Receptor (top)
11	Evaporator/chiller assembly
12	Temperature control
13	Waste line and precooler
14	Starting capacitor
15	Relay
16	Overload protector
17	Compressor (includes items 15 and 16)
18	Fan motor
19	Fan blade
20	Condenser
21	Dryer
23	Left-side panel

Figure 12–4 Parts lists for wall-mounted water cooler shown in Figure 5-3. (By permission of Halsey-Taylor Division, King-Seeley Thermos Company.)

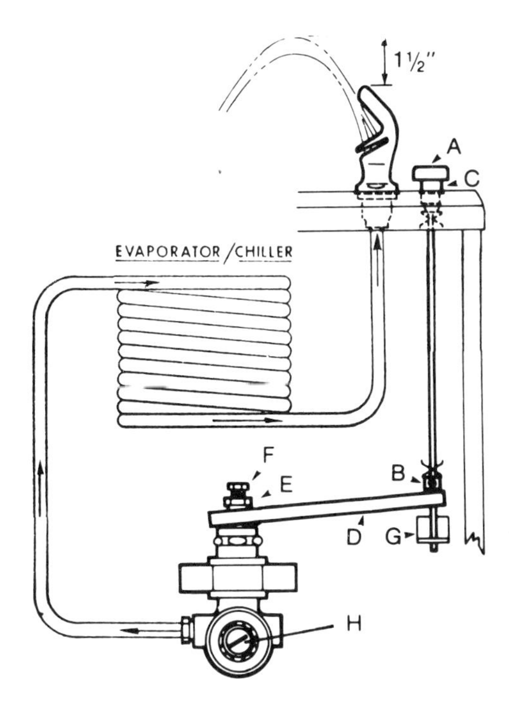

Figure 12–5 Details of a 2-Stream Projector. To adjust, loosen set screw F and locking collar B. Turn hex bushing E to right until snug against lever arm D and water does not trickle from projector. Hold cap A so that bottom of cap is in line with top of sleeve C. With locking collar B resting on lever D, tighten screw in collar securely on stem. Tighten set-screw F. Pushbutton stem must be in alignment with bracket G. Adjust screw H to regulate height of stream. (By permission of Halsey-Taylor Division, King-Seeley Thermos Company.)

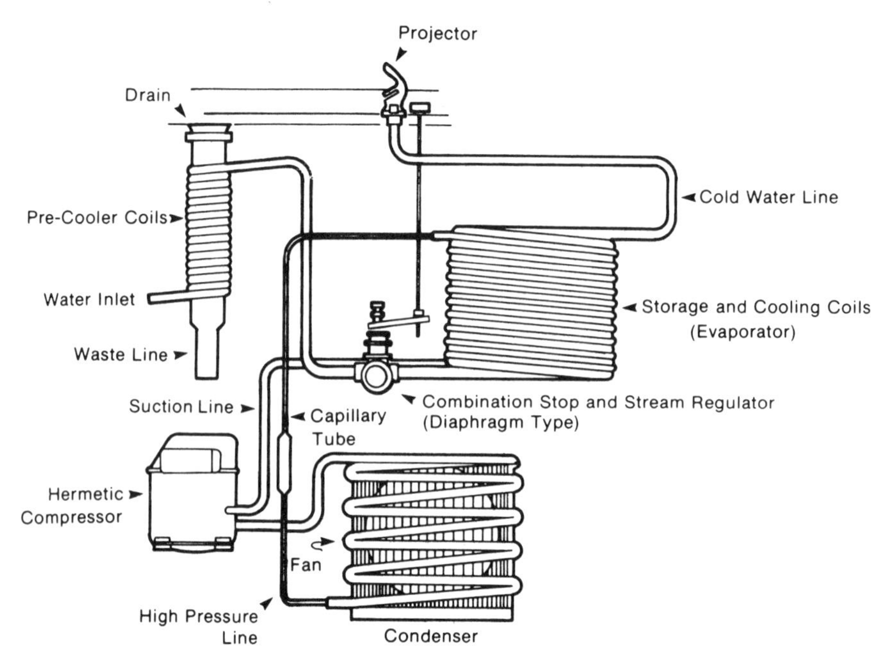

Figure 12–6 Schematic diagram of the water cooler refrigeration system. Note precooler coil and stream regulator. (By permission of Halsey-Taylor Division, King Seeley Thermos Company.)

Figure 12–7 Remote chiller for drinking water cooler. Thermostat for cooling coil and storage tank set for 50°F water, adjustable ±5°F. (By permission of Halsey-Taylor Division, King-Seeley Thermos Company.)

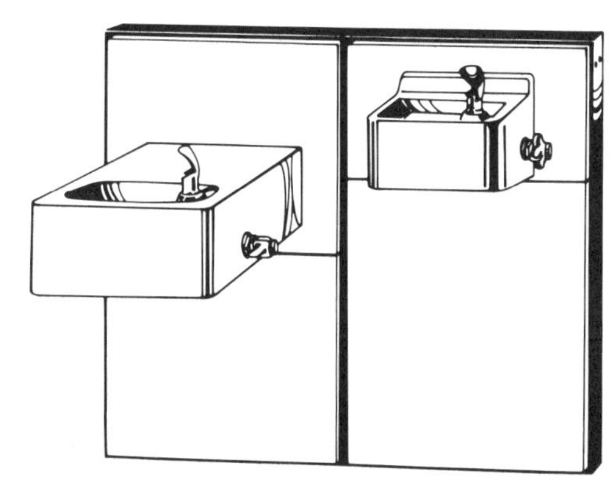

Bilevel Barrier-Free Fountain

Model: HBL-NR

Two handsome contour-formed receptors serve both the general public and the handicapped.

- Face-mounted, one-piece receptors are of 20-gauge stainless steel with rounded corners and edges.
- Two 2-stream, mound-building projectors with stream height regulator and antisquirt feature.
- Two removable access panels.
- Unit sizes: Left handicapped: 12½" W x 19" D x 6¼" H Right receptor: 10" W x 10" D x 7⅝" H.

Figure 12–8 Remote wall-mounted drinking fountain to be used with remote chiller (Figure 5–7), two-stream projector, and stream height regulator. (By permission of Halsey-Taylor Division, King-Seeley Thermos Company.)

By far the most popular model is the floor-mounted type (Figure 12–1). This unit is serviced from the front by removing the enclosure. In case of problems with the hermetic circuit, service can be performed at the location or the entire cooler removed and taken to the shop.

The wall-mounted cooler shown in Figure 12–2 has the advantage of leaving an open floor providing greater accessibility for cleaning.

Both types of coolers use ARI ratings which indicate the capacity of the cooler in gpm of water cooled from 80 °F inlet temperature to 50 °F drinking water temperature with an ambient temperature of 90 °F.

The precooler boosts the capacity by 60%. This unique device is double-walled construction with the incoming water tube coiled around and bonded to the drain line. The temperature control starts and stops the compressor and is adjustable plus or minus 5 °F.

The interior view (Figure 12–3) shows the arrangement of parts. A typical parts list is shown in Figure 12–4. The most common problem with drinking water can be a stopped-up drain or an incorrect stream adjustment. Both of these are maintenance items. Details of the stream adjustment are given with Figure 12–5.

The refrigeration system schematic is shown in Figure 12–6. Basically, it consists of a hermetically cooled system using a capillary tube type of metering device. Most self-contained drinking water coolers have air-cooled condensers.

Where it is desirable to place the refrigeration mechanism outside an air-conditioned space or to connect a number of fountains to the same cooler, a remote chiller is used (Figure 12–7). The remote chiller is used with a type of fountain shown in Figure 12–8. These coolers are less efficient since they cannot make use of the precooler.

The next group of equipment that will be discussed are the ice makers. It is evident that ice making is a large and important refrigeration industry.

AUTOMATIC ICE DISPENSERS

This is a large group of equipment, which includes ice cube makers, flake ice machines, and industrial ice equipment. Whereas years ago ice was made or stored at a central location and delivered to the user, now individual machines are located at the point of use.

This is a big business, and there are many types of equipment available which will be described.

Ice Cube Makers

Self-contained ice cube makers are available in a variety of sizes, ranging from 50 lb through 650 lb per day. They make clear ice (not clouded). This is done by forming the ice with running water to eliminate the air bubbles. The advantage of a clear cube is that it does not weaken the carbonation in a mixed drink. Various shapes of cubes are available, depending on the method of freezing. The equipment also includes a bin for storage, as shown in Figure 12–9. The most common refrigerant for small self-contained ice cube makers is R-22.

Units with remote condensing units are also available (Figure 12–10). The advantage of a remote condensing unit is the removal of noise and heat from the space. Remote

Figure 12–9 Self-contained cube ice maker. Produces up to 650 lb of hard ice in 24 hours. Provides choice between chips and cubes. Storage capacity can be arranged to fit needs up to 1200 lb. Horsepower of compressor ranges from ⅓ to 2 hp. (By permission of McQuay Group, McQuay-Perfex, Inc.)

Figure 12–10 Remote air-cooled condensing unit connected to ice cube maker. Can be located up to 40 ft from ice maker. Operates at ambient temperatures from −25 to 115 °F. Removes condenser heat from a conditioned space. (By permission of McQuay Group, McQuay-Perfex, Inc.)

Condenser Air Temperature °F	Incoming Water Temperatures		
	50°F	70°F	90°F
-25 thru 50	650	618	586
70	615	555	513
90	590	525	475
100	567	519	437
110	520	500	425

Figure 12–11 Typical ice-making capacity chart (pounds per 24 hours). Capacity decreases with increase in ambient temperature and rise in incoming water temperature. (By permission of McQuay Group, McQuay-Perfex, Inc.)

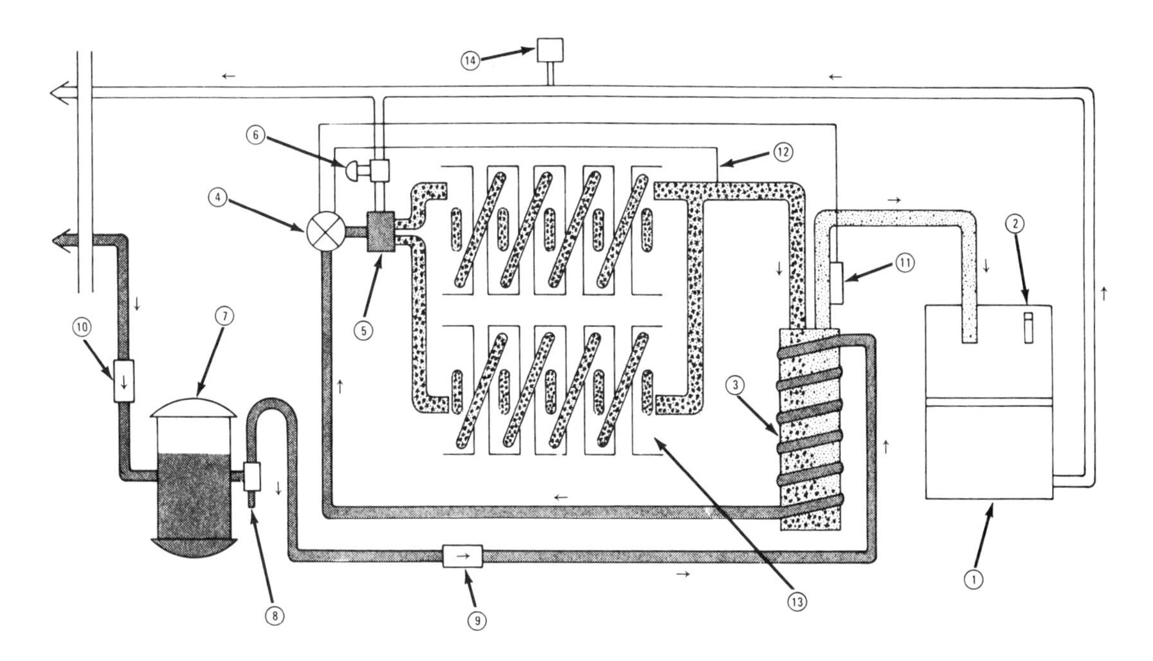

Figure 12–12 Schematic of refrigeration system for ice cube maker with air-cooled condensing unit. Note hot-gas valve, which is opened to defrost evaporator during harvest cycle. (By permission of McQuay Group, McQuay-Perfex, Inc.)

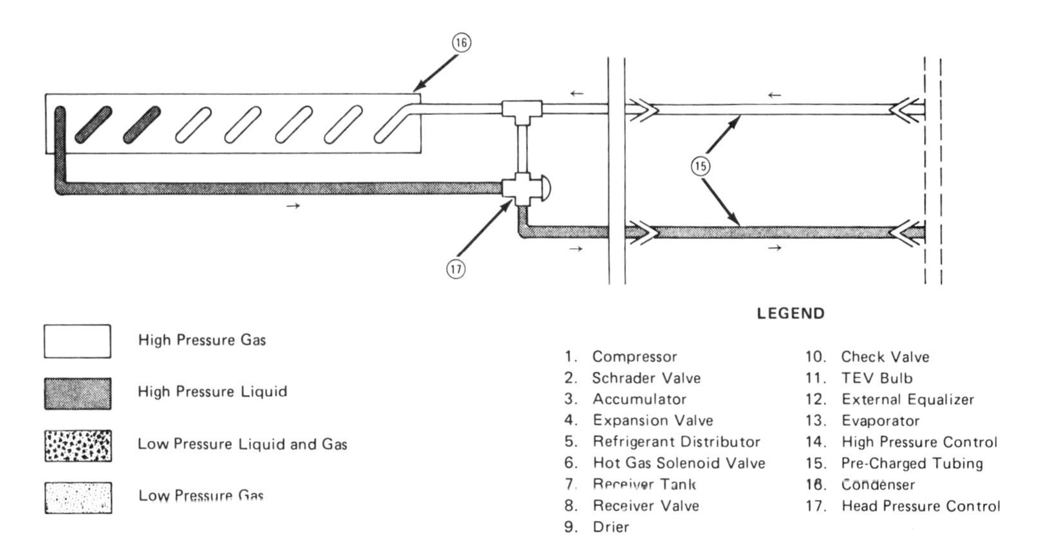

Figure 12–13 Air-cooled condenser section for ice cube maker. Note head pressure control for low-ambient conditions. (By permission of McQuay Group, McQuay-Perfex, Inc.)

air-cooled condensing units will operate in ambient temperatures from −25 to 115°F. A typical capacity table for one of these units is shown in Figure 12–11.

Note the schematic view of the refrigeration system for the ice cube maker in Figure 12–12. During the freezing cycle the hot refrigerant vapor flows to the condenser. During the harvest cycle or defrost cycle the hot refrigerant vapor flows through the evaporator due to the opening of the hot-gas valve. The accumulator is surrounded by a coil of hot liquid refrigerant to vaporize any liquid that falls into this tank.

The wiring diagram for the ice cube maker is shown in Figure 13-6 (page 000). There are two thermostats, the bin thermostat (TC1) and the ice-size thermostat (TC2). When TC1 calls for ice making, and the water pump switch (SW1) is in "auto" position, the water pump is started. The unit is then in the control of the TC2 thermostat. If the TC2 thermostat indicates that ice needs to be made, it operates the "freeze" cycle. If the TC2 thermostat indicates that the ice making is completed and defrosting should occur, it operates the "harvest" cycle. TC1 controls the operation of the compressor at the same time that it controls the water pump.

The air-cooled condenser section of the refrigeration system schematic is shown in Figure 12-13. The water system is shown in Figure 12-14. The level of the water in the pan is controlled by a float valve. Water is added only during the harvest cycle. The pump circulates the water from the sump to the top of the distribution pan. A blow-down cycle is incorporated to prevent the accumulation of solids in the sump. The sump is completely drained and filled with fresh water each time the bin thermostat (TC1) is satisfied (opened).

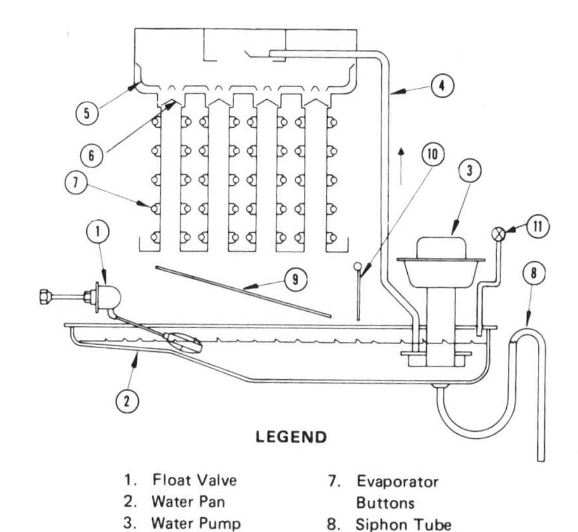

Figure 12-14 Typical water circuit for ice cube maker. Cubes are formed by water continually flowing through evaporator tubes. Blow-down system changes water in sump completely each time unit is shut down by the bin thermostat. (By permission of McQuay Group, McQuay-Perfex, Inc.)

Flake Ice Machines

Flake ice machines are made in sizes ranging from 310 lb per 24 hours to 3300 lb every 24 hours. Various-sized storage bins can be obtained. An illustration of a self-contained unit is shown in Figure 12-15 and a typical capacity table is shown in Figure 12-16.

Ice is made on the outside of an evaporator drum and is removed by a rotating cage assembly. When the ice builds up to the required thickness the knives on the cage assembly crack the ice, dropping it into a special conveyor which feeds it into the spout directed toward the bin. This prevents ice from dropping into the water reservoir.

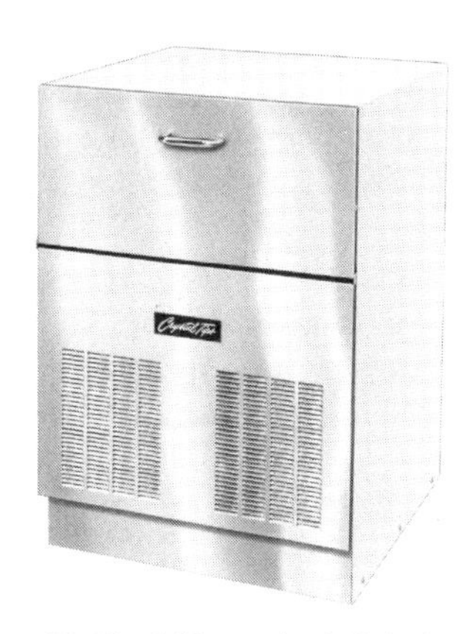

Figure 12–15 Self-contained flake ice machine. Uses flooded evaporator and capillary tube type of metering device. Capacity to 350 lb in 24 hours. Draws 11 A current with 115-V AC power supply. Uses R-12 or R-502 refrigerant depending on model. Standard bin holds 40 lb of ice. (By permission of McQuay Group, McQuay-Perfex, Inc.)

Room Temp. °F	Incoming Water Temperature — °F				
	50	**60**	70	80	90
60	353	318	293	276	264
70	342	309	285	269	258
80	331	300	278	263	252
90	322	292	271	257	247
100	295	270	250	237	228

Figure 12–16 Typical ice-making capacities (pounds per 24 hours) of self-contained flake ice machine. Capacities decrease with increases in room temperature or rise in incoming water temperature. (By permission of McQuay Group, McQuay-Perfex, Inc.)

Figure 12–17 Refrigeration circuit for a self-contained flake ice machine. Evaporator is vertical cylindrical tube with formed corrugations running horizontally for formation of ice. Ice is removed by cage assembly when the thickness builds up. 1, Compressor; 2, Schrader valve; 3, heater loop; 4, condenser; 5, drier; 6, evaporator; 7, evaporator housing tube; 8, O-ring seal; 9, capillary restrictor tube; 10, accumulator; 11, strainer. (By permission of McQuay Group, McQuay-Perfex, Inc.)

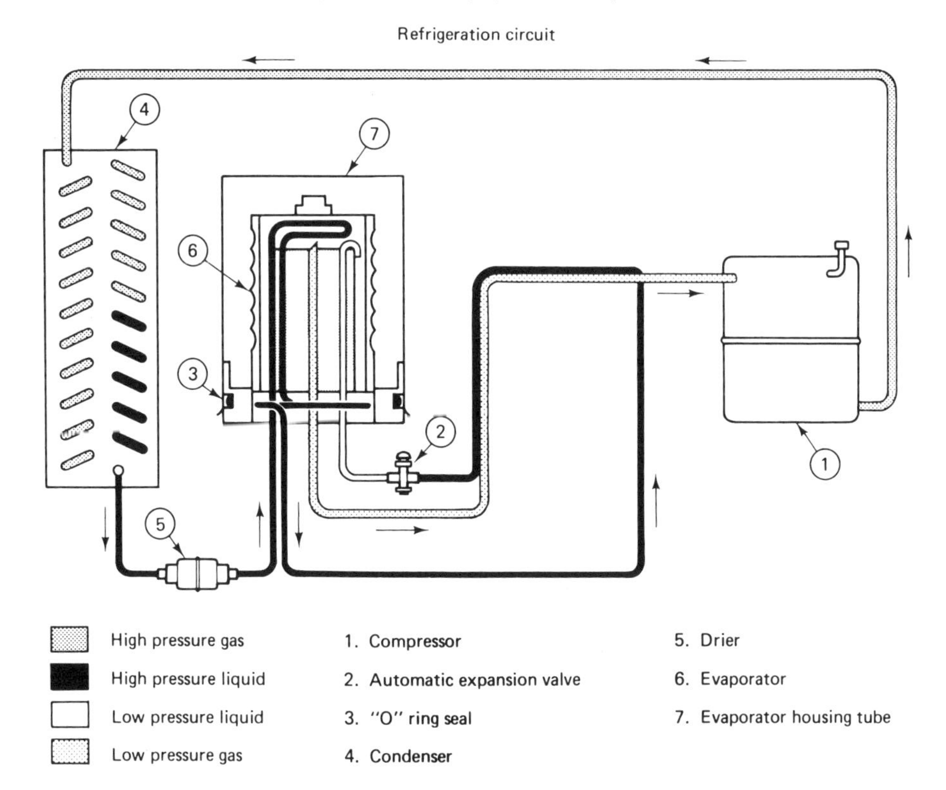

The gear motor assembly is mounted above the revolving drum. The bin thermostat regulates the operation of the compressor. No defrost cycle is required. Makeup water is fed into a float chamber or reservoir. Water from the reservoir is fed by gravity onto the refrigerated drum.

Note the illustration of the refrigeration cycle in Figure 12–17. The system employs a hermetically sealed compressor with a thermal expansion type of metering device. A heat loop is made by using the hot-liquid line to warm the top of the evaporator cylinder to prevent ice forming. A suction accumulator is used to protect the compressor.

Large Ice-Making Machines

Large ice-making machines have some interesting applications. They are used for hospitals and hotels. They are used by the fishing industry, the chemical industry, the poultry industry, and for making concrete. Various types of refrigerants are used: R-12, R-22, R-500, R-502, and R-717. Special equipment is required for ice handling in some of the larger systems.

Figure 12–18 shows the exterior and Figure 12–19 shows a cutaway view of the ice maker section. This machine is capable of making 32,000 lb of ice in 24 hours.

Referring to Figure 12–19, water is pumped to the top of the rotating cylinder. The cylinder is double walled with refrigerant flowing between the walls, freezing the ice on

Figure 12–18 Evaporator sections, industrial ice maker unit. This equipment can use standard refrigerants or brine, either flooded or pumped systems. Stationary evaporator. No defrosting cycles. No water loss, all water being converted into ice. (By permission of North Star Ice Equipment Corporation.)

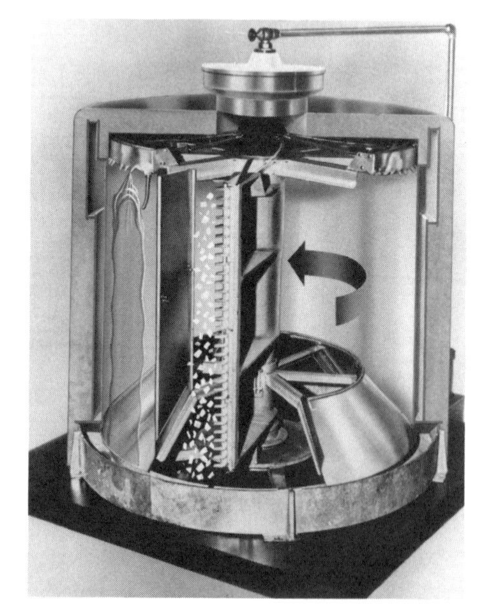

Figure 12–19 Cutaway view of ice-making evaporator. Ice forms on the inside surface of cylinder and is removed by rotor with adjustable ice-removing tools that do not touch the surface. Drip shield prevents water from dropping into brine. (By permission of North Star Ice Equipment Corporation.)

the surface. Excess water drops to a pan below the cylinder and is recirculated. A rotating cutter bar removes the ice from the cylinder without touching the surface. Water is separated from the ice. The thickness of the ice can be adjusted by regulating the position of the cutter bar. Ice can be delivered at temperatures of 0 °F (−18 °C) or lower.

Where conditions permit, evaporator and condensing unit can be combined on a single base. Figure 12–20 shows an evaporator with an air-cooled condensing unit. Figure 12–21 shows an evaporator with a water-cooled condensing unit.

Figure 12–20 Complete ice equipment package, evaporator, and air-cooled condensing unit mounted on same base. Ice bin must be directly below. (By permission of North Star Ice Equipment Corporation.)

Figure 12–21 Complete ice equipment package, evaporator, and water-cooled condensing unit mounted on same base. Ice bin must be directly below. (By permission of North Star Ice Equipment Corporation.)

Figure 12–22 shows the use of ice rakes to position ice in the bins. Figure 12–22a shows the bin before ice is delivered. Figure 12–22b shows the ice rake in use. Figure 12–23 shows a complete ice plant, which may be a desirable purchase for a company using over 5 tons of ice a day.

Figure 12–22 (a) Bin arrangement with ice rakes before ice is delivered to bin; (b) bin with ice and ice rakes in operation. (By permission of North Star Ice Equipment Corporation.)

(a)

(b)

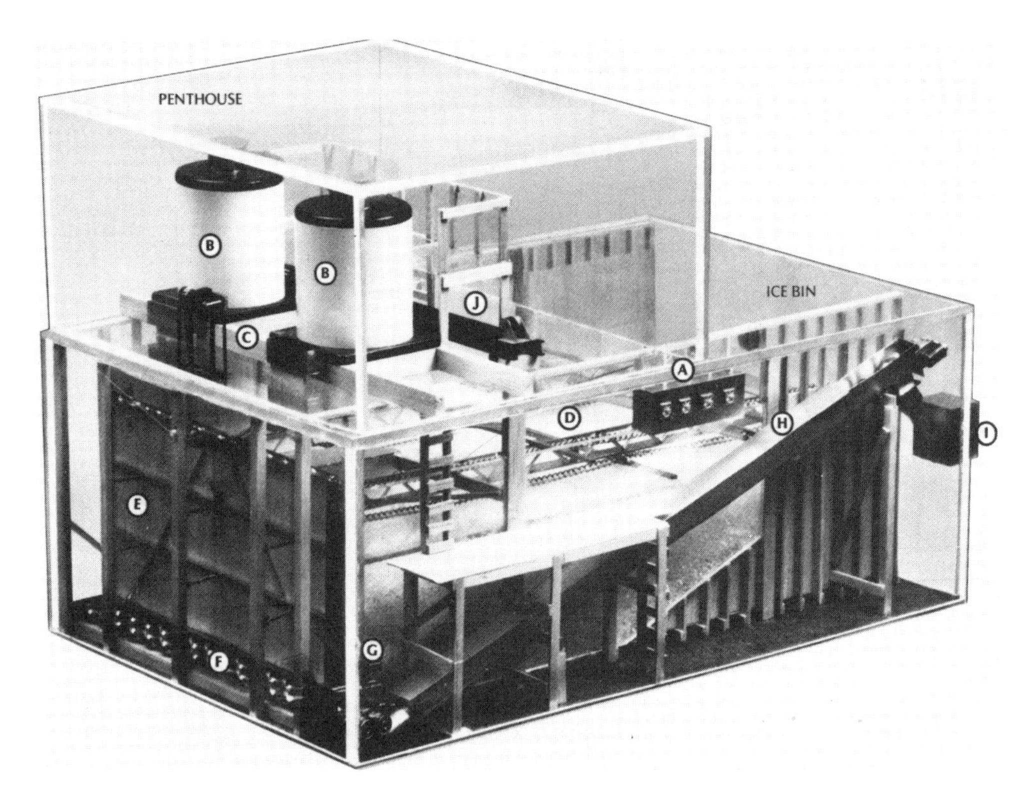

Figure 12–23 Complete ice plant. A, Bin cooling unit; B, ice maker—model and number as required to meet daily ice demand; C, motorized ice bin door operator; D, ice rake; E, bin door; F, twin screws; G, feed reservoir; H, elevating screw conveyor; I, motorized ice discharge door-screw conveyors beyond as required; J, ice rake hoist. (By permission of North Star Ice Equipment Corporation.)

Tube Ice Machines

A large-capacity machine that makes cylindrical ice or crushed ice is shown in Figure 12–24. Capacities are shown in Figure 12–25. Figure 12–26 shows the water and refrigerant circuits. Referring to the diagram, the compressor (3) delivers hot gas to the air-cooled condenser. Liquid refrigerant flows into the receiver (15). The refrigerant passes through the thawing chamber (16), the heat exchanger (13), the expansion valve (17), and into the freezer compartment. The suction gas leaves at the top of the freezer (2), passes through the heat exchanger (13), then back to the compressor (3).

Water is supplied to the reservoir (7) through the float valve (12) and is pumped (6) to the water header (8) at the top of the freezer (2). Water is continuously circulated by the pump (6) during the freezing cycle. Defrost is initiated by the pressure switch (36), which opens the hot-gas solenoid (18) and closes the refrigerant-liquid solenoid valve (20). During defrost the pump stops and the ice falls into the rotating cutter for sizing.

Figure 12–27 shows the tube ice maker with the front panel removed. The unit is placed in operation by the on/off thaw switch (3). Then the unit operates automatically in response to the ice bin thermostat. The type of ice produced is regulated by the ice

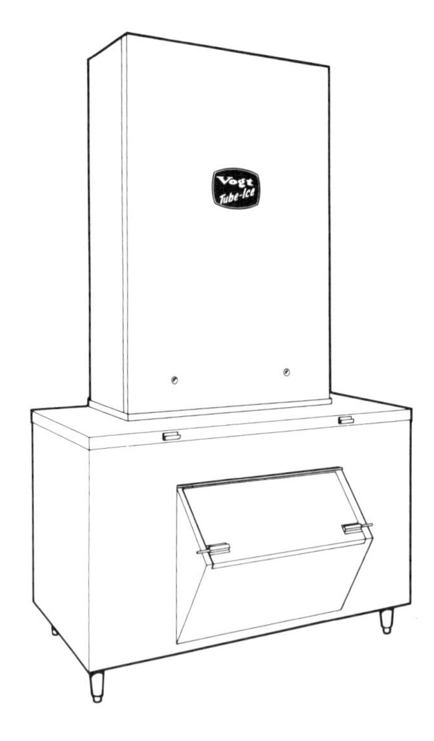

Figure 12–24 This machine produces tube ice with a normal size of ⅞ in. diameter and ¾ in. long. A switch is provided to shift from tube ice to crushed ice. Capacities range from 1350 to 4000 lb per 24 hours. Condensing unit horsepower ranges from 2 to 6.5 hp. (By permission of Henry Vogt Machine Company.)

Figure 12–25 Capacities are shown for three different-sized machines. Also shown are the power and water requirements. (By permission of Henry Vogt Machine Company.)

SPECIFICATIONS

	MODEL 1800						MODEL 3000						MODEL 4000					
	Capacity Lbs. Per 24 Hr.		KWH Per 100 Lbs.		Water Requirements		Capacity Lbs. Per 24 Hr.		KWH Per 100 Lbs.		Water Requirements		Capacity Lbs. Per 24 Hr.		KWH Per 100 Lbs.		Water Requirements	
Water Temperature	Hertz		Cylinder	Crushed	Ice Making GPH	Condenser GPM	Hertz		Cylinder	Crushed	Ice Making GPH	Condenser GPM	Hertz		Cylinder	Crushed	Ice Making GPH	Condenser GPM
	60	50					60	50					60	50				
85°	1350	1120	3.60	4.00	6.75	5.1	2350	1950	4.05	4.50	11.75	8.0	3200	2655	3.60	4.00	16.0	11.8
75°	1450	1205	3.50	3.90	7.25	2.5	2525	1995	3.90	4.35	12.65	4.0	3400	2820	3.50	3.90	17.0	5.9
65°	1550	1285	3.40	3.80	7.75	1.7	2650	2200	3.80	4.25	13.25	2.6	3600	2990	3.40	3.80	18.0	3.8
55°	1650	1370	3.25	3.65	8.25	1.2	2850	2365	3.65	4.10	14.25	2.0	3800	3155	3.25	3.65	19.0	3.0
45°	1750	1450	3.10	3.50	8.75	1.0	3000	2490	3.45	3.90	15.00	1.6	4000	3320	3.10	3.50	20.0	2.4
Weight	Net: 1105 lbs. — Shipping: 1170 lbs.						Net: 1195 lbs. — Shipping: 1260 lbs.						Net: 1480 lbs. — Shipping: 1545 lbs.					
Standard Electrical Requirements	15 ampere service						30 ampere service						40 ampere service					
	Either 208 volt, 3 phase, 60 hertz, or 230 volt, 3 phase, 60 hertz. (Optional electrical arrangements on request.)																	
Standard Condensing Unit	2 hp rating						5 hp rating						6.5 hp rating					
	Refrigerant 12 with accessible hermetic compressor and water cooled shell and coil condenser. Remote air cooled condenser available as optional extra.																	
Ice Size	Cylinder: ⅞" diameter x ¾" long (adjustable to 1½"). Crushed: 3/16" - ¼" thick.																	

NOTE: Capacity based on Ambient (Air) Temperature 90°F.
Water and KWH based on 60 Hertz operation.
Water for Ice Making does not include Blowdown.

NSF UL

selector switch (1). The ice-clean switch is always placed in the "ice" position during normal operation. The "clean" position of the switch is used only when the cleaning operation is to be performed. If it should be necessary to instantly stop the machine, the circuit breaker switch for either the cutter motor (4) or the pump motor (5) can be turned to the "off" position.

Figure 12–26 Description of how the tube ice machine operates is illustrated by this diagram and the one shown in Figure 12–27. See the text for a description. (By permission of Henry Vogt Machine Company.)

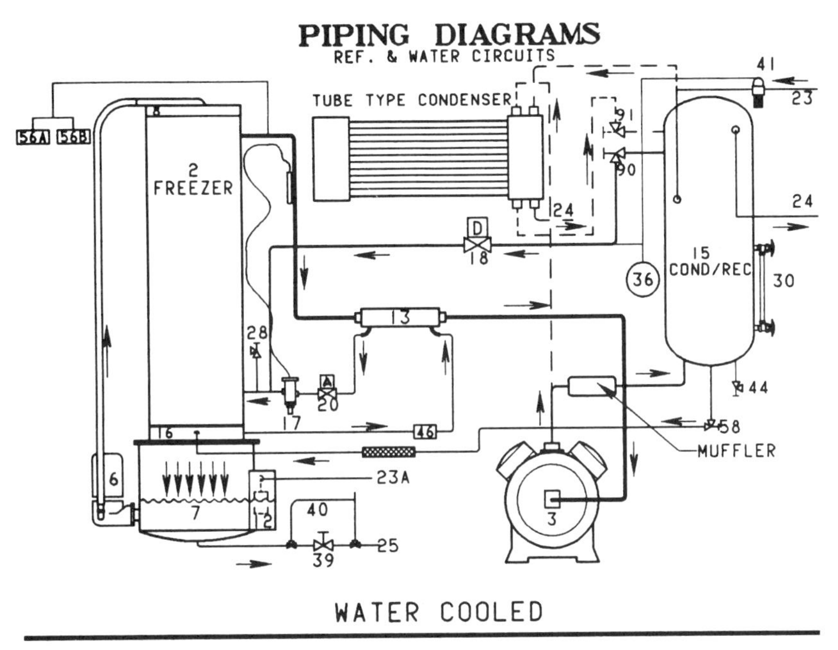

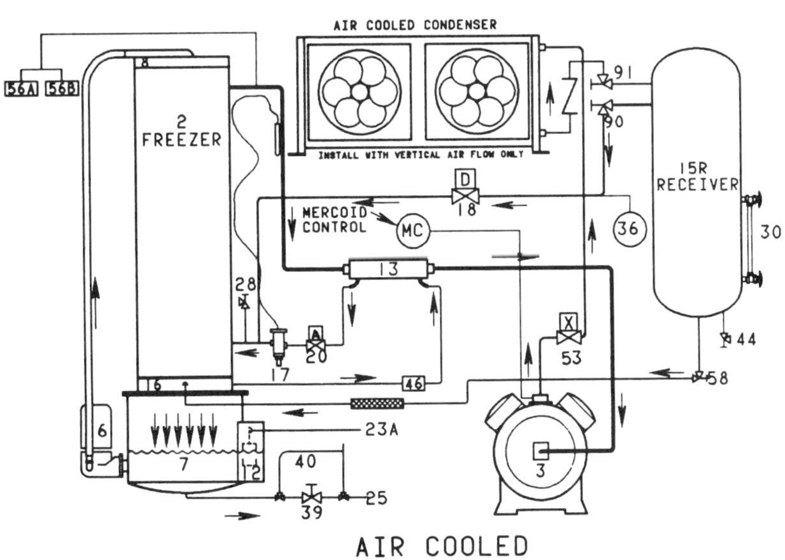

ICE BUILDERS

Figure 12–28 shows an ice builder. This equipment accumulates ice to store cooling capacity. A relatively small compressor can be used to build up this ice bank by running a sufficient number of hours. These units are used chiefly where an intermittent requirement for chilled water is needed and permits the use of a relatively small compressor to build up the storage capacity. They are also used where the compressor can be run on lower cost off-peak power.

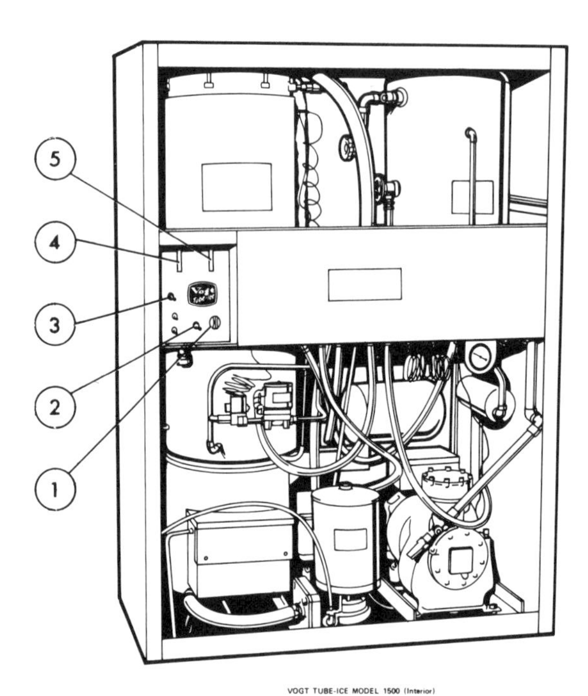

Figure 12–27 Interior construction of the tube ice machine. 1, Ice selector switch (crushed/auto/cylinder); 2, ice/clear switch; 3, on–off/thaw switch; 4, circuit breaker switch (for cutter motor); 5, circuit breaker switch (for pump motor); 6, condenser pressure gauge. (By permission of Henry Vogt Machine Company.)

Figure 12–28 This equipment stores refrigeration in the form of ice and releases it when required in the form of cooled water at temperatures of 32 to 33 °F. When peak loads occur for short periods a small compressor can be used to store a relatively large amount of cooling. Air agitation permits even accumulation of ice on the coils and greater efficiency of heat transfer. (By permission of Continental Equipment Corporation.)

Figure 12–29 shows the rebuilding time based on various suction temperatures and ice thicknesses. Units are designed to store 5000 to 100,000 lb of ice. The storage capacity ranges from 60 to 1200 tons/per hour.

Starting with a tank of water at 32 to 33 °F and using a 5 °F suction temperature, a 1-ton (12,000 Btuh) R-12 compressor will produce 1000 lb of ice in 12 hours. Water in the tank is agitated by air pressure to produce even freezing on the pipe coils.

Figure 12–29 Amount of rebuilding time required to build up ice storage capacity, based on full-flooded liquid overfeed system, 1¼-in. pipe coil, and U factor of 40. For example, with 0 °F evaporating temperature it would take 7.5 hours to obtain 100% of the rated capacity building up 2½ in. of ice on the coils. (By permission of Continental Equipment Corporation.

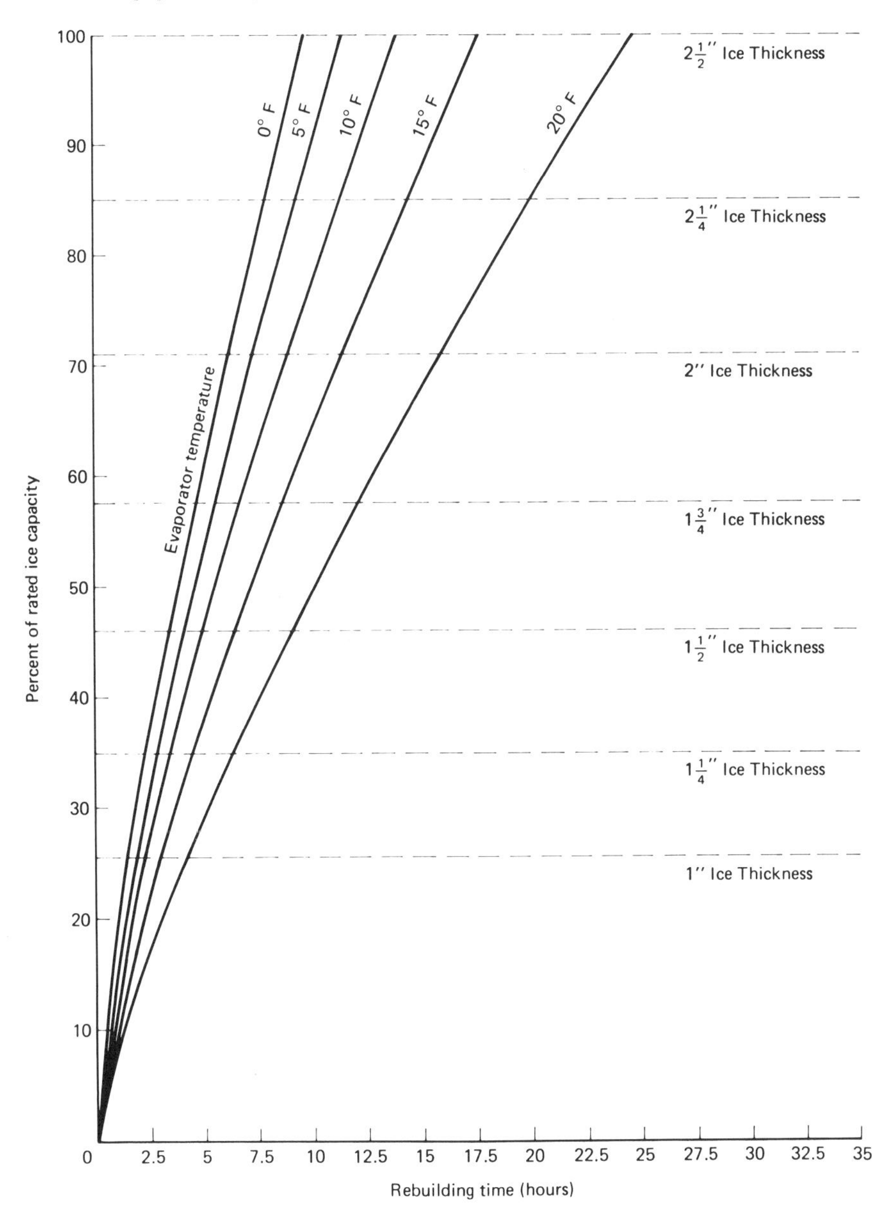

Figure 12–30 Prefabricated insulated refrigerated buildings use 4- or 5-in.-thick metal-clad insulated panels. Boxes are made suitable for temperatures as low as −30°F (−34.4°C). Urethane insulation foamed in place makes panels light and easy to handle. (By permission of Bally Case and Cooler, Inc.)

Figure 12–31 Key to building construction. 1, Standard-size insulated panels; 2, die-formed metal exterior; 3, 4- or 5-in.-thick urethane insulation; 4, sliding entrance door; 5, built-up insulated roof; 6, interior partitions; 7, horizontal steel struts; 8, open-web steel joist roof construction; 9, H-column support load-bearing beams; 10, lateral bracing; 11, vestibules using standard panels; 12, antisweat heaters around door openings; 13, slab urethane for built-in floors; 14, two-way swing vestibule doors; 15, self-contained refrigeration system; 16, self-closing entrance door; 17, patented joining mechanism for panels; 18, tongue-and-groove on panel edges; 19, color-coated panels. (By permission of Bally Case and Cooler, Inc.)

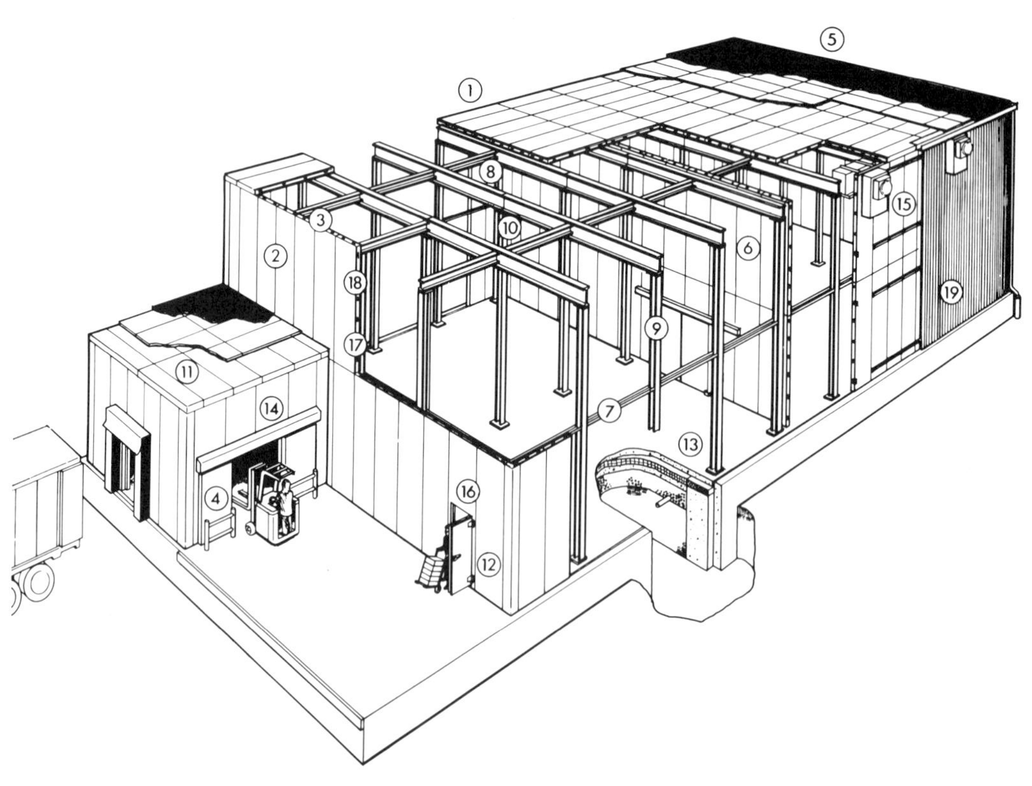

Figure 12-32 Buildings can be refrigerated from a central plant ammonia system. Illustration shows insulated ammonia piping to building. (By permission of Bally Case and Cooler, Inc.)

PREFABRICATED REFRIGERATED BUILDINGS

Refrigerated buildings are used by food, dairy, and beverage processors and distributors and by many other industry groups. Prefabricated metal-clad panels are available with 4- or 5-in.-thick urethane insulation, suitable for temperatures down to $-30\,^{\circ}F$ ($-34.4\,^{\circ}C$). These panels are so constructed that they can easily be joined together using suitable building framework to form complete buildings for outdoor use (Figure 12-30).

Details of the construction are shown in Figure 12-31. As shown in the illustration, a great deal of flexibility is provided in planning the building using standard components. Any width or length of the building can be arranged as well as building heights up to 36 ft and higher.

Figure 12-32 shows the installation of an ammonia refrigeration system adjacent to the building. Refrigeration can be tailored to fit the needs of the user, or certain types of packaged refrigeration units can be used as shown in Figure 12-31.

FOOD STORE EQUIPMENT

There is a large variety of food store equipment available. In general it serves two purposes: to preserve the food and to display it for sale. The largest volume of food cases is sold to supermarkets and, therefore, equipment is designed to make the best possible application for these large users. The types of units can be classified in about six groups: (1) walk-in coolers and freezers, (2) dairy product display cases, (3) frozen food and ice cream display cases, (4) meat display cases, (5) delicatessen display cases, and (6) produce display cases.

Walk-in Coolers and Freezers

Walk-in coolers and freezers are made with modular insulated metal-clad panels. Using corner sections and standard wall and ceiling panels, various-sized boxes can be made. Panels are assembled on the job using special eccentric cam fasteners. Insulated panels are used for the floors. Wall panels are usually clad with aluminum, floors with galvanized iron.

Figure 12–33 shows an assembled walk-in cooler. Figure 12–34 shows the sections of a modular walk-in cooler. Figure 12–35 shows the installation of an interior partition in a walk-in cooler. This makes it possible to include a cooler and a freezer in the same structure. By entering the cooler first and then the freezer, more economical operation is possible since a common wall is used.

Figure 12–33 Walk-in cooler/freezer can be assembled in any shape or size for indoor or outdoor use from standard modular panels. (By permission of Bally Case and Cooler, Inc.)

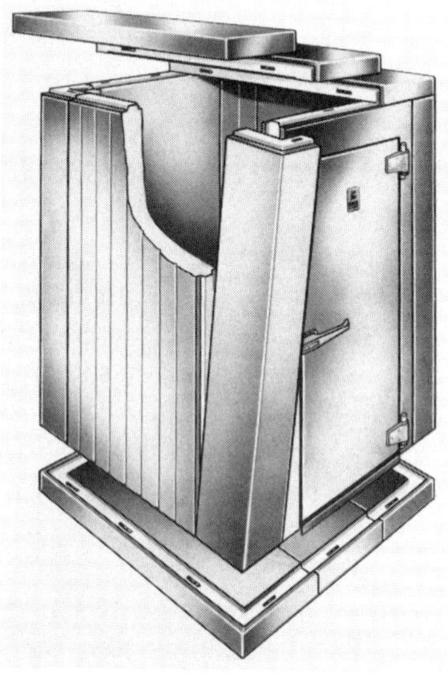

Figure 12–34 By the use of sectional panels and the patented joining mechanism, walk-in boxes are shipped knocked down and are quickly assembled at the job location. (By permission of Bally Case and Cooler, Inc.)

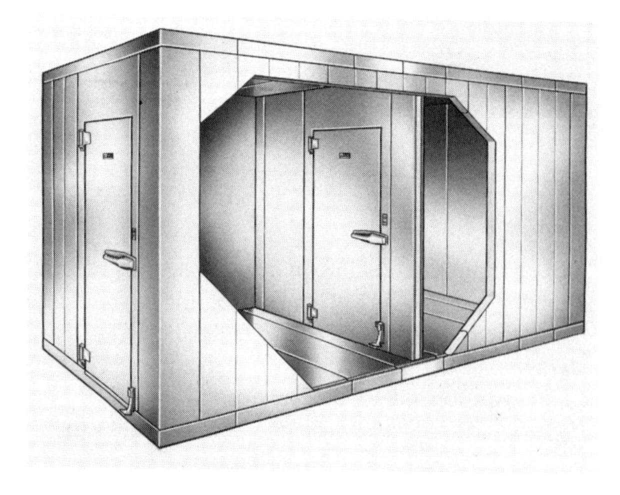

Figure 12–35 Walk-in cooler/freezer can be installed with two or more compartments by the use of partition panels. (By permission of Bally Case and Cooler, Inc.)

Figure 12-36 Storage carts. Used to store such items as dairy products in the walk-in refrigerator, to be moved directly into the refrigerated display case. (By permission of Warren/Sherer.)

Walk-in refrigerators, although essential to the food store operation, are not usually display units. These serve as storage areas to supply the needs of various display cases. Special racks or carts can be used to stack items in the cooler (Figure 12-36) such as milk, which are then transported directly to the display case.

Dairy Product Display Cases

Figure 12-37 shows the exterior of the dairy merchandiser and Figure 12-38, the cutaway view. This cabinet is arranged for easy entrance of storage carts to simplify loading the product. The refrigeration evaporator is in the top bunker with fans both in the top and the bottom of the air passage.

Figure 12-37 Display case for dairy products using storage carts. (By permission of Warren/Sherer.)

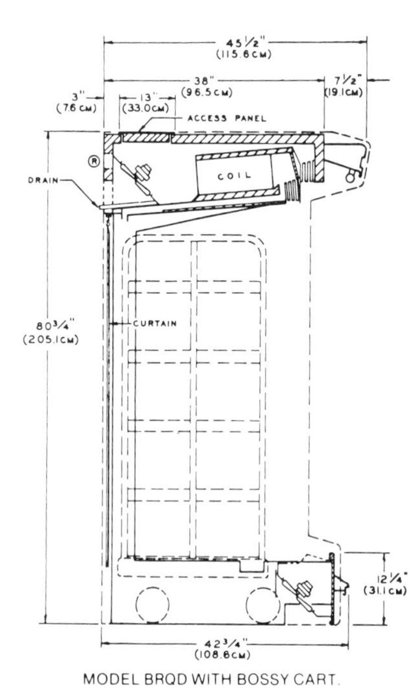

Figure 12-38 Type of dairy product merchandiser, showing cross-sectional view. (By permission of Warren/Sherer.)

Figure 12-39 Dairy-delicatessen merchandiser, using air defrost. (By permission of Warren/Sherer.)

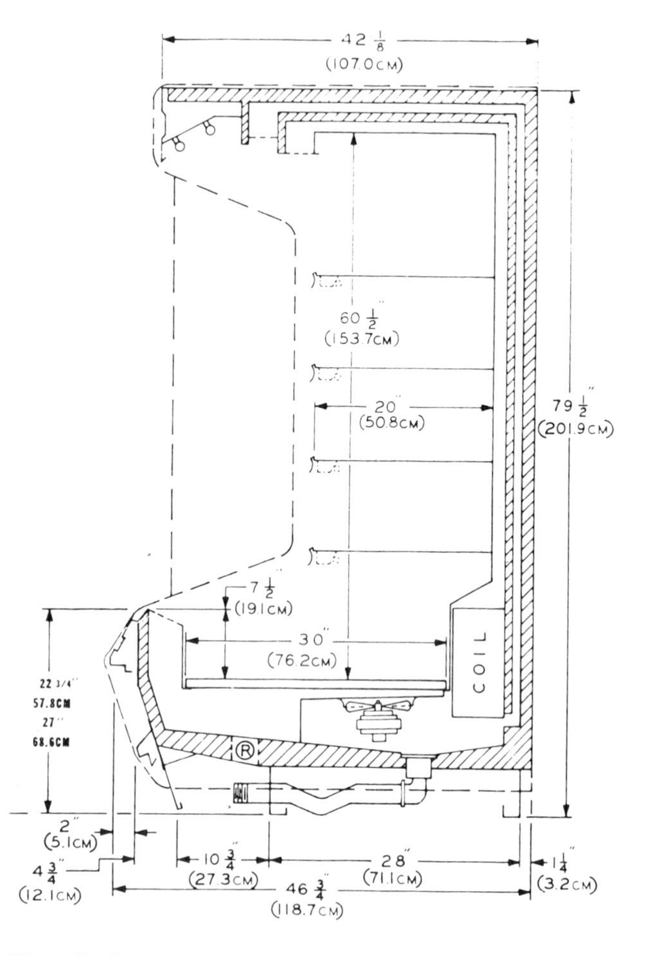

Figure 12-40 Dairy-delicatessen merchandiser, cutaway view, showing position of evaporator and fan. Note that this unit has two airstreams flowing across the front of the cabinet. The inner airstream is refrigerated. (By permission of Warren/Sherer.)

Figure 12–39 illustrates the dairy–delicatessen merchandiser, with cutaway view in Figure 12–40. Note the heavy loading of produce in this cabinet. The evaporator is located in the lower rear of the cabinet and a single fan provides air delivery for the two jet airstreams. This cabinet utilizes either electric or hot-gas defrost. Two plastic jets at the top front of the cabinet direct the two air curtains to minimize cold air in the aisles of the store.

Frozen Food and Ice Cream Display Cases

One of the most popular types of frozen-food display cases is the high-back type with an open front. This case has three jet airstreams to control temperature within the case and to isolate cold air from ambient air. One fan reverses for air defrost (see Figures 12–41 and 12–42).

Figure 12–41 Cabinet used to merchandise frozen food and ice cream. It has an open front with three jets of controlled air and air defrost. (By permission of Warren/Sherer.)

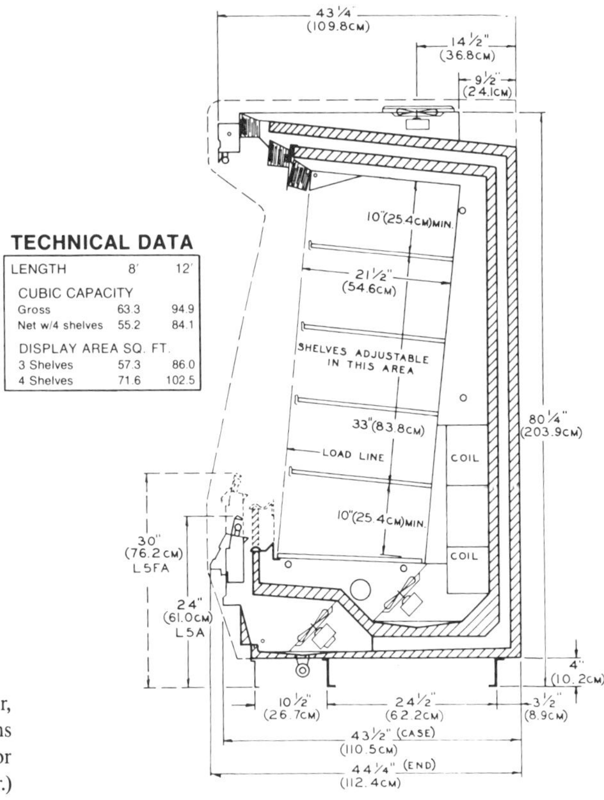

TECHNICAL DATA

LENGTH	8′	12′
CUBIC CAPACITY		
Gross	63.3	94.9
Net w/4 shelves	55.2	84.1
DISPLAY AREA SQ. FT.		
3 Shelves	57.3	86.0
4 Shelves	71.6	102.5

Figure 12–42 Frozen food/ice cream merchandiser, cutaway view. Note the three fans and three airstreams across the open front. The inner fan is reversible for the air defrost cycle. (By permission of Warren/Sherer.)

Meat Display Cases

Figure 12–43 shows the exterior view of the meat display case and Figure 12–44, the sectional view. This case has a full-length evaporator coil on the rear wall of the cabinet. The fans pressurize the flow of air through louvres on the rear of the cabinet over the top of the product to maintain even temperature. A hinged access door over the fan plenum permits easy access for cleaning. The system uses air defrost to conserve energy. Product racks are electrostatic epoxy coated.

Figure 12–43 Meat display case arranged for maximum visibility of the product. Air defrost is used. (By permission of Warren/Sherer).

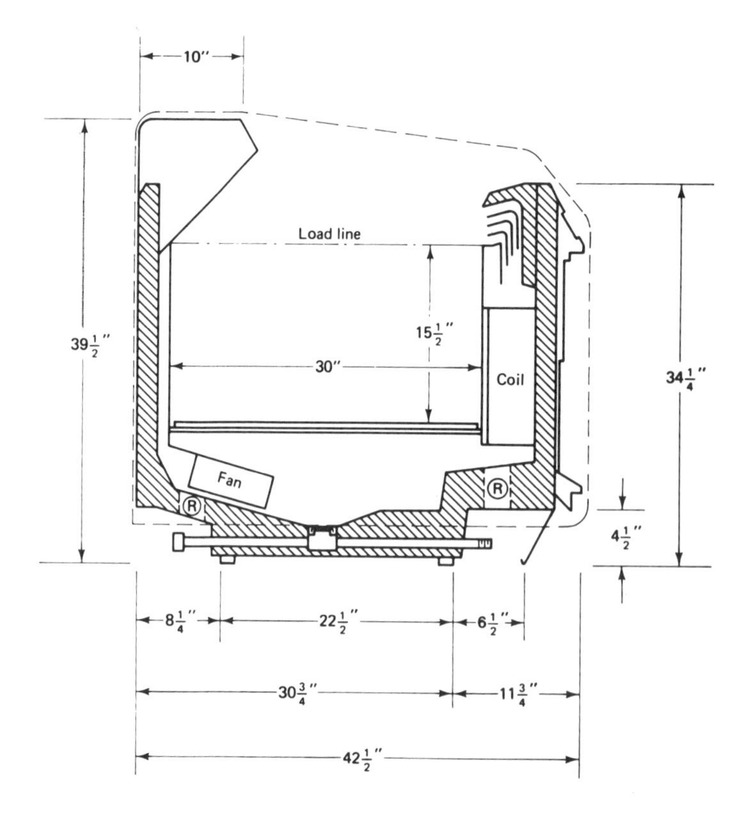

Figure 12–44 Meat display case, cross-sectional view, showing position of evaporator coil and fan. Uses pressure-charged compartment full length of case to provide proper air distribution over the full length of the display bed. (By permission of Warren/Sherer.)

Figure 12-45 Delicatessen merchandiser, totally enclosed. (By permission of Warren/Sherer.)

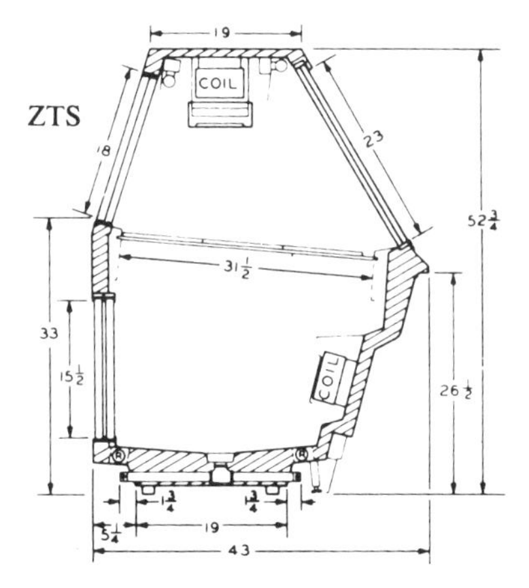

Figure 12-46 Delicatessen case, cutaway view. This equipment has two evaporator coils and gravity circulation. (By permission of Warren/Sherer.)

Delicatessen Display Cases

The exterior view is shown in Figure 12-45 and the cutaway vieww in Figure 12-46. This cabinet is completely enclosed to permit maintaining high humidity for exposed product. The evaporative coil is located in the top of the cabinet and cold air flows over the product by gravity action. Double or triple glass can be supplied. The cabinet is available for double duty, providing refrigerated storage area in the lower portion of the cabinet, with a separate evaporator coil.

Produce Display Cases

An exterior view of the produce merchandiser is shown in Figure 12-47 and a cutaway view in Figure 12-48. A large evaporator coil permits the use of low velocity air to cool the product without damage. A hinged door makes possible ready access to the coil and fan for cleaning.

Figure 12–47 Open-front produce display unit. This unit has low air velocity, top air discharge, and controlled air outlets from the rear. (By permission of Warren/Sherer.)

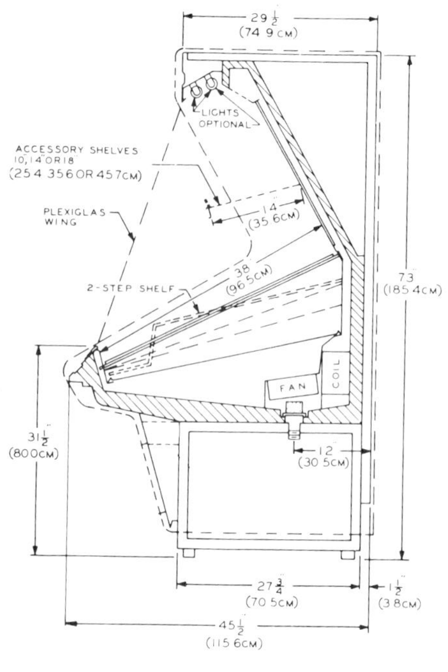

Figure 12–48 Produce display unit, cutaway view. Motor fan and coil plenum for controlling airflow over produce. (By permission of Warren/Sherer.)

Figure 12–49 Island produce display unit used center-mounted coil and fans on both wings, with water spray accessory. (By permission of Warren/Sherer.)

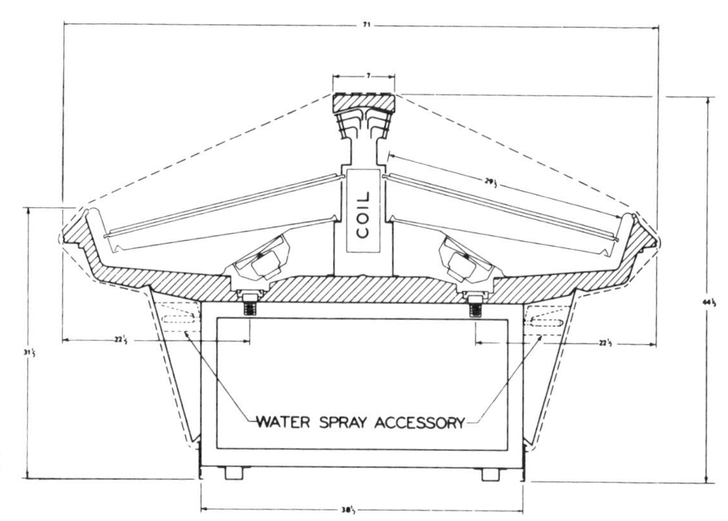

Figure 12–50 Island produce display unit, cutaway view. (By permission of Warren/Sherer.)

An island type shown in Figure 12–49, with cutaway in Figure 12–50, is available. Note the placement of the evaporator coil in the center section and multiple fans to direct the air over the product. A water spray system is also available.

Rack-Mounted Condensing Units

Rack-mounted condensing units, shown in Figure 12–51, are available to supply the refrigeration for individual cases. These units are air cooled and supplied with the necessary control and accessories. The racks are placed in a ventilated equipment room where the

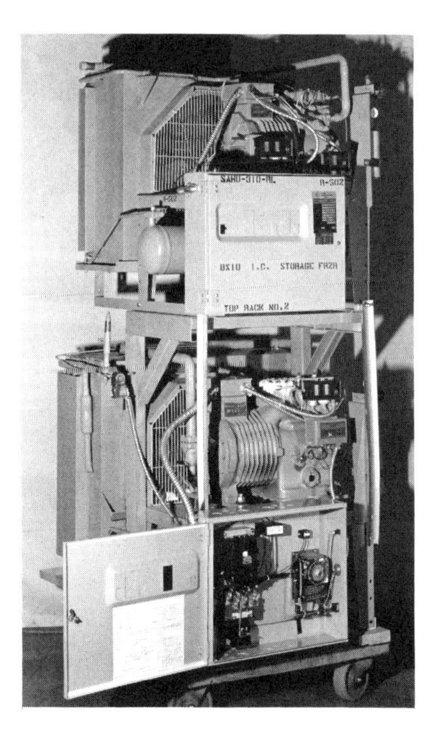

Figure 12–51 Rack-mounted condensing unit. Usually located in equipment room. One compressor is used for each case. (By permission of Warren/Sherer.)

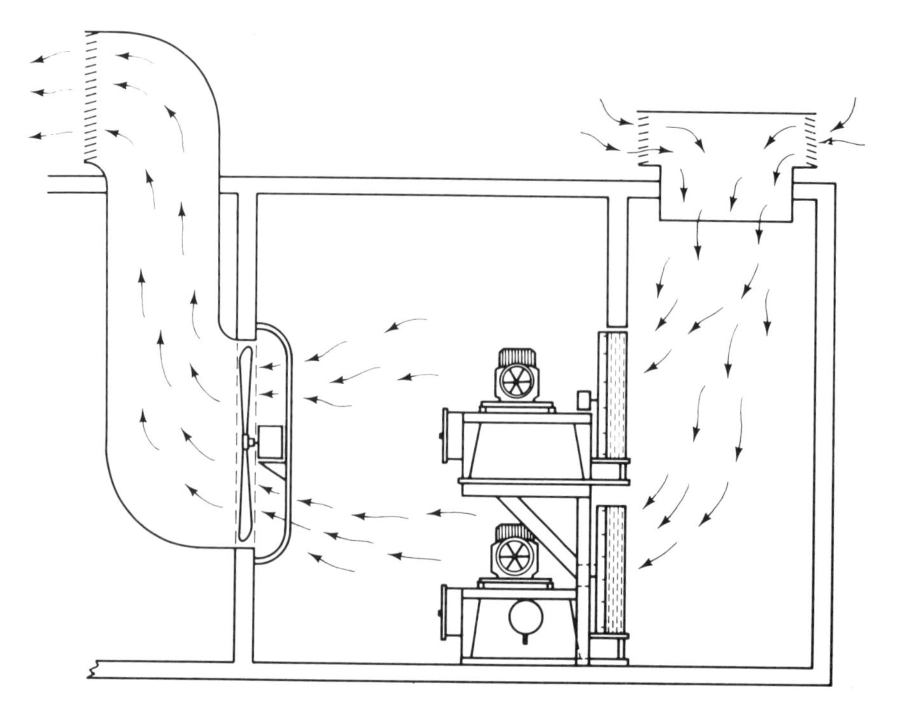

Figure 12–52 Ventilation is provided for the equipment room using rack-mounted condensing units. Fan cycles to maintain room temperatures. (By permission of Warren/Sherer.)

air temperature is controlled between 75 and 95 °F winter and summer. The receiver capacity is sufficient to hold the entire charge of refrigerant.

Figure 12–52 shows the ventilation arrangement for the equipment room. Condenser discharge can be directed into the store in winter for heating to conserve energy.

Multiple Compressor Units

This arrangement consists of two or more parallel compressors placed on a frame to supply refrigeration for a series of cases. EPR valves are used to maintain the required temperature in indiviudal cases (see Figure 12–53). The advantage of this arrangement is that standby capacity is available in case one compressor needs to be repaired. Also, extra capacity is available for pull-down. These multiple units are fully equipped with controls and accessories installed at the factory and tailored to fit the application. A novel arrangement is provided to permit the use of hot-gas defrost on one part of the system and to reuse the condensed refrigerant in another part of the system where normal operation is taking place.

Systems can be equipped with heat reclaimer coils, placed in the air-handling unit, to use condenser heat for space conditioning during winter months. Oil-level regulators are incorporated to adequately lubricate each compressor as required.

Figure 12–53 shows a three-compressor system. A fourth satellite compressor can be added for low-temperature operation (Figure 12–54). Figure 12–55 shows a piping layout

Figure 12–53 Three compressors in parallel. (By permission of Warren/Sherer.)

Figure 12–54 Three compressors with satellite compressor. (By permission of Warren/Sherer.)

for a two-compressor parallel system. Figure 12–56 shows a two-compressor piping layout with a satellite compressor. Figure 12–57 shows a heat reclaimer arrangement with a normal refrigeration cycle. Figure 12–59 shows a satellite system with hot-gas defrost. A suggested setting for pressure controls is shown in Figure 12–58. A typical defrost schedule is shown in Figure 12–60.

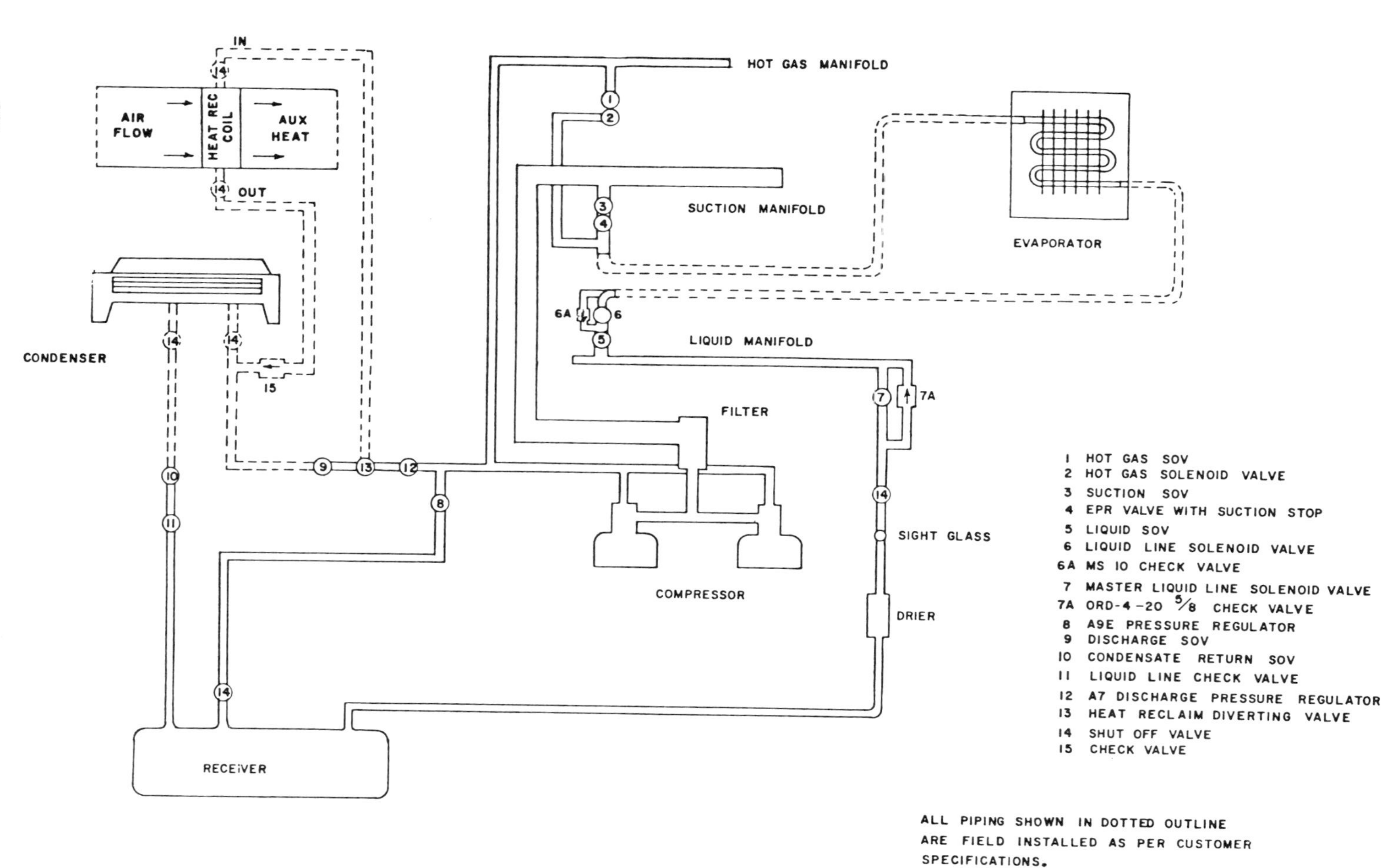

Figure 12–55 Layout of piping for two compressors in parallel system. (By permission of Warren/Sherer).

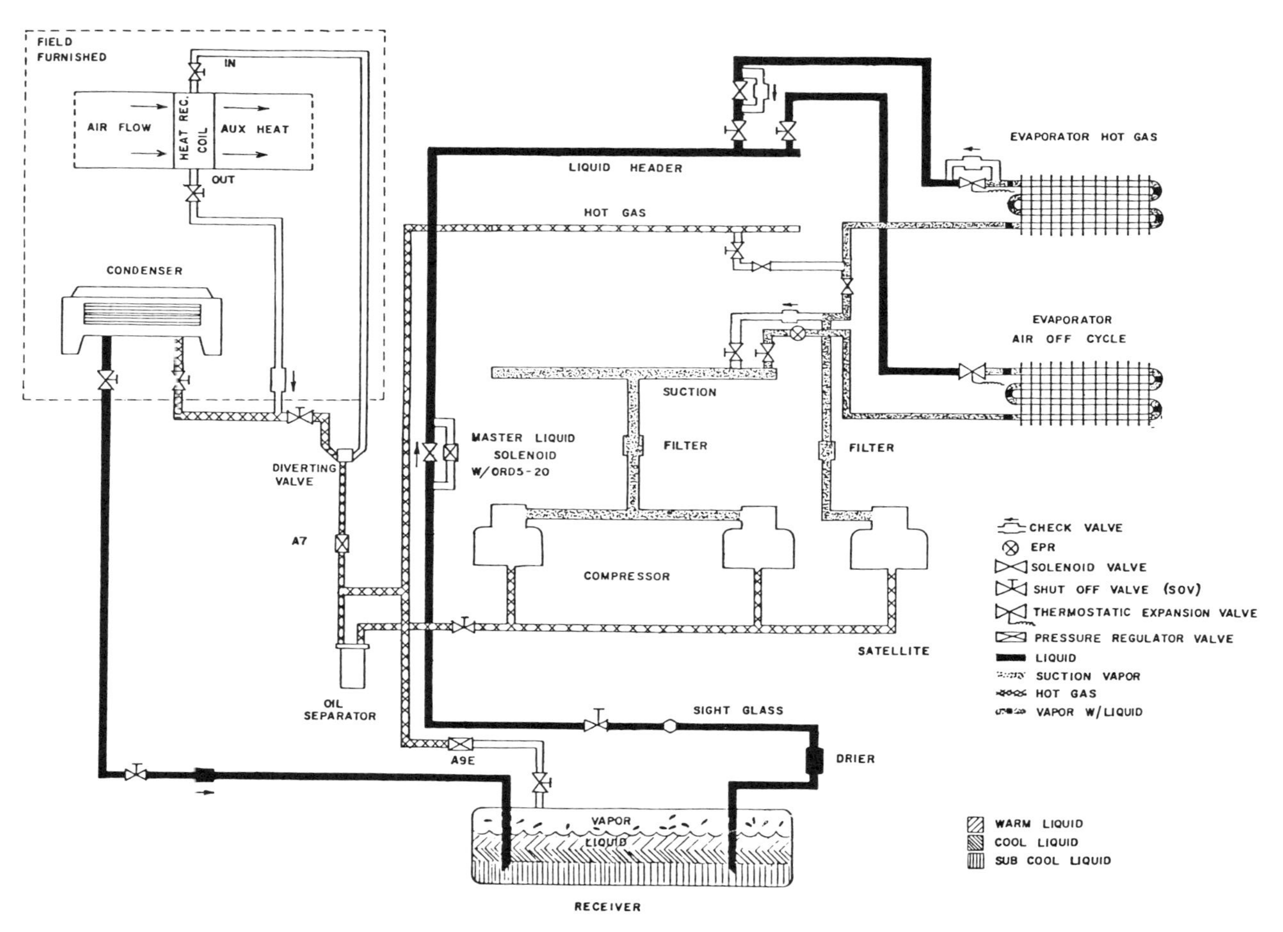

Figure 12–56 Layout of piping for two compressors in parallel system, with satellite compressor. (By permission of Warren/Sherer.)

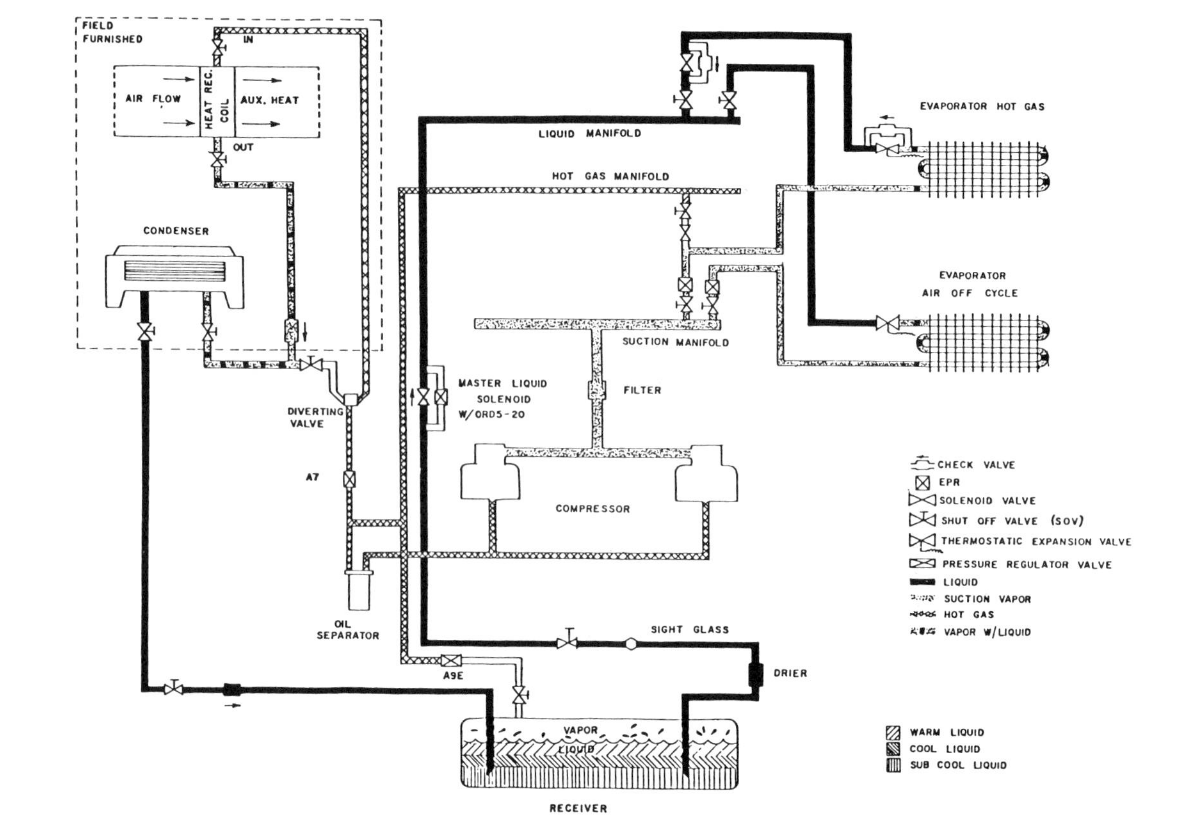

Figure 12–57 Heat reclaimed/normal refrigeration with a two-compressor system. (By permission of Warren/Sherer.)

Figure 12–58 Recommended control settings for parallel systems. (By permission of Warren/Sherer.)

A. Set discharge pressure regulator at 180 psig with R-502; 110 psig with R-12 and gauge on compressor discharge service valve.

B. Set receiver pressure regulator at 160 psig with R-502; 90 psig with R-12 and gauge on receiver outlet valve.

C. High-pressure controls:

R-502	350	Cutout
R-12	275	Cutout

D. Low-pressure controls:

1. Dual Metic without SSPC-2 option. Set low-pressure control as shown:

System	Compressor	Cut-out	Cut-in
R-502	Comp 1	1	9
LT	Comp 2	9	15
	Satellite (ice cream)	1	6
R-502	Comp 1	20	30
MT(+10°F)	Comp 2	29	38
R-502	Comp 1	30	41

Figure 12.58 (cont.)

System	Compressor	Cutout	Cutin
MT(+20°F)	Comp 2	41	48
R-12	Comp 1	3	10
MT(+10°F)	Comp 2	8	13
R-12	Comp 1	8	15
MT(+20°F)	Comp 2	15	19

2. Tri Metic and Dual Metic with SSPC-2 option: If the SSPC control is being used to control temperatures in one or more systems, set the SSPC pressure control to average approximate EPR valve settings. A pressure differential setting of 6 psig is recommended. If EPR valves are being used to control all systems, set the SSPC slightly lower than the lowest EPR valve setting. Again a 6-psig pressure differential setting is recommended. The compressor dual pressure controls should be set low enough that control is always by the SSPC. Any satellite compressor will be controlled separately, by means of the dual preessure control.

E. Adjustable time-delay controls: Dual Metric without SSPC control:
   1. First compressor: approximately 90 sec (optional)
   2. Second compressor: approximately 180 sec

Note: SSPC refers to Solid State Suction Pressure Controls.

Figure 12–59 Hot-gas defrost/normal refrigeration for a two-compressor system with a satellite. (By permission of Warren/Sherer.)

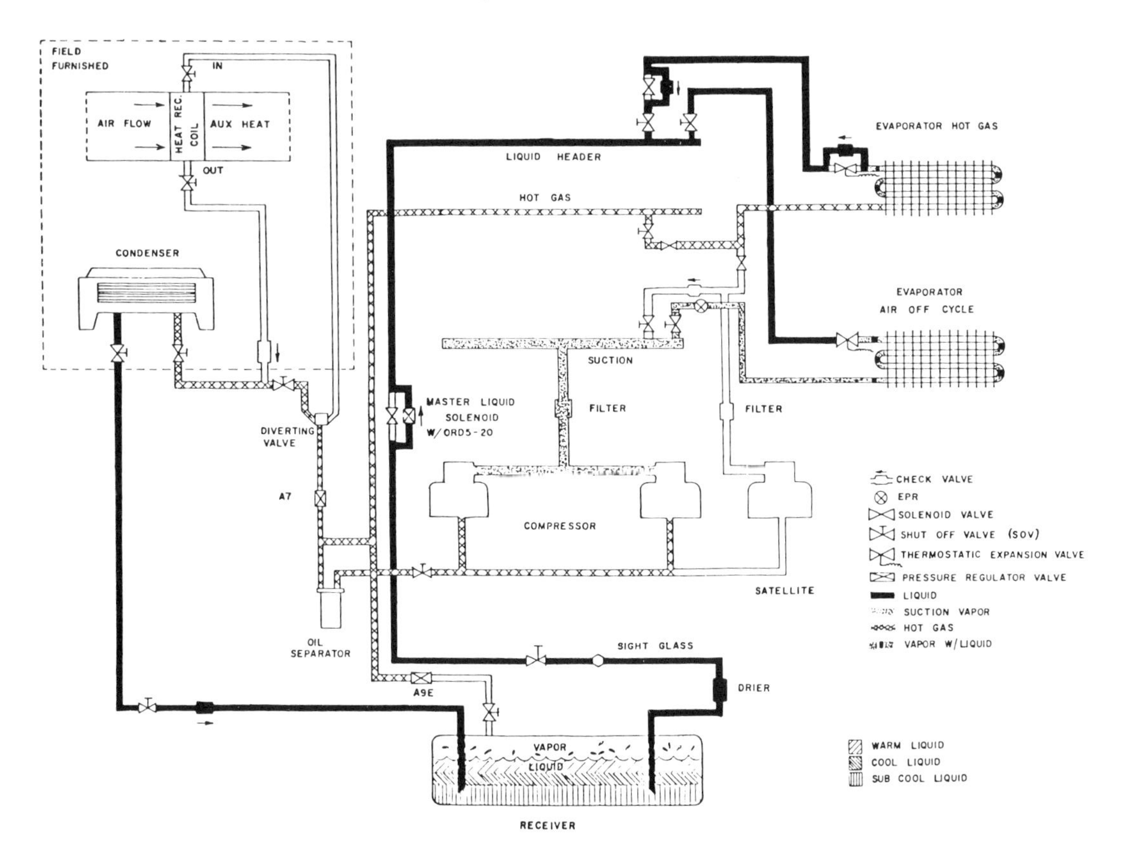

Fixture	Approximate defrost time (min)				Approximate defrost failsafe setting (min)			
	Air def.	Off cycle	Elec.	Hot gas	Air def.	Off cycle	Elec.	Hot gas
Meat and Deli								
M1A(G)	15	45	—	14	45	50	—	18
M4A(G)	20	45	—	14	45	50	—	18
JM, SJM	—	—	16	10	—	—	20	14
ZT ZTS 7900 (deli)	—	72	—	—	—	72	—	—
D6 (deli)	—	36	20	10	—	40	20	14
D5 (deli)	—	30	—	10	—	30	—	14
Dairy								
D6	—	36	20	10	—	40	22	14
D5	—	30	20	10	—	30	22	14
BQD BRQD JSQD	—	35	20	10	—	40	24	14
Produce								
HZV, ZV, TZP	—	30	—	—	—	32	—	—
Reach-ins								
L4H(A) (FF)	24	—	32	16	54	—	36	18
I4H(A) (IC)	28	—	36	16	54	—	40	18
D4H (bev.)	—	34	—	—	—	36	—	—
Frozen food								
XLA, ILA, BILA and EBILA	30	—	30	20	60	—	60	46
LM1A (G)	40	—	—	20	54	—	—	46
WTLA and EWTLA	35	—	30	20	40	—	40	36
L4 (A), L5 (A)	26	—	24	14	54	—	30	18
Ice cream								
XLA, ILA, BILA and EBILA	35	—	35	25	60	—	60	46
WTLA and EWTLA	40	—	35	25	60	—	60	36
I5F (A)		—	30	16		—	34	18
Walk-ins								
30° Meat	—	—	30	15	—	—	34	18
35° Dairy and produce	—	60	—	—	—	60	—	—
Poultry and deli	—	60	—	—	—	60	—	—
Meat prep. cutting room	—	60	—	—	—	60	—	—
Frozen-food storage	—	—	30	15	—	—	34	18
Ice cream storage	—	—	30	15	—	—	34	18

Note: The recommended settings are based on 75 °F–55% RH store conditions and properly loaded cases. Some adjustments may be required in case temperature or defrost frequency after initial opening dates and store settles down to usual traffic and environment.
General control recommendations:
[a]EPR valves should only be used on Dualmetics and Trimetics on cases requiring higher-temperature evaporators than system design level.

Figure 12–60 Defrost times for typical commercial refrigeration units. (By permission of Warren/Sherer.)

Defrost cycles per day	Refrigerant	EPR valve setting[a] (lb)	Discharge air temperature[b] (°F)	Approx. Pressure control setting[c] (lb)	
				Cutout	Cut-in
3	12	12	24 to 30	11	27
	502	38		37	63
3 to 6	12	12	26 to 30	11	27
	502	38		37	63
6	12	9	23 to 25	9	27
	502	32		32	63
1	12	18[d]	34 to 38	15	28
	502	48		42	65
4	12	14	24 to 30	12	28
	502	40		40	60
4	12	14	24 to 30	12	28
	502	40		40	60
	12	17	28 to 32	15	28
	502	47		46	60
4	12	15	28 to 32	5	25
	502	35		25	60
4	12	17	24 to 28	15	28
	502	47		46	60
4	12	18	38 to 42	20	35
	502	50		52	68
1	502	10	-5° to -12°	8	17
1 to 2	502	4	-15° to -22°	2	8
2	12	16	34 to 38	15	28
	502	42		43	63
1	502	12	-10° to 0°	9	16
4	502	12	-10° to 0°	9	16
1	502	12	-10° to 0°	9	16
4	502	12	-10° to 0°	9	16
1 to 2	502	5	-28° to -16°	2	8
1 to 2	502	5	-28° to -16°	2	8
6	502		-22° to -12°	5	12
2 to 4	12	19 to 23	28° to 32°	20	28
	502	48 to 56		51	65
2	12	22 to 26	35° to 39°	20	28
	502	54 to 61		51	65
2	12	21 to 25	33° to 38°	18	28
1	12	24 to 29	45° to 50°	—	—
2 to 4	502	15 to 18	-5° to -10°	—	—
2 to 4	502	10 to 12	-10° to -15°	—	—

[b]Thermostats are recommended as the primary control with Mastermetic units except on service meat cases (ZT and 7900) as noted in note 4.

[c]Low-pressure controls may require different settings if cases are controlled by thermostats.

[d]Service meat cases (ZT and 7900) should always have EPR as primary control and temperature thermostat and secondary control for peak performance.

Figure 12-60 (continued)

Reach-in Refrigerators

A wide variety of reach-in refrigerators and freezers are available, ranging in size from 6 to 40 ft^3 storage capacity. Some of the larger models are available with roll-in loading racks. Most refrigerators have self-contained refrigeration using a capillary tube type of metering device. Vaporizers are standard equipment on many models to evaporate the condensate that is removed from the coil during defrost. Many of the refrigerators have coils coated with a substance such as methaculate to prevent corrosion of the coils due to food acid. A fan-delay thermostat is used to prevent the fan from running until the coil reaches the proper temperature after defrost.

A picture of the undercounter refrigerator is shown in Figure 12–61. A roll-in type refrigerator is shown in Figure 12–62. A display refrigerator is shown in Figure 12–63. A sandwich bar refrigerator is shown in Figure 12–64, and a refrigerated base preparation table is shown in Figure 12–65.

Figure 12–61 Under-the-counter freezer. Evaporators are plasticized. This unit will maintain 0°F cabinet temperature with 100°F ambient temperatures. Circuit amperes, 12.5, with 115-V AC power. (By permission of Victory Manufacturing Company.)

Figure 12–62 Roll-in storage freezer. Refrigerating equipment is in the top. Air ducts are totally enclosed. Evaporators are plasticized. The unit will maintain 0°F with 100°F ambient. Current, 16.8 A with 115-V ac power. (By permission of Victory Manufacturing Company.)

Figure 12-63 Display refrigerator. This unit will maintain +40°F with 100°F ambient temperature. All condensate water disposed of automatically. Draws 25.8 A with 115-V ac power. (By permission of Victory Manufacturing Company.)

Figure 12-64 Sandwich bar refrigerator, self-contained. This unit has a $\frac{1}{8}$-hp condensing unit. (By permission of Star Metal Company.)

Figure 12-65 Prepration table, self-contained unit with a $\frac{1}{2}$-hp condensing unit. (By permission of Star Metal Company.)

Figure 12–66 Sectional floral refrigerator with remote condensing unit. A variety of combinations are available. This equipment maintains high humidity. (By permission of Buchbinder.)

Figure 12–66 shows a florist-type refrigerator. These are specially designed to maintain high humidity (usually about 80 percent relative humidity) with a gentle movement of refrigerated air. An ethylene purifier is often included in the package to remove excess ethylene gas.

PACKAGED LIQUID COOLERS

These coolers are designed specifically for industrial application primarily to cool cutting oil for the machine tool industry. The sizes range from ¾ to 60 hp, sometimes larger. The complete package can include not only the chiller equipment but also the coolant pump and temperature control.

A unique evaporator for a liquid cooler is shown in Figure 12–67. Note that the tubes can be cleaned by removing the two headers without disturbing the refrigeration cycle. Figure 12–68 shows the complete cooler with the panels removed, illustrating the use of multiple evaporators.

Evaporator sections are connected in parallel or in series depending on the cooler size to produce the desired temperature reduction, pressure drop, and the capacity required. Units are available with either one or two refrigeration circuits for capacity control as shown in Figure 12–69.

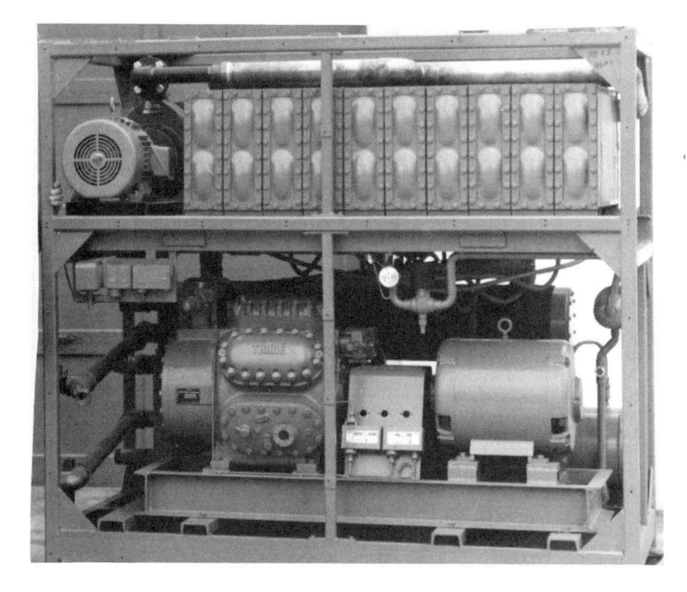

Figure 12-67 Liquid chiller unit, 40 hp. Capacity, 400,000 Btuh when cooling light oil for 100 °F to 70 °F. It has three stages of capacity reduction. Remote air-cooled condenser. (By permission of Hansen Refrigeration Machinery.)

Figure 12-68 Evaporator, multitube counterflow type. (By permission of Hansen Refrigeration Machinery.)

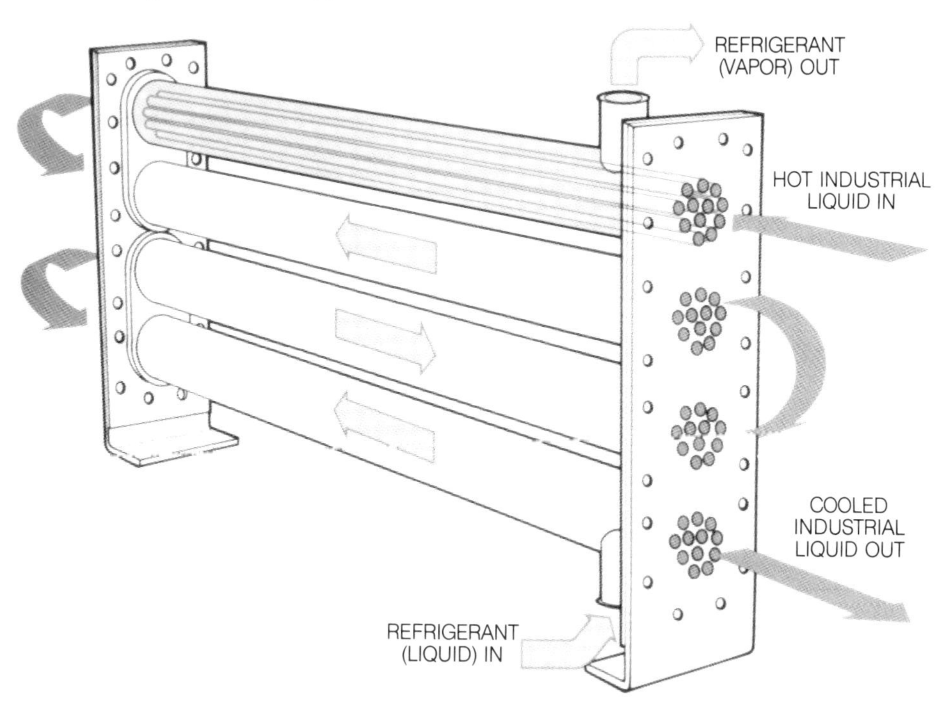

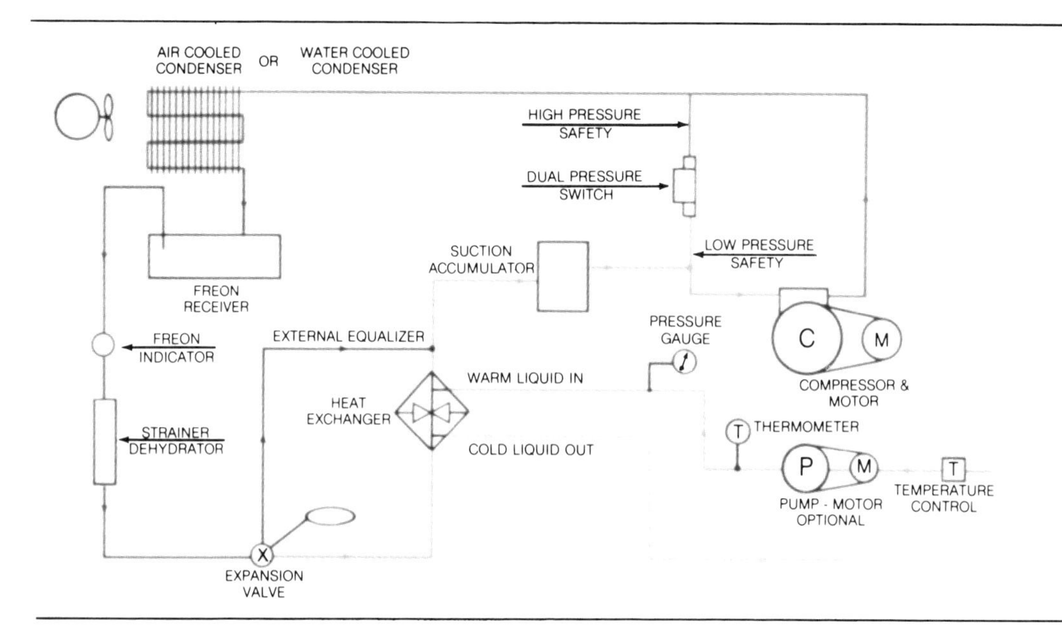

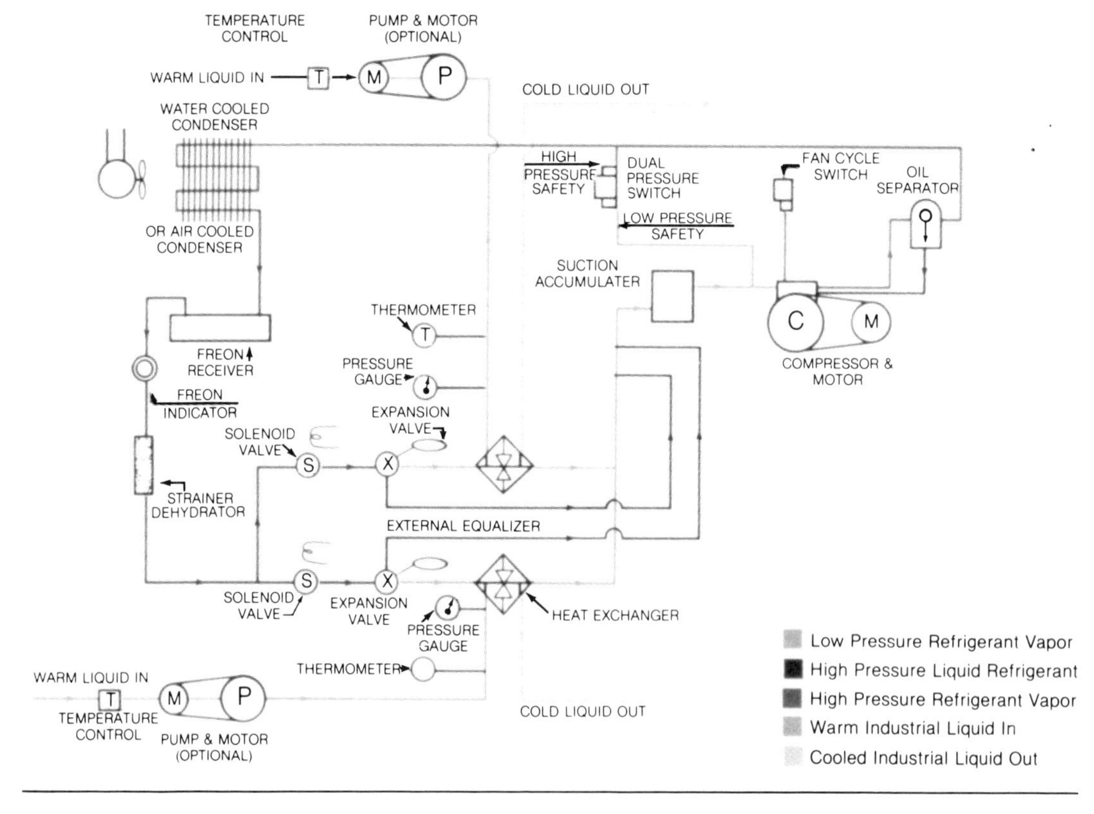

Figure 12-69 Refrigeration schematic. A, with one-circuit system. B, with two-circuit system. (By permission of Hansen Refrigeration Machinery.)

LARGE INDUSTRIAL PACKAGED CHILLERS

Large industrial-type chiller packages are manufactured using R-22 or R-717. Figure 12–70 shows a packaged unit used by a large petrochemical storage complex. The capacity is 50 tons of refrigeration at 0°F suction temperature using R-22.

Figure 12–71 shows a glycol chiller used for final milk cooling. This is a 60-ton package using ammonia refrigerant. The package includes condenser sump tank, glycol mixing tank, condenser pump, compressor, jacket water pump, glycol pump, and chiller.

Figure 12–70 Glycol chiller unit for final milk cooling in high-temperature short-time pasteurizing system. Has a 60-ton refrigeration capacity. (By permission of Crepaco, Inc.)

Figure 12–71 Packaged units for large petrochemical storage complex. Used for maintaining pressure in 5000-ton butadiene storage tank. Capacity is 50 tons of refrigeration at 0°F suction temperature, refrigerant R-22. (By permission of Crepaco, Inc.)

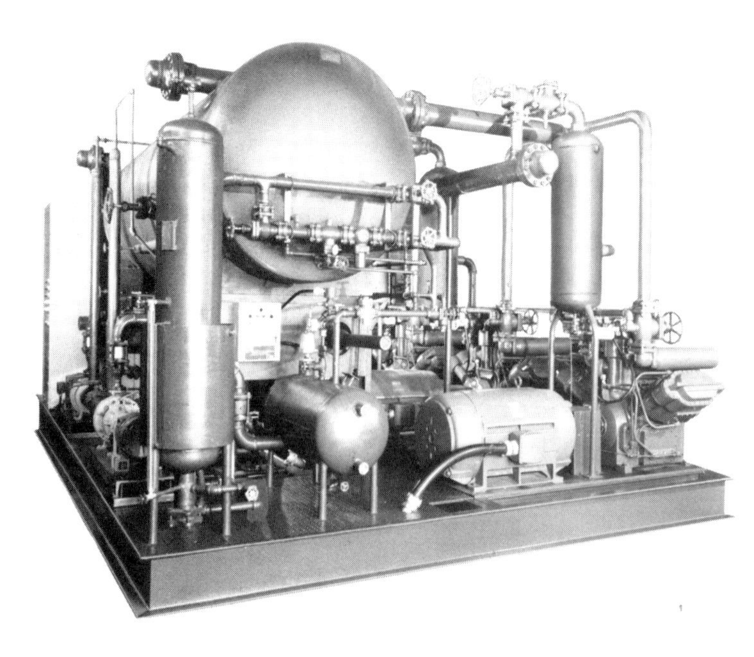

Figure 12–72 Packaged refrigeration unit, two-section, factory preassembled. Total capacity over 500 tons of refrigeration at 20°F suction, refrigerant—ammonia. Installed in a large cheese plant. (By permission of Crepaco, Inc.)

Figure 12–72 shows a two-section chiller package used for a large cheese plant. It has the capacity of 500 tons at 20°F suction and uses ammonia refrigerant.

TRANSPORT REFRIGERATION UNITS

There are basically two different types of transport refrigeration units: one type for trucks (see Figure 12–73) and the other type for trailers or "semi's" (see Figure 12–74). The most common type of trailer unit is mounted at the top front of the trailer (Figure 12–75). The undermount unit is shown in Figure 12–76.

Figure 12–73 Front-mount diesel-powered cooling/heating unit for large trucks. Continuous-run operation. High-speed or low-speed cooling or heating. Hot-gas defrost. Available with normal evaporator temperatures −20, 0, and 35°F. Capacities in 8500, 14,500, and 25,000 Btuh, respectively. Can also be operated with electric power. Capacities based on 100°F ambient temperature. (By permission of Thermo King Corporation.)

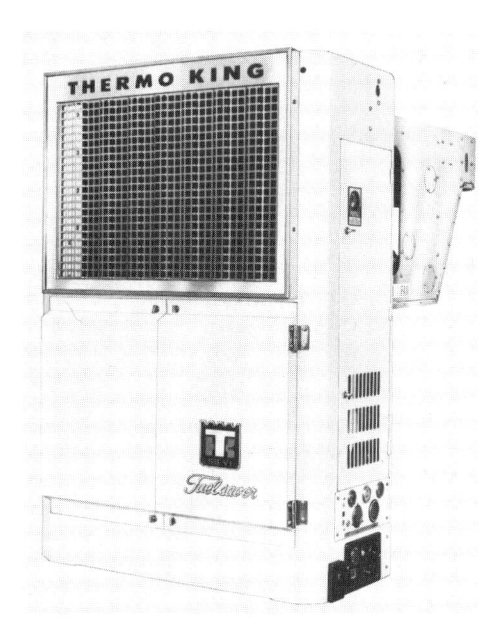

Figure 12–74 Fuelsaver heating and cooling, high speed and low speed. Runs continuously. Automatic or manual defrost. Capacity at −20°F evaporator, 11,000 Btuh; at 0°F evaporator, 19,500 Btuh; and at 35°F evaporator, 39,000 Btuh. Based on 100°F ambient temperature. (By permission of Thermo King Corporation.)

ıre 12–75 Front-mount cooling/heating : designed for application. Continuous ration with six-stage thermostat-conled system. (By permission of Thermo g Corporation.)

Figure 12–76 Undermount cooling/heating unit for trailers. Two-piece design: (1) evaporator and fan section installed in trailer and (2) engine-driven condensing unit mounted on underside of trailer. (By permission of Thermo King Corporation.)

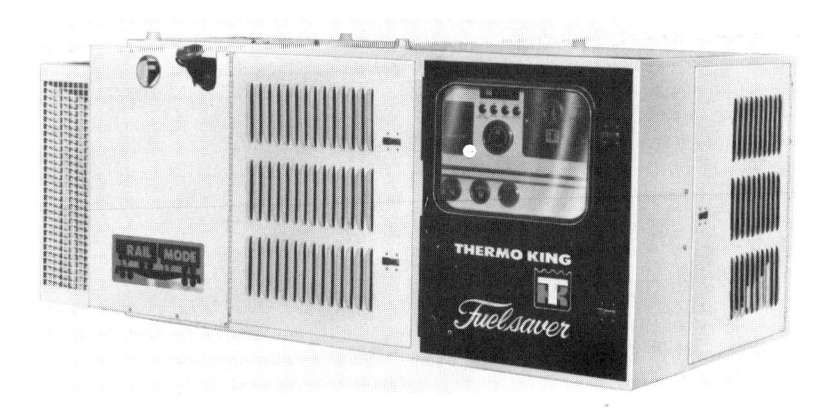

Component parts and features of the truck refrigeration unit are shown in Figure 12-77. Note that the complete unit is factory assembled ready for installation. All units are available with diesel engine drive, a complete refrigeration system for cooling and heating, a control system, and accessories. Optional electric motors are available for stationary operation.

The net cooling capacity depends on the temperature maintained. Capacities range from 4000 to 19,500 Btuh at 0 °F and 100 °F ambient. The fuel consumption varies from 0.13 to 5 gph, depending on the type of service.

The typical refrigeration cycle is shown in Figures 12-78 and 12-79, and the typical heating and defrost cycle is shown in Figures 12-80 and 12-81.

The units are factory assembled with engine compressor and condenser in the outside enclosure and the evaporator projecting inside the trailer. The engine runs continuously and the operator has the option of high-speed or low-speed cooling and high-speed or low-speed heating. When required, the unit will automatically defrost with hot gas. All units have an hour meter to determine service periods for the engine.

Selection of the proper unit is based on the length of the truck body, the temperature maintained inside the body, the amount of insulation, and the number of door closings per day. A typical guide for unit selection is shown in Figure 12-82.

Figure 12-77 Heating/cooling unit for trucks, cutaway view. Engine driven with electric standby power available. High- and low-speed heating and cooling. Capacity at 0 °F evaporator, 7000 Btuh; at 35 °F evaporator, 14,000 Btuh, based on engine drive. (By permission of Thermo King Corporation.)

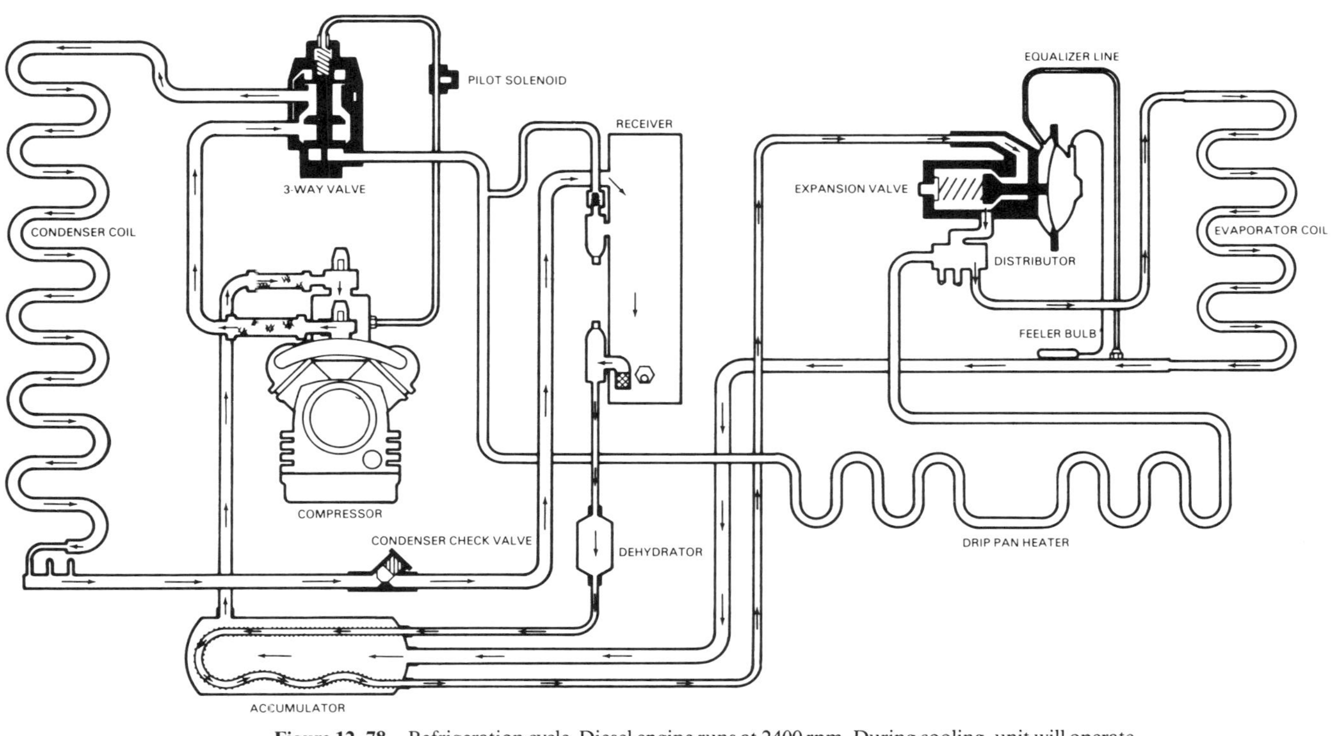

Figure 12–78 Refrigeration cycle. Diesel engine runs at 2400 rpm. During cooling, unit will operate on high speed until box temperature reaches 2°F above set point, then switch to low speed. Fuel saver switch locks unit into low speed. (By permission of Thermo King Corporation.)

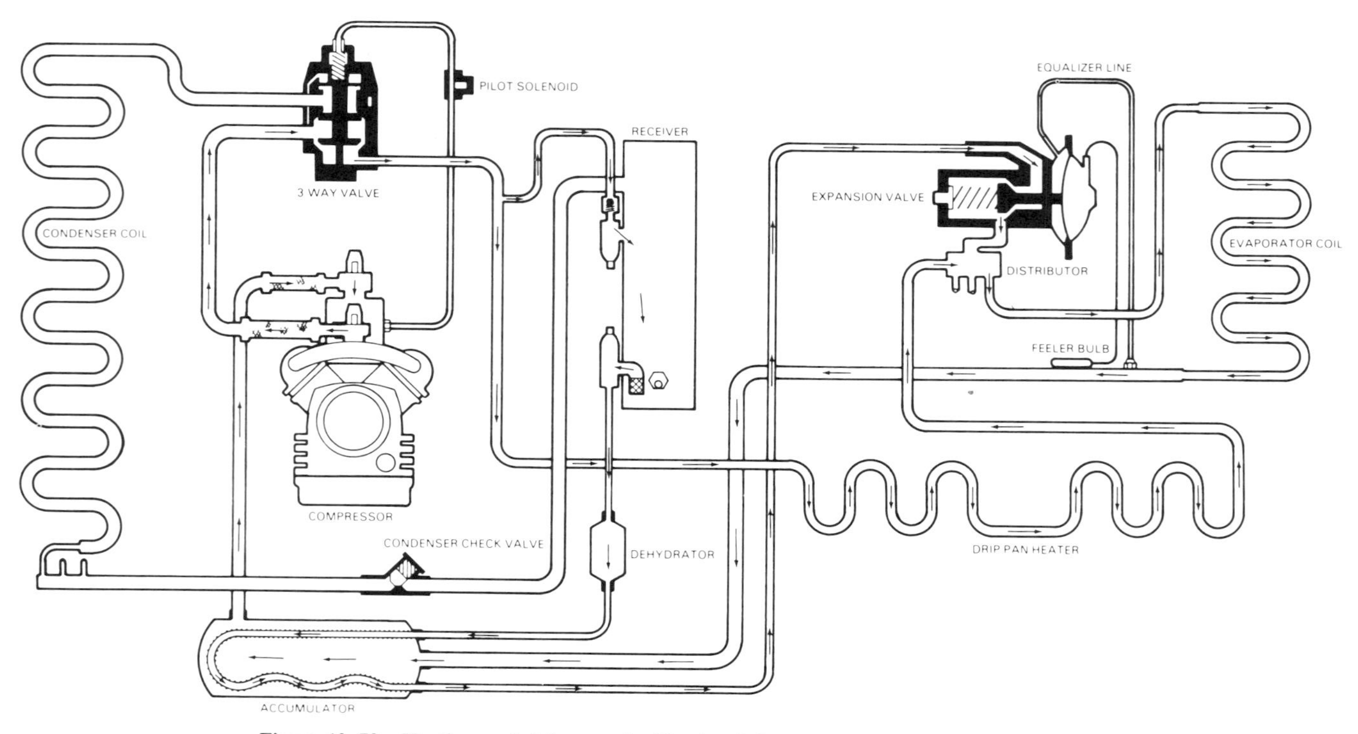

Figure 12–79 Heating and defrost cycle. Heating is by reverse-cycle principle using hot-gas discharge from compressor. Defrost is manual or automatic. (By permission of Thermo King Corporation.)

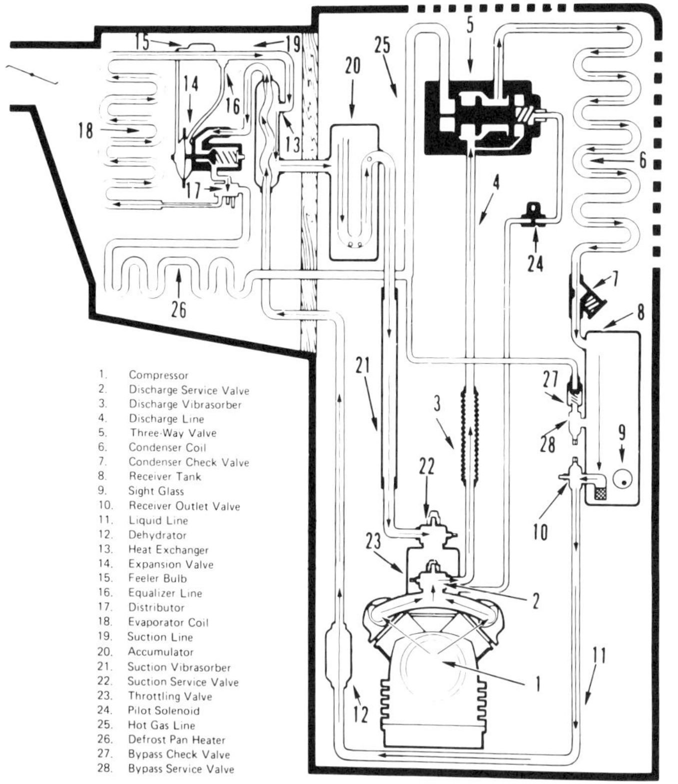

Figure 12-80 Refrigeration cycle. Operates on fuel saver principle. When thermostat calls for cooling, unit operates for 8 minutes at low speed. If high cooling is still required, unit automatically shifts to high speed. (By permission of Thermo King Corporation.)

Figure 12-81 Heating and defrost cycle. Unit defrosts automatically when air flows across coil is restricted. When thermostat calls for heating, unit operates at low speed for 8 minutes. If high heating is still required, unit shifts to high speed. This arrangement saves fuel. (By permission of Thermo King Corporation.)

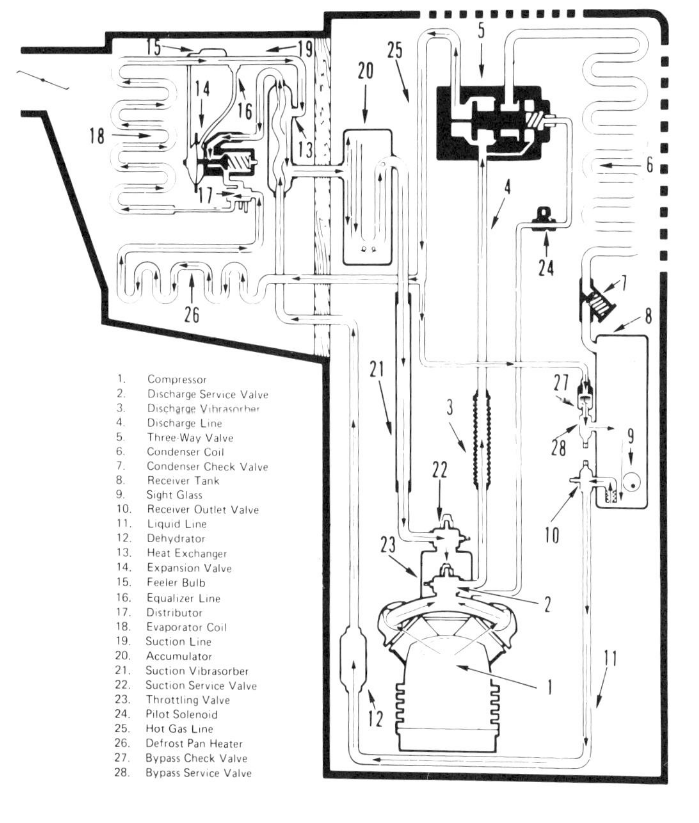

LENGTH OF TRUCK BODY – 22 FT (6.75 m)

Outside temp. 100°F/38°C Carrying temp. of precooled cargo	Average Number of Door Openings per day	AVERAGE POLYURETHANE INSULATION (for styrofoam insulation double polyurethane insulation thickness indicated) 2" / 51mm	3" / 76mm	4" /102mm
−20°F −29°C	1–6	suggest 4" / 102mm insulation	suggest 4" / 102mm of insulation	TWD XRWL
	7–12	suggest 5" / 127mm insulation See footnote No. 5		
	13–25			
0°F −18°C	1–6	TWD XRWL	TWD XRWL	TWD XRWL XKW
	7–12	suggest 3" / 76mm insulation	TWD XRWL	TWD XRWL XKW
	13–25	suggest 3" /76mm insulation	TWD XRWL	TWD XRWL XKW
35°F 2°C	1–6	**XKW** XMD XMT	**XKW** XMD XMT	**XKW** XMD XMT
	7–12	**XKW** TWD	**XKW** TWD	**XKW** TWD
	13–25	**XKW** TWD	**XKW** TWD	**XKW** TWD
55°F 13°C	1–6	**XMD** **XMT** XKW	**XMD** **XMT** XKW	**XMD** **XMT** XKW
	7–12	**XMD** **XMT** XKW	**XMD** **XMT** XKW	**XMD** **XMT** XKW
	13–25	**XMD** **XMT** XKW	**XMD** **XMT** XKW	**XMD** **XMT** XKW

Models in brackets have same refrigeration capacity

1. For optimal performance use unit set in **BOLD TYPE FACE.**
2. Tables assume product precooled to required carrying temperature.
3. Overlapping, shower-type curtains are recommended.
4. For XMT applications below 35°F/+1.7°C ambient, models 30, 50 are recommended for greatest efficiency on heat and defrost cycles.
5. For 5"/127mm insulation use TWD/XRWL.

Figure 12–82 Typical unit selection chart. Size selected is based on the length of the truck, box temperature, average door openings per day, and amount of insulation. (By permission of Thermo King Corporation.)

REVIEW EXERCISE

Select the letter representing your choice of the correct answer (there may be more than one).

12-1. The ARI rating standards for water coolers give the gph at the following conditions:

________ a. 80°F inlet, 50°F outlet, 90°

________ b. 70°F inlet, 50°F outlet, 90°F ambient

________ c. 70°F inlet, 50°F outlet, 80°F ambient

________ d. 80°F inlet, 50°F outlet, 80°F ambient

12-2. The refrigeration system for most types of drinking water coolers use what type of metering device?

________ a. Thermal expansion valve

________ b. High-pressure float

________ c. Capillary tube

________ d. Low-pressure float

12-3. How is clear ice made?

________ a. By the use of a chemical

________ b. By agitating the container

________ c. Using a syphon to remove the air

________ d. With running water

12-4. The most common refrigerant used for small self-contained ice cube makers is:

________ a. R-12

________ b. R-22

________ c. R-502

________ d. R-500

12-5. Commercial-type ice makers will deliver ice at temperatures:

________ a. 25°F and higher

________ b. 0°F and lower

________ c. 0°0C

________ d. −5°C and higher

12-6. Referring to 100 percent rated capacity for an ice builder, with an ice thickness of 2½-in. and a rebuilding time of 10 hours, the refrigerant suction temperature is:

________ a. 0°F

________ b. 5°F

________ c. 10°F

________ d. 15°F

12-7. The cut-in point of a low-pressure control for a reach-in beverage cooler using R-12 refrigerant should be:

________ a. 8 psig

________ b. 18 psig

________ c. 28 psig

________ d. 38 psig

12-8. Loading a refrigerated merchandiser in a supermarket (e.g., milk cartons) can be greatly speeded up by the use of:

________ a. Improving the refrigeration system

________ b. Cart stocking

________ c. Longer refrigerated cases

________ d. Using self-contained units

12-9. The most common location for the evaporator coil on open-type merchandiser cases is:

________ a. In the top of the case

________ b. Inside of the case

________ c. In the bottom of the case

________ d. In some remote location

12-10. A heat reclaimer coil can be used to:

________ a. Utilize condenser heat to heat the building

________ b. Pick up heat from the outdoors

________ c. Act as a heat exchanger for the refrigeration system

________ d. Reclaim heat from inside the building

13 Wiring Diagrams and Control Circuits

OBJECTIVES

After studying this chapter, the reader will be able to:

- Determine the sequence of system operation from the information in the wiring diagram.
- Use the information supplied in wiring diagrams for installation and service of various types of commercial, industrial, and institutional refrigeration equipment.
- Modify manufacturers' wiring diagrams where necessary to add supplementary controls as needed.

WIRING DIAGRAM PRACTICES

There is considerable variation in the way manufacturers make wiring diagrams. It is therefore the purpose of this text to present a number of different wiring diagrams so that the reader can become familiar with the practices that occur.

TYPES OF WIRING DIAGRAMS

Wiring diagrams are principally of three different types: (1) field or installation, (2) connection or pictorial, and (3) schematic.

The *installation wiring diagram* indicates the electrical connections that must be made in the field. Usually these relate to the power requirement, but they may also indicate the wiring for a remote thermostat or separate blower unit, anything outside the main prewired unit that needs to be wired in the field.

The *connection diagram* shows the exact manner of wiring the unit, where each wire is connected to the electrical devices.

In the case of packaged equipment, much of this wiring is done at the factory. The diagram usually shows the actual position of each control and the wiring to each connected terminal. If alternate wiring arrangements can be made using auxiliary controls, these wiring arrangements are shown on the connection diagram.

The most important diagram is the *schematic*. This shows not only the separate circuits included in the wiring but also indicates their function in the complete system. Some manufacturers furnish only the schematic diagram and omit the connection diagram. Actual field wiring can be done from the schematic diagram.

Since the schematic diagram is the most important and most useful, most of this discussion will relate to schematic wiring diagrams.

A schematic wiring diagram is often called a *ladder diagram,* which serves to describe its configuration. The power supply lines are usually drawn vertically, with horizontal lines, representing the individual circuits. Each circuit has a load and one or more switches.

TYPICAL WIRING DIAGRAMS

The following are examples of actual wiring diagrams using various types of refrigerated equipment. They serve to indicate the industry practice.

Drinking Water Coolers

Figure 13–1 shows the wiring diagram for a drinking water cooler. This diagram serves as both a schematic and a connection diagram. There are three loads: the compressor, the fan motor, and the current relay coil. There are three switches: a thermostat, a current relay switch (NO), and an overload switch (NC).

When the thermostat calls for cooling, current flows through the relay coil and the run winding of the compressor. This initial rush of current (LRA) energizes the relay coil, closing the relay switch, causing current to flow through the start winding of the compressor. When the compressor reaches about three-fourths full speed, the current through the run winding drops off (FLA), deenergizing the relay coil, opening the relay switch, and cutting off current to the start winding of the compressor. The start capacitor is in series with the start winding of the compressor to increase the starting torque. The start capacitor is removed from the circuit at the same time the compressor start winding is deenergized. The condenser fan motor is wired in parallel with the compressor and runs when the compressor runs.

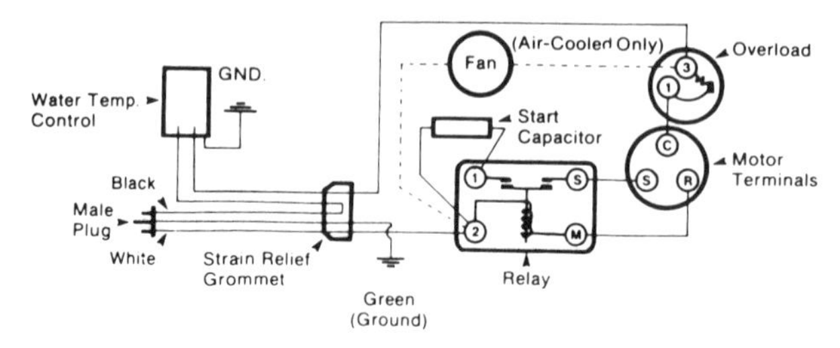

Figure 13–1 Drinking water cooler, wiring diagram. The thermostat controls operation of the compressor. The compressor uses capacitor start motor and current-type relay. The condenser fan is operated at the same time as the compressor. See Figure 12–3 for description of cooler parts and Figure 12–4 for refrigeration cycle. (By permission of Halsey-Taylor Division, King-Seeley Thermos Company.)

Electric Defrost Systems for Walk-in Freezers

The simplest type of electric defrost system has application in a walk-in freezer installation. The refrigeration system consists of a condensing unit and a blower-type evaporator unit.

The wiring diagram for the system is shown in Figure 13–2. This system uses a single-phase compressor motor and an evaporator with three fan motors. The thermostat operates a solenoid valve in the liquid refrigerant line indirectly operating the compressor on a pump-down cycle. The compressor is shut down by the low-pressure control. The defrost cycle is time initiated and temperature terminated. Power connections are made to N and 1 in the time clock.

The sequence of operation is as follows:

1. For normal operation the system operates in response to the thermostat on a pump-down cycle arrangement.
2. The timer initiates the defrost cycle, usually one to three defrost cycles per day. The timer opens switch A in the time clock, which stops the compressor and evaporator fan motors, and closes switch B, energizing the defrost heaters.
3. When the coil warms up to 55 °F, the defrost termination thermostat closes, energizing the holding coil in the timer, which closes switch A and opens switch B. The compressor runs. The evaporator fans do not run since the fan delay thermostat is open. The safety heater thermostat would open only if the coil temperature reaches 75 °F and the defrost termination thermostat fails to close. Holding the evaporator fans off when the compressor starts is good practice since it prevents blowing warm air.
4. When the temperature of the coil reaches 35 °F the fan control switch closes and the evaporator fan runs.

Figure 13–2 Electric defrost wiring diagram. The defrost system is time initiated and temperature terminated. Compressor operates on pump-down cycle. The terminating thermostat provides for a fan delay at the end of the defrost cycle. (By permission of Dunham-Bush, Inc.)

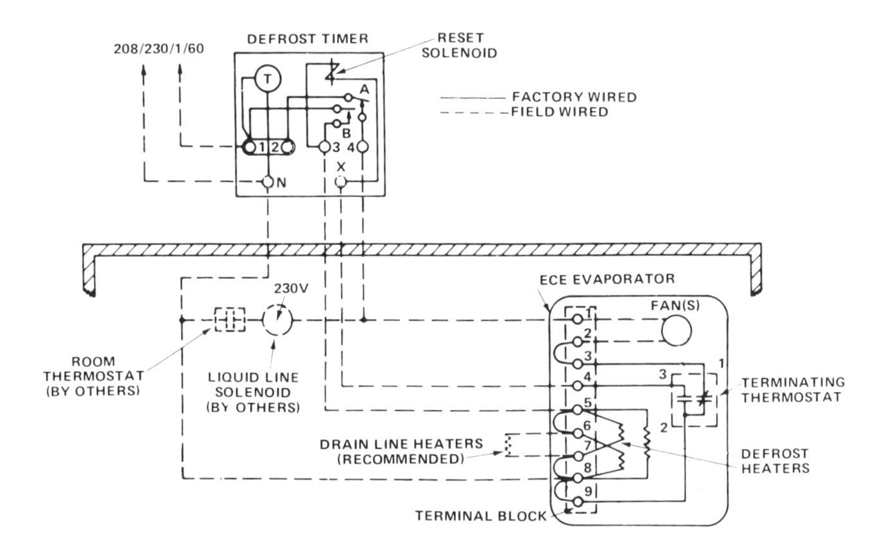

Hot Gas Defrost Systems for Walk-in Freezers

One of the most effective gas defrost systems is the Thermobank system. The electrical diagram is shown in Figure 13–3. The system operates on a pump-down cycle with the thermostat opening the liquid solenoid valve and the compressor starting when the contacts make on the low-pressure control. During the normal cycle the hot gas from the compressor warms the storage tank water before going to the condenser.

Figure 13–3 Thermobank wiring diagram shows the controls used to operate the normal refrigeration and the defrost cycle (see Chapter 8 for refrigeration diagrams). (By permission of Kramer-Trenton Company.)

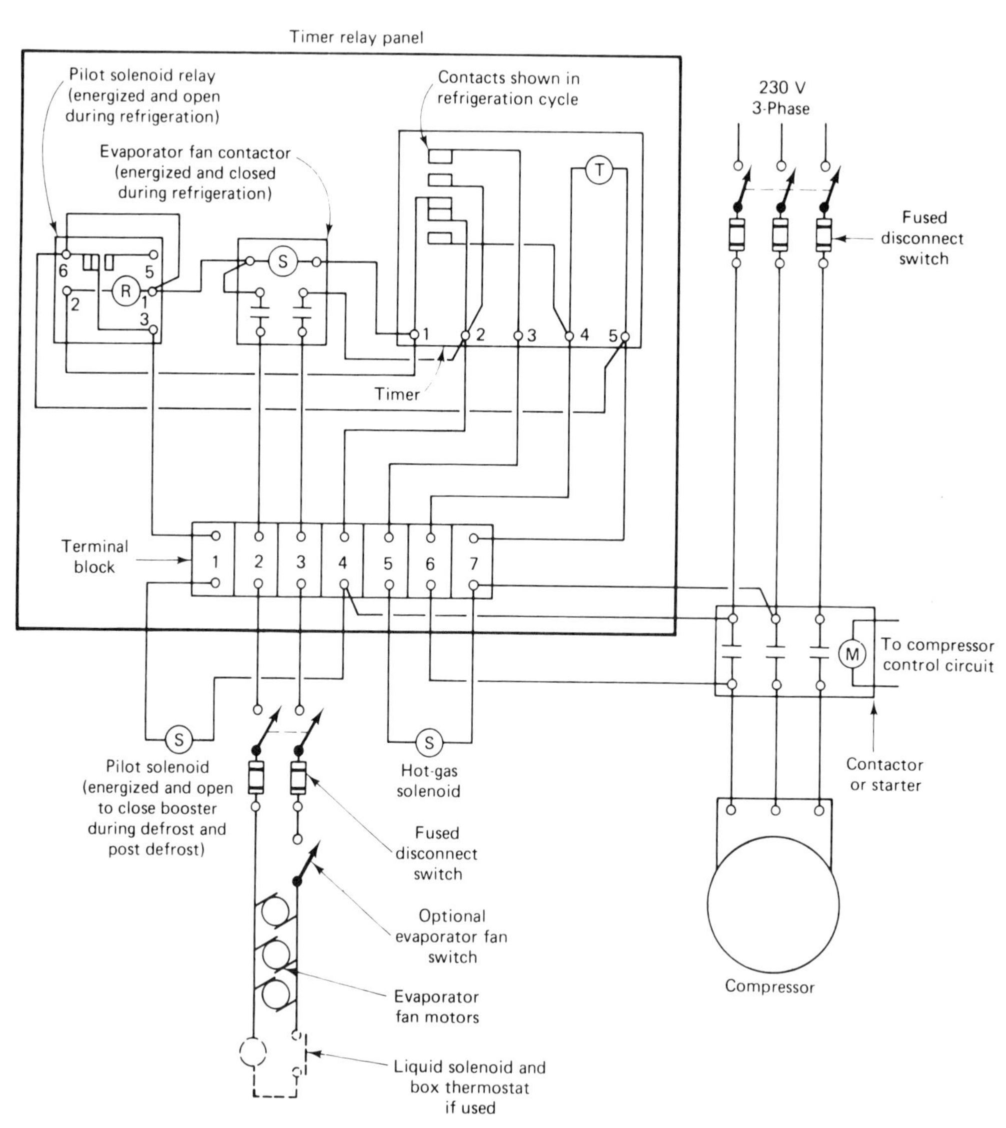

When the time clock calls for defrost, the evaporator motor stops, the suction-line solenoid valve closes, and the hot-gas solenoid opens. Suction gas passes through the hold-back valve and into the storage tank. This warms the suction gas before it reaches the compressor. The compressor discharge passes through the evaporator coil, removing the frost. The defrost cycle is usually terminated in about 7 minutes.

At the termination of the defrost cycle the hot-gas solenoid is deenergized and the valve closes. The compressor operates to lower the suction pressure and, after 3 minutes, the system is restored to normal operation. The evaporator fans start when the suction-line solenoid is opened.

This type of hot-gas defrost system is especially useful where a single freezer room is involved. The tank of water stores extra heat required for defrost; otherwise, the heat of compression would be the sole source.

Ice Cube Makers

Figure 13-4 shows the wiring diagram for an ice cube maker. The on/off *bin thermostat* (SPST) controls the starting and stopping of the unit. If the capillary senses 34 °F, the unit will shut off. When the capillary senses 38 °F, the unit will start again.

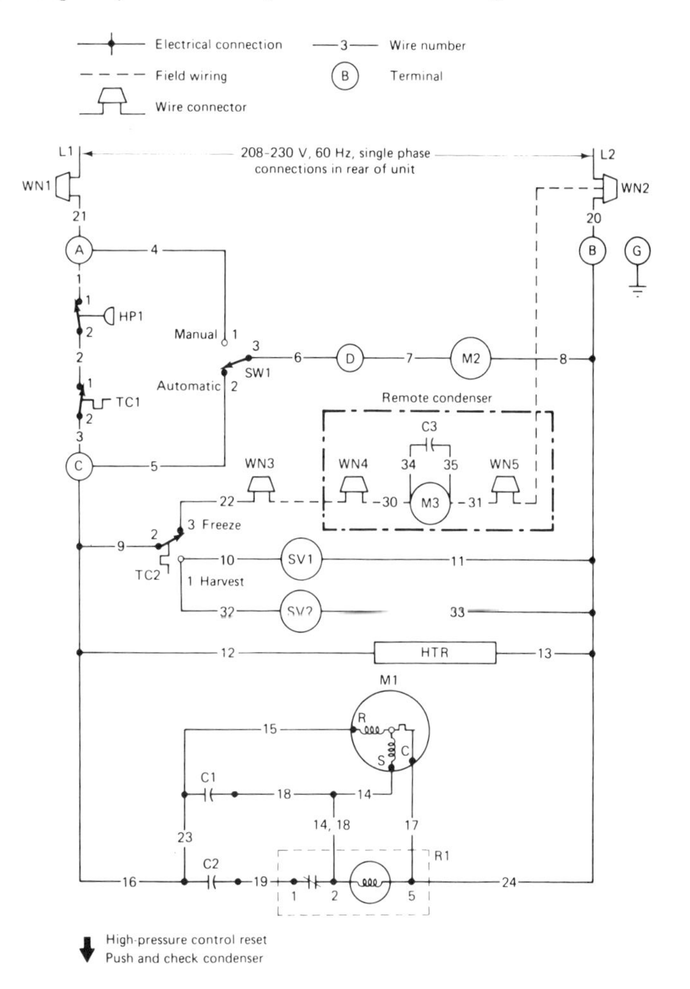

Figure 13-4 Packaged ice cube maker, wiring diagram. Bin thermostat (TC1) ice-size thermostat controls the complete operation of the unit (see description of sequence of operation in Chapter 12). SV1, Hot-gas solenoid valve; SV2, water solenoid valve; M1, compressor motor; M2, water pump motor; M3, fan motor; HP1, high-pressure control; TC1, bin thermostat; TC2, ice-size thermostat; SW1, water pump switch; HTR, ice-size heater; C1, run capacitor; C2, start capacitor; R1, compressor relay; C3, fan motor capacitor. (By permission of McQuay Group, McQuay-Perfex, Inc.)

When placed in the "auto" position, the *water pump switch* (SPDT) will cause the pump to run whenever the complete unit runs. The *ice-size thermostat* (SPDT) slides into a sheath that mounts in the ice-making area of the evaporator. When the ice increases in size, the temperature of the sensor drops to 38 °F and the unit switches to the harvest cycle. When the evaporator warms up to 48 °F, indicating an end of the harvesting, the thermostat switches back to the freeze cycle.

The *ice-size heater* is a wire-wound resistor located under the front cover of the evaporator casing and slides over the sheath containing the ice-size thermostat capillary. It prevents the capillary from dropping below 38 °F prematurely during the freeze cycle before the ice forms to contact it. This heater is on continuously since it is not effective when surrounded by ice, due to the rapid heat dissipation.

The *hot-gas solenoid* (NC) is actuated by the ice-size thermostat during the harvest cycle. The *water solenoid* will add water to the pan only during the harvest cycle. Water rises to the point of overflow and impurities are bled off through the siphon tubes.

Ice Dispensers

Figure 13–5 shows the wiring diagram for an ice dispenser unit. Note in the diagram that each circuit is indicated by a letter and the circuits are named from A to G. The power

Figure 13–5 Schematic wiring diagram for an A, compressor fan motor line; B, series controls and gear motor circuit; C, water solenoid and switch; D, ice portion control circuit; E, step-down transformer; F, spout switch reset line circuit; G, full ice bin switches. Level control energizes relay A. Relay A energizes compressor contactor and auger drive unit. (By permission of Scotsman.)

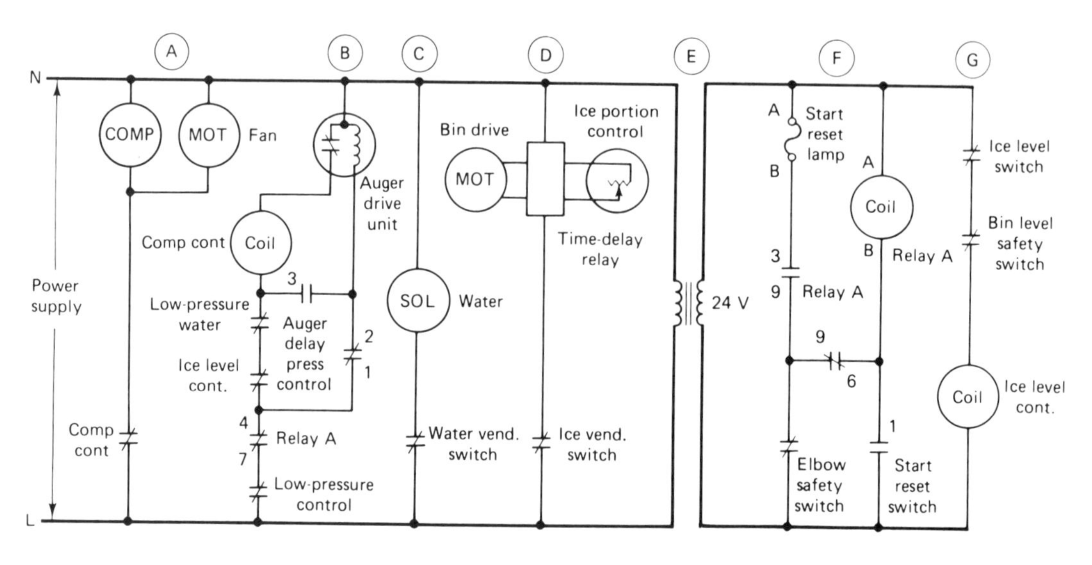

lines in this diagram are horizontal and the circuit lines are vertical. The principal loads are:

Load	Circuit location
1. Compressor motor	A
2. Condenser fan motor	A
3. Compressor contactor coil	B
4. Auger drive unit	B
5. Solenoid water valve	C
6. Ice portion motor	D
7. Control circuit transformer	E
8. Relay A, three switches	F
9. Ice-level control, one switch	G

The equipment has a number of operating switches:

Operating switch	Circuit location
1. Ice-level control	B
2. Water vend switch	C
3. Ice vend switch	D

There are also a number of safety switches:

Safety switch	Circuit location
1. Low-pressure water	B
2. Low refrigerant pressure	B
3. Elbow safety switch	F
4. Bin-level safety switch	G
5. Start reset switch	F
6. Auger-delay pressure control	B

During the normal ice-making mode the ice-level control operates the compressor and the auger drive unit. Ice is held in the bin until the ice vending switch is made which operates the bin drive motor. The water vend switch operates the water solenoid valve.

Relay A is primarily used in the safety control circuit to prevent the operation of the compressor and auger drive unit when the elbow safety switch is activated. The auger delay switch will permit the auger to clear the evaporator of ice in case the ice-level control stops the compressor.

Flake Ice Machines

A wiring diagram for a flake ice machine is shown in Figure 13–6. The unit is started by manually closing the on/off switch. Assuming that the bin is not full of ice, the gear motor is started through the gear motor relay and the main relay is energized. The main relay has two NO switches. One switch forms a holding circuit for the main relay. The

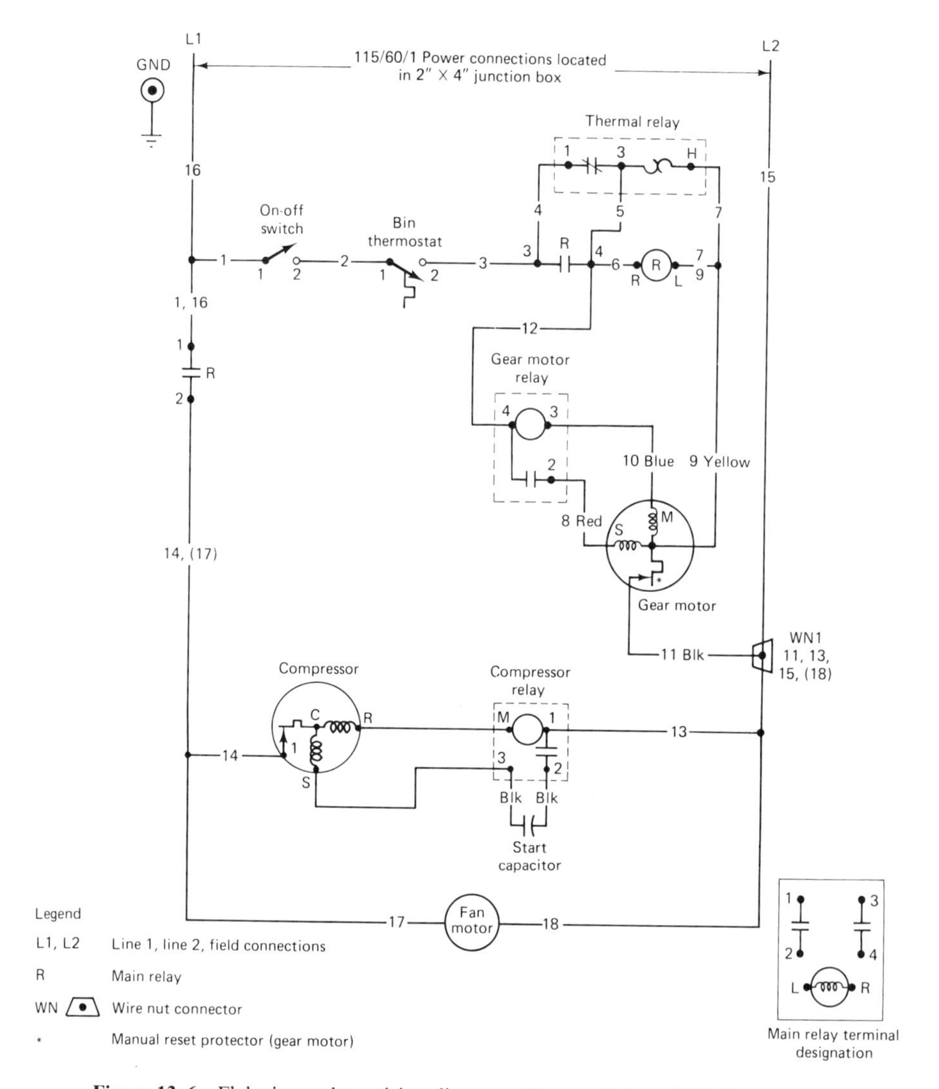

Figure 13-6 Flake ice maker, wiring diagram. (By permission of McQuay Group, McQuay-Perfex, Inc.)

other switch starts the compressor and fan. A safety circuit is provided by the thermal relay. When the unit starts to make ice, the thermal relay contacts open and remains open until the unit is stopped by the bin thermostat. After a delay of from 2 to 5 minutes, and ice is no longer on the evaporator, the thermal relay closes, permitting the unit to restart by the action of the bin thermostat.

Large-Capacity Flake Ice Equipment

Figure 13-7 shows the wiring diagram and the refrigeration piping diagram for a large flake ice machine. This unit uses a flooded evaporator and can be adapted for either halocarbon refrigerants or ammonia. Referring to the wiring diagram for halocarbon

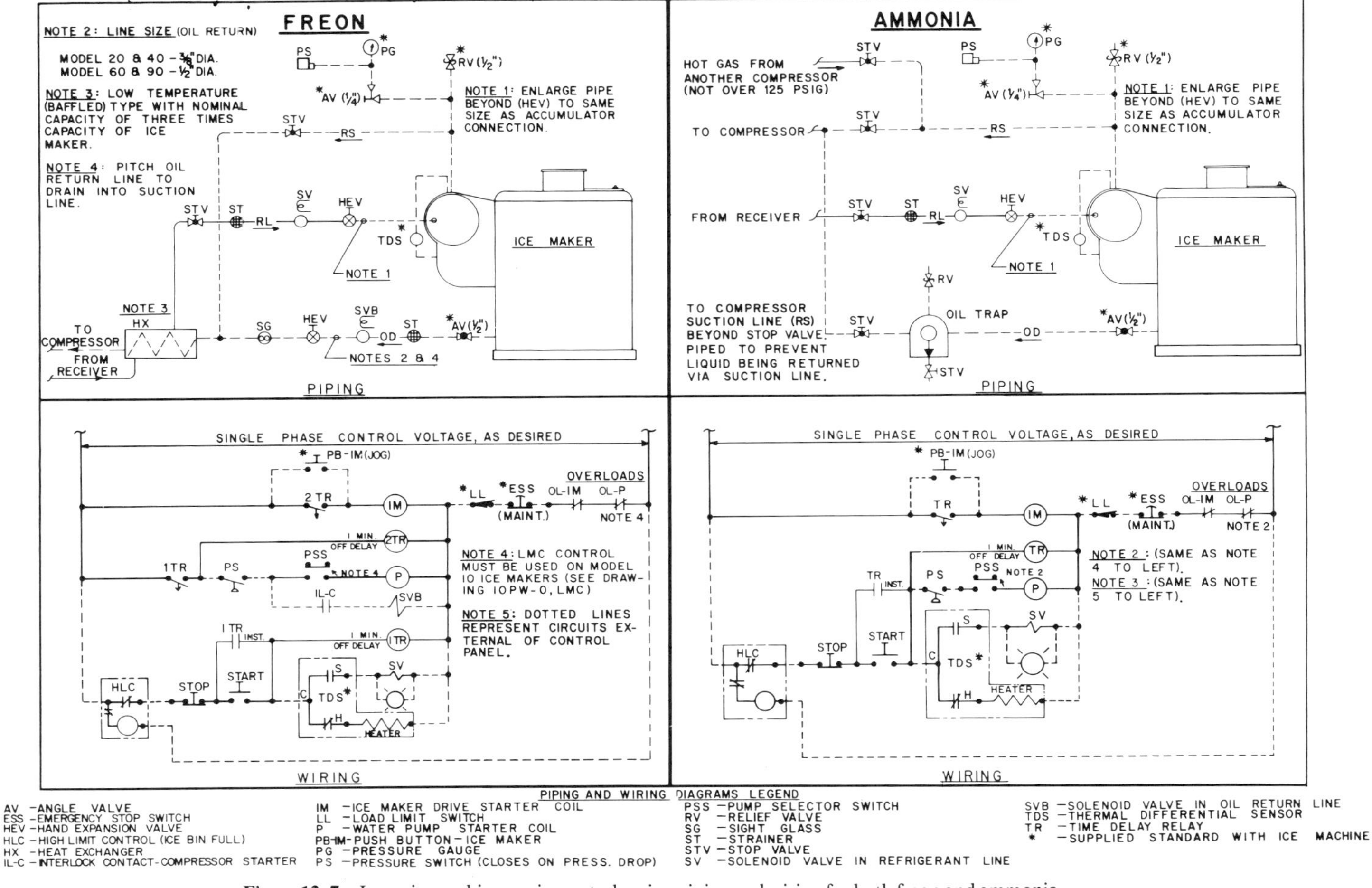

Figure 13–7 Large ice-making equipment, showing piping and wiring for both freon and ammonia systems. The system is manually started by pushbutton start switch. Refrigerant solenoid opens to start the compressor and time-delay switch (1TR) is energized to start the water pump (P). After 1-minute delay the second time-delay switch (2TR) starts the drive motor. Ice is made continuously until the equipment is stopped. (By permission of North Star Ice Equipment Company.)

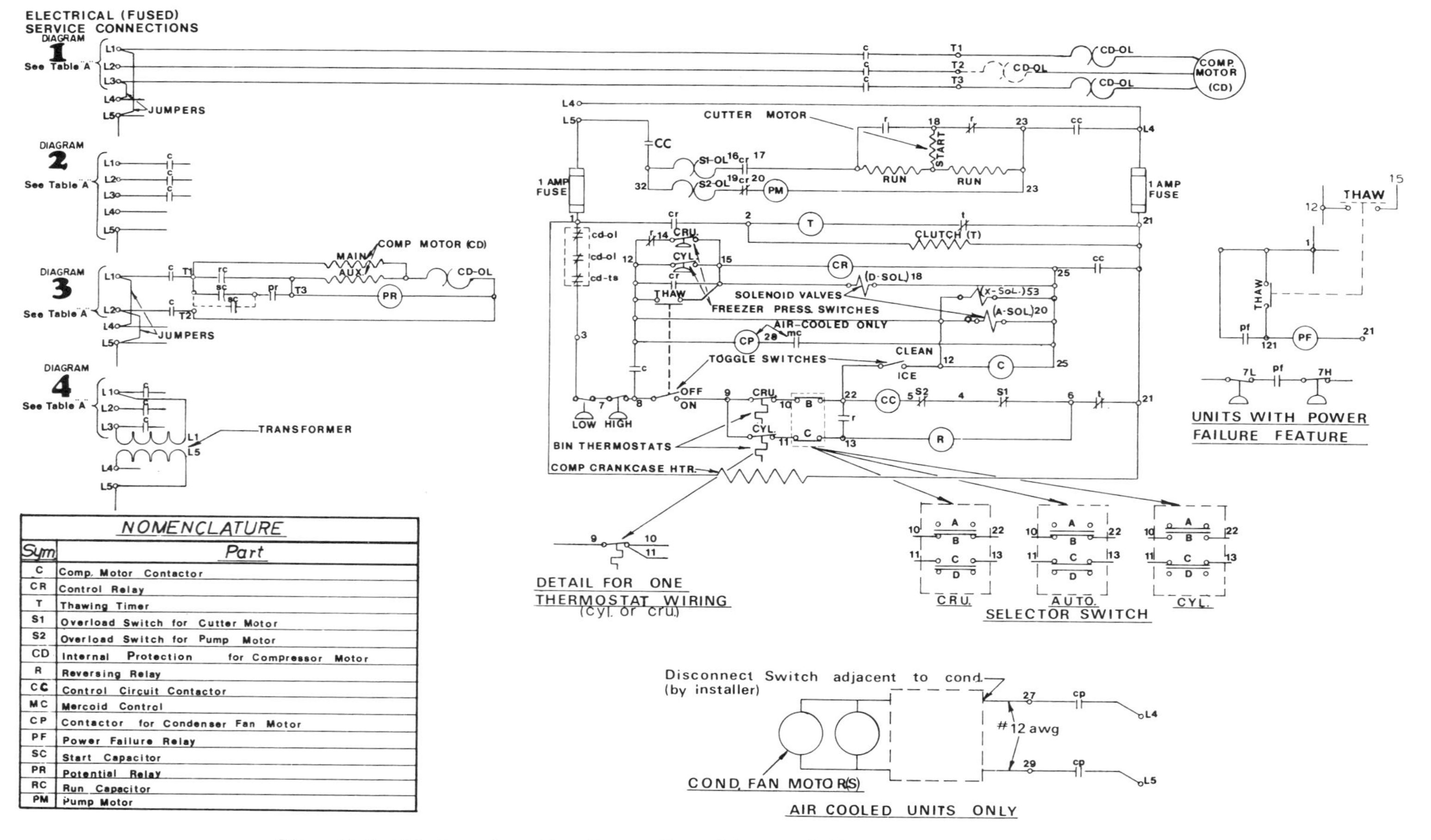

NOMENCLATURE

Sym	Part
C	Comp. Motor Contactor
CR	Control Relay
T	Thawing Timer
S1	Overload Switch for Cutter Motor
S2	Overload Switch for Pump Motor
CD	Internal Protection for Compressor Motor
R	Reversing Relay
CC	Control Circuit Contactor
MC	Mercoid Control
CP	Contactor for Condenser Fan Motor
PF	Power Failure Relay
SC	Start Capacitor
PR	Potential Relay
RC	Run Capacitor
PM	Pump Motor

Figure 13–8 Tube ice maker, wiring diagram. The unit is controlled by the on-off "thaw" switch. The selector switch determines the type of ice produced. An ice bin thermostat can be used to operate the machine automatically. During freezing, water is continuously circulated by the pump (PM). During freezing, hot gas solenoid (D-sol) is closed. At the end of the freezing cycle the hot gas solenoid (D-sol) is opened by the pressure switch. The water pump (P) is stopped and the ice crusher is started. The clean ice switch is always set on the "ice" position unless the unit is being cleaned. (By permission of Henry Vogt Machine Company.)

refrigerants, note that the unit is started manually. There is a safety arrangement to shut off the machine with a high-limit control (HLC) when the ice bin is full.

When the start switch is operated, the first time-delay relay (lTR) is energized and the accumulator heater is turned on. After a 1-minute delay the relay (1TR) switch closes, energizing the second time-delay relay (2TR), and if the compressor is running (operated by separate controls), the interlock (IL-C) will close, energizing the solenoid valve in the oil return line.

After 1 minute the relay (2TR) switch will close, energizing the ice-maker drive. Note that the solenoid valve in the refrigerant liquid line is energized after the compressor is running and a temperature differential is sensed between the vapor and liquid in the accumulator. When the liquid solenoid opens, the accumulator heater is turned off by the thermal differential senser (TDS).

Note that there are four safety controls in series with one side of the power supply, a load limit switch (LL), an emergency stop switch (ESS), an overload for the ice-maker drive, and an overload for the pump motor.

Tube Ice Machines

The unit is started by action of the thaw switch. However, the unit will not start unless the low-pressure switch is closed. Also, the bin thermostat must be closed, indicating the need for ice (see Figure 13–8).

Power is then supplied to the control circuit contactor (which has three switches), the compressor contactor (which has four switches), and the liquid-line solenoid valve. The compressor and the water pump motor (PM) start and ice is made in the evaporator.

The freezing period for the production of ice is controlled by the freezer pressure switch. When the pressure drops to close this switch the control relay (CR) is energized, operating four switches. One of these energizes the thawing timer (T), which governs the ice discharging period. A second control relay switch opens the hot-gas solenoid (D). The two additional control relay switches stop the water pump and start the cutter motor.

Refrigeration Compressors

Figure 13–9 shows the wiring diagram for a rotary screw compressor. This compressor starts 100% unloaded so that a normal torque motor can be used. It is designed with a water-cooled oil chiller, water-cooled cylinder block, and water-cooled seal bearings. Water flow is controlled by solenoid valves. An optional arrangement is provided for return of oil from the oil separator using a solenoid valve.

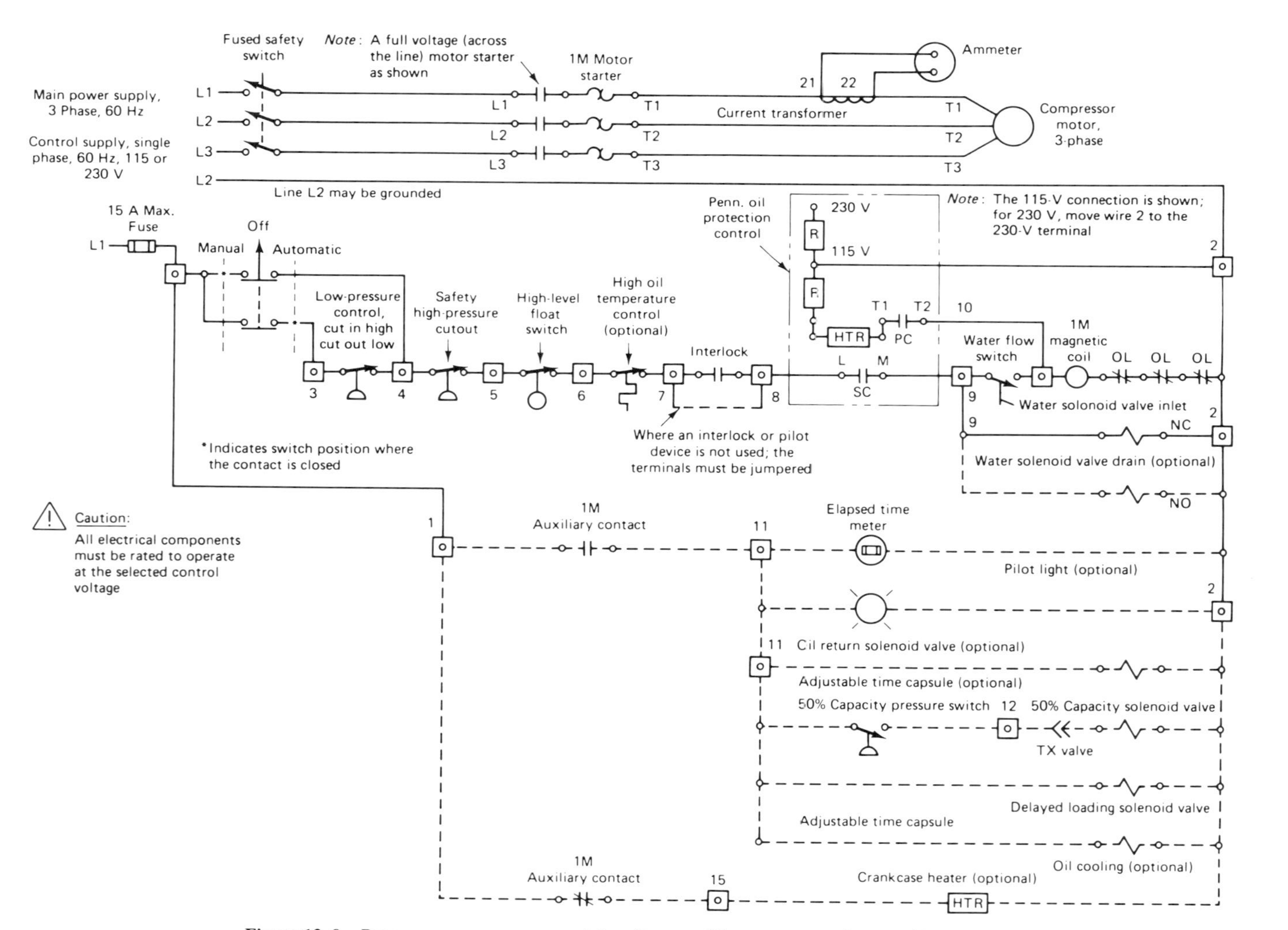

Figure 13–9 Rotary screw compressor wiring diagram. The compressor is started by a manual pushbutton switch and the auxiliaries are started by an extra set of contacts in the compressor contactor. Most of the safety controls are wired in series with the control circuit of the compressor starter. (By permission of Crepaco, Inc.)

Supermarket Systems

The earlier supermarket systems used a series of rack-mounted air-cooled condensing units installed in a compressor room. The room was ventilated in winter to maintain a minimum temperature of 64°F for the air-cooled condensers.

The modern approach to supermarket refrigeration is the use of groups of parallel-connected compressors. Banks of compressor systems are selected to meet the needs of fixture selection for a specific store. These compressors are controlled by a common suction pressure switch which is set for the lowest-temperature fixture on the system. EPR valves are used on individual cases where higher suction temperatures are required. Individual cases can use air defrost, electric defrost, or hot-gas defrost.

A typical wiring diagram for the three-compressor system is shown in Figure 13–10. The control cirucit for the defrost system is shown in Figure 13–11.

Referring to Figure 13–10, the compressor is controlled by a pressure switch. Each compressor is equipped with an oil pressure failure control, a cooling fan for the compressor, and a crankcase heater. When the compressor starts, the crankcase heater is turned off and the cooling fan comes on. A red light shows when the oil pressure switch cuts out the compressor.

A solid-state pressure control is available to replace the mechanical switches (see Figure 13–12). The compressor wiring diagram shows the solid-state control connected (SSPC). This control has one set of cut-in and cut-out suction pressure settings and two time-delay relays. The solid-state unit will stage the compressor to run only the number of machines required to maintain the suction pressure. It will also rotate compressor usage so that all compressors will receive equal wear. To prevent compressor short cycling one relay prevents adding compressors in less than 3 minute intervals and another prevents stopping compressors in less than 3 minute intervals.

Visual information is provided by the solid-state control to show suction pressure, cut-in setting, and cutout setting. Indicator lights are provided to show "defrost in process." The control also provides for at least one compressor to operate during a defrost cycle even though the suction pressure is below the cut-in setting.

The timing controllers regulate the defrost cycles. Defrosting periods are staggered so that all systems do not defrost at once. This provides an adequate supply of hot gas for defrost since most of the machines are operating on their normal cycle. As shown in Figure 13–11, each timer controls the solenoid valves and the fans required for the defrost cycle.

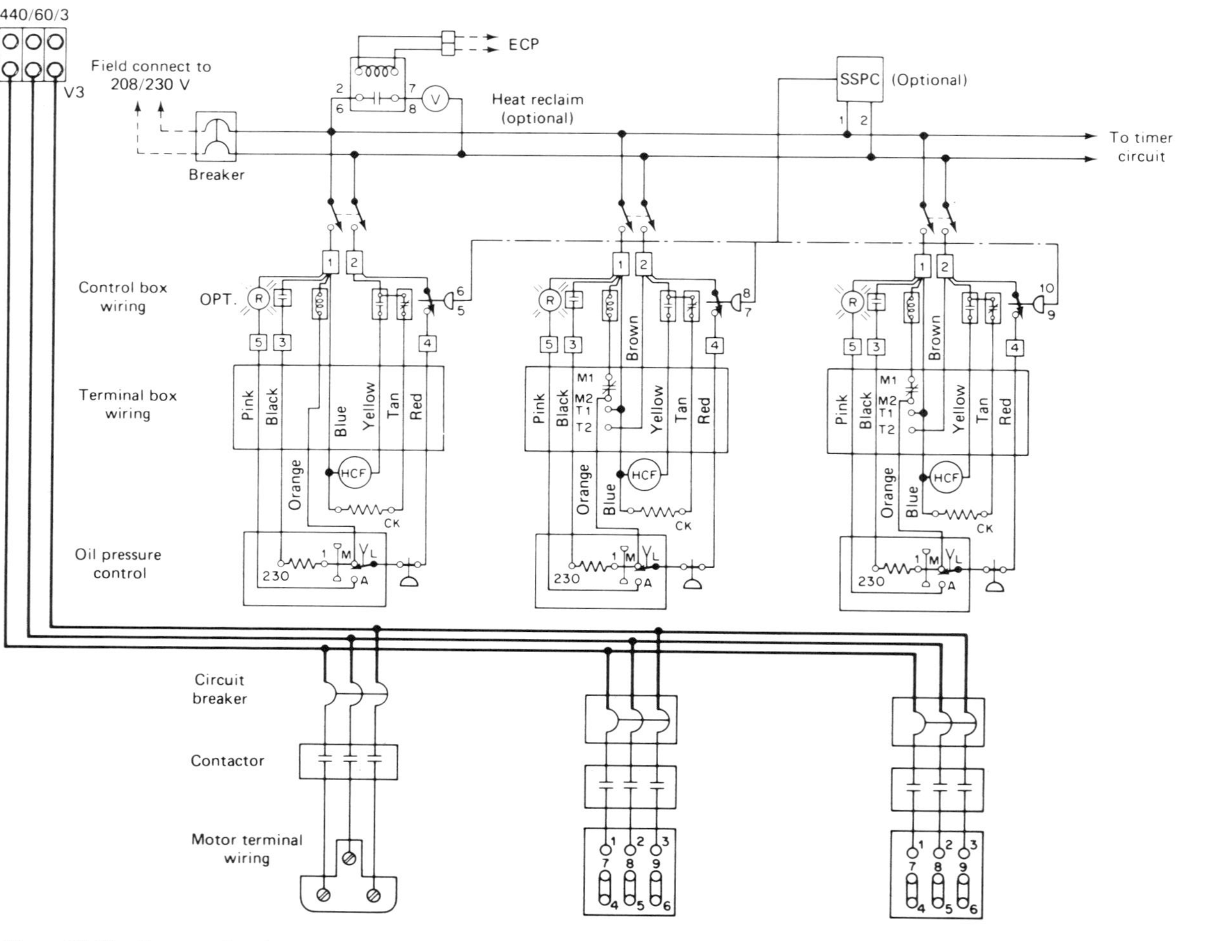

Figure 13–10 Supermarket three-compressor unit, wiring diagram. The individual units operate in response to the suction pressure control. The crankcase heater is energized only when the compressor is off. With a call for cooling, the holding coil (HC) is energized. The oil pressure failure control can stop the compressor in case of low oil pressure. BRKR, Breaker; CK, crankcase heater; CKT, circuit; D, defrost termination; ECP, environmental control panel; G, hot-gas solenoid; L, liquid-line solenoid; ML, master liquid solenoid; S, suction stop solenoid; SOL, solenoid; SSPC, solid-state pressure control; T, thermostat; TM, timer motor; V, three-way diverting valve. (By permission of Warren/Sherer.)

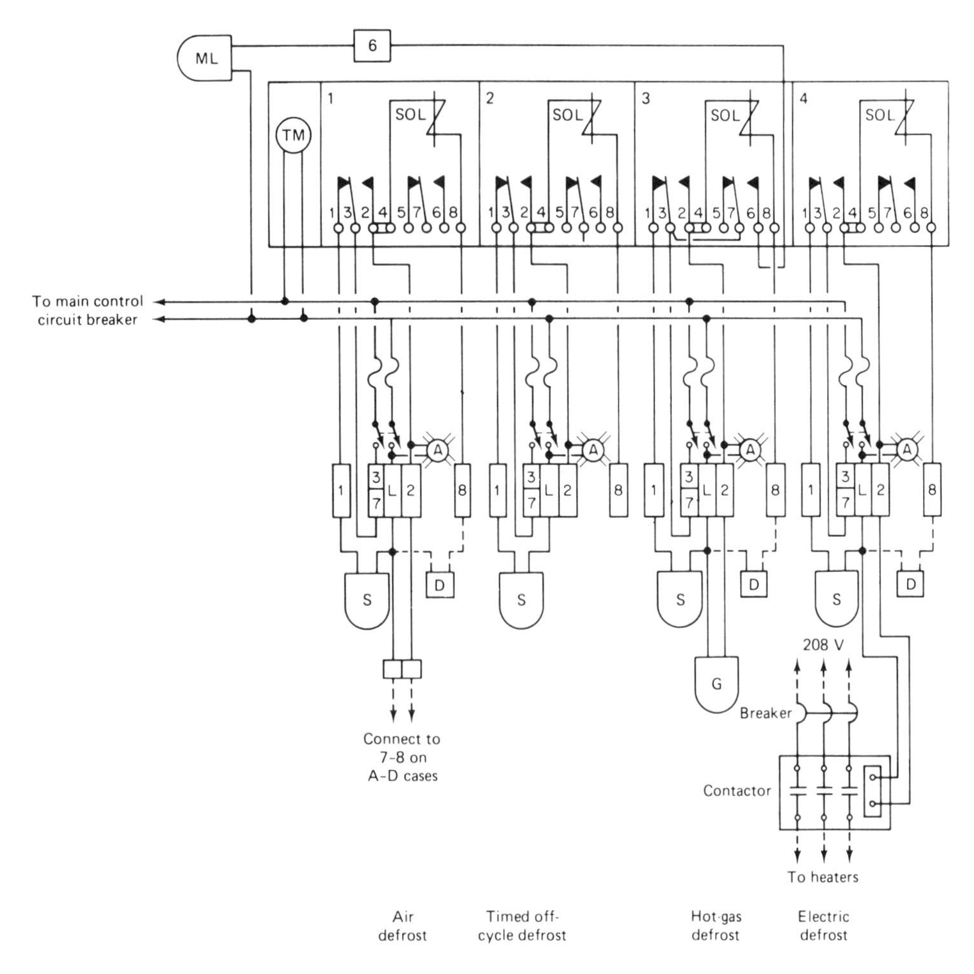

Figure 13–11 Supermarket compressors defrost system. The diagram shows typical timing system and defrost controls for three types of defrost systems. Defrost systems are timed to "stagger" defrosting. See Figure 13–10 for legend. (By permission of Warren/Sherer.)

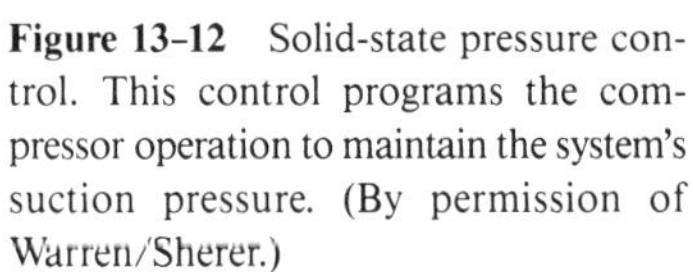

Figure 13–12 Solid-state pressure control. This control programs the compressor operation to maintain the system's suction pressure. (By permission of Warren/Sherer.)

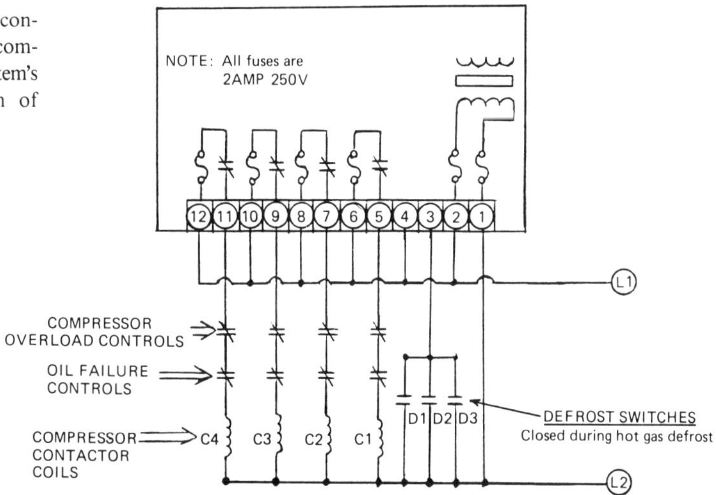

Diagrams for the refrigeration piping, showing the operation of the system for cooling, heating, and defrost, are shown in Figures 12–82 through 12–84. Refer to these piping diagrams for details on how the system operates during the three modes of use.

The schematic wiring diagram is shown in Figure 13–13 and a convection diagram for the same unit is shown in Figure 13–14. Each of these drawings requires considerable study for a full understanding. Each one is complete and fully describes the electrical wiring.

The following information will assist the reader in determining the sequence of operation. Referring to the schematic drawing (Figure 13–13), a number of aspects of the electrical system are indicated.

Operation. The primary controls consist of a number of switches:

1. *Unit on/off switch.* This switch is placed in the "on" position to start the unit. In the "on" position the electrical system receives 12-V dc power from the battery. The "off" position deenergizes the electrical system, closes the fuel solenoid, and stops the engine.
2. *Preheat/start switch (PHS).* In the preheat position the glow plugs (GP) are energized. In the "start" position the glow plugs and the starting motor (SM) are energized.
3. *Diesel-electric switch (DE).* This switch is placed in "diesel" position to operate with a diesel engine and in "electric" position to operate the compressor with an electric motor.
4. *Manual defrost switch (MD).* The unit defrosts automatically at coil temperatures below 32 °F. The manual switch can be used at any time to defrost the evaporator.
5. *Thermostat (CM).* The thermostat is set at the desired box temperature.

Indicator lights

PL1: white, cooling

PL2: amber, heating

PL3: tan, defrost

PL4: amber, fuel saver; compressor operating partly unloaded

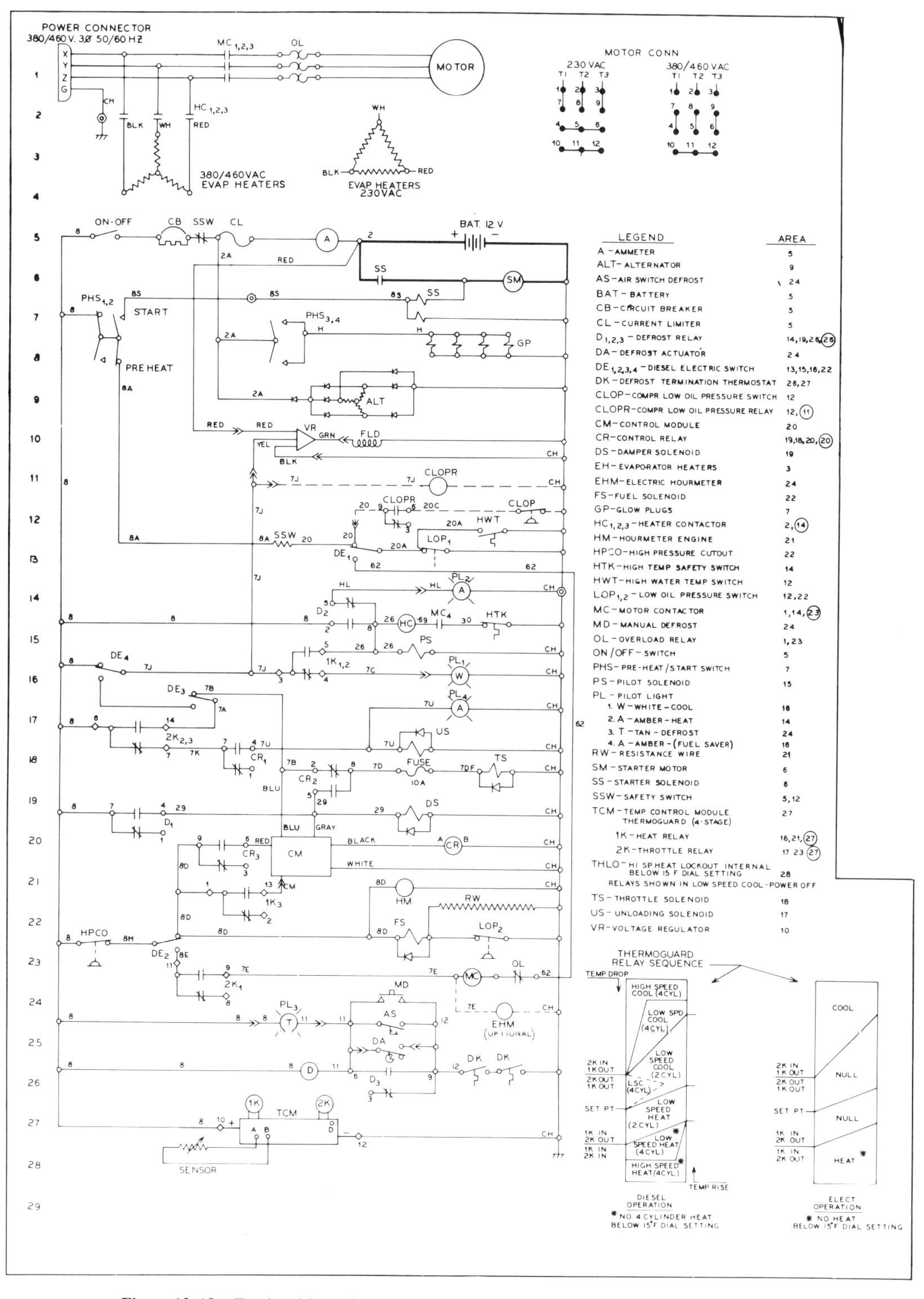

Figure 13–13 Truck refrigeration system, schematic wiring diagram (for diagrams of the refrigeration, heating, and defrost systems, see Figures 12–82, 12–83, and 12–84). (By permission of Thermo-King Corporation.)

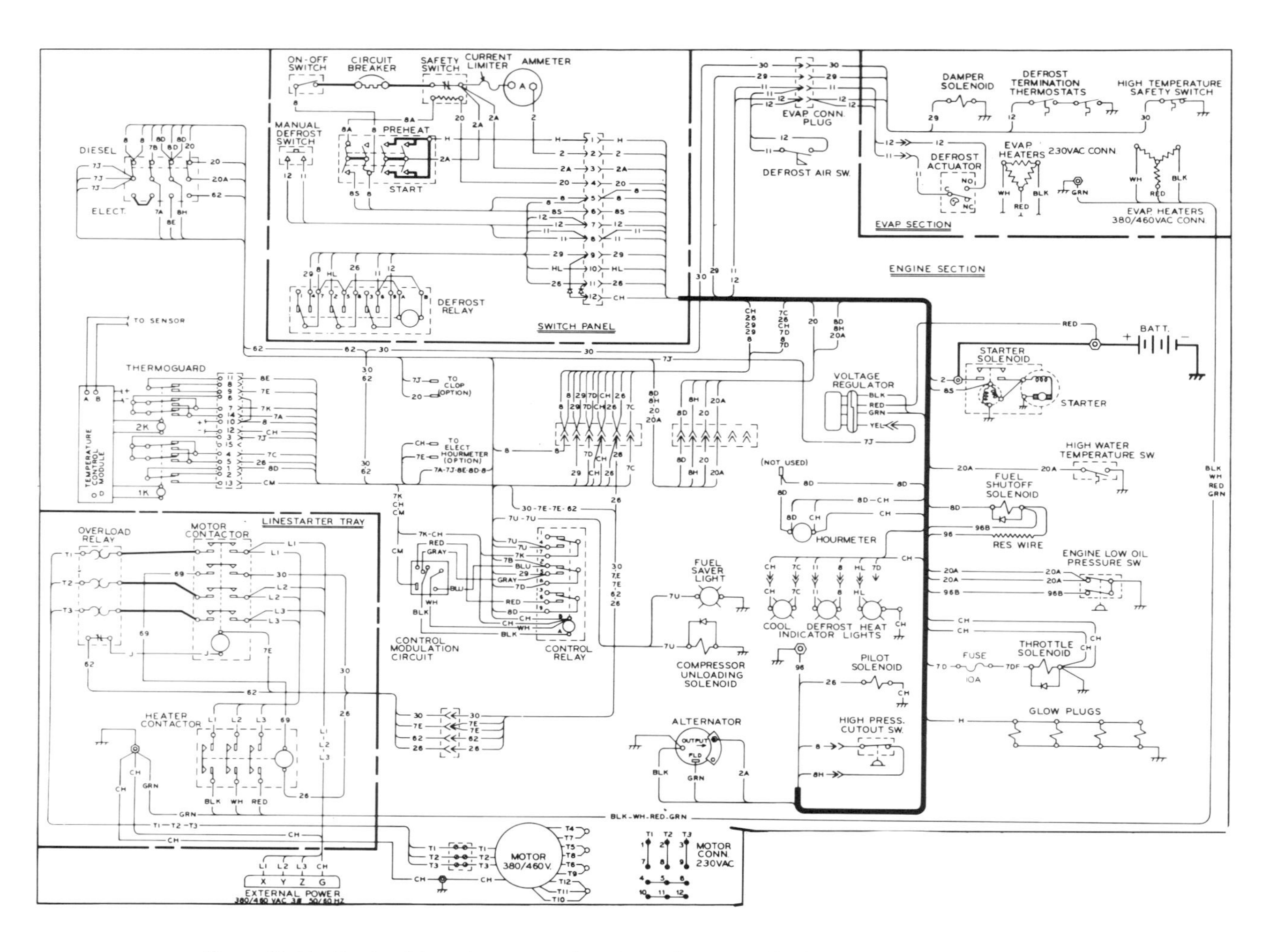

Figure 13–14 Truck refrigeration, connection wiring diagram. This shows the actual position of the various switches and loads with the connecting numbered wires. Note the use of "wiring harness" diagramming. This connection diagram can be used with the schematic, Figure 13–13. (By permission of ThermoKing Corporation.)

Cooling and heating sequence of controls. The entire operation of the system is controlled by the temperature control module (TCM) located in line 27 of the schematic diagram. The TCM operates in response to the temperature senser. The modes of operation are shown in the diagram at the lower right of the schematic diagram.

Condition	Relay position
1. High-speed cool (four cycle)	1K out, 2K in
2. Low-speed cool (two cycle)	1K out, 2K out
3. Low-speed heat (two cycle)	1K in, 2K out
4. High-speed heat (four cycle)	1K in, 2K out

The unit runs continuously and modulates from full cool to full heat to maintain the proper temperature. Note the use of the 1K and 2K relays to operate the system:

Relay	Control operated
1K1, 2	Pilot solenoid (PS) to control the defrost cycle
1K3	Control module (CM), which operates the control relay (CR), which has three switches: CR_1, CR_2, and CR_3
2K1	Operates the motor contactor (MC)
2K2, 3	Operates the unloader solenoid (US)

Another important relay that has three switches is the defrost relay (D): D_1, D_2, and D_3. Thus, using the 1K, 2K, CR, and D relays, most of the system is controlled from the thermostat.

There are a number of important solenoids in the system: the pilot solenoid (PS), the unloading solenoid (US), the damper solenoid (DS), the fuel solenoid (FS), the throttle solenoid (TS), and the defrost solenoid (D). There are two contactors: the motor contactor (MC) and the heater contactor (HC). All of these represent loads operated by the relays.

REVIEW EXERCISE 1

Select the letter representing your choice of the correct answer (there may be more than one).

13–1. The type of wiring diagram that is most useful in analyzing how a system operates is:

________ a. Installation
________ b. Pictorial
________ c. Schematic
________ d. Power circuit

13–2. The single circuits or ladders of a schematic wiring diagram can be drawn:

________ a. Horizontally
________ b. Vertically

13–3. The legend shows:

________ a. The meaning of the symbols
________ b. A history of the equipment
________ c. The manufacturer's name
________ d. How the system operates

13–4. The control that initiates the electric defrost system is:

________ a. Pressure control
________ b. Temperature control
________ c. Thermostat
________ d. Timer

13–5. Which type of defrost system is generally used for supermarket refrigeration?

________ a. Hot-gas defrost
________ b. Electric defrost

13–6. A Thermobank is what type of defrost system?

________ a. Electric
________ b. Hot gas
________ c. Air
________ d. Glycol

13–7. On a bin-type self-contained ice maker the control that starts and stops the ice-making equipment (not including safety controls) is:

________ a. Suction pressure control
________ b. Timer
________ c. Pressure regulator
________ d. Bin thermostat

13–8. In a supermarket equipment room ventilation is provided in winter to keep the room temperature above:

________ a. 60°F
________ b. 64°F
________ c. 68°F
________ d. 72°F

13-9. On Figure 13-26, giving the wiring diagram for a flake ice maker, what is the purpose of the thermal relay?

________ a. To provide a time delay before restarting the unit

________ b. To operate the pump for the water system

________ c. To provide temperature-limit protections

________ d. To terminate the harvest cycle

13-10. On an electric defrost system during the defrost cycle, do the evaporator fans continue to run?

________ a. Yes

________ b. No

REVIEW EXERCISE 2

Figure 13-2 is a wiring diagram for an electric defrost unit. The solenoid valve is used to provide a pump-down capacity for the compressor. The electric defrost unit is time initiated and temperature terminated. A time-delay relay is provided, which permits the coil to cool down before the fan is started, preventing water from blowing into the room.

Referring to Figure 13-2, answer the following questions:

1. How many switches are there in the time-delay relay? ________
2. When is the fan off? ________

3. How many switches are there in the time clock? ________
4. What does the thermostat do? ________

5. Is the thermostat line voltage or low voltage? ________
6. How many electric heat circuits are used? ________
7. How many electrical connections need to be made to the evaporator? ________
8. How many electrical controls need to be field wired? ________
9. Name them:

10. Are the evaporator fans field wired? ________

14 Installation and Operation

OBJECTIVES

After studying this chapter, the reader will be able to:

- Identify the various types of equipment used for installation of refrigeration systems.
- Describe various procedures required to install a refrigeration system.
- Describe the various procedures required to check, test, and start the system.

INSTALLATION ARRANGEMENT

The installation arrangement is dependent on a number of factors: the use of the equipment, the size of the job, the type of refrigerant, the type of refrigerant condenser, and unique features of the application. The installation must provide for a number of essential conditions, including:

1. Proper location for the equipment
2. The piping arrangement
3. The electrical connections
4. Check, test, and start

The Safety Code for Mechanical Refrigeration (ANSI/ASHRAE 15-74; see Appendix III) should be referred to for all matters pertaining to the safe use of refrigerants. Most local codes incorporate this information as part of their code requirements. All installations must be constructed to comply with local code regulations.

SELECTING A LOCATION FOR THE EQUIPMENT

In selecting the best location for equipment, a number of factors should be considered:

1. Ventilation or provision for dissipation of heat
2. Effect of weather conditions, such as susceptibility to freeze-up
3. Space for installation and service
4. Isolation of noise and vibration
5. Availability of required services such as electricity, water, and drains.

Ventilation

All mechanical equipment produces some heat. Unless provision is made to dissipate it, heat can build up to a point that the equipment will not function properly. If equipment is placed in an enclosed space, adequate mechanical ventilation is usually necessary.

Air-cooled condensers, evaporative condensers, and cooling towers should be located outside the building whenever possible. Even if outdoor locations are used, adequate provision for airflow around the unit must be provided without interference from obstructions.

Weather Conditions

Weather conditions must be considered not only in the location of the equipment, but also in the selection of equipment to operate under the prevailing weather. Special care needs to be exercised in locating water-cooled equipment or water-circulating equipment to prevent freeze-up during the winter months. An interior room may not offer the required protection without the addition of supplementary heat.

In locating air-cooled units outside the building, provision needs to be made for low-ambient control during winter operation. Many options are available, depending on the weather conditions encountered.

Space for Installation and Service

If equipment for a retrofit installation needs to be brought into a building, the contractor must be certain that adequate space is available through doorways and past obstructions.

Provision needs to be made for space not only for installation but also for service. If shell-and-tube condensers are installed in an equipment room, space needs to be provided for cleaning tubes and even replacing them if such a condition becomes necessary.

Equipment should be located with the proper minimum clearance from walls and obstructions to provide space for repairs and service. Normally, manufacturers supply this information together with other essential installation information.

Isolation of Noise and Vibration

Noise and vibration can be real troublemakers unless provision is made to handle them properly. When a choice of locations is available, problems from noise and vibration

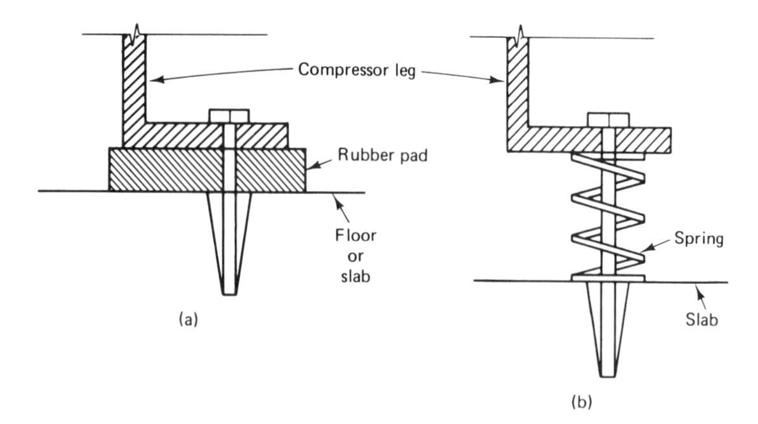

Figure 14–1 Vibration isolators. Most compressors need some type of vibration isolation to prevent sound and movement to other parts of the building. (a) One method, often used on the ground floor, is to bolt the compressor down dolidly to a concrete pad. (b) Another method, used when the compressor is on upper floors, is to mount the base of the compressor on isolation springs. Manufacturer's recommendations should be followed.

can often be minimized by utilizing certain locations in preference to others. For example, if a compressor can be located on the ground floor on an isolation base, some of the problems related to vibration and sound transmission are solved. It is much easier to solve these problems at the time of the original installation than to go back in answer to complaints.

One method of minimizing vibration is to bolt a refrigeration compressor solidly to an isolated concrete pad on the ground. When mounting the equipment on the building structure, floating-spring vibration isolators can also be used under the base of a unit to prevent or reduce sound and vibration transmission (see Figure 14–1). Vibration isolators such as those shown in Figure 7–1 can be used in the piping connections to the compressor to reduce vibration. These isolators should be located to permit linear movement but not compression of the isolator. Piping suspended from the ceiling can use spring-loaded hangers. Vibration loops in the piping also act to prevent vibration transmission.

A compressor discharge line to an air-cooled condenser located on the roof should have a muffler installed in the discharge piping to prevent vibration. Discharge lines should be securely bolted to solid areas of the structure.

A massive concrete wall separating a compressor from the surrounding building will materially reduce sound transmission. Many large refrigeration systems are located in separate buildings to isolate the sound.

Availability of Utility Services

The location of water, drain, and electrical service must be considered. Any location of equipment that simplifies the connection to existing facilities is helpful. New services may be expensive and in some instances not justified. Local utility companies should be consulted in case there is any question on the service available in the area.

REFRIGERANT PIPING

Reference should be made to Chapters 2 and 7 for information on refrigerant piping. For halocarbon refrigerants, the preference is for copper piping. Soft copper with flared compression fittings is used for piping near small units. Only types K and L copper tubing is approved for refrigeration work.

Hard copper with silver-brazed joints is used for the major part of the installation in accordance with code regulations. A typical oxacetylene torch suitable for most brazing operations is shown in Figure 14–2. A review of the procedure for silver brazing is indicated below:

1. Thoroughly clean the joints.
2. Fit the joint and support it.
3. Apply a silver brazing flux.
4. Apply heat evenly at the proper temperature.
5. Melt the silver brazing alloy so that it enters the joint completely by capillary action.
6. Cool the joint and remove the unused flux.

To prevent the formation of scale on the inside of the piping, flow dry nitrogen gas through the piping during brazing. Nitrogen in cylinders has a pressure as high as 1500 psi. Therefore, the pressure must be reduced to cause a gentle flow through the pipe. Care must be exercised in handling nitrogen due to its extremely high pressure.

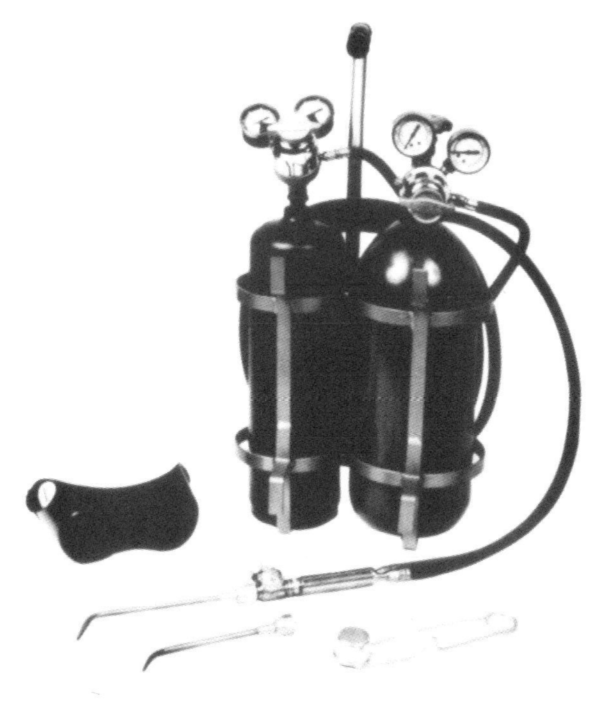

Figure 14–2 Brazing torches. Acetylene torches are available for mixing air and acetylene or oxygen and acetylene. The hottest flame is obtained using oxygen with acetylene and is required for most silver brazing. (By permission of National Cylinder Gas Company.)

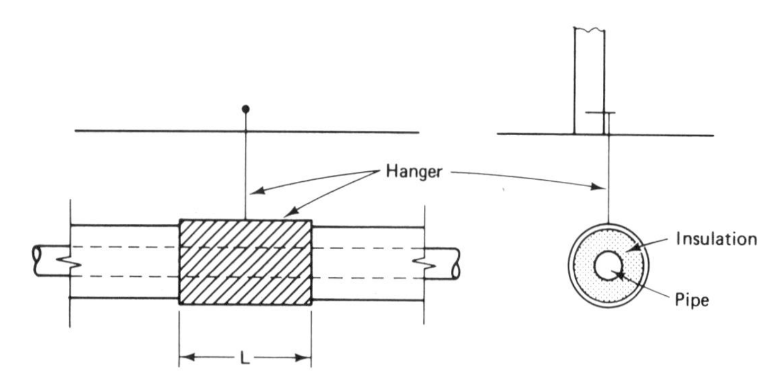

Figure 14-3 Insulation pipe hangers. Support the insulation without compressing it. A sleeve will usually serve this purpose.

When brazing near a valve, it is advisable, in order not to damage the mechanism by excessive heat, to remove the internal working parts or to wrap the valve with a wet rag. There are also certain compounds on the market that can be applied to copper piping to isolate the heating effect.

Suction piping should be insulated when condensate water dripping from the pipe can cause damage or discomfort. When insulation is applied, the metal hanger should surround the insulation to prevent transmission of the cold to the uninsulated material (see Figure 14-3).

The most common method of joining iron or steel piping is by welding, although valves and some accessories are connected by matching flanges. Pipe sizes 1½ in. and smaller are usually socket welded, whereas pipes 2 in. and larger are butt welded.

All piping must be tested for leaks using pressures higher than working pressures. These requirements are included in the safety code for mechanical refrigeration, ANSI/ASHRAE 15-74.

ELECTRICAL CONNECTIONS

Power and control wiring are discussed in Chapters 10 and 11. All wiring should comply with local and national electrical codes. A fused disconnect is provided to isolate each major piece of equipment and permit electrically isolating that equipment in case service is required. All wiring connections should be made with a good grade of commercial connector to provide good contact and prevent possible separation.

DEHYDRATION AND EVACUATION

The presence of moisture or water in a refrigeration system can cause serious damage to the equipment. Water reacts with different types of refrigerants in different manners.

The hydrocarbon group, such as ethane, will absorb almost no water. It is heavier than the refrigerant and sinks to the bottom. The halogenated hydrocarbon group, such as R-12 and R-22, will absorb small amounts of water.

Refrigerants such as R-717 (ammonia) and R-764 (sulfur dioxide) will absorb water in any amount. For R-717, moisture becomes a problem only when excessive amounts are absorbed. Generally, the problem is corrected by replacing the refrigerant.

In R-764 systems moisture can be serious. A drier placed in the line will absorb small amounts. If the R-764 contains considerable moisture, the refrigerant should be replaced.

Referring to the halocarbon refrigerants, moisture can cause a number of undesirable effects: (1) copper plating, (2) formation of sludges that are highly corrosive, (3) freeze-up at the expansion valve, and (4) damage to motor windings on suction gas-cooled hermetic units.

A good way to determine the presence of moisture in the refrigerant is to use a moisture-indicating sight glass in the liquid line (see Figure 7–32). A color change indicates a wet system.

There are two principal ways of removing the moisture: (1) the use of chemical driers and (2) evacuation. Chemical driers can be the sealed type, used on smaller jobs, or the cartridge type with replacable cores used for larger jobs. The desiccant used in chemical driers is usually alumina, silica gel, or calcium sulfate.

Evacuation

A vacuum of at least 29.8 in. of mercury should be drawn on the system before charging with refrigerant. With larger systems it is advisable to evacuate for 8 to 12 hours to be sure to remove any moisture. It is also advisable to make a final leak detection test by turning off the vacuum pump (Figure 14–4) and seeing if the system holds its vacuum.

Low vacuums, used in evacuation, can best be measured in microns (Figure 14–5). A micron is one-millionth of a meter. There are about 25,000 microns to an inch. A vacuum of 29 in. of mercury would equal 25,000 microns. It is desirable for a good vacuum pump to be capable of drawing a vacuum of 29.8 in. of mercury or 5000 microns. A micron gauge is available at most refrigeration supply stores.

One reason for reaching a low vacuum during evacuation is that it permits water to be evaporated and removed. Figure 14–6 gives the boiling point of water for low pressures (psia). Unless a boiling temperature below ambient temperature is reached, the water in the system will not be removed.

Figure 14–4 Vacuum pump. Made in single- or two-stage models for evacuating the system. (By permission of Robinaire Division of Sealed Power Corp.)

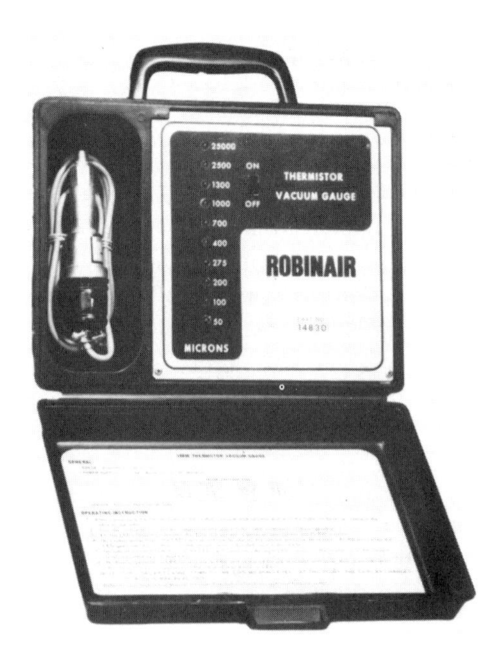

Figure 14–5 Low vacuum measuring instrument. (By permission of Robinaire Division of Sealed Power Corp.)

Figure 14–6 Effects of pressure on the boiling point of water. By creating a low pressure in the system with a vacuum pump, moisture in the piping can be boiled off and removed. Note that the water would start to boil at 60°F if the pressure is reduced to 0.522 in. mercury absolute. Moisture in the piping can be boiled off and removed.

Absolute pressure (in. mercury)	temperature (°F)
7.569	150
5.881	140
4,525	130
3.446	120
2.596	110
1.933	100
1.422	90
1.032	80
0.739	70
0.522	60
0.363	50
0.248	40
0.180	32

A convenient way to evacuate the system and later charge it with refrigerant is to utilize the gauge manifold (Chapter 15). By properly adjusting the gauge manifold valves, an arrangement for evacuating can easily be changed to charging without the danger of air entering the system.

Where the vacuum pump is not capable of drawing a deep vacuum, double or triple evacuation is recommended. This means evacuating at as low a pressure as the pump will go, passing a dry gas like dry nitrogen back into the vessel or system, and repeating the process one or two times. The use of heat to facilitate the boiling process is also recommended.

Charging the System with Refrigerant

After evacuation, the proper amount of refrigerant is added by either the vapor method or the liquid method. For larger systems the liquid method is advisable to save time. A drier should be installed in the system liquid line to remove any residual moisture that may have entered during charging. A moisture-indicating sight glass is helpful in measuring the dryness of the system.

The position of the drum for both charging methods is shown in Figure 14–7. In charging with vapor it is usually necessary to supply additional heat to vaporize the refrigerant. The proper method of doing this is shown in Figure 14–8. Direct heat from

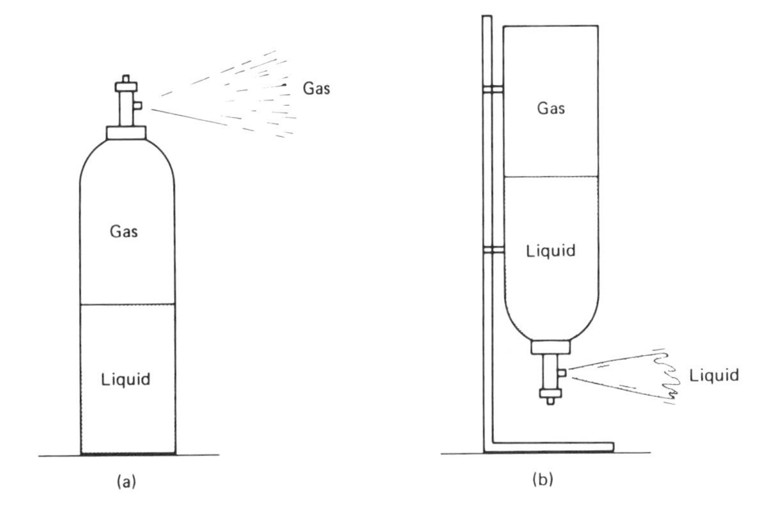

Figure 14–7 Charging refrigerant: (a) position of cylinder when charging with vapor; (b) position of cylinder when charging with liquid.

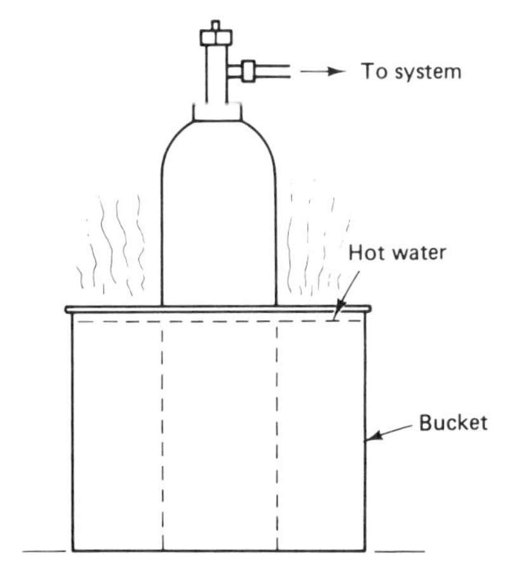

Figure 14–8 Method of vaporizing refrigerant. In charging vapor refrigerant into the system heat must be added to the liquid refrigerant. The approved method is to place the cylinder in warm water (120 °F). The cylinder valve must always be open. An open flame should *never* be used.

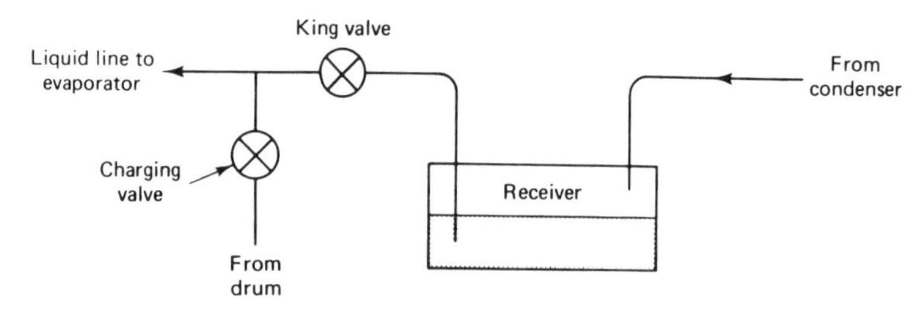

Figure 14–9 Liquid charging. To charge with liquid, enter the system between the receiver and metering devices. Close the king valve. Open the charging valve. The compressor will draw the liquid from the drum.

a torch should never be used, due to the danger of explosion. The refrigerant drum should be placed in warm water (120 °F) to control the vaporization properly.

If vapor is being charged, the drum should be connected on the downstream side of the expansion valve. If the system is being charged with liquid, the drum should be connected on the upstream side of the expansion valve.

A convenient place to charge vapor is directly into the compressor through the gauge port connection of the suction service valve. This connection lends itself to the use of the gauge manifold.

A convenient place to charge liquid into the system is through a charging valve located in the liquid line (see Figure 14–9).

Correct amount of refrigerant. The correct amount of refrigerant can be determined in two principal ways, depending on the type of system. Where a critical charge is required, the amount of the charge is indicated on the nameplate. This charge can be measured in a Dial-a-Charge (Figure 14–10) cylinder or weighted on a suitable scale.

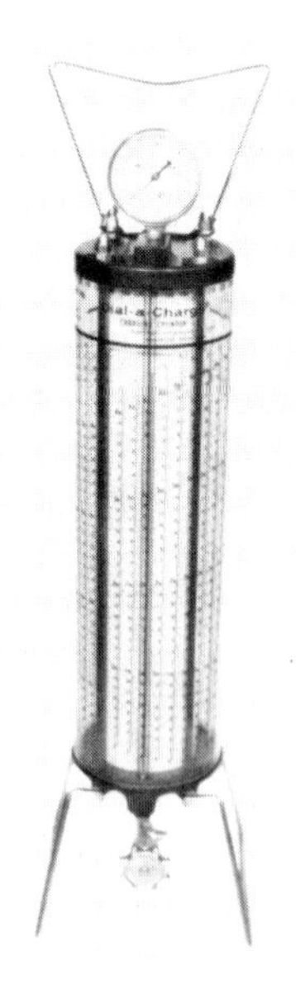

Figure 14–10 Dial-a-Charge for accurately measuring the refrigerant charge. (By permission of Robinaire Division of Sealed Power Corp.)

Figure 14-11 Electronic sightglass. This device operates using ultrasonic principles and emits an audible signal when the system reaches full charge. (By permission of TFI Instruments, Inc.)

For most built-up systems the correct charge is determined by use of a liquid sight glass (see Figure 7-32). Refrigerant is added until the bubbles disappear from the sight glass. It is advisable to place almost the full charge into the system, then run it to permit the pressures to stabilize. If bubbles are still present after the stabilization period, more refrigerant is added until the bubbles disappear.

In most systems where there is a properly sized receiver, a small overcharge will not create a problem. An undercharge is not desirable, however, since it limits the capacity of the system.

An electronic sight glass has been developed that operates by using ultrasonic principles, emitting an audible signal whenever the refrigerant system is fully charged. This device is particularly useful in determining the refrigerant charge in critically charged systems (see Figure 14-11).

Purging Noncondensables

If the suction pressure in a system is below atmospheric pressure, a leak in the piping will cause air to enter the system. This is evidenced by too high a head pressure in the condenser, where a mixture of liquid and vapor exists. The head pressure does not correspond to the saturated temperature of the refrigerant.

Air is usually the principal noncondensable. On a halocarbon system air can usually best be removed from the receiver or top of the condenser. The system is stopped and kept above atmospheric pressure. The air tends to collect at the high points. Properly located valves can be opened to the atmosphere for a brief period to expel the air, with a minimum loss of refrigerant and oil. In purging it is advisable to use a gauge manifold so that pressures can be monitored. In purging a water-cooled condenser, care must be exercised to prevent freeze-up. Purging may reduce the pressure sufficiently to cause the refrigerant remaining in the vessel to boil at below-freezing temperatures.

CHECK, TEST, AND START

A number of important procedures are necessary before the equipment can be started:

1. Test for refrigerant leaks.
2. Remove or loosen the hold-down bolts on the compressor.
3. Properly position all valves.
4. Check all lubrication, including the compressor oil.
5. Check the belt tension.
6. Check the water and drain system.
7. Check the electrical system.
8. Supply customer information.

Testing for Leaks

There are a number of ways to test for leaks on a halocarbon system:

1. Use a halide torch.
2. Use an electronic leak detector.
3. Use soap bubbles.

The type of test will depend on conditions on the job. In larger systems it is usually advantageous to charge the system with a mixture of a small amount of refrigerant with dry nitrogen for testing purposes. Dry nitrogen is relatively inexpensive and makes a good medium for the refrigerant. Due to the high pressure in the nitrogen drum, a regulator must be used. Usually, the test pressure is specified by the code. After testing with nitrogen it should be purged before evacuation of the system and final charging.

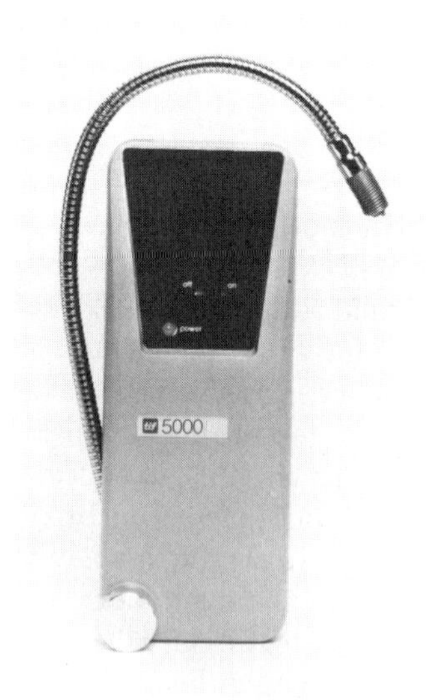

Figure 14-12 Refrigerant leak detector. This equipment uses computerized circuitry. It produces a beeping signal which increases in frequency as it approaches the refrigerant leak. It is battery operated. (By permission of TFI Instruments, Inc.)

Using a halide torch, air is drawn over a heated copper wire. If the air contains refrigerant, the color of the flame will change.

The electronic detector can be used to detect small leaks. The presence of refrigerant will cause the instrument to emit a loud noise (see Figure 14–12).

The soap bubble test can be used for any refrigerant. The outward pressure of the refrigerant leak will form bubbles at the location of the leak when the bubble solution is applied.

If the refrigerant is ammonia, the location of the leak can be detected by burning a sulfur candle and passing the flame into the area of the suspected leak. A safer method is the application of hydrochloric acid vapor used in the same manner. In either case, white smoke will form in the presence of ammonia. If ammonia has leaked into water or a brine solution, Nessler's solution can be used to detect the presence of the refrigerant. Ammonia causes the solution to turn yellow.

Loosening the Hold-Down Bolts

If any hold-down bolts are used for the compressor or motor, these should be loosened or removed. Hold-down bolts are useful for shipping but restrict the operation of the equipment. In most cases they should be saved in case they are needed later.

Positioning the Valves

Compressor valves should be opened and other valves positioned for operating conditions. All three-way valves should be checked to see that they are in the proper position. Receiver valves need to be checked to see that the flow of refrigerant is not restricted.

When the system is started, the initial pull-down may overload the compressor. The suction valve of the compressor can be manually controlled to maintain as nearly normal an operating pressure as possible. Suction pressure should not be allowed to drop too low or to rise too high during pull-down.

Checking the Lubrication

The amount of oil in the compressor crankcase should be checked for proper level. In larger systems some oil will remain in the system even though there are no improper traps. Some oil may, therefore, need to be added to the compressor crankcase after the system has been operated for a short time.

All motors that do not have permanent oil supplies should be checked for lubrication. The manufacturers' recommendations are helpful in determining the proper oil to use and the periods of lubrication.

Checking the Compressor Oil

Larger compressors are usually equipment with a sight glass in the crankcase to indicate the oil level. On small hermetics the original oil is usually supplied by the manufacturer and, unless there is a major service problem, no additional oil needs to be added.

The paper test can often be advantageously used to check the oil on halocarbon systems. A refrigerant connection is slightly opened and refrigerant oil mixture sprayed on a piece of paper. A very slight stain on the paper when the refrigerant has evaporated is an indication of proper oil charge.

On larger halocarbon systems, when the compressor is idle for prolonged periods, a crankcase heater should be used to prevent refrigerant from diluting the oil.

On a new system the compressor can be charged with oil before the system is evacuated. To add oil after the system is in operation requires a siphon arrangement. An oil supply connection is made to the crankcase. A pressure below atmospheric pressure in the crankcase will pull the oil in. Also, a hand oil pump can be used to force oil into the crankcase against a higher atmospheric pressure. It is important in adding oil to an operating system to add only oil and no air or moisture.

Somewhat similar arrangements can be used for removing oil. A pump can be used or a siphon arrangement.

Checking the Belt Tension

Refer to Figure 14–13. Belt tension should be checked prior to startup. Alignment of pulleys can be checked with a straightedge. Belt tension can be adjusted so that the belts do not slip. This usually allows about a 1-in. depression when squeezed with the fingers, as shown in the illustration. If belts are too tight, they can overload the motor.

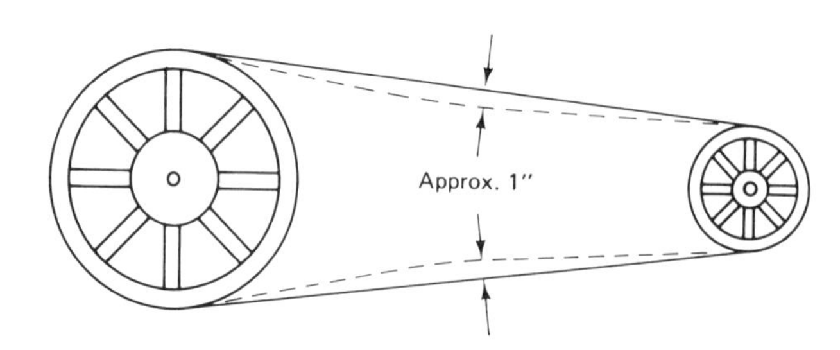

Figure 14–13 Belt tension. Both alignment and belt tension on all belt-driven types of equipment must be checked. A straight edge can be used to check alignment. About a 1-in. deflection in the belt is the allowance considered good practice.

Checking the Water System

If a water-cooled condenser is used, all manual valves should be open and drain systems properly installed. Any condensate drains should be open and ready to use. Water pressure should be checked to assure proper flow. If the medium being cooled is water or brine, the manual valves should be checked for proper position. All safety controls, such as relief valves, should be checked to see that they operate properly.

Checking the Electrical System

Use meters to check the power supply, which should agree with the voltage on the motor nameplate. After the system is started, check to see that the ampere draw of the motors is within the limits indicated by the manufacturer(s).

The control system should be checked to see that proper voltages are available and that the control system operates as designed. Thermostats must be properly set and safety controls tested for proper function.

After the system is operated for a period of time, usually a day or two, controls should be rechecked to see that they are properly set and have maintained their calibration.

Supplying Customer Information

All information essential for operation and maintenance of the system should be left on the job. The owner and/or assigned employees should be properly instructed on operating and maintenance procedures. All instructions for obtaining service when needed should be plainly posted on the job.

REVIEW EXERCISE

Select the letter representing your choice of the correct answer (there may be more than one).

14-1. Information pertaining to the safe use of refrigerants is included in:

________ a. ANSI/ASHRAE 15-74 code
________ b. The *Mechanical Engineers' Handbook*
________ c. Printed on the outside of a refrigerant drum
________ d. On the label attached to each piece of refrigeration equipment

14-2. Nitrogen cylinders can have a pressure as high as:

________ a. 1500 psi
________ b. 2000 psi
________ c. 2500 psi
________ d. 3000 psi

14-3. Which of the following refrigerants will absorb any amount of water?

________ a. R-12
________ b. R-22
________ c. R-502
________ d. R-717

14-4. The following condition is *not* caused by moisture in a halogenated carbon refrigerant:

________ a. Copper plating
________ b. High-pressure drop
________ c. Formation of sludges
________ d. Freeze-up at the expansion valve

14-5. The following material is a type of refrigerant desiccant:

________ a. Sodium carbonate
________ b. Porous rock
________ c. Silica gel
________ d. Ethylene glycol

14-6. What sizes of steel pipe are usually butt-welded?

________ a. 1 inch and larger
________ b. 1½ inches and larger
________ c. 2 inches and larger
________ d. 3 inches and larger

14-7. The high side of a refrigeration system should be tested for leaks using an internal pressure of not less than:

________ a. 150 psig
________ b. 200 psig
________ c. 250 psig
________ d. refer to the safety code

14-8. Which type of leak detection method can be used for all refrigerants?

________ a. Sulfur stick

________ b. Halide torch

________ c. Electronic

________ d. Soap bubble

14-9. In evacuating a system using a vacuum pump, how low a vacuum should be drawn?

________ a. 29.6 in. of mercury

________ b. 29.7 in. of mercury

________ c. 29.8 in. of mercury

________ d. 29.9 in. of mercury

14-10. How many microns are there in an inch?

________ a. 250,000

________ b. 25,000

________ c. 2500

________ d. 250

15 Service

OBJECTIVES

After studying this chapter, the reader will be able to:

- Select the proper equipment to perform standard service requirements.
- Describe the methods of servicing common types of refrigeration equipment and components.

THE SERVICE BUSINESS

Servicing commercial refrigeration equipment is an essential business. In many cases, these systems provide an important function, such as preserving food or medicines. The equipment must be kept in operation or a severe loss occurs. Well-trained refrigeration service people are in demand. Although much can be learned by study, the practical use of these principles is important in becoming proficient in the servicing business.

USING THE GAUGE MANIFOLD

One of the most useful tools of the refrigeration mechanic is the gauge manifold (Figure 15-1). It can be used to:

1. Read pressures
2. Charge refrigerant vapor through the compressor
3. Purge the receiver
4. Charge refrigerant liquid into the high side of the system
5. Build up system pressures to test for leaks
6. Charge oil into the compressor

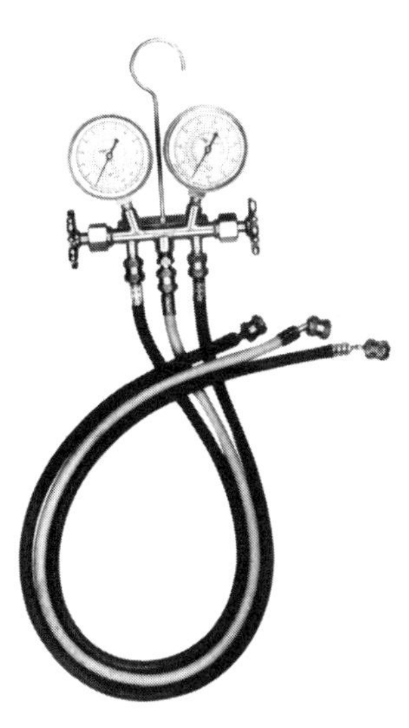

Figure 15–1 Gauge manifold. The left-hand gauge is a compound gauge for reading suction pressure. Its range is usually from 30 in. mercury vacuum to 120 psi. The right-hand gauge is a high-pressure gauge for reading discharge pressure. Its range is usually from 0 to 500 psi. The hand wheels shut off valves to the service connections. System pressures can be read with both handwheel valves closed.

Typical connections to the compressor service valves are shown in Figure 15–2. The gauge manifold has three connections as shown. Usually, one hose connects to the compressor suction valve, another hose connects to the compressor discharge valve, and the center hose is used for service functions such as supplying refrigerant or connecting to a vacuum pump.

Each valve on the manifold has two positions, open and closed. The service valves on the compressor are three-position valves: front seat, back seat, and midposition. Each compressor valve has three connections: one to the compressor, one to the gauge port, and one to the refrigeration system.

Figure 15–2 Gauge manifold. Figure on the left shows refrigeration gauge manifold connected to compressor service valves. Valves on gauge manifold are two-position valves. Valves on compressor are three-position valves. The four figures on the right show various uses of the manifold and the corresponding position of the valves.

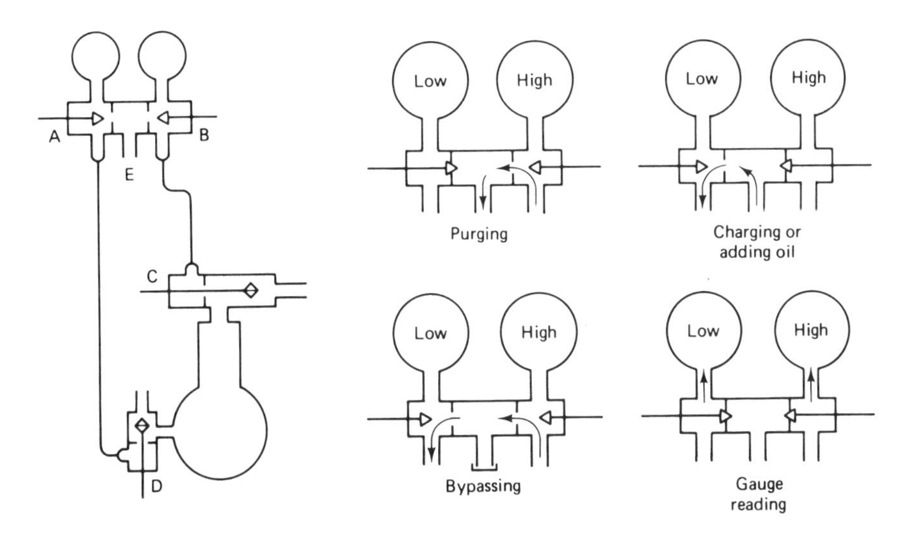

The position of the four valves for the various servicing functions is indicated below. In Figure 15–2 the gauge manifold low-side valve is marked A, the gauge manifold high-side valve is marked B, the compressor low-side valve is marked D, and the compressor high-side valve is marked C. The service connection is marked E.

1. To test operating pressures:
   a. Close A.
   b. Close B.
   c. Crack open back seat of C.
   d. Crack open back seat of D.
2. To charge vapor refrigerant through the compressor:
   a. Connect refrigerant drum to E.
   b. Open A.
   c. Close B.
   d. Crack open back seat of C.
   e. Close front seat of D.
3. To purge air from the receiver:
   a. Close A.
   b. Open B.
   c. Crack open back seat of C.
4. To charge liquid refrigerant into the high side:
   a. Close A.
   b. Open B.
   c. Place C in midposition.
5. To build up pressure in the system for testing:
   a. Seal E with cap.
   b. Open A.
   c. Open B.
   d. Crack open back seat of C.
   e. Put D in midposition.
6. To charge oil into the compressor:
   a. Connect oil supply to E.
   b. Open A.
   c. Close B.
   d. Back seat C.
   e. Front seat D.

Connecting the Gauge Manifold

One of the advantages of using the gauge manifold for service is that when it is properly connected, a number of service operations can be performed without air or other contaminants entering the system. To make this possible, the lines must be purged before the compressor valves are opened.

To illustrate how this is done, referring to Figure 15–2, assume that a third hose is connected between the center part of the manifold (E) and a refrigerant cylinder. All hose connections are made finger-tight. In case there is any leak, the sealing gasket needs to be replaced.

Place the refrigerant cylinder in an upright position, to permit the use of only vapor refrigerant. Close valves A and B on the manifold. Be sure that compressor valves C and D are in the back-seat position. Crack open but do not disconnect the hose connections to the compressor valves. Open the valve on the top of the refrigerant cylinder. Crack open manifold valve A until the refrigerant vapor purges the line connected to compressor valve D. While the refrigerant is gently flowing from the line, tighten the connection to compressor valve D. Perform the same operation using valves B and C. Close valves A and B and read the system pressures by cracking open the back seat on valves C and D.

Any time the hoses are changed and opened to the air, the purging operation needs to be performed. Many service personnel use a "tee" at the center connection of the manifold. This permits connecting a vacuum pump and the charging cylinder at the same time. With this arrangement the system can be evacuated and charged without opening the refrigerant lines.

After the manifold is connected to the system, and before service work is performed, the connections should be leak-tested. This is particularly important if the suction is run below atmospheric pressure or if tests are to be made after evacuating to determine whether or not the system holds the vacuum. Be sure that there are no leaks in the hose connections.

After servicing is complete, the manifold hoses are removed without loss of refrigerant from the system. This is done by closing valves C and D, closing drum valve to E, and removing the hoses from the compressor valves. Caps on the compressor valves should be replaced.

USING AND CARING FOR SERVICE VALVES

As previously indicated, service valves permit direct entrance of the refrigerant into the compressor, receiver, or water-cooled condenser. The valve stem is adjustable. In the running position the valve is *back seated* to permit refrigerant to move through the piping system. The back-seat position closes the gauge port.

In the *front-seated* position the flow of refrigerant is shut off to the system. Special care must be exercised not to operate the compressor with the discharge valve closed. Excessive head pressures can occur and cause an explosion. In shipping compressors the manufacturer often installs a holding charge. Most compressors are shipped with the valves closed. After the compressor is connected to the system, the valves are opened.

The *midposition* of the valve is useful in reading pressures while the system is in operation. Before the gauges are removed the valves are back-seated to close the gauge ports.

On some compressors the valve stem is packed and the tightening nut must be loosened before adjusting a compressor valve. At the completion of work the packing nut is retightened to prevent leakage.

The packing material is usually asbestos, lead, or graphite. The packing must provide a leaktight seal on the stem when the packing nut is tightened. Packing should be replaced if the seal leaks. The body of the valve is usually brass (on halocarbon equipment), and the stem, steel. Stems can rust unless kept properly lubricated with refrigerant oil.

If the packing around the compressor valve stem is dry, refrigerant oil should be used to lubricate it. There is always a danger of "freezing" the stem if the packing is dry. To relieve a frozen stem, apply heat from a butane torch.

A ratchet-type valve wrench should be used except to crack the valve. When the valve is cracked open, a fixed wrench is preferred so that the valve can be quickly closed if necessary.

SERVICING COMPRESSORS AND CONDENSING UNITS

The type of service for compressors and condensing units depends on the type of equipment installed. In general, hermetically sealed compressors are not serviced in the field. It is important to test the hermetic unit properly before replacing it. The service mechanic must be certain that the relay, capacitors, and overloads are all functioning. The resistance of each of the windings can be tested for possible shorts or grounds. The "acid" test, however is whether the compressor will or will not run if supplied with proper power and proper accessories.

If the compressor is semihermetic, the valve plate can be removed for service. If the compressor is an open type, either direct drive or belt driven, the unit offers complete opportunity to service any part of it.

Testing Open-Type Compressors

The two most common service problems with open-type compressors are leaking valves and leaking seals. A test of the pumping ability of the compressor will determine the condition of the valves. One way to do this is to close the suction valve momentarily while the compressor is running. Under these conditions the machine should develop a normal head pressure, with a suction pressure of 20 in. of mercury or better. If the compressor cannot produce a suction pressure of 20 in. of mercury, it should be overhauled. A good machine should produce a suction of 28 in. of mercury or better.

If a compressor connected in a system is stopped after the pressures stabilize, a good compressor will hold its pressures provided that the metering device is a TEV type. If the suction pressure creeps up, there is a leaky suction valve. If the head pressure drops down to atmospheric pressure, there is a leaky discharge valve. If needed, the valve plate assembly usually can easily be replaced on an open-type unit.

A refrigerant leak at the seal can usually be detected by an oil trace around the seal or with a suitable leak detector. A leaking seal causes loss of refrigerant and/or air entering the system. A leaking seal must be replaced.

Removing Compressors

Before removing the compressor, the refrigerant should be pumped into the receiver and the receiver valves closed. Be certain that the receiver is not filled to more than 85 percent of its volume, to allow for expansion. Let the compressor continue to operate and close the suction valve to permit the suction to operate in a vacuum. Shut off the power. Close the discharge valve. Bleed refrigerant into the suction side of the compressor until the gauge reads 0 or 1 psig, and close the suction valve. The compressor is ready for removal. When the compressor is removed, close all openings with cork or rubber stoppers to exclude air and moisture. In general, compressor overhauls are best performed by companies specializing in this work.

Replacing Crankshaft Seals

On an open-type compressor where the motor is external to the compressor, the shaft must be sealed to prevent leakage of refrigerant. The seal provides a rubbing surface so smooth that leakage does not occur. The rubbing surface on one side is hardened tool steel and on the other, bronze or carbon. Spring or bellow pressure holds the seal in place. Some modern refrigeration compressors use a synthetic rubber or a neoprene seal.

Replacing a worn seal was a laborious process on some of the older machines. The crankshaft surface and the surface of the seal were carefully lapped (ground) and polished to provide a leakproof connection. Modern compressors provide a seal kit for replacement in which both rubbing surfaces are part of the seal assembly. The following is a suggested procedure for replacing the seal:

1. Open the electrical disconnect switch.
2. Close the discharge and suction valves.
3. Allow the refrigerant gas to escape by loosening the gauge plugs.
4. Remove the shutoff valves from the compressor.
5. Remove the compressor, flywheel, and seal assembly.
6. Wash the replacement parts in solvent. The seal surfaces should be carefully washed with a clean chamois.
7. Check the shaft. If it is corroded or marred in any way, polish it with fine emory cloth.
8. Apply a light coat of refrigerant oil to the clean shaft and seal parts.
9. Replace the seal assembly, assemble the compressor, and place it back in operating position. Use new gaskets. Cap screws should be tightened, using no torque higher than:
   a. ¼-in. cap screws: 100 in.-lb
   b. ⁵⁄₁₆-in. cap screws: 275 in.-lb
   c. ⅜-in. cap screws: 400 in.-lb
   d. ½-in. cap screws: 500 in.-lb
10. Recharge the compressor with oil.
11. Attach the flywheel.
12. Evacuate, open valves, and test for leaks.

Replacing Valve Plates

Leaking valves are one of the most common service problems. They cause loss of capacity and poor performance. In modern machines it is common practice to replace the entire valve plate assembly. The procedure is as follows:

1. Pump down the compressor to 2 to 5 psig suction pressure.
2. Open the electrical disconnect switch.
3. Close the compressor valves.
4. Bleed off the refrigerant by cracking the discharge valve gauge plug.

5. Remove the compressor head and valve plate.
6. Inspect the valve plate and the internal parts of the compressor. Look for possible corrosion or carbonized oil. Clean the system if necessary and replace the oil charge.
7. Remove all the old gasket material from the compressor head and valve plate.
8. Wash the new valve plate in solvent and remove with high-pressure air if possible.
9. Apply a light coat of refrigerant oil to the new gaskets and valve plate.
10. Reassemble the valve plate and cylinder head.
11. Tighten the cylinder head bolts evenly (use maximum torque as covered under seal replacement).
12. Purge some refrigerant through the compressor crankcase and cylinder head or evacuate with good vacuum pump.
13. Open compressor valves, test for leaks, and place back in operation.

SERVICING BURNED-OUT HERMETICS

Burnouts are usually caused by moisture in the system, producing a breakdown of the electrical windings of the motor. A burnout can be detected by testing for acid in the refrigerant. Acid test kits are convenient to use for this purpose.

Whenever a compressor burns out, it must be replaced. In many systems the refrigerant can be saved. The system is pumped down. The compressor valves are closed and the compressor is removed. The system is flushed out with R-11, using CO_2 or a liquid pump to circulate the cleaning fluid (see Figure 15–6).

After cleaning, the new compressor is installed. A suction-line filter is placed in the suction line. The system is evacuated, charged, and tested for leaks. Driers should be replaced until all traces of the acid are removed.

SERVICING CONDENSERS

The condenser is the essential part of the refrigeration system that disposes of the heat that has been absorbed by the evaporator. Unless the heat is properly expelled from the system, the unit cannot operate properly. Some of the common problems of condensers are:

1. Noncondensable in the system
2. Undercharge or overcharge
3. Dirty condensers
4. Problems associated with the heat-removal medium, such as fans on air-cooled condensers or pumps on water-cooled systems
5. Excess oil in the condenser

Purging Noncondensables

The chief noncondensables that get into the system are air and associated moisture. Moisture can cause burnouts unless it is removed by a drier. Air can usually be purged from the system as described in Chapter 14. Air is usually collected in the condenser or receiver and can usually be purged out through the compressor discharge valve port with the system shut off.

Air in the system can cause excessive head pressure. To check the head pressure, add 30 to 35 °F to the ambient temperature for air-cooled condensers or 10 to 15 °F to the leaving water temperature for water-cooled condensers. Check the equivalent head pressure in the appropriate refrigerant table to obtain the normal head pressure.

On large ammonia systems operating at suction pressures below atmospheric pressure, it is common practice to use an automatic air purge unit. This removes any noncondensable material from the system.

Correcting Undercharge or Overcharge

An undercharge of refrigerant can reduce condensing capacity by reducing the amount of effective condensing surface. The proper refrigerant charge can be checked by reference to the liquid sight glass or the superheat condition of the evaporator.

Refrigerant can be added or removed in accordance with instructions described in Chapter 14. Use of the gauge manifold is recommended as described in the early part of this chapter.

Cleaning Condensers

On an *air-cooled condenser* the outside surface must be clean to permit proper airflow. A good brush, compressed air, or a suitable cleaning fluid can be used to clean the air-cooled condenser. On some types of *water-cooled condensers* the water passages can be cleaned by power-driven tools. Shell-and-coil condensers must be cleaned using acid. A circulating system using a dilute solution of hydrochloric acid is shown in Figure 15-3.

Figure 15-3 Apparatus used for cleaning a shell and coil water-cooled condenser. This equipment uses an acid solution. At the conclusion of the cleaning process the condenser must be thoroughly flushed with clean water to remove all traces of acid.

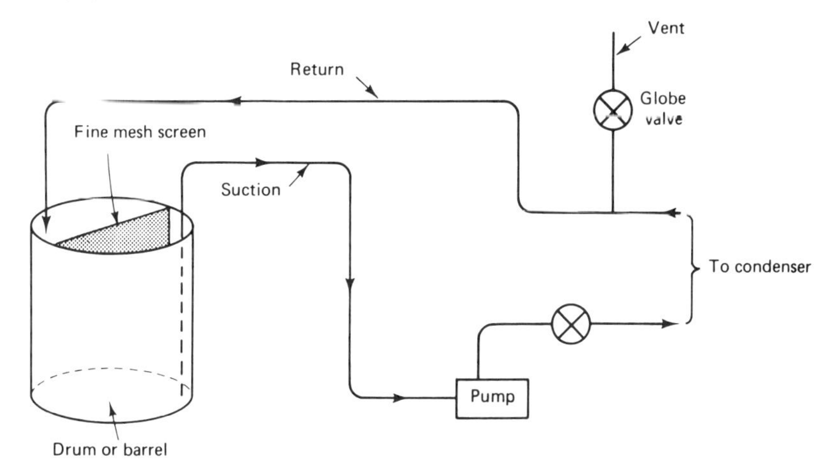

When acid systems are used for cleaning condensers, the system must be thoroughly flushed with clear water to remove any trace of acid before the system is put into use. Testing the water in the condenser with litmus paper or some other acid indicator is essential after the system has been cleaned.

The water in a cooling-tower system is usually treated with a scale inhibitor to prevent buildup of deposits in the condenser tubes.

SERVICING EVAPORATORS

The problems with evaporators depend to some extent on the type, although there are some common problems. Types of service problems include:

1. Refrigerant controls not functioning properly
2. Poor performance due to a dirty evaporator, poor air distribution, a defective fan, a defective pump motor, or ice accumulation
3. Incorrect selection of evaporator, such as too high a pressure drop, inadequate capacity, or the system not matching the load
4. Refrigerant leaks

The most important first step is to see that the evaporator is clean. If any contamination is blocking the surface, this condition should be corrected first.

The refrigerant control should be checked to determine if the superheat is properly adjusted. Usually, 10 °F superheat is normal.

The temperature drop of the cooling medium over the evaporator should be checked. In a direct-expansion coil, the air temperature drop should be on the order of 15 °F.

On a direct-expansion coil the refrigerant pressure drop should be 2 or 3 psig for a standard thermal expansion valve (TEV) control. Higher-pressure drops require a TEV with an external equalizer. Larger evaporator coils with multiple circuits require a refrigerant distributor at the outlet of the valve.

Leaking evaporators must be removed for repair. The system is pumped down and shut off. The refrigerant is allowed to bleed slightly through the coil connections when the coil is removed. The connections are capped as soon as the coil is disconnected. The repaired or replacement coil is leak tested before and after replacement.

SERVICING REFRIGERANT CONTROLS

In most refrigeration systems, in addition to the compressor, the only other items of mechanical equipment that require service are the refrigeration flow controls. This discussion will include: (1) thermal expansion valves, (2) low-side float valves, (3) solenoid valves, (4) pressure regulators.

Thermal Expansion Valves

A cutaway diagram of a typical TEV assembly is shown in Figure 15-4. Before servicing the thermal expansion valve the service person should check the entire system to be certain that the cause of the problem is not from some other source. Install a gauge manifold and read the pressures. Check the sight glass to make certain that the charge is adequate and the refrigerant is dry. Check the electrical system. Refer to Chapter 10 for adjustment of the superheat.

Two conditions that warrant special attention are:

1. Frost back, such that the suction line is frosted
2. A starved coil, indicated by uneven frosting of the coil, causing poor cabinet temperatures

The frosting of the suction line is due to liquid refrigerant passing completely through the coil and evaporating in the suction line. This may be due to:

1. TEV may be too large.
2. Needle valve may be leaking.
3. Valve is adjusted for too low superheat.
4. Thermal bulb may not be properly attached to the suction line.
5. Thermal bulb may be located in too warm a position.
6. External equalizer may be clogged.
7. Needle valve may be held open by dirt.

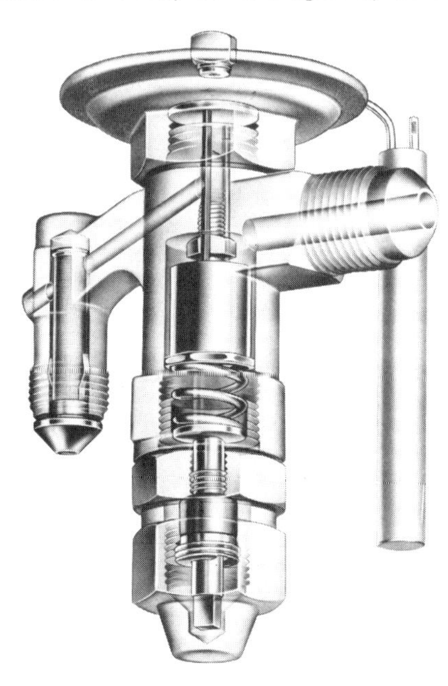

Figure 15-4 Internal parts of thermal expansion valve. Note the enclosed adjustment shaft at the bottom of the valve. Using a ratchet-type valve wrench, the spring tension or superheat can be changed as required. Note the pins that contact the bottom of the diaphram to transmit the spring tension. (By permission of Sporlan Valve Company.)

These problems are best solved by a process of elimination. The size of the valve required can be checked with the load. The superheat can be checked with suitable thermometers. The valve must be designed for use with the system refrigerant. The pressure drop through the coil can be checked. If it exceeds 3 psig, a properly operating external equalizer line must be used. If the valve does not close or shut down, the needle is leaking or clogged open.

A starved coil can be caused by any of the following:

1. The screen on the expansion valve may be clogged.
2. Moisture in the system may cause the orifice to freeze shut. This is usually an intermittent condition.
3. The valve may be too small.
4. The use of improper refrigerant oil may cause the formation of wax in the orifice.
5. Rupture of the capillary tube to the thermal bulb may cause erratic operation.

Again, these possible problems can be checked by testing until the proper solution is found.

In many cases the thermal expansion valve must be removed for service. This requires pumping down the system and valving off the flow of refrigerant. The pressure at the valve is reduced to 0 psig or a few pounds above atmospheric pressure. The valve is removed and connecting lines are sealed. When the valve is replaced, it is purged with refrigerant, tightened in place, leaktested, and put into operation. If the cause of the problem is external to the valve, the condition is corrected before the valve is put back into service.

Low-Side Float Systems

Some of the problems that occur on low-side float systems are:

1. Leaking needle valves
2. Low refrigerant charge
3. Evaporator coil not level
4. Leaking float ball
5. Clogged strainer screen

A leaking needle valve will produce a hissing sound. Further, the suction line may sweat back as far as the compressor. Usually, a leaking needle valve will have to be replaced. Some manufacturers recondition float valves.

A flooded evaporator coil that is not level can be corrected by adjusting the mounting. If the refrigerant charge is low, the needle valve will not shut off properly under a no-load condition. This can easily be corrected by adding more refrigerant. Usually, a leaking float ball must be replaced.

If the strainer is clogged, it must be removed and cleaned. One method of cleaning the screen is to heat it carefully and then submerge it in water. A weak solution of hydrochloric acid can also be used for cleaning the screen. All the acid must be removed before the screen is reused.

Solenoid Valves

Service for solenoid valves includes correcting such problems as:

1. Coil burnouts
2. Failure to open
3. Failure to close
4. Alternating current hum

Coil burnouts can be caused by improper electrical characteristics, under- or overvoltage, or mechanical interference with the operation of the plunger. *Failure to open* can be caused by coil burnout, an open circuit to the coil connections, improper electrical characteristics, or in pilot-operated valves, dirt, scale, or sludge may prevent the piston from lifting. *Failure to close* can be caused by a valve that is held open by the manual lift stem; in a pilot-operated valve, dirt, scale, sludge, or a damaged pilot port may prevent closing. *Alternating-current hum* can be caused by a loose coil housing.

Pressure Regulators

In servicing pressure regulators, it is essential to measure the existing operating temperatures and pressures to diagnose the problem properly. A comparison needs to be made between what the valve is doing and what it is supposed to do. After the exact malfunction is determined, corrective action can be applied.

There are two basic problems that occur with pressure regulating valves: failure to open and failure to close. Most valves of this type can be disassembled. The normal service for most valves is to disassemble the valve and clean it thoroughly. If this does not solve the problem, the element can usually be replaced.

USING FILTERS AND DRIERS

Moisture is a deadly enemy of a refrigeration system. Therefore, the use of driers is standard practice in all commercial halocarbon systems.

The size of the drier is determined by the type of refrigerant used, the horsepower of the system, and to some extent, the type of system. Manufacturers' recommendations are usually followed. If the liquid line is warm on one side of the drier and cold on the other, some reexpansion is taking place and the drier should be replaced. A moisture-indicating sight glass is helpful in determining when the system is dry. Driers/strainers collect loose material in the system and protect the compressor. Driers can easily be replaced by pumping down the system and allowing 2 to 5 psig pressure in the lines for purging during a quick replacement.

REMOVING THE REFRIGERANT

One of the most effective ways to remove halocarbon refrigerant from a system for reuse is the method shown in Figure 15-5. A separate condensing system is used to pump vapor refrigerant from the system. The tank is cooled to prevent the condenser from becoming liquid bound. Special precaution should always be exercised in filling a pressure vessel with refrigerant. Liquid must not occupy more than 85% of the available space. A cylinder without proper vapor space can develop an excessive hydrostatic pressure and explode.

As mentioned earlier in our discussion of purging, when refrigerant is removed from the system, evaporating pressures must be kept above freezing temperatures where purging is performed from condensers or chillers. Water in the pressure vessel can freeze and break tubes. One way to prevent this is to keep the water flowing during the purging operation. Keeping the pump running will usually keep the water temperature above freezing even though the surrounding area is below freezing temperatures.

Figure 15-5 Removing refrigerant from the system. There are two methods: (1) by migration, using a cold cylinder (which is slow) and (2) by use of a compressor. The left-hand view shows the migration method. The right-hand view shows the use of the compressor. In either case, a drier should be used in the connecting piping.

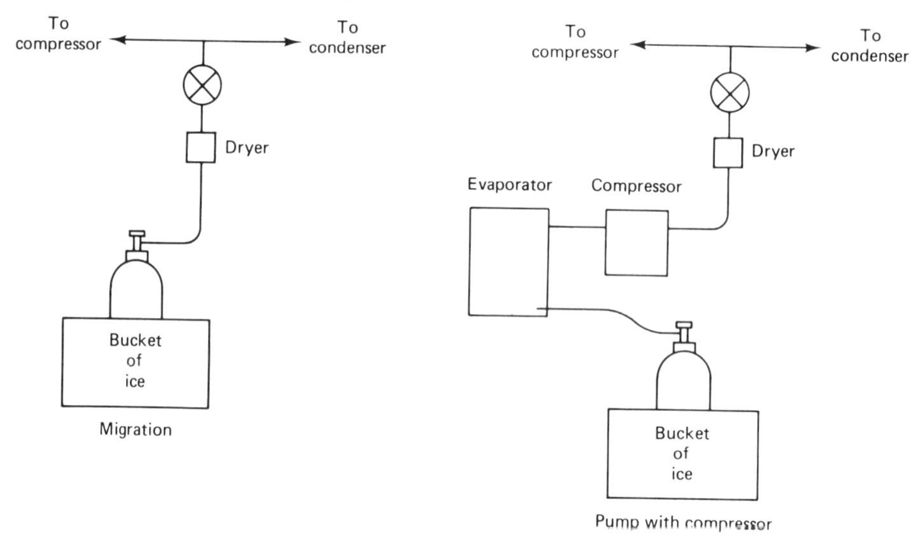

CLEANING THE SYSTEM WITH LIQUID REFRIGERANT

Following a severe hermetic compressor burnout, the system should be thoroughly cleaned. One way to do this is to use the apparatus shown in Figure 15-6. The refrigerant used for the cleaning process is liquid F-11. This refrigerant has a boiling point of about 75 °F at atmospheric pressure and therefore can readily be handled in liquid form, in most locations. The refrigerant is circulated through the system by a separate pump and filtered. An accessory that can stop the flow, such as an expansion valve, is bypassed during the cleaning process, as shown in Figure 15-6.

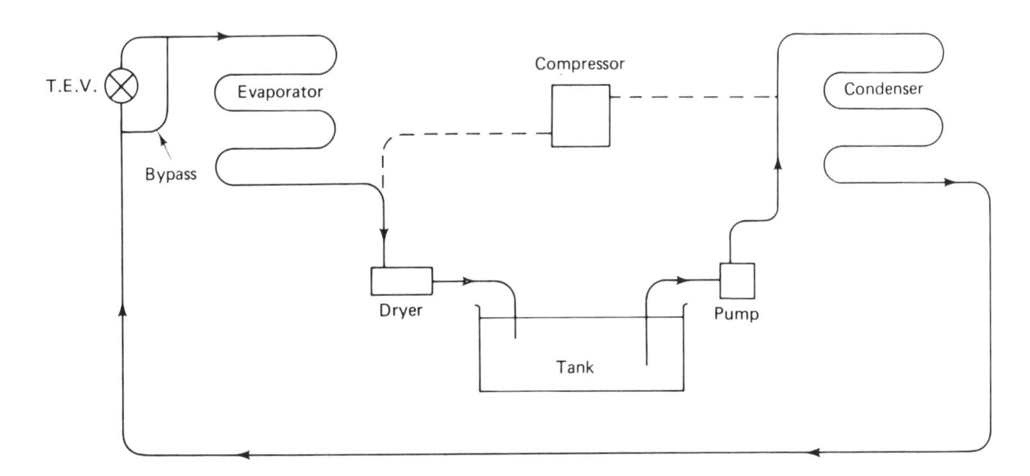

Figure 15-6 Using a liquid refrigerant flush to clean out the piping after a burn out. R-11 is often used for this purpose. The compressor is disconnected and the flushing equipment connected to the piping as shown. The thermal expansion valve is bypassed.

After the system is cleaned, the R-11 is removed, a drier is inserted in the liquid line and the unit is charged with the operating refrigerant.

A second method is to use the compressor to circulate the operating refrigerant through a "system cleaner" inserted in the suction line. This system cleaner is actually a large filter-drier, a cylinder containing a strainer and a desiccant cell. The cylinder cleaner is usually heated by a coil from the hot-gas line to prevent trapping liquid. The desiccant is changed periodically until the system is clean. This method is preferred by some service people since it requires less interruption of the system operation.

SUMMARY OF HALOCARBON REFRIGERATION SYSTEM SERVICE

1. Liquid refrigerant must be removed from equipment or piping being serviced.
2. Pressures in the system must be reduced to 2 to 5 psig to remove a part in the refrigeration system being replaced.
3. All refrigerant connections should be sealed when the system is opened.
4. Always install new gaskets when parts require them.
5. When the system is placed back in service, all air must be purged from the system.
6. Always replace the drier when the system is opened.
7. Use a vacuum pump capable of producing a vacuum of 28.8 in. of mercury or better to evacuate the system before charging.
8. Always test for leaks before the full charge is added.
9. After the system is put back into service, controls should be properly adjusted.
10. A check back after the system has been in operation for a few days is highly recommended.

REVIEW EXERCISE

Select the letter representing your choice of the correct answer (there may be more than one).

15-1. The following is *not* the proper use of a gauge manifold:
- ________ a. Read refrigerant pressures
- ________ b. Read water pressures
- ________ c. To purge a receiver
- ________ d. Charge oil into the compressor

15-2. A compressor service valve has how many positions?
- ________ a. Two
- ________ b. Three
- ________ c. Four
- ________ d. Five

15-3. To use a gauge manifold to test the pressures in a system, what position should the manifold valves be in?
- ________ a. Opened
- ________ b. Closed

15-4. What type of wrench should be used in completely opening or closing the compressor service valves?
- ________ a. Ratchet-type valve wrench
- ________ b. Crescent wrench
- ________ c. Torque wrench
- ________ d. Open-end refrigeration wrench

15-5. On which type of compressor can the valve plate be readily removed?
- ________ a. Hermetically sealed reciprocating
- ________ b. Semihermetic reciprocating
- ________ c. Hermetic rotary
- ________ d. Helical screw

15-6. To determine the normal operating condensing temperature of a water-cooled condenser, *how many °F* should be added to the leaving-water temperature?
- ________ a. 10 to 15°F
- ________ b. 20 to 25°F
- ________ c. 30 to 35°F
- ________ d. 40 to 45°F

15-7. To determine the normal operating condensing temperature for an air-cooled condenser, *how many* °F should be added to the ambient air temperature?
- ________ a. 10 to 15°F
- ________ b. 20 to 25°F
- ________ c. 30 to 35°F
- ________ d. 40 to 45°F

15-8. Which of the following types of water-cooled condensers must be cleaned with an acid solution?

________ a. Shell-and-tube

________ b. Tube-in-tube

________ c. Cascade

________ d. Shell-and-coil

15-9. The normal superheat setting of a thermal expansion valve is *how many °F?*

________ a. 4°F

________ b. 6°F

________ c. 8°F

________ d. 10°F

15-10. How many pressures are required to operate a thermal expansion valve?

________ a. One

________ b. Two

________ c. Three

________ d. Four

16 Troubleshooting

OBJECTIVES

After studying this chapter, the reader will be able to:

- State the procedures required for a competent troubleshooter.
- Perform the operations necessary to effect a "quick fix."
- Use troubleshooting aids.

TROUBLESHOOTING SKILL

Troubleshooting is an important skill that a refrigeration service person needs to perform. There are two parts to troubleshooting. The first is finding the cause of the problem, and the second, correcting it. The service person, who finds the problem, needs to know a great deal about the equipment. He or she needs to know what things are most likely to go wrong. Then he or she, or someone possibly less skillful, can perform the necessary repairs. It is important to put the equipment back on line as quickly as possible.

EIGHT STEPS OF TROUBLESHOOTING PROCEDURE

A competent troubleshooter should do the following things:

1. *Interviewing the operator.* Whoever is responsible for the equipment can usually provide helpful information to assist in finding the problem. They may relate such conditions as:

"It ran fine until we had a bad electrical storm."

"The motor starter has been giving me a lot of trouble."

Talking to the operator usually saves a lot of time in diagnosing the problem.

2. *Verifying the symptoms.* This action seems to uncover operator problems. To the uninitiated there are many operator-induced problems. Suppose that a switch or a valve was left in the wrong position.

By operating the equipment the troubleshooter can usually spot a symptom caused by improper operation or locate the cause of the symptom. For example, the operator may state that he must constantly press the reset buttom on the high limit to keep the machine operating. By operating the unit, the troubleshooter finds, for example, that the system grossly overloads. Good troubleshooters always verify the symptoms before digging into the equipment itself.

3. *Attempting a quick fix.* Competent troubleshooters attempt quick fixes even before the trouble is located. They know the malfunctions that are most likely to occur. These things are relatively easy to try, and they either solve the problem quickly or provide valuable information. The troubleshooter checks fuses, adjusts controls, cleans contacts, checks interlocks, and so on. If the quick fix works, valuable time is saved. If not, only a small amount of time has been lost and a certain number of causes can be eliminated.

Experienced troubleshooters know that certain quick fixes are a form of preventive maintanance. By that is meant general servicing of the equipment to prevent breakdown and to extend its life. Many service problems are caused by inadequate maintenance or a total lack of it. Many of these quick fixes are considered necessary for continued functioning of the equipment.

Some problems are more likely to occur than others. For example, if a lamp will not light, the first thing the troubleshooter does is to test the bulb. If a car will not start, the first thing to check may be the battery. The probability is high that these are the causes of the trouble. Troubleshooters must be armed with information on the most frequent causes of trouble and their solutions.

4. *Reviewing troubleshooting aids.* Many manufacturers provide troubleshooting aids to assist in locating service problems. If talking to the operator and trying a quick fix does not solve the problem, these aids can be very useful.

In the simplest form, troubleshooting aids may be an if/then chart: *if* a certain condition exists, *then* the suggested action is indicated. These charts usually list the common problems and their solution. The manufacturer may furnish a series of warning lights with a troubleshooting chart as a key to the needed action when specific lights are either on or off.

Some equipment may require a testing board, which is a device furnished by the manufacturer to diagnose problems. This test equipment can greatly reduce the time required to find the difficulty and prevent unnecessary changing of parts.

5. *Following a step-by-step search.* When other procedures fail, a troubleshooter turns to a step-by-step analysis of the equipment to find the problem. Some manufacturers furnish a flowchart that can be followed to check all aspects of the equipment. Each part is tested; if it is found to be satisfactory the service person proceeds to check the next part, and so on. This procedure is slow, but it will eventually lead to the problem.

It also requires a large amount of skill. The mechanic needs to be able to read diagrams, use test equipment, and interpret information as well as locate components and test points.

6. *Correcting the problem.* After the trouble is found, either the troubleshooter or someone else corrects the problem. Some manufacturers furnish factory personnel to solve difficult service problems. After the trouble is diagnosed, the owner's mechanic is often called upon to change the defective part or perform the required service. Diagnosing the trouble may require years of experience and technical expertise. Different skills are required to correct the trouble after it is found.

7. *Making final checks.* After the equipment is put back on line it is important to determine that the unit is operating properly. Certain parts may need to be worn-in or, after running for a short period, adjusted. Even a phone call may provide a valuable check to determine if the equipment is operating properly.

8. *Completing the project by* carrying out the following steps:

a. *Keep a written log of the services performed.* This is a valuable tool for analyzing future service problems.
b. *Recommend preventive maintenance where needed.* This prevents reoccurrence of the problem.
c. *Instruct the owner.* Advise him or her of what has been done and of ways to prevent reoccurrence of the problem. Advise the owner of any unique features of the equipment and of any improved methods for operating the equipment to give better performance.

TROUBLESHOOTING AIDS

Indicated above is a proven procedure for troubleshooting. Figures 16-1 and 16-2 summarize the various troubleshooting aids. In all cases, reference to the manufacturer's information is an essential step for approaching a problem on specific equipment.

Figure 16-1 Examples of quick fix procedures.

Condition	Check for quick fix
System does not run at all	Check electrical system fuses, thermostat settings, reset buttons, and safety switches.
Only part of the system operates	Make electrical check on defective load device. Do they have power? If so, is the motor inoperative?
System operates but does not have capacity	Install gauge manifold and read suction and discharge pressures. Analysis is as follows:[a]

Figure 16–1 Examples of quick fix procedures (continued).

Condition	Check for quick fix		
	Discharge pressure	Suction pressure	Problem
	L	L	Undercharge
	L	H	Defective compressor
	H	H	Overload or overcharge
	H	L	Restriction

[a]L, lower-than-normal pressure; H, higher-than-normal pressure.

Figure 16–2 Troubleshooting symptoms, problems, and solutions.

Symptom	Problem	Solution
	Air-cooled condensers	
1. High head pressure	Dirty condenser	Clean condenser
	Restricted airflow	Remove restriction
	Incorrect fan rotation	Reverse motor rotation
	Fan too slow	Replace fan motor
	Motor out on overload	Reset; check volts, amperes; adjust speed
	Prevailing winds	Install wind deflector
	Air short cycling	Install baffles
2. Pressure not corresponding to ambient temperature on shutdown	Noncondensables in refrigerant	Purge or recharge system
3. Temperature-drop at liquid level on return bends; bleed-off reduces head pressure but does not cause bubbles in sight glass	Overcharge	Reduce the refrigerant charge to normal level
4. Unit fails to start during low-ambient conditions	LPC has the unit shut off	Install low ambient-control arrangement
	Water-cooled condensers	
1. High water quantity, low water temperature rise	Scaled condenser tubes	Clean condenser tubes (see Chapter 5)
	Overcharge	Bleed refrigerant to normal level
2. High head pressure	Water valve fully open	Clean condenser

Figure 16–2 Troubleshooting symptoms, problems, and solutions (continued).

Symptom	Problem	Solution
	Evaporative condensers	
1. System shuts down on high head pressure or compressor overload	Low air quantity	Increase fan speed or replace fan motor
	Pump failure	Replace/rebuild pump
	Dirty water screen	Clean water screen
	Scaled coil tubes	Use wire brush or spray with acid (see Chapter 15)
	Evaporators	
1. Lower-than-normal suction pressure	Low refrigerant charge	Add refrigerant
2. Temperature drop in liquid line	Plugged drier	Replace drier
3. High expansion valve superheat	Poor expansion valve adjustment	Adjust TEV
	Defective expansion valve	Replace TEV
4. Some circuits of coil sweating, some dry (multicircuit coil)	Poor refrigerant distribution	Clean plugged tubes in refrigerant distributor or replace distributor
	Water chillers	
1. High water temperature drop through chiller	Shortage of circulating water	Increase water flow through chiller by adjustment or pump replacement
2. Low suction temperature	Shortage of refrigerant	Add refrigerant
	Restriction	Replace plugged drier
	Compressors	
1. Seizure	Lack of lubricating oil due to: (a) oil traps in piping,	Repipe to remove oil traps
	(b) flooded start due to oil accumulation in compressor during shutdown, or	Install check valve in discharge piping (see Chapter 7)
	(c) low refrigerant piping velocity causing oil not to return properly	Install double risers (see Chapter 7)
2. Excessive noise	Lack of proper lubrication	Check oil pressure and crankcase oil level
	Slugging on start-up	Repair or replace crankcase heater
3. Lack of capacity	Poor pumping ability	Repair or replace compressor valves
		Rebuild or replace compressor

Figure 16-2 Troubleshooting symptoms, problems, and solutions (continued).

Symptom	Problem	Solution
4. Overheating	High or low voltage	Check power input
	Shortage of refrigerant	Add refrigerant
	High head pressure	Clean condenser Check for overcharge
	Excessive compression ratio	Replace with two-stage equipment
	Metering devices	
A. Expansion valves		
1. Overfeeding or flooding	Improper superheating	Adjust superheat on expansion valve
	Improper bulb location	Relocate bulb (see Chapter 6)
	Wrong refrigerant in the system	Replace with proper refrigerant
	Light or low load	Replace expansion valve having a better range
	Excessive oil in the system	Remove oil to maintain normal level
2. Underfeeding or starving	Wrong type of valve	Replace expansion valve
	Shortage of refrigerant	Add refrigerant
	Improper superheat	Adjust superheat on expansion valve
	Plugged drier	Replace drier
	Improper bulb location	Relocate bulb (see Chapter 6)
	Plugged equalizer line	Repipe equalizer line
3. Hunting	Oversized expansion valve	Replace expansion valve
	Light load	Replace expansion valve having a greater range
	Refrigerant circuit too long	Replace coil with proper circuiting
	Rapid load changes	Replace coil with multi-circuit arrangement
	Intermittent bubbles in the liquid line	Check the refrigerant charge
B. Capillaries		
1. Low suction pressure and starved evaporation	Shortage of refrigerant	Repair leak and recharge system
	Too much oil in circulation	Recharge system with oil and refrigerant
	Plugged drier	Replace drier
	Low head pressure	Check refrigerant charge
	Plugged capillary tube	Clean capillary or replace

Figure 16–2 Troubleshooting symptoms, problems, and solutions (continued).

Symptom	Problem	Solution
2. Flooding evaporator and increased suction pressure	High head pressure	Clean condenser
	Excessive subcooling	Check refrigerant charge
	Overcharge of refrigerant	Recharge system
C. Low- and high-side floats		
1. Starving the coil	Float valve sticking or leaking	Repair or replace valve
2. Flooding the coil	Float valve sticking or leaking	Repair or replace valve
	Low- and high-pressure switches	
1. Nuisance trip on startup (low pressure control)	Expansion valve slow to adjust to proper feed	Readjust LPC[a]
2. Nuisance trip on startup (high pressure control)	Heat buildup on an air-cooled condenser exposed to direct sun during "off" cycle	Readjust HPC[a]
	Solenoid valves	
1. Humming sound	Low voltage, loose connection, or sticking plunger	Check voltage; repair or replace defective coil or valve
2. Temperature drop across closed valve	Valve leakage	Repair or replace valve
	Water regulating valves	
1. Valve fails to close off when the system shuts down	Leaking water valve	Head pressure must be reduced 10-20 lb at shut-down to cause the valve to close; adjust or replace valve
	Strainer-driers	
1. Freeze-up of expansion valve	Strainer-drier releasing moisture on system temperature rise	Replace strainer-drier
2. Excessive pressure drops through drier causing temperature drop in liquid line	Restriction in the drier causing reexpansion of liquid	Replace strainer-drier
	Sight glasses	
1. Bubbles in sight glass	Shortage of refrigerant	Add refrigerant
	Restriction	Remove restriction
	Excessive liquid lift	Increase liquid subcooling
	Undersized liquid line	Correct liquid line piping

Figure 16–2 Troubleshooting symptoms, problems, and solutions (continued).

Symptom	Problem	Solution
2. Moisture indicating sight glass shows moisture in the system	Moisture in system	Install or replace drier
	Defective moisture indicator	Test moisture indicator using a fresh drum of refrigerant

[a]If a reset-type pressure switch is used, a good test is to place a small fuse across the pressure switch terminals. The fuse will blow in case of an overload and identify whether the LPC or HPC is causing the problem.

REVIEW EXERCISE

Select the letter representing your choice of the correct answer (there may be more than one).

16-1. What is the first thing a troubleshooter should do?

________ a. Talk to the operator
________ b. Verify the symptoms
________ c. Attempt quick fixes
________ d. Review troubleshooting aids

16-2. What is the final thing a troubleshooter should do?

________ a. Make final checks
________ b. Keep a written log
________ c. Recommend preventive maintenance
________ d. Instruct the owner

16-3. Indicate an example of a "quick fix":

________ a. Purge the system
________ b. Check electrical fuses
________ c. Add refrigerant
________ d. Remove refrigerant

16-4. If a unit with air-cooled condenser fails to start during low-ambient conditions, what is likely to be the problem?

________ a. Overcharge
________ b. Expansion valve out of adjustment
________ c. LPC has unit shut off
________ d. Noncondensables in the system

16-5. A water-cooled condensing unit uses a high quantity of water and has a low water temperature rise. What is likely to be the cause?

________ a. Defective water valve
________ b. Scaled condenser tubes
________ c. Shortage of refrigerant
________ d. Defective electrical system

16-6. If the gauges indicate lower than normal suction and discharge pressure, what condition is likely to be the cause?

________ a. Poor refrigerant distribution
________ b. Poor expansion valve adjustment
________ c. Low refrigerant charge
________ d. Plugged drier

16-7. What is likely to be the cause of high water temperature drop through a water chilling unit?

________ a. Shortage of circulating water
________ b. Shortage of refrigerant
________ c. Defective compressor
________ d. Defective refrigerant piping

16-8. What is a likely cause of compressor seizure?

________ a. Low refrigerant charge

________ b. Lack of lubricating oils

________ c. Overcharge of refrigerant

________ d. A dirty condenser

16-9. What condition could cause a compressor to be unusually noisy?

________ a. Undercharge of refrigerant

________ b. Defective valves

________ c. Lack of proper lubrication

________ d. Plugged drier

16-10. What condition could cause high suction pressure and low discharge pressure in a system?

________ a. Undercharge

________ b. Defective compressor

________ c. Overload or overcharge

________ d. Restriction

17 Applications

OBJECTIVES

After studying this chapter, the reader will be able to:

- Describe the proper refrigeration conditions for a wide variety of products.
- Determine the proper refrigeration process for many common applications.
- Determine the relative merits of various refrigeration systems used for cooling or freezing a specific product.

CLASSIFICATION OF APPLICATIONS

There are many important applications for commercial refrigeration systems. In this section the topics discussed include:

1. Food preservation
2. Transport refrigeration
3. Low-temperature applications
4. Industrial applications

The topic of *food refrigeration* includes many areas. Some of them are:

1. Commercial freezing of foods
2. Precooling of fruits and vegetables
3. Refrigeration of meat products
4. Refrigeration of poultry products
5. Refrigeration of fishery products

6. Refrigeration of dairy products
7. Refrigeration of fruits and fruit-juice concentrates
8. Refrigeration of vegetables
9. Refrigeration of precooled and prepared foods
10. Refrigeration of bakery products
11. Refrigeration of candies, nuts, and dried fruits
12. Refrigeration of beverages
13. Refrigeration of eggs and egg products
14. Refrigeration warehouses and storage facilities
15. Retail food store refrigeration

The distribution of chilled and frozen foods includes:

1. Trucks, trailers, and containers
2. Railroad car refrigeration
3. Marine refrigeration
4. Air transport

The low-temperature refrigeration applications include:

1. Environmental test equipment
2. Cryogenics
3. Low-temperature metallurgy
4. Biomedical applications

The industrial applications include:

1. Ice manufacture
2. Ice rinks
3. Concrete dams, subsurface soil, and foundations
4. Refrigeration in the chemical industry

Each of these applications has its special considerations insofar as design and equipment selection are concerned. In order for the mechanic to properly service these systems he or she must have a thorough understanding of how the equipment should function. In this text only a few of these applications will be covered in detail. An effort will be made to cover the most important areas.

FOOD PRESERVATION

The perishable food industry is one of the largest, if not the largest industry in the country today. An industry of this size is extremely important. Proper refrigeration is an important factor in the success of this business.

Perishable foods can be classified into six groups:

1. Meats
2. Poultry
3. Seafood
4. Fruits
5. Vegetables
6. Dairy products

These perishable foods can be divided into two groups. The first three—meats, poultry, and seafood—are animal products. They are preserved and kept palatable. The exceptions, of course, are lobsters, oysters, and some other types of shellfish that are transported, then either cooked or eaten alive. Fruits and vegetables are as much alive while they are being transported as they are while they are growing but require an entirely different set of preservation conditions.

The principal causes of spoilage in foods are:

1. *Microbiological.* These include bacteria, molds, and fungi.
2. *Enzymes.* These are chemical in nature and do not deteriorate.
3. *Oxidation changes.* These are caused by atmospheric oxygen coming in contact with the food, producing discoloration and rancidity.
4. *Surface dehydration.* In freezing this is called "freezer burn."
5. *Wilting.* This applies to leafy vegetables that lose their crispness.
6. *Suffocation.* Certain fresh vegetables must have air. When packed in cellophane bags, the bags must have holes or the vegetable "suffocates."

The types of preservation include salting, smoking, pickling, pasturizing, canning, dehydration, cold storage above freezing, and storage below freezing. The last two of these require refrigeration.

Keep in mind that in the storage of meats, poultry, and fish that preservation depends on protecting the product against the effects of death, whereas fruits and vegetables must be kept alive without harmful changes. Each principal group of products requires separate treatment.

Meats

This product deteriorates through the harmful action of bacteria. The enzymes serve to tenderize the meat. Aging is the process of utilizing the good effects of the enzymes without the harmful effects of the bacteria. Sanitation is the most important factor in controlling bacteria. Air has many forms of bacteria present. One of the best ways of controlling this infection is through the use of germicidal or ultraviolet lamps. Oxidation is detrimental to meats, causing discolorization and deterioration of the flavor. Dehydration can be controlled to a large extent by maintaining high humidity in the storage room. High humidity also protects against moisture loss, which lowers the number of pounds that are available for sale.

Good practice requires that pork trimming be rapidly cooled after it is cut. This prevents destructive enzymatic action, which causes discoloration, rancidity, and poor flavor. Figure 17–1 shows a chill cabinet where trucks loaded with trimmmings can be unloaded into the chiller room through side-opening doors. Fans circulate the chilled air over the meat at low velocity. Air temperatures are in the 0 °F range with a TD of 10 °F. A continuous trimming chiller is shown in Figure 17–2. Spray-type coolers are used.

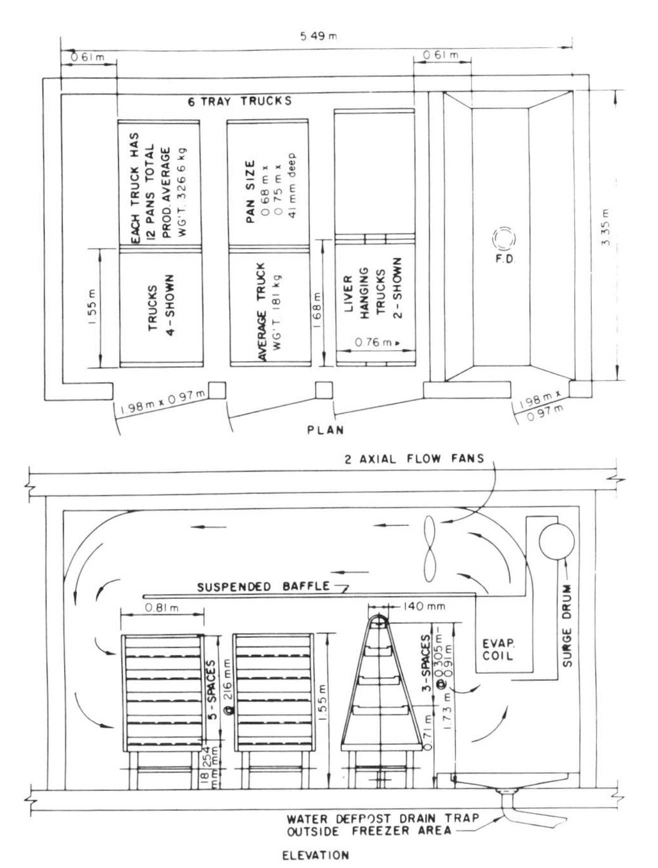

Figure 17–1 Chilling room for variety meats. Trimmings are placed on the shelf trucks in pans. Air temperatures vary from −29 to −17.8 °C. The desired internal meat temperature is −1.7 to −1.1 °C. (By permission of *ASHRAE Handbook.*)

Figure 17–2 Continuous trimming chiller. The air is chilled by two brine spray unit coolers. A temperature difference of 5.56 °C is maintained between room temperature and the refrigerant evaporating temperature. Coils are sprayed with antifreeze such as propylene glycol to remove the frost. (By permission of *ASHRAE Handbook.*)

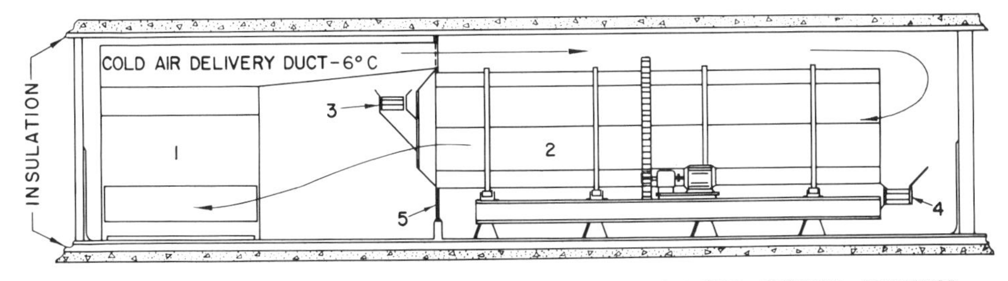

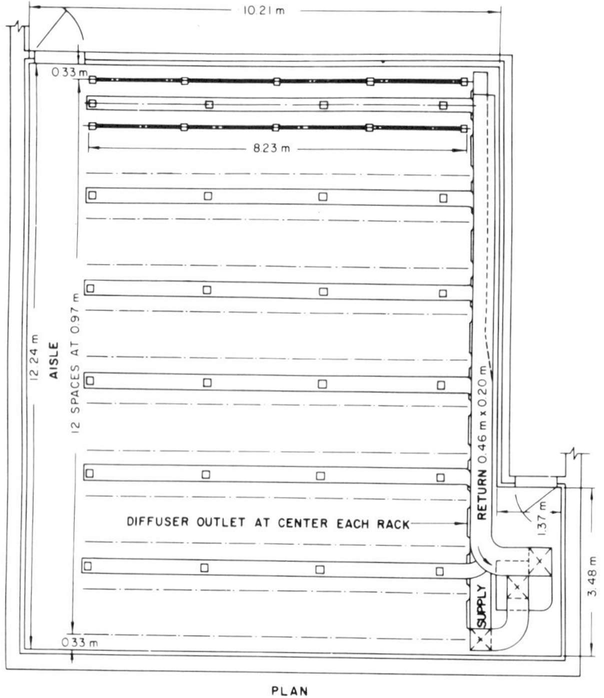

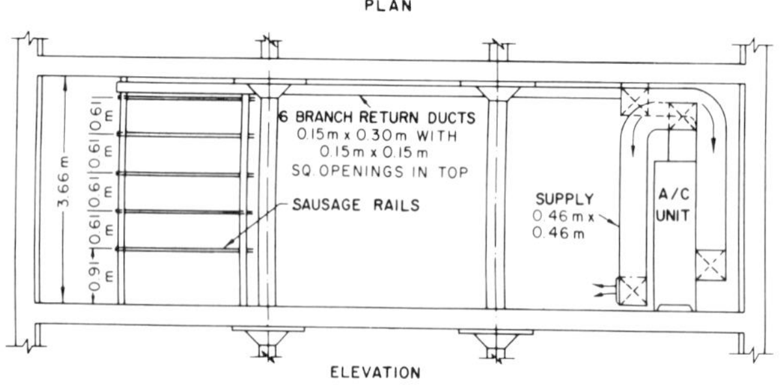

Figure 17-3 Sausage dry room. Conditions in this room remove moisture from the sausage. The keeping quality of sausage is dependent on the use of curing ingredients and spices, and removal of moisture. Thirty percent of the moisture is removed. Typical conditions in the room are 7.2 to 12.8 °C and 70 to 75% relative humidity. (By permission of *ASHRAE Handbook*.)

The layout for a sausage dry room is shown in Figure 17-3. The purpose of this room is to remove about 30 percent of the moisture, to a point where the sausage will keep for a long time virtually without refrigeration. This process is used as an alternative to the smoking process. The U.S. Department of Agriculture (USDA) requires that this room be maintained at temperatures above 45 °F, and the length of time in the room depends on the diameter and method of preparation of the product.

Poultry

Problems associated with the preservation of poultry are similar to those of meat in many respects except that poultry spoils much faster. However, poultry can be precooled by the use of cold water without detrimental effects. This is a relatively simple and effective process and therefore quite generally used. Bacteria and enzyme action is useful only in preserving game birds, as such action has a tendency to enhance the "game flavor."

Seafood

This product is the most perishable of all the animal foods. Yet there is a vast difference in the keeping quality of different kinds of fish. For example, swordfish can be kept refrigerated for 24 days and be in a more palatable condition than mackerel refrigerated for 24 hours. Commercial fish are usually refrigerated with ice.

Fruits and Vegetables

The problem with fruits and vegetables is that they are still alive after they are picked. They grow, breathe, and ripen. Most fruits and vegetables are picked in an unripened condition. The best-tasting products are vine ripened. However, it is not practical to ship and store vine-ripened fruits and vegetables. The purpose of refrigeration is to slow down the ripening process so that these can reach consumers before they reach the spoiling stage.

Vegetables quickly lose their vitamin content when surface drying takes place. It is interesting to note that products shipped from California to Chicago that have been properly iced after harvest will be fresher than produce supplied from Illinois farms and shipped to a Chicago market without being iced.

Another way to improve the product when it reaches the user is to package the product. This cuts down the surface drying. Packages are usually made of cellophane or some similar product. These containers must have holes so that the product can breathe (exchange oxygen and CO_2). Otherwise, the product will die, and a dead product will spoil rapidly.

A number of products require special treatment: bananas, for example. These are picked green and must be ripened for marketing. Banana ripening is initiated by the introduction of ethylene gas. For this to be effective banana rooms must be airtight. Refrigeration is provided using R-12 refrigerant since an ammonia leak can be destructive to the fruit. Rooms such as those shown in Figure 17–4 are cooled using 45 to 65 °F air. A design temperature difference of 15 °F and a refrigerant temperature of 40 °F are considered good practice.

An interesting application of refrigeration is the processing of fruit-juice concentrate (Figure 17–5). Hot-gas discharge from the compressor is used to supply heat for the evaporation of the juices. The water vapor is condensed by evaporating liquid refrigerant in a shell-in-tube condenser. Water vapor is used to superheat the suction gas. This type of apparatus provides a continuous process.

Citrus Fruits, Bananas, and Subtropical Fruits

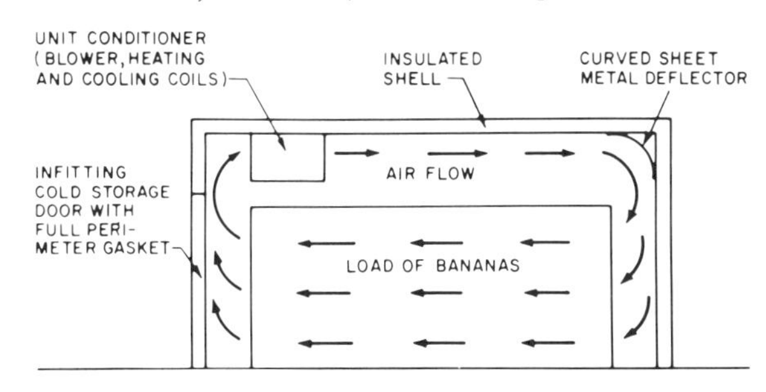

Figure 17–4 Banana room. Ripening is initiated by exposure to ethylene gas. Rooms must be airtight. Exposure time is usually 24 hours. Floor drain must be trapped to prevent gas leakage. R-12 is recommended. Air temperature range is 7.2 to 18.3 °C. Recommended design temperature difference is 8.3 °C with 4.4 °C evaporating temperature. (By permission of *ASHRAE Handbook*.)

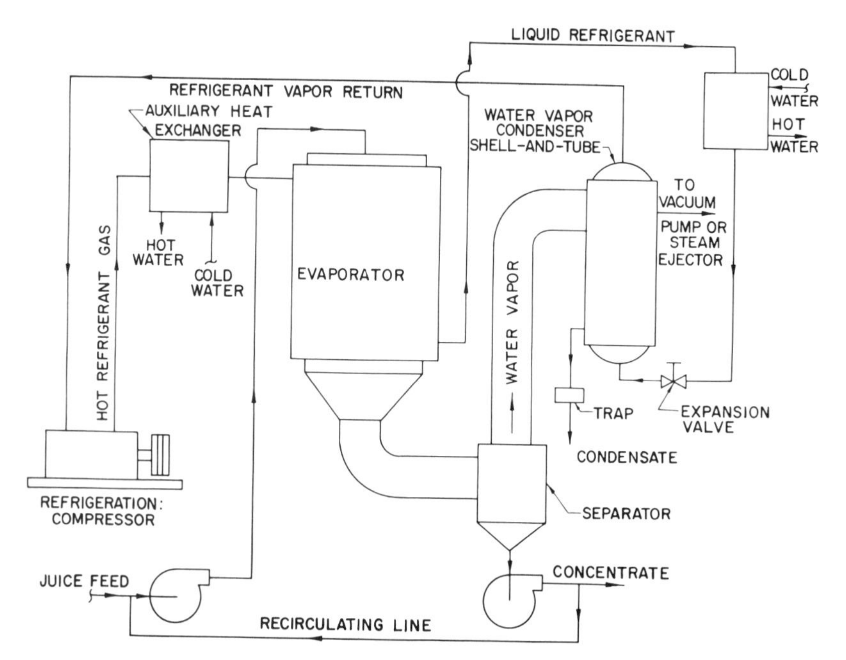

Figure 17–5 Equipment for fruit juice concentrates. Hot refrigerant gas is used to supply heat for evaporation of juices. Water vapor is condensed by evaporating liquid refrigerant. Vapor and concentrate are usually separated by an arrangement of cyclones and baffles. Pressure in the system is maintained by steam ejector or vacuum pump. (By permission of *ASHRAE Handbook*.)

Dairy Products and Eggs

Sanitation is extremely important in all aspects of handling milk. The bacteria content must be controlled. Certain limits are set for the number of bacteria (the bacteria count) for milk supplied by the producer. Souring takes place due to the formation of lactic acid. The best means of controlling the growth and multiplication of bacteria is by refrigeration.

Butter is primarily fat that is extracted from sour milk. The fat oxidizes unless it is refrigerated, which prevents it from becoming rancid and unpalatable.

Cheeses are refrigerated to prevent too rapid mold growth. The surface must be kept moist or the cheese will become hard and brittle. The moisture facilitates mold. Some mold enhances the flavor. However, too much mold creates waste because it usually must be removed for sale. Refrigeration retards the growth of mold.

One example of the use of mechanized equipment to handle food products is shown in Figure 17–6. This figure shows the steps in egg product processing. The various products are shown to the right in the drawing. The U.S. Department of Agriculture requires certain holding temperatures for egg products. Egg products are extremely perishable and should be kept at 34 °F or lower at all times before use.

For long-term storage of fresh foods the product is kept as close to freezing temperature as possible (e.g., at 34 °F) and at as high a humidity as practically possible (usually 70 to 90% RH). Long-term storage also includes some special treatments, such as the control of CO_2 in the air. These special applications will receive separate treatment.

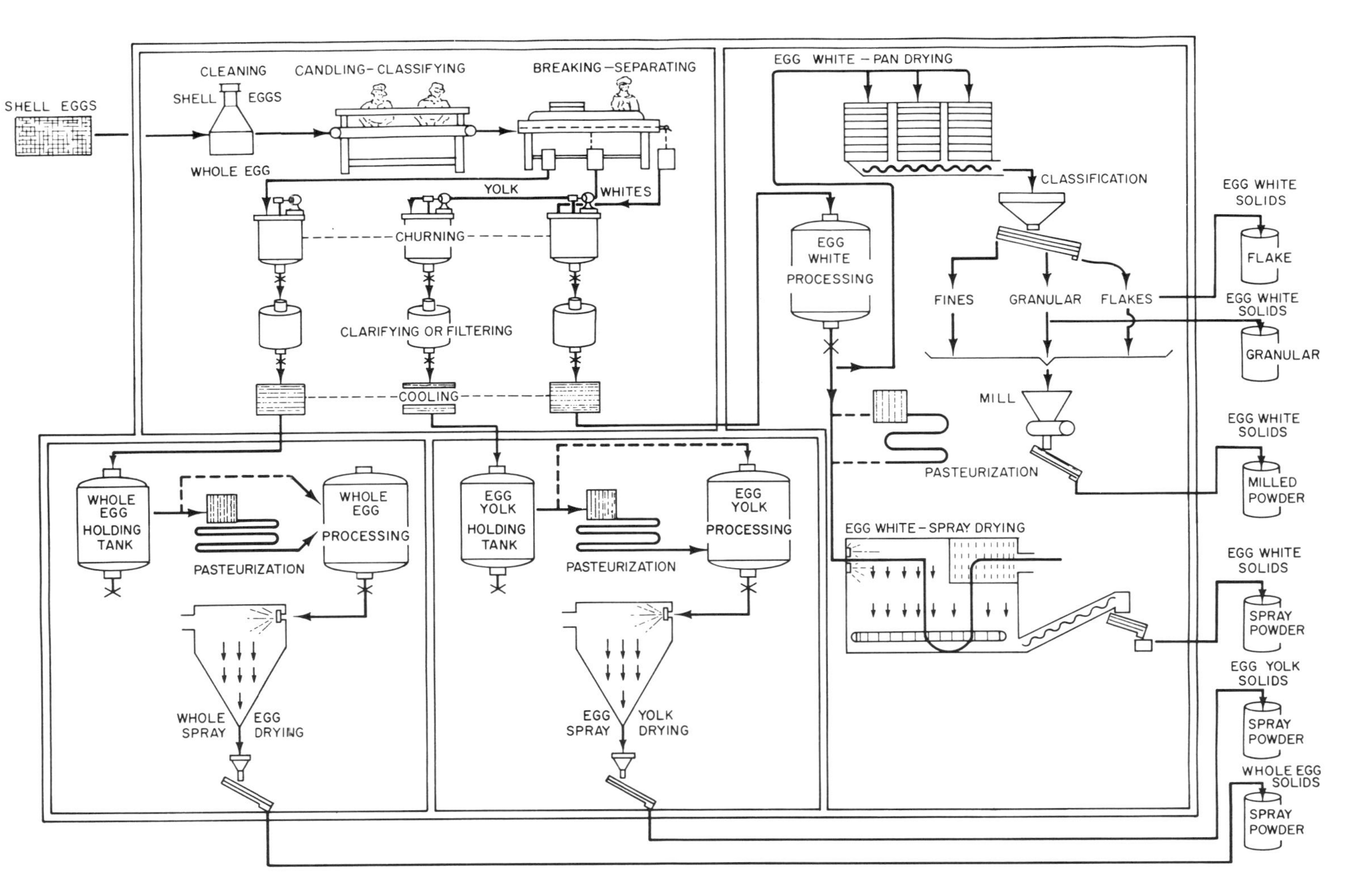

Figure 17–6 Egg product processing. Various unit processes are involved in the manufacture of egg solids or dried egg products. Practically all egg products are required by law to be pasteurized. The USDA requires certain holding temperatures for egg products ranging frm 4.2 to 18.3 °C. (By permission of *ASHRAE Handbook*.)

Frozen Foods

The quality of frozen foods requires controlling many factors: the types of seeds, the method of growing and harvesting, treatment of the product when it is picked, proper processing and packaging, proper storage, and proper cooking by the user. At any one of these steps the product can be damaged.

Due to the importance of the frozen-food business and its dependence on proper refrigeration, a number of topics will be discussed: (1) quick freezing, (2) refreezing, (3) storage conditions, (4) transportation, (5) retail food stores, (6) storing frozen cooked foods, and (7) using frozen foods.

Quick freezing. *Quick freezing* is essential. It produces small ice crystals, which are less damaging to the product. Produce should be frozen immediately after harvest—the sooner the better. Small packages are better to freeze than large packages because the interior freezes more quickly. Speed in freezing is important.

Small packages may be frozen on or between refrigerated plates or in a "blast" freezer. Foods are frozen at temperatures between -5 and $-20\,^{\circ}$F. Freezer burn should be avoided. This is the condition of surface oxidation that causes discolorization of the product. It is prevented by packaging in airtight containers or by waxing or glazing the product. Ice glazing is used to prevent surface drying of fish.

Vegetables must be blanched before freezing. This consists of placing the product in boiling water to kill bacteria. Fruits are often glazed with a sugar syrup to prevent oxidation. Air is removed from citrus juice before freezing. Meats become about 20 percent more tender through freezing.

Commercial Freezing Systems. Commercial freezing systems can be divided into five groups:

1. Air-blast freezers
2. Contact freezers
3. Immersion freezers
4. Cryogenic freezers
5. Liquid-refrigerant freezers

The air-blast freezer (Figure 17–7) using a stationary tunnel, produces satisfactory results for a number of products. Products are placed in trays which are held in racks. Air-blast freezers can be mechanized to provide a continuous process, as shown in Figures 17–8 and 17–9. These freezers are used primarily for packaged products.

Modern belt-type freezers use vertical airflow and greatly improve the contact between air and product. Two types are shown in Figures 17–10 and 17–11.

The fluidation principle is illustrated in Figure 17–12. Solid particles are floated by upward streams of air. In freezing, each particle is separated from the other and free to move. Product moves through the bed without the use of a belt (Figure 17–13).

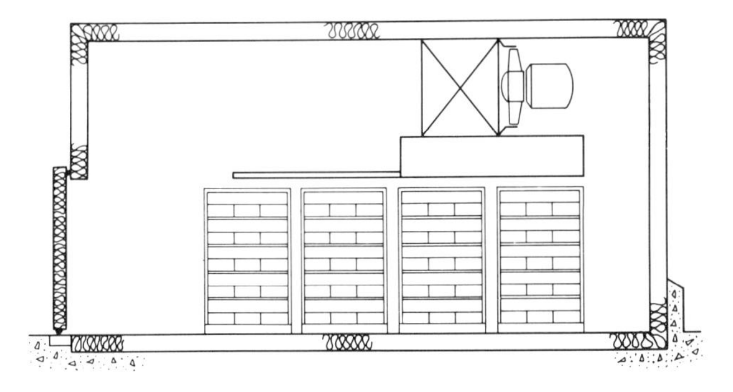

Figure 17-7 Stationary freezing tunnel. Practically all products can be frozen in a stationary tunnel. Conditions are determined in relation to the product being frozen. (By permission of *ASHRAE Handbook.*)

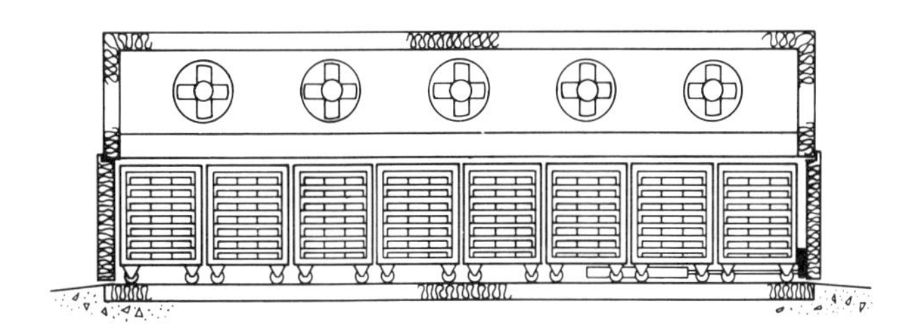

Figure 17-8 Push-through tunnel. This type of freezer offers a convenient means of placing and removing freezing racks. It can be used for a variety of products. Conditions depend on the product being frozen. (By permission of *ASHRAE Handbook.*)

Figure 17-9 Carrier freezer. This freezer arrangement is similar to two push-through tunnels on top of one another. In the top section carriers are pushed forward; in the lower section they are returned. In both ends there are elevating mechanisms. (By permission of *ASHRAE Handbook.*)

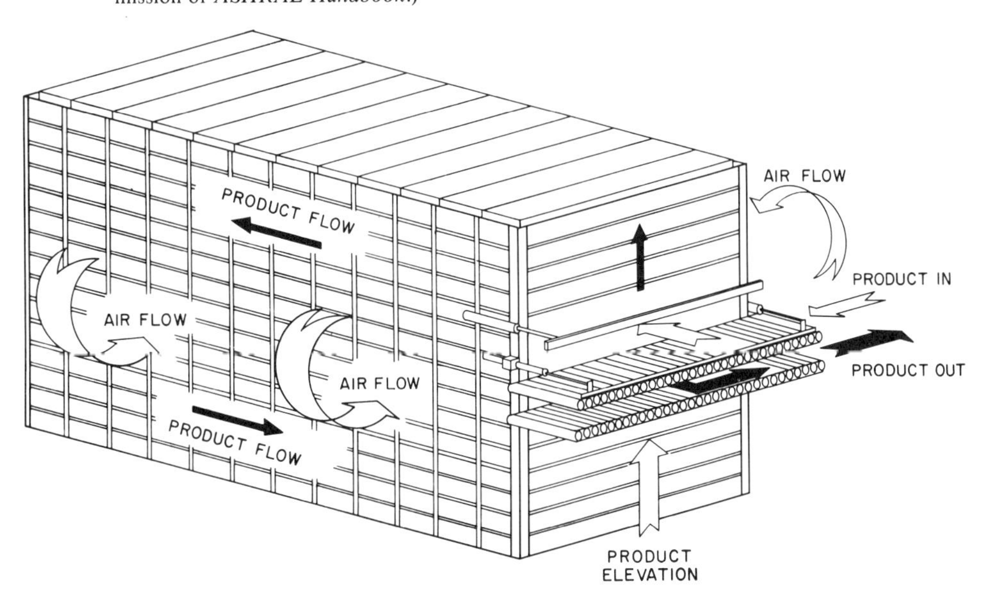

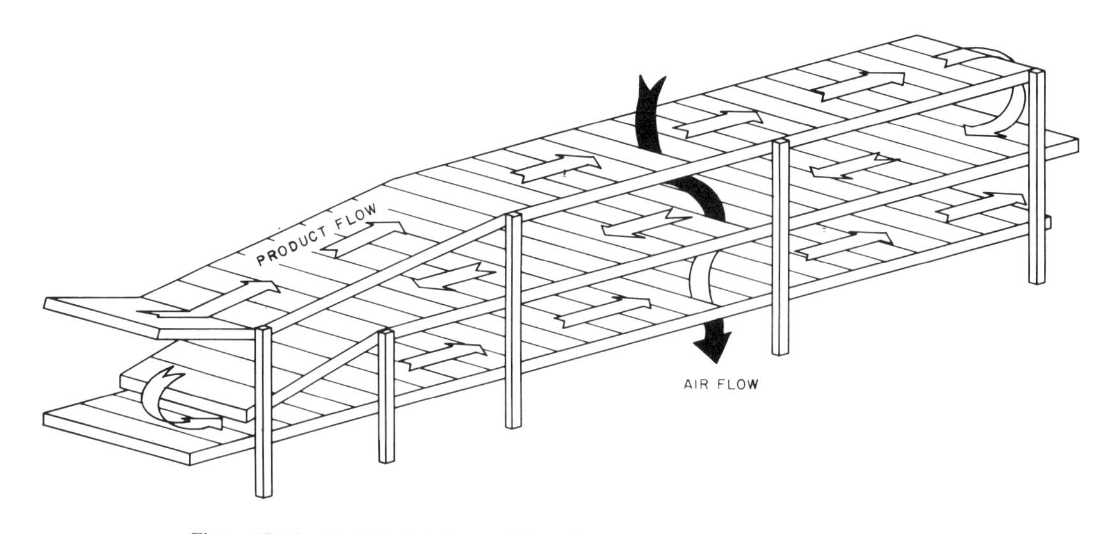

Figure 17-10 Multiple-belt freezer. These freezers are suitable for individual quick freezing of fried fish sticks, fish portions, bakery items, and similar products. (By permission of *ASHRAE Handbook.*)

Figure 17-11 Spiral belt freezer. This freezer is designed to save floor space. It is placed in a refrigerated room and the freezing time is regulated by the speed of the belt. It is suitable for a wide variety of unpackaged meat products. (By permission of *ASHRAE Handbook.*)

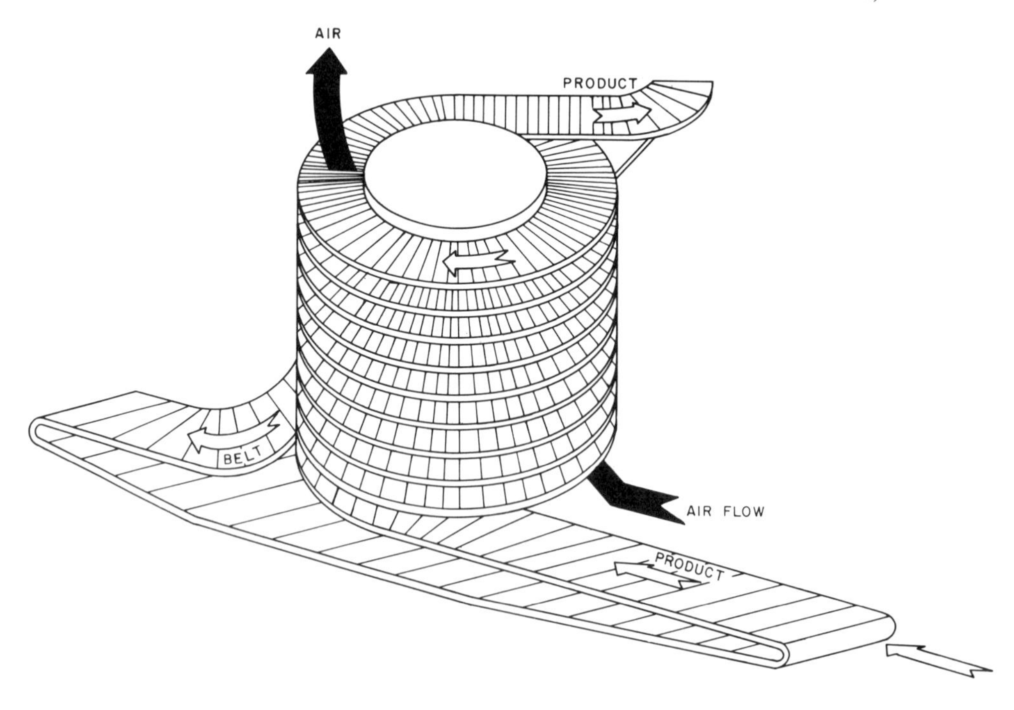

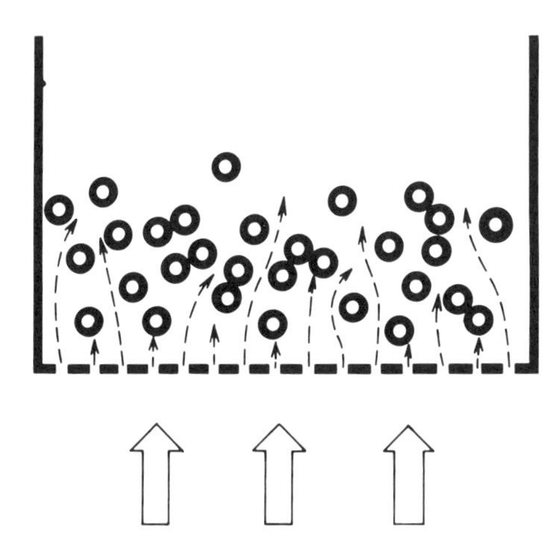

Figure 17–12 Fluidation principle. Fluidization is the process of freezing small items such as peas by an upward flow of freezing air. At a certain air velocity the particles being frozen will float and maintain separation. The principle can be used in a continuous freezer as shown in Figure 17–13. (By permission of *ASHRAE Handbook*.)

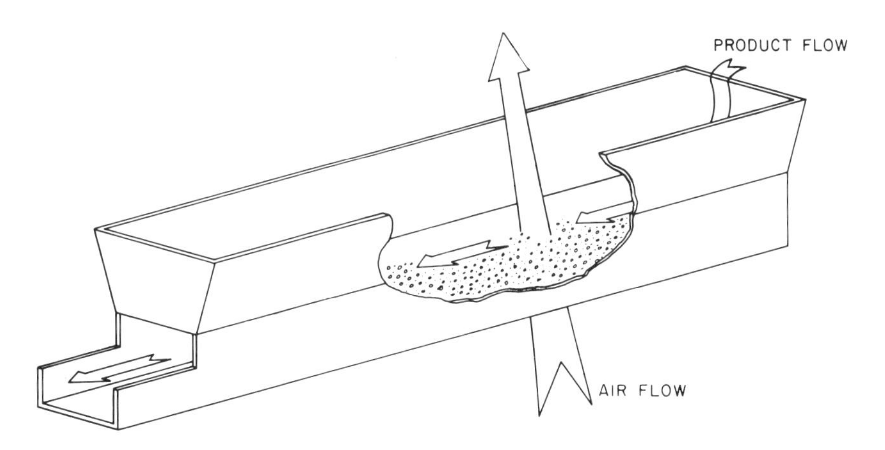

Figure 17–13 Fluidized-bed freezer. This is an automatic continuous process freezer using the fluidization principle. It is suitable for small vegetable pieces, french fried potatoes, cooked meatballs, and many similar products. (By permission of *ASHRAE Handbook*.)

Relatively thin packages of produce can be quickly frozen between freezing plates. A typical freezer of this type is shown in Figure 17–14.

Cryogenic freezers use liquid nitrogen (LN_2) or liquid carbon dioxide (LCO_2) as refrigerants. The boiling liquid comes in direct contact with the product. After use, the refrigerant is wasted to the atmosphere. A freezer of this type is shown in Figure 17–15.

Some products, such as shrimp, are frozen by immersion in a boiling, highly purified R-12 refrigerant. In these systems the refrigerant is recovered by condensing on the surface of refrigerated coils. A liquid refrigerant freezing system is shown in Figure 17–16.

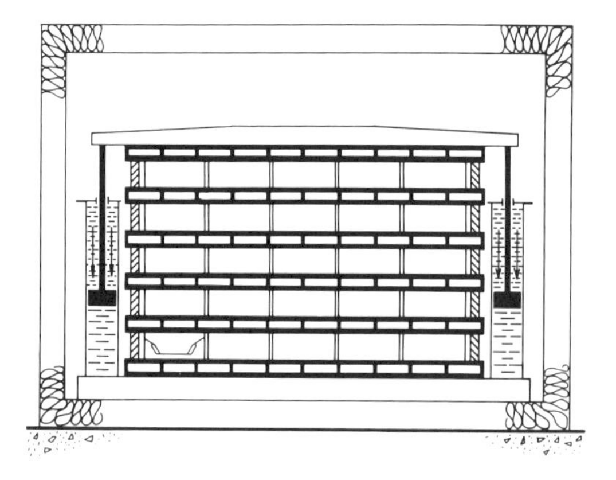

Figure 17–14 Plate freezer. The product to be frozen is firmly pressed between two refrigerated plates. This process is limited to products with good transfer rate and thickness not to exceed 50 mm. (By permission of *ASHRAE Handbook*.)

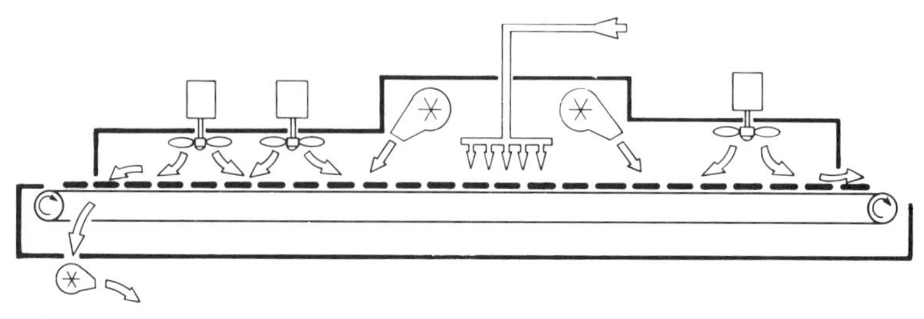

Figure 17–15 Cryogenic freezer. This freezer uses a liquid nitrogen spray (−196 °C). The product passes through the freezer on a moving belt. Vapor refrigerant is discharged at −30 to −100 °C. This process provides fast freezing. Vaporized nitrogen is wasted. (By permission of *ASHRAE Handbook*.)

Figure 17–16 Liquid refrigerant freezing system. The product comes into direct contact with boiling refrigerant. R-12 (−30 °C), highly purified, is used. Vaporized refrigerant is recovered by condensation on refrigerated coils. This process is commonly used to freeze shrimp. Excellent heat transfer ratio. (By permission of *ASHRAE Handbook*.)

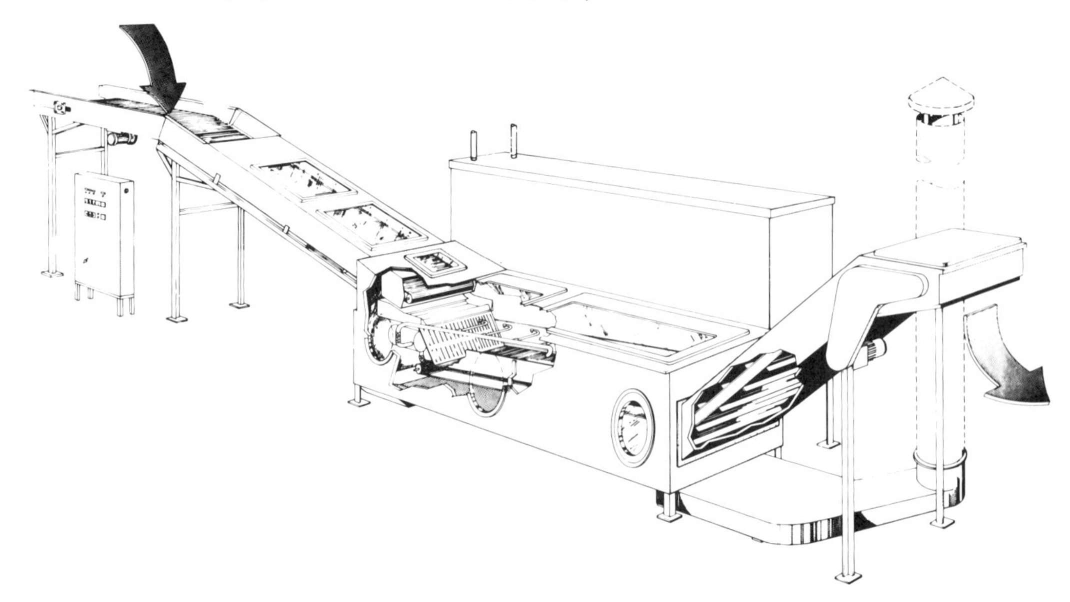

Refreezing. When a vegetable is in a frozen state, it is dead. It may come to life when thawed out and is then subject to spoiling. Some of the quality of the product is lost when it is refrozen, and this practice should be avoided if at all possible. Some canneries freeze products to prevent spoilage until they can schedule the final canning process.

Storage. There are certain parts of the structure of foods that thaw out at temperatures below −20 °F. A good, general storage temperature is −5 °F with only 1 °F fluctuation up or down from this temperature. Wide fluctuations of temperature cause changes in the ice crystal formation in the product, damaging the cells.

If foods do not contain fatty acids which can become rancid and the food is properly protected against surface dehydration, the food can be stored almost indefinitely.

Most modern refrigerated warehouses are one-story structures. A typical floor-plan layout is shown in Figure 17–17. The building construction is shown in Figure 17–18. Note that the outer walls are independent of the rest of the building. It is usually convenient to use penthouse refrigeration equipment rooms (see Figure 17–19).

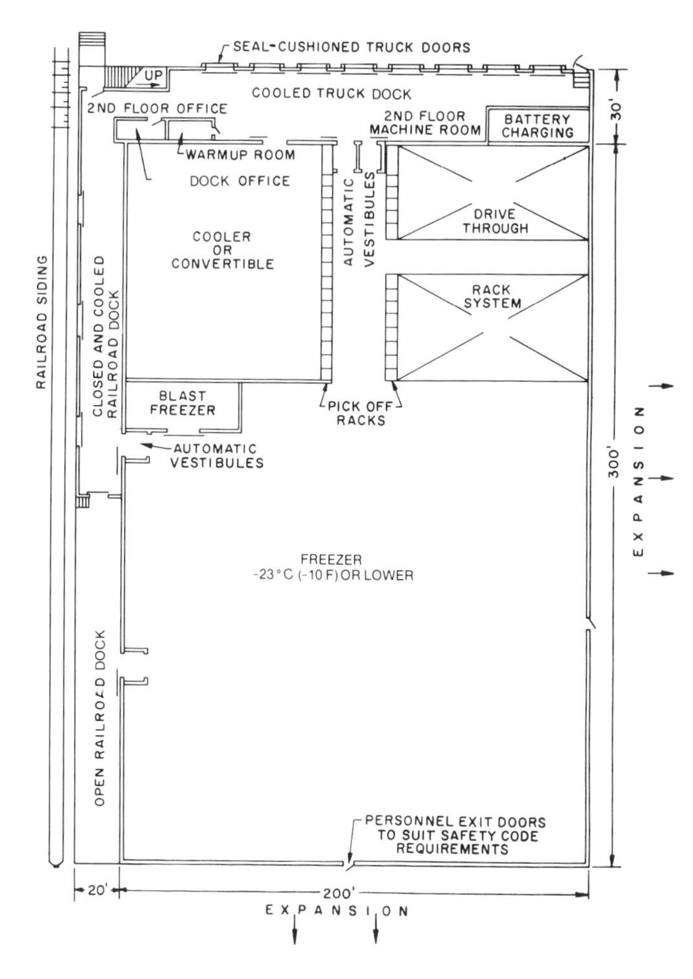

Figure 17–17 One-story warehouse. In the layout, note refrigerated shipping platforms; conditioned vestibules; first-in, first-out rack arrangement; blast freezer; office; and shop. (By permission of *ASHRAE Handbook*.)

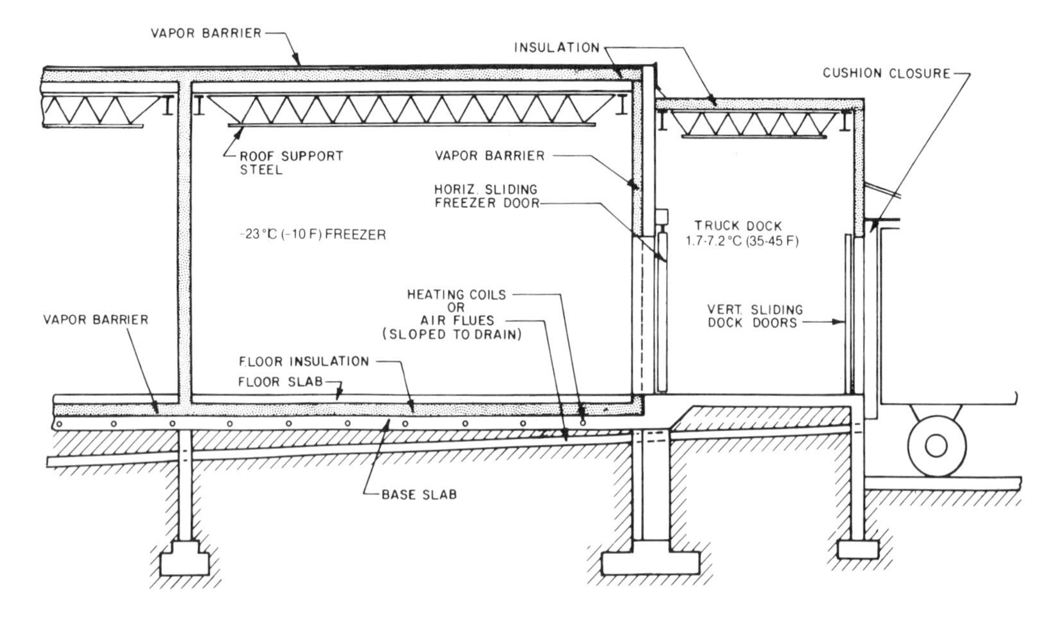

Figure 17–18 Typical one-story construction with hung insulated ceiling and underfloor warming pipes. (By permission of *ASHRAE Handbook*.)

Figure 17–19 Penthouse application of cooling units. (By permission of *ASHRAE Handbook*.)

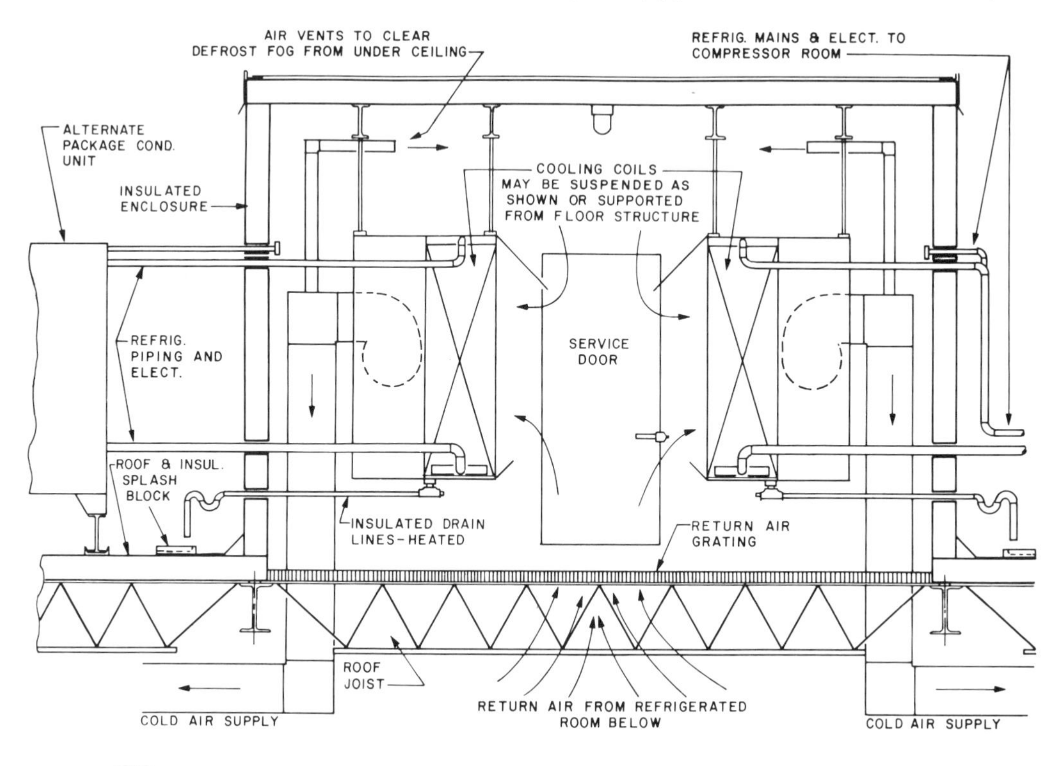

Transportation. Good-quality refrigeration systems are now available to transport either fresh or frozen foods. For fresh vegetables the preference is still for ice. However, high-humidity refrigeration systems are available. Piggyback-type "semis" with refrigeration systems can be loaded on flat cars for rail transportation. Twenty-four-hour service is available in most major cities to service these refrigerated systems.

Retail food stores. Excellent facilities are now available in retail stores (or supermarkets) for the storage and sale of fresh vegetables and frozen foods. Even with good refrigeration, fresh vegetables have a relatively short shelf life. Frozen foods can be stored for months if necessary and still maintain their original goodness.

Storing frozen cooked foods. This business has been built on the merits of good refrigeration systems. One modern development has been the display freezer. Frozen cooked foods, such as TV dinners, can be displayed at eye level. The customer reaches through an "air curtain" to obtain the product. The use of microwave ovens has made possible quick use of these prepared foods by the purchaser.

Using frozen foods. Although it is desirable to quick-freeze foods, it is equally desirable to slow-thaw the foods. This causes the least disturbance to the food cells and to the flavor. The one exception is the use of the microwave oven, which causes heat to penetrate the frozen product and thaw it uniformly. The use of microwave ovens for thawing frozen foods is a great asset for providing quick use.

TRANSPORTATION REFRIGERATION

The transporation of refrigerated products is a large and important aspect of commercial refrigeration. Many truck bodies are well insulated, and reliable refrigeration units are available to maintain the required temperatures. There are a number of types of refrigerated trucks and trailers. There is some difference between those used for long and short hauls, as well as the different types of products being transported.

Long-Haul Systems

Long-distance hauls are usually handled by trailer trucks or "semis." These trailers can be detached from the engine or "tractor" portion of the assembly. The trailer is usually equipped with a stand-alone refrigeration system. An illustration of a typical insulated trailer is shown in Figure 17–20. Notice the bulkhead in the front for return air coming

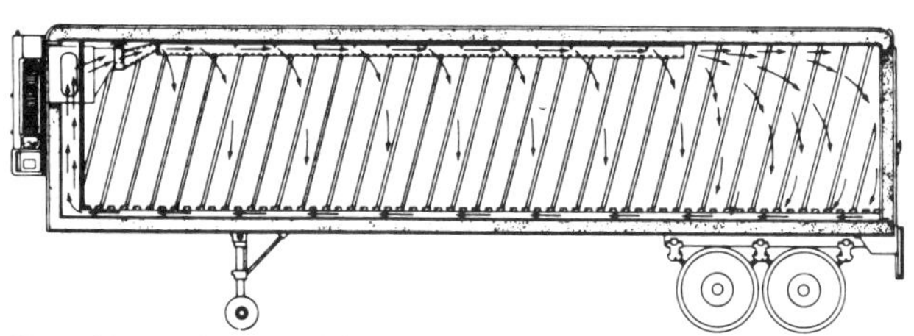

[a] Note false bulkhead which forms return air plenum taking air from floor.

Figure 17–20 Section of trailer showing air circulation. (By permission of *ASHRAE Handbook*.)

back to the evaporator through a raised floor. This illustration shows a "piggyback" or "clip-on" refrigeration unit.

Some trailers use refrigeration consisting of the evaporation of liquid nitrogen (LN_2) or liquid carbon dioxide (LCO_2). In these systems the evaporated refrigerant is wasted to the outdoors.

The "semi" trailer with stand-alone refrigeration can be placed on a railroad flat car and transported. Local delivery is accomplished by connecting a tractor and making the local delivery.

There are four different temperature conditions maintained in refrigerated transport, depending on the type of product being handled. These are:

Type of service	Temperature °F
1. Air conditioning for floral products, candy, etc.	55 to 70
2. Medium temperature for perishable foods	32 to 40
3. Fresh meats	28 to 32
4. Frozen foods	0 to –5

Some systems that are used both winter and summer have heating equipment as well as cooling equipment. Reverse-cycle refrigeration is popular for these applications. Reverse-cycle refrigeration can also be used for defrosting evaporators where conditions cause the accumulation of ice on the coils.

Local Delivery Equipment

Where smaller trucks are used for local deliveries, refrigeration systems with "eutectic plates" are advantageously used. Eutectic plates are constructed with a provision for refrigeration storage capacity. Plates are made with an interior volume that holds a special liquid called a eutectic solution. This liquid stores the cold by changing the state of the solution. Just as ice freezes at 32°F, these eutectic solutions freeze at various temperatures depending on the desired conditions in the truck. Plates are available for operating temperatures of −58, −29, −14, −12, −10, −8, −6, +18, +23, and +26°F.

The plates are usually connected to the refrigeration compressor at night, when the truck is not being used. The refrigeration capacity of the plates is sufficient to carry the load during the truck's next day's use. Some trucks carry their own compressor so that they can easily be plugged into electric service wherever they may be. Illustrations of the eutectic plates are shown in Figure 17–21. Another arrangement for ice cream or frozen-food service consists of equipping the truck with auxiliary drive or an electric motor to operate the compressor while the truck is in service.

Figure 17–21 Plate-type evaporator with entectic solution for holdover cooling capacity. (By permission of Dole Refrigerating Company.)

Mechanical Refrigeration

There are many types of systems available for furnishing the power to drive the compressor:

1. Independent gasoline (or propane or butane) engines
2. Power takeoffs from vehicle engines
3. Electric motors

The compressor may be either on board the vehicle or at the garage for out-of-service use only. Evaporations are of many types: finned and pipe coils for forced or gravity circulation, standard plates, or eutectic plates.

Equipment must be designed to withstand severe motion, shock, and vibration. Short, rigid lines are subject to "cold working. " They become hard and brittle through continual motion. Thus, flexible hose connections are often used. Air-cooled condensers need to be designed so that normal airflow due to movement of the truck does not prevent adequate air circulation. Units with independent engines require batteries for cranking, fuel tanks, and pumps.

LOW-TEMPERATURE APPLICATIONS

Making Ice Cream

There are two general types of equipment used for freezing ice cream:

1. Self-contained equipment for freezing "soft" ice cream or custard. This is a fresh product, quickly frozen for immediate use.

2. Large central plant systems. Here refrigeration is a large part of the cost of the product. Ammonia is the principal refrigerant used.

Due to the constituents of ice cream, there is no one freezing point for the entire product. The freezing point varies from +28 to −68 °F. Freezing commences at about 28 °F and at about 27 °F about 90% of the original water is frozen out as ice. The products used for soft ice cream are usually delivered to the freezer in a precooled condition, 35 °F and the frozen product is cooled by the freezer to 24 °F.

The same process is used in freezing ice cream in larger plants with larger equipment except that the soft ice cream is placed in hardening rooms where the storage temperature is down to −20 °F. Product temperatures at these levels require a refrigerant evaporating temperature as low as −30 to −40 °F. For these temperatures, two-stage ammonia equipment is the most common type of refrigerant used.

Bakery Refrigeration

Proper refrigeration is an essential part of bakery production. Refrigeration is used primarily in three areas:

1. Storage of the ingredients prior to use
2. Controlling the temperature of the dough during the mixing process
3. Providing refrigerated storage and freezing facilities for the product, to coordinate production with the marketing arrangement

Storing ingredients. Certain items, such as corn syrup, liquid sugar, lard, and vegetable oil, are stored at 125 °F. Specialized vinegars and shortenings are stored at 85 °F to prevent mold and rancidity. Yeast should always be stored below 45 °F. Cocoa, milk products, spices, and other raw materials subject to insect infestation should be stored at the same temperature as yeast.

Mixing. During mixing, heat is produced in two ways:

1. *Heat of friction:* The electrical energy of the mixer is converted into heat.
2. *Heat of hydration:* This heat is produced when the material in the mixer absorbs water.

The temperature rise in the dough is dependent not only on the heat produced in mixing but also on the specific heat of the mix.

The temperature at which yeast is most active is quite critical. Yeast is dormant at temperatures below 45 °F. It is extremely active when mixed with water and certain sugars in the range 80 to 100 °F. The cells of the yeast die at temperatures of about 140 °F.

To absorb this excess heat the mixture is supplied with water at about 36 to 40 °F and the mix is also cooled by a water jacket surrounding the dough. The final dough temperature is about 100 to 105 °F.

For many years the princpial refrigerant used in bakeries was ammonia. Many systems using R-12 or R-22 refrigerants are now in use, providing chilled water to the mixer at the desired temperature.

Fermentation. There are two processes for preparing the dough: the straight dough process and the sponge dough process.

In the straight dough process the liquid sponge (made of yeast, flour, and malt) is made up in an unjacketed tank at a temperature of 69 to 81 °F and allowed to ferment for 1 hour. The second hour the fermentation takes place in a jacketed tank at about 84 °F. A continuous process is provided by allowing the brew to be drawn from one tank while the other is fermenting. In the sponge dough process, 3½ to 5 hours is required for fermenting at a sponge temperature of about 81 °F.

In the straight dough process, all the ingredients are mixed at once. In the sponge dough process, only part of the amount of flour and water are mixed with the yeast. After the fermenting takes place, the remaining flour and water are added.

Bread cooling and processing. Bread is cooled to 90 to 95 °F for handling and slicing. This prevents condensation within the wrapper, which can cause mold to develop.

Refrigeration is required for a bread-wraping machine. Evaporator temperature is usually in the range of 10 °F. The plate surface temperature must be held to about 16 °F. Refrigeration is not required if bags are used.

Bread is frozen at temperatures of from 16 to 20 °F. Freezing rooms with temperatures from 0 to −30 °F are used for freezing bread. To prevent moisture loss and crystallization of the starch, frozen bread should be stored at temperatures below 0 °F. This usually requires two-stage refrigeration equipment.

Ice Rinks

The category of "ice rinks" includes all types of ice sheets, indoors or outdoors, used for any purpose—skating, ice hockey, figure skating, curling, and so on. The rinks are usually made by circulating a secondary refrigerant (brine)—glycol, methanol, or a calcium chloride solution—through a series of pipe coils located below the surface of the ice. The central chilling system usually uses R-22, although other refrigerants have been used. A hockey rink is usually 85 × 200 ft (26 × 61 m), a curling rink 14 × 146 ft (4.3 × 45 m), a figure skating rink 16 × 40 ft (5 × 12 m), and a speed-skating rink 1400 ft (400 m) oval and 35 ft wide with 392 ft (112 m) straightaways.

The amount of refrigeration is usually dependent on the location of the rink (indoors or outdoors), the use of the rink, and the conditions of the building (for indoor rinks) (see Figure 17-22).

Ice rink conditions. Indoor rinks are operated from 6 to 11 months a year, depending mainly on profitability. Rooms for heated rinks are usually maintained at 50 to 60 °F with relative humidities ranging from 60 to 90%.

It is customary to control the temperature of the ice closely, by controlling either the brine temperature or the ice surface temperature. A 1-in. thickness of ice is considered

Four to Five Winter Months, above 37-deg Latitude		
	ft²/ton	m²/kW
Outdoors, unshaded	85 to 300	2.2 to 7.8
Outdoors, covered	125 to 200	3.3 to 5.2
Indoors, uncontrolled atmosphere	175 to 300	4.6 to 7.8
Indoors, controlled atmosphere	150 to 350	4.0 to 9.1
Curling rinks, indoors	200 to 400	5.2 to 10.4
Year-Round (Indoors), (Controlled Atmosphere)		
	ft²/ton	m²/kW
Sports arena	100 to 150	2.6 to 4.0
Sports arena, accelerated ice making	50 to 100	1.3 to 2.6
Ice recreation center	130 to 175	3.4 to 4.6
Figure skating clubs and studios	135 to 185	3.5 to 4.8
Curling rinks	150 to 225	4.0 to 5.9
Ice shows	75 to 130	2.0 to 3.4

Figure 17–22 Ice rink refrigeration requirements for a variety of applications. (By permission of *ASHRAE Handbook.*)

good practice. The temperature of ice for hockey is 22 °F, for figure skating 26 °F, and for recreational skating 26 to 28 °F. Average brine temperatures are usually 10 °F lower than the required ice temperature. Usually, two compressors are used, two being required for pull-down, one for normal operation. The temperature differential between supply and return brine is usually 2 °F.

Construction and equipment. Brine systems usually use 1-in. steel or polyethylene pipe 4 in. on centers. Pipes are set dead level, with sand fill around and over the pipes. A balanced flow distribution system is shown in Figure 17–23. An expansion tank must be installed in the piping.

The construction of the space below the rink is very important. There must be proper drainage to prevent "heaving" caused by water freezing below the rink. Many new rinks install heater cables below the floor of the rink or circulate warm brine to prevent heaving. A "header trench" is located at one end of the rink to house the piping header. A snow-melting pipe is usually located at the opposite end of the rink for melting the scraped ice, removed by scraping the surface of the rink (see Figure 17–24).

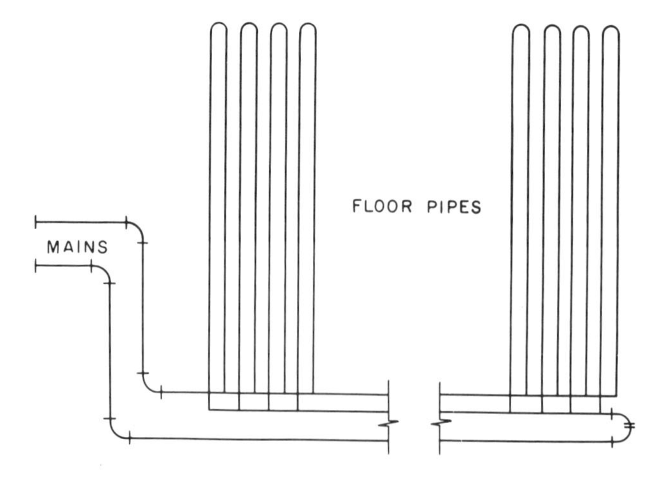

Figure 17–23 Piping diagram for ice rink to provide balanced flow. (By permission of *ASHRAE Handbook.*)

To obtain a smooth surface, after scraping, the ice is sprayed with a fine mist. Some automatic resurfacing units supply this needed surfacing water. Maintaining the 1-in. ice thickness is extremely important.

It is usually imporatnt to install a defogging air conditioner near the ceiling of an indoor rink to prevent condensation (ceiling dripping) and the formation of fog during certain operating conditions.

Figure 17-24 Types of ice rink surface floors. (By permission of *ASHRAE Handbook*.)

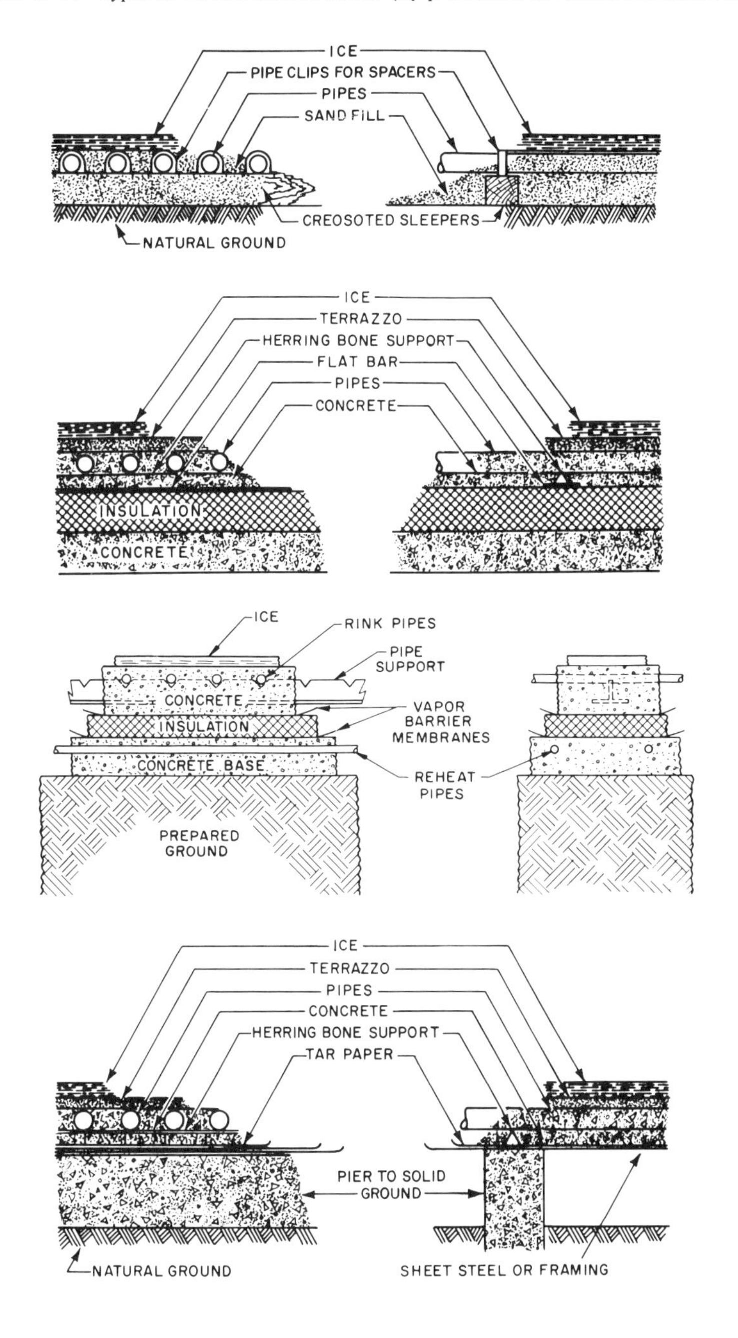

REVIEW EXERCISE

Select the letter representing your choice of the correct answer (there may be more than one).

17-1. Which type of food products are kept alive while they are being refrigerated?

_______ a. Meats
_______ b. Poultry
_______ c. Seafood
_______ d. Dairy products

17-2. Bacteria molds and fungi are examples of which type of food spoilage?

_______ a. Microbiological
_______ b. Enzymes
_______ c. Oxidation
_______ d. Dehydration

17-3. Which food spoils faster?

_______ a. Meat
_______ b. Poultry

17-4. Which food product requires ethylene gas to initiate the ripening process?

_______ a. Lettuce
_______ b. Apples
_______ c. Grapes
_______ d. Bananas

17-5. Why is cheese refrigerated?

_______ a. To kill all bacteria
_______ b. To prevent too rapid growth of mold
_______ c. To prevent oxidation
_______ d. To substitute for pasteurization

17-6. A fluidized-bed freezer can be used for the following food products:

_______ a. Berries
_______ b. Steaks
_______ c. Bananas
_______ d. Chicken

17-7. Which is a better storage temperature for most food products?

_______ a. $-5\,^\circ F$
_______ b. $-20\,^\circ F$

17-8. For what type of food product is ice the preferred form of cooling?

_______ a. Corn
_______ b. Potatoes
_______ c. Fresh meats
_______ d Fresh vegetables

17-9. If a microwave oven is not being used, which is the superior method of thawing frozen food?

________ a. Rapid-thaw

________ b. Slow-thaw

17-10. Which is the proper temperature range for transporting fresh meats?

________ a. 55 to 70°F

________ b. 32 to 40°F

________ c. 28 to 32°F

________ d. 0 to −5°F

18 Calculating the Refrigeration Load and Selecting the Equipment

OBJECTIVES

After studying this chapter, the reader will be able to:

- Determine the load imposed on the refrigeration equipment.
- Make recommendations on construction, equipment, or operation to reduce energy consumption.
- Determine the best selection of equipment and refrigerant to meet the load requirements.

REFRIGERATION LOAD CALCULATION IS UNIQUE

Calculating the load for refrigeration is unique in many respects and therefore requires special treatment. Some of the elements that differ from other types of cooling load calculation are as follows:

1. Loads are calculated on a 24-hour basis.
2. Product loads are usually greater than the transmission losses. Special consideration needs to be given to the amount of product brought into the room, how often it is brought in, the temperature reduction, whether or not the product needs to be frozen in the room, and so on. The product load must be accurately determined.
3. The hourly load is affected by the running time of the compressor. For example, if the total 24-hour load is 160,000 Btu and the compressor runs 16 hours per

day, the load on an hourly basis is

$$\frac{160{,}000\ \text{Btu}}{16\ \text{hours of operation}} = 10{,}000\ \text{Btuh}$$

This hourly load is used in selecting the refrigerating equipment for the project.

Sufficient idle time needs to be provided for the system to allow for the defrost period.

Three main topics are discussed in the following sections:

1. *The survey:* includes information that needs to be secured in order to calculate the load.
2. *The load calculation:* uses the data from the survey and the tables provided to calculate the refrigeration load. The method of determining the product load for rooms "above freezing temperatures" differs from rooms "below freezing temperatures."
3. *Selection of the equipment:* involves the selection of the condensing unit and evaporators from the calculated load requirements.

The forms for the survey, load calculation, and selection of equipment, including examples, are given in Figures 18-1 through 18-5.

THE SURVEY

All pertinent data relative to the condition of the project are recorded on the survey sheet (Figure 18-1). The following data are particularly important:

1. The design inside and outside conditions
2. The humidity requirements of the product
3. The dimension of the space, construction materials, thickness and type of insulation, and exposures. Note that outside dimensions are used.
4. The nature of the product load:
   a. Storage temperature above or below freezing.
   b. Type of product: meat, vegetables, etc.
   c. Condition of the product when it is loaded in the room: temperature, frozen or not frozen.
   d. How much product is loaded in the refrigerator each day.
   e. Miscellaneous loads: lights, motors, people, etc.
5. The kinds of services available, such as electricity, water, drains, etc.

Note that certain charts and tables are necessary in order to completely fill out the survey sheet. This information is used later in making the calculations. Figures 18-6 through 18-22 provide the tables and charts to be used with the survey sheet.

REF.	Load Survey and Estimating Data
9, 9A	Design ambient: ______°FDB, ______°FWB, ______%RH, ______°F summer ground temp. (use 55 °F for insulated freeze floor slab)
16A, B, C, D	Room design: ______°FDB, ______°FWB, ______%RH ______°F winter design ambient
	Access area: ______°FDB, ______°FWB, ______%RH, (Ante-RM/loading dock/other)
	Room dim. outside: ______ ft W ______ ft L ______ ft H ______ Total ft² (outside surface)

	Insulation				Wall	Adj.	Effective	Sun		Overall wall heat gain
	Type	Thick (inches)	K Factor	U* Factor	thick-ness	area °F	wall TD	effect (°F)	Total TD	(Btu/24 hr/ft²)
N. Wall								■		
S. Wall										
E. Wall (10)										
W. Wall										
Ceiling (11)										
Floor (12)								■		

$$*\text{U Factor} = \frac{K}{\text{Insul. Thickness (in.)}}$$

Refrig. door(s): ______ Vent fan(s): ______

Room int. vol: ______ W X ______ L X ______ H = ______ ft³

(inside room dimension = outside dimension–wall thicknesses)

Floor area: ______ W X ______ L = ______ ft²

Electrical power: ______ V ______ Ph ______ Hz: Control: ______ V

Type control: ______

Product data and class of product: ______

16A–D

	Amount of Product			Product temp (°F)		Specific heat					
Type product	Amount storage	Daily turn-over	Freezing or cooling	Enter-ing	Final	Above freeze	Below freeze	Lat.ht. freeze (Btu/lb)	Highest product freeze temp.	Heat or respir'n (Btu/lb) 24 hr)	( ) Pull-down ( ) Freezing time (hr)

Evap. TD ______ ; Type defrost: ☐ Air, ☐ Hot gas, ☐ Electric

Class product ______ : ______ ;

No. of defrosts–Total time per 24 hr; ______ hr.

Compressor running time: ______ hr.

Box usage: ☐ Average, ☐ Heavy, ☐ Extra heavy

Product load and additional information ______

20

21A & B Packaging ______ Containers ______ Wgt. ______ Sp. ht. ______ (container)

22 Pallets: No. ______ Size ______ Wgt. ea. ______ Sp. ht. ______

Product racks: No. ______ Mat'l ______ Wgt. ea. ______ Sp. ht. ______

21A & B

Estimating Product Loading Capacity of Room

Estimated product loading = 0.40 X ______ ft³ X ______ lb./ft³ = ______ lb:

(room volume) (loading density)

Miscellaneous loads

People: No. ______ hr. ______

Motors (other than evap. fan)

Use: ______ , ______ hp. ______ hr. ______

______ , ______ hp. ______ hr. ______

Fork lifts: ______ No. ______ hp ______ hr./day, Other ______

Lights ______ W/ft²

Figure 18–1 Refrigeration load survey (for rooms above and below freezing. (By permission of Dunham-Bush, Inc.)

REF.	I. Wall Loss (Transmission Load)				
	Surface	TD	Area of surface	Wall heat gain factor	Btu/24 hr
12	N. Wall		____ ft L X ____ ft H = ____ ft^2 X	____ =	
	S. Wall		____ ft L X ____ ft H = ____ ft^2 X	____ =	
	E. Wall		____ ft L X ____ ft H = ____ ft^2 X	____ =	
	W. Wall		____ ft L X ____ ft H = ____ ft^2 X	____ =	
	Ceiling		____ ft L X ____ ft W = ____ ft^2 X	____ =	
	Floor		____ ft L X ____ ft W = ____ ft^2 X	____ =	
	Box		Total surface = ____ ft^2 X	____ =	
			I	Total wall transmission load =	
13, 14 15	II. (Long Form) Infiltration (Air Change Load) ____ ft^3 X ____ air changes/24hr X ____ service factor X ____ Btu/ft^3 =				
			II	Total infiltration load =	
	III. Product Load				
16A, B, C	Product temp. reduction above freezing (sensible heat) ____ *lb/day X ____ °F temp. reduction X ____ sp. ht. =				
	Product freezing (latent heat load) ____ *lb/day X ____ Btu/lb/latent heat ____ =				
17	Product temp. reduction below freezing (sensible heat) ____ *lb/day X ____ °F temp. reduction X ____ sp. ht. =				
19	Heat of respiration ____ lb product (storage) X ____ Btu/lb/24 hr =				
20	Miscellaneous product loads (1) containers (2) pallets (3) other ____ lb/day X ____ °F temp. reduction X ____ sp. ht. = ____ lb/day X ____ °F temp. reduction X ____ sp. ht. =				
			III	Total product load =	
	IV. Miscellaneous loads (a) Lights ____ ft^2 floor area X ____ W/ft^2 X 3.41 Btu/W X ____ hr/24 hr = (1 to 1½ W/ft^2 in storage areas and 2 to 3 for work areas) (b) Occupancy ____ No. of people X ____ Btu/hr X ____ hr = (c) Motors ____ Btu/hp/hr X ____ hp X ____ hr/24 hr =				
23	____ Btu/hr X ____ hp X ____ hr/24 hr = (d) Material handling				
24	____ Forklift(s) X ____ eqiv. hp X 3100 Btu/hr/hp X ____ hr operation = Other ____ =				
	*If the product pull down is accomplished in less than 24 hr the daily product will be: Pounds product X $\frac{24\ hr}{\text{Pull-down hours}}$		IV	Total miscellaneous loads =	
				Total Btu load I to IV =	
				Add 10% safety factor =	
	Total Btu/24 hr with safety factor (not including evap. fan or defrost heat loads) 24-hr base refrigeration load =				

Figure 18–2 Refrigeration load calculations form. (By permission of Dunham-Bush, Inc.)

REF.	Equipment Selection from Load Calculation Form
6, 7	1. Determine evap. TD required for class of product and room temp. ______°F (TD) (from load survey data)
	2. Determine compressor running time based on operating temperatures and defrost requirements ______ hr (from load survey data)
	3. Evaporator temp. °F = __________ − __________ = ________°F (room temp) [evap. TD (from load survey data)]
25	4. Comp. suct. temp. °F = ______________ − ______________ = __________°F (evap. suct. temp) (suct. line loss)
	Btu/24 hr base refrigeration load with safety factor = ____________________ (not including evaporator fan or defrost heat) Preliminary hourly load = Btu/24 hr (base load) / hr/day (comp. running time) = __________ Btu/hr Fan heat load estimate Btu/hr = __________ Qty. X ______ Watts ea. X 3.41 Btu/W X ______ hr = ________ Btu/24 hr (MOTORS) (INPUT) or __________ Qty. X ______ hp ea. X ______ Btu/hp/hr X ____ hr = ________ Btu/24 hr (MOTORS)
24	Defrost heat load estimate Btu/hr = _____ Qty. evaps. X ______ W ea. X ______ hr X 3.41 Btu/W X ______ Defrost load factor* = ________ Btu/hr *Use 0.50 for electric defrost, 0.40 for hot gas defrost Btu/24 hr total load = __________ + __________ + __________ = __________ Btu/24 hr (base load) (fan heat) (defrost heat) or __________ X __________ = __________ Btu/hr (base load) (base load mult.) Actual hourly load = Btu/24 hr (total load) / hr/day (comp. running time) = __________ Btu/24 hr

Equipment Selection

	Compressor units		Condensing units		Evaporators		Condensers	
Model no.								
Quantity								
Capacity (ea.) Btu/hr)								
Air volume (ea.) (cfm)								
	Design	Actual	Design	Actual	Design	Actual	Design	Actual
Evaporator temp. (°F)								
Evap. TD (°F)								
Suction temp. (°F)								
Condensing temp. (°F)								
Design ambient temp. (°F)								
Min. oper. ambient temp. (°F)								

Figure 18–3 Refrigeration equipment selection from load calculation form. (By permission of Dunham-Bush, Inc.)

Location: Atlanta, Ga.
Room use: Storage of mixed vegetables
Outside dimensions: 12 × 14 × 9
Room temperature: 35°F

Insulation: 3 in. molded polystyrene
Total room load: 8000 lb
Daily delivery: 2000 lb precooled to 45°F
Containers: Paper cartons, total weight 200 lb (a)

REF.	Load Survey and Estimating Data

9, 9A — Design ambient: 95 °FDB, ____ °FWB, 50 %RH, ____ °F summer ground temp. (use 55 °F for insulated freeze floor slab)

16A, B, C, D — Room design: 35 °FDB, ____ °FWB, 80 %RH ____ °F winter design ambient

Access area: 95 °FDB, ____ °FWB, 50 %RH, (Ante-RM/loading dock/other)

Room dim. outside: 12 ft W 14 ft L 9 ft H 804 Total ft² (outside surface)

	Insulation Type	Thick (inches)	K Factor	U* Factor	Wall thickness	Adj. area °F	Effective wall TD	Sun effect (°F)	Total TD	Overall wall heat gain (Btu/24 hr/ft²)
N. Wall	POLY.	3"	.20	.067	5"	95	60		60	96
S. Wall										
E. Wall (10)										
W. Wall										
Ceiling (11)										
Floor (12)										

$$\text{*U Factor} = \frac{K}{\text{Insul. Thickness (in.)}}$$

Refrig. door(s): (1) 7 x 4 Vent fan(s): ____

Room int. vol: 11 W X 13 L X 8 H = 1144 ft³
(inside room dimension = outside dimension–wall thicknesses)

Floor area: 11 W X 13 L = 143 ft²

Electrical power: 240 V 3 Ph 60 Hz: Control: 120 V

Type control: ____

Product data and class of product: ____

16A–D

Type product	Amount of Product: Amount storage	Daily turn-over	Freezing or cooling	Product temp (°F): Enter-ing	Final	Specific heat: Above freeze	Below freeze	Lat.ht. freeze (Btu/lb)	Highest product freeze temp.	Heat or respir'n (Btu/lb) 24 hr)	( ) Pull-down ( ) Freezing time (hr)
MIX. VEG.	8000	2000	COOLING	45	35	.9				2.0	

Evap. TD 10 ; Type defrost: ☐ Air, ☐ Hot gas, ☐ Electric

Class product 2 : ____ ;

No. of defrosts–Total time per 24 hr; ____ hr.

Compressor running time: 16 hr.

Box usage: ☐ Average, ☐ Heavy, ☐ Extra heavy

Product load and additional information 2000 lb. mixed vegetables per day, entering temperature 45°F

20 — Packaging Paper cartons Containers ____ Wgt. 200 lbs Sp. ht. .32 (container)

21A & B — Pallets: No. ____ Size ____ Wgt. ea. ____ Sp. ht. ____

22 — Product racks: No. ____ Mat'l ____ Wgt. ea. ____ Sp. ht. ____

21A & B — Estimating Product Loading Capacity of Room

Estimated product loading = 0.40 X ____ ft³ X ____ lb./ft³ = ____ lb:
(room volume) (loading density)

Miscellaneous loads
People: No. ____ hr. ____

Motors (other than evap. fan)
Use: ____, ____ hp. ____ hr. ____
____, ____ hp. ____ hr. ____

Fork lifts: ____ No. ____ hp ____ hr./day, Other ____

Lights ____ W/ft²

Figure 18-4 Sample calculation showing load data for equipment selection, for *above-freezing conditions:* (a) data; (b) completed forms. (By permission of Dunham-Bush, Inc.)

REF.	I. Wall Loss (Transmission Load)				
	Surface	**TD**	**Area of surface**	**Wall heat gain factor**	**Btu/24 hr**
12	N. Wall		______ ft L X ______ ft H = ______ ft²	X ______ =	
	S. Wall		______ ft L X ______ ft H = ______ ft²	X ______ =	
	E. Wall		______ ft L X ______ ft H = ______ ft²	X ______ =	
	W. Wall		______ ft L X ______ ft H = ______ ft²	X ______ =	
	Ceiling		______ ft L X ______ ft W = ______ ft²	X ______ =	
	Floor		______ ft L X ______ ft W = ______ ft²	X ______ =	
	Box	60	Total surface = 804 ft²	X 96 =	77,184
			I Total wall transmission load	=	77,184

REF.		Btu/24 hr
13, 14, 15	II. (Long Form) Infiltration (Air Change Load) 1144 ft³ X 17.5 air changes/24hr X 1 service factor X 2.49 Btu/ft³ =	49,850
	II Total infiltration load =	49,850
16A, B, C	III. Product Load Product temp. reduction above freezing (sensible heat) 2000 *lb/day X 10 °F temp. reduction X .9 sp. ht. =	18,000
	Product freezing (latent heat load) ______ *lb/day X ______ Btu/lb/latent heat =	
17	Product temp. reduction below freezing (sensible heat) ______ *lb/day X ______ °F temp. reduction X ______ sp. ht. =	
19	Heat of respiration 8000 lb product (storage) X 2.0 Btu/lb/24 hr =	16,000
20	Miscellaneous product loads (1) containers (2) pallets (3) other 200 lb/day X 10 °F temp. reduction X .32 sp. ht. =	640
	______ lb/day X ______ °F temp. reduction X ______ sp. ht. =	
	III Total product load =	34,640
	IV. Miscellaneous loads (a) Lights ______ ft² floor area X ______ W/ft² X 3.41 Btu/W X ______ hr/24 hr = (1 to 1½ W/ft² in storage areas and 2 to 3 for work areas)	
	(b) Occupancy ______ No. of people X ______ Btu/hr X ______ hr =	
	(c) Motors ______ Btu/hp/hr X ______ hp X ______ hr/24 hr =	
23	______ Btu/hr X ______ hp X ______ hr/24 hr =	
24	(d) Material handling ___ Forklift(s) X ______ eqiv. hp X 3100 Btu/hr/hp X ______ hr operation =	
	Other ______ =	
	IV Total miscellaneous loads =	
	Total Btu load I to IV =	161,674
	Add 10% safety factor =	16,167
	Total Btu/24 hr with safety factor (not including evap. fan or defrost heat loads) 24-hr base refrigeration load =	177,841

*If the product pull down is accomplished in less than 24 hr the daily product will be:

$$\text{Pounds product} \times \frac{24 \text{ hr}}{\text{Pull-down hours}}$$

(b)

Figure 18–4 (Cont.)

REF.	Equipment Selection from Load Calculation Form
6, 7	1. Determine evap. TD required for class of product and room temp. 10 °F (TD) (from load survey data)
	2. Determine compressor running time based on operating temperatures and defrost requirements 16 hr (from load survey data)
	3. Evaporator temp. °F = 35 (room temp) – 10 [evap. TD = 25 °F (from load survey data)]
25	4. Comp. suct. temp. °F = 25 (evap. suct. temp) – 2 (suct. line loss) = 23 °F
24	Btu/24 hr base refrigeration load with safety factor = 177,641 (not including evaporator fan or defrost heat) Preliminary hourly load = Btu/24 hr (base load) / hr/day (comp. running time) = ______ Btu/hr Fan heat load estimate Btu/hr = ______ Qty. (MOTORS) X ______ Watts ea. (INPUT) X 3.41 Btu/W X ______ hr = ______ Btu/24 hr or 1 Qty. (MOTORS) X 1/6 hp ea. X 4350 Btu/hp/hr X 24 hr = 17,400 Btu/24 hr Defrost heat load estimate Btu/hr = ______ Qty. evaps. X ______ W ea. X ______ hr X 3.41 Btu/W X ______ Defrost load factor* = ______ Btu/hr *Use 0.50 for electric defrost, 0.40 for hot gas defrost Btu/24 hr total load = 177,641 (base load) + 17,400 (fan heat) + ______ (defrost heat) = 195,041 Btu/24 hr or ______ (base load) X ______ (base load mult.) = ______ Btu/hr Actual hourly load = 195041 / 16 Btu/24 hr (total load) / hr/day (comp. running time) = 12,190 Btu/24 hr

Equipment Selection

	Compressor units		Condensing units		Evaporators		Condensers	
Model no.			154		WJ 120			
Quantity			1		1			
Capacity (ea.) Btu/hr)			12,600		12,000			
Air volume (ea.) (cfm)								
	Design	Actual	Design	Actual	Design	Actual	Design	Actual
Evaporator temp. (°F)					25			
Evap. TD (°F)								
Suction temp. (°F)			25					
Condensing temp. (°F)								
Design ambient temp. (°F)			100					
Min. oper. ambient temp. (°F)								

(b)

Figure 18-4 (Cont.)

Location: Atlanta, Ga.
Room use: Storage of frozen beef
Outside dimensions: 10 × 12 × 9
Room temperature: 0°F

Insulation: 5 in. molded polystyrene
Total room load: 2000 lb
Daily delivery: 300 lb precooled to 40°F
Containers: Paper cartons, total weight 200 lb
Type of defrost: electric

(a)

REF.	Load Survey and Estimating Data

9, 9A — Design ambient: 95 °FDB, ___ °FWB, 50 %RH, ___ °F summer ground temp. (use 55 °F for insulated freeze floor slab)

16A, B, C, D — Room design: 0 °FDB, ___ °FWB, ___ %RH ___ °F winter design ambient

Access area: 95 °FDB, ___ °FWB, 50 %RH, (Ante-RM/loading dock/other)

Room dim. outside: 10 ft W 12 ft L 9 ft H 636 Total ft² (outside surface)

REF.		Insulation Type	Insulation Thick (inches)	Insulation K Factor	Insulation U* Factor	Wall thick-ness	Adj. area °F	Effective wall TD	Sun effect (°F)	Total TD	Overall wall heat gain (Btu/24 hr/ft²)
	N. Wall	Polystyrene	5"	.20	.04	6	95	95		95	91
	S. Wall										
10	E. Wall										
	W. Wall										
11	Ceiling										
12	Floor										

$$^*\text{U Factor} = \frac{K}{\text{Insul. Thickness (in.)}}$$

Refrig. door(s): (1) – 7 × 4 Vent fan(s): ___

Room int. vol: 9 W × 11 L × 8 H = 792 ft³
(inside room dimension = outside dimension–wall thicknesses)

Floor area: 9 W × 11 L = 99 ft²

Electrical power: 240 V 3 Ph 60 Hz: Control: 120 V

Type control: ___

Product data and class of product: ___

REF.	Type product	Amount of Product: Amount storage	Daily turn-over	Freezing or cooling	Product temp (°F): Enter-ing	Final	Specific heat: Above freeze	Below freeze	Lat.ht. freeze (Btu/lb)	Highest product freeze temp.	Heat or respir'n (Btu/lb) 24 hr)	( ) Pull-down ( ) Freezing time (hr)
16A–D	Beef	2000	300	Freezing	40	0	.80	.40	100	28		

Evap. TD 10 ; Type defrost: ☐ Air, ☐ Hot gas, ☐ Electric

Class product II : ___ ;

No. of defrosts—Total time per 24 hr; 6 hr.

Compressor running time: 18 hr.

Box usage: ☑ Average, ☐ Heavy, ☐ Extra heavy

Product load and additional information ___

20 —

21A & B — Packaging ___ Containers ___ Wgt. 200 Sp. ht. .32 (container)

22 — Pallets: No. ___ Size ___ Wgt. ea. ___ Sp. ht. ___

Product racks: No. ___ Mat'l ___ Wgt. ea. ___ Sp. ht. ___

21A & B — Estimating Product Loading Capacity of Room

Estimated product loading = 0.40 × ___ ft³ × ___ lb./ft³ = ___ lb:
(room volume) (loading density)

Miscellaneous loads

People: No. ___ hr. ___

Motors (other than evap. fan)

Use: ___ , ___ hp. ___ hr. ___
___ , ___ hp. ___ hr. ___

Fork lifts: ___ No. ___ hp ___ hr./day, Other ___

Lights ___ W/ft²

(b)

Figure 18-5 Sample calculations, using load data for equipment selection, for *below-freezing conditions:* (a) data; (b) completed forms. (By permission of Dunham-Bush, Inc.)

REF.	I. Wall Loss (Transmission Load)				
	Surface	TD	Area of surface	Wall heat gain factor	Btu/24 hr
12	N. Wall		____ ft L X ____ ft H = ____ ft² X	____ =	
	S. Wall		____ ft L X ____ ft H = ____ ft² X	____ =	
	E. Wall		____ ft L X ____ ft H = ____ ft² X	____ =	
	W. Wall		____ ft L X ____ ft H = ____ ft² X	____ =	
	Ceiling		____ ft L X ____ ft W = ____ ft² X	____ =	
	Floor		____ ft L X ____ ft W = ____ ft² X	____ =	
	Box	95	Total surface = 636 ft² X	91 =	57,876
			I Total wall transmission load	=	57,876
13, 14 15	II. (Long Form) Infiltration (Air Change Load) 792 ft³ X 15.3 air changes/24hr X 1 service factor X 3.28 Btu/ft³ =				39,745
			II Total infiltration load	=	39,745
	III. Product Load				
16A, B, C	Product temp. reduction above freezing (sensible heat) 300 *lb/day X 12 °F temp. reduction X .80 sp. ht. =				2880
	Product freezing (latent heat load) 300 *lb/day X 100 Btu/lb/latent heat =				30,000
17	Product temp. reduction below freezing (sensible heat) 300 *lb/day X 28 °F temp. reduction X .40 sp. ht. =				3360
19	Heat of respiration ____ lb product (storage) X ____ Btu/lb/24 hr =				
20	Miscellaneous product loads (1) containers (2) pallets (3) other 200 lb/day X 40 °F temp. reduction X .32 sp. ht. =				2560
	____ lb/day X ____ °F temp. reduction X ____ sp. ht. =				
			III Total product load	=	38,800
	IV. Miscellaneous loads				
	(a) Lights ____ ft² floor area X ____ W/ft² X 3.41 Btu/W X ____ hr/24 hr = (1 to 1½ W/ft² in storage areas and 2 to 3 for work areas)				
	(b) Occupancy ____ No. of people X ____ Btu/hr X ____ hr =				
	(c) Motors ____ Btu/hp/hr X ____ hp X ____ hr/24 hr =				
23	____ Btu/hr X ____ hp X ____ hr/24 hr =				
	(d) Material handling				
24	___ Forklift(s) X ____ eqiv. hp X 3100 Btu/hr/hp X ____ hr operation =				
	Other ____ =				
	*If the product pull down is accomplished in less than 24 hr the daily product will be: Pounds product X 24 hr / Pull-down hours		IV Total miscellaneous loads	=	
			Total Btu load I to IV	=	136,421
			Add 10% safety factor	=	13642
			Total Btu/24 hr with safety factor (not including evap. fan or defrost heat loads) 24-hr base refrigeration load	=	150,063

(b)

Figure 18–5 (Cont.)

REF.	Equipment Selection from Load Calculation Form
6, 7	1. Determine evap. TD required for class of product and room temp. 10 °F (TD) (from load survey data)
	2. Determine compressor running time based on operating temperatures and defrost requirements 18 hr (from load survey data)
	3. Evaporator temp. °F = 0 (room temp) − 10 [evap. TD = −10 °F (from load survey data)]
25	4. Comp. suct. temp. °F = −10 (evap. suct. temp) − 3 (suct. line loss) = −13 °F
24	Btu/24 hr base refrigeration load with safety factor = ____ (not including evaporator fan or defrost heat) Preliminary hourly load = Btu/24 hr (base load) / hr/day (comp. running time) = ____ Btu/hr Fan heat load estimate Btu/hr = 1 Qty. (MOTORS) X 210 Watts ea. (INPUT) X 3.41 Btu/W X 24 hr = 17 186 Btu/24 hr or ____ Qty. (MOTORS) X ____ hp ea. X ____ Btu/hp/hr X ____ hr = ____ Btu/24 hr Defrost heat load estimate Btu/hr = 1 Qty. evaps. X 3380 W ea. X 1½ hr X 3.41 Btu/W X .5 Defrost load factor* = 8644 Btu/hr *Use 0.50 for electric defrost, 0.40 for hot gas defrost Btu/24 hr total load = 150,063 (base load) + 17,186 (fan heat) + 8644 (defrost heat) = 175,893 Btu/24 hr or ____ (base load) X ____ (base load mult.) = ____ Btu/hr Actual hourly load = 175,893 Btu/24 hr (total load) / 18 hr/day (comp. running time) = 9771 Btu/24 hr

Equipment Selection

	Compressor units		Condensing units		Evaporators		Condensers	
Model no.			32 C		NDE 105			
Quantity			1		1			
Capacity (ea.) Btu/hr)			11,500		10,500			
Air volume (ea.) (cfm)					1420			
	Design	Actual	Design	Actual	Design	Actual	Design	Actual
Evaporator temp. (°F)					−10			
Evap. TD (°F)								
Suction temp. (°F)			−10					
Condensing temp. (°F)								
Design ambient temp. (°F)			100					
Min. oper. ambient temp. (°F)								

(b)

Figure 18-5 (Cont.)

Guidelines for Room Conditions

As a guide for storage temperatures and humidity conditions, see Figures 18–6, 18–7, and 18–9. The various classes of food used in these charts are described as follows:

Class 1. Such products as eggs, unpacked butter and cheese, and most vegetables held for comparatively long periods of time. These products require very high relative humidity because it is necessary to effect a minimum of moisture evaporation during storage.

Class 2. Such foods as cut meats, fruits, and similar products. These require high relative humidities but not as high as class 1.

Class 3. Carcass meats and fruit such as melons, which have tough skins. These products require only moderate relative humidities because they have surfaces whose rate of moisture evaporation is moderate.

Class 4. Canned goods, bottled goods, and other products which have a protective covering. These products need only low relative humidities or are not affected by relative humidity. Products from whose surfaces there is a very low rate of moisture evaporation or none at all fall into this class.

Type Coils	Class 1	Class 2	Class 3	Class 4
Forced Air Coils	6 - 9°F	9 - 12°F	12 - 20°F	Above 20°F
Gravity Coils	14 - 18°F	18 - 22°F	21 - 28°F	27 - 37°F

*Temperature difference is defined as average fixture temperature minus average refrigerant temperature.

Figure 18–6 Temperature differences for four classes of foods and two types of coils. Temperature difference is defined as the average fixture temperature minus the average refrigerant temperature. (By permission of Dunham-Bush, Inc.)

Room Temp. °F	Evap. Temp. °F	Defrosting		Compressor Running Time Hrs.	Rel. Hum. %	Evap. °F TD
		Type	No./24 Hrs.			
Over 35	Over 30	None	–	18 - 20 & up	90 85 80 75	8 - 10 10 - 12 12 - 15 16 - 20
35 & up		Ambient	4	16		
35 to 25	Below 30	Elec.	4	18 - (20*)		
10 & less		or Hot Gas	6	(18*) - 20	-	8 - 10

*Preferred compressor running time.

Figure 18–7 Temperature-difference guidelines for evaporator temperatures above and below freezing. (By permission of Dunham-Bush, Inc.)

State	City	Summer Dry Bulb °F	Summer Wet Bulb °F	Winter Dry Bulb °F	Elev. Ft.
Alabama	Birmingham	97	79	19	610
	Mobile	96	80	26	211
Alaska	Fairbanks	82	64	-53	436
	Juneau	75	66	- 7	17
Arizona	Phoenix	108	77	31	1117
	Tucson	105	74	29	2584
Arkansas	Fort Smith	101	79	15	449
	Little Rock	99	80	19	257
California	Bakersfield	103	72	31	495
	Fresno	101	73	28	326
	Los Angeles	94	72	42	312
	Oakland	85	65	35	3
	San Francisco	80	64	32	8
Colorado	Denver	92	65	- 2	5283
Connecticut	Hartford	90	77	1	15
	New Haven	88	77	5	6
Delaware	Dover	93	79	13	38
	Wilmington	93	79	12	78
District of Columbia	Washington	94	78	15	14
Florida	Jacksonville	96	80	29	24
	Key West	90	80	55	6
	Miami	92	80	44	9
	Tampa	92	81	36	19
Georgia	Atlanta	95	78	15	1005
	Savannah	95	81	24	42
Hawaii	Honolulu	87	75	60	7
Idaho	Boise	96	68	4	2842
Illinois	Chicago	94	78	- 3	594
	Peoria	94	78	- 2	652
	Springfield	95	79	- 5	587
Indiana	Evansville	96	79	6	381
	Fort wayne	93	77	0	791
	Indianapolis	93	78	0	793
	Terre Haute	95	79	3	601
Iowa	Des Moines	95	79	- 7	948
	Sioux City	96	79	-10	1095
Kansas	Dodge City	99	74	3	2594
	Topeka	99	79	3	877
	Wichita	102	77	5	1321
Kentucky	Louisville	96	79	8	474
Louisiana	New Orleans	93	81	32	3
	Shreveport	99	81	22	252
Maine	Bangor	88	75	- 8	162
	Portland	88	75	- 5	61
Maryland	Baltimore	94	79	16	14
Massachusetts	Boston	91	76	6	15
	Springfield	86	74	- 3	247
Michigan	Detroit	92	76	4	633
	Grand Rapids	91	76	2	601
	Lansing	89	76	2	652
Minnesota	Duluth	85	73	-19	1426
	Minneapolis	92	77	-14	822
Mississippi	Vicksburg	97	80	23	234
Missouri	Kansas City	100	79	4	742
	St. Louis	96	79	7	465
Montana	Billings	94	68	-10	3367
	Helena	90	65	-17	3893
Nebraska	Lincoln	100	78	- 4	1150
	Omana	97	79	- 5	978
Nevada	Reno	95	64	12	4490
New Hampshire	Concord	91	75	-11	339

State	City	Summer Dry Bulb °F	Summer Wet Bulb °F	Winter Dry Bulb °F	Elev. Ft.
New Jersey	Atlantic City	91	78	14	11
	Newark	94	77	11	11
	Trenton	92	78	12	144
New Mexico	Albuquerque	96	66	14	5310
	Sante Fe	90	65	7	7045
New York	Albany	91	76	1	19
	Buffalo	88	75	3	705
	New York	93	77	11	132
North Carolina	Asheville	91	75	13	2770
	Charlotte	96	78	18	735
	Raleigh	95	79	16	433
	Wilmington	93	52	23	30
North Dakota	Bismarck	95	74	-24	1647
Ohio	Cincinnati	94	78	8	761
	Cleveland	91	76	2	777
	Columbus	92	77	2	212
	Dayton	92	77	0	997
	Toledo	92	72	0	900
Oklahoma	Oklahoma City	102	78	11	1280
	Tulsa	102	79	12	650
Oregon	Portland	91	69	26	57
Pennsylvania	Erie	88	76	7	732
	Philadelphia	93	78	11	7
	Pittsburgh	90	75	7	749
	Scranton	89	75	2	940
Rhode Island	Providence	89	76	6	55
South Carolina	Charleston	95	81	26	9
	Greenville	95	77	19	957
South Dakota	Huron	97	77	-16	1282
	Rapid City	96	72	- 9	3165
	Sioux Falls	95	77	-14	1430
Tennessee	Chattanooga	97	78	75	670
	Knxoville	95	77	13	980
	Memphis	98	80	17	263
	Nashville	97	79	12	577
Texas	Amarillo	98	72	8	3607
	Corpus Christi	95	32	32	43
	Dallas	101	79	19	481
	El Paso	100	70	21	3918
	Galveston	91	32	32	5
	Houston	96	80	29	158
	San Antonio	99	77	25	792
Utah	Salt Lake City	97	67	5	4220
Vermont	Burlington	88	74	-12	331
Virginia	Norfolk	94	79	20	26
	Richmond	96	79	14	152
	Roanoke	94	76	15	1174
Washington	Seattle	81	69	28	14
	Spokane	93	66	- 2	2357
West Virginia	Charleston	92	76	9	939
	Parkersburg	93	77	8	615
Wisconsin	Green Bay	88	·75	-12	683
	Madison	92	77	- 9	858
	Milwaukee	90	77	- 5	672
Wyoming	Cheyenne	89	63	- 6	6128
Death Valley - 280' Elev.					

Summer deisgn DB & WB temperatures equal to or exceeded 1% of 4 summer months. (about 30 hours), and winter DB temperatures 99% of 3 winter months (about 22 hours).

Ground temperatures (GT) for cold storage room calculations may be approximated for range -30° to +30°F. Winter design Dry Bulb (WDB) by GT, °F = 65 + WDB/2.

Province	Sta. City	Summer Dry Bulb °F	Summer Wet Bulb °F	Winter Dry Bulb °F	Elev. Ft.
Alberta	Calgary	87	66	-29	3540
British Columbia	Vancouver	80	68	15	60
Manitoba	Winnipeg	90	78	-28	786
Newfoundland	Gander	85	69	- 5	482
Northwest Territories	Fort Smith	85	67	-49	665
Nova Scotia	Halifax	83	69	0	136
Ontario	Toronto	90	77	- 3	578
Prince Edward Island	Charlottetown	84	71	- 6	186
Quebec	Montreal	88	75	-16	98
	Quebec	86	75	-19	245
Saskatchewan	Regina	92	73	-34	1884
Yukon	White Horse	78	62	-45	2289

ASHRAE 1972 Fundamentals - Reprinted by permission

Figure 18–8 Design temperatures in United States and Canada. (By permission of *ASHRAE 1972 Fundamentals.*)

Product	Short-Time Storage		24 - 72 Hr. Storage	
	°F	Moisture	°F	Moisture
Vegetables	36 - 42	Class 2	32 - 36	Class 1
Fruits	36 - 42	Class 2	32 - 36	Class 2
Meats (cut)	34 - 38	Class 2	32 - 36	Class 2
Meats (carcass)	34 - 38	Class 3	32 - 36	Class 3
Poultry	32 - 36	Class 2	30 - 35	Class 1 - 2‡
Fish	35 - 40	Class 1	35 - 40	Class 1
Eggs	36 - 42	Class 2	31 - 35	Class 1
Butter, Cheese	38 - 45	Class 1*	35 - 40	Class 2 - 3†
Bottled beverages	35 - 45	Class 4	40 - 45	Class 4
Frozen Foods			0	Class 2

ASRE 1959 Data Book and ASHRAE 1962 Guide and Data book
- Reprinted by permission

Temperatures and moisture conditions shown above give customary restaurant practice. Some variations may be expected due to different varieties of fruits or vegetables and the growing conditions in different geographical locations. The USDA Handbook No. 66 is another authoritative source for conditions for the long-time preservation of fruits and vegetables.

* If not packaged.

† If packaged

‡ Freeze and hold at 0°F or below if held for more than 72 hours.

Figure 18-9 Storage temperature and humidity conditions for selected food products. These will serve as a guide for determining the required compressor and evaporator operation. (By permission of *ASRE 1959 Data Book* and *ASHRAE 1962 Guidebook*.)

Storage Temperature °F	Cork or equivalent thickness (Inches)	
	Northern U.S.	Southern U.S.
50 to 60	2	3
40 to 50	3	4
25 to 40	4	5
15 to 25	5	6
0 to 15	6	7
0 to -15	7	8
-15 to -40	9	10

ASHRAE 1972 - Reprinted by permission

Figure 18-10 Minimum insulation thickness for storage temperatures of −40 to +60°F. (By permission of *ASHRAE 1972 Fundamentals*.)

Figure 18-11 Temperature allowance for sun effect on various outside surfaces. Add the appropriate degrees Farenheit to the normal temperature difference for heat leakage calculations to compensate. (By permission of *ASHRAE 1972 Fundamentals*.)

Type of Surface	East Wall	South Wall	West Wall	Flat Roof
Dark colored surfaces such as: Slate roofing, Tar roofing, Black paints	8	5	8	20
Medium colored surfaces, such as: Unpainted wood, Brick, Red Tile, Dark cement, Red, gray, or green paint	6	4	6	15
Light colored surfaces, such as: White stone, Light colored cement, White paint	4	2	4	9

ASHRAE 1972 Fundamentals - Reprinted by permission

CALCULATING THE REFRIGERATION LOAD

The total load is made up of:

1. Transmission load (wall load)
2. Infiltration load (air-change load)
3. Product load
4. Miscellaneous loads
5. Safety factor

Transmission Load

This is the load due to the heat loss of the structure. Six exposures are included. The area of each exposure is multiplied by the wall heat gain factor found in Figure 18-12.

To obtain this factor from the table in Figure 18-12, it is necessary to know the thickness of the insulation, the temperature difference (between the inside and the outside of the room), and the type of insulation.

For example, with molded polystyrene the K factor is 0.20. The K factor is the Btuh/ft^2 of surface/°F temperature difference for 1-in. thickness of the material. With 3 in. of molded polystyrene and a temperature difference of 55 °F, referring to Figure 18-12, the factor would be 88. Assuming the area of the north wall to be 100 ft^2, the heat loss through the wall in 24 hours would be 8800 Btu.

Infiltration Load

The infiltration factors are given in Figure 18-13 for above-freezing conditions and in Figure 18-14 for below-freezing conditions. These factors are largely based on experience, and relate to the volume of the room. They are given in terms of air changes in 24 hours. The amount of heat required to cool the infiltration air is given in Figure 18-15. The *infiltration load* is calculated as follows:

$$H = V \times \mathrm{AC} \times \mathrm{HR}$$

where
V = volume of the room, ft^3
AC = air changes in 24 hours (see Figure 18-13 or 18-14)
HR = heat removed in cooling the air, Btu/ft^3 (see Figure 18-15)

Assuming that the volume of the room is 1000 ft^3 and, referring to Figure 18-13, the air changes per 24 hours are 17.5. Referring to Figure 18-15, with a room temperature of 40 °F and an outside condition of 95 °F DB and 50% RH, the heat removed in cooling outside air to refrigerated room conditions is 2.31 Btu/ft^3. By substituting these values in the formula above, the infiltration load is determined as follows:

$$H = 1000\ \mathrm{ft}^3 \times 1.75\ \text{air changes} \times 2.31\ \mathrm{Btu/ft}^3$$
$$= 40{,}000\ \mathrm{Btu/24\ hours}$$

INSULATION		TEMPERATURE DIFFERENCE IN °F (AMBIENT TEMPERATURE - STORAGE TEMPERATURE)																					
Thickness Inches	k Factor	1	10	20	30	40	45	50	55	60	65	70	75	80	85	90	95	100	105	110	115	120	125
1	.30	7.2	72	144	216	288	324																
	.25	6.0	60	120	180	240	270	300	330														
	.20	4.8	48	96	144	192	216	240	264	288	312												
	.16	3.84	38	77	115	154	173	192	211	230	250	269	288	307									
	.14	3.36	34	67	101	134	151	168	185	202	218	235	252	269	286	302							
2	.30	3.6	36	72	108	144	162	180	198	216	234	252	270	288	306								
	.25	3.0	30	60	90	120	135	150	165	180	195	210	225	240	255	270	285	300					
	.20	2.4	24	48	72	96	108	120	132	144	136	168	180	192	204	216	228	240	252	264	276	288	300
	.16	1.92	19	38	58	77	86	96	106	115	125	134	144	154	163	173	182	192	202	211	221	230	240
	.14	1.68	17	34	50	67	76	84	92	101	109	118	126	134	143	151	160	168	176	185	193	202	210
3	.30	2.4	24	48	72	96	108	120	132	144	156	168	180	192	204	216	228	240	252	264	276	288	300
	.25	2.0	20	40	60	80	90	100	110	120	130	140	150	160	170	180	190	200	210	220	230	240	250
	.20	1.6	16	32	48	64	72	80	88	96	104	112	120	128	136	144	152	160	168	176	184	192	200
	.16	1.28	13	26	38	51	58	64	70	77	83	90	96	102	109	115	122	128	134	141	147	154	160
	.14	1.13	11	23	34	45	50	56	62	67	73	78	84	90	95	101	106	112	118	123	129	134	140
4	.30	1.8	18	36	54	72	81	90	99	108	117	126	135	144	153	162	171	180	189	198	207	216	225
	.25	1.5	15	30	45	60	68	75	83	90	98	105	113	120	128	135	143	150	158	165	173	180	188
	.20	1.2	12	24	36	48	54	60	66	72	78	84	90	96	102	108	114	120	126	132	138	144	150
	.16	0.96	10	19	29	38	43	48	53	58	62	68	72	77	82	87	91	96	101	106	111	115	120
	.14	0.84	9	17	25	34	38	42	46	50	55	59	63	68	71	75	80	84	88	92	97	101	105
5	.30	1.44	14	29	42	58	65	72	79	87	94	101	108	115	122	130	137	144	151	159	166	173	180
	.25	1.2	12	24	36	48	54	60	66	72	78	84	90	96	102	108	114	120	126	132	138	144	150
	.20	0.96	10	19	29	38	43	48	53	58	62	67	72	77	82	96	91	96	101	106	110	115	120
	.16	0.76	8	15	23	31	35	38	42	46	50	54	58	61	65	69	73	77	81	84	88	92	96
	.14	0.67	7	13	20*	27	30	34	37	40	44	47	50	54	57	60	64	67	71	74	77	81	84
6	.30	1.2	12	24	36	48	54	60	66	72	78	84	90	96	102	108	114	120	126	132	138	144	160
	.25	1.0	10	20	30	40	45	50	55	60	65	70	75	80	85	90	95	100	105	110	115	120	125
	.20	0.8	8	16	24	32	36	40	44	48	52	56	60	64	68	72	76	80	84	88	92	96	100
	.16	0.64	6	13	19	26	29	32	35	38	42	45	48	51	54	58	61	64	67	70	74	77	80
	.14	0.56	6	11	17	22	25	28	31	34	36	39	42	45	48	50	53	56	59	62	64	67	70
7	.30	1.02	10	20	30	41	46	52	57	62	67	72	77	82	88	93	98	103	108	113	118	124	129
	.25	0.85	9	17	26	34	39	43	47	51	56	60	64	51	73	77	81	86	90	94	99	103	107
	.20	0.68	7	14	21	27	31	34	38	41	45	48	51	55	58	62	65	69	72	75	•79	82	86
8	.30	0.90	9	18	27	36	41	45	50	54	69	63	68	72	77	81	86	90	95	99	104	108	113
	.25	0.75	8	15	23	30	34	38	41	45	49	53	56	60	64	68	71	75	79	83	86	90	94
	.20	0.60	6	12	18	24	27	30	33	36	39	42	45	48	51	54	57	60	63	66	69	72	75
9	.30	0.80	8	16	24	32	36	40	44	48	52	56	60	64	68	72	76	80	84	88	92	96	100
	.25	0.67	7	13	20	27	30	34	37	40	44	47	50	54	57	60	64	67	70	74	77	80	84
10	.30	0.72	7	14	21	29	32	36	40	43	47	50	54	58	61	65	68	72	76	79	83	86	90
	.25	0.60	6	12	18	24	27	30	33	36	39	42	45	48	51	54	57	60	63	66	69	72	75
11	.30	0.65	6.5	13	19.5	26	30	33	36	40	43	46	50	53	56	60	63	66	69	73	76	79	82
	.25	0.55	5.5	11	17	22	25	28	30	33	36	39	41	44	47	50	52	55	58	61	63	66	69
12	.30	0.60	6	12	18	24	27	30	33	36	39	42	45	48	51	54	57	60	63	66	69	72	75
	.25	0.50	5	10	15	20	23	25	28	30	33	35	38	40	43	45	48	50	53	55	58	60	63
Single Glass		27	270	540	810	1080	1220	1350	1490	1620													
Double Glass		11	110	220	330	440	500	500	610	660	715	770	825	880	936	990	1050	1100	1160	1210	1270	1320	1375
Triple Glass		7	70	140	210	280	320	350	390	420	454	490	525	560	595	630	665	700	740	770	810	840	875
Concrete Floor Slab on the ground (See Note)		For 6 or 8 inch concrete floor on the ground, not insulated use the TD between the average summer ground temp. and box temp. (Note: Not for freezers or for coolers operated close to freezing. Insulated floors recommended for all coolers and freezers (use insul. thickness and TD room to slab temperature).																					
Floor under 144 sq. ft.		6	60	120	180	240	270																
Floor over 144 sq. ft.		4.5	45	90	135	180	203																

Insulation • k = 0.30 Corkboard, Mineral Wool
0.25 Fiberglass Styrofoam, Expanded Polystyrene
0.20 Moulded Polystyrene
0.16 Sprayed Urethane, Foam Urethane Sheets, Slabs and Panels
0.14 Urethane Panels (Foamed in Place.

•Check with supplier for applicable k factor

k Factor in Btu/Hr./Sq. Ft./Deg. F/Inch

ASHRAE 1977 Fundamentals - Reprinted by permission

Figure 18–12 Wall heat gain table. Btu/24 hr./Sq. ft. of outside surface. (By permission of *ASHRAE 1977 Fundamentals.)*

Volume Cu. Ft.	Air Changes Per 24 Hrs.	Volume Cu. Ft.	Air Changes Per 24 Hrs.	Volume Cu. Ft.	Air Changes Per 24 Hrs.	Volume Cu. Ft.	Air Changes Per 24 Hrs.
200	44.0	800	20.0	5000	7.2	25,000	3.0
250	38.0	1000	17.5	6000	6.5	30,000	2.7
300	34.5	1500	14.0	8000	5.5	40,000	2.3
400	29.5	2000	12.0	10,000	4.9	50,000	2.0
500	26.0	3000	9.5	15,000	3.9	75,000	1.6
600	23.0	4000	8.2	20,000	3.5	100,000	1.4
						350,000	1.13*
						700,000	0.97*

NOTE: For heavy usage multiply the above values by a service factor of 2.
For long storage multiply the above values by 0.6.

* = Extrapolated

ASHRAE 1967 Fundamentals - Reprinted by permission

Figure 18-13 Average air changes per 24 hours for storage rooms *above freezing,* due to door opening and infiltration. (By permission of *ASHRAE 1967 Fundamentals.*)

Volume Cu. Ft.	Air Changes Per 24 Hrs.	Volume Cu. Ft.	Air Changes Per 24 Hrs.	Volume Cu. Ft.	Air Changes Per 24 Hrs.	Volume Cu. Ft.	Air Changes Per 24 Hrs.
200	33.5	800	15.3	5000	5.6	25,000	2.3
250	29.0	1000	13.5	6000	5.0	30,000	2.1
300	26.2	1500	11.0	8000	4.3	40,000	1.8
400	22.5	2000	9.3	10,000	3.8	50,000	1.6
500	20.0	3000	7.4	15,000	3.0	75,000	1.3
600	18.0	4000	6.3	20,000	2.6	100,000	1.1
						150,000	0.88*
						200,000	.77

NOTE:
- For heavy usage multiply the above values by a service factor of 2.
  For long storage multiply the above values by 0.6.
- For 2 doors in same wall multiply by 1.25.
- For 2 doors in opposite walls multiply by 2.5, but two open doors on adjacent or opposite walls should not be tolerated.

* = Extrapolated

ASHRAE 1967 Fundamentals - Reprinted by permission

Figure 18-14 Average air changes per 24 hours for storage rooms *below freezing,* due to door opening and infiltration. (By permission of *ASHRAE 1967 Fundamentals.*)

Storage Room Temp. F	Temperature of Outside Air, F							
	85		90		95		100	
	Relative Humidity, %							
	50	60	·50	60	50	60	50	60
65	0.65	0.85	0.93	1.17	1.24	1.54	1.58	1.95
60	0.85	1.03	1.13	1.37	1.44	1.74	1.78	2.15
55	1.12	1.34	1.41	1.66	1.72	2.01	2.06	2.44
50	1.32	1.54	1.62	1.87	1.93	2.22	2.28	2.65
45	1.50	1.73	1.80	2.06	2.12	2.42	2.47	2.85
40	1.69	1.92	2.00	2.26	2.31	2.62	2.67	3.06
35	1.86	2.09	2.17	2.43	2.49	2.79	2.85	3.24
30	2.00	2.24	2.26	2.53	2.64	2.94	2.95	3.35

From Chapter 27 "ASHRAE Guide and Data Book, 1961". Used by permission.

Storage Room Temp. F	Temperature of Outside Air, F							
	40		50		90		100	
	Relative Humidity, %							
	70	80	70	80	50	60	50	60
30	0.24	0.29	0.58	0.66	2.26	2.53	2.95	3.35
25	0.41	0.45	0.75	0.83	2.44	2.71	3.14	3.54
20	0.56	0.61	0.91	0.99	2.62	2.90	3.33	3.73
15	0.71	0.75	1.06	1.14	2.80	3.07	3.51	3.92
10	0.85	0.89	1.19	1.27	2.93	3.20	3.64	4.04
5	0.98	1.03	1.34	1.42	3.12	3.40	3.84	4.27
0	1.12	1.17	1.48	1.56	3.28	3.56	4.01	4.43
-5	1.23	1.28	1.59	1.67	3.41	3.69	4.15	4.57
-10	1.35	1.41	1.73	1.81	3.56	3.85	4.31	4.74
-15	1.50	1.53	1.85	1.92	3.67	3.96	4.42	4.86
-20	1.63	1.68	2.01	2.09	3.88	4.18	4.66	5.10
-25	1.77	1.80	2.12	2.21	4.00	4.30	4.78	5.21
-30	1.90	1.95	2.29	2.38	4.21	4.51	4.90	5.44

Figure 18-15 Heat removed in cooling air to refrigerator conditions. Btu per cu. ft. (By permission of *ASHRAE Guide and Data Book*, 1961.)

Product Load

Product load, above freezing. Any product brought into the cooler above room temperatures must be cooled. This product load adds to the refrigeration requirements.

For products stored at above-freezing temperatures the product load includes:

1. Lowering the temperature of the product to room temperature.
2. Determining the heat of respiration.

The heat of respiration is the continuous heat given off by a product stored above freezing temperatures due to oxidation.

Sample Product Load Calculations, Above Freezing. A refrigerated room is used to cool mixed vegetables to a temperature of 40°F. The room has the capacity to hold 10,000 lb. The loading arrangement provides for placing 1000 lb of mixed vegetables, precooled to 50°F, in the room per day.

1. The *temperature reduction* load is calculated as follows:

$$H = W \times (T_2 - T_1) \times \mathrm{SH}$$

where
H = load, Btu
W = weight of product, lb
T_2 = temperature (°F) of product entering cooler
T_1 = room temperature (°F)
SH = specific heat of product (see Figure 18–16)

Substituting the values from the example gives us

$$H = 1000 \text{ lb} \times (50\,^\circ\text{F} - 40\,^\circ\text{F}) \times 0.90 = 9000 \text{ Btu/24 hours}$$

2. *Heat of respiration* is calculated as follows:

$$H = W \times \mathrm{HR}$$

where
W = total weight of product in storage, lb
HR = heat of respiration, Btu/lb/24 hours (see Figure 18–19)

Substituting the values in the example:

$$H = 10{,}000 \text{ lb} \times 2.0 \text{ Btu/lb/24 hours} = 20{,}000 \text{ Btu/24 hours}$$

Total Product Load.

1. Temperature reduction	9,000 Btu/24 hours
2. Heat of respiration	20,000 Btu/24 hours
Total	29,000 Btu/24 hours

Commodity	Retail Storage °F	% RH	Storage Temp. °F	Relative Humidity %	Approximate Storage Life	Water Content %	Highest Freezing Point °F	Specific Heat Above Freezing Btu/lb/F	Specific Heat Below Freezing Btu/lb/F	Latent Heat (Calculated) Btu/lb
					FRUITS					
Apples*	35	85 - 88	30 - 40	90	3 - 8 months	84.1	29.3	0.87	0.45	121
Apricots	35	80 - 85	31 - 32	90	1 - 2 weeks	85.4	30.1	0.88	0.46	122
Avocados	50	85 - 90	45 - 55	85 - 90	2 - 4 weeks	65.4	31.5	0.72	0.40	94
Bananas*	56	90 - 95	56	85 - 95	8 days	74.8	30.6	0.80	0.42	108
Blackberries			31 - 32	95	3 days	84.8	30.5	0.88	0.46	122
Blueberries			31 - 32	90 - 95	2 weeks	82.3	29.7	0.86	0.45	118
Cherries	35	80 - 85	31 - 32	90	10 - 14 days	83.0	28.8	0.87	0.45	120
Coconuts	35	80 - 85	32 - 35	80 - 85	1 - 2 months	46.9	30.4	0.58	0.34	67
Cranberries	40	85 - 90	36 - 40	90 - 95	2 - 4 months	87.4	30.4	0.90	0.46	124
Currants	36	85 - 90	31 - 32	90 - 95	10 - 14 days	84.7	30.2	0.88	0.45	120
Dates (fresh)			0 - 10			78	27.1	.82	.43	112
Dewberries			31 - 32	90 - 95	3 days	84.5	29.7	.88		
Dried fruits*	35	50 - 60	32	50 - 60	9 to 12 months	14.0 - 26.0		0.31 - 0.41	0.26	20 - 37
Figs, fresh	40	65 - 75	31 - 32	85 - 90	7 - 10 days	78.0	27.6	0.82	0.43	112
Frozen-pack fruits			-10 - 0		6 - 12 months					
Gooseberries			31 - 32	90 - 95	2 - 4 weeks	88.9	30.0	0.90	0.46	126
Grapefruit *	45	85 - 90	50 - 60	85 - 90	4 - 6 weeks	88.8	30.0	0.91	0.46	126
Grapes (Chapter 29)										
American type	35	80 - 90	31 - 32	85 - 90	2 - 8 weeks	81.9	29.7	0.86	0.44	116
European type			30 - 31	90 - 95	3 - 6 months	81.6	28.1	0.86	0.44	116
Lemons*	55	85 - 90	32 or 50 - 58	85 - 90	1 - 6 months	89.3	29.4	0.91	0.46	127
Limes	45	85 - 90	48 - 50	85 - 90	6 - 8 weeks	86.0	29.1	0.80	0.46	122
Logan blackberries			31 - 32	85 - 90	5 - 7 days	8.29	29.7	0.86	0.45	118
Mangoes			55	85 - 90	2 - 3 weeks	81.4	30.3	0.85	0.44	117
Melons, Cantaloupe	45	85 - 90	36 - 40	90 - 95	5 - 15 days	92.0	29.9	0.93	0.48	132
Persian			45 - 50	90 - 95	2 weeks	92.7	30.5	0.94	0.48	132
Honeydew and Honey Ball			45 - 50	90 - 95	3 - 4 weeks	92.6	30.3	0.94	0.48	132
Casaba			45 - 50	85 - 90	4 - 6 weeks	92.7	30.1	0.94	0.48	132
Watermelons			40 - 50	80 - 90	2 - 3 weeks	92.1	31.3	0.97	0.48	132
Oranges*	40	85 - 90	32 - 48	85 - 90	3 - 12 weeks	87.2	30.6	0.90	0.46	124
Orange juice, chilled			30 - 35		3 - 6 weeks	89.0		0.91	0.47	128
Papayas			45	85 - 90	1 - 3 weeks	90.8	30.4	0.82	0.47	130
Peaches and nectarines	35	80 - 85	31 - 32	90	2 - 4 weeks	89.1	30.3	0.90	0.46	124
Pears	35	90 - 95	29 - 31	90 - 95	2 - 7 months	82.7	29.2	0.86	0.45	118
Pears, green			32	90 - 95	1 - 2 weeks	74.3	30.9	0.79	0.42	106
Persimmons			30	90	3 - 4 months	78.2	28.1	0.84	0.43	112
Pineapples										
Mature green	50	85 - 90	50 - 55	85 - 90	3 - 4 weeks		30.2			
Ripe	40	85 - 90	45	85 - 90	2 - 4 weeks	85.3	30.0	0.88	0.45	122
Plums, including fresh prunes	40	80 - 85	31 - 32	90 - 95	2 - 4 weeks	82.3	30.5	0.88	0.45	122
Pomegranates			32	90	2 - 4 weeks		26.6			
Quinces	35	80 - 85	31 - 32	90	2 -3 months	85.3	28.4	0.88	0.45	122
Raspberries	31	85 - 90								
Black			31 - 32	90 - 95	2 - 3 days	80.6	30.0	0.84	0.44	122
Red			31 - 32	90 - 95	2 - 3 days	84.1	30.9	0.87	0.45	121
Frozen (red or black)			-10 - 0		1 year					
Strawberries										
Fresh	31	85 - 90	31 - 32	90 - 95	5 - 7 days	89.9	30.6	0.92		129
Frozen*			-10 - 0		1 year	72.0			0.42	103
Tangerines	40	85 - 90	32 - 38	85 - 90	2 - 4 weeks	87.3	30.1	0.90	0.46	125

Estimating Specific and Latent Heats:
Sp. Ht. above freezing = 0.20 + 0.008 x % water.
Sp. Ht. below freezing = 0.20 + 0.003 x % water.
Latent Heat = 143.4 x % water.

Figure 18-16 Storage requirements and properties of perishable products: (a) fruits; (b) vegetables; (c) meats and seafood; (d) miscellaneous. (By permission of *ASHRAE 1974 Fundamentals* and *ASHRAE 1972 Fundamentals*.)

Commodity	Retail Storage °F % RH	Storage Temp. °F	Relative Humidity %	Approximate Storage Life	Water Content %	Highest Freezing Point °F	Specific Heat Above Freezing Btu/lb/F	Specific Heat Below Freezing Btu/lb/F	Latent Heat (Calculated) Btu/lb
VEGETABLES									
Artichokes (Globe)	40 90 - 95	31 - 32	95	2 weeks	83.7	29.9	0.87	0.45	120
Jerusalem		31 - 32	90 - 95	5 months	79.5	27.5	0.83	0.44	114
Asparagus	40 85 - 90	32 - 36	90 - 95	2 - 3 weeks	93.0	30.9	0.94	0.48	134
Beans (Green or snap)	45 85 - 90	45	90 - 95	7 - 10 days	88.9	30.7	0.91	0.47	128
Lima	40 85 - 90	32 - 40	90	7 days	66.5	31.0	0.73	0.40	94
Beets	40 85 - 90								
Bunch		32	95	10 - 14 days		31.3			
Topped		32	95 - 100	4 - 6 months	87.6	30.1	0.90	0.46	126
Broccoli, sprouting	40 90 - 95	32	95	10 - 14 days	89.9	30.9	0.92	0.47	130
Brussells sprouts	40 90 - 95	32	90 - 95	3 - 4 weeks	84.9	30.5	0.88	0.46	122
Cabbage, late	35 90 - 95	32	95 - 100	3 - 4 months	92.4	30.4	0.94	0.47	132
Carrots	40 85 - 90								
Prepackaged		32	80 - 90	3 - 4 weeks	88.2	29.5	0.90	0.46	126
Topped		32	90 - 95	4 - 5 months	88.2	29.5	0.90	0.46	126
Cauliflower	35 85 - 90	32	95	2 - 4 weeks	91.7	30.6	0.93	0.47	132
Celeriac		32	95 - 100	3 - 4 months	88.3	30.3	0.91	0.46	126
Celery	35 85 - 90	32	95	1 - 2 months	93.7	31.1	0.95	0.48	135
Collards		32	95	10 - 14 days	86.9	30.6	0.90		
Corn, sweet	35 85 - 90	32	95	4 - 8 days	73.9	30.9	0.79	0.42	106
dried							0.28	0.23	15
Cucumbers	50 85 - 95	50 - 55	90 - 95	10 - 14 days	96.1	31.1	0.97	0.49	137
Eggplant	50 85 - 95	45 - 50	90 - 95	7 - 10 days	92.7	30.6	0.94	0.48	132
Endive (escarole)	35 90 - 95	32	95	2 - 3 weeks	93.3	31.9	0.94	0.48	132
Frozen-pack vegetables		-10 - 0		6 - 12 months					
Garlic, dry	35 85 - 90	32	65 - 70	6 - 7 months	61.3	30.5	0.69	0.40	89
Horseradish	45	30 - 32	95 - 100	10 - 12 months	73.4	28.7	0.78	0.42	104
Kale		32	95	3 - 4 weeks	86.6	31.1	0.89	0.46	124
Kohlrabi		32	95	2 - 4 weeks	90.1	30.2	0.92	0.47	128
Locks, green	35 90 - 95	32	95	1 - 3 months	85.4	30.7	0.88	0.46	126
Lettuce, head	35 90 - 95	32 - 34	95 - 100	2 - 3 weeks	94.8	31.7	0.96	0.48	136
Mushrooms	40 75 - 80	32	90	3 - 4 days	91.1	30.4	0.93	0.47	130
Olives, fresh		45 - 50	85 - 90	4 - 6 weeks	75.2	29.4	0.80	0.42	108
Onions and onion sets	50 70 - 75	32	65 - 70	1 - 8 months	87.5	30.6	0.90	0.46	124
Parsley		32	95	1 - 2 months	85.1	30.0	0.88	0.45	122
Parsnips	35 90 - 95	32	98 - 100	4 - 6 months	78.6	30.4	0.84	0.44	112
Peas, green	35 85 - 90	32	95	1 - 3 weeks	74.3	30.9	0.79	0.42	106
dried					9.5		0.28	0.22	14
Peppers, sweet	50 90 - 95	45 - 50	90 - 95	2 - 3 weeks	92.4	30.7	0.94	0.47	132
Peppers, chili (dry)		32 - 50	60 - 70	6 months	12.0		0.30	0.24	17
Popcorn, unpopped	40 85	32 - 40	85	4 - 6 months	13.5		0.31	0.24	19
Potatoes	50 85 - 90								
Early crop		50 - 55	90		81.2	30.9	0.85	0.44	116
Late crop		38 - 50	90		77.8	30.9	0.82	0.43	111
Sweet potatoes	55 85 - 90	55 - 60	85 - 90	4 - 7 months	68.5	29.7	0.75	0.40	97
Pumpkins	55 70 - 75	50 - 55	70 - 75	2 - 3 months	90.5	30.5	0.92	0.47	130
Radishes-spring, prepackaged	35 90 - 95	32	95	3 - 4 weeks	93.6	30.7	0.95	0.48	134
winter		32	95 - 100	2 - 4 months	93.6		0.95	0.48	134
Rhubarb	35 95	32	95	2 - 4 weeks	94.9	30.3	0.96	0.48	134
Rutabagas	35 95	32	98 - 100	4 - 6 months	89.1	30.1	0.91	0.47	127
Salsify		32	98 - 100	2 - 4 months	79.1	30.0	0.83	0.44	113
Sauerkraut	45 75 - 80				8.9	2.6	0.92	0.47	129
Spinach	35 90 - 95	32	95	10 - 14 days	92.7	31.5	0.94	0.48	132
Squash	50 70 - 75								
acorn	50 85 - 95	45 - 50	70 - 75	5 - 8 weeks		30.5			
summer	50 85 - 95	32 - 50	85 - 95	5 - 14 days	94.0	31.1	0.95		135
winter	55 85 - 95	50 - 55	70 - 75	4 - 6 months	88.6	30.3	0.91		127
Tomatoes	55 85 - 90								
Mature green	50 85 - 90	57 - 70	85 - 90	1 - 3 weeks	93.0	31.0	0.95	0.48	134
Firm ripe		45 - 50	85 - 90	4 - 7 days	94.1	31.1	0.94	0.48	134
Turnips, roots	35 90 - 95	32	95	4 - 5 months	91.5	30.1	0.93	0.47	130
Vegetables, mixed		32 - 40	90 - 95	1 - 4 weeks	90.0		0.90	0.45	130
Yams		60	85 - 90	3 - 6 months	73.5	28.5	0.79	0.40	130

Estimating Specific and Latent Heats:
Sp. Ht. above freezing = 0.20 + 0.008 x % water.
Sp. Ht. below freezing = 0.20 + 0.003 x % water.
Latent Heat = 143.4 x % water.

Figure 18–16 (b) (Cont.)

Commodity	Retail Storage °F	% RH	Storage Temp. °F	Relative Humidity %	Approximate Storage Life	Water Content %	Highest Freezing Point °F	Specific Heat Above Freezing Btu/lb/F	Specific Heat Below Freezing Btu/lb/F	Latent Heat (Calculated) Btu/lb
MEATS - FISH - SHELLFISH										
Bacon						20		0.50	0.30	29
Cured, Farm Style	55	55 - 65	60 - 65	85	4 - 6 months	13 - 29		.3 - .43	.24 - .29	18 - 41
Packer Style			34 - 40	85	2 - 6 weeks					
Frozen			-10 - 0	90 - 95	4 - 6 months					
Fat Backs			34 - 36	85 - 90	0 - 3 months	6 - 12		.25 - .30	.22 - .24	9 - 17
Hams & Shoulders Fresh	34	85 - 88	32 - 34	85 - 90	7 - 12 days	47 - 54	28 - 29	.58 - .63	.34 - .36	67 - 77
Cured	55	55 - 65	60 - 65	50 - 60	0 - 3 years	40 - 45		.52 - .56	.32 - .33	57 - 64
Frozen			-10 - 0	90 - 95	6 -8 months				.34 - .36	
Pork Fresh	34	85 - 90	32 - 34	85 - 90	3 - 7 days	32 - 44	28 - 29	.46 - 55		46 - 63
Fresh						60	28	68	.38	86.5
Frozen			-10 - 0	90 - 95	4 - 6 months				.3 - .33	
Smoked	55	55 - 65				57		.60	.32	
Sausage Casings			40 - 45	85 - 90				0.6		
Drying						65.5	26	0.89	.56	93
Franks	35	80 - 90				60	29	.86	.56	86
Fresh	35	85 - 90				65	26	.89	.56	93
Smoked	40	80 - 85	40 - 45	85 - 90	6 months	60	25	.68	.38	86
Beef, Dried						5 - 15		.22 - 34	.19 - .26	7 - 22
Fresh-Lean	34	85 - 90	32 - 34	88 - 92	1 - 6 weeks	62 - 67	28 - 29	.70 - .84	.40	89 - 110
Fresh-Fat							28	.6	.35	79
Brined	40	80 - 85						.75		
Frozen			-10 - 0	90 - 95	9 - 12 months				.38 - .43	
Veal	34	85 - 90	32 - 34	90 - 95	5 - 10 days	64 - 70	28 - 29	.71 - .76	.30	92 - 100
Cut Meats	34	85 - 90	30 - 34	88 - 92		65	29	.72	.40	95
Lamb, Fresh	34	85 - 90	32 - 34	85 - 90	5 - 12 days	60 - 70	28 - 29	.68 - .76	.38 - .51	86 - 100
Frozen			-10 - 0	90 - 95	8 - 10 months				.38 - .51	
Rabbits, Fresh	34	85 - 90	32 - 34	90 - 95	1 - 5 days	68.0		.74	.40	98
Frozen			-10 - 0	90 - 95	0 - 6 months					
Livers, Fresh	34	85 - 90	32 - 34	80 - 85	1 - 6 weeks	65.5	29	.72	.40	93.3
Frozen			-10 - 0	90 - 95	3 - 4 months	70			.41	100
Poultry, Fresh	28	85 - 90	32	85 - 90	1 week	74	27	.79		106
*Frozen (eviscerated)			0 or below	90 - 95	8 - 12 months				.42	
Fish, Fresh (Red)	30	80 - 95	33 - 35	90 - 95	5 - 15 days	62 - 85	28	.70 - .86		89 - 122
Dried			40 - 50	50 - 60	6 - 8 months			.56	.34	65
Frozen			-20 - 0	90 - 95	6 - 12 months	62 - 85			.38 - .45	89 - 122
Smoked			40 - 50	50 - 60	6 - 8 months			.70	.39	92
Brine Salted			40 - 50	90 - 95	10 - 12 months			.76	.41	100
Mild Cured			28 - 35	75 - 90	4 - 8 months			.76	.41	100
Shellfish, Fresh	32	70 - 75	30 - 33	85 - 95	3 - 7 days	80 - 87	28	.83 - .90		113 - 125
Frozen			-20 - 0	90 - 95	3 - 8 months				.44 - .46	113 - 125
Oysters & Clams, Shell			33	90 - 95		80.4	27	.83	.44	116
Shucked	32	70 - 75	33	90 - 95		87	27	.90	.46	125
Scallops	32	70 - 75	33	90 - 95		80.3	28	.89	.48	116
Shrimp	32	70 - 75	33	90 - 95		70.8	28	.83	.45	119
Lobsters, Crabs, Boiled	32	80 - 90	25	89 - 90	10 days	79		.85	.44	113

Estimating Specific and Latent Heats:
Sp. Ht. above freezing = 0.20 + 0.008 x % water.
Sp. Ht. below freezing = 0.20 + 0.003 x % water.
Latent Heat = 143.4 x % water.

Figure 18-16 (c) (Cont.)

Commodity	Retail Storage °F % RH	Storage Temp. °F	Relative Humidity %	Approximate Storage Life	Water Content %	Highest Freezing Point °F	Specific Heat Above Freezing Btu/lb/F	Specific Heat Below Freezing Btu/lb/F	Latent Heat (Calculated) Btu/lb
MISCELLANEOUS									
Beer (Barrels, metal keg	40	35 - 40		3 - 9 weeks	90.2	28	.92		129
wood keg	40 85 - 90	35 - 40	85 - 90	3 - 9 weeks	90.2	28	.92		129
Bread	0	0		3 wks - 3 mos.	32 - 37	20	.7	.34	46 - 53
Bread Dough	35 85 -90	35 - 40	85 - 90	1 - 3 days	46 - 58		.6 - .75	.34 - .37	66 - 83
Butter	40 75 - 80	40	75 - 85	1 month	16	31	.5	.34	23
Candy	34 40 - 50	0 - 34	40 - 65				.93		
Cavier (Tub)			85 - 90		55	20			
Cheese, American	40 75 - 80	40 - 45	65 - 70		60	17	.64	.36	79
Camembert	40 80 - 85	30 - 34			60	18	.70	.40	86
Limberger	40 80 - 85	30 - 34			55	19	.70	.40	86
Roquefort	45 75 - 80	30 - 34			55	3	.65	.32	79
Swiss	40 75 - 80	30 - 34			55	15	.64	36	79
Chocolate (Coating)	65 40 - 50	60	40 - 50	6 months	55	95 - 85	.55	.30	40
Cream 40%	35	33		7 weeks	73	28	.85	.4	90
Cream (sweetened)		-15		sev. months	72.5		.78	.42	104
Coffee, green	37 80 - 85	35 - 37	80 - 85	2 - 4 months	10 - 15		.30	.24	14 - 21
Eggs, shell	40 80 - 85	29 - 31	80 - 85	5 - 6 months	66	28	.73	.40	96
frozen, whole		0 - less		1 year +	74			.42	106
frozen, yolk		0 - less		1 year +	55			.36	79
frozen, white		0 - less		1 year +	88			.46	126
Whole Egg Solids	40	35 - 40	low	6 - 12 months	2 - 4		.22	.21	4
Yolk Solids		35 - 40	low	6 - 12 months	3 - 5		.23	.21	6
Flake Albumen		room	low	1 year +	12 - 16		.31	.24	20
Dried Spray Albumen		room	low	1 year +	5 - 8		.26	.22	11
Flour	82 60 - 65	78	60 - 65		13.5		.38	.28	
Flowers, cut (see table)		32							
Furs-Woolens	40 45 - 55	34 - 40	45 - 55	sev. years			.40		
Furs-to-shock		15							
Honey	40 65 - 70			1 year +	18		.35	.26	26
Hops	32 50 - 60	29 - 32	50 - 60	sev. months					
Ice Cream	-15	-20 - 15		3 - 12 months	58 - 63	21	.66 - .7	.37 - .39	86
Ice		25 - 28			100		1.00	0.5	144
Lard	45 75 - 80	45	90 - 95	4 - 8 months	0		.52		
Malt		0	90 - 95	12 - 14 months					
Maple Sugar	45 65 - 70	31	65 - 70	1 year +	5		.24	.21	7
Maple Syrup	45 65 - 70	31	65 - 70	1 year +	35.5		.48	.31	51
Milk	35 65 - 75	33			87.5	31	.93	.49	124
Nuts (Dried)	40 65 - 75	32 - 50	65 - 75	6 - 10 months	3 - 10		.21 - .29	.19 - .24	4.3 - 14
Oleomargarine	45 75 - 80	35	60 - 70	1 year +	15.5		.32	.25	22
Tobacco Hogsheads		50 - 65	50 - 55	1 year					
Bales		35 - 40	70 - 85	1 - 2 years					
Cigarettes		35 - 46	50 - 55	6 months					
Cigars		35 - 50	60 - 65	2 months					
Veg. Seed	50 55 - 65	32 - 50	50 - 65		7.0 - 15.0		.29	.23	16
Yeast	35 80 - 85	31 - 32			70.9		.77	.41	102

Estimating Specific and Latent Heats:
Sp. Ht. above freezing = 0.20 + 0.008 x % water.
Sp. Ht. below freezing = 0.20 + 0.003 x % water.
Latent Heat = 143.4 x % water.

Figure 18–16 (d) (Cont.)

Product load, below freezing. If the product is stored below freezing, the product load is figured in three steps:

1. Cooling required to bring the product down to the freezing temperature
2. Cooling required to freeze the product
3. Cooling required to cool the frozen product to storage temperatures

Sample Product Load Calculations, below Freezing. A refrigerated room is used to store frozen-fresh lean beef at a storage temperature of 0°F. One thousand pounds of beef is delivered to the room per day at a temperature of 50°F.

1. The *temperature reduction load* is calculated as follows:

$$H = W \times (T_2 - T_F) \times \text{SH}$$

where W = weight of product delivered per 24 hours, lb
T_2 = °F temperature of product entering the room
T_F = °F freezing temperature of product (see Figure 18–16)
SH = specific heat of product, above freezing (see Figure 18–16)

Substituting the values from the example gives us

$$\begin{aligned} H &= 1000 \text{ lb} \times (50\,°\text{F} - 38\,°\text{F}) \times 0.73 \text{ Btu/lb/°F} \\ &= 8760 \text{ Btu/24 hours} \end{aligned}$$

2. *Product freezing* is calculated as follows:

$$H = W \times \text{LH}$$

where W = weight of frozen product, lb
LH = latent heat of freezing (see Figure 18–16)

Substituting the values from the example gives us

$$\begin{aligned} \text{H} &= 1000 \text{ lb} \times 100 \text{ Btu/lb} \\ &= 100{,}000 \text{ Btu/24 hours} \end{aligned}$$

3. *Product temperature reduction below freezing* is calculated as follows:

$$H = W \times (T_F - T_1) \times \text{SH}$$

where W = weight of product, lb
T_F = °F freezing temperature of product
T_1 = °F storage temperature
SH = specific heat of product below freezing (see Figure 18–16)

Figure 18–17 Average water content and specific and latent heats of various fishery products. (By permission of *ASHRAE 1974 Applications.*)

Fish	Water Content %	Average Freezing Point, F	Specific Heat[a] Above Freezing, Btu/(Lb) (F deg)	Specific Heat[b] Below Freezing, Btu/(Lb) (F deg)	Latent Heat[b] Btu/Lb
WHOLE FISH					
Haddock, Cod	78	28	0.82	0.43	112
Halibut	75	28	0.80	0.43	108
Tuna	70	28	0.76	0.41	100
Herring (kippered)	70	28	0.76	0.41	100
Herring (smoked)	64	28	0.71	0.39	92
Salmon	64	28	0.71	0.39	92
Menhaden	62	28	0.70	0.38	89
FISH FILLETS OR STEAKS					
Haddock, Cod, Ocean perch	80	28	0.84	0.44	115
Hake, Whiting	82	28	0.86	0.45	118
Pollock	79	28	0.83	0.44	113
Mackerel	57	28	0.66	0.37	82
SHELL FISH					
Scallop meat	80	28	0.84	0.44	115
Shrimp	83	28	0.86	0.45	119
American lobster	79	28	0.83	0.44	113
Oysters and Clams (meat and liquor)	87	28	0.90	0.46	125

[a]Calculated by Siebel's formula, for values above freezing $S = 0.008a + 0.20$; for values below freezing $S = 0.003a + 0.20$ where a = water content in percent; 0.20 = specific heat of solid constituents of the substance.
[b]Values for latent heat (latent heat of fusion) in Btu per lb, calculated by multiplying the percentage of water content by the latent heat of fusion of water 143.4 Btu.

Substituting the values from the example gives us

$$H = 1000 \text{ lb} \times (28\,^\circ\text{F} - 0\,^\circ\text{F}) \times 0.40 \text{ Btu/lb/}^\circ\text{F})$$
$$= 11{,}200 \text{ Btu/24 hours}$$

Total Product Load

1. Reduction to freezing	8,760 Btu/24 hours
2. Latent heat	100,000 Btu/24 hours
3. Reduction below freezing	11,200 Btu/24 hours
Total	119,960 Btu/24 hours

Figure 18–18 Storage conditions for cut flowers and nursery stock. (By permission of *ASHRAE 1974 Fundamentals.*)

Commodity	Storage Temperature F	Relative Humidity, %	Approximate Storage Life	Method of Holding	Highest Freezing Point, F
CUT FLOWERS:					
Calla Lily	40	90 - 95	1 week	Dry pack	-
Camellia	45	90 - 95	3 - 6 days	Dry pack	30.6
Carnation	31 - 32	90 - 95	2 - 4 weeks	Dry pack	30.8
Chrysanthemum	31 - 32	90 - 95	2 - 4 weeks	Dry pack	30.5
Daffodil (Narcissus)	32 - 33	90 - 95	1 - 3 weeks	Dry pack	31.8
Dahlia	40	90 - 95	3 - 5 days	Dry pack	-
Gardenia	32 - 33	90 - 95	2 weeks	Dry pack	31.0
Gladiolus	40 - 42	90 - 95	1 week	Dry pack	31.4
Iris, tight buds	31 - 32	90 - 95	2 weeks	Dry pack	30.6
Lily, Easter	32 - 35	90 - 95	2 - 3 weeks	Dry pack	31.1
Lily-of-the-Valley	31 - 32	90 - 95	2 - 3 weeks	Dry pack	-
Orchid	55	90 - 95	1 - 2 weeks	Water	31.4
Peony, tight buds	32 - 35	90 - 95	4 - 6 weeks	Dry pack	30.1
Rose, tight buds	32	90 - 95	1 - 2 weeks	Dry pack	31.2
Snapdragon	40 - 42	90 - 95	1 - 2 weeks	Dry pack	30.4
Sweet peas	31 - 32	90 - 95	2 weeks	Dry pack	30.4
Tulips	31 - 32	90 - 95	4 - 8 weeks	Dry pack	-
GREENS:					
Asparagus (plumosus)	32 - 40	90 - 95	4 - 5 months	Polylined cases	26.0
Fern, dagger and wood	30 - 32	90 - 95	2 - 3 months	Dry pack	28.9
Holly	32	90 - 95	4 - 5 weeks	Dry pack	27.0
Huckleberry	32	90 - 95	1 - 4 weeks	Dry pack	26.7
Laurel	32	90 - 95	1 - 4 weeks	Dry pack	27.6
Magnoli	35 - 40	90 - 95	1 - 4 weeks	Dry pack	27.0
Rhododendron	32	90 - 95	1 - 4 weeks	Dry pack	27.6
Salal	32	90 - 95	1 - 4 weeks	Dry pack	26.8
BULBS:					
Amaryllis	38 - 45	70 - 75	5 months	Dry	30.8
Caladium	70	70 - 75	2 - 4 months	-	29.7
Crocus	48 - 63	-	2 - 3 months	-	-
Dahlia	40 - 45	70 - 75	5 months	Dry	28.7
Gladiolus	38 - 50	70 - 75	8 months	Dry	28.2
Hyacinth	55 - 70	-	2 - 5 months	-	29.3
Iris, Dutch, Spanish	80 - 85	70 - 75	4 months	Dry	-
Lily					
Gloriosa	63	70 - 75	3 - 4 months	Poly liner	-
Candidum	31 - 33	70 - 75	1 - 6 months	Poly liner & peat	-
Croft	31 - 33	70 - 75	1 - 6 months	Poly liner & peat	-
Longiflorum	31 - 33	70 - 75	1 - 10 months	Poly liner & peat	28.9
Speciosum	31 - 33	70 - 75	1 - 6 months	Poly liner & peat	-
Peony	33 - 35	70 - 75	5 months	Dry	-
Tuberose	40 - 45	70 - 75	4 months	Dry	-
Tulip	31 - 32	70 - 75	5 - 6 months	Dry	27.6
NURSERY STOCK:					
Trees and Shrubs	32 - 36	80 - 85	4 - 5 months	†	-
Rose Bushes	32	85 - 95	4 - 5 months	Bare rooted with poly liner	-
Strawberry Plants	30 - 32	80 - 85	8 - 10 months	Bare rooted with poly liner	29.9
Rooted Cuttings	33 - 40	85 - 95	-	Poly wrap	-
Herbaceous Perennials	27 - 28 or 33 - 35†	80 - 85	-		-
Christmas Trees	22 - 32	80 - 85	6 - 7 weeks	-	-

*Data from USDA Handbook No. 66 and bulletin by Post and Fischer.
†For details for various trees, shrubs, and perennials, see bulletin by Mahistede and Fletcher.

Product	Storage temperature (°F)			
	32 °F	40 °F	60 °F	°F Other
		Fruits		
Apples	0.25 - 0.450	0.55 - 0.80	1.5 - 3.4	
Apricots	0.55 - 0.63	0.70 - 1.0	2.33 - 3.74	
Avacados	–	–	6.6 - 15.35	
Bananas	–	–	2.3 - 2.75	@ 68° 4.2 - 4.6
Blackberries	1.70 - 2.52	5.91 - 5.00	7.71 - 15.97	
Blueberries	0.65 - 1.10	1.0 - 1.35	3.75 - 6.5	@ 70° 5.7 - 7.5
Cherries	0.65 - 0.90	1.4 - 1.45	5.5 - 6.6	
Cherries, sour	0.63 - 1.44	1.41 - 1.45	3.0 - 5.49	
Cranberries	0.30 - 0.35	0.45 - 0.50	–	
Figs, mission	–	1.18 - 1.45	2.37 - 3.52	
Gooseberries	0.74 - 0.96	1.33 - 1.48	2.37 - 3.52	
Grapefruit	0.20 - 0.50	0.35 - 0.65	1.1 - 2	
Grapes				
American	0.30	0.60	1.75	
European	0.15 - 0.20	0.35 - 0.65	1.10 - 1.30	
Lemons	0.25 - 0.45	0.30 - 0.95	1.15 - 2.50	
Limes	–	0.405	1.485	
Melons				
Cantaloupes	0.55 - 0.63	0.96 - 1.11	3.70 - 4.22	
Honey dew	–	0.45 - 0.55	1.2 - 1.65	
Oranges	0.20 - 0.50	0.65 - 0.8	1.85 - 2.6	
Peaches	0.45 - 0.70	0.70 - 1.0	3.65 - 4.65	
Pears	0.35 - 0.45	–	4.40 - 6.60	
Plums	0.20 - 0.35	0.45 - 0.75	1.20 - 1.40	
Raspberries	1.95 - 2.75	3.40 - 4.25	9.05 - 11.15	
Strawberries	1.35 - 1.90	1.80 - 3.40	7.80 - 10.15	
Tangerines	1.63	2.93	–	
		Vegetables		
Artichokes (globe)	2.48 - 4.93	3.48 - 6.56	8.49 - 15.90	
Asparagus	2.95 - 6.60	5.85 - 11.55	11.0 - 25.75	
Beans				
green or snap	–	4.60 - 5.7	16.05 - 22.05	
lima	1.15 - 1.6	2.15 - 3.05	11.0 - 13.7	
Beets, topped	1.35	2.05	3.60	
Broccoli	3.75	5.50 - 8.80	16.9 - 25.0	
Brussels sprouts	1.65 - 4.15	3.30 - 5.50	6.60 - 13.75	
Cabbage	0.60	0.85	2.05	
Carrots, topped	1.05	1.75	4.05	
Cauliflower	1.80 - 2.10	2.10 - 2.40	4.70 - 5.40	
Celery	0.80	1.20	4.10	
Corn, sweet	3.60 - 5.65	5.30 - 6.60	19.20	
Cucumbers	–	–	1.65 - 3.65	
Garlic	0.33 - 1.19	0.63 - 1.08	1.18 - 3.0	
Horseradish	0.89	1.19	3.59	
Kohlrabi	1.11	1.78	5.37	
Leeks	1.04 - 1.78	2.15 - 3.19	9.08 - 12.82	
Lettuce				
Head	1.15	1.35	3.95	
Leaf	2.25	3.20	7.20	
Mushrooms	3.10 - 4.80	7.80	–	@ 50° 11.0
Okra	–	6.05	15.8	
Olives	–	–	2.37 - 4.26	
Onions				
Dry	0.35 - 0.55	0.90	1.20	
Green	1.15 - 2.45	1.90 - 7.50	7.25 - 10.70	

Figure 18-19 Heats of respiration (approximate, Btu/lb/24 hr) of fruits, vegetables, and other perishable foods. All fruits and vegetables are living and give off heat in storage. If heat of respiration is not given, an approximate value or average value should be used. (By permission of *ASHRAE 1978 Handbook*.)

Product	Storage temperature (°F)			
	32 °F	40 °F	60 °F	°F Other
		Vegetables (continued)		
Peas, green	4.10 - 4.20	6.60 - 8.0	19.65 - 22.25	
Peppers, sweet	1.35	2.35	4.25	
Potatoes				
Immature	–	1.30	1.45 - 3.4	
Mature	–	0.65 - 0.90	0.75 - 1.30	
Sweet	–	0.85	2.15 - 3.15	
Radishes				
With tops	1.59 - 1.89	2.11 - 2.30	7.67 - 8.5	
Topped	0.59 - 0.63	0.85 - 0.89	3.04 - 3.59	
Rhubarb, topped	0.89 - 1.44	1.19 - 2.0	3.41 - 4.97	
Spinach	2.10 - 2.45	3.95 - 5.60	18.45 - 19.0	
Squash, yellow	1.3 - 1.41	1.55 - 2.04	8.23 - 9.97	
Tomatoes,				
Mature green	–	0.55	3.10	
Ripe	0.50	0.65	2.8	
Turnips	0.95	1.10	2.65	
Vegetables, mixed	2.0	–	–	
		Miscellaneous		
Caviar, tub	–	–	1.91	
Cheese				
American	–	–	2.34	
Camembert	–	–	2.46	
Limburger	–	–	2.46	
Roquefort	–	–	–	@ 45° 2.0
Swiss	–	–	2.33	
Flowers, cut		0.24 Btu/24 hr/ft^2 floor area		
Honey	–	0.71	–	
Hops	–	–	–	@ 35° 0.75
Malt	–	–	–	@ 50° 0.75
Maple sugar	–	–	–	@ 45° 0.71
Maple syrup	–	–	–	@ 45° 0.71
Nuts	0.074	0.185	0.37	
Nuts, dried	–	–	–	@ 35° 0.50

Figure 18–19 (Cont.)

Material description	Specific heat [Btu/(lb) (°F)]	Density (lb/ft³)	Thermal conductivity [Btuh/(ft²) (°F/ft)]
Aluminum (allow 1100)	0.214	171	128
Aluminum bronze (76% Cu, 22% Zn, 2% Al)	0.09	517	58
Asbestos			
Fiber	0.25	150	0.097
Insulation	0.20	36	0.092
Asphalt	0.22	132	0.43
Bakelite	0.35	81	9.7
Brick, building	0.2	123	0.4
Brass			
red (85% Cu, 15% Zn)	0.09	548	87
yellow (65% Cu, 35% Zn)	0.09	519	69
Bronze	0.104	530	17(32)
Cellulose	0.32	3.4	0.033
Cement (portland clinker)	0.16	120	0.017
Clay	0.22	63	
Coal	0.3	90	0.098(32)
Concrete (stone)	0.156(392)	144	0.54
Copper (electrolytic)	0.092	556	227
Earth (dry and packed)		95	0.037
Fireclay brick	0.198(212)	112	0.58(392)
Glass:			
Crown (soda-lime)	0.18	154	0.59(200)
Flint (lead)	0.117	267	0.79
Pyrex	0.20	139	0.59(200)
Gypsum	0.259	78	0.25
Ice			
32°F	0.487	57.5	1.3
−4°F	0.465		1.41*
Iron			
Cast	0.12(212)	450	27.6(129)
Wrought		485	34.9
Lead	0.0309	707	20.1
Limestone	0.217	105	0.54
Marble	0.21	162	1.5
Paper	0.32	58	0.075
Paraffin	0.69	56	0.14(32)
Plaster		132	0.43(167)
Porcelain	0.18	162	1.3
Rock salt	0.219	136	
Rubber			
Vulcanized (soft)	0.48	68.6	0.08
Hard		74.3	0.092
Sand	0.191	94.6	0.19
Snow			
Freshly fallen		7	0.34
At 32°F		31	1.3
Steel (mild)	0.12	489	26.2
Stone (quarried)	0.2	95	
Tar			
Pitch	0.59	67	0.51
Bituminous		75	0.41
Woods			
Hardwoods: maple, oak, etc.	0.45/0.65	23/70	0.066/0.148
Softwoods: fir, pine, etc.	0.65/0.67	23/46	0.061/0.093

*For thermal conductivity k in Btuh/(ft²) (°F/in.), multiply tabular values by 12.

Figure 18-20 Properties of solids. Values are for room temperature unless otherwise noted in brackets. (By permission of *ASHRAE 1972 Applications*.)

Commodity	Type of Package	Outside Dimensions of Package, In.	Avg Gross Wt of Pkg, Lb	Avg Net Wt of Mdse, Lb	Avg Gross Wt Density, Lb per Cu Ft	Avg Net Wt Density, Lb per Cu Ft
Apples	Wood Box Northwestern	$19\frac{1}{2} \times 11 \times 12\frac{3}{16}$	50	42	33.1	27.8
	Fiber Tray Carton	$20\frac{1}{2} \times 12\frac{1}{2} \times 13\frac{1}{4}$	$46\frac{3}{4}$	43	23.8	21.9
	Fiber Master Carton	$22\frac{1}{2} \times 12\frac{1}{2} \times 13$	$44\frac{3}{4}$	41	21.2	19.4
	Fiber Bulk Carton	$19 \times 12\frac{1}{2} \times 13$	$44\frac{3}{4}$	41	25.0	22.9
	Pallet Box	$47 \times 47 \times 30$	1030	900	26.9	23.5
Beef						
Boneless	Fiber Carton	$28 \times 18 \times 6$	146	140	83.4	80.0
Fores	Loose	—	—	—	—	22.2
Hinds	Loose	—	—	—	—	22.2
Celery	Wirebound Crates	$20\frac{1}{4} \times 16 \times 9\frac{3}{4}$	60	55	32.8	30.0
	Fiber Carton	$16 \times 11 \times 10$	36	32	35.4	31.4
Cheese	Hoops	$16 \times 16 \times 13$	84	78	43.6	40.5
	Wood, Export	$17 \times 17 \times 14$	87	76	37.1	32.5
Cheese, Swiss	Wheels	$32\frac{1}{2} \times 32\frac{1}{2} \times 7$	—	171	—	40.0
Chili Peppers	Bags	$45 \times 21 \times 26$	234	229	16.5	16.1
Citrus Fruits						
Oranges	Box	$12\frac{1}{8} \times 13\frac{1}{4} \times 26\frac{1}{4}$	77	69	31.5	28.3
	Bruce Box	$13 \times 11 \times 26\frac{1}{4}$	88	83	40.5	38.2
	Pallet, 40 Cartons	$40 \times 48 \times 58\frac{1}{2}$	1690	1480	26.0	22.8
California Oranges	Fiber Carton	$16\frac{3}{8} \times 10\frac{1}{16} \times 10\frac{1}{2}$	40	37	38.0	35.2
Florida Oranges	Fiber Carton	$19\frac{1}{4} \times 12\frac{1}{4} \times 8$	45	37	41.3	33.9
Lemons	Fiber Carton	$16\frac{3}{8} \times 10\frac{1}{16} \times 10\frac{1}{2}$	40	37	40.0	37.0
Grapefruit	Fiber Carton	$19\frac{1}{4} \times 12\frac{1}{4} \times 8$	40	38	36.7	34.9
Coconut, Shredded	Bags	$38 \times 18\frac{1}{2} \times 8$	101	100	31.0	30.7
Cranberries	Fiber Carton	$15\frac{3}{4} \times 11\frac{1}{4} \times 10\frac{1}{2}$	26	24	24.1	22.2
Cream	Tins	$12 \times 12 \times 14$	$52\frac{3}{4}$	50	45.2	42.9
Dried Fruit						
Dates	Wood Box	$15\frac{1}{2} \times 10 \times 6\frac{1}{2}$	$26\frac{1}{2}$	25	45.4	42.9
	Fiber Carton	$14 \times 14 \times 11$	32	30	25.7	24.0
Raisins, prunes, figs, peaches	Fiber Carton	$15 \times 11 \times 7$	32	30	47.9	44.9
Eggs, Shell	Wood Cases	$26 \times 12 \times 13$	55	45	23.4	19.1
Eggs, Frozen	Cans	$10 \times 10 \times 12\frac{1}{2}$	32	30	44.2	41.5
Frozen Fishery Products						
Blocks	4/13½ lb Carton	$20\frac{3}{4} \times 12\frac{1}{8} \times 6\frac{3}{4}$	56	54	57.0	55.0
	4/16½ lb Carton	$19\frac{3}{4} \times 10\frac{3}{4} \times 11\frac{1}{4}$	68	66	49.2	47.8
Fillets	12/16 oz Carton	$12\frac{3}{4} \times 8\frac{5}{8} \times 3\frac{13}{16}$	13.5	12	55.8	49.6
	10/5 lb Carton	$14\frac{1}{2} \times 10 \times 14$	52.25	50	44.6	42.7
	5/10 lb Carton	$14\frac{1}{2} \times 10 \times 14$	52.2	50	44.5	42.7
Fish Sticks	12/8 oz Carton	$11 \times 8\frac{3}{8} \times 3\frac{7}{8}$	6.9	6	33.6	29.3
	24/8 oz Carton	$16\frac{7}{16} \times 8\frac{5}{16} \times 4\frac{5}{8}$	13.8	12	37.8	32.9
Panned Fish	None, Glazed	Wooden Boxes	—	—	—	35.0
Portions	2, 3, 5, and 6 lb Carton	Custom Packing	—	—	—	29–33
Round Ground Fish	None, Glazed	Stacked Loose	—	—	—	33–35
Round Halibut	None, Glazed	Wooden Box, Loose	—	—	—	30–35
Round Salmon	None, Glazed	Stacked Loose	—	—	—	33–35
Shrimp	2½ and 5 lb Cartons	Custom Packing	—	—	—	35.0
Steaks	1, 5, or 10 lb Packages	Custom Packing	—	—	—	50–60
Frozen Fruits, Juices and Vegetables						
Asparagus	24/12 oz Carton	$13\frac{1}{2} \times 11\frac{3}{4} \times 82$	21	18	27.	23.8
Beans, Green	36/10 oz Carton	$12\frac{1}{2} \times 11 \times 8$	$25\frac{1}{2}$	$22\frac{1}{2}$	40.1	35.3
Blueberries	24/12 oz Carton	$12 \times 11\frac{1}{2} \times 8$	20	18	31.3	28.2
Broccoli	24/10 oz Carton	$12\frac{1}{2} \times 11\frac{1}{2} \times 8\frac{1}{2}$	$18\frac{1}{2}$	15	26.2	21.2
Citrus Concentrates	Fiber Carton 48/6 oz	$13 \times 8\frac{3}{4} \times 7\frac{1}{2}$	27	26	54.7	52.7
Peaches	24/1 lb Carton	$13\frac{1}{2} \times 11\frac{1}{4} \times 7\frac{1}{2}$	27	24	41.0	36.4
Peas	6/5 lb Carton	$17 \times 11 \times 9\frac{1}{2}$	32	30	31.1	28.2
	48/12 oz Carton	$21\frac{1}{2} \times 8\frac{1}{2} \times 12\frac{1}{2}$	38	36	28.7	27.2
Potatoes, French Fries	12/16 oz Carton	—	—	—	—	28.6
	24/9 oz Carton	—	—	—	—	24.0
Spinach	24/14 oz Carton	$12\frac{1}{2} \times 11 \times 8\frac{1}{2}$	24	21	35.5	31.0
Strawberries	30 lb Can	$12\frac{1}{2} \times 10 \times 10$	32	30	44.2	41.5
	24/1 lb Carton	$13 \times 11 \times 8$	28	24	42.3	36.2
	450 lb Barrel	$35 \times 25 \times 25$	—	450	—	35.5

Figure 18-21 Space, weight, and density data for commodities stored in refrigerated warehouses. (By permission of *ASHRAE 1974 Applications.*)

Commodity	Type of Package	Outside Dimensions of Package, In.	Avg Gross Wt of Pkg, Lb	Avg Net Wt of Mdse, Lb	Avg. Gross Wt Density, Lb per Cu Ft	Avg. Net Wt Density, Lb per Cu Ft
Grapes, California	Wood Lug Box	6½ × 15 × 18	31	28	32.4	29.2
Lamb, Boneless	Fiber Box	20 × 15 × 5	57	53	65.7	61.0
Lard (2/28 lb)	Wood Export Box	18 × 13¼ × 7¾	64	56	59.8	52.5
Lettuce, head	Fiber Carton	20½ × 13½ × 9½	37½	35	24.7	—
	Fiber Carton	20½ × 14¼ × 10½	45–55	42–52	26.9	25.2
	Pallet, 30 Cartons	42 × 50 × 66	1350	1170	16.8	14.6
Milk, Condensed	Barrels	35 × 25½ × 25½	670	600	50.9	45.6
Nuts						
Almonds, in Shell	Sacks	24 × 15 × 33	91½	90	13.3	13.1
Almonds, Shelled	Cases	6¾ × 23½ × 11	32	28	31.7	27.7
English Walnuts, in Shell	Sacks	25 × 11 × 31	103	100	20.9	20.3
English Walnuts, Shelled	Fiber Carton	14 × 14 × 10	27	25	23.8	22.0
Peanuts, Shelled	Burlap Bag	35 × 10 × 15	127	125	39.2	38.6
Pecans, in Shell	Burlap Bag	35 × 22 × 12	126½	125	23.7	23.4
Pecans, Shelled	Fiber Carton	13 × 13 × 11	32	30	29.8	27.9
Peaches	¾ Bushel Baskets	16⅞ top dia.	41	38	43.9	40.7
	½ Bushel Baskets	14½ top dia.	28	25	45.0	40.2
	Wirebound Crate	19 × 11¾ × 11⅛	42	38	29.2	26.4
	Wood Lug Box	18⅛ × 11½ × 5¾	26	23	38.0	33.1
Pears	Wood Box	8½ × 11½ × 18	52	48	51.0	47.1
Pears, place pack	Fiber Carton	18½ × 12 × 10	52	46	40.5	35.6
Pork						
Bundle Bellies	Bundles	23½ × 10½ × 7	57	57	57.0	57.0
Loins (Regular)	Wood Box	28 × 10 × 10	60	54	37.0	33.3
Loins (Boneless)	Fiber Box	20 × 15 × 5	57	52	65.7	59.9
Potatoes	Sack	33 × 17½ × 11	101	100	27.5	27.2
Poultry, Fresh (Eviscerated)						
Fryers, Whole, 24–30 to Pkg.	Wirebound Crate	24 × 10 × 7	65	60	27.5	25.4
Fryer Parts	Wirebound Crate	17¾ × 10 × 12½	54	50	42.1	38.9
Poultry, Frozen (Eviscerated)						
Ducks, 6 to Pkg.	Fiber Carton	22 × 16 × 4	32½	31	39.9	38.0
Fowl, 6 to Pkg.	Fiber Carton	20¾ × 18 × 5½	33½	31	28.2	26.1
Fryers, cut up, 12 to Pkg.	Fiber Carton	17¼ × 15¾ × 4¼	30½	28	45.4	41.7
Roasters, 8 to Pkg.	Fiber Carton	20¾ × 18 × 5½	32½	30	27.3	25.2
Turkeys,						
3–6 lb., 6 to Pkg.	Fiber Carton	21 × 17 × 6½	30	27	22.5	20.1
6–10 lb, 6 to Pkg.	Fiber Carton	26 × 21½ × 7	52½	48	23.3	21.2
10–13 lb, 4 to Pkg.	Fiber Carton	26½ × 16 × 7½	50	46	27.2	25.0
13–16 lb, 4 to Pkg.	Fiber Carton	29 × 18½ × 9	67½	62	24.2	22.2
16–20 lb, 2 to Pkg.	Fiber Carton	17 × 16 × 9	39	36	27.7	25.4
20–24 lb, 2 to Pkg.	Fiber Carton	19 × 16½ × 9½	47½	44	27.6	25.5
Tomatoes						
Florida	Fiber Carton	19 × 10⅞ × 10¾	43	40	33.3	31.0
	Wirebound Crate	18¾ × 11 15/16 × 11 15/16	64	60	41.3	38.7
California	Wood Lug Box	17½ × 14 × 7¾	34	30	30.9	27.3
Texas	Wood Lug Box	17½ × 14 × 6⅝	34	30	36.2	31.9
Veal (Boneless)	Fiber Carton	20 × 15 × 5	57	53	65.7	61.0

Figure 18–21 (Cont.)

Figure 18–22 Piling heights using pallets without auxiliary support. (By permission of *ASHRAE 1974 Applications.*)

Commodity	Units High in Pile	Pallets In Pile	Overall Height Ft
Eggs, wood cases	12	3	14½
Apples, boxes	12	3	13⅔
Grapes, lugs	21	3	14¼
Canned goods, cartons		3–4	15–18
Tierces, on end	4	3–4	13
Cans, 30 lb	12	3	14¼
Butter, 68 lb cubes	12	3	14¼

Commodity	Units High in Pile	Pallets in Pile	Overall Height Ft.
Butter, 68 lb cubes	12	6	15
Frozen fruits & vegetables, cartons		2–4	10–20
Frozen juice concentrate, cartons		4–5	18
Turkeys, boxes approx. 80 lb	21	3	18
Boned meat, fiber cartons, if flat		3–4	13–15

Cooler Temperature, F	Heat Equivalent Per Person, Btuh
50	720
40	840
30	950
20	1050
10	1200
0	1300
-10	1400

When people enter a refrigerated space for short durations, they carry with them a considerable amount of heat over and above that listed in Fig. 18–23. Some allowance must be made if traffic load of this type is heavy.

ASHRAE 1977 Fundamentals—Reprinted by permission

Figure 18-23 Heat equivalent of occupancy. (By permission of *ASHRAE 1977 Fundamentals.*)

Evaporator Fan Motor HP	Connected Load In Refrigerated Space Btu/HP/Hr.	Motor Losses Outside Refrig. Space Btu/HP/Hr.	Connected Load Outside Refrig. Space Btu/HP/Hr.
1/20 HP	6400	2545	
1/15 HP	5700	2545	
1/12 HP	5300	2545	
1/10 HP	4950	2545	
1/8 HP	4650	2545	
1/6 HP	4350	2545	
1/4 HP	4000	2545	1455
1/3 HP	3850	2545	1305
1/2 HP	3700	2545	1155
3/4 HP	3600	2545	1055
1 HP	3500	2545	955
2 HP	3300	2545	755
3	3200	2545	655
5 HP	3100	2545	555
7½ HP	3050	2545	505
10 to 20	3000	2545	455

*For motors rated in Watts (input) multiply Watts by 3.41 Btu/W/Hr.

At a given fan speed the fan will deliver approximately the same air quantity through a given coil at any air temperature.

The lower the air temperature the greater the weight of air handled by the fan at a given fan speed. The fan horsepower will vary directly with the air density or inversely with the absolute temperature of the air (°F + 460° or °C + 273°)

The increased heat dissipation of the motor at the lower temperatures offsets the greater HP requirement and increased amp draw.

For altitudes above and below sea level, the weight of air handled by the fan and the fan HP wll varry directly with the change in air density.

ASHRAE 1977 Fundamentals - Reprinted by permission

Figure 18-24 Heat equivalent of electric motors. (By permission of *ASHRAE 1977 Fundamentals.*)

Miscellaneous Loads

Miscellaneous loads consist of people who occupy the space and/or lights and motors in use in the space for extended periods of time (see Figures 18-23 and 18-24).

Safety Factor

It is recommended that a 10 percent safety factor be added to the total load calculation to allow for possible variations in the product load. Good judgment will temper this amount depending on the operating conditions. Evaporation fan and defrost loads are not included in the normal calculations and should be added when they apply.

SELECTING THE EQUIPMENT

The following information is needed to properly select the compressor or condensing unit and the evaporator:

1. Load (Btu/24 hr)
2. Running time for the compressor (hours/per day)
3. Load (Btu/hr)
4. Refrigerator box temperature (°F)
5. Refrigerator box relative humidity (%)
6. Refrigerant evaporating temperature (°F)
7. Temperature difference between box temperature and refrigerant evaporating temperature (°F)
8. Suction-line temperature rise (see Figure 18–25) due to pressure drop from evaporator to compressor (°F) (used only when long runs are encountered)

With the foregoing data and the manufacturers' data sheets on compressors (or condensing units) and evaporators, the proper selections can be made. The compressor selected must match the evaporator selected to give the design refrigerant evaporator temperature.

Equipment Selection Example

A refrigerated box is used to cool mixed vegetables using an R-12 refrigerating system. The system uses air defrost. A study of the information on the survey sheet and the load calculation sheet indicates the following essential requirements:

Load	10,474 Btuh
Box temperature	35 °F
Evaporating temperature	25 °F

A model AH15H compressor is selected from the manufacturer's data (Figure 18–26). Using 100 °F ambient temperature and 25 °F suction temperature, this machine has a capacity of 12,600 Btuh. A model WJ120 evaporator is selected from the manufacturer's data given in Figure 18–27. This unit has a capacity of 12,000 Btuh at 10 °F TD.

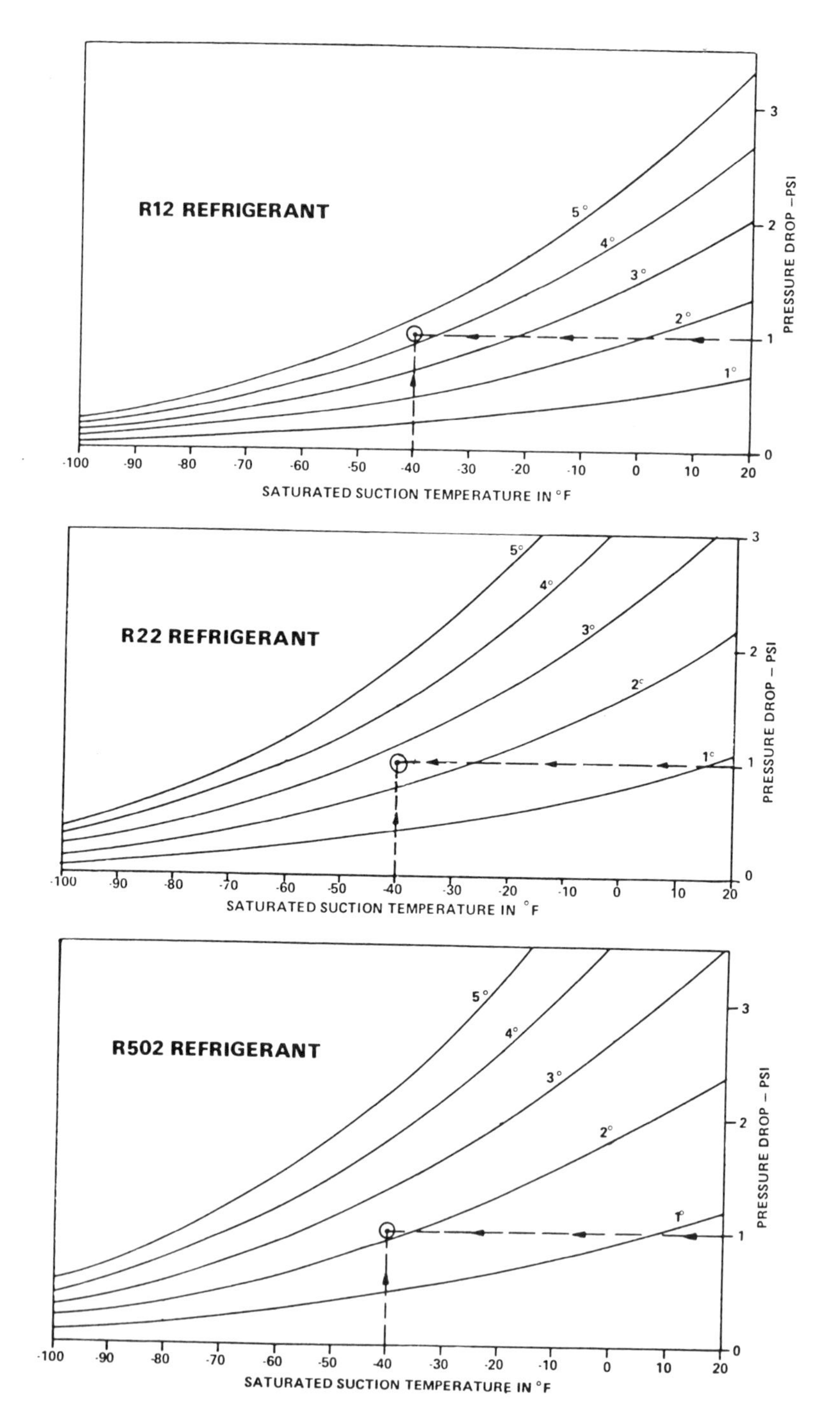

Figure 18–25 Temperature loss due to drop in suction lines, for R-12, R-22, and R-502. (By permission of Dunham-Bush, Inc.)

CAPACITY DATA R12

1750 RPM FOR 60 HERTZ AND 1450 RPM FOR 50 HERTZ
AIR COOLED ('AH'), AIR/WATER ('AWH') CONDENSING UNITS
CAPACITY IN BTUH

HP	60 HZ. AH/AWH	50 HZ. AH/AWH	Ambient Air Temp. °F	LOW TEMP.			COMMERCIAL TEMP.					HIGH TEMP.		
			SATURATED SUCTION TEMPERATURE °F	-40°F	-30°F	-20°F	-10°F	0°F	10°F	20°F	25°F	30°F	40°F	50°F
			SUCTION PRESSURE - PSIG OR *VACUUM INCHES OF MERCURY	10.9*	5.5*	0.6	4.5	9.2	14.64	21.04	24.61	28.45	36.97	46.70
				BTUH	BTUH	BTUH	BTUH	BTUH	BTUH	BTUH	BTUH	BTUH	BTUH	BTUH
½	Δ 5HCL		90 100	- -	1,370 1,250	1,900 1,750	2,600 2,350	3,350 3,050	4,150 3,800	5,000 4,650	5,400 5,100	6,000 5,800	6,900 6,700	7,800 7,600
¾	Δ 8HCL		90 100	- -	1,550 1,400	2,200 2,000	3,000 2,700	3,800 3,500	4,700 4,400	5,700 5,400	6,250 5,950	6,700 6,500	7,900 7,700	9,500 9,000
	Δ 9HCL		90 100	- -	1,950 1,750	2,750 2,500	3,700 3,400	4,700 4,350	5,800 5,400	7,050 6,500	7,800 7,200	8,400 7,700	9,700 8,900	11,100 10,200
1	Δ 10HCL		90 100	- -	2,250 2,050	3,200 2,850	4,350 3,900	5,550 5,100	6,900 6,450	8,300 7,850	9,150 8,700	9,800 9,400	11,400 11,000	13,000 12,600
	Δ 11HCL**		90 100	2,070 1,850	2,900 2,650	4,150 3,800	5,600 5,100	7,100 6,600	8,800 8,400	10,600 10,000	11,700 11,100	12,500 12,000	14,600 14,100	16,600 16,100
1½	15H	15H5	90 100	- -	- -	- -	- -	- -	11,800 9,700	12,200 11,600	13,400 12,600	14,600 13,800	16,800 15,900	18,900 18,000
	15C	15C5	90 100	- -	- -	- -	7,100 6,500	9,300 8,700	11,500 10,900	14,500 13,600	16,000 15,300	- -	- -	- -
	15L	15L5	90 100	2,800 2,500	4,200 3,700	5,800 5,400	7,600 7,000	9,600 8,900	- -	- -	- -	- -	- -	- -
2	20H	20H5	90 100	- -	- -	- -	- -	- -	14,500 12,000	16,500 14,500	17,500 15,500	19,000 17,000	22,000 20,000	26,000 23,000
	20C	20C5	90 100	- -	- -	- -	8,600 8,100	11,300 10,600	15,400 14,000	17,500 16,200	19,400 18,000	- -	- -	- -
	20L	22L5	90 100	3,600 3,400	5,500 5,000	7,500 7,000	9,800 9,100	12,200 11,400	- -	- -	- -	- -	- -	- -
3	30H	32H5	90 100	- -	- -	- -	- -	- -	18,000 17,000	23,000 21,500	25,500 24,000	28,000 26,000	32,500 30,500	38,000 35,400
	30C	30C5	90 100	- -	- -	- -	- -	18,600 17,300	23,000 21,500	28,700 27,000	32,000 30,000	- -	- -	- -
	32C		90 100	- -	- -	- -	12,500 11,500	16,200 15,000	19,900 18,500	25,400 24,000	28,200 26,400	- -	- -	- -
	30L	30L5	90 100	6,000 5,400	8,500 8,000	12,000 11,000	16,000 15,000	20,500 19,000	- -	- -	- -	- -	- -	- -
		41PL5	90 100	6,560 6,070	9,700 9,000	13,900 12,400	18,900 17,500	24,100 22,500	- -	- -	- -	- -	- -	- -
5	50H	50H5	90 100	- -	- -	- -	- -	- -	30,000 29,300	37,500 36,000	41,000 39,000	45,000 42,700	52,500 50,000	61,500 58,500
	50C	51PC5	90 100	- -	- -	- -	21,300 18,600	27,400 24,600	33,500 30,600	41,800 38,300	46,300 43,000	- -	- -	- -
	51PL		90 100	8,000 7,400	11,800 11,000	17,000 15,100	23,000 21,300	29,400 27,400	- -	- -	- -	- -	- -	- -
		51PH5	90 100	- -	- -	- -	- -	- -	37,200 36,300	46,500 44,600	50,800 48,400	55,800 53,000	65,100 62,000	76,300 72,500
		62PC5	90 100	- -	- -	- -	26,200 22,900	33,700 30,200	41,200 37,600	51,400 47,100	57,000 52,900	- -	- -	- -
		61PL5	90 100	9,800 9,100	14,500 13,500	20,900 18,600	28,300 26,200	36,200 33,700	- -	- -	- -	- -	- -	- -
7½	76PH/C		90 100	- -	- -	- -	26,500 23,900	34,000 32,400	41,500 38,000	54,000 50,500	60,000 56,000	67,000 62,500	84,500 78,500	105,000 96,000
	77PC	77PC5	90 100	- -	- -	- -	32,000 30,000	41,000 39,000	54,000 50,000	67,000 63,500	75,000 70,000	- -	- -	- -
	76PL	75PL5	90 100	10,700 9,700	16,600 15,500	25,400 23,300	35,400 32,600	47,100 43,500	- -	- -	- -	- -	- -	- -
		77PH5	90 100	- -	- -	- -	- -	- -	50,400 46,700	66,400 62,100	73,800 68,900	82,400 76,900	104,000 96,600	129,000 118,000
10	101PH	101PH5	90 100	- -	- -	- -	- -	- -	68,900 64,600	90,600 84,700	100,700 93,900	112,300 104,800	141,700 131,700	176,100 161,000
	101PC	101PC5	90 100	- -	- -	- -	40,000 37,500	51,300 48,800	67,500 62,500	83,800 79,400	93,800 87,500	- -	- -	- -
	104PL		90 100	12,500 11,750	20,000 18,800	30,000 28,200	46,000 43,300	61,000 57,400	- -	- -	- -	- -	- -	- -
15	D154PH		90 100	- -	- -	- -	- -	- -	74,800 70,600	101,200 92,700	106,400 99,300	119,000 111,000	147,600 137,000	182,000 166,200
	154PC†		90 100	- -	- -	- -	42,700 39,400	56,800 52,500	71,200 67,200	95,900 89,600	- -	- -	- -	- -

NOTES: † This unit will not operate at suction temperatures greater than 20°F.

Δ AH units only.

* Inches of mercury below one atmosphere.

** Not available for +30°F through +50°F SST 208V/1Φ applications.

Figure 18-26 Capacity data for R-12 condensing units. (By permission of Dunham-Bush, Inc.)

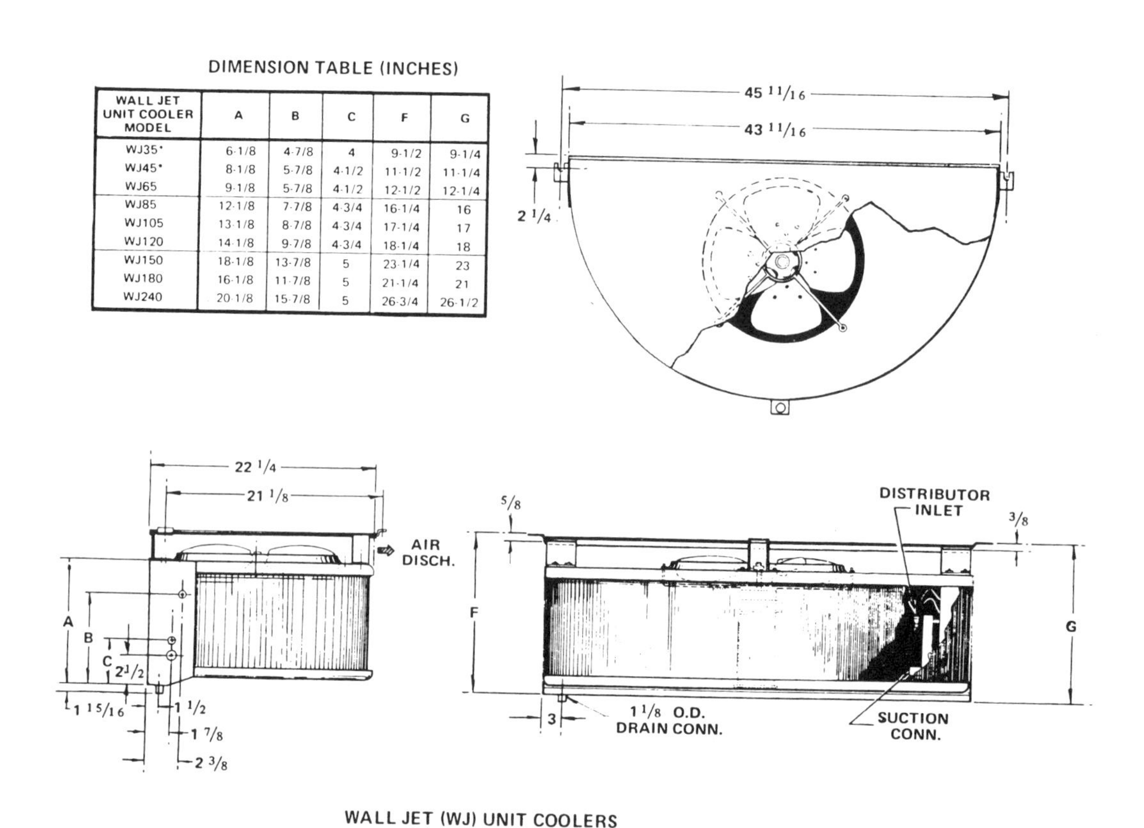
DIMENSION TABLE (INCHES)

WALL JET UNIT COOLER MODEL	A	B	C	F	G
WJ35*	6-1/8	4-7/8	4	9-1/2	9-1/4
WJ45*	8-1/8	5-7/8	4-1/2	11-1/2	11-1/4
WJ65	9-1/8	5-7/8	4-1/2	12-1/2	12-1/4
WJ85	12-1/8	7-7/8	4-3/4	16-1/4	16
WJ105	13-1/8	8-7/8	4-3/4	17-1/4	17
WJ120	14-1/8	9-7/8	4-3/4	18-1/4	18
WJ150	18-1/8	13-7/8	5	23-1/4	23
WJ180	16-1/8	11-7/8	5	21-1/4	21
WJ240	20-1/8	15-7/8	5	26-3/4	26-1/2

WALL JET (WJ) UNIT COOLERS
SERIES "C"

WALL JET UNIT COOLER MODEL	UNIT CAPACITY BTU/HR @ 10°F TD	UNIT CAPACITY BTU/HR @ 15°F TD	FAN CFM	FAN SIZE INCH.	MOTOR HP 120/1/60†	TOTAL FAN MTR. AMPS 120V	OPTIONAL HEAT EXCHANGER RECOMMENDED	CONNECTIONS** COIL INLET	CONNECTIONS** SUCTION O.D.	CONNECTIONS** DRAIN O.D.	APPROX. SHIPPING WEIGHT LBS.
WJ35*	3,500	5,250	625	10	1/20	.82	B-25XS	1/2 FL.	1/2	1-1/8	40
WJ45*	4,500	6,750	725	10	1/20	.82	B-25XS	1/2 FL.	1/2	1-1/8	46
WJ65	6,500	9,750	1050	12	1/10	1.70	B-50XS	1/2 FL.	1/2	1-1/8	52
WJ85	8,500	12,750	1300	12	1/10	1.70	B-75XS	1/2 FL.	3/4	1-1/8	60
WJ105	10,500	15,750	1600	14	1/6	1.70	B-75XS	1/2 FL.	3/4	1-1/8	66
WJ120	12,000	18,000	1800	14	1/6	1.70	B-75XS	1/2 FL.	3/4	1-1/8	72
WJ150	15,000	22,500	2375	16	1/6	1.70	B-120XS	1/2 FL.	7/8	1-1/8	82
WJ180	18,000	27,000	2900	16	1/6	1.70	B-120XS	1/2 FL.	7/8	1-1/8	88
WJ240	24,000	36,000	4000	16	1/2	3.00	B-200XS	1/2 FL.	1-1/8	1-1/8	108

NOTES: * ALL COILS EXCEPT WJ35 AND WJ45 ARE EQUIPPED WITH DISTRIBUTORS AND EQUALIZING TUBES.
** ALL REFRIGERANT CONNECTIONS ARE MADE FROM THE LEFT HAND SIDE WHEN FACING INTO THE AIR DISCHARGE.
† OPTIONAL 208/240/1/60 FAN MOTORS AVAILABLE ON SPECIAL ORDER. STOCK UNITS ARE 120/1/60.
AIR THROW IS 15 TO 25 FT. DEPENDING UPON THE INSTALLATION HEIGHT, TEMPERATURE DIFFERENCE AND UNIT MODEL

Figure 18–27 Capacity data for wall-jet unit coolers. (By permission of Dunham-Bush, Inc.)

Checking the Balancing-Out Point

The performance of the compressor and evaporator can be plotted as shown in Figure 18–28 for the selections made to determine the balancing-out point. Note that the intersection of the two curves shows the combined performance to produce 12,600 Btuh at 24.5 °F refrigerant evaporating temperature.

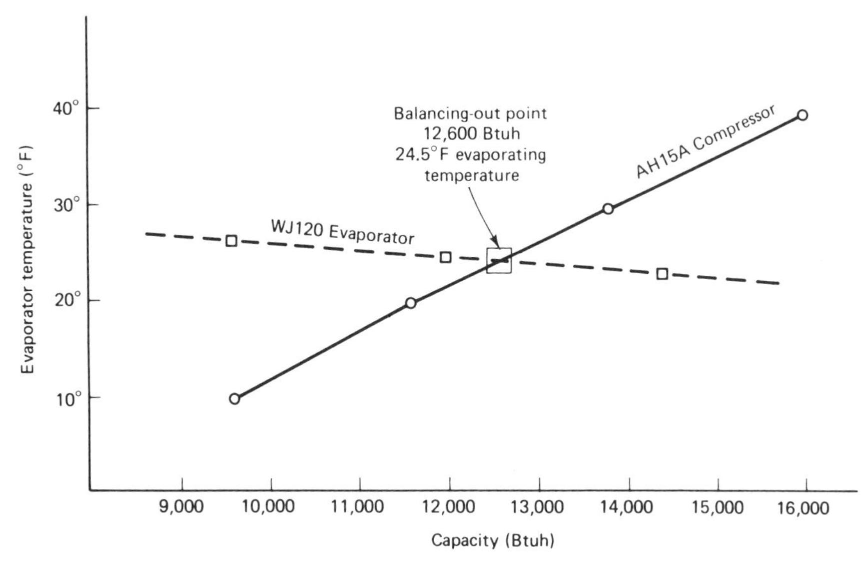

Figure 18–28 Balancing-out point of evaporator and compressor selected for the sample calculation (above freezing).

An alternative selection can be studied as shown in Figure 18–29. A commercial temperature compressor is selected and the balancing-out point determined for three coil selections. Note that sufficient capacity may be produced with smaller coils but the problem here is low humidity due to a high TD.

In determining the balancing-out point between coil and compressor, the information given in Figure 18–25 for the temperature drop in the suction line may be useful where long runs are encountered.

Figure 18–29 Intersection of the performance curves of three evaporators with one compressor.

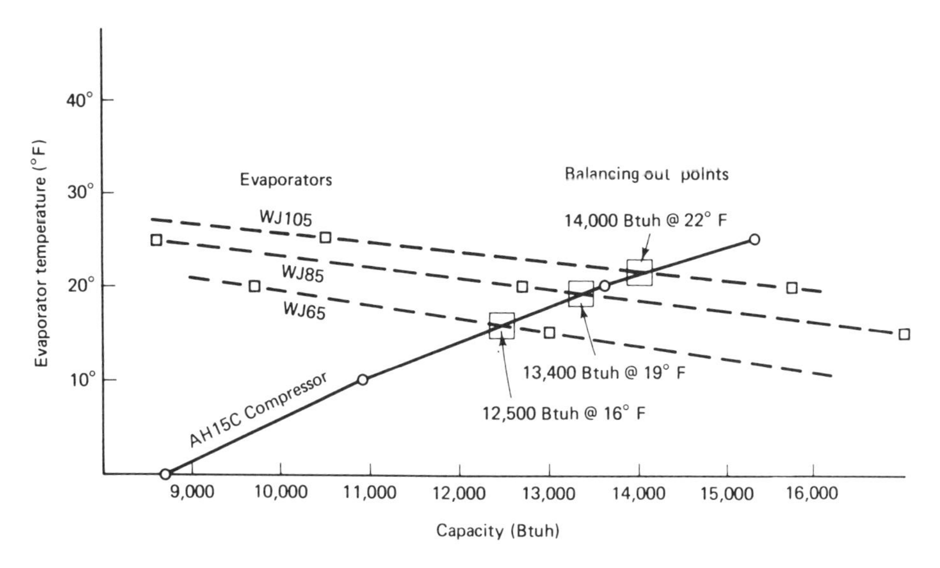

REVIEW EXERCISE

Using the formulas provided in this chapter, calculate the load and select the equipment for a refrigeration system with the following specifications. A walk-in cooler, with dimensions 10 ft wide × 12 ft long × 9 ft high (outside), located in Atlanta, Georgia, is used to cool mixed vegetables to 40 °F. The wall, floor, and ceiling of the room have 3 in. of molded polystyrene insulation. The room holds 3800 lb of the product and 1000 lb is delivered to the room daily, at a precooled temperature of 50 °F, in 25 paper cartons. Unloaded, each paper carton weighs 4 lb.

19 Using Tables and Charts

OBJECTIVES

After studying this chapter, the reader will be able to:

- Utilize the performance information contained in a refrigerant table.
- Calculate the operating characteristics of a refrigeration cycle.
- Predict the performance of a refrigeration system using a pressure-enthalpy diagram.

READING THE REFRIGERANT TABLE

Refrigerant tables are useful in determining the performance of the refrigeration cycle. A copy of the table for R-12, showing the properties of liquid and saturated vapor is shown in Figure 1–5. An example of the use of this chart in determining the effect of adding heat to the refrigerant follows.

Note that Figure 1–5 gives the conditions of the refrigerant from − 150 to + 233 °F. Reading from left to right, for 0 °F refrigerant:

Absolute pressure	= 23.849 psia
Gauge pressure	= 9.153 psig
Volume of the vapor	= 1.6089 ft^3/lb
Density of the liquid	= 90.659 lb/ft^3
Enthalpy (total heat) of liquid	= 8.5207 Btu/lb
Enthalpy of the vapor	= 77.271 Btu/lb
Entropy of the liquid	= 0.019323 Btu/lb/ °R
Entropy of the vapor	= 0.16888 Btu/lb/ °R

Entropy is much like specific heat. It is the amount of heat required to raise 1 pound of the refrigerant 1 degree Rankine. Rankine temperatures are absolute temperatures and start from absolute zero. Thus, 0 °R is equal to −460 °F.

Adding Heat to the Refrigerant

Referring to Figure 1–7, there are two parts of the refrigeration cycle where heat is added: in the evaporator and through the compressor. Referring to the data in Figure 1–5, the condition of the refrigerant before and after passing through the compressor can be observed. Assume that the evaporating temperature of the refrigerant is 0 °F and the condensing temperature is 100 °F. From the table it can be determined that under these saturated conditions the low-side pressure of the system is 23.85 psia (9.15 psig) and the high-side pressure is 131.9 psia (117.2 psig). Other characteristics are shown in the following tabulation:

System characteristics	Before entering the compressor	After entering the compressor
Pressure	9.15 psig	117.2 psig
Temperature	80 °F	200.4 °F
Volume	1.948 ft^3/lb[a]	0.4022 ft^3/lb[a]
Heat content	89.000 Btu/lb[a]	104.804 Btu/lb[a]

[a] Data from "Thermodynamic Properties of Refrigerants," *ASHRAE Handbook.*

The heat of compression is, therefore, 104.804 Btu/lb minus 89.000 Btu/lb, or 15.804 Btu/lb.

Note that in this example it is assumed that the suction gas is superheated 80 °F and the discharge gas is superheated 100 °F. *These tabulations show that the compressor adds 15.804 Btu/lb to the refrigerant.* This is due to the fact that work is done on the refrigerant and this mechanical work is changed into heat. Energy is never destroyed, but its form can be changed.

The changes that take place in the refrigerant in passing through the evaporator can also be observed. For example, the amount of heat that needs to be removed from the liquid refrigerant in passing through the metering device is as follows:

Enthalpy (total heat) of liquid at 100 °F	= 31.100 Btu/lb	
Enthalpy of liquid at 0 °F	= 8.521 Btu/lb	
Heat to be removed from liquid:	= 22.579 Btu/lb	(A)
Enthalpy of saturated vapor at 0 °F	= 77.271 But/lb	
Enthalpy of liquid at 0 °F	= 8.521 Btu/lb	
Latent heat of vaporization at 0 °F	= 68.750 Btu/lb	(B)

The net refrigerating effect is the difference between the heat added by vaporization of the refrigerant (B) and the heat removed from the liquid (A):

Net refrigerating effect = B - A
= 68.750 Btu/lb - 22.579 Btu/lb
= 46.171 Btu/lb

Calculating from the Tables and Charts

Many other calculations can be made by using data from these tables. Using the data given above and applying them to 1 ton of refrigeration, the following characteristics of the refrigeration system can be determined:

1. Total refrigerating effect
2. Pounds of refrigerant circulated
3. Heat of compression
4. Horsepower of compression
5. Compressor displacement
6. Compression ratio

Note that the tables are all based on processing 1 lb of refrigerant. In order for the information to be more useful, it needs to be converted to a *rate* of refrigeration. The ton of refrigeration expresses a rate: a ton of refrigeration is the quantity of cooling or heating required to remove 12,000 Btuh or 200 Btu/min.

From the tables the necessary figures can be derived to determine that quantity of refrigeration required to produce 1 ton of refrigeration under the conditions specified. This may be done through the following calculations:

1. Determine the *total refrigerating effect* due to:
   a. Evaporation (previously calculated above: B - A)
   b. Superheat in the evaporator.

   Calculate as follows, allowing 20°F superheat in the evaporator:*

Enthalpy of R-12 superheated to 20°F, 9.15 psig	=	80.161 Btu/lb	
Enthalpy of R–12 saturated at 0°F, 9.15 psig	=	77.271 Btu/lb	
Heat added by superheat in the evaporator	=	2.89 Btu/lb	(C)
Net refrigerating effect due to evaporation	=	46.17 Btu/lb	(B–A)
Total refrigerating effect, (B–A) + C	=	49.06 Btu/lb	(D)

*Data from "Thermodynamic Properties of Refrigerants," *ASHRAE Handbook.*

2. Determine the amount of *refrigerant circulated:*

$$\frac{\text{Ton of refrigeration}}{\text{Refrigerating effect (D)}} = \text{refrigerant circulated} \quad \text{(E)}$$

$$\frac{200 \text{ Btu/min}}{49.06 \text{ Btu/lb} \quad \text{(D)}} = 4.077 \text{ refrigerant, lb/min.} \quad \text{(E)}$$

3. Determine the *heat of compression* per ton of refrigeration (refer to page 430 for heat of compression per lb:

   Heat of compression per lb × lb of refrigerant circulated =
   = 15.804 Btu/lb × 4.077 lb/min (E) = 64.433 Btu/min

4. Determine the *horsepower per ton of refrigeration* (given that the heat equivalent of 1 hp = 42.42 Btu/min):

$$\frac{\text{Heat of compression}}{\text{Heat equivalent of 1 hp}}$$

$$= \frac{64.433 \text{ Btu/min} \quad (F)}{42.42 \text{ Btu/min}} = 1.519 \text{ hp} \quad (G)$$

5. Determine the *compressor displacement* (refer to the table on page 430 for the entering volume of the refrigerant):

$$\frac{\text{Total refrigerating effect}}{\text{Entering volume of refrigerant per pound}}$$

$$= \frac{49.06 \text{ Btu/lb} \quad (D)}{1.94 \text{ ft}^3\text{/lb}} = 25.3 \text{ Btu/ft}^3 (H)$$

   To compute compressor displacement per ton:

$$\frac{200 \text{ Btu/min}}{25.3 \text{ Btu/ft}^3 \quad (H)} = 7.91 \text{ ft}^3\text{/min} \quad (J)$$

6. Determine the *compression ratio.* The compression ratio is a very useful calculation since it determines the "stress" placed on the compressor in performing its function in the system. Each compressor has its limits insofar as the compression ratio is concerned. If these limits are exceeded, the compressor cannot do the job required. The compression ratio is calculated by dividing the absolute pressure on the high side of the system by the absolute pressure on the low side of the system. Thus, in the system being studied here, using system characteristics given in the table on page 430:

$$\frac{131.9 \text{ psia}}{23.85 \text{ psia}} = 5.5 \text{ (ratio)}$$

   It is usually desirable to have the compression ratio as low as possible to obtain the greatest efficiency from the compressor. The 5.5 ratio in the example given would be considered average, whereas a ratio of 8 would be high and a ratio approaching zero would be unacceptably low. The compression ratio affects the volumetric efficiency of the machine, which will be discussed further in the next section.

VOLUMETRIC EFFICIENCY

In a reciprocating compressor when the piston rises to its full height in the cylinder (Figure 19-1) and the vapor is compressed, there is a small volume between the top of the piston and the top of the cylinder where the piston does not travel. This space is

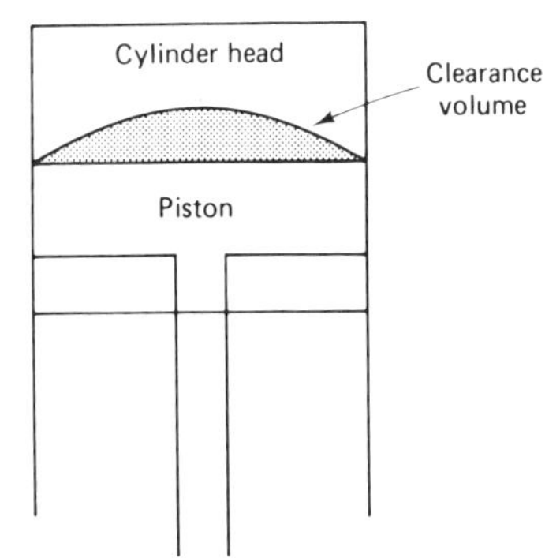

Figure 19-1 Clearance volume of piston cylinder head. The space at the top of the compression cylinder is not included in the travel of the piston.

necessary from a mechanical standpoint to allow space for the valve and to prevent the piston from touching the top of the cylinder during its stroke. The volume of vapor in this space is not discharged from the cylinder during the compressor stroke and represents an inefficiency as far as the performance of the compressor is concerned. This characteristic relates to the efficiency of the compressor and causes a loss of power in the compressor. The ratio of the loss of power to the usable power is called *volumetric efficiency*. This quantity is usually found experimentally; it is not a constant but varies with the discharge pressure, the suction pressure, the type of refrigerant used, and the valve action. The volumetric efficiency can vary from 65 to 95%. Probably the most important influencing factor is the compression ratio. The higher the compression ratio, the lower the volumetric efficiency. This factor helps to explain the reason for lower capacities at high head pressures and low suction pressures. In selecting or operating compressor equipment it is, therefore, important to keep the compression ratio within reasonable limits so that the effects of low volumetric efficiency can be avoided.

OPERATING CONDITIONS THAT INCREASE THE CAPACITY OF THE SYSTEM

Assuming that there is no change in the mechanical equipment, what can be done to increase the capacity of a system? Every operating engineer and owner is interested in securing the most refrigerating effect with the least amount of power expended. From the study of the refrigeration tables as they related to the cycle of operation, a number of things can be done to increase the capacity within the limits of the facilities available. The following are some of the things that will produce results:

1. *Increase the low-side pressure.* This means increasing the evaporating temperature. Other functions remaining the same, this reduces the compression ratio and increases the capacity of the compressor.
2. *Reduce the high-side pressure.* This means reducing the condensing temperature. Other functions remaining the same, if the compression ratio is reduced, the compressor is more efficient.
3. *Lower the temperature of the vapor entering the compressor.* If saturated vapor

rather than superheated vapor can enter the compressor without the danger of liquid "slugging," the capacity of the compressor will be increased. This is due to the reduction in cubic feet of vapor per pound entering the compressor. Thus, more pounds of refrigerant are pumped by the compressor.

When one change is made in the operating conditions of a system, other conditions may be altered to an extent that makes the change impractical. For example, increasing the capacity of the compressor may create the need for a larger compressor motor. Thus, each change that is made in the operating characteristics of the system must be viewed from the standpoint of total effect.

Plotting System Performance on a Pressure–Enthalpy Diagram

The pressure–enthalpy (P-E) diagram (Figure 19–2) shows the relationships of six different properties of a refrigerant. These properties are:

1. Temperature, normal degrees Fahrenheit
2. Pressure, psi absolute (psia)
3. Volume, cubic feet per pound
4. Enthalpy, Btu per pound
5. Entropy, Btu per pound per degrees Rankine
6. Quality, percent vapor by weight

Absolute pressure is shown on the horizontal lines of the chart. Enthalpy is shown on the vertical lines. Constant-temperature lines are shown on vertical lines extending upward to the left. Constant-volume lines go horizontally upward to the right. Constant-entropy lines go diagonally upward to the right. Constant-quality lines are inside the mixture area and go upward to the right.

The "inverted U"-shaped area represents the mixture of liquid and vapor refrigerant. The area to the left represents subcooled liquid. The area to the right represents superheated vapor. The critical temperature of the refrigerant, 233.6 °F for R-12, is shown in the center of the inverted U. The critical temperature is the highest temperature at which liquid can exist. Above this temperature no amount of pressure can liquefy the refrigerant. Calculations can be made from the numbers found in the chart. However, for most accurate results, the tables are used.

Figure 19–3 shows a plot on the P-E diagram of a typical refrigeration cycle having an evaporating temperature of 0 °F and a condensing temperature of 120 °F. For the sake of simplicity, no subcooling of the liquid or superheating of the suction gas is shown.

Starting at the metering device (A), the action in the metering device is shown by the vertical line *AB*. The action in the evaporator is shown by line *BC*. The action in the compressor is shown by line *CD*. The action in the condenser is shown by line *DA*.

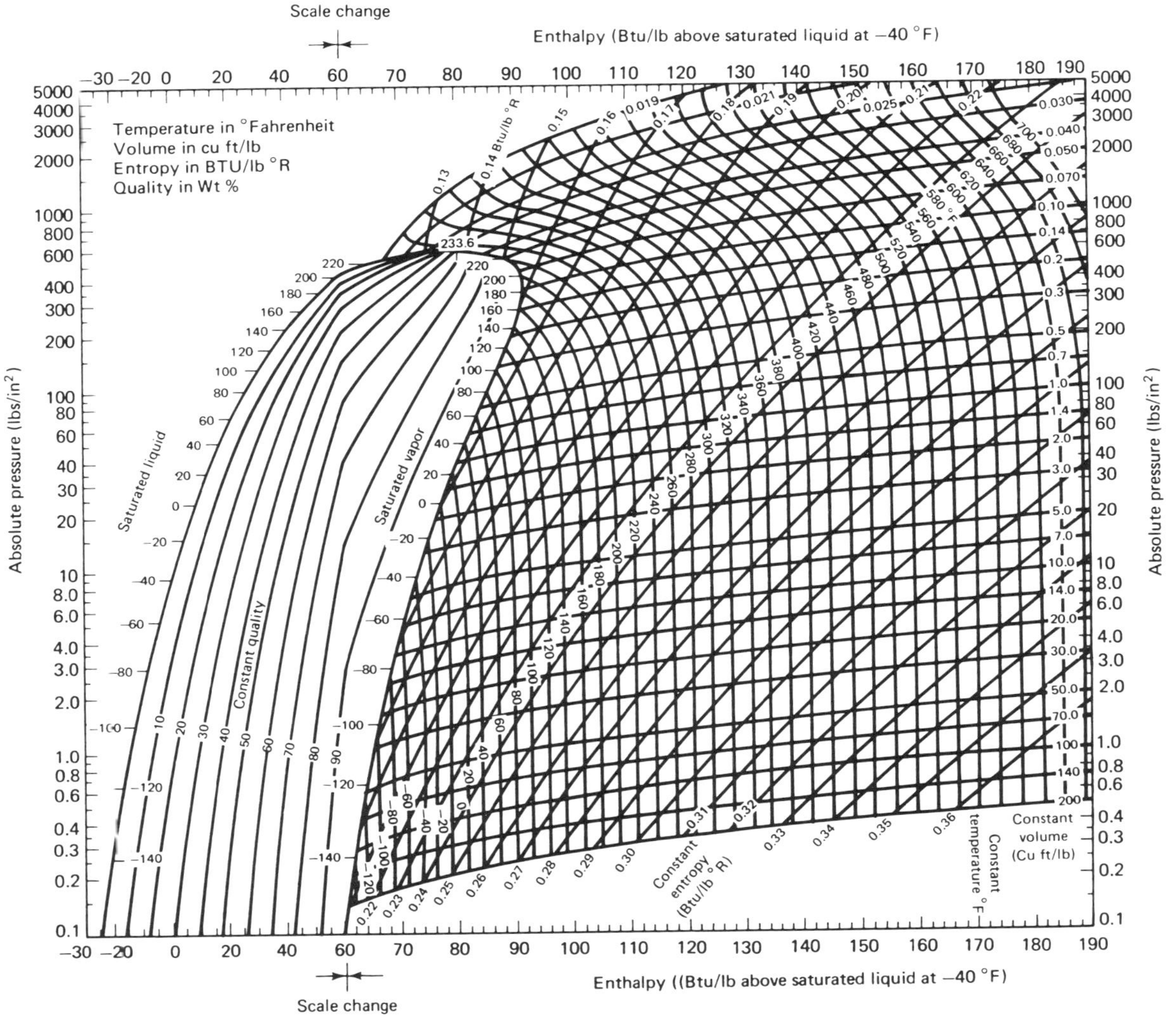

Figure 19–2 Pressure–enthalpy diagram for R-12. The diagram shows change in heat content of the refrigerant for various conditions of the refrigerant. (By permission of E.I. du Pont de Nemours & Company, Inc.)

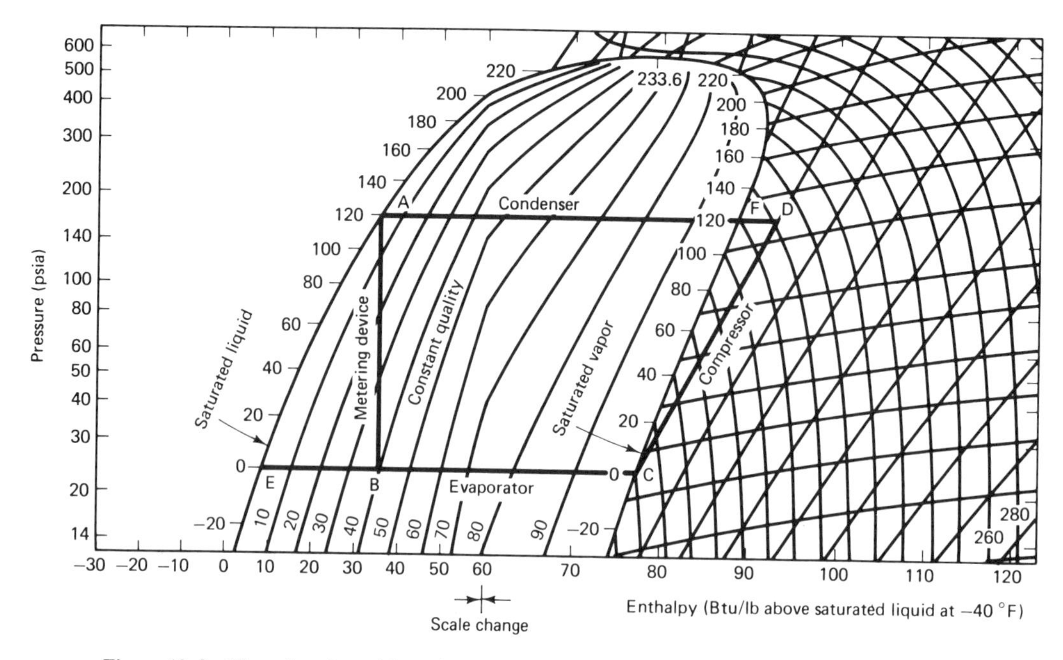

Figure 19-3 Plot of entire refrigeration cycle with 120°F condensing temperature and 0°F evaporating temperature. No superheat or subcooling. (By permission of E.I. du Pont de Nemours & Company, Inc.)

Reading the P-E Chart

For an example of reading the chart, refer to Figure 19-3:

Point A:

Condensing temperature	120°F
Condensing pressure	172.3 psia
Total heat	36.0 Btu/lb
Condition of refrigerant	100% liquid

Point B:

Evaporating temperature	0°F
Evaporating pressure	23.3 psia
Total heat	36.0 Btu/lb
Condition of refrigerant	40% vapor

Point C:

Evaporating temperature	0°F
Evaporating pressure	23.8 psia
Total heat	77.3 Btu/lb
Condition of refrigerant	100% vapor
Volume of vapor	1.6 ft³/lb
Suction vapor temperature	0°F
Entropy	0.16 Btu/lb/°R

Point D:

Condensing temperature	120 °F
Condensing pressure	172.3 psia
Total heat	92.5 Btu/lb
Condition of refrigerant	Vapor superheated 20 °F
Volume of vapor	0.28 ft^3/lb
Discharge vapor temperature	140 °F
Entropy	0.18 Btu/lb/°R

Point E:

Evaporating temperature	0 °F
Evaporating pressure	172.3 psia
Total heat	8.5 Btu/lb
Condition of refrigerant	100% liquid

Point F:

Condensing temperature	120 °F
Condensing pressure	172.3 psia
Total heat	88.6 Btu/lb
Condition of refrigerant	100% vapor

Latent heat of vaporization. The latent heat of R-12 at 0 °F evaporating temperature is shown in Figure 19-4. Reading this value on the entropy scale, latent heat

Figure 19-4 Pressure-enthalpy diagram for R-12, Heat of vaporization at 0 °F temperature. (By permission of E.I. du Pont de Nemours & Company, Inc.)

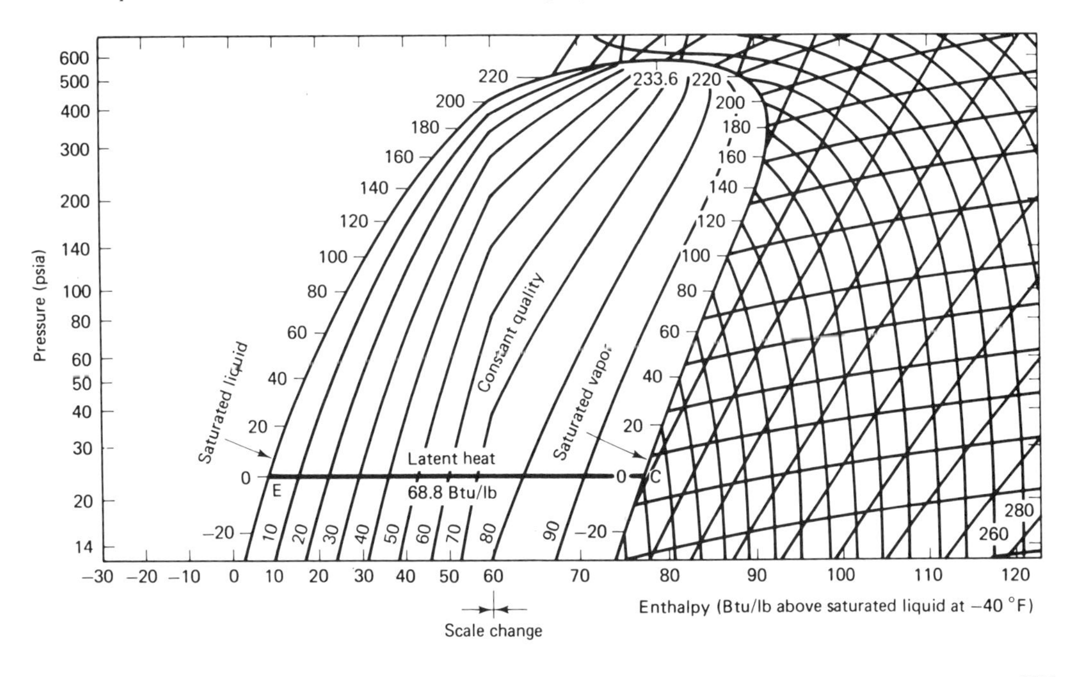

is 68.8 Btu/lb (77.3 Btu/lb – 8.5 Btu/lb (E–C)). This is the quantity of heat required to vaporize 1 lb of R-12 at 0 °F evaporating temperatures.

Note that as the evaporating temperature rises, the latent heat of vaporization decreases up to the critical point (233.6 °F) at which no amount of pressure can condense the refrigerant.

Net refrigerating effect. Referring to Figure 19–5, note that on the vertical line passing from *A* to *B* there is no change in the total heat content of the refrigerant in passing through the metering device. However, a sizable portion of the refrigerant has been evaporated to cool the refrigerant from 120 °F to 0 °F. This amount of heat is 27.5 Btu/lb (36.0 Btu/lb – 8.5 Btu/lb (B-E)). At point *B* where the refrigerant enters the evaporator, about 40 percent of the refrigerant has been evaporated. This vapor is called *flash gas.*

The amount of heat picked up in the evaporator by the vaporization of the remaining refrigerant is 41.3 Btu/lb (77.3 Btu/lb – 36.0 Btu/lb (B-C)). This is the net refrigerating effect.

Compressor heat. The action of the compressor is shown in Figure 19–6. The suction gas enters the compressors at 0 °F and is compressed and superheated to 140 °F. The amount of superheat added by the compressor is 15.2 Btu/lb (92.5 Btu/lb – 77.3 Btu/lb (D-C)). The increase in pressure is from 23.8 psia (9.2 psig) to 172.3 psia (157.6 psig). In practice the condensing temperature may be somewhat higher due to the design of the compressor, the friction in the machine producing heat, and the ability of the compressor to operate along constant-entropy lines.

Figure 19–5 Pressure–enthalpy diagram for R-12, showing net refrigerating effect at conditions indicated. (By permission of E.I. du Pont de Nemours & Company, Inc.)

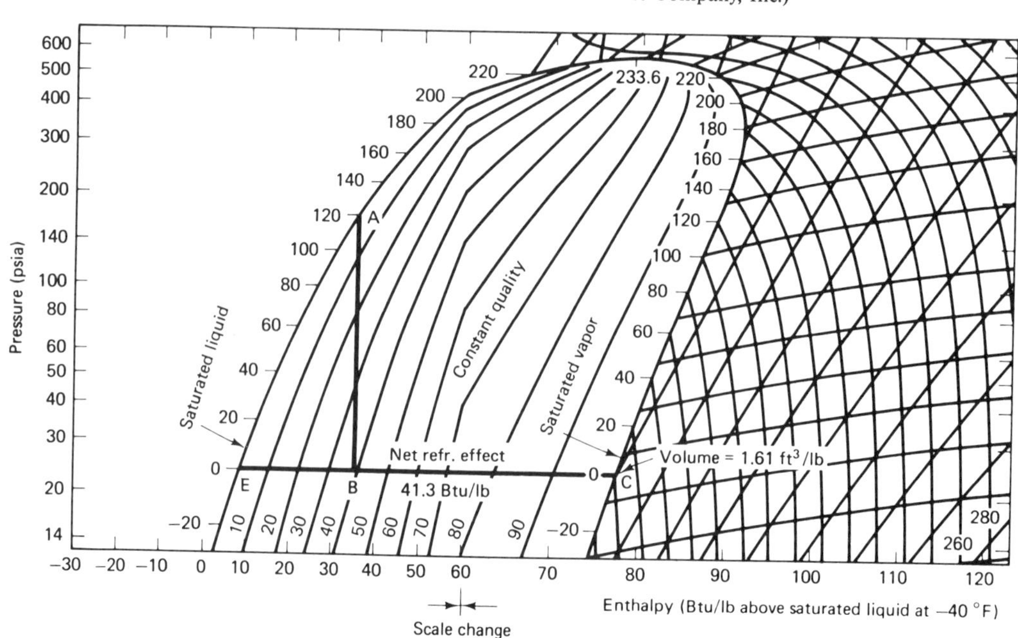

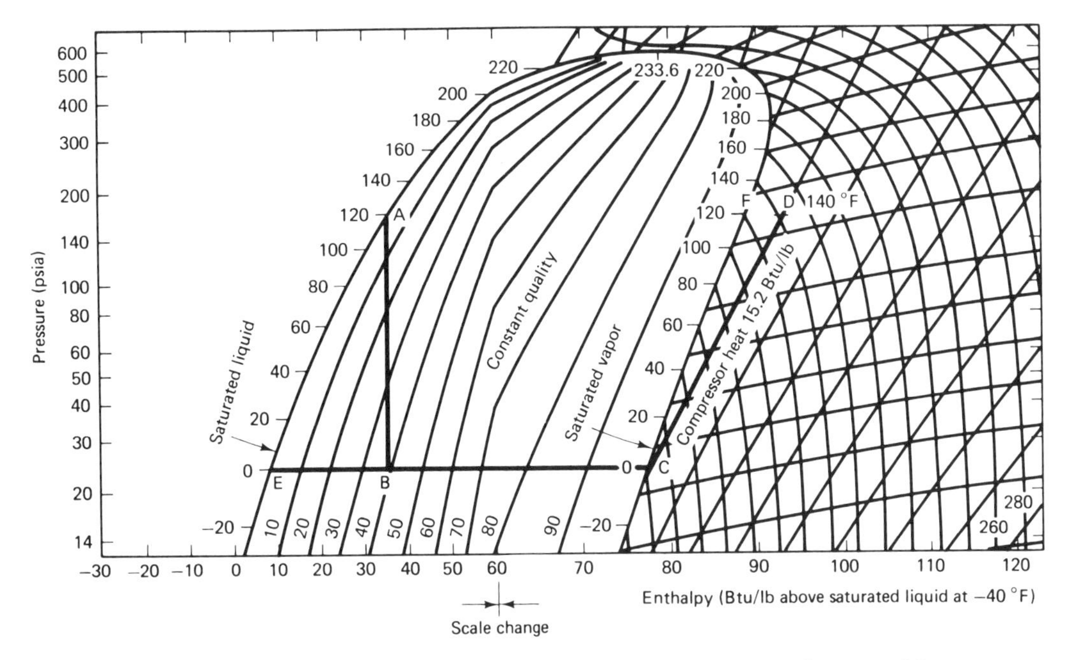

Figure 19-6 Pressure-enthalpy diagram for R-12, showing heat of compression at conditions indicated. (By permission of E.I. du Pont de Nemours & Company, Inc.)

Heat removed by the condenser. The heat removed by the condenser is shown in Figure 19–7. Note that before the condensing action can take place, 3.9 Btu/lb (92.5 Btu/lb - 88.6 Btu/lb (D-F)) of heat must be removed in the form of superheat. After the superheat is removed, the condenser must absorb 52.6 Btu/lb (88.6 Btu/lb - 36.0 Btu/lb (F-A) of heat to condense the refrigerant. The action of the condenser is shown in the diagram from point *D* to point *A*.

The effect of reducing the condensing temperature. The effect of reducing the condensing temperature is shown by the dashed lines in Figure 19–8. This figure shows the condensing temperature reduced to 100 °F. Note that under these conditions the condenser must remove more heat, 59.2 Btu/lb (55.9 + 3.3 Btu/lb). However, the net refrigerating effect is increased to 46.2 Btu/lb. Less flash gas enters the evaporator from the metering device. The system operates more efficiently since the compressor adds a reduced amount of heat, 13.0 Btu/lb.

Effect of subcooling the liquid. Figure 19–9 shows the effect of subcooling the liquid refrigerant. This requires additional surface in the condenser to cool the liquid refrigerant after it has been condensed, to a temperature below the condensing temperature. Ten degrees of subcooling is considered good minimum practice.

Figure 19–9 shows subcooling of 20 °F (from 120 °F to 100 °F). The amount of heat removed by subcooling is 4.9 Btu/lb. The net refrigerating effect is increased to 46.2 Btu/lb. The amount of flash gas entering the evaporator has been reduced to 32.8 percent.

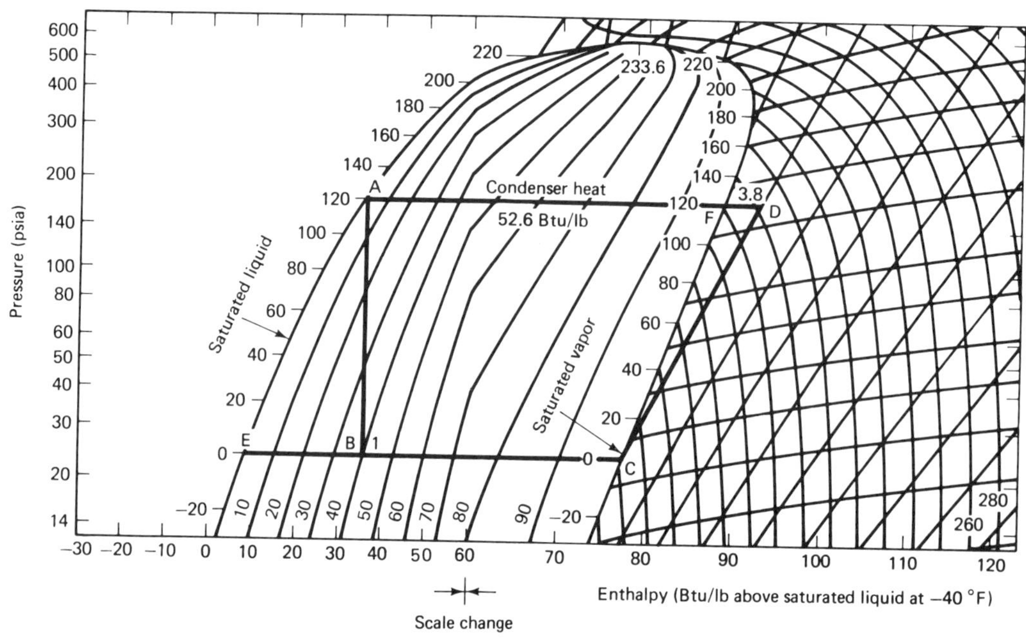

Figure 19-7 Pressure-enthalpy diagram for R-12, showing heat dissipated by the condenser at conditions indicated. (By permission of E.I. du Pont de Nemours & Company, Inc.)

Figure 19-8 Pressure-enthalpy diagram for R-12, showing the effect of lowering condensing temperature. (By permission of E.I. du Pont de Nemours & Company, Inc.)

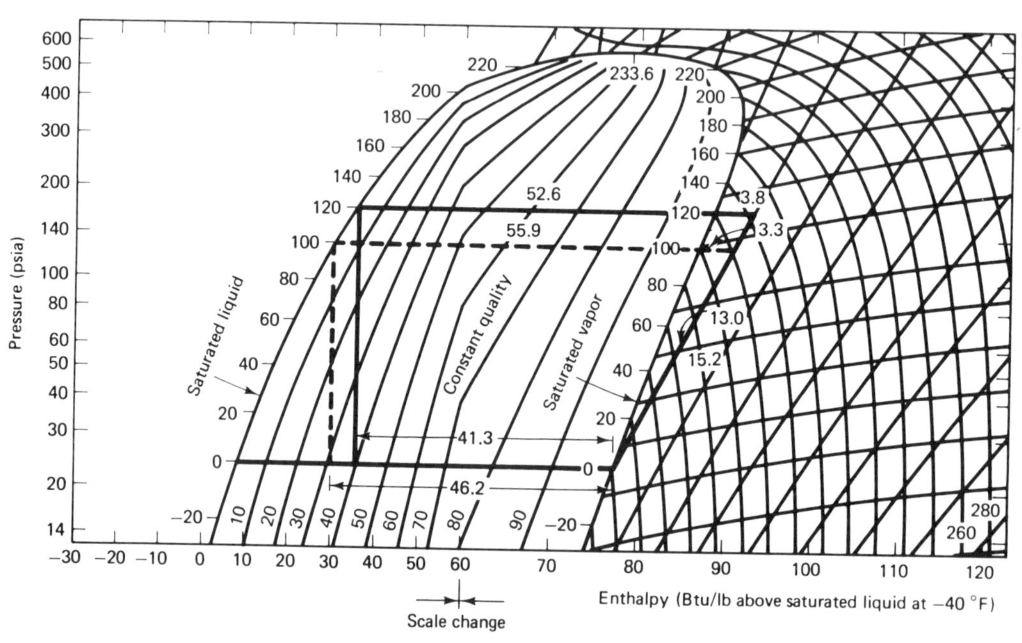

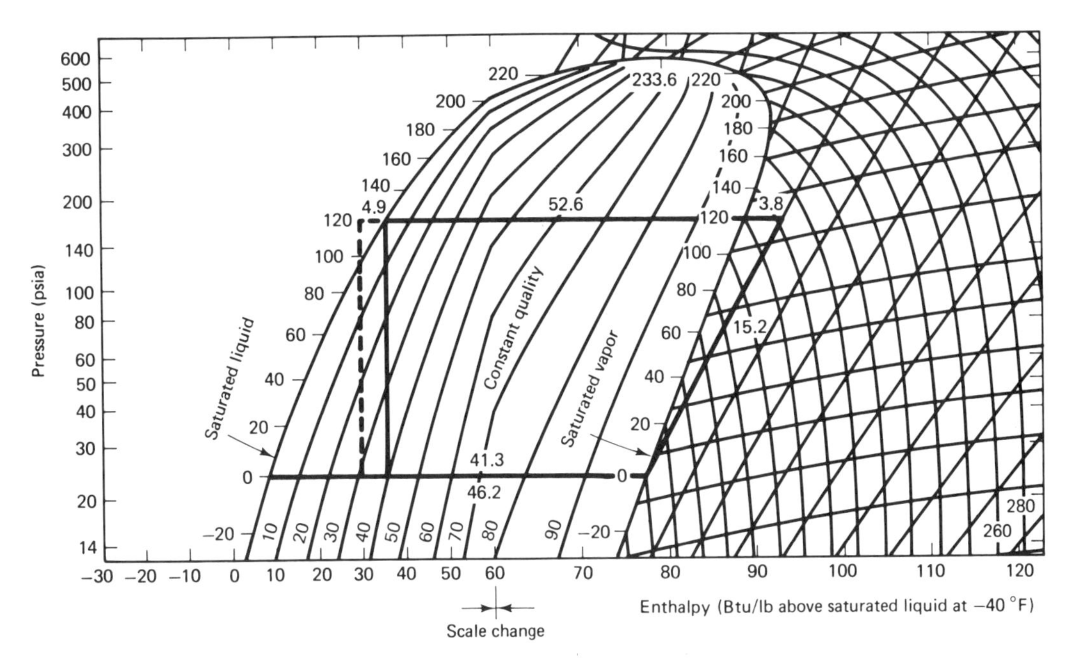

Figure 19-9 Pressure-enthalpy diagram for R-12, showing the effect of liquid subcooling. (By permission of E.I. du Pont de Nemours & Company, Inc.)

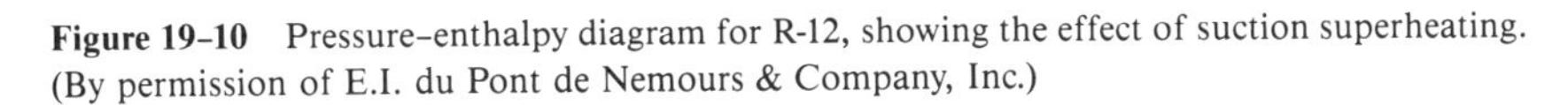
Figure 19-10 Pressure–enthalpy diagram for R-12, showing the effect of suction superheating. (By permission of E.I. du Pont de Nemours & Company, Inc.)

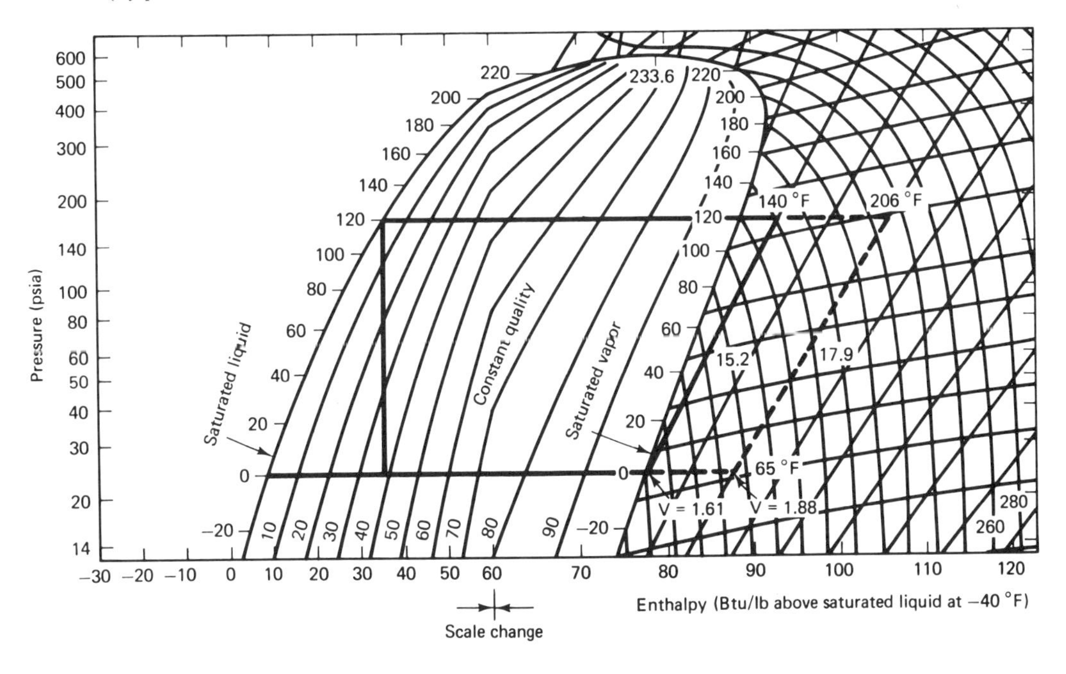

Effect of superheating the suction gas. Figure 19–10 shows the effect of superheating the suction gas. If this is done in the evaporator, it adds to the net refrigerating effect. If superheat is added to the suction line after it leaves the evaporator, it decreases the efficiency of the refrigeration cycle. Assume that in Figure 19–10 the suction-line refrigerant is superheated to 65 °F after it leaves the evaporator. The discharge temperature is raised to 206 °F and additional condensing surface must be used in the condenser to remove more superheat. More horsepower is required by the compressor to increase its work load. Insulating the suction line may be one solution in reducing the suction-line superheat.

Pressure–Enthalpy Chart in SI Units

Referring to the information given in Chapter 1 on measurement, the pressure–enthalpy chart, Figure 19–2 for R-12, is shown in English units. To compare the measurement terms for SI units, a similar chart is shown in Figure 19–11 using SI units. A comparison of the terms used is shown in the following table:

Condition	English units	SI units
Pressure	psia	MPa
Enthalpy	Btu/lb	kJ/kg
Temperature	°F	°K
Density	lb/ft^3	kg/m^3
Entropy	Btu/lb –°R	kJ/kg –°K

The meaning of the SI units is as follows:

MPa = megapascals, a measure of pressure ("Mega" means 1,000,000)

kJ/kg = kilojoules per kilogram, a measure of enthalpy ("Kilo" means 1000)

°K = degrees Kelvin, a measure of temperature in Celsius degrees starting from absolute zero

kg/m^3 = kilograms per cubic meter, a measure of density

kJ/kg–°K = kilojoules per kilogram per degree Kelvin, a measure of entropy or specific heat

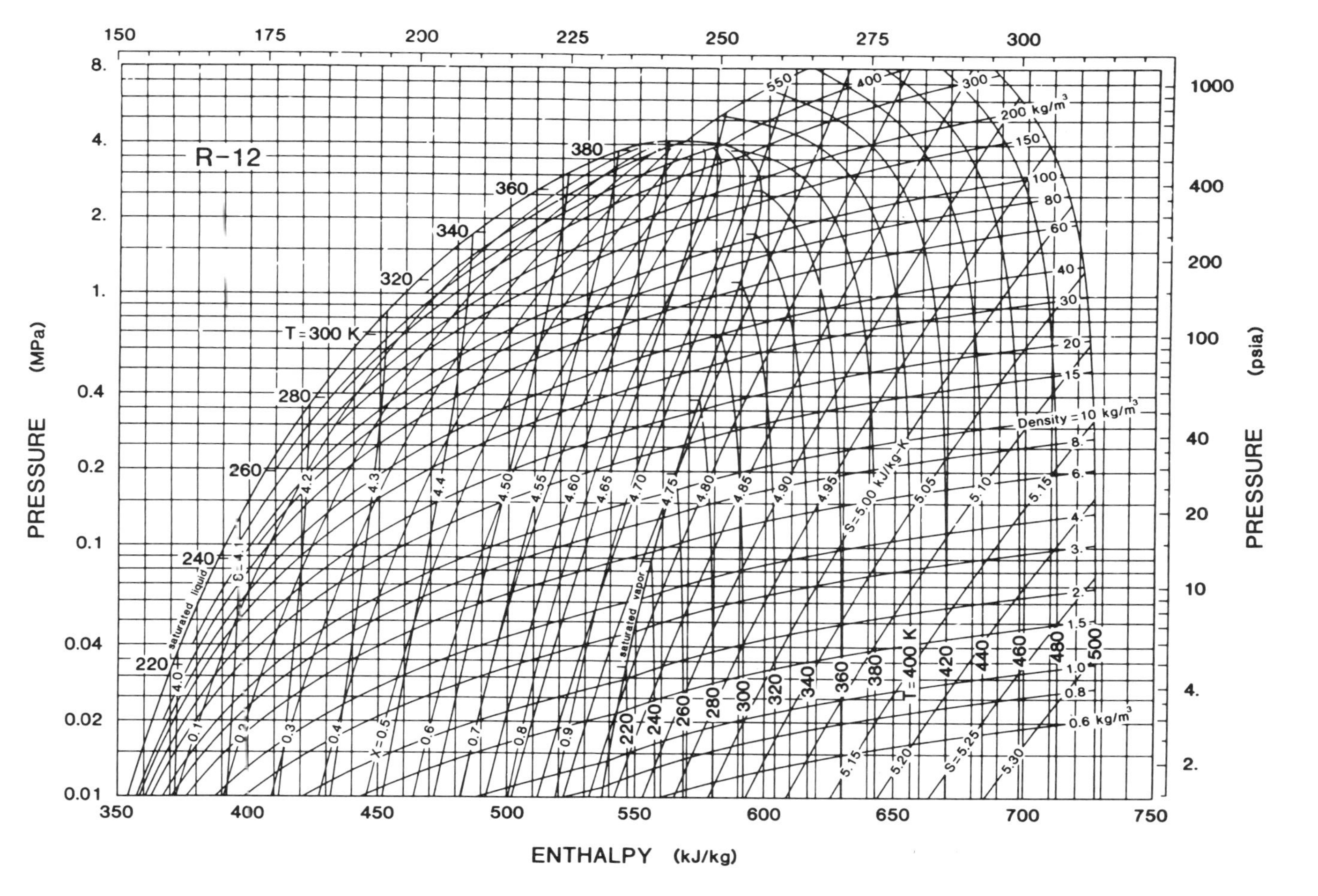

Figure 19–11 (a) Pressure–enthalpy diagram for R-12, showing the use of SI units; (b) R-12, properties of saturated liquid and vapor, showing use of SI units. (By permission of *ASHRAE Handbook*.)

Temp K	Pressure MPa	Volume Vapor m³/kg	Density Liquid kg/m³	Enthalpy Liquid kJ/kg	Enthalpy Vapor kJ/kg	Entropy Liquid kJ/kg · K	Entropy Vapor kJ/kg · K
170	0.000867	13.460	1686.1	328.51	523.56	3.7732	4.9205
175	0.001395	8.6113	1673.2	332.73	525.75	3.7977	4.9006
180	0.002175	5.6757	1660.1	336.96	527.97	3.8215	4.8826
185	0.003298	3.8433	1647.0	341.18	530.21	3.8446	4.8664
190	0.004875	2.6673	1633.8	345.40	532.48	3.8671	4.8517
195	0.007039	1.8930	1620.5	349.63	534.77	3.8891	4.8385
200	0.009948	1.3713	1607.2	353.87	537.07	3.9105	4.8265
205	0.013787	1.0121	1593.7	358.11	539.39	3.9315	4.8157
210	0.018765	0.75991	1580.1	362.37	541.72	3.9520	4.8060
215	0.025118	0.57957	1566.4	366.64	544.05	3.9721	4.7972
220	0.033110	0.44844	1552.6	370.94	546.40	3.9918	4.7893
222	0.036829	0.40624	1547.1	372.66	547.33	3.9996	4.7864
224	0.040876	0.36876	1541.5	374.39	548.27	4.0073	4.7836
226	0.045271	0.33540	1535.9	376.12	549.21	4.0150	4.7809
228	0.050035	0.30565	1530.3	377.85	550.15	4.0226	4.7783
230	0.055189	0.27905	1524.7	379.59	551.09	4.0302	4.7758
232	0.060756	0.25522	1519.0	381.34	552.02	4.0377	4.7734
234	0.066758	0.23383	1513.3	383.08	552.96	4.0452	4.7712
236	0.073218	0.21459	1507.6	384.84	553.90	4.0526	4.7690
238	0.080160	0.19726	1501.9	386.59	554.83	4.0600	4.7669
240	0.087609	0.18161	1496.1	388.36	555.77	4.0674	4.7649
242	0.095589	0.16745	1490.3	390.13	556.70	4.0747	4.7630
243.36	0.101325	0.15861	1486.3	391.33	557.33	4.0797	4.7618
244	0.10413	0.15463	1484.5	391.90	557.63	4.0820	4.7612
246	0.11324	0.14299	1478.6	393.68	558.56	4.0892	4.7595
248	0.12297	0.13241	1472.7	395.46	559.49	4.0964	4.7578
250	0.13334	0.12278	1466.8	397.25	560.42	4.1036	4.7562
252	0.14436	0.11399	1460.9	399.05	561.34	4.1107	4.7547
254	0.15608	0.10597	1454.9	400.86	562.26	4.1178	4.7533
256	0.16852	0.09863	1448.9	402.67	563.18	4.1249	4.7519
258	0.18170	0.09191	1442.8	404.48	564.10	4.1319	4.7506
260	0.19566	0.08574	1436.7	406.31	565.01	4.1389	4.7493
262	0.21042	0.08007	1430.6	408.14	565.92	4.1459	4.7481
264	0.22602	0.07486	1424.4	409.97	566.82	4.1528	4.7470
266	0.24248	0.07005	1418.2	411.82	567.73	4.1598	4.7459
268	0.25983	0.06563	1412.0	413.67	568.62	4.1667	4.7448

Temp K	Pressure MPa	Volume Vapor m³/kg	Density Liquid kg/m³	Enthalpy Liquid kJ/kg	Enthalpy Vapor kJ/kg	Entropy Liquid kJ/kg · K	Entropy Vapor kJ/kg · K
270	0.27811	0.06154	1405.7	415.53	569.52	4.1735	4.7438
272	0.29735	0.05775	1399.3	417.40	570.41	4.1804	4.7429
274	0.31757	0.05425	1392.9	419.27	571.29	4.1872	4.7420
276	0.33881	0.05101	1386.5	421.16	572.17	4.1940	4.7411
278	0.36110	0.04800	1380.0	423.05	573.05	4.2007	4.7403
280	0.38448	0.04520	1373.4	424.95	573.91	4.2075	4.7395
282	0.40896	0.04260	1366.8	426.86	574.78	4.2142	4.7388
284	0.43459	0.04018	1360.1	428.77	575.63	4.2209	4.7380
286	0.46140	0.03793	1353.4	430.70	576.48	4.2276	4.7373
288	0.48943	0.03583	1346.6	432.63	577.33	4.2343	4.7367
290	0.51870	0.03386	1339.7	434.58	578.16	4.2409	4.7360
292	0.54924	0.03203	1332.7	436.53	578.99	4.2475	4.7354
294	0.58111	0.03031	1325.7	438.49	579.81	4.2542	4.7348
296	0.61431	0.02870	1318.6	440.46	580.62	4.2608	4.7343
298	0.64890	0.02719	1311.4	442.44	581.42	4.2673	4.7337
300	0.68491	0.02578	1304.2	444.43	582.21	4.2739	4.7332
302	0.72236	0.02445	1296.8	446.43	582.99	4.2804	4.7326
304	0.76131	0.02320	1289.3	448.44	583.76	4.2870	4.7321
306	0.80177	0.02203	1281.8	450.46	584.53	4.2935	4.7316
308	0.84380	0.02092	1274.1	452.49	585.27	4.3000	4.7311
310	0.88742	0.01988	1266.4	454.53	586.01	4.3065	4.7306
315	1.0037	0.01753	1246.5	459.68	587.79	4.3227	4.7294
320	1.1308	0.01549	1225.8	464.91	589.49	4.3388	4.7281
325	1.2693	0.01371	1204.1	470.21	591.07	4.3549	4.7268
330	1.4198	0.01214	1181.5	475.61	592.54	4.3710	4.7253
335	1.5832	0.01077	1157.6	481.10	593.86	4.3871	4.7237
340	1.7600	0.009549	1132.3	486.71	595.02	4.4033	4.7218
345	1.9510	0.008465	1105.4	492.45	595.98	4.4195	4.7196
350	2.1572	0.007493	1076.5	498.36	596.71	4.4360	4.7170
355	2.3794	0.006616	1045.3	504.47	597.14	4.4527	4.7138
360	2.6188	0.005819	1011.1	510.84	597.20	4.4699	4.7098
365	2.8765	0.005085	972.99	517.57	596.77	4.4878	4.7047
370	3.1541	0.004398	929.67	524.81	595.62	4.5066	4.6980
375	3.4532	0.003735	878.34	532.83	593.34	4.5273	4.6887
380	3.7764	0.003048	811.63	542.34	588.78	4.5515	4.6737
*384.95	4.125	0.001792	558.0	566.9	566.9	4.614	4.614

*Critical point

Figure 19–11 (b) (Cont.)

Temp K	Viscosity, μPa · s Sat. Liquid	Sat. Vapor	Gas (1 Atm.)	Thermal Conductivity, mW/m · K Sat. Liquid	Sat. Vapor	Gas (1 Atm.)	Specific Heats, kJ/kg · K Sat. Liquid C_p	Sat. Liquid C_v	Sat. Vapor C_p	Sat. Vapor C_v	Gas (0 Atm.) C_p	Gas (0 Atm.) C_v	Velocity of Sound, m/s Sat. Liquid	Sat. Vapor	Gas (1 Atm.)
170	1210		—	116.3		—	0.825		0.444		0.443				—
180	969		—	112.6		—	0.835		0.459		0.457				—
190	794		—	108.9		—	0.844		0.475		0.472				
200	664	8.68	—	105.2	5.0	—	0.854		0.489		0.485				—
210	565	9.02	—	101.5	5.4	—	0.863		0.507		0.499				—
220	488	9.38	—	97.8	5.83	—	0.872		0.524		0.512				
230	426	9.76	—	94.1	6.27	—	0.882		0.548		0.525				—
240	377	10.16	—	90.4	6.72	—	0.892		0.572		0.537				
243.4[a]	396	10.30	10.30	88.1	6.88		0.895		0.580		0.541				
250	337	10.58	10.57	86.8	7.18	7.19	0.902		0.595		0.549				
260	303	11.02	10.98	83.2	7.67	7.69	0.913		0.615		0.560				
270	275	11.48	11,40	79.6	8.18	8.19	0.926		0.637		0.571				
280	252	11.96	11.80	76.0	8.70	9.19	0.942		0.663		0.582				
290	231	12.46	12.21	72.4	9.23	9.19	0.960		0.690		0.592				
300	213.9	13.00	12.60	68.7	9.77	9.70	0.979		0.725		0.602				
310	198.9	13.56	13.00	65.0	10.30	10.21	1.00		0.77		0.612				
320	185.7	14.15	13.39	61.3	10.86	10.73	1.04		0.82		0.621				
330	174.1	14.80	13.78	57.6	11.45	11.25	1.09		0.88		0.630				
340	160.0	15.50	14.17	53.9	12.09	11.78	1.16		0.95		0.639				
350	144.5	16.40	14.54	50.2	12.8	12.31	1.26		1.09		0.647				
360	127.5	17.46	14.92	46.3	13.6	12.31	1.26		1.31		0.655				
370	105.5	19.0	15.29	41.3	15.6	13.41	1.55		1.76		0.663				
380	75.0	22.2	15.66			13.96	∞	d	∞	d	0.675				
385.1[b]	31.0	31.0	15.85	∞	∞	14.24	∞	d	∞	d	0.675			0	0
390	—	—	16.03	—	—	14.51	—	—	—	—	0.678			—	—
400	—	—	16.39	—	—	15.06	—	—	—	—	0.685			—	—
410	—	—	16.74	—	—	15.62	—	—	—	—	0.692			—	—
420	—	—	17.10	—	—	16.19	—	—	—	—	0.698			—	—
430	—	—	17.45	—	—	16.77	—	—	—	—	0.704			—	—
440	—	—	17.80	—	—	17.35	—	—	—	—	0.710			—	—
450	—	—	18.14	—	—	17.94	—	—	—	—	0.716			—	—
460	—	—	18.49	—	—	17.94	—	—	—	—	0.721			—	—
470	—	—	18.83	—	—	19.12	—	—	—	—	0.726			—	—
480	—	—	19.16	—	—	19.71	—	—	—	—	0.731			—	—
490	—	—	19.49	—	—	20.30	—	—	—	—	0.736			—	—
500	—	—	19.83	—	—	20.90	—	—	—	—	0.741			—	—

[a]Normal boiling point. [b]Critical point. [c]Very large. [d]Large. [e]Small.

Figure 19–11 (b) (Cont.)

The conversion factors are:

To convert from:	To:	Multiply by:
psi	Pa	6.894×10^3
Btu/lb	J/kg	2.326×103
°F	°K	$\frac{°F + 459.67}{1.8}$
lb/ft³	kg/m³	1.601×10
Btu/lb−°R	kJ/kg−°K	No change in unit; same for both systems

For example,

$$\begin{aligned} 14.7 \text{ psi} &= 14.7 \times 6.894 \times 10^3 \\ &= 14.7 \times 6894 \\ &= 101{,}342 \text{ pascal} \\ &= 0.101 \text{ MPa} \end{aligned}$$

Summary

The P-E diagram is useful in supplying a picture of the action that takes place in the refrigeration cycle. Plotting performance conditions on the P-E chart gives theoretical solutions to problems that must be tempered by actual considerations. Power losses in the motor, volumetric efficiency of the compressor, and pressure drop in piping will affect the actual performance of the system. The P-E chart does indicate trends and points to areas to explore for improving the efficiency of the system.

REVIEW EXERCISE

Using the pressure–enthalpy diagram for R-12, plot the refrigeration cycle for the following set of conditions (see Figure 19–12)

Condensing temperature	130 °F
Evaporating temperature	20 °F
Superheat in evaporator	10 °F
Subcooling in the condenser	10 °

Note that when the compressor performance is plotted, it follows the constant-entropy line. From the diagram, determine the following:

1. Latent heat of evaporation ________________ Btu/lb
2. Superheat in the evaporator ________________ Btu/lb
3. Net refrigerating effect ________________ Btu/lb
4. Heat added by the compressor ________________ Btu/lb
5. Superheat removed by the condenser ________________ Btu/lb
6. Subcooling in the condenser ________________ Btu/lb
7. Heat removed in the condenser by condensation ________________ Btu/lb
8. Total heat removed by the condenser ________________ Btu/lb
9. The compression ratio ________________

*10. The pounds of refrigerant pumped per ton ________________ lb/min

*11. The compressor displacement ________________ ft^3/min

*12. The theoretical horsepower of the compressor ________________ hp

*13. Capacity of compressor, in Btuh's ________________ Btuh

*Use tables if necessary for more accurate results.

Figure 19–12 Pressure-enthalpy diagram for sample calculations. R-12 refrigerant. (By permission of E.I. du Pont de Nemours & Company, Inc.)

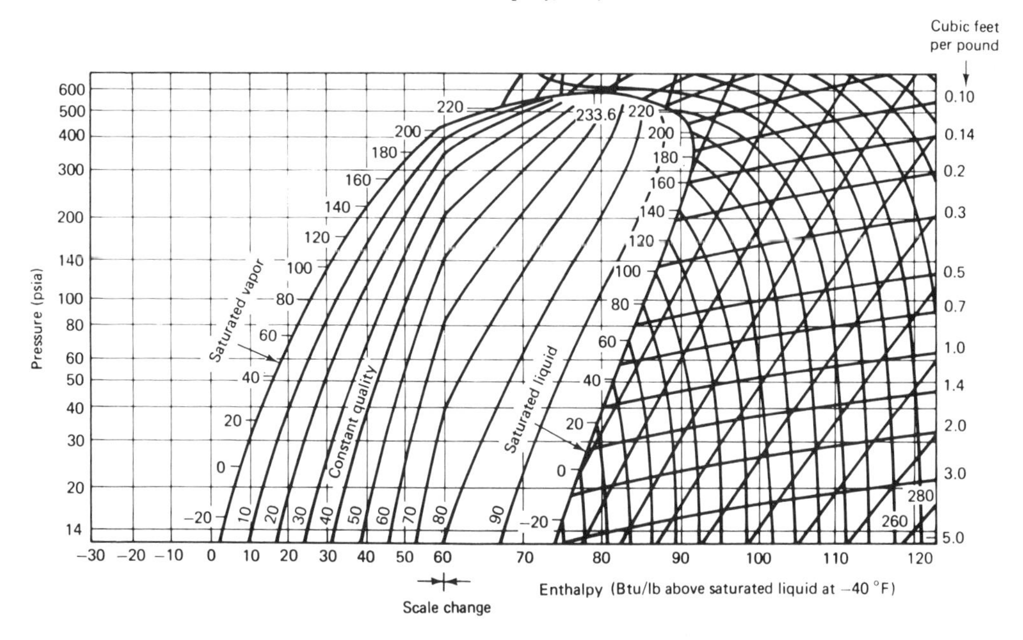

Appendix

I SI Units and Conversion Factors

SI UNITS

The most familiar SI unit is probably "Celsius," used to measure temperature throughout the world. "SI" stands for International System of units. The United States has been slow in putting these measurement units into practice. There is so much now being written using SI units that it is important for the refrigeration technician to be familiar with converting SI units into British units. One of the best sources of information is the ASHRAE publication, *1981 Fundamentals Handbook,* Chapter 37.

Since temperature, pressure, and heat are the most frequently used measurements referred to in the text, a brief discussion of SI units and conversion factors is useful.

Temperature

In the British system if we start with absolute zero, the temperature readings are in degrees Rankine (°R). If we use SI units, the readings are in degrees Kelvin (°K). Rankine and Kelvin temperatures are not used as commonly as Fahrenheit and Celsius which have already been discussed. The conversion formulas are

$$^\circ C = \frac{^\circ F - 32}{1.8} \quad \text{and} \quad ^\circ F = 1.8\,(^\circ C) + 32$$

Pressure

In the British system pressure per square inch (gauge or absolute) is commonly used and sometimes inches of mercury vacuum. In the SI system the base unit is the pascal. A pascal is a force of 1 newton applied to an area of 1 square meter. Usually, pounds per square inch is converted directly to newtons per square meter. The conversion factor is

0.1129. Thus, 10 psi = 1.129 n/m². A newton is a force that imparts to a mass of 1 kilogram an acceleration of 1 meter per second per second.

Heat

In the British system the commonly used term is Btu or Btuh. In the SI system the most common term is joule. A joule is the work done by 1 newton acting through a distance of 1 meter. The conversion factor is Btu = 1.055 kJ (kilojoules or 1000 joule).

CONVERSION FACTORS

The conversion factors given in Figure A–1 can be used to convert SI units to English units, or vice versa. The conversion table uses the "E" notation, which is convenient for writing both small and large numbers. The numbers are expressed as factors of the appropriate power of 10. Examples are shown as follows:

Number	Scientific notation	E Notation
3,159,000,000	3.159×10^{9}	3.159E+09
0.3159	3.159×10^{-1}	3.159E−01

An asterisk (*) indicates that the conversion factor is exact. All other conversion factors have been rounded to the figures given.

Figure A–1 SI units and conversion factors. (By permission of *ASHRAE Handbook*.)

To convert from:	To:	Multiply by:
	Acceleration	
ft/s^2	meter per second squared (m/s^2)	2.048 000*E − 01
$in./s^2$	meter per second squared (m/s^2)	2.540 000*E − 02
	Angle	
degree (angle)	radian (rad)	1.745 329 E − 02
minute (angle)	radian (rad)	2.908 882 E − 04
second (angle)	radian (rad)	4.848 137 E − 06
	Area	
acre	square meter (m^2)	4.046 973 E + 03
are	square meter (m^2)	1.000 000*E + 02
circular mil	square meter (m^2)	5.067 075 E − 10
ft^2	square meter (m^2)	9.290 304*E − 02
hectare	square meter (m^2)	1.000 000*E + 04
$in.^2$	square meter (m^2)	6.451 600*E − 04
mi^2 (international)	square meter (m^2)	2.589 988 E + 06

To convert from:	To:	Multiply by:
mi^2 (U.S. statute)	square meter (m^2)	2.589 998 E+06
yd^2	square meter (m^2)	8.361 274 E−01
	Bending Moment or Torque	
dyne · cm	newton meter (N · m)	1.000 000*E−07
kg_f · m	newton meter (N · m)	9.806 650*E+00
oz_f · in.	newton meter (N · m)	7.061 552 E−03
lb_f · in.	newton meter (N · m)	1.129 848 E−01
lb_f · ft.	newton meter (N · m)	1.355 818 E+00
	Bending Moment or Torque per Unit Length	
lb_f · ft/in.	newton meter per meter (N · m/m)	5.337 866 R+01
lb_f in./in.	newton meter per meter (N · m/m)	4.448 222 E+00
	Capacity (see Volume)	
	Density (see Mass per Unit Volume)	
	Electricity and Magnetism	
abampere	ampere (A)	1.000 000*E+01
abcoulomb	coulomb (C)	1.000 000*E+01
abfarad	farad (F)	1.000 000*E+09
abhenry	henry (H)	1.000 000*E−09
abmho	siemens (S)	1.000 000*E+09
abohm	ohm (Ω)	1.000 000*E−09
abvolt	volt (V)	1.000 000*E−08
ampere hour	coulomb (C)	3.600 000*E+03
EMU of capacitance	farad (F)	1.000 000*E+09
EMU of current	ampere (A)	1.000 000*E+01
EMU of electric potential	volt (V)	1.000 000*E−08
EMU of inductance	henry (H)	1.000 000*E−09
EMU of resistance	ohm (Ω)	1.000 000*E−09
ESU of capacitance	farad (F)	1.112 650 E−12
ESU of current	ampere (A)	3.335 6 E−10
ESU of electric potential	volt (V)	2.997 9 E+02
ESU of inductance	henry (H)	8.987 554 E+11
ESU of resistance	ohm (Ω)	8.987 554 E+11
faraday (based on carbon-12)	coulomb (C)	9.648 70 E+04
faraday (chemical)	coulomb (C)	9.649 57 E+04
faraday (physical)	coulomb (C)	9.652 19 E+04
gamma	tesla (T)	1.000 000*E−09
gauss	tesla (T)	1.000 000*E−04
gilbert	ampere (A)	7.957 747 E−01
maxwell	weber (Wb)	1.000 000*E−08
mho	siemens (S)	1.000 000*E+00
oersted	ampere per meter (A/m)	7.957 747 E+01
ohm centimeter	ohm meter (Ω · m)	1.000 000*E−02
ohm circular-mil per foot	ohm meter (Ω · m)	1.662 426 E−09
statampere	ampere (A)	3.335 640 E−10

Figure A–1 (Cont.)

To convert from:	To:	Multiply by:
statcoulomb	coulomb (C)	3.335 640 E − 10
statfarad	farad (F)	1.112 650 E − 12
stathenry	henry (H)	8.987 554 E + 11
statmho	siemens (S)	1.112 650 E − 12
statohm	ohm (Ω)	8.987 554 E + 11
statvolt	volt (V)	2.997 925 E + 02
unit pole	weber (Wv)	1.256 637 E − 07
	Energy (Includes Work)	
British thermal unit (IT)[2]	joule (J)	1.055 056 E + 03
British thermal unit (39F)	joule (J)	1.059 67 E + 03
British thermal unit (59F)	joule (J)	1.054 80 E + 30
British thermal unit (60F)	joule (J)	1.054 68 E + 03
calorie (IT)	joule (J)	4.186 800*E + 00
calorie (15 °C)	joule (J)	4.185 80 E + 00
calorie (20 °C)	joule (J)	4.181 90 E + 00
calorie, kilogram (IT)	joule (J)	4.186 800*E + 03
electronvolt	joule (J)	1.602 19 E − 19
erg	joule (J)	1.000 000*E − 07
ft · lb_f	joule (J)	1.355 818 E + 00
ft-poundal	joule (J)	4.214 011 E − 02
kilocalorie (IT)	joule (J)	4.186 800*E + 03
kW · h	joule (J)	3.600 000*E + 06
therm	joule (J)	1.055 056 E + 08
W · h	joule (J)	3.600 000*E + 03
W · s	joule (J)	1.000 000*E + 00
	Energy per Unit Area Time	
erg/(cm^2 · s)	watt per (W/m^2)	1.000 000*E − 03
W/cm^2	square (W/m^2)	1.000 000*E + 04
$W/in.^2$	meter (W/m^2)	1.550 003 E + 03
	Flow (see Mass per Unit Time or Force Volume per Unit Time)	
dyne	newton (N)	1.000 000*E − 05
dilogram-force	newton (N)	9.806 650*E + 00
kilopond	newton (N)	9.806 650*E + 00
kip (1000 lb_f)	newton (N)	4.448 222 E + 03
ounce-force	newton (N)	2.780 139 E − 01
pound-force (lb_f)	newton (N)	4.448 222 E + 00
poundal	newton (N)	1.382 550 E − 01
ton-force (2000 lb_f)	newton (N)	8.896 444 E + 03
	Force per Unit Area (see Pressure)	
	Force per Unit Length	
lb_f/ft	newton per meter (N/m)	1.459 390 E + 01
$lb_f/in.$	newton per meter (N/m)	1.751 268 E + 02

Figure A-1 (Cont.)

To convert from:	To:	Multiply by:
	Heat	
Btu (IT) · ft/h · ft² – F	watt per meter kelvin (W/m · K)	1.730 735 E+00
Btu (IT) · in./h · ft² · F	watt per meter kelvin (W/m · K)	1.442 279 E–01
Btu (IT)/ft²	joule per square meter (J/m²)	1.135 653 E+04
Btu (IT)/h · ft²	watt per square meter (W/m²)	3.154 591 E+00
Btu (IT)/h · ft² · F	watt per square meter kelvin (W/m² · K)	5.678 264 E+00
Btu (IT)/h · ft² · F	watt per square meter kelvin (W/m² · K)	5.678 264 E+00
Btu (IT)/lb	joule per kilogram (J/kg)	2.326 000*E+03
Btu (IT)/lb · F	joule per kilogram kelvin (J/kg · K)	4.186 800*E+03
Btu (IT)/ft³	joule per cubic meter (J/m³)	3.725 895 E+04
cal (IT)/g	joule per kilogram (J/kg)	4.186 800*E+03
cal (IT)/g · °C	joule per kilogram kelvin (J/kg · K)	4.186 800*E+03
clo	kelvin square meter per watt (K · m²/W)	2.003 712 E–01
F · ft² · h/Btu (IT)	kelvin square meter per watt (K · m²/W)	1.761 102 E–01
F · ft² · h/Btu (IT) · in.	kelvin meter per watt (K · m/W)	6.933 471 E+00
	Length	
angstrom	meter (m)	1.000 000*E–10
chain	meter (m)	2.011 684 E+01
fathom	meter (m)	1.828 804 E+00
foot	meter (m)	3.048 000*E–01
foot (U.S. survey)	meter (m)	3.048 006 E–01
inch	meter (m)	2.540 000*E–02
light year	meter (m)	9.460 55 E+15
microinch	meter (m)	2.540 000*E–08
micron	meter (m)	1.000 000*E–06
mil	meter (m)	2.540 000*E–05
mile (international nautical)	meter (m)	1.852 000*E+03
mile (U.S. nautical)	meter (m)	1.852 000*E+03
mile (international)	meter (m)	1.609 344*E+03
mile (U.S. statute)	meter (m)	1.609 347 E+03
rod	meter (m)	5.029 210 E+00
yard	meter (m)	9.144 000*E–01
	Light	
cd/in.²	candela per square meter (cd/m²)	1.550 003 E+03
footcandle	lux (lx)	1.076 391 E+01
footlambert	candela per square meter (cd/m²)	3.426 259 E+00
lambert	candela per square meter (cd/m²)	3.183 099 E+03
	Mass	
grain	kilogram (kg)	6.479 891*E–05
gram	kilogram (kg)	1.000 000*E–03

Figure A–1 (Cont.)

To convert from:	To:	Multiply by:
hundredweight (long)	kilogram (kg)	5.080 235 E+01
hundredweight (short)	kilogram (kg)	4.535 924 E+01
ounce (avoirdupois)	kilogram (kg)	2.834 952 E−02
ounce (troy or apothecary)	kilogram (kg)	3.110 348 E−02
pound (lb avoirdupois)	kilogram (kg)	4.535 924 E−01
pound (troy or apothecary)	kilogram (kg)	3.732 417 E−01
slug	kilogram (kg)	1.459 390 E+01
ton (assay)	kilogram (kg)	2.916 667 E−02
ton (long, 2240 lb)	kilogram (kg)	1.016 047 E+03
ton (metric)	kilogram (kg)	1.000 000*E+03
ton (short, 2000 lb)	kilogram (kg)	9.071 847 E+02
tonne	kilogram (kg)	1.000 000*E+03
	Mass per Unit Area	
oz/ft^2	kilogram (kg/m^2)	3.051 517 E−01
oz/yd^2	per square (kg/m^2)	3.390 575 E−02
lb/ft^2	meter (kg/m^2)	4.882 428 E+00
	Mass per Unit Capacity (see Mass per Unit Volume)	
	Mass per Unit Length	
lb/ft	kilogram per meter (kg/m)	1.488 164 E+00
lb/in.	kilogram per meter (kg/m)	1.785 797 E+01
	Mass Per Unit Time (Includes Flow)	
perm (0°C)	kilogram per pascal second square meter (kg/Pa · s · m^2)	5.721 35 E−11
perm (23°C)	kilogram per pascal second square meter (kg/Pa · s · m^2)	5.745 25 E−11
perm. in. (0°C)	kilogram per pascal second meter (kg/Pa · s · m)	1.453 22 E−12
perm. in. (23°C)	kilogram per pascal second meter (kg/Pa · s · m)	1.459 29 E−12
lb/h	kilogram per second (kg/s)	1.259 979 E−04
lb/min	kilogram per second (kg/s)	7.559 873 E−03
lb/s	kilogram per second (kg/s)	4.535 924 E−01
lb/hp · h	kilogram per joule (kg/J)	1.689 659 E−07
ton (short)/h	kilogram per second (kg/s)	2.519 958 E−01
	Mass per Unit Volume (Includes Density and Mass Capacity) (Note: kg/m^3 = kilogram per cubic meter)	
grain/gal (U.S. liquid)	(kg/m^3)	1.711 806 E−02
g/cm^3	(kg/m^3)	1.000 000*E+03
oz (avoirdupois)/gal (U.K. liquid)	(kg/m^3)	6.236 021 E+00
oz (avoirdupois)/gal (U.S. liquid)	(kg/m^3)	7.489 152 E+00
oz (avoirdupois)/in^3	(kg/m^3)	1.729 994 E+03
lb/ft^3	(kg/m^3)	1.601 846 E+01

Figure A-1 (Cont.)

To convert from:	To:	Multiply by:
lb/in.^{3}	(kg/m^3)	2.767 990 E+04
lb/gal (U.K. liquid)	(kg/m^3)	9.977 633 E+01
lb/gal (U.S. liquid)	(kg/m^3)	1.198 264 E+02
lb/yd^3	(kg/m^3)	5.932 764 E−01
slug/ft^3	(kg/m^3)	5.153 788 E+02
ton (long)/yd^3	(kg/m^3)	1.328 939 E+03
ton (short)/yd^3	(kg/m^3)	1.186 553 E+03
	Power	
Btu (IT)/h	watt (W)	2.930 722 E−01
Btu (IT)/min	watt (W)	1.758 427 E+01
Btu (IT)/s	watt (W)	1.055 056 E+03
cal (IT)/s	watt (W)	4.186 800*E+00
erg/s	watt (W)	1.000 000*E−07
ft · lb_f/h	watt (W)	3.766 161 E−04
ft · lb_f/min	watt (W)	2.259 697 E−02
ft · lb_f/s	watt (W)	1.355 818 E+00
horsepower (550 ft · lb_f/s)	watt (W)	7.456 999 E+02
horsepower (boiler)	watt (W)	9.809 50 E+03
horsepower (electric)	watt (W)	7.460 000*E+02
kilocalorie (IT)/h	watt (W)	1.163 000*E+00
ton (refrigeration)	watt (W)	3.516 853 E+03
	Pressure of Stress (Force per Unit Area)	
atmosphere (standard)	pascal (Pa)	1.013 250*E+05
atmosphere (technical = 1 kg_f/cm^2)	pascal (Pa)	9.806 650*E+04
bar	pascal (Pa)	1.000 000*E+05
centimeter of mercury (0 °C)	pascal (Pa)	1.333 22 E+03
centimeter of water (4 °C)	pascal (Pa)	9.806 38 E+01
dyne/cm^2	pascal (Pa)	1.000 000*E−01
foot of water (39.2 F)	pascal (Pa)	2.988 98 E+03
g_f/cm^2	pascal (Pa)	9.806 650*E+01
inch of mercury (32 F)	pascal (Pa)	3.386 38 E+03
inch of mercury (60 F)	pascal (Pa)	3.376 85 E+03
inch of water (39.2 F)	pascal (Pa)	2.490 82 E+02
inch of water (60 F)	pascal (Pa)	2.488 4 E+02
kg_f/cm^2	pascal (Pa)	9.806 650*E+04
kg_fm^2	pascal (Pa)	9.806 650*E+00
kg_fmm^2	pascal (Pa)	9.806 650*E+06
kip/in.^{2} (ksi)	pascal (Pa)	6.894 757 E+06
millibar	pascal (Pa)	1.000 000*E+02
millimeter of mercury (0 °C)	pascal (Pa)	1.333 22 E+02
poundal/ft^2	pascal (Pa)	1.488 164 E+00
lb_f/ft^2	pascal (Pa)	4.788 026 E+01
lb_f/in^2 (psi)	pascal (Pa)	6.894 757 E+03
psi	pascal (Pa)	6.894 757 E+03
torr (mm Hg, 0 °C)	pascal (Pa)	1.333 22 E+02

Figure A–1 (Cont.)

To convert from:	To:	Multiply by:
	Speed (see Velocity)	
	Temperature	
degree Celsius	kelvin (K)	$T_k = t\,°C + 273.15$
degree Fahrenheit	degree Celsius (°C)	$T\,°C = (t\,°F - 32)/1.8$
degree Fahrenheit	kelvin (K)	$T_k = (t\,°F + 459.67)/1.8)$
degree Rankine	kelvin (K)	$TK = T\,°R/1.8$
kelvin	degree Celsius (°C)	$t\,°C = TK - 273.15$
	Time	
day	second (s)	8.640 000*E + 04
day (sidereal)	second (s)	8.616 409 E + 04
hour	second (s)	3.600 000*E + 03
hour (sidereal)	second (s)	3.590 170 E + 03
minute	second (s)	6.000 000*E + 01
minute (sidereal)	second (s)	5.983 617 E + 01
second (sidereal)	second (s)	9.972 696 E − 01
year (365 days)	second (s)	3.153 600*E + 07
year (sidereal)	second (s)	3.155 815 E + 07
year (tropical)	second (s)	3.155 693 E + 07
	Torque (see Bending Moment)	
	Velocity (Includes Speed) (*Note:* m/s = meter per second)	
ft/h	(m/s)	8.466 667 E − 05
ft/min	(m/s)	5.080 000*E − 03
ft/s	(m/s)	3.048 000*E + 01
in./s	(m/s)	2.540 000*E − 02
km/h	(m/s)	2.777 778 E − 01
knot (internatnional)	(m/s)	5.144 444 E − 01
mi/h (international)	(m/s)	4.470 400*E − 01
mi/min (international)	(m/s)	2.682 240*E + 01
mi/s (international)	(m/s)	1.609 344*E + 03
	Viscosity	
centipoise	pascal second (Pa · s)	1.000 000*E − 03
centistokes	square meter per second (m^2s)	1.000 000*E − 06
ft^2/s	square meter per second (m^2/s)	9.290 304*E − 02
poise	pascal second (Pa · s)	1.000 000*E − 01
poundal · s/ft^2	pascal second (Pa · s)	1.488 164 E + 00
lb/ft · h	pascal second (Pa · s)	4.133 789 E − 04
lb/ft · s	pascal second (Pa · s)	1.488 164 E + 00
lb_f · s/ft^2	pascal second (Pa · s)	4.788 026 E + 01
lb_f · $s/in.^2$	pascal second (Pa · s)	6.894 757 E + 03
rhe	L per pascal second (L/Pa · s)	1.000 000*E + 01
slug/ft · s	pascal second (Pa · s)	4.788 026 E + 01
stokes	square meter per second (m^2/s)	1.000 000*E − 04

Figure A–1 (Cont.)

To convert from:	To:	Multiply by:
	Volume (Includes Capacity)	
acre-foot	cubic meter (m^3)	1.233 489 E + 03
barrel (oil, 42 gal)	cubic meter (m^3)	1.589 873 E − 01
board foot	cubic meter (m^3)	2.359 737 E − 03
bushel (U.S.)	cubic meter (m^3)	3.523 907 E − 02
cup	cubic meter (m^3)	2.365 882 E − 04
fluid ounce (U.S.)	cubic meter (m^3)	2.957 353 E − 05
ft^3	cubic meter (m^3)	2.831 685 E − 02
gallon (Canadian liquid)	cubic meter (m^3)	4.546 090 E − 03
gallon (U.K. liquid)	cubic meter (m^3)	4.546 092 E − 03
gallon (U.S. dry)	cubic meter (m^3)	4.404 884 E − 03
gallon (U.S. liquid)	cubic meter (m^3)	3.785 412 E − 03
gill (U.K.)	cubic meter (m^3)	1.420 654 E − 04
gill (U.S.)	cubic meter (m^3)	1.182 941 E − 04
$in.^3$	cubic meter (m^3)	1.638 706 E − 05
liter	cubic meter (m^3)	1.000 000*E − 03
ounce (U.K. fluid)	cubic meter (m^3)	2.841 307 E − 05
ounce (U.S. fluid)	cubic meter (m^3)	2.957 353 E − 05
peck (U.S.)	cubic meter (m^3)	8.809 768 E − 03
pint (U.S. dry)	cubic meter (m^3)	5.506 105 E − 04
pint (U.S. liquid)	cubic meter (m^3)	4.731 765 E − 04
quart (U.S. dry)	cubic meter (m^3)	1.101 221 E − 03
quart (U.S. liquid)	cubic meter (m^3)	9.463 529 E − 04
stere	cubic meter (m^3)	1.000 000*E + 00
tablespoon	cubic meter (m^3)	1.478 676 E − 05
teaspoon	cubic meter (m^3)	4.928 922 E − 06
ton (register)	cubic meter (m^3)	2.831 685 E + 00
yd^3	cubic meter (m^3)	7.645 549 E − 01
	Volume per Unit Time (Includes Flow)	
ft^3/min	cubic meter per second (m^3/s)	4.719 474 E − 04
ft^3/s	cubic meter per second (m^3/s)	2.831 685 E − 02
gallon (U.S. liquid)/hp · h)	cubic meter per joule (m^3/J)	1.410 089 E − 09
$in.^3/min.$	cubic meter per second (m^3/s)	2.731 177 E − 07
yd^3/min	cubic meter per second (m^3/s)	1.274 258 E − 02
gallon (U.S. liquid) per day	cubic meter per second (m^3/s)	4.381 264 E − 08
gallon (U.S. liquid) per minute	cubic meter per second (m^3/s)	6.309 020 E − 05
	Work (see Energy)	

Figure A-1 (Cont.)

Appendix

II Psychometric Charts

Two psychometric charts are shown in Figures A–2 and A–3. The use of these charts in this text is primarily to determine relative humidity, having measured wet- and dry-bulb temperatures. For example, for 40° dry bulb (4.4°C) and 37.5° wet bulb (3.05°C), the relative humidity is 80%.

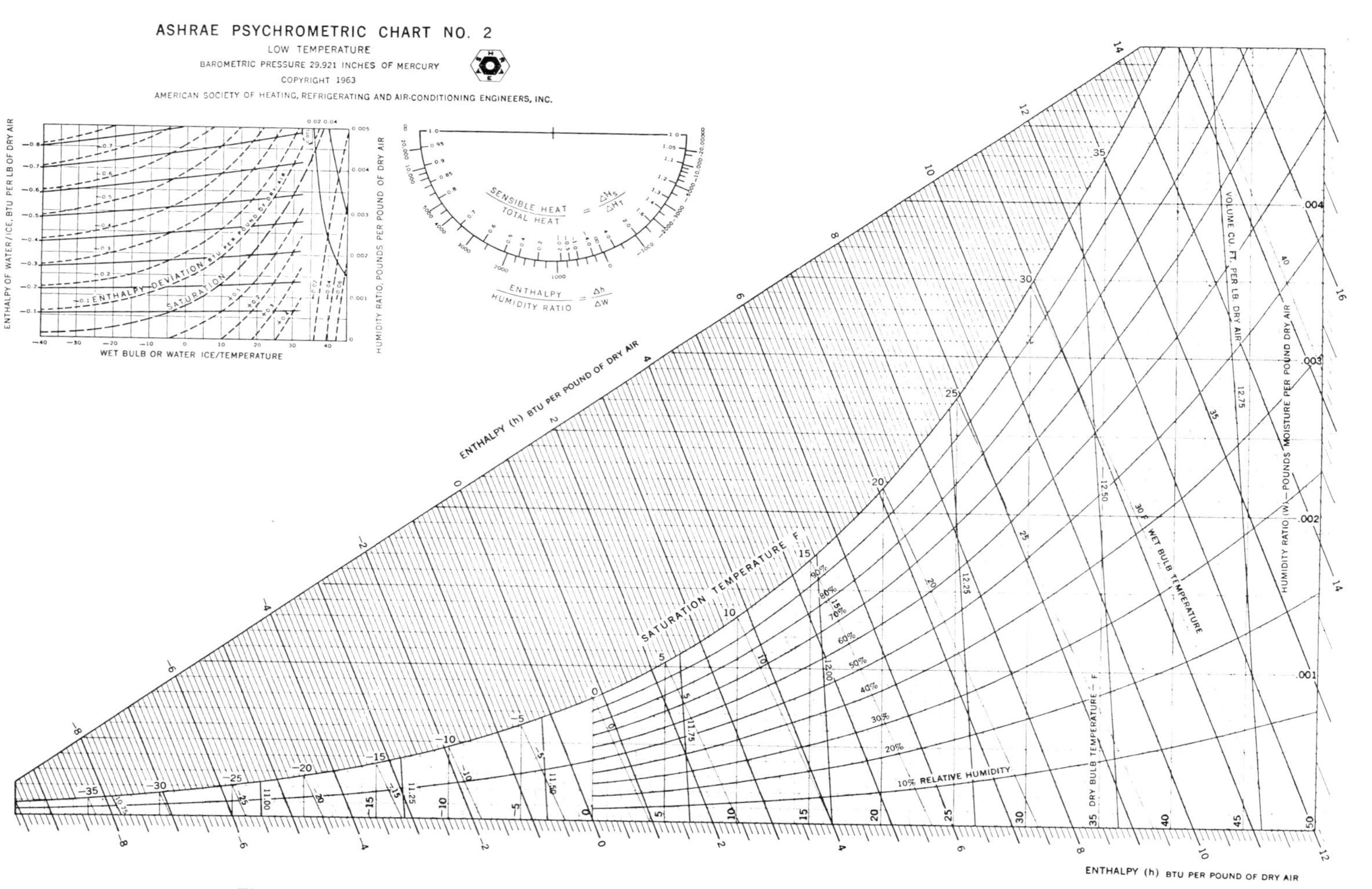

Figure A–2 Psychometric chart: low temperatures, English units. (By permission of *ASHRAE Handbook*.)

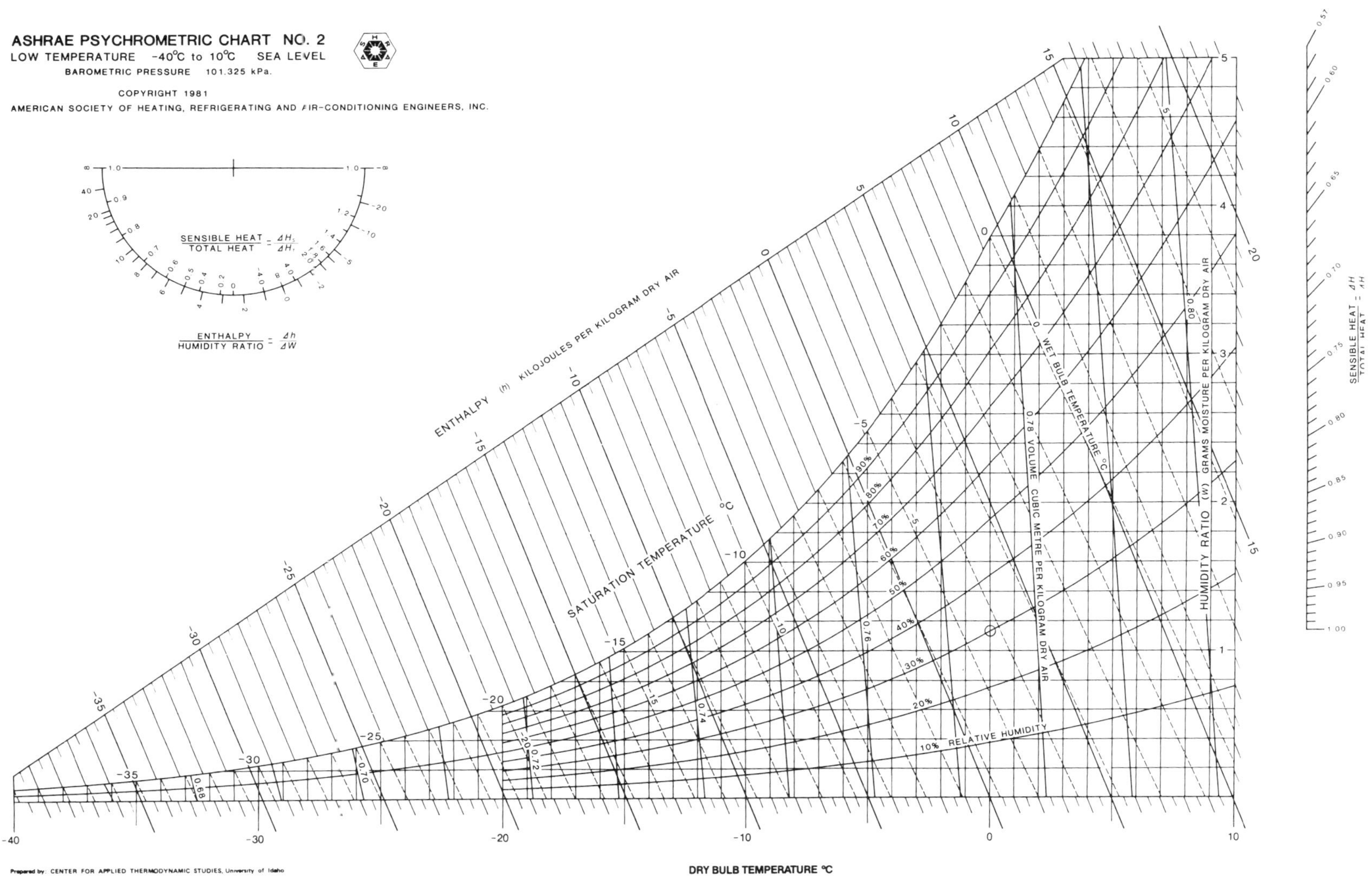

Figure A–3 Psychometric chart: low temperatures, SI units. (By permission of *ASHRAE Handbook*.)

Appendix

Safety Code of Mechanical Refrigeration

Anyone working in the refrigeration industry must be familiar with the information included in the Safety Code of Mechanical Refrigeration. Due to the importance of this material, it is included in this text for ready reference (Figure A–4).

CONTENTS

Section		Page
1.	Scope and Purpose	1
2.	Definitions	1
3.	Occupancy Classification	3
4.	Refrigerating System Classification by Type	4
5.	Refrigerant Classification	4
6.	Requirements for Institutional, Public Assembly, Residential, and Commercial Occupancies	6
7.	Requirements for Industrial Occupancies	8
8.	Design and Construction of Equipment	9
9.	Pressure-Limiting Devices	12
10.	Pressure-Relief Protection	12
11.	Installation Requirements	16
12.	Field Tests	17
13.	General Requirements	18
14.	Titles and Sources of Reference Standards	19

1. SCOPE AND PURPOSE

1.1 Scope. The application of this Code is intended to assure the safe design, construction, installation, operation, and inspection of every *refrigerating system* employing a fluid which normally is vaporized and liquefied in its refrigerating cycle, when employed under the occupancy classifications listed in Section 3. The provisions of this Code are not intended to apply to the use of water or air as a *refrigerant,* nor to gas bulk storage tanks that are not permanently connected to a *refrigerating system,* nor to *refrigerating systems* installed on railroad cars, motor vehicles, motor drawn vehicles or on shipboard. (For shipboard installations see ANSI/ASHRAE 26-1978)

1.2 Purpose. This Code is intended to establish reasonable safeguards to life, limb, health, and property; to define certain practices which are inconsistent with safety; and to prescribe standards of safety which will properly influence future progress and developments in *refrigerating systems.*

1.3 Application. This Code *shall* apply to *refrigerating systems* installed subsequent to its adoption and to parts replaced or added to systems installed prior to or subsequent to its adoption. In cases of practical difficulty or unnecessary hardship, the authority having jurisdiction may grant exceptions from the literal requirements of this Code or permit the use of other devices, materials or methods, but only when it is clearly evident that equivalent protection is thereby secured.

Equipment *listed* by *an approved nationally recognized testing laboratory* is deemed to meet the design, manufacture, and factory test requirements section of this code or equivalent, for the *refrigerant* or *refrigerants* for which such equipment is designed. *Listed refrigerating systems* are not required to be field tested to comply with Section 12 of this Code.

2. DEFINITIONS

2.1 ***Absorber (Adsorber)*** is that part of the *low side* of an *absorption system* used for absorbing (adsorbing) vapor *refrigerant.*

2.2 ***Absorption (Adsorption) System*** is a *refrigerating system* in which the gas evolved in the *evaporator* is taken up by an *absorber (adsorber).*

2.3 ***Approved*** means acceptable to the authorities having jurisdiction.

2.4 ***An Approved Nationally Recognized Testing Laboratory*** is one acceptable to the authorities having jurisdiction, that provides uniform testing and examination procedures and standards for meeting the design, manufacture and factory test requirements of this code, is properly organized, equipped, and qualified for testing, and has a follow-up inspection service of the current production of the listed products.

2.5 ***Brazed Joint*** is a gas-tight joint obtained by the joining of metal parts with alloys which melt at temperatures higher than 800F (426.5 °C) but less than the melting temperatures of the joined parts.

2.6 ***Brine*** is any liquid, used for the transmission of heat without a change in its state, having no flash point or a flash point above 150F (65.5 °C) determined by American Society for Testing and Materials method D93. (See 14.1.)

2.7 ***Companion or Block Valves*** are pairs of mating *stop valves,* valving off sections of systems and arranged so that these sections may be joined before opening these valves or separated after closing them.

2.8 ***Compressor*** is a specific machine, with or without accessories, for compressing a given *refrigerant* vapor.

2.9 ***Compressor Unit*** is a compressor with its prime mover.

2.10 ***Condenser*** is that part of the system designed to liquefy *refrigerant* vapor by removal of heat.

2.11 ***Condenser Coil*** is a *condenser* constructed of pipe or tubing other than a shell and tube or shell and coil type.

2.12 ***Condensing Unit*** is a specific refrigerating machine combination for a given *refrigerant,* consisting of one or more power-driven *compressors, condensers,*

Figure A–4 Safety code for mechanical refrigeration, ANSI/ASHRAE 15-1978. (By permission of *ASHRAE Handbook.*)

liquid receivers (when required), and the regularly furnished accessories.

2.13 ***Container*** is a cylinder for the transportation of *refrigerant*. (See 14.2)

2.14 ***Critical Pressure, Critical Temperature and Critical Volume*** are the terms given to the state points of a substance at which liquid and vapor have identical properties. Above the *critical pressure* or *critical temperature* there is no line of demarcation between liquid and gaseous phases.

2.15 ***Department Store*** is the entire space occupied by one *tenant* or more than one *tenant* in an individual store where more than 100 persons commonly assemble on other than the street-level floor for the purpose of buying personal wearables and other merchandise.

2.16 ***Design Pressure*** is the maximum allowable working pressure, psig (Pa), for which a specific part of a system is designed.

2.17 ***Direct System*** (See 4.2)

2.18 ***Double Direct System*** (See 4.3)

2.19 ***Double Indirect Vented Open-Spray System*** (See 4.4.4)

2.20 ***Duct*** is a tube or conduit used for conveying or encasing purposes as specifically defined below:

a) *Air Duct* is a tube or conduit used for conveying air. (The air passages of *self-contained systems* are not to be construed as *air ducts*.)

b) *Pipe Duct* is a tube or conduit used for encasing pipe.

c) *Wire Duct* is a tube or conduit used for encasing either moving or stationary wire, rope, etc.

2.21 ***Entrance*** is a confined passageway immediately adjacent to the door through which people enter a building.

2.22 ***Evaporator*** is that part of the system designed to vaporize liquid *refrigerant* to produce refrigeration.

2.23 ***Evaporator Coil*** is an *evaporator* constructed of pipe or tubing other than a shell and tube or shell and coil type.

2.24 ***Exit*** is a confined passageway immediately adjacent to the door through which people leave a building.

2.25 ***Field Test*** is a test performed in the field to prove system tightness.

2.26 ***Fusible Plug*** is a device having a predetermined-temperature fusible member for the relief of pressure.

2.27 ***Generator*** is a device equipped with a means of heating used in an *absorption system* to drive *refrigerant* out of solution.

2.28 ***Hallway*** is a corridor for the passage of people.

2.29 ***High Side*** means the parts of a *refrigerating system* subjected to *condenser* pressure.

2.30 ***Humanly Occupied Space*** is a space normally frequented or occupied by people but excluding *machinery rooms* and walk-in coolers used primarily for refrigerated storage.

2.31 ***Indirect Closed-Surface System*** (See 4.4.2)

2.32 ***Indirect Open-Spray System*** (See 4.4.1)

2.33 ***Indirect System*** (See 4.4)

2.34 ***Indirect Vented Closed-Surface System*** (See 4.4.3)

2.35 ***Internal Gross Volume*** is the volume as determined from internal dimensions of the *container* with no allowance for volume of internal parts.

2.36 ***Limited Charge System*** is a system in which, with the *compressor* idle, the internal volume and total *refrigerant* charge are such that the *design pressure* will not be exceeded by complete evaporation of the *refrigerant* charge.

2.37 ***Listed*** means equipment that has been tested and is identified as acceptable by *an approved nationally recognized testing laboratory*.

2.38 ***Liquid Receiver*** is a vessel permanently connected to a system by inlet and outlet pipes for storage of a liquid *refrigerant*.

2.39 ***Lobby*** is a waiting room or large *hallway* serving as a waiting room.

2.40 ***Low Side*** means the parts of a *refrigerating system* subjected to *evaporator* pressure.

2.41 ***Machinery*** is the refrigerating equipment forming a part of the *refrigerating system* including any or all of the following: *compressor, condenser, generator, absorber (adsorber), liquid receiver,* connecting *piping,* or *evaporator*.

2.42 ***Machinery Room*** is a room in which a *refrigerating system* is permanently installed and operated but not including *evaporators* located in a cold storage room, refrigerator box, air cooled space, or other enclosed space. Closets solely contained within, and opening only into, a room shall not be considered *machinery rooms* but *shall* be considered a part of the *machinery room* in which they are contained or open into. It is not the intent of this definition to cause the space in which a *self-contained system* is located to be classified as a *machinery room*. (See 11.13).

2.43 ***Machinery Room, Class T,*** is a *machinery room* with specific restrictions and requirements. (See 11.13.5)

2.44 ***Manufacturer*** is the company or organization which evidences its responsibility by affixing its name or nationally registered trademark or trade name to the refrigeration equipment concerned.

2.45 ***Mechanical Joint*** is a gas-tight joint, obtained by the joining of metal parts through a positive-holding mechanical construction.

2.46 ***Nonpositive Displacement Compressor*** is a *compressor* in which increase in vapor pressure is attain-

Figure A–4 (Cont.)

ed without changing the internal volume of the compression chamber.

2.47 ***Piping*** means the pipe or tube mains for interconnecting the various parts of a *refrigerating system*. *Piping* includes pipe, flanges, bolting, gaskets, valves, fittings, the pressure containing parts of other components such as expansion joints, strainers, and devices which serve such purposes as mixing, separating, muffling, snubbing, distributing, metering or controlling flow, pipe supporting fixtures and structural attachments.

2.48 ***Positive Displacement Compressor*** is a *compressor* in which increase in pressure is attained by changing the internal volume of the compression chamber.

2.49 ***Premises*** are the buildings and that part of the grounds of one property, where an installation would affect the safety of those buildings or adjacent property, and the occupants thereof.

2.50 ***Pressure-Imposing Element*** is any device or portion of the equipment used for the purpose of increasing the *refrigerant* vapor pressure.

2.51 ***Pressure-Limiting Device*** is a pressure-responsive mechanism designed to automatically stop the operation of the *pressure-imposing element* at a predetermined pressure.

2.52 ***Pressure-Relief Device*** is a pressure actuated valve or *rupture member* designed to automatically relieve excessive pressure. (Not temperature actuated.)

2.53 ***Pressure-Relief Valve*** is a pressure-actuated valve held closed by a spring or other means and designed to automatically relieve pressure in excess of its setting.

2.54 ***Pressure Vessel*** is any refrigerant-containing receptacle of a *refrigerating system* other than *evaporators* (each separate section of which does not exceed 1/2 cu ft (.014m^3) of refrigerant containing volume), *evaporator coils, compressors, condenser coils,* controls, headers, pumps, and *piping*.

2.55 ***Refrigerant*** is a substance used to produce refrigeration by its expansion or vaporization.

2.56 ***Refrigerating System*** is a combination of interconnected refrigerant-containing parts constituting one closed *refrigerant* circuit in which a *refrigerant* is circulated for the purpose of extracting heat. (See Section 4 for classification of *refrigerating systems* by type.)

2.57 ***Rupture Member*** is a device that will rupture at a predetermined pressure.

2.58 ***Saturation Pressure*** of a *refrigerant* is the pressure at which there is stable coexistence of the vapor and liquid or the vapor and solid phase.

2.59 ***Sealed Absorption System*** is a unit system for Group 2 *refrigerants* only in which all refrigerant- containing parts are made permanently tight by welding or brazing against *refrigerant* loss.

2.60 ***Self-Contained System*** is a complete factory-made and factory-tested system in a suitable frame or enclosure which is fabricated and shipped in one or more sections and in which no refrigerant-containing parts are connected in the field other than by *companion* or *block valves*.

2.61 ***Shall*** Where *"shall"* or *"shall not"* is used for a provision specified, that provision is intended to be mandatory.

2.62 ***Should*** *"Should"* or *"it is recommended"* is used to indicate provisions which are not mandatory but which are recommended good practice.

2.63 ***Soldered Joint*** is a gas-tight joint obtained by the joining of metal parts with metallic mixtures or alloys which melt at temperatures not exceeding 800F (426.5 °C) and above 400F (204.5 °C).

2.64 ***Stop Valve*** is a device to shut off the flow of *refrigerant*.

2.65 ***Tenant*** *shall* be construed as a person, firm, or corporation possessed with the legal right to occupy *premises*.

2.66 ***Ultimate Strength*** is the highest stress level which the component can tolerate without rupture.

2.67 ***Unprotected Tubing*** is tubing which is not protected by enclosure or suitable location so that it is exposed to crushing, abrasion, puncture or similar mechanical damage under installed conditions.

2.68 ***Unit System*** is a *self-contained system* which has been assembled and tested prior to its installation and which is installed without connecting any refirgerant-containing parts. A *unit system* may include factory-assembled *companion or block valves*.

2.69 ***Welded Joint*** is a gas-tight joint, obtained by the joining of metal parts in the plastic or molten state.

3. OCCUPANCY CLASSIFICATION

3.1 Locations Governed by This Code in which *refrigerating systems* may be placed are grouped by occupancy as follows:

3.1.1 Institutional Occupancy *shall* apply to that portion of the *premises* in which persons are confined to receive medical, charitable, educational, or other care or treatment, or in which persons are held or detained by reason of public or civic duty, including among others, hospitals, nursing homes, asylums, sanitariums, police stations, jails, courthouses with cells, and similar occupancies.

3.1.2 Public Assembly Occupancy *shall* apply to that portion of the *premises* in which persons congregate for civic, political, educational, religious, social, or recreational purposes; including among others, armories, assembly rooms, auditoriums, ballrooms, bath houses, bus terminals, broadcasting studios, churches, colleges, courthouses without cells, dance halls, *department stores,* exhibition halls, fraternity halls, libraries,

Figure A–4 (Cont.)

lodge rooms, mortuary chapels, museums, passenger depots, schools, skating rinks, subway stations, theaters, enclosed portions of arenas, race tracks and stadiums, and similar occupancies.

3.1.3 Residential Occupancy *shall* apply to that portion of the *premises* in which sleeping accommodations are provided, including among others, clubhouses, convents, dormitories, hotels, lodging houses, multiple story apartments, residences, studios, tenements, and similar occupancies.

3.1.4 Commercial Occupancy *shall* apply to that portion of the *premises* used for the transaction of business; for the rendering of professional services; for the supplying of food, drink, or other bodily needs and comforts; for manufacturing purposes or for the performance of work or labor (except as included under 3.1.5 Industrial Occupancy) including among others, bake shops, fur storage, laboratories, loft buildings, markets, office buildings, professional buildings, restaurants, stores other than *department stores,* and similar occupancies.

3.1.5 Industrial Occupancy *shall* apply to an entire building or *premises* or to that portion of a building usedfor manufacturing, processing, or storage of materials or products, including among others, chemical, food, candy and ice cream factories, ice- making plants, meat packing plants, refineries, perishable food warehouses and similar occupanices. In an Industrial Occupancy, when the number of persons in a refrigerated space, served by a *direct system,* on any floor above the first floor (ground level or deck level) exceeds one person per 100 sq ft (9.29 sq m) of floor area, the requirements of Commercial Occupancy *shall* apply unless that refrigerated space containing more than one person per 100 sq ft (9.29 sq m) of floor area, above the first floor is provided with the required number of doors opening directly into *approved* building exits. Such refrigerated space shall be cut off from the rest of the building by tight construction with tight-fitting doors. (See 14.17).

NOTE: The above does not prohibit openings for the passage of products from one refrigerated space to another refrigerated space.

3.1.6 Mixed Occupancy *shall* apply to a building occupied or used for different purposes in different parts. When the occupancies are cut off from the rest of the building by tight partitions, floors, and ceilings and protected by self-closing doors, the requirements for each type of occupancy *shall* apply for its portion of the building or *premises.* For example, the cold storage spaces in retail frozen food lockers, hotels, and *department stores* might be classified under Industrial Occupancy, whereas other portions of the building would be classified under other occupancies. When the occupancies are not so separated, the occupancy carrying the more stringent requirements *shall* govern.

3.2 Adjacent Locations. Equipment other than *piping* installed in locations adjacent to areas outlined in 3.1.1 through 3.1.6 and located outside of, but less than 20 ft (6.10 m) from any building opening, shall be governed by the occupancy classification of the building. Equipment installed in a non-adjacent location, such as equipment in a separate building located 20 ft (6.10 m) or more from an opening in any other building, *shall* be governed by the provisions of 3.1.

4. REFRIGERATING SYSTEM CLASSIFICATION BY TYPE

4.1 Refrigerating Systems are classified by the method employed for extracting heat as follows:

4.2 Direct System is one in which the *evaporator* is in direct contact with the material or space refrigerated or is located in air-circulating passages communicating with such spaces.

4.3 Double Direct System is one in which an evaporative *refrigerant* is used in a secondary circuit to condense or cool a *refrigerant* in a primary circuit. For the purpose of this Code, each system enclosing a separate body of an evaporative *refrigerant shall* be considered as a separate *direct system.*

4.4 Indirect System is one in which a *brine* cooled by the *refrigerant,* is circulated to the material or space refrigerated or is used to cool air so circulated. *Indirect systems* which are distinguished by the type or method of application are as given in the following paragraphs:

4.4.1 *Indirect Open-Spray System* is one in which a *brine* cooled by an *evaporator* located in an enclosure external to a cooling chamber is circulated to such cooling chamber, and is sprayed therein.

4.4.2 *Indirect Closed-Surface System* is one in which a *brine* cooled by an *evaporator* located in an enclosure external to a cooling chamber, is circulated to and through such a cooling chamber in pipes or other closed circuits.

4.4.3 *Indirect Vented Closed-Surface System* is one in which a *brine* cooled by an *evaporator* located in a vented enclosure external to a cooling chamber, is circulated to and through such cooling chamber in pipes or other closed circuits.

4.4.4 *Double Indirect Vented Open-Spray System* is one in which a *brine* cooled by an *evaporator* located in a vented enclosure, is circulated through a closed circuit to a second enclosure where it cools another supply of a *brine* and this liquid in turn is circulated to a cooling chamber and is sprayed therein.

5. REFRIGERANT CLASSIFICATION

5.1 General. *Refrigerants shall* be classified by their toxicity or flammability and accordingly are divided in-

4

ASHRAE STANDARD 15-78

Figure A–4 (Cont.)

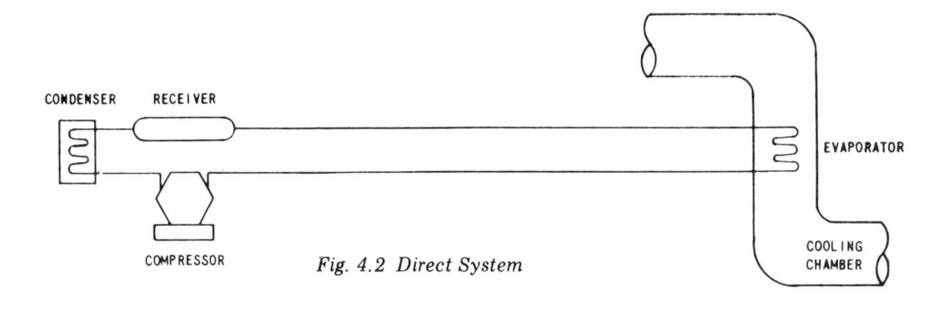

Fig. 4.2 Direct System

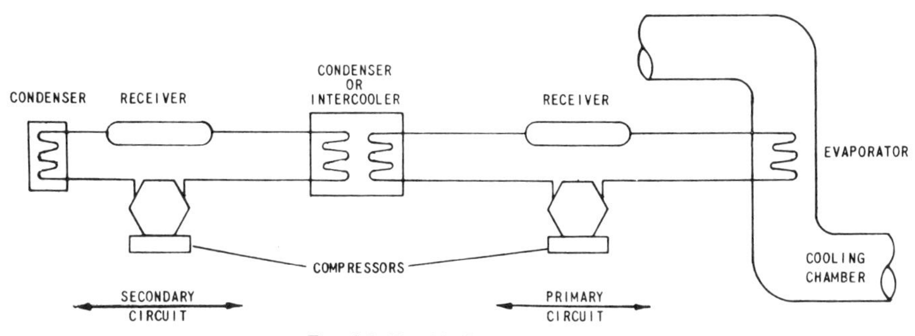

Fig. 4.3 Double Direct System

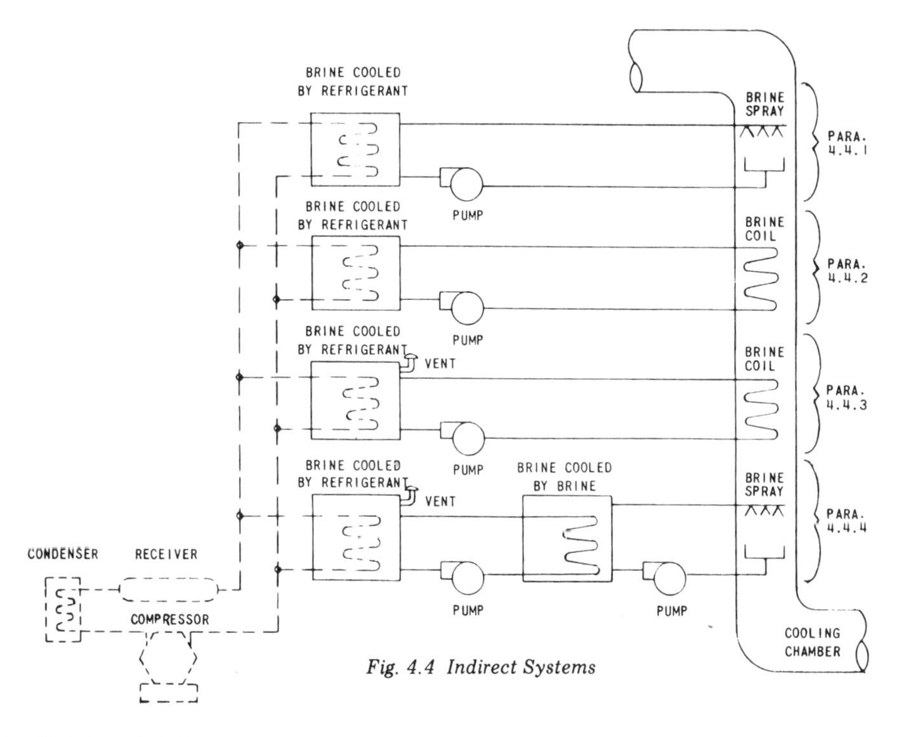

Fig. 4.4 Indirect Systems

ASHRAE STANDARD 15-78 5

Figure A-4 (Cont.)

to groups as follows:

Refrigerant*	Name	Chemical Formula
5.1.1		
Group 1		
R-11	Trichlorofluoromethane	CCl_3F
R-12	Dichlorodifluoromethane	CCl_2F_2
R-13	Chlorotrifluoromethane	$CClF_3$
R-13B1	Bromotrifluoromethane	$CBrF_3$
R-14	Tetrafluoromethane	CF_4
R-21	Dichlorofluoromethane	$CHCl_2F$
R-22	Chlorodifluoromethane	$CHClF_2$
R-30	Dichloromethane (Methylene chloride)	CH_2Cl_2
R-113	Trichlorotrifluoroethane	CCl_2FCClF_2
R-114	Dichlorotetrafluoroethane	$CClF_2CClF_2$
R-115	Chloropentafluoroethane	$CClF_2CF_3$
R-C318	Octafluorocyclobutane	C_4F_8
R-500	Dichlorodifluoromethane, 73.8% and Ethylidene Difluoride, 26.2%	CCl_2F_2/CH_3CHF_2
R-502	Chlorodifluoromethane, 48.8% and Chloropentafluoroethane, 51.2%	$CHClF_2/CClF_2CF_3$
R-503	Trifluoromethane 40.1%, and Chlorotrifluoromethane 59.9%	$CHF_3/CClF_3$
R-744	Carbon dioxide	CO_2
5.1.2		
Group 2		
R-40	Methyl chloride	CH_3Cl
R-611	Methyl formate	$HCOOCH_3$
R-717	Ammonia	NH_3
R-764	Sulphur dioxide	SO_2
5.1.3		
Group 3		
R-170	Ethane	C_2H_6
R-290	Propane	C_3H_8
R-600	Butane	C_4H_{10}
R-600a	Isobutane	$CH(CH_3)_3$
R-1150	Ethylene	C_2H_4
R-1270	Propylene	C_3H_6

*Group Classification-See 14.19

6. REQUIREMENTS FOR INSTITUTIONAL, PUBLIC ASSEMBLY, RESIDENTIAL, AND COMMERCIAL OCCUPANCIES

6.1 General

6.1.1 Public stairway, stair landing, *entrance* or *exit*. No portion of a *refrigerating system shall* be installed in or on a public stairway, stair landing, *entrance*, or *exit*.

6.1.2 Public *hallway* or *lobby*. No portion of a *refrigerating system shall* interfere with free passage through these areas. No portion of a *refrigerating system* containing a Group 2 *refrigerant shall* be permitted in public *hallways* or *lobbies* of Institutional or Public Assembly Occupancies. *Refrigerating systems* installed in a public *hallway* or *lobby shall* be limited to:

a) *Unit Systems* containing not more than the quantities of a Group 1 *refrigerant* specified in Table 1, or

b) *Sealed absorption systems* containing not more than 3 lbs (1.36 kg.) of a Group 2 *refrigerant* when in Residential and Commercial Occupancies.

6.1.3 When the refrigerant-containing parts of a system are located in one or more enclosed spaces, the cubical content of the smallest enclosed *humanly occupied space* other than the *machinery room, shall* be used to determine the permissible quantity of *refrigerant* in the system. Where a *refrigerating system* has *evaporator* coils serving individual stories of a building, the story having the smallest volume *shall* be used to determine the maximum quantity of *refrigerant* in the entire system.

6.1.4 When the *evaporator* is located in an *air duct* system, the cubical content of the smallest *humanly occupied* enclosed space served by the *air duct* system *shall* be used to determine the permissible quantity of *refrigerant* in the system. If the air flow to any enclosed space served by the *air duct* system cannot be shut off or reduced below one-quarter of its maximum, the cubical contents of the entire space served by the *air duct* system may be used to determine the permissible quantity of *refrigerant* in the system.

6.1.5 Where the return air space above a suspended ceiling is one continuous space and not an enclosed *air duct* in which the return air is confined, this space may be included in calculating the cubical content of the *humanly occupied space*.

6.1.6 In Institutional and Public Assembly Occupancies, direct *expansion coils* or *evaporators* used for air conditioning and located downstream from, and in proximity to, a heating coil, or located upstream within 18 in. (0.46m) of a heating coil, *shall* be fitted with a *pressure relief device* discharging to the outside of the building in an *approved* manner; except that such a relief device *shall not* be required on *unit* or *self-contained systems* if the internal volume of the *low side* of the system which may be shut off by valves, divided by the total weight of *refrigerant* in the system less the weight of *refrigerant* vapor contained in the other parts of the system at 110F (43.5 °C) exceeds the specific volume of the *refrigerant* at critical conditions of temperature and pressure.

NOTE: The above exemption is also stated in formula form below.

$V_1/W_1 - W_2$ *shall* be more than V_{gc}

where V_1 = low side volume, cu ft (cu m)

V_{gc} = specific volume at critical conditions of temperature and pressure, cu ft per lb (cu m per kg)
W_1 = total weight of refrigerant in system, lb (kg)
V_2 = total volume of system less V_1 cu ft (cu m.)
V_{gt} = specific volume of refrigerant vapor at 110F (43.5 °C), ft^3/lb (cu m/kg)
$W_2 = V_2/V_{gt}$ = weight of refrigerant vapor in V_2 at 110F (43.5°C)

6.2 Group 1 Refrigerants

6.2.1 ***Direct Systems.*** The maximum permissible quantity of a Group 1 *refrigerant* in a *direct system shall* be as specified in Table 1 except for additional limitations specified in 6.2.1.1 for Institutional Occupancies.

6.2.1.1 ***Direct Systems*** in Institutional Occupancies *shall* be limited to systems each containing not more

6

ASHRAE STANDARD 15-78

Figure A-4 (Cont.)

TABLE 1 MAXIMUM PERMISSIBLE QUANTITIES OF GROUP 1 REFRIGERANTS FOR DIRECT SYSTEMS

Refrigerant		Maximum Quantity in lb per 1000 cu ft of Humanly Occupied Space*	Maximum Quantity in kg per 100 cubic meters of Humanly Occupied Space
R-11	Trichlorofluoromethane	35	56.2
R-12	Dichlorodifluoromethane	31	49.8
R-13	Chlorotrifluoromethane	27	43.1
R-13B1	Bromotrifluoromethane	38	60.8
R-14	Tetrafluoromethane	23	36.7
R-21	Dichlorofluoromethane	13	20.8
R-22	Chlorodifluoromethane	22	35.3
R-30	Dichloromethane (Methylene chloride)	6	9.5
R-113	Trichlorotrifluoroethane	24	38.5
R-114	Dichlorotetrafluoroethane	44	70.7
R-115	Chloropentafluoroethane	40	64.0
R-C318	Octafluorocyclobutane	50	80.2
R-500	Dichlorodifluoromethane 73.8% and Ethylidene Difluoride, 26.2%	26	41.7
R-502	Chlorodifluoromethane, 48.8% and Chloropentafluoroethane, 51.2%	30	48.0
R-503	Trifluoromethane, 40.1% and Chlorotrifluoromethane 59.9%	22	35.3
R-744	Carbon dioxide	11	17.7

*Volatile Charge in a control shall not be considered as refrigerant

TABLE 2 MAXIMUM PERMISSIBLE QUANTITIES OF GROUP 2 REFRIGERANTS FOR DIRECT SYSTEMS

Type of refrigerating system	Maximum pounds for various occupancies (kg) Institutional	Public assembly	Residential	Commercial
Sealed Absorption Systems:				
a) In public *hallways* or *lobbies*	0 (0)	0 (0)	3 (1.4)	3 (1.4)
b) In other than public *hallways* or *lobbies*	0*(0)	6 (2.7)	6 (2.7)	20 (9.1)
Self-Contained or *Unit Systems:*				
a) In public *hallways* or *lobbies*	0 (0)	0 (0)	0 (0)	0 (0)
b) In other than public *hallways* or *lobbies*	0 (0)	0*(0)	6 (2.7)	20 (9.1)

*6 lbs (2.72 kg) allowed when installed in kitchens, laboratories and mortuaries.

than 50% of the permissible quantities of Group 1 *refrigerants* specified in Table 1 except in kitchens, laboratories and mortuaries.

6.2.2 ***Indirect Systems.*** A system containing more than the quantity of a Group 1 *refrigerant* allowed in Table 1 *shall* be of the *indirect* type. All refrigerant-containing parts except *piping shall* be installed in a *machinery room* or installed outside the building. *Piping shall* be installed in accordance with 11.12.3. The *machinery room shall* be used for mechanical equipment only.

6.2.3 Open flames in *machinery rooms.* No open flames or apparatus to produce an open flame *shall* be installed in a *machinery room* where any *refrigerant* other than carbon dioxide is used unless the flame is enclosed so that the products of combustion are vented to the open air. The use of matches, cigarette lighters, halide leak detectors, and similar devices *shall not* be considered a violation of this paragraph or of 6.2.4.

6.2.4 Open flames in institutional occupancies. In Institutional Occupancies where more than 1 lb (0.45 kg.) of a Group 1 *refrigerant*, other than carbon dioxide is used in a system, any portion of which is in a room where there is an apparatus for producing an open flame, then such *refrigerant shall* be classed in Group 2, unless the flame-producing apparatus is provided with a hood and flue capable of removing the products of combustion to the open air.

6.3 Group 2 Refrigerants

6.3.1 ***Direct Systems.*** Direct systems containing Group 2 *refrigerants shall not* be used for air conditioning for human comfort. For other applications, the maximum permissible quantity of Group 2 *refrigerants* in a *direct system shall* be as specified in Table 2.

6.3.2 ***Indirect Systems.*** Except as provided in 6.3.2.1 or 6.3.2.2.1, the maximum permissible quantity of a Group 2 *refrigerant* in any *indirect system shall* be

Figure A–4 (Cont.)

as specified in Table 3. Except as provided in 6.3.2.1, such systems *shall* be of the following type:

a) Institutional and Public Assembly Occupancies - *indirect vented closed-surface,* or *double indirect vented open-spray.*

b) Residential and Commercial Occupancies - *indirect closed-surface, indirect vented closed-surface,* or *double indirect vented open-spray,* or secondary circuit of *double direct* type.

TABLE 3 MAXIMUM PERMISSIBLE QUANTITIES OF GROUP 2 REFRIGERANTS FOR INDIRECT SYSTEMS

Occupancy	Class T Machinery Rooms Max Lb (kg)
Institutional	500 (226.8)
Public Assembly	1000 (453.6)
Residential	No Limit
Commercial	No Limit

6.3.2.1 *Indirect Systems* using Group 2 *refrigerants* and conforming with the provisions of 6.3.1 for *direct systems* shall be permitted.

6.3.2.2 *Indirect Systems* using Group 2 *refrigerants* not in excess of the quantities shown in Table 3, other than such systems conforming with the provisions of 6.3.2.1 or 6.3.2.2.1, *shall* have all refrigerant-containing parts, except *piping,* installed in Class T *machinery room* in accordance with 11.13.5, except that an air-cooled or evaporative *condenser* may be installed outside the building. *Piping shall* be installed in accordance with 11.12.3. *The Class T machinery room shall* be used for mechanical equipment only.

6.3.2.2.1 A sealed ammonia-water absorption *unit system* containing not more than 20 lbs. (9.07 kg) of ammonia and installed outdoors adjacent to a commercial or residential occupancy *shall not* be required to conform with provisions of 6.3.2.2.

6.3.2.3 Flame-producing devices, hot surfaces, and electrical equipment in *Class T machinery rooms* (see 11.13.5). Where a *Class T machinery room* is provided to comply with 6.3.2.2 to house a *refrigerating system* containing any Group 2 *refrigerant* other than sulphur dioxide, no flame-producing device or hot surface above 800F (426.5 °C) *shall* be permitted in such room and all electrical equipment in the room *shall* conform to the requirements of Hazardous Locations Class I, Division 2, of the National Electrical Code (see 14.6). The use of matches, cigarette lighters, halide leak detectors, combustion engines, gas turbines and similar devices *shall not* be considered a violation of this paragraph, provided the quantities of Group 2 *refrigerants* in Table 4 are not exceeded.

6.3.2.4 Group 2 *refrigerants* listed in Table 4 *shall not* be used in a *refrigerating system* in excess of 1000 lbs (454 kg) unless approved by the authority having jurisdiction.

6.4 Group 3 Refrigerants

6.4.1 Group 3 *refrigerants shall not* be used in Institutional, Public Assembly, Residential, or Commercial Occupancies except in laboratories for Commercial Occupancies. In such laboratory installations only *unit systems* containing not more than 6 lbs (2.72 kg) *shall* be used unless the number of persons does not exceed one person per 100 sq ft (9.29 sq m) of laboratory floor area, in which case the requirements of Section 7 shall apply.

7. REQUIREMENTS FOR INDUSTRIAL OCCUPANCIES

7.1 General. There *shall* be no restrictions on the quantity or kind of *refrigerant* used in an Industrial Occupancy, except as specified in 7.2, 7.3, 7.4, and 7.5.

7.2 Open Flames. When the quantity of flammable *refrigerant* in any one *refrigerating system* exceeds the amount given in Table 4 for each 1000 cu. ft. (28.3 cu m) of room volume in which the system, or any part thereof is installed, then no flame producing device or hot surface above 800F (426.5 °C) *shall* be permitted and such room *shall* be considered a Class I, Division 2 Location, and all electrical equipment in the room *shall* conform to the requirements of Class I, Division 2, of the National Electrical Code. (see 14.6)

7.3 Flammable Refrigerants as listed in Table 4, *shall not* be used in a *refrigerating system* in excess of 1000 lbs (454 kg) unless approved by the authority having jurisdiction.

7.4 Ammonia Systems

7.4.1 *Machinery Room.* When ammonia is used in a *refrigerating system,* all refrigerant-containing parts, except *piping* and *evaporators,* and except refrigerant-containing components installed outside the building, *shall* be installed in a *machinery room* (see 11.13) under

TABLE 4 MAXIMUM PERMISSIBLE QUANTITIES OF FLAMMABLE REFRIGERANTS

Refrigerant	Group	Name	Maximum Quantity in lb per 1000 cu ft of Room Volume	Maximum Quantity in kg per 100 cu meters of Room Volume
R-40	2	Methylchloride	10	15.9
R-611	2	Methylformate	7	11.3
R-170	3	Ethane	2 1/2	3.9
R-290	3	Propane	2 1/2	3.9
R-600	3	Butane	2 1/2	3.9
R-600a	3	Isobutane	2 1/2	3.9
R-1150	3	Ethylene	2	3.2
R-1270	3	Propylene	2	3.2

Figure A-4 (Cont.)

the conditions of 7.4.1.1 or 7.4.1.2 or 7.4.1.3, and such room *shall* have no flame-producing apparatus permanently installed and operated.

7.4.1.1 The room *shall* be provided with a continuously operated, independent mechanical ventilation system, and this room *shall* be considered a "Non-Hazardous (Unclassified) Location" in accordance with the National Electrical Code (see 14.6). Failure of the mechanical ventilation system *shall* also initiate a supervised alarm so corrective action can be initiated.

7.4.1.2 The room *shall* be provided with an independent mechanical ventilation system actuated automatically by a vapor detector(s) when the concentration of ammonia in the room exceeds 40,000 parts per million (25% of LEL), and also operable manually, and this room *shall* be considered a "Non-Hazardous (Unclassified) Location" in accordance with the National Electrical Code (see 14.6). The vapor detector(s) *shall* also initiate a supervised alarm so corrective action can be initiated. Periodic tests of detector(s)/alarm/mechanical ventilation system *shall* be performed.

7.4.1.3 Where mechanical ventilation is not provided in accordance with 7.4.1.1 or 7.4.1.2, the room *shall* be considered a Class I, Division 2 Location, and all electrical equipment in the room *shall* conform to the requirements for a Class I, Division 2 Location, of the latest edition of the National Electrical Code (see 14.6).

7.5 Refrigerated Storage Areas and Work Areas

7.5.1 Refrigerated Storage Areas. There *shall* be no additional requirements for refrigerant-containing components in refrigerated storage areas for any *refrigerant,* except as specified in 7.2 and 7.3.

7.5.2 Refrigerated Work Areas. In refrigerated work areas where *refrigerants* in Table 4 are used, the area *shall* be considered a Class I, Division 2 Location, and all electrical equipment in the room *shall* conform to the requirements for a Class I, Division 2 Location of the National Electrical Code (See 14.6).

7.5.2.1 When any *refrigerant* other than those listed in Table 4 is used, the area *shall* be considered a "Non-Hazardous (Unclassified) Location" in accordance with the National Electrical Code (see 14.6).

7.5.2.2 When any Group 2 or Group 3 *refrigerant* is used, means *shall* be taken to adequately safeguard *piping*, controls, and other refrigeration equipment in working areas to minimize the possibility of accidental damage or rupture from external sources.

7.5.2.3 Areas through which *piping* for a Group 2 or Group 3 *refrigerant* is run *shall* be considered a refrigerated work area.

8. DESIGN AND CONSTRUCTION OF EQUIPMENT

8.1 Materials

8.1.1 All materials used in the construction and installation of *refrigerating systems shall* be suitable for conveying the *refrigerant* used. Some *refrigerants* are corrosive to the usual materials when moisture or air, or both, are present. No material *shall* be used that will deteriorate because of the *refrigerant,* or the oil, or the combination of both.

8.1.2 Aluminum, Zinc, or Magnesium *shall not* be used in contact with methyl chloride in a *refrigerating system.* Magnesium alloys *shall not* be used in contact with any halogenated *refrigerant.*

8.1.3 Copper and its alloys *shall not* be used in contact with ammonia except as a component of bronze alloys for compressor parts or similar uses where compatibility has been established.

8.2 Design Pressure

8.2.1 *Design pressures shall* be selected high enough for all operating and standby conditions. When selecting the *design pressure,* consideration *should* be given to allowances for setting *pressure-limiting devices* and *pressure-relief devices* sufficiently above operating conditions to avoid nuisance shutdowns and for shipping conditions.

Minimum *design pressure shall not* be less than 15 psig (103.4 kPa gage) and except as noted in 8.2.2, 8.2.3, and 8.2.4, *shall not* be less than the saturation pressure of the *refrigerant* at the following temperatures:

a) *low sides* of all systems +80F (26.5 °C)
b) *high side* of water or evaporatively cooled systems +105F (40.5 °C)
c) *high sides* of air-cooled systems +125F (51.5 °C)

Corresponding pressures are given in Table 5 for the *refrigerants* in common use.

8.2.1.1 All operating conditions are intended to cover maximum pressures attained under any anticipated normal operating conditions.

8.2.1.2 Standby conditions are intended to include conditions which may be attained in the system when not operating, including pressure developed in the *low side* of the system resulting from equalization after the system has stopped.

8.2.1.3 Shipping conditions are intended to include consideration of maximum pressures attainable due to anticipated environmental conditions during transit.

8.2.2 The *design pressure* for either the *high* or *low side* need not exceed the *critical pressure* of the *refrigerant* unless the system is intended to operate at these conditions.

8.2.3 When a part of a *limited charged system* is protected by a *pressure-relief device,* the *design pressure* of that part need not exceed the setting of the *pressure-relief device.*

8.2.4 When a *compressor* is used as a booster to obtain a low pressure and discharges into the suction side of another system, the booster *compressor* is considered

Figure A-4 (Cont.)

TABLE 5
MINIMUM* DESIGN PRESSURES
***Selection of Higher Design Pressures May be Required to Satisfy Actual Shipping, Operating or Standby Conditions**

			Minimum Design Pressures, PSIG (k Pa gage)	
			High Side	
Refrigerant	Name	Low Side	Evap. or Water Cooled	Air Cooled
R-11	Trichlorofluoromethane	15 (103)	15 (103)	21 (145)
R-12	Dichlorodifluoromethane	85 (586)	127 (876)	169 (1165)
R-13	Chlorotrifluoromethane	521 (3592)	547 (3772)	547 (3772)
R-13B1	Bromotrifluoromethane	230 (1586)	321 (2213)	410 (2827)
R-14	Tetrafluoromethane	544 (3751)	544 (3751)	544 (3751)
R-21	Dichlorofluoromethane	15 (103)	29 (200)	46 (317)
R-22	Chlorodifluoromethane	144 (993)	211 (1455)	278 (1917)
R-30	Methylene Chloride	15 (103)	15 (103)	15 (103)
R-40	Methyl Chloride	72 (496)	112 (772)	151 (1041)
R-113	Trichlorotrifluoroethane	15 (103)	15 (103)	15 (103)
R-114	Dichlorotetrafluoroethane	18 (124)	35 (241)	53 (365)
R-115	Chloropentafluoroethane	152 (1048)	194 (1338)	252 (1737)
R-170	Ethane	616 (4247)	709 (4888)	709 (4888)
R-290	Propane	129 (889)	188 (1296)	244 (1682)
R-C318	Octafluorocyclobutane	34 (234)	59 (407)	85 (586)
R-500	Dichlorodifluoromethane, 73.8% and Ethylidene Difluoride, 26.2%	102 (703)	153 (1055)	203 (1400)
R-502	Chlorodifluoromethane, 48.8% and Chloro-pentafluoromethane 51.2%	162 (1117)	232 (1600)	302 (2082)
R-503	Trifluoromethane 40.1%, Chlorotri-fluoromethane 59.9%	617 (4254)	617 (4254)	617 (4254)
R-600	N-Butane	23 (158)	42 (290)	61 (420)
R-600a	Isobutane	39 (269)	63 (434)	88 (607)
R-611	Methyl Formate	15 (103)	15 (103)	15 (103)
R-717	Ammonia	139 (958)	215 (1482)	293 (2020)
R-744	Carbon Dioxide	955 (6585)	1058 (7295)	1058 (7295)
R-764	Sulfur Dioxide	45 (310)	78 (538)	115 (793)
R-1150	Ethylene	732 (5074)	732 (5074)	732 (5074)
R-1270	Propylene	160 (1103)	228 (1572)	294 (2027)

a part of the *low side* provided that a low pressure stage *compressor* of the *positive displacement* type is protected by a *pressure-relief device*. (See 10.5).

8.2.5 Any components connected to *pressure vessels shall* have a *design pressure* equal to or greater than the *pressure vessels*.

8.3 Refrigerant-Containing Pressure Vessels

8.3.1 *Pressure vessels* exceeding 6 in. (152 mm) Inside Diameter and having maximum internal or external *design pressure* greater than 15 psig (103.4 kPa gage), *shall* comply with the rules of Section VIII of the ASME Boiler and Pressure Vessel Code (See 14.15) covering the requirements for the design, fabrication, inspection and testing during construction of unfired *pressure vessels*.

8.3.2 *Pressure vessels* not exceeding 6 in. (152 mm) Inside Diameter, except those having a maximum internal or external *design pressure* of 15 psig (103.4 kPa gage) or less, *shall* be *listed* either individually or as part of refrigeration equipment, by *an approved nationally recognized testing laboratory* or *shall* meet the design, fabrication, and testing requirements of Section VIII of the ASME Boiler and Pressure Vessel Code. (See 14.15).

8.3.3 *Pressure vessels* having a maximum internal or external *design pressure* of 15 psig (103.4 kPa gage) or less, except as noted in 8.3.4 and 8.3.5, *shall* have an *ultimate strength* to withstand at least three times the *design pressure* and *shall* be tested to at least 1-1/3 times the *design pressure* for which they are rated.

8.3.4 If a *pressure-relief device* is used to protect a *pressure vessel* not exceeding 6 in. (152 mm) Inside Diameter, the *ultimate strength* of the *pressure vessel* so protected *shall* be sufficient to withstand at least 2-1/2 times the pressure setting of the *pressure-relief device*.

8.3.5 If a *fusible plug* is used to protect a *pressure vessel* not exceeding 6 in. (152 mm) Inside Diameter, the *ultimate strength* of the *pressure vessel* so protected *shall* be sufficient to withstand at least 2-1/2 times the *refrigerant saturation pressure* corresponding to the stamped temperature on the *fusible plug,* or at least 2-1/2 times the *critical pressure* of the *refrigerant* used, whichever is smaller.

Figure A-4 (Cont.)

8.4 Refrigerant Piping, Valves, Fittings, and Related Parts

8.4.1 Refrigerating *piping,* valves, fittings, and related parts, having a maximum internal or external *design pressure* greater than 15 psig (103.4 kPa gage) *shall* be *listed* either individually or as part of refrigeration equipment by *an approved nationally recognized laboratory;* or *shall* comply with the ANSI Code for Refrigeration Piping B31.5 where applicable (See 14.10). In either case, the following additional requirements apply:

8.4.2 Specific minimum requirements for *unprotected refrigerant* pipe or tubing.

8.4.2.1 *Unprotected* seamless drawn temper copper water tube used for *refrigerant piping* erected on the *premises, shall* conform to ASTM Specification B88 Types K or L (see 14.13) for specifications, dimensions and tolerances.

8.4.2.2 Unprotected soft annealed copper tubing used for *refrigerant piping* erected on the *premises, shall* conform to, and *shall* be limited to the sizes and wall thicknesses and tolerances in ASTM Specification B280 (See 14.14).

8.4.3 Metal enclosures or *pipe ducts* for soft copper tubing.

Rigid or flexible metal enclosures *shall* be provided for soft, annealed copper tubing used for *refrigerant piping* erected on the *premises* and containing Group 2 or 3 *refrigerants,* except that no enclosures *shall* be required for connections between a *condensing unit* and the nearest protected (see 2.67) riser, provided such connections do not exceed 6 ft (1.83 m) in length.

8.4.4 Joints on copper tubing which are made by the addition of filler metal *shall* be *brazed* in *refrigerating systems* containing Group 2 or Group 3 *refrigerants,* and may be *brazed* or *soldered* in *refrigerating systems* containing Group 1 *refrigerants.*

8.5 Components Other than Pressure Vessels and Piping

8.5.1 Every pressure-containing component of a *refrigerating system* other than *pressure vessels, piping,* pressure gages and control mechanisms, *shall* be *listed* either individually or as part of refrigeration equipment by *an approved nationally recognized testing laboratory* or *shall* be designed, constructed and assembled to have an *ultimate strength* sufficient to withstand at least three times the *design pressure* for which it is rated.

8.5.2 Liquid level gage glasses, except those of the bull's-eye or reflex type, *shall* have automatic closing shut-off valves, and such glasses *shall* be protected against damage.

8.5.3 Dial of a pressure gage, when the gage is permanently installed on the *high side* of a *refrigerating system, shall* be graduated to at least 1.2 times the *design pressure.*

8.5.4 *Liquid receivers* or parts of a system designed to receive the *refrigerant* charge during pumpdown, *shall* have sufficient capacity to receive the charge without the liquid occupying more than 90% of the volume when the temperature of the *refrigerant* is 90 F (32°C).

8.5.5 Butt welded steel, other than open hearth type which is not rephosphorized, or wrought iron pipe *shall not* be used for *evaporator* or *condenser coils.*

8.6 Service Provisions

8.6.1 General. All systems *shall* have provisions to handle safely the *refrigerant* charge for service purposes. Properly located *stop valves,* separate storage tanks or adequate venting for safe disposal are satisfactory for this purpose.

8.6.2 Systems containing more than 6 lbs (2.72 kg) of a Group 2 or 3 *refrigerant,* other than systems utilizing *nonpositive displacement compressors, shall* have *stop valves* installed as follows:

a) Each inlet of each *compressor, compressor unit,* or *condensing unit;*

b) Each discharge outlet of each *compressor, compressor unit,* or *condensing unit,* and of each *liquid receiver.*

8.6.3 Systems containing 100 lbs (45.4 kg) or more of *refrigerant.* All systems containing 100 lbs (45.4 kg) or more of *refrigerant,* other than systems utilizing *nonpositive displacement compressors, should* have *stop valves* at the locations specified in 8.6.2, and on each inlet of each *liquid receiver,* except on the inlet of a receiver in a *condensing unit* or on the inlet of a receiver which is an integral part of a *condenser.*

8.6.4 *Stop valves* used with soft annealed copper tubing or hard drawn copper tubing 7/8 in. (22.2 mm) outside diameter or smaller *shall* be securely mounted, independent of tubing fastenings or supports.

8.6.5 *Stop valves shall* be suitably labeled if it is not obvious what they control. Numbers may be used to label the valves provided a key to the numbers is located near the valves.

8.7 Factory Tests

8.7.1 Every refrigerant-containing part of every system *shall* be tested and proved tight by the manufacturer at not less than the *design pressure* for which it is rated, except as noted in 8.7.2, 8.7.3 and 8.7.3.1.

8.7.2 *Pressure vessels shall* be tested in accordance with Section 8.3 of this code,

8.7.3 The test pressure applied to the *high side* of each factory-assembled *refrigerating system shall* be at least equal to the *design pressure* of the component in the *high side* which has the lowest rated *design pressure.* The test pressure applied to the *low side* of each factory-assembled *refrigerating system shall* be at least equal to the *design pressure* of the component in the *low side* which has the lowest rated *design pressure.*

Figure A-4 (Cont.)

8.7.3.1 In testing systems using *nonpositive displacement compressors,* the entire system *shall* be considered for test purposes as the *low side* pressure.

8.8 Nameplate

Each *unit system* and each separate *condensing unit, compressor* or *compressor unit* sold for field assembly in a *refrigerating system shall* carry a nameplate marked with the *manufacturer's* name, nationally registered trademark or trade name, identification number, the *design pressures,* and the *refrigerant* for which it is designed. The *refrigerant shall* be designated according to ANSI/ASHRAE 34-1978 Number Designation of Refrigerants.

9. PRESSURE-LIMITING DEVICES

9.1 *Pressure-limiting devices shall* be provided on all systems containing more than 20 lbs (9.1 kg) of *refrigerant* and operating above atmospheric pressure and on all water-cooled systems so constructed that the *compressor* or *generator* is capable of producing a pressure in excess of the *high side design pressure* except water-cooled *unit systems* containing not more than 3 lbs (1.36 kg) of a Group 1 *refrigerant* providing the operating pressure developed in the system with the water supply shut off does not exceed one-fifth the *ultimate strength* of the system, or providing an overload device will stop the action of the *compressor* before the pressure exceeds one-fifth the *ultimate strength* of the system.

9.2 When required by 9.1, the maximum setting to which a *pressure-limiting device* may readily be set by use of the adjusting means provided *shall not* exceed the *design pressure* of the *high side* of a system which is not protected by a *pressure relief device* or 90% of the setting of the *pressure-relief device* installed on the *high side* of a system except as provided in 9.2.1. The *pressure-limiting device shall* stop the action of the *pressure imposing element* at a pressure no higher than this maximum setting.

9.2.1 On systems using *nonpositive displacement compressors,* the *pressure-limiting device* may be set at the *design pressure* of the *high side* of the system provided the *pressure-relief device* is, (1) located in the *low side,* (2) subject to *low side* pressure and, (3) there is a permanent (unvalved) relief path between the *high side* and the *low side* of the system.

9.3 *Pressure-limiting devices,* when required by 9.1, *shall* be connected between the *pressure-imposing element* and any *stop valve* on the discharge side with no intervening *stop valves* in the line leading to the *pressure-limiting device.*

10. PRESSURE-RELIEF PROTECTION

10.1 General

Every *refrigerating system shall* be protected by a *pressure-relief device* or some other means designed to safely relieve pressure due to fire or other abnormal conditions.

10.1.1 In addition, all *pressure vessels shall* be protected in accordance with the requirements of 10.4.

10.1.2 All *pressure-relief devices* (not *fusible plugs) shall* be direct pressure actuated. Each part of a *refrigerating system* which can be valved off and which contains one or more *pressure vessels* having internal diameters greater than 6 in. (152 mm) and containing liquid *refrigerant shall* be protected by a *pressure-relief device.*

10.1.3 *Stop valves shall not* be located between the means of pressure relief and the part or parts of the system protected thereby, except when the parallel relief devices mentioned in 10.4.2 are so arranged that only one can be rendered inoperative at a time for testing or repair purposes.

10.1.4 All *pressure-relief devices* and *fusible plugs shall* be connected as nearly as practicable directly to the *pressure vessel* or other parts of the system protected thereby, above the liquid *refrigerant* level, and installed so that they are readily accessible for inspection and repair and so that they cannot be readily rendered inoperative. *Fusible plugs* may be located above or below the liquid *refrigerant* level except on the *low side.*

10.1.5 The seats and discs of *pressure-relief devices shall* be constructed of suitable material to resist *refrigerant* corrosion or other chemical action caused by the *refrigerant.* Seats or discs of cast iron *shall not* be used.

10.2 Setting of Pressure-Relief Devices.

10.2.1 *Pressure-relief valve* setting. All *pressure-relief valves shall* be set to start to function at a pressure not to exceed the *design pressure* of the parts of the system protected.

10.2.2 *Rupture member* setting. All *rupture members* used in lieu of, or in series with, a relief valve *shall* have a nominal rated rupture pressure not to exceed the *design pressure* of the parts of the systems protected.

The conditions of application *shall* conform to the requirements of Section VIII, Division 1, of the ASME Boiler and Pressure Vessel Code.

Rupture members installed ahead of relief valves need not be larger, but *shall not* be smaller, than the relief valve inlet.

10.3 Marking of Relief Devices and Fusible Plugs

10.3.1 All *pressure-relief valves* for refrigerant-containing components *shall* be set and sealed by the *manufacturer* or an assembler as defined in paragraph UG-136(c)(4) of Section VIII, Division I, of the ASME Boiler and Pressure Vessel Code. Each *pressure-relief valve shall* be marked by the *manufacturer* or assembler with the data required in Paragraph UG-129 (a) of Section VIII, Division 1, of the ASME Boiler and Pressure Vessel Code except relief valves for systems with *design pressures* of 15 psig (103.4 kPa gage) or less may be marked by the *manufacturer,* with the pressure setting and capacity.

Figure A–4 (Cont.)

10.3.2 Each *rupture member* for refrigerant-containing *pressure vessels shall* be marked with the data required in Paragraph UG-129 (d) of Section VIII, Division 1, of the ASME Boiler and Pressure Vessel Code.

10.3.3 *Fusible plugs shall* be marked with the melting temperature in degrees F (°C).

10.4 Pressure Vessel Protection

10.4.1 General

Pressure vessels shall be provided with pressure relief protection in accordance with rules given in Paragraphs UG-125 to UG-134 inclusive, of Section VIII, Division 1, of the ASME Boiler and Pressure Vessel Code, with such additional modifications as are necessary for control of *refrigerants.*

10.4.2 *Pressure vessels* over 3 cu ft ($8.5 \times 10^{-2} m^3$). Each *pressure vessel* containing liquid *refrigerant* with internal gross volume exceeding 3 cu ft ($8.5 \times 10^{-2} m^3$) except as specified in 10.4.4, and which may be shut off by valves from all other parts of a *refrigerating system, shall* be protected by a *pressure-relief device,* having sufficient capacity to prevent the pressure in the *pressure vessel* from rising more than 10% above the setting of the *pressure-relief device.*

10.4.2.1 *Pressure vessels* over 3 cu ft ($8.5 \times 10^{-2} m^3$) but less than 10 cu ft ($28.3 \times 10^{-2} m^3$). Under conditions specified in 10.4.2, a single *pressure-relief device* may be used on *pressure vessels* having less than 10 cu ft ($28.3 \times 10^{-2} m^3$) *internal gross volume.*

10.4.2.2 *Pressure vessels* of 10 cu ft ($28.3 \times 10^{-2} m^3$) *internal gross volume* or over. Under conditions specified in 10.4.2, if a *pressure-relief valve* is used, a relief device system consisting of *pressure-relief valve* in parallel with a second *pressure-relief valve* as described in Paragraph 10.1.3 *shall* be provided on *pressure vessels* having *internal gross volume* of 10 cu ft ($28.3 \times 10^{-2} m^3$) or over. Each *pressure-relief valve shall* have sufficient capacity to prevent the pressure in the *pressure vessel* from rising more than 10% above the setting of the *pressure-relief valve.*

10.4.2.2.1 *Pressure-relief valves* discharging into *low side* of the system. Under conditions permitted in 10.4.8.1, a single relief valve (not *rupture member)* of the required relieving capacity may be used on vessels of 10 cu ft ($28.3 \times 10^{-2} m^3$) or over.

10.4.2.3 *Pressure-relief devices* in parallel on large vessels. In cases where large *pressure vessels* containing liquid *refrigerant* except as specified in 10.4.4, require the use of two or more *pressure-relief devices* in parallel to obtain the capacity required, the battery of *pressure-relief devices shall* be considered as a unit, and therefore as one *pressure-relief device.*

10.4.3 *Pressure vessels* with *internal gross volume* of 3 cu ft ($8.5 \times 10^{-2} m^3$) or less. Each *pressure vessel* having an *internal gross volume* of 3 cu ft. ($8.5 \times 10^{-2} m^3$) or less, containing liquid *refrigerant,* except as specified in 10.4.4, and which may be shut off by valves from all other parts of a *refrigerating system shall* be protected by a *pressure-relief device,* or *fusible plug. Pressure vessels* of less than 3 in. (76.2 mm) inside diameter are exempt from the requirements.

10.4.4 *Pressure-relief device* for *pressure vessels* used as, or as part of *evaporator. Pressure vessels* having internal diameters greater than 6 in. (152 mm) used as, or as part of, *evaporators* insulated or installed in insulated space, and which may be shut off by valves from all other parts of a *refrigerating system shall* be protected by a *pressure-relief device* in accordance with the provisions of Paragraph 10.4.2, and 10.4.3 except that the provisions of Paragraph 10.4.2.2, requiring a second parallel *pressure-relief valve, shall not* apply. *Pressure vessels* used as evaporators, having internal diameters of 6 in. (152 mm) or less are exempt from *pressure relief device* requirements.

10.4.5 Required capacity. The minimum required discharge capacity of the *pressure-relief device* or *fusible plug* for each *pressure vessel shall* be determined by the following:

$$C = fDL$$

where C = minimum required discharge capacity of the relief device in lb of air per min
D = outside diameter of the vessel in ft
l = length of the vessel in ft
f = factor dependent upon kind of *refrigerant*

$$C = 4.88fDL$$

where C = minimum required discharge capacity of the relief device in kg per min. of air
D = outside diameter of the vessel in metres.
L = length of the vessel in metres
f = factor dependent upon kind of *refrigerant,* as follows:

Kind of Refrigerant	Value of f
Ammonia (*Refrigerant* 717)	0.5
Refrigerants 12,22 and 500	1.6
Refrigerant 502, 503 and *Refrigerants* 13, 13B1 and 14 when on cascaded systems	2.5
All other *refrigerants*	1.0

When one *pressure-relief device* or *fusible plug* is used to protect more than one *pressure vessel* the required capacity *shall* be the sum of the capacities required for each *pressure vessel.*

10.4.6 The rated discharge capacity of a *pressure-relief valve* expressed in pounds of air per minute (kilograms of air per minute), *shall* be determined in accordance with Paragraph UG-131 Section VIII, of the ASME Boiler and Pressure Vessel Code. (See 14.15). All pipe and fittings between the *pressure-relief valve* and the parts of the system it protects *shall* have at least the area of *pressure-relief valve* inlet.

10.4.7 The rated discharge capacity of a *rupture member* or *fusible plug* discharging to atmosphere under critical flow conditions in pounds (kg) of air per

Figure A-4 (Cont.)

minute *shall* be determined by the following formulas:

$$C = 0.8\,P_1\,d^2$$

$$d = 1.12\sqrt{C/P_1}$$

Where C = rated discharge capacity in lbs of air per min.
d = smallest of the internal diameter of the inlet pipe, retaining flanges, *fusible plug*, or *rupture member* in inches.

Where for *rupture members*

P_1 = (Rated pressure PSIG × 1.10) + 14.7, PSIA

for *fusible plugs*

P_1 = absolute saturation pressure, corresponding to the stamped temperature melting point of the *fusible plug* or the *critical pressure* of the *refrigerant* used, whichever is smaller, PSIA

$$C = 4.08 \times 10^{-5}\,P_1 d^2$$

$$d = 156.6\sqrt{C/P_1}$$

Where C = rated discharge capacity in kg of air per min.
d = smallest of the internal diameter of the inlet pipe, retaining flanges, *fusible plug* or *rupture member* in millimeters.

Where for *rupture members*

P_1 = (Rated pressure kPa × 1.10) + 101.33, kPa

for *fusible plugs*

P_1 = Absolute saturation pressure, corresponding to the stamped temperature melting point of *fusible plug* or the *critical pressure* of the *refrigerant* used, whichever is smaller, kPa.

10.4.8 *Pressure relief devices* and *fusible plugs* on any system containing a Group 3 *refrigerant*, on any system containing more than 6 lbs (2.72 kg) of a Group 2 *refrigerant* (except as indicated in 10.4.8.2), and on any system containing more than 100 lbs. (45.36 kg) of a Group 1 *refrigerant, shall* discharge to the atmosphere at a location not less than 15 ft (4.57 m) above the adjoining ground level and not less than 20 ft (6.1 m) from any window, ventilation opening, or exit in any building. Discharge *piping* connected to the discharge side of a *fusible plug* or *rupture member shall* have provisions to prevent plugging the *piping* in the event the *fusible plug* or *rupture member* functions.

10.4.8.1 *Pressure-relief valves* may discharge into the *low side* of the system, provided the *pressure-relief devices* are of a type not appreciably affected by back pressures and provided the *low side* of the system is equipped with *pressure-relief devices*. The relief devices on the *low side* of the system *shall* have sufficient capacity to protect the *pressure vessels* that are relieved into the *low side* of the system, or to protect all *pressure vessels* on the *low side* of the system, whichever relieving capacity is the largest, as computed by the formula in 10.4.5. Such *low side pressure-relief device shall* be set in accordance with 10.2.1 and vented to the outside of the building in accordance with 10.4.8.

10.4.8.2 Ammonia Discharge. Where ammonia is used, the discharge may be into a tank of water which *shall* be used for no purpose except ammonia absorption. At least one gallon of fresh water *shall* be provided for each pound (1m³ for each 120 kg) of ammonia in the system. The water used *shall* be prevented from freezing without the use of salt or chemicals. The tank *shall* be substantially constructed of not less than 1/8 in (3.2 mm) or No. 11 U.S. gage iron or steel. No horizontal dimension of the tank *shall* be greater than one half the height. The tank *shall* have hinged cover, or, if of the enclosed type, *shall* have a vent hole at the top. All pipe connections *shall* be through the top of the tank only. The discharge pipe from the *pressure-relief valves shall* discharge the ammonia in the center of the tank near the bottom. An indirect ammonia-water absorption *unit system* installed outdoors adjacent to a single family residence is not required to comply with 10.4.8 provided the discharge is shielded and dispersed.

10.4.8.3 Sulphur Dioxide Discharge. Where sulphur dioxide is used, the discharge may be into a tank of absorptive *brine* which *shall* be used for no purpose except sulphur dioxide absorption. There *shall* be one gallon (1m³) of standard dichromate *brine* (2 1/2 lb sodium dichromate per gallon of water or 300 kg per m³) for each pound (120 kg) of sulphur dioxide in the system. *Brines* made with caustic soda or soda ash may be used in place of sodium dichromate provided the quantity and strength give the equivalent sulphur dioxide absorbing power. The tank *shall* be substantially constructed of not less than 1/8 in. (3.2 mm) or No. 11 U.S. gage iron or steel. The tank *shall* have a hinged cover, or, if of the enclosed type, *shall* have a vent hole at the top. All pipe connections *shall* be through the top of the tank only. The discharge pipe from the *pressure-relief valve shall* discharge the sulphur dioxide in the center of the tank near the bottom.

10.4.8.4 The size of the discharge pipe from the *pressure-relief device* or *fusible plug shall not* be less than the size of the *pressure-relief device* or *fusible plug* outlet. The discharge from more than one relief device or *fusible plug* may be run into a common header, the area of which *shall* be not less than the sum of the areas of the pipes connected thereto.

10.4.8.5 The length of the discharge *piping* permitted to be installed on the outlet of a *pressure-relief device* or *fusible plug shall* be determined as follows:

$$C = (3Pd^{5/2})/(L^{1/2})$$

$$d = [(C^2L)/(9P^2)]^{1/5}$$

Where C = minimum required discharge capacity in lb. of air per min.
d = internal diameter of pipe in inches.
L = length of discharge pipe in ft.
P = 0.25P. (P. is defined under 10.4.7).
(See Table 6 for computations derived from the preceding formula)

$$C = (3.35 \times 10^{-5}Pd^{5/2})/(L^{1/2})$$

$$d = [(C^2L)/(1.12 \times 10^{-9} \times P^2)]^{1/5}$$

Where C = minimum required discharge capacity in kg of air per min.
d = internal diameter of pipe in millimeters
L = length of discharge pipe in meters
P = 0.25P. (P. is defined under 10.4.7).
(See Table 6 for computations derived from the preceding formula)

Figure A-4 (Cont.)

TABLE 6 LENGTH OF DISCHARGE PIPING FOR PRESSURE-RELIEF DEVICES OF VARIOUS DISCHARGE CAPACITIES

Equiv. Length of discharge pipe, ft (meters) (L)	Discharge capacity in lb of air per min (C) Standard wall iron pipe sizes, in. (Discharge capacity in kg of air per min (C) Standard wall iron pipe sizes, mm.)							
	1/2 (12.7)	3/4 (19.0)	1 (25.4)	1 1/4 (31.8)	1 1/2 (38.1)	2 (50.8)	2 1/2 (63.5)	3 (76.2)
RELIEF DEVICE SET AT 25 PSIA (P_1)—(172.4 kPa absolute)								
50(15.2)	0.81(.37)	1.6(.72)	2.9(1.32)	5.9(2.68)	8.7(3.95)	16.3(7.39)	25.3(11.5)	43.8(19.9)
75(22.9)	0.67(.30)	1.4(.64)	2.4(1.09)	4.9(2.22)	7.2(3.26)	13.3(6.03)	20.9(9.48)	35.8(16.2)
100(30.5)	0.58(.26)	1.2(.54)	2.1(.95)	4.2(1.90)	6.2(2.81)	11.5(5.22)	18.0(8.16)	30.9(14.0)
150(45.7)	0.47(.21)	0.95(.43)	1.7(.77)	3.4(1.54)	5.0(2.27)	9.4(4.26)	14.6(6.62)	25.3(11.5)
200(61.0)	0.41(.18)	0.8(.36)	1.5(.68)	2.9(1.32)	4.4(2.00)	8.1(3.67)	12.6(5.72)	21.8(9.89)
300(91.4)	0.33(.15)	0.67(.30)	1.2(.54)	2.4(1.09)	3.6(1.63)	6.6(2.99)	10.5(4.76)	17.9(8.12)
RELIEF DEVICE SET AT 50 PSIA (P_1)—(344.8 kPa absolute)								
50(15.2)	1.6 (.72)	3.3(1.50)	5.9(2.68)	11.9(5.40)	17.4(7.89)	32.5(14.7)	50.6(22.9)	87.6(39.7)
75(22.9)	1.3 (.59)	2.7(1.22)	4.9(2.22)	9.7(4.40)	14.3(6.49)	26.5(12.0)	41.8(19.0)	71.5(32.4)
100(30.5)	1.2 (.54)	2.3(1.04)	4.2(1.90)	8.4(3.81)	12.3(5.58)	23.0(10.4)	36.0(16.3)	61.7(28.0)
150(45.7)	0.94(.43)	1.9(.86)	3.5(1.59)	6.9(3.13)	10.0(4.54)	18.7(8.48)	29.2(13.2)	50.6(23.0)
200(61.0)	0.81(.37)	1.6(.72)	2.9(1.32)	5.9(2.68)	8.7(3.95)	16.3(7.39)	25.3(11.5)	43.7(19.8)
300(91.4)	0.66(.30)	1.3(.59)	2.5(1.13)	4.9(2.22)	7.1(3.22)	13.3(6.03)	21.0(9.52)	35.7(16.2)
RELIEF DEVICE SET AT 75 PSIA (P_1)—(517.1 kPa absolute)								
50(15.2)	2.4(1.09)	4.9(2.22)	8.9(4.04)	17.9(8.12)	26.1(11.8)	48.7(22.1)	75.9(34.4)	131.5(59.6)
75(22.9)	2.0(.91)	4.1(1.86)	7.3(3.31)	14.6(6.62)	21.4(9.71)	39.8(18.0)	62.6(28.4)	107.0(48.5)
100(30.5)	1.7(.77)	3.5(1.59)	6.4(2.90)	12.6(5.72)	18.5(8.39)	34.4(15.6)	54.0(24.5)	92.6(42.0)
150(45.7)	1.4(.64)	2.8(1.27)	5.2(2.36)	10.3(4.67)	15.0(6.80)	28.0(12.7)	43.8(19.9)	75.9(34.4)
200(61.0)	1.2(.54)	2.5(1.13)	4.4(2.00)	8.9(4.04)	13.1(5.94)	24.4(11.1)	37.9(17.2)	65.6(29.8)
300(91.4)	0.9(.41)	2.0(.91)	3.7(1.68)	7.3(3.31)	10.7(4.85)	19.9(9.03)	31.5(14.3)	53.5(24.3)
RELIEF DEVICF SET AT 100 PSIA (P_1)—(689.5 kPa absolute)								
50(15.2)	3.2(1.45)	6.6(2.99)	11.9(5.40)	23.8(10.8)	34.8(15.8)	65.0(29.5)	101.2(45.9)	175.2(79.5)
75(22.9)	2.7(1.22)	5.4(2.45)	9.7(4.40)	19.4(8.80)	28.6(13.0)	53.0(24.0)	83.6(37.9)	143.0(64.9)
100(30.5)	2.3(1.04)	4.6(2.09)	8.5(3.86)	16.8(7.62)	24.6(11.2)	45.9(20.8)	72.0(32.6)	123.6(56.1)
150(45.7)	1.9(.86)	3.8(1.72)	6.9(3.13)	13.7(6.21)	20.0(9.07)	37.4(17.0)	58.4(26.5)	101.2(45.9)
200(61.0)	1.6(.72)	3.3(1.50)	5.9(2.68)	11.9(5.40)	17.5(7.94)	32.5(14.7)	50.6(23.0)	87.6(39.7)
300(91.4)	1.3(.59)	2.7(1.22)	4.9(2.22)	9.7(4.40)	14.2(6.44)	26.5(12.0)	42.0(19.0)	71.4(32.2)
RELIEF DEVICE SET AT 150 PSIA (P_1)—(1034 kPa absolute)								
50(15.2)	4.9(2.22)	9.9(4.49)	17.9(8.12)	35.7(16.2)	52.3(23.7)	97.5(44.2)	151.8(68.8)	262.8(119)
75(22.9)	4.0(1.81)	8.1(3.67)	14.6(6.62)	29.2(13.2)	42.9(19.4)	79.5(36.1)	125.4(56.9)	214.5(97.3)
100(30.5)	3.5(1.59)	6.9(3.13)	12.7(5.76)	25.2(11.4)	36.9(16.7)	68.9(31.2)	108.0(49.0)	185.4(84.1)
150(45.7)	2.8(1.27)	5.7(2.58)	10.4(4.72)	20.6(9.34)	30.0(13.6)	56.1(25.4)	87.6(39.7)	151.8(68.8)
200(61.0)	2.4(1.09)	4.9(2.22)	8.9(4.04)	17.8(8.07)	26.2(11.9)	48.7(22.1)	75.9(34.4)	131.4(59.6)
300(91.4)	1.9(.86)	4.0(1.81)	7.4(3.36)	14.6(6.62)	21.1(9.57)	39.7(18.0)	63.0(28.6)	107.1(48.6)
RELIEF DEVICE SET AT 200 PSIA (P_1)—(1379 kPa absolute)								
50(15.2)	6.5(2.95)	13.2(5.99)	23.8(10.8)	47.6(21.6)	69.7(31.6)	130.0(59.0)	202.4(91.8)	350.4(159)
75(22.9)	5.3(2.40)	10.8(4.90)	19.4(8.80)	38.9(17.6)	57.2(25.9)	106.0(48.1)	167.2(75.8)	286.0(130)
100(30.5)	4.6(2.09)	9.2(4.17)	16.9(7.66)	33.6(15.2)	49.2(22.3)	91.8(41.6)	144.0(65.3)	247.2(112)
150(45.7)	3.8(1.72)	7.6(3.45)	13.8(6.26)	27.4(12.4)	40.0(18.1)	74.8(33.9)	116.8(53.0)	202.4(91.8)
200(61.0)	3.2(1.45)	6.5(2.95)	11.8(5.35)	23.8(10.8)	34.9(15.8)	64.9(29.4)	101.2(45.9)	175.2(79.5)
300(91.4)	2.6(1.18)	5.3(2.40)	9.8(4.44)	19.4(8.80)	28.4(12.9)	52.9(24.0)	84.0(38.1)	142.8(64.8)
RELIEF DEVICE SET AT 250 PSIA (P_1)—(1723 kPa absolute)								
50(15.2)	8.1(3.67)	16.5(7.48)	29.8(13.5)	59.5(27.0)	87.1(39.5)	162.5(73.7)	253.0(115)	437.0(198)
75(22.9)	6.7(3.04)	13.5(6.12)	24.3(11.0)	48.6(22.0)	71.5(32.4)	132.5(60.1)	209.0(94.8)	357.5(162)
100(30.5)	5.8(2.63)	11.6(5.26)	21.2(9.62)	42.0(19.0)	61.6(27.9)	114.8(52.1)	180.0(81.6)	309.0(140)
150(45.7)	4.7(2.13)	9.5(4.31)	17.3(7.85)	34.3(15.6)	50.0(22.7)	93.5(42.4)	146.0(66.2)	253.0(115)
200(61.0)	4.1(1.86)	8.2(3.72)	14.8(6.71)	29.7(13.5)	43.7(19.8)	81.2(36.8)	126.5(57.4)	219.0(99.3)
300(91.4)	3.3(1.50)	6.7(3.04)	12.3(5.58)	24.3(11.0)	35.5(16.1)	66.2(30.0)	105.0(47.6)	178.5(81.0)
RELIEF DEVICE SET AT 300 PSIA (P_1)—(2068 kPa absolute)								
50(15.2)	9.7(4.40)	19.8(8.98)	35.7(16.2)	71.4(32.4)	104.5(47.4)	195.0(88.4)	303.6(138)	525.6(238)
75(22.9)	7.9(3.58)	16.2(7.35)	29.1(13.2)	58.3(26.4)	85.8(38.9)	159.0(72.1)	250.8(114)	429.0(195)
100(30.5)	6.9(3.13)	13.9(6.30)	25.4(11.5)	50.4(22.9)	73.9(33.5)	137.7(62.5)	261.0(98.0)	370.8(168)
150(45.7)	5.6(2.54)	11.3(5.12)	20.7(9.39)	41.1(18.6)	60.0(27.2)	112.2(50.9)	175.2(79.5)	303.6(138)
200(61.0)	4.9(2.22)	9.8(4.44)	17.8(8.07)	35.6(16.1)	52.4(23.8)	97.4(44.2)	151.8(68.8)	262.8(119)
300(91.4)	3.9(1.77)	7.9(3.58)	14.7(6.67)	29.1(13.2)	42.6(19.3)	79.4(36.0)	126.0(57.2)	214.2(97.2)

ASHRAE STANDARD 15-78 15

Figure A-4 (Cont.)

10.5 Positive Displacement Compressor Protection

Every Group 1 *refrigerant positive displacement compressor* operating above 15 psig (103.4 kPa) and having a displacement exceeding 50 cfm (1.42 m^3/min), and every Group 2 or Group 3 *refrigerant positive displacement compressor, shall* be equipped by the *manufacturer* with a *pressure relief device* of adequate size and pressure setting to prevent rupture of the *compressor,* located between the *compressor* and *stop valve* on the discharge side. The *pressure-relief device shall* discharge into the low pressure side of the system, or to the atmosphere at a location not less than 15 ft (4.57 m) above the adjoining ground level and not less than 20 ft (6.1 m) from any window, ventilator opening, or *entrance* of any building.

11. INSTALLATION REQUIREMENTS

11.1 Foundations and supports for *condensing units* or *compressor units shall* be of substantial and noncombustible construction when more than 6 in. (152 mm) high.

11.2 Moving *machinery shall* be guarded in accordance with *approved* safety standards. (See 14.3 and 14.22)

11.3 Clear space adequate for inspection and servicing of *condensing units* or *compressor unit shall* be provided.

11.4 *Condensing units* or *compressor units* with enclosures *shall* be readily accessible for servicing and inspection.

11.5 Water supply and discharge connections shall be made in accordance with *approved* safety and health standards. (See 14.4)

11.5.1 Discharge water lines *shall not* be directly connected to the waste or sewer systems. The waste or discharge from such equipment *shall* be through an *approved* air gap and trap.

11.6 Illumination adequate for inspection and servicing of *condensing units* or *compressor units shall* be provided. (See 14.5)

11.7 Electrical equipment and wiring *shall* be installed in accordance with *approved* safety standards. (See 14.6)

11.8 Gas fuel devices and equipment used with *refrigerating systems shall* be installed in accordance with *approved* safety standards. (See 14.7 and 14.20.)

11.9 Air duct systems of air-conditioning equipment for human comfort using mechanical refrigeration *shall* be installed in accordance with *approved* safety standards. (See 14.8 and 14.9)

11.9.1 *Air ducts* passing through a *Class T machinery room shall* be of tight construction and *shall* have no openings in such rooms.

11.10 Joints and refrigerant-containing parts in *air ducts.* Joints and all refrigerant-containing parts of a *refrigerating system* located in an *air duct* carrying conditioned air to and from a *humanly occupied space shall* be constructed to withstand a temperature of 700 F (353.3 °C) without leakage into the air stream.

11.11 Exposure of *refrigerant* pipe joints. *Refrigerant* pipe joints erected on the *premises shall* be exposed to view for visual inspection prior to being covered or enclosed.

11.12 Location of Refrigerant Piping.

11.12.1 *Refrigerant piping* crossing an open space which affords passageway in any building *shall* be not less than 7-1/2 ft (2.29m) above the floor unless against the ceiling of such space.

11.12.2 Free passageway *shall not* be obstructed by *refrigerant piping. Refrigerant piping shall not* be placed in any elevator, dumbwaiter, or other shaft containing a moving object, or in any shaft which has openings to living quarters or to main *exit hallways. Refrigerant piping shall not* be placed in public *hallways, lobbies,* or *stairways,* except that such *refrigerant piping* may pass across a public *hallway,* if there are no joints in the section in the public *hallway,* and provided nonferrous tubing of 1-1/8 in. (28.6 mm) outside diameter and smaller be contained in a rigid metal pipe.

11.12.3 *Refrigerant piping shall not* be installed vertically through floors from one story to another except as follows:

a) It may be installed from the basement to the first floor, from the top floor to a *machinery* penthouse or to the roof, or between adjacent floors served by the *refrigerating system.*

b) For the purpose of interconnecting separate pieces of equipment not located as described by 11.12.3(a), the *piping* may be carried in an *approved* rigid and tight continuous fire resisting *pipe duct* or shaft having no openings into floors not served by the *refrigerating system* or it may be carried on the outer wall of the building provided it is not located in an air shaft, closed court or in other similar spaces enclosed within the outer walls of the building. The *pipe duct* or shaft *shall* be vented to the outside.

c) *Piping* of *direct systems* containing Group 1 *refrigerants* as governed by 6.2.1, need not be enclosed where it passes through space served by that system.

11.12.4 *Refrigerant piping* may be installed horizontally in closed floors or in open joist spaces. *Piping* installed in concrete floors *shall* be encased in *pipe duct.*

11.13 Machinery Room Requirements.

11.13.1 Each refrigerating *machinery room shall* be provided with a tight-fitting door or doors and have no partitions or openings that will permit the passage of escaping *refrigerant* to other parts of the building.

11.13.2 Each refrigerating *machinery room shall* be provided with means for ventilation to the outer air. The ventilation *shall* consist of windows or doors open-

Figure A-4 (Cont.)

TABLE 7 MINIMUM AIR DUCT AREAS AND OPENINGS

Weight of refrigerant in system, lb (kg)	Mechanical discharge of air, cfm (M^3/min)	Duct area, sq ft (M^2)	Open areas of windows and doors sq ft (M^2)
up to 20 (9.07)	150 (4.2)	1/4 (2.3×10^{-2})	4 (37.2×10^{-2})
50 (22.7)	250 (7.1)	1/3 (3.1×10^{-2})	6 (55.7×10^{-2})
100 (45.4)	400 (11.3)	1/2 (4.6×10^{-2})	10 (92.9×10^{-2})
150 (68.0)	550 (15.6)	2/3 (6.2×10^{-2})	12 1/2 (1.16)
200 (90.7)	680 (19.2)	2/3 (6.2×10^{-2})	14 (1.30)
250 (113)	800 (22.6)	1 (9.3×10^{-2})	15 (1.39)
300 (136)	900 (25.5)	1 (9.3×10^{-2})	17 (1.58)
400 (181)	1,100 (31.2)	1 1/4 (11.6×10^{-2})	20 (1.86)
500 (227)	1,275 (36.1)	1 1/4 (11.6×10^{-2})	22 (2.04)
600 (272)	1,450 (41.1)	1 1/2 (13.9×10^{-2})	24 (2.23)
700 (318)	1,630 (46.2)	1 1/2 (13.9×10^{-2})	26 (2.42)
800 (363)	1,800 (51.0)	2 (18.6×10^{-2})	28 (2.60)
900 (408)	1,950 (55.2)	2 (18.6×10^{-2})	30 (2.79)
1,000 (454)	2,050 (58.0)	2 (18.6×10^{-2})	31 (2.88)
1,250 (567)	2,250 (63.7)	2 1/4 (20.9×10^{-2})	33 (3.06)
1,500 (680)	2,500 (70.8)	2 1/4 (20.9×10^{-2})	37 (3.44)
1,750 (794)	2,700 (76.5)	2 1/4 (20.9×10^{-2})	38 (3.53)
2,000 (907)	2,900 (82.1)	2 1/4 (20.9×10^{-2})	40 (3.72)
2,500 (1134)	3,300 (93.4)	2 1/2 (23.2×10^{-2})	43 (4.00)
3,000 (1361)	3,700 (105)	3 (27.9×10^{-2})	48 (4.46)
4,000 (1814)	4,600 (130)	3 3/4 (34.8×10^{-2})	55 (5.11)
5,000 (2268)	5,500 (156)	4 1/2 (41.8×10^{-2})	62 (5.76)
6,000 (2722)	6,300 (178)	5 (46.4×10^{-2})	68 (6.32)
7,000 (3175)	7,200 (204)	5 1/2 (51.1×10^{-2})	74 (6.87)
8,000 (3629)	8,000 (226)	5 3/4 (53.4×10^{-2})	80 (7.43)
9,000 (4082)	8,700 (246)	6 1/4 (58.1×10^{-2})	85 (7.90)
10,000 (4536)	9,500 (269)	6 1/2 (60.4×10^{-2})	90 (8.36)
12,000 (5443)	10,900 (309)	7 (65.0×10^{-2})	100 (9.29)
14,000 (6350)	12,200 (345)	7 1/2 (69.7×10^{-2})	109 (10.1)
16,000 (7258)	13,300 (377)	7 3/4 (72.0×10^{-2})	118 (11.0)
18,000 (8165)	14,300 (405)	8 (74.3×10^{-2})	125 (11.6)
20,000 (9072)	15,200 (430)	8 1/4 (76.6×10^{-2})	130 (12.1)
25,000 (11340)	17,000 (481)	8 3/4 (81.3×10^{-2})	140 (13.0)
30,000 (13608)	18,200 (515)	9 (83.6×10^{-2})	145 (13.5)
35,000 (15876)	19,400 (549)	9 1/4 (85.9×10^{-2})	150 (13.9)
40,000 (18144)	20,500 (580)	9 1/2 (88.2×10^{-2})	155 (14.4)
45,000 (20412)	21,500 (609)	9 3/4 (90.6×10^{-2})	160 (14.9)

ing to the outer air, of the size shown in Table 7, or of mechanical means capable of removing the air from the room in accordance with Table 7. The amount of ventilation for *refrigerant* removal purposes *shall* be determined by the *refrigerant* content of the largest system in the *machinery room.*

11.13.3 Air supply and return *ducts* used for *machinery room* ventilation *shall* serve no other area.

11.13.4 Mechanical ventilation, when used, *shall* consist of one or more power-driven exhaust fans, which *shall* be capable of removing from the refrigerating *machinery room* the amount of air specified in Table 7. The inlet to the fan, or fans, or *air duct* connection shall be located near the refrigerating equipment. The outlet from the fan, or fans, or *air duct* connections *shall* terminate outside of the building in an *approved* manner. When *air ducts* are used either on the inlet or discharge side of the fan, or fans, they *shall* have an area not less than specified in Table 7. Provision *shall* be made for the inlet of air to replace that being exhausted.

11.13.5 *Machinery room, Class T* is a *machinery room* having no flame-producing apparatus permanently installed and operated and also conforming to the following:

a) Any doors, communicating with the building, *shall* be *approved* self-closing, tight-fitting fire doors.

b) Walls, floor, and ceiling *shall* be tight and of not less than one-hour fire-resistive construction.

c) It *shall* have an *exit* door which opens directly to the outer air or through a vestibule-type *exit* equipped with self-closing, tight-fitting doors.

d) Exterior openings, if present, *shall not* be under any fire escape or any open stairway.

e) All pipes piercing the interior walls, ceiling, or floor of such room *shall* be tightly sealed to the walls, ceiling, or floor through which they pass.

f) Emergency remote controls to stop the action of the *refrigerant compressor shall* be provided and located immediately outside the *machinery room.*

g) An independent mechanical ventilation system *shall* be provided and operated continuously.

h) Emergency remote controls for the mechanical means of ventilation *shall* be provided and located outside the *machinery room.*

12. FIELD TESTS

12.1 General

Every refrigerant-containing part of every system that is erected on the *premises,* except *compressors, con-*

Figure A-4 (Cont.)

densers, evaporators, safety devices, pressure gages, control mechanisms and systems that are factory tested, *shall* be tested and proved tight after complete installation, and before operation.

The *high* and *low side* of each system *shall* be tested and proved tight at not less than the lower of the *design pressure* or the setting of the *pressure-relief device* protecting the *high* or *low side* of the system, respectively, except as noted in 12.1.1.

12.1.1 Systems erected on the *premises* using Group 1 *refrigerant* and with copper tubing not exceeding 5/8 in. (15.9 mm) outside diameter may be tested by means of the *refrigerant* charged into the system at the saturated vapor pressure of the *refrigerant* at 70F (21 °C) minimum.

12.2 Test Medium

Oxygen or any combustible gas or combustible mixture of gases *shall not* be used within the system for testing.

12.2.1 The means used to build up the test pressure *shall* have either a *pressure-limiting device* or a pressure-reducing device with a *pressure-relief device* and a gage on the outlet side. The *pressure-relief device shall* be set above the test pressure but low enough to prevent permanent deformation of the system components.

12.3 A dated declaration of test *should* be provided for all systems containing 50 lb (22.68 kg) or more of *refrigerant*. The declaration should give the name of the *refrigerant* and the *field test* pressure applied to the *high side* and the *low side* of the system. The declaration of test *should* be signed by the installer and, if an inspector is present at the tests, he *should* also sign the declaration. When requested, copies of this declaration *shall* be furnished to the enforcing authority.

13. GENERAL REQUIREMENTS

13.1 Signs

Each *refrigerating system* erected on the *premises shall* be provided with an easily legible permanent sign securely attached and easily accessible, indicating thereon the name and address of the installer, the kind and total number of pounds (kg) of *refrigerant* required in the system for normal operations, and the *field test* pressure applied.

13.2 Metal signs for systems containing more than 100 lbs (45.36 kg) of *refrigerant*. Systems containing more than 100 lbs (45.36 kg) of *refrigerant shall* be provided with metal signs having letters not less than 1/2 in. (12.7 mm) in height designating the main shutoff valves to each vessel, main steam or electrical control, remote control switch, and *pressure-limiting device*. On all exposed high pressure and low pressure *piping* in each room where installed outside the *machinery room, shall* be signs, as specified above, with the name of the *refrigerant* and the letters "HP" or "LP".

13.3 New sign for changed *refrigerant*. When the kind of *refrigerant* is changed as provided in 13.4 (substitution of *refrigerant*), there *shall* be a new sign, of the same type as specified in 13.2, indicating clearly that a substitution has been made, and stating the same information for the new *refrigerant* as was stated in the original.

13.4 Substitution of kind of *refrigerant* in a system *shall not* be made without the permission of the approving authority, the user, and the makers of the original equipment, and due observance of safety requirements.

13.5 Charging and discharging *refrigerants*. When *refrigerant* is added to a system, except a *unit system* requiring less than 6 lbs (2.72 kg) of *refrigerant* it *shall* be charged into the low pressure side of the system. Any point on the downstream side of the main liquid line *stop valve shall* be considered as part of the low pressure side when operating with said *stop valve* in the closed position. No service *container shall* be left connected to a system except while charging or withdrawing *refrigerant*.

13.6 *Refrigerants* withdrawn from *refrigerating systems shall* be transferred to *approved containers* only. (See 14.2.) No *refrigerant shall* be discharged to a sewer.

13.7 *Containers* used for *refrigerants* withdrawn from a *refrigerating system shall* be carefully weighed each time they are used for this purpose, and the *containers shall not* be filled in excess of the permissible filling weight for such *containers* and such *refrigerants* as are prescribed in the pertinent regulations of the Department of Transportation. (See 14.2)

13.8 *Refrigerant* stored in a *machinery room shall* be not more than 300 lbs (136 kg), in addition to the charge in the system and the *refrigerant* stored in a permanently attached *receiver,* and then only in *approved* storage *containers*. (See 14.2)

13.9 Masks or helmets. At least two masks or helmets *shall* be provided at a location convenient to the *machinery room* when an amount of a Group 2 *refrigerant* exceeds 100 lbs. (45.36 kg).

13.9.1 Only complete helmets or masks marked as *approved* by the Bureau of Mines of the United States Department of the Interior and suitable for the *refrigerant* employed *shall* be used and they *shall* be kept in a suitable cabinet immediately outside the *machinery room* or other *approved* accessible location.

13.9.2 Canisters or cartridges of helmets or masks *shall* be renewed immediately after having been used or the seal broken and, if unused, the canisters *shall* be renewed not later than the date noted on the canister labels.

13.10 Maintenance. All *refrigerating systems shall* be maintained by the user in a clean condition, free from accumulations of oily dirt, waste and other debris, and *shall* be kept readily accessible at all times.

13.11 Responsibility as to operation of the system. It *shall* be the duty of the person in charge of the *premises*

Figure A-4 (Cont.)

on which a *refrigerating system* containing more than 50 lbs (22.68 kg) of *refrigerant* is installed, to place a card conspicuously as near as practicable to the *refrigerant compressor* giving directions for the operation of the system, including precautions to be observed in case of a breakdown or leak as follows:

a) Instruction for shutting down the system in case of emergency;

b) The name, address, and day and night telephone numbers for obtaining service;

c) The name, address, and telephone number of the municipal inspection department having jurisdiction, and instructions to notify said department immediately in case of emergency.

13.12 Pressure gages *should* be checked for accuracy prior to test and immediately after every occasion of unusually high pressure, equal to full scale reading either by comparison with master gages or by setting the pointer as determined by a dead weight pressure gage tester.

14. TITLES AND SOURCES OF REFERENCE STANDARDS

The following listing clearly identifies published standards by name of organization, standard number, year of issue and title to which reference is made in this code.

14.1 American National Standard Z11.7-1973, Method of Test for Flash Point by Means of the Pensky-Martens Closed Tester. (ASTM D93-72) 1,2

14.2 U.S. Department of Transportation Code of Federal Regulations 49 TRANSPORTATION Parts 100-199 "Regulations for Transportation of Explosive and other Dangerous Articles by Land and Water in Rail Freight Service and by Motor Vehicle (Highway) and Water, Including Specifications for Shipping Containers", Agent R.M. Grazianos Tariff No. 31, Revised as of October 1, 1975.

14.3 American National Standard B15.1-1972, Safety Code for Mechanical Power-Transmission Apparatus.

14.4 American National Standard A40.8-1955, National Plumbing Code.[1]

14.5 American National Standard A11.1-1973, for Industrial Lighting.[1]

14.6 National Fire Protection Association Standard 70, National Electrical Code.[1,4] American National Standard ANSI/NFPA 70-1978.

14.7 National Fire Protection Association Standard 54, Installation of Gas Appliance and Gas Piping, American National Standard-Z223.1 1974.[1,4]

14.8 National Fire Protection Association Standard ANSI/NFPA 90A-1976, Air Conditioning and Ventilating Systems.[4]

14.9 National Fire Protection Association Standard ANSI/NFPA 90B-1976, Warm Air Heating and Air Conditioning Systems.[4]

14.10 American National Standard B31.5-1974, Code for Pressure Piping: Refrigerant Piping.[1]

14.11 American National Standard B36.10-1970, Wrought-Steel and Wrought-Iron Pipe.[1]

14.12 American National Standard H26.1-1967, Specifications for Seamless Copper Pipe, Standard Sizes. (ANSI/ASTM B42-78). 1,2

14.13 American National Standard H23.1-1973, Specifications for Seamless Copper Water Tube. (ANSI/ASTM B88-78). 1,2

14.14 American National Standard H23.5-1967, Specifications for Seamless Copper Tube for Refrigeration Field Service. (ANSI/ASTM B280-77). 1,2

14.15 ASME Boiler and Pressure Vessel Code, Section VIII 1974 Edition, known as ASME Pressure Vessel Code, Division 1.[5]

14.16 For reference to items not found in the body of the Code, the ASHRAE HANDBOOK AND PRODUCT DIRECTORY, current editions, are recommended.[6]

14.17 National Fire Protection Association Standard ANSI/NFPA 101-1976, Building Exits Code for Life Safety from Fire.[4]

14.18 American National Standard H27.1-1973, Specification for Seamless Red Brass Pipe, Standard Sizes. (ANSI/ASTM B43-76) 1,2

14.19 American National Standard and ASHRAE Standard (ANSI/ASHRAE 34-1978), Number Designation of Refrigerants (formerly ANSI B79.1-1968) 1,6

14.20 National Fire Protection Association Standard ANSI/NFPA 37-1975, Installation and Use of Combustion Engines and Gas Turbines.[4]

14.21 American National Standard and ASHRAE Standard (ANSI/ASHRAE 26-1978) Recommended Practice for Mechanical Refrigeration Installations on Shipboard (formerly ANSI B59-1-1964) 1,6

14.22 American National Standard Z262.1-1974, Refrigeration and Air-Conditioning Condensing and Compressor Units. 1

REFERENCE SOURCES

1. American National Standards Institute (ANSI)
   1430 Broadway
   New York, NY 10018
2. American Society for Testing and
   Materials (ASTM)
   1916 Race Street
   Philadelphia, PA 19103
3. U.S. Department of Transportation (DOT)
   Materials Transportation Bureau
   2100 Second Avenue, S.W.
   Washington, DC 20590
4. National Fire Protection Association (NFPA)
   470 Atlantic Avenue
   Boston, MA 02210
5. American Society of Mechanical Engineers (ASME)
   United Engineering Center
   345 East 47th Street
   New York, NY 10017
6. American Society of Heating, Refrigerating and
   Air-Conditioning Engineers, Inc. (ASHRAE)
   United Engineering Center
   345 East 47th Street
   New York, NY 10017

Figure A-4 (Cont.)

Pressure-enthalpy Diagrams and Tables

Pressure–enthalpy diagrams and tables for R-22, R-502, and R-717.

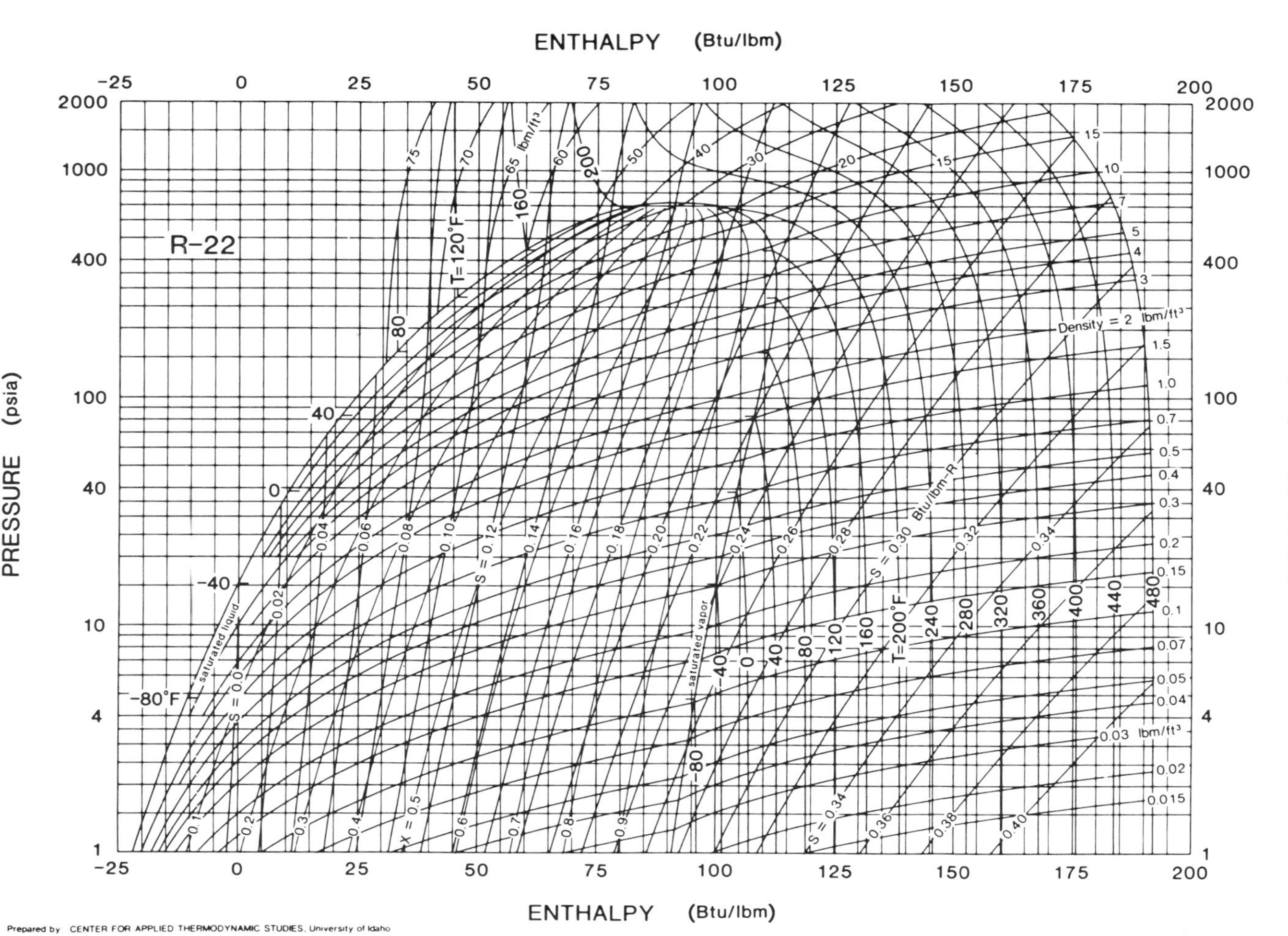

Prepared by CENTER FOR APPLIED THERMODYNAMIC STUDIES, University of Idaho

Figure A–5 P-E diagram for R-22. (By permission of *ASHRAE Handbook*.)

Temp F	Pressure psia	Pressure psig	Volume cu ft/lb Vapor v_g	Density lb/cu ft Liquid $1/v_f$	Enthalpy** Btu/lb Liquid h_f	Enthalpy** Btu/lb Vapor h_g	Entropy** Btu/lb · R Liquid s_f	Entropy** Btu/lb · R Vapor s_g
−150	0.27163	29.36816*	141.23	98.236	−25.974	87.521	−0.07147	0.29501
−140	0.44692	29.01126*	88.532	97.363	−23.725	88.681	−0.06432	0.28729
−130	0.71060	28.47441*	57.356	96.480	−21.463	89.848	−0.05736	0.28027
−120	1.0954	27.6910*	38.280	95.587	−19.185	91.020	−0.05055	0.27388
−110	1.6417	26.5788*	26.242	94.684	−16.886	92.196	−0.04389	0.26805
−100	2.3983	25.0383*	18.433	93.770	−14.564	93.371	−0.03734	0.26274
−90	3.4229	22.9522*	13.235	92.843	−12.216	94.544	−0.03091	0.25787
−80	4.7822	20.1846*	9.6949	91.905	−9.838	95.710	−0.02457	0.25342
−70	6.5522	16.5809*	7.2318	90.952	−7.429	96.868	−0.01832	0.24932
−60	8.8180	11.9677*	5.4844	89.986	−4.987	98.014	−0.01214	0.24556
−50	11.674	6.154*	4.2224	89.004	−2.511	99.144	−0.00604	0.24209
−40	15.222	0.526	3.2957	88.006	0.000	100.257	0.00000	0.23888
−30	19.573	4.877	3.6049	86.991	2.547	101.348	0.00598	0.23591
−20	24.845	10.149	2.0826	85.956	5.131	102.415	0.01189	0.23315
−18	26.020	11.324	1.9940	85.747	5.652	102.626	0.01307	0.23262
−16	27.239	12.543	1.9099	85.537	6.175	102.835	0.01425	0.23210
−14	28.501	13.805	1.8302	85.326	6.699	103.043	0.01542	0.23159
−12	29.809	15.113	1.7544	85.114	7.224	103.250	0.01659	0.23108
−10	31.162	16.466	1.6825	84.901	7.751	103.455	0.01776	0.23058
−8	32.563	17.867	1.6141	84.688	8.280	103.660	0.01892	0.23008
−6	34.011	19.315	1.5491	84.473	8.810	103.863	0.02009	0.22960
−4	35.509	20.813	1.4872	84.258	9.341	104.065	0.02125	0.22912
-2	37.057	22.361	1.4283	84.042	9.874	104.266	0.00241	0.22864
0	38.657	23.961	1.3723	83.825	10.409	104.465	0.02357	0.22817
2	40.309	25.613	1.3189	83.606	10.945	104.663	0.02472	0.22771
4	42.014	27.318	1.2680	83.387	11.483	104.860	0.02587	0.22725
†5	42.888	28.192	1.2434	83.277	11.752	104.958	0.02645	0.22703
6	43.775	29.079	1.2195	83.167	12.022	105.056	0.02703	0.22680
8	45.591	30.895	1.1732	82.946	12.562	105.250	0.02818	0.22636
10	47.464	32.768	1.1290	82.724	13.104	105.442	0.02932	0.22592
12	49.396	34.700	1.0869	82.501	13.648	105.633	0.03047	0.22548
14	51.387	36.691	1.0466	82.276	14.193	105.823	0.03161	0.22505
16	53.438	38.742	1.0082	82.051	14.739	106.011	0.03275	0.22463
18	55.551	40.855	0.97144	81.825	15.288	106.198	0.03389	0.22421
20	57.727	43.031	0.93631	81.597	15.837	106.383	0.03503	0.22379
22	59.967	45.271	0.90270	81.368	16.389	106.566	0.03617	0.22338
24	62.272	47.576	0.87055	81.138	16.942	106.748	0.03730	0.22297
26	64.644	49.948	0.83978	80.907	17.496	106.928	0.03844	0.22257
28	67.083	52.387	0.81031	80.675	18.052	107.107	0.03958	0.22217
30	69.591	54.895	0.78208	80.441	18.609	107.284	0.04070	0.22178
32	72.169	57.473	0.75503	80.207	19.169	107.459	0.04182	0.22139
34	74.818	60.122	0.72911	79.971	19.729	107.632	0.04295	0.22100
36	77.540	62.844	0.70425	79.733	20.292	107.804	0.04407	0.22062
38	80.336	65.640	0.68041	79.495	20.856	107.974	0.04520	0.22024
40	83.206	68.510	0.65753	79.255	21.422	108.142	0.04632	0.21986
42	86.153	71.457	0.63557	79.013	21.989	108.308	0.04744	0.21949
44	89.177	74.481	0.61448	78.770	22.558	108.472	0.04855	0.21912
46	92.280	77.584	0.59422	78.526	23.129	108.634	0.04967	0.21876
48	95.463	80.767	0.57476	78.280	23.701	108.795	0.05079	0.21839
50	98.727	84.031	0.55606	78.033	24.275	108.953	0.05190	0.21803
52	102.07	87.38	0.53808	77.784	24.851	109.109	0.05301	0.21768
54	105.50	90.81	0.52078	77.534	25.429	109.263	0.05412	0.21732
56	109.02	94.32	0.50414	77.282	26.008	109.415	0.05523	0.21697
58	112.62	97.93	0.48813	77.028	26.589	109.564	0.05634	0.21662
60	116.31	101.62	0.47272	76.773	27.172	109.712	0.05745	0.21627
62	120.09	105.39	0.45788	76.515	27.757	109.857	0.05855	0.21592
64	123.96	109.26	0.44358	76.257	28.344	110.000	0.05966	0.21558
66	127.92	113.22	0.42981	75.996	28.932	110.140	0.06076	0.21524
68	131.97	117.28	0.41653	75.733	29.523	110.278	0.06186	0.21490
70	136.12	121.43	0.40373	75.469	30.116	110.414	0.06296	0.21456
72	140.37	125.67	0.39139	75.202	30.710	110.547	0.06406	0.21422
74	144.71	130.0[illegible]	0.37949	74.934	31.307	110.677	0.06516	0.21388
76	149.15	134.45	0.36800	74.664	31.906	110.805	0.06626	0.21355
78	153.69	138.99	0.35691	74.391	32.506	110.930	0.06736	0.21321
80	158.33	143.63	0.34621	74.116	33.109	111.052	0.06846	0.21288
82	163.07	148.37	0.33587	73.839	33.714	111.171	0.06956	0.21255
84	167.92	153.22	0.32588	73.560	34.322	111.288	0.07065	0.21222
†86	172.87	158.17	0.31623	73.278	34.931	111.401	0.07175	0.21188
88	177.93	163.23	0.30690	72.994	35.543	111.512	0.07285	0.21155
90	183.09	168.40	0.29789	72.708	36.158	111.619	0.07394	0.21122
92	188.37	173.67	0.28917	72.419	36.774	111.723	0.07504	0.21089
94	193.76	179.06	0.28073	72.127	37.394	111.824	0.07613	0.21056
96	199.26	184.56	0.27257	71.833	38.016	111.921	0.07723	0.21023
98	204.87	190.18	0.26467	71.536	38.640	112.015	0.07832	0.20989
100	210.60	195.91	0.25702	71.236	39.267	112.105	0.07942	0.20956
102	216.45	201.76	0.24962	70.933	39.897	112.192	0.08052	0.20923
104	222.42	207.72	0.24244	70.626	40.530	112.274	0.08161	0.20889
106	228.50	213.81	0.23549	70.317	41.166	112.353	0.08271	0.20855
108	234.71	220.02	0.22875	70.005	41.804	112.427	0.08381	0.20821
110	241.04	226.35	0.22222	69.689	42.446	112.498	0.08491	0.20787
112	247.50	232.80	0.21589	69.369	43.091	112.564	0.08601	0.20753
114	254.08	239.38	0.20974	69.046	43.739	112.626	0.08711	0.20718
116	260.79	246.10	0.20378	68.719	44.391	112.682	0.08821	0.20684
118	267.63	252.94	0.19800	68.388	45.046	112.735	0.08932	0.20649
120	274.60	259.91	0.19238	68.054	45.705	112.782	0.09042	0.20613
122	281.71	267.01	0.18692	67.714	46.368	112.824	0.09153	0.20578
124	288.95	274.25	0.18163	67.371	47.034	112.860	0.09264	0.20542
126	296.33	281.63	0.17648	67.023	47.705	112.891	0.09375	0.20505
128	303.84	289.14	0.17147	66.670	48.380	112.917	0.09487	0.20468
130	311.50	296.80	0.16661	66.312	49.059	112.936	0.09598	0.20431
140	351.94	337.25	0.14418	64.440	52.528	112.931	0.10163	0.20235
150	396.20	381.50	0.12448	62.402	56.143	112.728	0.10739	0.20020
160	444.53	429.83	0.10701	60.145	59.948	112.263	0.11334	0.19776
170	497.26	482.56	0.091279	57.581	64.019	111.438	0.11959	0.19490
180	554.78	540.09	0.076790	54.549	68.498	110.068	0.12635	0.19133
190	617.59	602.89	0.062837	50.677	73.711	107.734	0.13409	0.18646
200	686.36	671.66	0.047438	44.571	80.862	102.853	0.14460	0.17794
204.81	721.91	707.21	0.030525	32.760	91.329	91.329	0.16016	0.16016

**Based on 0 for the saturated liquid at −40 F.
†Standard cycle Temperatures.

*From published data (1964) of E. I. du Pont de Nemours & Co., Inc. Used by permission.
*Inches of mercury below one standard atmosphere.

Figure A-6 R-22: properties of liquid and saturated vapor. (By permission of *ASHRAE Handbook.*)

Temp, F	Viscosity, $lb_m/ft \cdot h$			Thermal Conductivity, $Btu/h \cdot ft \cdot F$			Specific Heat, c_p, $Btu/lb_m \cdot F$				Temp, F
	Sat. Liquid	Sat. Vapor	Gas, P=1 atm $\times 10^{-2}$‡	Sat. Liquid	Sat. Vapor	Gas, P=1 atm $\times 10^{-3}$‡	Sat. Liquid	Sat. Vapor	Gas $(c_p)_{0\ atm}$	Gas $(c_p)_{1\ atm}$	
−100	1.167			0.0789			0.255		0.1260		−100
− 80	1.014			0.0757			0.256		0.1292		− 80
− 60	0.894			0.0725			0.259	0.139	0.1324		− 60
− 40	0.798	0.0245	2.45	0.0693	0.0040	4.04	0.262	0.146	0.1356		− 40
− 20	0.719	0.0257	2.57	0.0661	0.0044	4.43	0.266	0.152	0.1388		− 20
0	0.654	0.0269	2.68	0.0630	0.0048	4.81	0.271	0.158	0.1420		0
20	0.599	0.0282	2.80	0.0598	0.0052	5.20	0.276	0.165	0.1452		20
40	0.553	0.0295	2.91	0.0566	0.0056	5.58	0.283	0.175	0.1484		40
60	0.513	0.0309	3.03	0.0534	0.0060	5.97	0.291	0.187	0.1515		60
80	0.480	0.0325	3.14	0.0502	0.0064	6.35	0.300	0.204	0.1546		80
100	0.449	0.0343	3.25	0.0471	0.0068	6.74	0.313	0.226	0.1577		100
120	0.427	0.0362	3.37	0.0439	0.0072	7.12	0.332	0.253	0.1608		120
140	0.392	0.0383	3.48	0.0407	0.0077	7.51	0.357	0.288	0.1638		140
160	0.344	0.0411	3.59	0.0371	0.0084	7.90	0.390	0.332	0.1668		160
180	0.285	0.045	3.70	0.0318	0.0105	8.28	0.433		0.1697		180
190	0.244	0.049	3.75	0.0288	0.0119	8.48			0.1712		190
200	0.182	0.058	3.81	0.0238	0.0140	8.67			0.1726		200
205*	0.074	0.074	3.83	0.0177	0.0177	8.76			0.1733		205*
220			3.92			9.05			0.1754		220
240			4.02			9.44			0.1782		240
300			4.34			10.59			0.1863		300
400			4.86			12.5			0.1983		400
440			5.06			13.3			0.2026		440

*Critical Temperature. Tabulated properties ignore critical region effects.
‡Actual value = (Table value) × (Indicated multiplier).

Figure A–6 (Cont.)

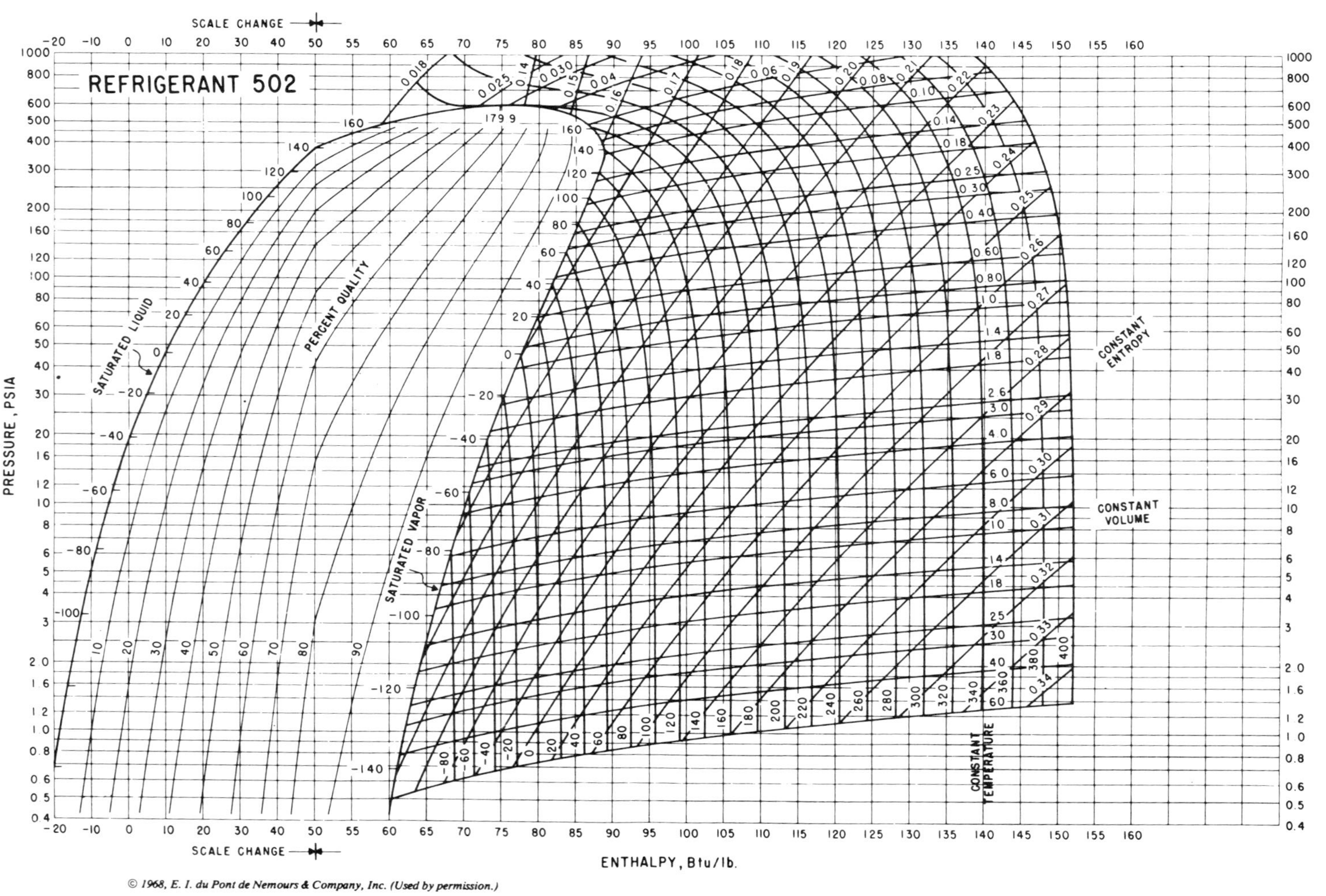

© 1968, E. I. du Pont de Nemours & Company, Inc. (Used by permission.)

Figure A–7 P-E Diagram for R-502. (By permission of *ASHRAE Handbook*.)

Temp F	Pressure psia	Pressure psig	Volume cu ft/lb Vapor v_g	Density lb/cu ft Liquid $1/v_f$	Enthalpy** Btu/lb Liquid h_f	Enthalpy** Btu/lb Vapor h_g	Entropy** Btu/lb · R Liquid s_f	Entropy** Btu/lb · R Vapor s_g
−150	0.4267	29.05*	69.573	102.87	−21.37	59.91	−0.0585	0.2040
−140	0.677	28.54*	45.200	101.89	−19.71	61.08	−0.0532	0.1996
−130	1.043	27.80*	30.222	100.89	−18.00	62.27	−0.0479	0.1956
−120	1.562	26.74*	20.739	99.896	−16.24	63.47	−0.0427	0.1920
−110	2.283	25.27*	14.572	98.882	−14.42	64.67	−0.0374	0.1888
−100	3.261	23.28*	10.461	97.857	−12.55	65.89	−0.0321	0.1860
−90	4.5612	20.63*	7.6591	96.82	−10.61	67.10	−0.0268	0.1834
−80	6.258	17.18*	5.7081	95.77	− 8.62	68.31	−0.0215	0.1811
−70	8.434	12.75*	4.3241	94.70	− 6.56	69.53	−0.0162	0.1791
−60	11.18	7.15*	3.3248	93.62	− 4.44	70.73	−0.0108	0.1773
−50	14.60	0.19*	2.5915	92.51	− 2.25	71.93	−0.0054	0.1757
−40	18.80	4.11	2.0453	91.39	0.00	73.11	0.0000	0.1742
−36	20.73	6.03	1.8666	90.94	0.92	73.59	0.0022	0.1737
−32	22.80	8.10	1.7065	90.48	1.85	74.05	0.0043	0.1732
−28	25.03	10.34	1.5628	90.02	2.79	74.52	0.0065	0.1727
−24	27.43	12.74	1.4336	89.56	3.73	74.98	0.0087	0.1722
−20	30.01	15.31	1.3171	89.09	4.69	75.44	0.0109	0.1718
−16	32.76	18.07	1.2120	88.62	5.66	75.90	0.0131	0.1714
−12	35.71	21.01	1.1168	88.14	6.64	76.35	0.0152	0.1710
−8	38.85	24.16	1 0307	87.66	7.63	76.80	0.0174	0.1706
−4	42.21	27.51	0.9525	87.17	8.63	77.25	0.0196	0.1702
0	45.78	31.08	0.3814	86.68	9.63	77.69	0.0218	0.1698
2	47.64	32.95	0.8482	86.43	10.14	77.91	0.0229	0.1697
4	49.57	34.87	0.8166	86.19	10.65	78.13	0.0240	0.1695
6	51.55	36.86	0.7864	85.94	11.16	78.35	0.0251	0.1693
8	53.59	38.90	0.7574	85.69	11.68	78.56	0.0262	0.1692
10	55.70	41.00	0.7299	85.43	12.19	78.78	0.0272	0.1690
12	57.86	43.17	0.7035	85.18	12.71	78.99	0.0283	0.1689
14	60.09	45.39	0.673	84.93	13.23	79.20	0.0294	0.1687
16	62.38	47.68	0.6541	84.67	13.76	79.42	0.0305	0.1686
18	64.73	50.04	0.6310	84.41	14.28	79.63	0.0316	0.1684
20	67.16	52.46	0.6088	84.15	14.81	79.84	0.0327	0.1683
22	69.64	54.95	0.5876	83.89	15.34	80.04	0.0338	0.1681
24	72.20	57.50	0.5673	83.63	15.87	80.25	0.0349	0.1680
26	74.82	60.13	0.5478	83.37	16.41	80.46	0.0360	0.1679
28	77.52	62.82	0.5291	83.10	16.94	80.66	0.0371	0.1677
30	80.29	65.59	0.5112	82.83	17.48	80.86	0.0382	0.1676
32	83.13	68.43	0.4940	82.56	18.02	81.06	0.0392	0.1675
34	86.04	71.34	0.4775	82.29	18.56	81.26	0.0403	0.1673
36	89.03	74.33	0.4616	82.02	19.11	81.46	0.0414	0.1672
38	92.09	77.39	0.4464	81.75	19.65	81.66	0.0425	0.1671
40	95.23	80.53	0.4318	81.47	20.20	81.85	0.0436	0.1670
42	98.45	83.75	0.4177	81.19	20.75	82.04	0.0447	0.1668
44	101.74	87.04	0.4041	80.91	21.31	82.24	0.0458	0.1667
46	105.12	90.42	0.3911	80.62	21.86	82.43	0.0469	0.1666
48	108.58	93.88	0.3786	80.35	22.42	82.61	0.0479	0.1665
50	112.12	97.42	0.3666	80.06	22.98	82.80	0.0490	0.1664
52	115.74	101.05	0.3550	79.77	23.54	82.98	0.0501	0.1663
54	119.45	104.75	0.3438	79.48	24.10	83.17	0.0512	0.1662
56	123.25	108.55	0.3330	79.19	24.66	83.35	0.0523	0.1661
58	127.13	112.43	0.3226	78.89	25.23	83.53	0.0533	0.1659
60	131.10	116.40	0.3126	78.59	25.80	83.70	0.0544	0.1658
64	139.31	124.61	0.2937	77.99	26.94	84.05	0.0566	0.1656
68	147.89	133.19	0.2761	77.37	28.09	84.39	0.0587	0.1654
72	156.84	142.15	0.2596	76.75	29.25	84.72	0.0609	0.1652
76	166.18	151.49	0.2443	76.11	30.41	85.04	0.0630	0.1650
80	175.92	161.22	0.2300	75.46	31.59	85.35	0.0651	0.1647
84	186.06	171.36	0.2165	74.79	32.77	85.64	0.0673	0.1645
88	196.62	181.92	0.2040	74.11	33.95	85.93	0.0694	0.1643
92	207.60	192.90	0.1922	73.41	35.15	86.20	0.0715	0.1641
96	219.02	204.32	0.1812	72.70	36.35	86.47	0.0736	0.1638
100	230.89	216.19	0.1708	71.97	37.56	86.71	0.0758	0.1636
110	262.61	247.91	0.1474	70.04	40.63	87.26	0.0810	0.1629
120	297.41	282.71	0.1271	67.96	43.77	87.68	0.0863	0.1621
130	335.54	320.84	0.1093	65.66	46.98	87.95	0.0917	0.1611
140	377.29	362.60	0.09359	63.08	50.32	88.00	0.0971	0.1599
150	423.06	408.35	0.07934	60.09	53.85	87.76	0.1027	0.1583
160	473.38	458.69	0.06604	56.43	57.73	87.01	0.1087	0.1560
170	529.11	514.41	0.05271	51.39	62.45	85.25	0.1160	0.1522
179.889	591.00	576.30	0.02857	35.00	74.65	74.65	0.1348	0.1348

[a]From published data of E. I. du Pont de Nemours & Co. (1969). Used by permission. *Inches of mercury below standard atmosphere. **Based on 0 for the saturated liquid at −40 F.

Figure A-8 R-502: properties of liquid and saturated vapor. (By permission of *ASHRAE Handbook*.)

Temp, F	Viscosity, $lb_m/ft \cdot h$			Thermal Conductivity, $Btu/h \cdot ft \cdot F$			Specific Heat, c_p, $Btu/lb_m \cdot F$				Temp, F
	Sat. Liquid	Sat. Vapor	Gas, $P = 1$ atm $\times 10^{-2}$‡	Sat. Liquid	Sat. Vapor	Gas, $P = 1$ atm $\times 10^{-3}$‡	Sat. Liquid	Sat. Vapor	Gas $(c_p)_{0\ atm}$	Gas $(c_p)_{1\ atm}$	
−100	1.39			0.0595			0.244				−100
− 80	1.16			0.0570			0.248				− 80
− 60	1.00	0.0228		0.0545			0.253	0.138			− 60
− 40	0.86	0.0244	2.42	0.0519	0.0046	4.58	0.259	0.149		0.148	− 40
− 20	0.76	0.0258	2.54	0.0494	0.0050	4.95	0.264	0.155		0.151	− 20
0	0.67	0.0270	2.66	0.0469	0.0053	5.31	0.271	0.160		0.154	0
20	0.60	0.0283	2.77	0.0444	0.0057	5.67	0.277	0.164		0.157	20
40	0.54	0.0295	2.89	0.0419	0.0060	6.03	0.285	0.171		0.160	40
60	0.487	0.0310	3.01	0.0394	0.0064	6.39	0.292	0.180		0.164	60
80	0.433	0.0327	3.12	0.0369	0.0068	6.76	0.300	0.195		0.167	80
100	0.380	0.0348	3.23	0.0344	0.0071	7.14	0.308	0.218		0.170	100
120	0.329	0.0373	3.34	0.0314	0.0075	7.52	0.316	0.249		0.173	120
140	0.284	0.039	3.45	0.0281	0.0083	7.91	0.326	0.310		0.176	140
160	0.243	0.045	3.56	0.0237	0.0090	8.31	0.335			0.178	160
170	0.207	0.053	3.62	0.021	0.0103	8.52	0.345			0.179	170
180*	0.074	0.074	3.67	0.014	0.014	8.73				0.181	180*
190			3.72			8.94				0.182	190
200			3.78			9.16				0.183	200
220			3.88			9.60				0.186	220
240			3.99			10.07				0.188	240
260			4.10			10.5				0.190	260
280			4.20			11.0				0.192	280
300			4.29			11.6				0.193	300
320						12.1					320
340						12.7					340
360						13.3					360
380						13.9					380
440						16.0					440
460						16.7					460
500						18.4					500

*Critical Temperature. Tabulated properties ignore critical region effects.
‡Actual value = (Table value) × (Indicated multiplier).

Figure A–8 (Cont.)

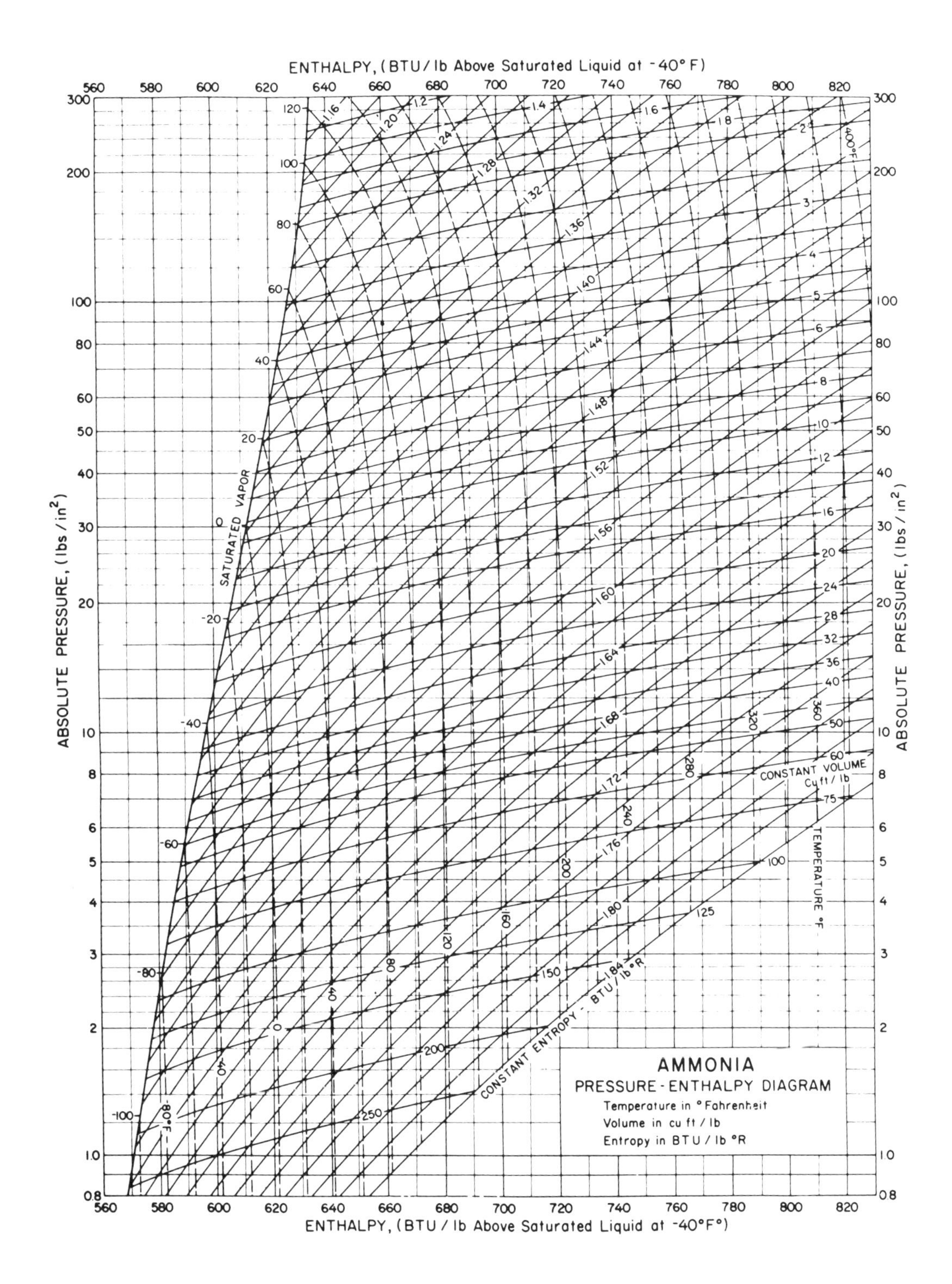

Figure A–9 P-E Diagram for R-717. (By permission of *ASHRAE Handbook*.)

Temp F	Pressure psia	Pressure psig	Volume cu ft/lb Vapor v_g	Density lb/cu ft Liquid $1/v_f$	Enthalpy** Btu/lb Liquid h_f	Enthalpy** Btu/lb Vapor h_g	Entropy** Btu/lb · R Liquid s_f	Entropy** Btu/lb · R Vapor s_g
−105	0.996	27.9*	223.2	45.71	−68.5	570.3	−0.1774	1.6243
−100	1.24	27.4*	182.4	45.52	−63.3	572.5	−0.1626	1.6055
−90	1.86	26.1*	124.1	45.12	−52.8	576.9	−0.1338	1.5699
−80	2.74	24.3*	86.50	44.73	−42.2	581.2	−0.1057	1.5368
−70	3.94	21.9*	61.60	44.32	−31.7	585.5	−0.0784	1.5059
−60	5.55	18.6*	44.73	43.91	−21.2	589.6	−0.0517	1.4769
−50	7.67	14.3*	33.08	43.49	−10.6	593.7	−0.0256	1.4497
−40	10.41	8.7*	24.86	43.08	0.0	597.6	0.0000	1.4242
−38	11.04	7.4*	23.53	42.99	2.1	598.3	0.0051	1.4193
−36	11.71	6.1*	22.27	42.90	4.3	599.1	0.0101	1.4144
−34	12.41	4.7*	21.10	42.82	6.4	599.9	0.0151	1.4096
−32	13.14	3.2*	20.00	42.73	8.5	600.6	0.0201	1.4048
−30	13.90	1.6*	18.97	42.65	10.7	601.4	0.0250	1.4001
−28	14.71	0.0	18.00	42.57	12.8	602.1	0.0300	1.3955
−26	15.55	0.8	17.09	42.48	14.9	602.8	0.0350	1.3909
−24	16.24	1.7	16.24	42.40	17.1	603.6	0.0399	1.3863
−22	17.34	2.6	15.43	42.31	19.2	604.3	0.0448	1.3818
−20	18.30	3.6	14.68	42.22	21.4	605.0	0.0497	1.3774
−18	19.30	4.6	13.97	42.13	23.5	605.7	0.0545	1.3729
−16	20.34	5.6	13.29	42.04	25.6	606.4	0.0594	1.3686
−14	21.43	6.7	12.66	41.96	27.8	607.1	0.0642	1.3643
−12	22.56	7.9	12.06	41.87	30.0	607.8	0.0690	1.3600
−10	23.74	9.0	11.50	41.78	32.1	608.5	0.0738	1.3558
−8	24.97	10.3	10.97	41.69	34.3	609.2	0.0786	1.3516
−6	26.26	11.6	10.47	41.60	36.4	609.8	0.0833	1.3474
−4	27.59	12.9	9.991	41.52	38.6	610.5	0.0880	1.3433
−2	28.98	14.3	9.541	41.43	40.7	611.1	0.0928	1.3393
0	30.42	15.7	9.116	41.34	42.9	611.8	0.0975	1.3352
2	31.92	17.2	8.714	41.25	45.1	612.4	0.1022	1.3312
4	33.47	18.8	8.333	41.16	47.2	613.0	0.1069	1.3273
6	35.09	20.4	7.971	41.07	49.4	613.6	0.1115	1.3234
8	36.77	22.1	7.629	40.98	51.6	614.3	0.1162	1.3195
10	38.51	23.8	7.304	40.89	53.8	614.9	0.1208	1.3157
12	40.31	25.6	6.996	40.80	56.0	615.5	0.1254	1.3118
14	42.18	27.5	6.703	40.71	58.2	616.1	0.1300	1.3081
16	44.12	29.4	6.425	40.61	60.3	616.6	0.1346	1.3043
18	46.13	31.4	6.161	40.52	62.5	617.2	0.1392	1.3006
20	48.21	33.5	5.910	40.43	64.7	617.8	0.1437	1.2969
22	50.36	35.7	5.671	40.34	66.9	618.3	0.1483	1.2933
24	52.59	37.9	5.443	40.25	69.1	618.9	0.1528	1.2897
26	54.90	40.2	5.227	40.15	71.3	619.4	0.1573	1.2861
28	57.28	42.6	5.021	40.06	73.5	619.9	0.1618	1.2825
30	59.74	45.0	4.825	39.96	75.7	620.5	0.1663	1.2790
34	64.91	50.2	4.459	39.77	80.1	621.5	0.1753	1.2721
38	70.43	55.7	4.126	39.58	84.6	622.5	0.1841	1.2652
42	76.31	61.6	3.823	39.39	89.0	623.4	0.1930	1.2585
46	82.55	67.9	3.547	39.19	93.5	624.4	0.2018	1.2519
50	89.19	74.5	3.294	39.00	97.9	625.2	0.2105	1.2453
54	96.23	81.5	3.063	38.80	102.4	626.1	0.2192	1.2389
58	103.7	89.0	2.851	38.60	106.9	626.9	0.2279	1.2325
62	111.6	96.9	2.656	38.40	111.5	627.7	0.2365	1.2262
66	120.0	105.3	2.477	38.20	116.0	628.4	0.2451	1.2201
70	128.8	114.1	2.312	38.00	120.5	629.1	0.2537	1.2140
72	133.4	118.7	2.235	37.90	122.8	629.4	0.2579	1.2110
74	138.1	123.4	2.161	37.79	125.1	629.8	0.2622	1.2080
76	143.0	128.3	2.089	37.69	127.4	630.1	0.2664	1.2050
78	147.9	133.2	2.021	37.58	129.7	630.4	0.2706	1.2020
80	153.0	138.3	1.955	37.48	132.0	630.7	0.2749	1.1991
82	158.3	143.6	1.892	37.37	134.3	631.0	0.2791	1.1962
84	163.7	149.0	1.831	37.26	136.6	631.3	0.2833	1.1933
†86	169.2	154.5	1.772	37.16	138.9	631.5	0.2875	1.1904
88	174.8	160.1	1.716	37.05	141.2	631.8	0.2917	1.1875
90	180.6	165.9	1.661	36.95	143.5	632.0	0.2958	1.1846
92	186.6	171.9	1.609	36.84	145.8	632.2	0.3000	1.1818
94	192.7	178.0	1.559	36.73	148.2	632.5	0.3041	1.1789
96	198.9	184.2	1.510	36.62	150.5	632.6	0.3083	1.1761
98	205.3	190.6	1.464	36.51	152.9	632.9	0.3125	1.1733
100	211.9	197.2	1.419	36.40	155.2	633.0	0.3166	1.1705
110	247.0	232.3	1.217	35.84	167.0	633.7	0.3372	1.1566
120	286.4	271.7	1.047	25.26	179.0	634.0	0.3576	1.1427
125	307.8	293.1	0.973	34.96	185.1	634.0	0.3679	1.1358

[a]From National Bureau of Standards *Circular* No. 142 (1945) and *Circular* No. 472 (1948).
*Inches of mercury below one standard atmosphere. **Based on 0 for the saturated liquid at −40 F. †Standard cycle temperatures.

Figure A–10 R-717: properties of liquid and saturated vapor. (By permission of *ASHRAE Handbook.*)

Temp, F	Viscosity, $lb_m/ft \cdot h$			Thermal Conductivity, $Btu/h \cdot ft \cdot F$			Specific Heat, c_p, $Btu/lb_m \cdot F$				Temp, F
	Sat. Liquid	Sat. Vapor	Gas, $P = 1$ atm $\times 10^{-2}$‡	Sat. Liquid	Sat. Vapor	Gas, $P = 1$ atm $\times 10^{-3}$‡	Sat. Liquid	Sat. Vapor	Gas $(c_p)_{0\ atm}$	Gas $(c_p)_{1\ atm}$	
−100				0.410			1.030				−100
− 80				0.395			1.044				− 80
− 60				0.380			1.056	0.523	0.477		− 60
− 40				0.365			1.066	0.543	0.480		− 40
− 20	0.629	0.0227	2.02	0.350	0.011	11.1	1.075	0.565	0.483	0.547	− 20
0	0.558	0.0237	2.11	0.335	0.012	11.7	1.083	0.590	0.486	0.536	0
20	0.494	0.0246	2.20	0.321	0.012	12.2	1.092	0.620	0.489	0.528	20
40	0.437	0.0256	2.29	0.306	0.013	12.9	1.103	0.655	0.493	0.522	40
60	0.386	0.0266	2.39	0.291	0.015	13.5	1.118	0.698	0.496	0.519	60
80	0.341	0.0277	2.48	0.276	0.016	14.2	1.135	0.750	0.500	0.517	80
100	0.301	0.0288	2.58	0.261	0.018	14.9	1.158	0.814	0.504	0.517	100
120	0.268	0.0299	2.67	0.246	0.020	15.6	1.187	0.866	0.508	0.518	120
140	0.238	0.0312	2.77	0.231	0.022	16.3	1.222		0.512	0.520	140
160	0.213	0.0325	2.87	0.216	0.025	17.1	1.265		0.517	0.523	160
180	0.190	0.0340	2.97	0.201	0.029	17.8	1.317		0.521	0.527	180
200	0.171	0.0358	3.06	0.186	0.034	18.6	1.379		0.526	0.532	200
220	0.154	0.0380	3.16	0.169	0.039	19.4	1.452		0.530	0.536	220
240	0.124	0.0411	3.26	0.149	0.044	20.2	1.536		0.535	0.542	240
250	0.110	0.043	3.31	0.137	0.049	20.7	1.59		0.537	0.544	250
260	0.094	0.046	3.36	0.120	0.060	21.1	1.65		0.540	0.547	260
270			3.41	0.102	0.081	21.5			0.542	0.550	270
271*	0.060	0.060	3.42	0.087	0.087	21.5			0.542	0.550	271*
280			3.46			21.9			0.544	0.552	280
290			3.51			22.3			0.547	0.555	290
300			3.56			22.8			0.549	0.557	300
320			3.66			23.7			0.554	0.562	320
360			3.86			25.4			0.564	0.572	360
400			4.06			27.3			0.574	0.581	400
440			4.27			29.1			0.584	0.590	440
500			4.57			32.0			0.599	0.602	500

*Critical Temperature. Tabulated properties ignore critical region effects.
‡Actual value = (Table value) × (Indicated multiplier).

Figure A-10 (Cont.)

Appendix

V Answers to Review Exercises

CHAPTER 1

1. Vapor
2. Vapor
3. 126.6 psig; 105 °F
4. 126.6 psig; liquid
5. 126.6 psig; 95 °F; liquid
6. 24.6 psig; 25 °F; mixture
7. 24.6 psig; 25 °F; mixture

CHAPTER 2

2–1. b
2–2. c
2–3. a
2–4. a
2–5. b
2–6. d
2–7. b
2–8. a
2–9. b
2–10. d

CHAPTER 3

3–1. a, b, c
3–2. b
3–3. a
3–4. b
3–5. c
3–6. d
3–7. a
3–8. b
3–9. a, c
3–10. d

CHAPTER 4

4–1. a
4–2. a
4–3. b
4–4. c

4-5. a
4-6. a
4-7. a
4-8. All
4-9. c
4-10. b

CHAPTER 5

5-1. c
5-2. b
5-3. d
5-4. b
5-5. a
5-6. b
5-7. b
5-8. c
5-9. b
5-10. d

CHAPTER 6

Review Exercise 1

6-1. a
6-2. c
6-3. b
6-4. d
6-5. b
6-6. a
6-7. a
6-8. b
6-9. a
6-10. d

Review Exercise 2

	Capillary tube	Direct expansion	Low-side float	Reversed cycle
Valve 1	Closed	Open	Closed	Closed
Valve 2	Closed	Closed	Closed	Closed
Valve 3	Open	Closed	Closed	Open
Valve 4	Closed	Closed	Open	Closed
Valve 5	Closed	Open	Open	Closed
Valve 6				
to float	Closed	Closed	Open	Closed
to evaporator	Open	Open	Closed	Open

CHAPTER 7

7-1. d
7-2. a, d
7-3. c
7-4. b
7-5. d
7-6. b
7-7. a
7-8. d
7-9. c
7-10. c

CHAPTER 8

8–1. d	8–6. a
8–2. a	8–7. c
8–3. b	8–8. d
8–4. c	8–9. c
8–5. b	8–10. a

CHAPTER 9

9–1. a	9–6. a
9–2. c	9–7. b
9–3. e	9–8. d
9–4. b	9–9. a
9–5. b	9–10. d

CHAPTER 10

10–1. c	10–6. c
10–2. b	10–7. a
10–3. d	10–8. c
10–4. b	10–9. c
10–5. a	10–10. d

CHAPTER 11

11–1. d	11–6. d
11–2. b	11–7. b
11–3. b	11–8. b
11–4. a	11–9. c
11–5. c	11–10. a

CHAPTER 12

12–1. a	12–6 a
12–2. c	12–7. c
12–3. d	12–8. b
12–4. b	12–9. c
12–5. b	12–10. a

CHAPTER 13

Review Exercise 1

13-1.	c	13-6.	b
13-2.	a, b	13-7.	d
13-3.	a	13-8.	b
13-4.	d	13-9.	a
13-5.	a	13-10.	b

Review Exercise 2

1. 2
2. During defrosting and for a short period after the compressor starts
3. 2
4. Energizes and deenergizes solenoid valve
5. Line voltage
6. 4
7. 6, including drain line heater
8. 3
9. Timer, thermostat, and solenoid valve
10. Yes

CHAPTER 14

14-1.	a	14-6.	c
14-2.	a	14-7.	d
14-3.	d	14-8.	d
14-4.	b	14-9.	c
14-5.	c	14-10.	b

CHAPTER 15

15-1.	b	15-6.	a
15-2.	b	15-7.	c
15-3.	b	15-8.	d
15-4.	a	15-9.	d
15-5.	a	15-10.	c

CHAPTER 16

16–1. a
16–2. d
16–3. b
16–4. c
16–5. b
16–6. c
16–7. a
16–8. b
16–9. c
16–10. d

CHAPTER 17

17–1. c, d
17–2. a
17–3. b
17–4. d
17–5. b
17–6. a
17–7. a
17–8. d
17–9. b
17–10. c

CHAPTER 18

See Figure A–11

CHAPTER 19

See Figure A–12

1. 66.5
2. 1.5
3. 45.0
4. 12.5
5. 4.0
6. 2.0
7. 51.5
8. 57.5
9. 5.42
10. 4.44
11. 5.33
12. 1.31
13. 11.988

REF.	Load Survey and Estimating Data
9, 9A	Design ambient: 95 °FDB, ______ °FWB, 50 %RH, ______ °F summer ground temp. (use 55 °F for insulated freeze floor slab)
16A, B, C, D	Room design: 40 °FDB, ______ °FWB, 80 %RH ______ °F winter design ambient
	Access area: 95 °FDB, ______ °FWB, 50 %RH, (Ante-RM/loading dock/other)
	Room dim. outside: 10 ft W 12 ft L 9 ft H 636 Total ft² (outside surface)

	Insulation				Wall thick-ness	Adj. area °F	Effective wall TD	Sun effect (°F)	Total TD	Overall wall heat gain (Btu/24 hr/ft²)
	Type	Thick (inches)	K Factor	U* Factor						
N. Wall	POLY.	3"	.20	.067	5"	95	55		55	88
S. Wall										
E. Wall (10)										
W. Wall										
Ceiling (11)										
Floor (12)										

$$^*\text{U Factor} = \frac{K}{\text{Insul. Thickness (in.)}}$$

Refrig. door(s): (1) 7 × 4 Vent fan(s): ______

Room int. vol: 9 W × 11 L × 8 H = 792 ft³
(inside room dimension = outside dimension–wall thicknesses)

Floor area: 9 W × 11 L = 99 ft²

Electrical power: 230 V 3 Ph 60 Hz: Control: 120 V

Type control: ______

Product data and class of product: ______

16A–D

Type product	Amount of Product			Product temp (°F)		Specific heat		Lat.ht. freeze (Btu/lb)	Highest product freeze temp.	Heat or respir'n (Btu/lb) 24 hr)	( ) Pull-down ( ) Freezing time (hr)
	Amount storage	Daily turn-over	Freezing or cooling	Enter-ing	Final	Above freeze	Below freeze				
MIX. VEG.	3800	1000	COOLING	50	40	.9					

Evap. TD 10; Type defrost: ☐ Air, ☐ Hot gas, ☐ Electric

Class product 2 : ______;

No. of defrosts–Total time per 24 hr; ______ hr.

Compressor running time: 16 hr.

Box usage: ☑ Average, ☐ Heavy, ☐ Extra heavy

Product load and additional information 1000 lbs mixed vegetables per day entering temperature 50°F

20 / 21A & B — Packaging Paper cartons Containers ______ Wgt. 100 lb Sp. ht. .32 (container)

22 — Pallets: No. ______ Size ______ Wgt. ea. ______ Sp. ht. ______

Product racks: No. ______ Mat'l ______ Wgt. ea. ______ Sp. ht. ______

21A & B — Estimating Product Loading Capacity of Room

Estimated product loading = 0.40 × ______ ft³ × ______ lb./ft³ = ______ lb:
(room volume) (loading density)

Miscellaneous loads

People: No. ______ hr. ______

Motors (other than evap. fan)

Use: ______, ______ hp. ______ hr. ______
______, ______ hp. ______ hr. ______

Fork lifts: ______ No. ______ hp ______ hr./day, Other ______

Lights ______ W/ft²

Figure A-11

REF.	I. Wall Loss (Transmission Load)				
	Surface	TD	Area of surface	Wall heat gain factor	Btu/24 hr
12	N. Wall		____ ft L X ____ ft H = ____ ft^2 X	____ =	
	S. Wall		____ ft L X ____ ft H = ____ ft^2 X	____ =	
	E. Wall		____ ft L X ____ ft H = ____ ft^2 X	____ =	
	W. Wall		____ ft L X ____ ft H = ____ ft^2 X	____ =	
	Ceiling		____ ft L X ____ ft W = ____ ft^2 X	____ =	
	Floor		____ ft L X ____ ft W = ____ ft^2 X	____ =	
	Box	55	Total surface = 636 ft^2 X	88 =	55,968
			I Total wall transmission load	=	55,968

REF.		Btu/24 hr
13, 14 15	II. (Long Form) Infiltration (Air Change Load) 792 ft^3 X 20 air changes/24hr X 1 service factor X 2.31 Btu/ft^3 =	36,590
	II Total infiltration load =	36,590
	III. Product Load	
16A, B, C	Product temp. reduction above freezing (sensible heat) 1000 *lb/day X 10 °F temp. reduction X .9 sp. ht. =	9000
	Product freezing (latent heat load) ____ *lb/day X ____ Btu/lb/latent heat =	
17	Product temp. reduction below freezing (sensible heat) ____ *lb/day X ____ °F temp. reduction X ____ sp. ht. =	
19	Heat of respiration ____ lb product (storage) X ____ Btu/lb/24 hr =	
20	Miscellaneous product loads (1) containers (2) pallets (3) other ____ lb/day X ____ °F temp. reduction X ____ sp. ht. =	
	____ lb/day X ____ °F temp. reduction X ____ sp. ht. =	9000
	III Total product load =	9000
	IV. Miscellaneous loads	
	(a) Lights ____ ft^2 floor area X ____ W/ft^2 X 3.41 Btu/W X ____ hr/24 hr = (1 to 1½ W/ft^2 in storage areas and 2 to 3 for work areas)	
	(b) Occupancy ____ No. of people X ____ Btu/hr X ____ hr =	
	(c) Motors ____ Btu/hp/hr X ____ hp X ____ hr/24 hr =	
23	____ Btu/hr X ____ hp X ____ hr/24 hr =	
	(d) Material handling	
24	___ Forklift(s) X ____ eqiv. hp X 3100 Btu/hr/hp X ____ hr operation =	
	Other ____ =	
	IV Total miscellaneous loads =	
	Total Btu load I to IV =	101,558
	Add 10% safety factor =	10,156
	Total Btu/24 hr with safety factor (not including evap. fan or defrost heat loads) 24-hr base refrigeration load =	111,714

*If the product pull down is accomplished in less than 24 hr the daily product will be:

$$\text{Pounds product} \times \frac{24 \text{ hr}}{\text{Pull-down hours}}$$

Figure A–11 (Cont.)

REF.	Equipment Selection from Load Calculation Form
6, 7	1. Determine evap. TD required for class of product and room temp. __10__ °F (TD) (from load survey data)
	2. Determine compressor running time based on operating temperatures and defrost requirements __16__ hr (from load survey data)
	3. Evaporator temp. °F = __40__ (room temp) − __10__ [evap. TD = __30__ °F (from load survey data)]
25	4. Comp. suct. temp. °F = __30__ (evap. suct. temp) − __2__ (suct. line loss) = __28__ °F
24	Btu/24 hr base refrigeration load with safety factor = __111,714__ (not including evaporator fan or defrost heat) Preliminary hourly load = __111,714__ Btu/24 hr (base load) / hr/day (comp. running time) = __6982__ Btu/hr Fan heat load estimate Btu/hr = ____ Qty. (MOTORS) X ____ Watts ea. (INPUT) X 3.41 Btu/W X ____ hr = ____ Btu/24 hr or __1__ Qty. (MOTORS) X __1/4__ hp ea. X __4350__ Btu/hp/hr X __24__ hr = __17,400__ Btu/24 hr Defrost heat load estimate Btu/hr = ____ Qty. evaps. X ____ W ea. X ____ hr X 3.41 Btu/W X ____ Defrost load factor* = ____ Btu/hr *Use 0.50 for electric defrost, 0.40 for hot gas defrost Btu/24 hr total load = __111,714__ (base load) + __17,400__ (fan heat) + ____ (defrost heat) = __129,114__ Btu/24 hr or ____ (base load) X ____ (base load mult.) = ____ Btu/hr Actual hourly load = __129,114__ Btu/24 hr (total load) / __16__ hr/day (comp. running time) = __8069__ Btu/24 hr

Equipment Selection

	Compressor units		Condensing units		Evaporators		Condensers	
Model no.			10 HCL		WJ 85			
Quantity			1		1			
Capacity (ea.) Btu/hr)			9400		8500			
Air volume (ea.) (cfm)					1300			
	Design	Actual	Design	Actual	Design	Actual	Design	Actual
Evaporator temp. (°F)					10			
Evap. TD (°F)								
Suction temp. (°F)			-30					
Condensing temp. (°F)								
Design ambient temp. (°F)			100					
Min. oper. ambient temp. (°F)								

Figure A-11 (Cont.)

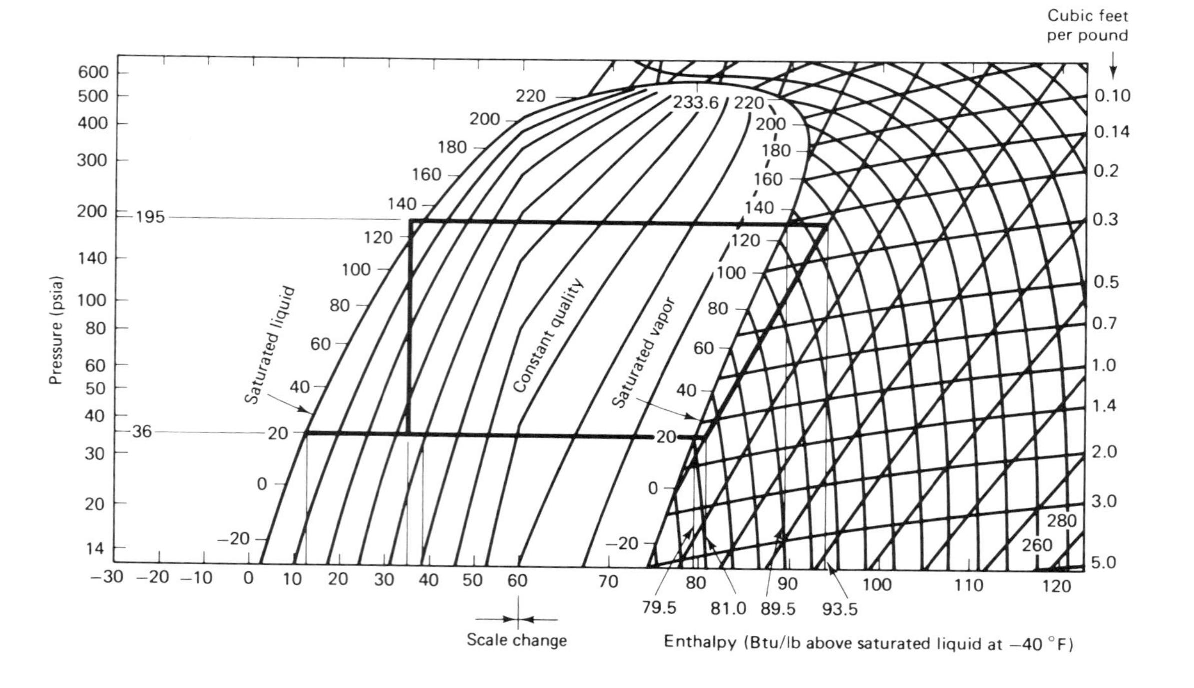
Cubic feet per pound
0.10
0.14
0.2
0.3
0.5
0.7
1.0
1.4
2.0
3.0
5.0
Pressure (psia)
600
500
400
300
200
140
100
80
60
50
40
30
20
14
195
36
Saturated liquid
Constant quality
Saturated vapor
233.6
220
200
180
160
140
120
100
80
60
40
20
0
−20
280
260
−30 −20 −10 0 10 20 30 40 50 60 70 80 90 100 110 120
79.5
81.0
89.5
93.5
Scale change
Enthalpy (Btu/lb above saturated liquid at −40 °F)

Figure A-12

Index

Accessories (halocarbon refrigerants):
compressor valves, 160
dehydrators, 161
heat exchangers, 159
liquid indicators, 158
pressure relief valves, 158
receiver tank valves, 160
shutoff valves, 160
vibration eliminators, 157
Answers to review questions, 490–94
Application, 368–89
classification, 368–69
dairy products and eggs, 374–75
food preservation, 369–70
fruits and vegetables, 373
meats, 370–72
poultry, 372
seafood, 373
frozen foods:
quick freezing, 376–80
Carrier freezer, 377
fluidation bed freezer, 379
freezing tunnel, 377
liquid refrigerant freezer, 380
multiple belt freezer, 378
plate freezer, 380
push-through tunnel, 377
spiral belt freezer, 378
refreezing, 381
storage, 381–82
transportation, 383
low temperature applications, 385
bakery refrigeration, 386–87
ice rinks, 387–89
making ice cream, 385
transportation refrigeration, 383
local delivery equipment, 384
long-haul systems, 383
mechanical refrigeration, 385

Bakery refrigeration, 386–87
Belt tension, 338
Brazing:
equipment, 25, 26
flowing silver alloy, 26
British thermal unit (Btu), definition, 1
Buildings (refrigerated), 272–73
Burn-outs, 348

Capacitors, 333–34
Cascade systems, 181–83
Change of state, 4
Charging (refrigerant), 333–35
Chillers (liquid), 292–96
Baudelot, 83
double-pipe, 82
selection, 83, 84
shell and coil, 81
shell and tube, 81
tank type, 82
Cleaning (condensers), 349–50
Cold, definition, 1
Compound systems, 179–81

Compressors, 36-61
  ammonia, 52, 54-57
  centrifugal, 57-61
  commercial temperature, 48
  helical rotary, 52, 53
  hermetically sealed, 37, 38
  high temperature, 48
  low temperature, 48
  open, 39, 40
  reciprocating, 37, 55
  rotary, 51
  rotary booster, 51, 52
  semi-hermetic, 38, 39
  two-stage, 55
  types, 36
  unloaders, 49, 50
Condensers, 87-101, 112, 113
  air cooled, 88-94
  controls, 111
  evaporative, 99-101
  heat rejection, 87, 88
  heat transfer, 87
  maintenance, 112
  water-cooled, 94-99
Condensers (air cooled), 88-94
  head pressure control, 92-94
  remote, 89
Condensers (evaporative), 99-101
  capacities, 100
  construction, 100
  schematic, 101
Condensers (water cooled), 94-99
  atmospheric, 99
  capacity, 96
  classification, 94
  double tube, 98
  non-condensables, 96
  series and parallel flow, 95
  shell and coil, 97
  shell and tube, 96
Condensing temperature reduction, 439-40
Condensing units, 40-48
  air and water cooled, 41
  air cooled, 40, 42, 44, 46
  compressor-receiver units, 41
  selection from manufacturers' data, 49
  water cooled, 41, 43, 45, 47
Contactors, 231, 232
Conversion factors, 446, 448-56
Cooling towers, 101-7
  capacities, 102-5
  freeze protection, 106
  pump selection, 107
  schematics, 106
Crankshaft seals, 347
Cryogenic systems, 178-83
  cascade systems, 181-83
  compound systems, 179-81
  intercoolers, 180, 182
Customer information, 339

Defrosting (evaporators):
  air, 67
  electric, 67
  hot gas:
    glycol, 72
    re-evaporation, 69
    re-use of refrigerant, 69
    reverse cycle, 69-71
    Thermobank, 69, 71, 72
Dehydration, 330-33
Drinking water cooler, 255-59, 306

Electrical combination devices:
  contactors, 231-32
  dual-circuit overloads, 233
  relays, 229-30
  starters, 233
Electrical components:
  capacitors, 233-34
  combination devices, 229-33
  loads, 207-19
  switches, 219-25
  types, 205
Electrical loads:
  heaters, 219
  lights, 219
  motors, 207
  solenoids, 208-16
    heat-reclaim, 214
    hot gas, 211
    industrial, 215
    liquid line, 209-10, 212
    pilot operated, 213
    reversing valves, 216
    suction line, 211
  transformers, 216-18
Electrical switches, 219-28
  fuses and circuit breakers, 228
  oil lubrication safety switch, 226-28
  pressurestats, 221, 222
    dual pressure control (HLPC), 222
    high pressure control (HPC), 221
    low pressure control (LPC), 221
  temperature controls, 223-25
    defrost control, 223-24
    fan controls, 223-24
    freeze protection, 225

general purpose, 223-24
time clocks, 225
types, 220
Electrical symbols:
capacitors, 233
combination devices, 229
overloads, 233
relays, 229
loads, 206
heaters, 219
lights, 219
motors, 207
solenoids, 209
transformers, 216
switches, 220
fuses and circuit breakers, 228
pressure, 220
temperature, 220
Electricity:
circuits, 191, 195-97
grounded, 198
open, 198
parallel, 196, 197
series, 195, 196
short, 198
conductance, 195
current (a.c. and d.c.), 194
electromagnetism, 194
horsepower, 193
Ohm's Law, 192
power (watts), 192
power factor, 193
power supplies, 197-98
Delta, 198
hertz, 198
phase, 198
Wye, 198
Equipment, 254-302
Automatic ice dispensers, 259
flake ice machines, 262-64
ice building machines, 270-71
ice cube makers, 260-62
large ice making machines, 264-67
tube ice makers, 267-69
classification, 254-55
drinking water coolers, 255-59
food store equipment, 273
dairy product display cases, 274-77
delicatessen display cases, 279
frozen food and ice cream cases, 277
meat display cases, 278-79
multiple compressor units, 282-83
defrost times, 288-89
heat reclaim, 286
hot gas defrost, 287
piping layout, 284-85
produce display cases, 279-81
rack mounted condensing units, 281-82
walk-in coolers and freezers, 274-75
industrial chillers (large), 295-96
liquid coolers (packaged), 292-94
reach-in refrigerators, 290-91
transport refrigeration units, 296-302
heating and defrost cycle, 300-301
refrigeration cycle, 299-301
selection chart, 302
types, 296-98
Evacuation, 330-33
Evaporative water cooler, 107
Evaporators, 64-84
bare tube, 65
classification, 64, 66
dry expansion, 65
finned tube, 65, 66
fin spacing, 80
flooded, 64
humidity control, 79
industrial product cooler, 67
matching coil and compressor, 78
performance, 79
plate, 65
refrigerator, 66
selecting a large product cooler, 73-77
selecting a small unit cooler, 78
temperature difference, 79

Filters and driers:
servicing, 353
types, 161
Fittings:
pipe and tubing, 19
special service, 19
wrought copper, 19-20
Float valves:
servicing, 252-53
types, 126-28
Flow controls (ammonia):
hot gas bypass valves, 136
pressure regulators, 134, 135
shutoff valves, 132
solenoid valves, 134
strainers, 132
Flow controls (halocarbons):
pressure regulating valves, 129
crankcase pressure regulator, 130, 131
evaporator pressure regulator, 129, 130
head pressure regulator, 130
hot gas bypass valves, 131
suction pressure regulator, 129, 130

Food preservation, 369-70, 376-83
Food store equipment, 273-89, 317-19
Frozen foods, 376-83
Fuses, 228

Gauge manifolds, 342-45
Gauges:
  compound, 28
  manifold, 29
  pressure, 28

Heat:
  enthalpy, 5, 6, 8
  heat of fusion, 7
  heat of vaporization, 7
  latent heat, 7
  measurement, 1-4
  movement of, 4
  specific heat, 2, 3, 8
  total heat, 14

Ice cream making, 385
Ice makers, 259-71, 309-15
Ice rinks, 387-89
Installation:
  charging with refrigerant, 333-34
  check, test, and start, 336
  checking belt tension, 338
  checking lubrication, 337
  checking water system, 338
  dehydration and evacuation, 330-33
  determining correct charge, 334-35
  electrical connections, 330
  location of equipment:
    noise and vibration, 327-28
    space requirement, 327
    ventilation, 327
    weather conditions, 327
  loosening hold-down bolts, 337
  piping, 329-30
  positioning valves, 337
  purging non-condensables, 335
  supplying customer information, 339
  testing for leaks, 336-37
  utilities, 328
Instruments (for measurement), 13

Leaks, testing for, 336-37
Load calculation, 392-427
  equipment selection form, 396
  load calculation form, 395
  miscellaneous loads, 422
  procedures, 406
  product loads, 410, 414-16
  safety factors, 423
  sample calculations:
    above freezing, 397-99
    below freezing, 400-401
  survey form, 394
  tables, 403-22
Load calculation tables:
  air changes, 408
  design outside temperature, 404
  heat of motors, 422
  heat of occupancy, 422
  heat of respiration, 417
  heat removed from air, 409
  insulation, 405
  piling heights, 421
  property of solids, 419
  space, weight, density of stored
      products, 420-21
  storage conditions, 403
  storage requirements, 411-14, 416
  sun effect, 405
  temperature difference, 403
  wall heat gain, 407
  water content of foods, 415

Locating equipment, 327-28
Lubrication, 29-30, 337

Metering devices, ammonia:
  expansion valves, 133
  float valves, 133-34
Metering devices, halocarbon refrigerants:
  automatic expansion valves (AEV), 116-17
  capillary tubes, 124-26
  hand expansion valves, 116
  high side floats, 127-28
  low side floats, 126-27
  thermal expansion valves (TEV), 117-25
  types, 115
Meters, electrical, 200-202
  ammeters, 201
  capacitance meters, 202
  ohmmeters, 200
  voltmeters, 201
  wattmeters, 202
Motors, electrical:
  hermetic compressor motors, 233-34, 243-44
    capacitors, 233-34
    overloads, 244
    relays for, 243

manufacturers' information, 198
motor speeds, 242
name plates, 198
FLA, 198
LRA, 198
service factors, 198
temperature rise, 198
power supplies, 237
single phase, 238–41
capacitor start (CS), 239
capacitor start, capacitor run (CSR), 240
permanent split capacitor (PSC), 240
shaded pole, 241
split phase, 239
wound rotor, 241
three phase, 242
squirrel cage, 242
synchronous, 242
wound rotor, 242
Motor starters:
across-the-line, 245
interlocking circuits, 251
reduced voltage, 245–50
auto transformer, 247, 250
part winding, 247–48
primary resistance, 247–49
Wye-Delta, 245–47

Oil safety switches, 226–28
Oils, lubricating:
reaction with refrigerant, 30
refrigeration types, 29
viscosity, 29
Open compressors, 346
Overloads, 233

Piping, ammonia refrigeration:
fitting losses, 147–48
flow rates, 146
joints, 141
materials, 140–41
system practices, 161–67
air blowers, 164–66
compressor piping, 161
condenser-receiver piping, 162
evaporative condenser, 163–65
parallel condensers, 163–64
suction traps, 164–67
Piping, halocarbon refrigerants:
fitting losses, 147–48
flow rates, 146
functions, 140
joints, 141
materials, 17–18, 140–41
sizing, 142–45
R-12, 142
R-22, 143
R-502, 144
R-717, 145
supports, 141
system practices, 148–61
condenser piping, 156–57
evaporator piping, 154–56
hot gas lines, 150–51, 153
liquid lines, 152
suction lines, 148–49, 153
Pressure:
absolute, 8
atmospheric, 8
gauge, 8
Pressure enthalpy diagrams, 436–40
R-12, 436–40
R-22, 481
R-502, 484
R-717, 487
Pressure regulators:
servicing, 353
types, 129–31
Pressurestats, 221–22
Pressure-temperature chart, 9
Psychrometric charts:
low temperature, 458–59
English units, 458
SI units, 459
Purging refrigerant, 335, 349

Refrigerants:
best system, 48
charging, 349, 354–55
comparison of industrial refrigerants, 56
containers, 31–32
cryogenic fluids, 31
grouping and classifying, 31
heat absorbing rates, 30
properties of R-12, 5, 6
properties of R-17, 131–32
qualities, 30
removal, 354
saturated conditions, 5–7
secondary, 83
subcooling, 8
supercooling, 8
Refrigerating effect, 10
Refrigeration:
definition, 1, 3
pump-down cycle, 209

Refrigeration (*cont.*)
refrigeration effect, 10
tons, 7
Refrigeration calculations, 431–32
Refrigeration cycle:
changes, 11–13
components, 10
Refrigeration tables and charts:
adding heat to the refrigerant, 430
calculations, 431–32
compression ratio, 432
compressor displacement, 432
heat of compression, 432
horsepower of compression, 432
pounds of refrigerant circulated, 432
total refrigerating effect, 431
conversion factors, 446
effects of reducing condensing temperature, 439–40
effects of subcooling liquid, 439, 441
effects of suction superheat, 441–42
example of P/E chart in SI units, 442–43
example of refrigeration table in SI units, 444–45
plotting performance on a P/E diagram, 436–40
compressor heat, 439
condenser heat, 439–40
latent heat of vaporization, 437
net refrigeration effect, 438
reading a refrigeration table, 429
summary, 446
volumetric efficiency, 432–33
Refrigerators:
reach-in, 290–91
walk-in, 307–9
Relays, 229–30

Safety Code of Mechanical Refrigeration, 400–479
Selecting equipment, walk-in coolers, 423–27
balancing evaporator capacity with compressor capacity, 427
condensing unit capacities, R-12, 425
temperature loss in suction lines, 424
unit cooler capacities, 426
Service, 242–55
caring for service valves, 345
charging system with liquid refrigerant, 354–55
cleaning condensers, 349–50
correcting refrigerant charge, 349
purging non-condensables, 349
removing refrigerant, 354
servicing burned out hermetics, 348
servicing compressors and condensing units, 346–50
removing compressors, 346
replacing crankshaft seals, 347
replacing valve plates, 347–48
testing open compressors, 346
servicing condensers, 348
servicing evaporators, 350
servicing refrigeration controls, 351–53
low side floats, 252–53
pressure regulators, 353
solenoid valves, 353
thermal expansion valves, 351–52
summary of halocarbon refrigeration service, 355
using filters and driers, 353
using gauge manifolds, 342–45
Service valves, 345
SI units, 442–46
conversion factors, 448–56
Soldering, soft, 24
Solenoid valves, 208–16
servicing, 353
Solvents and cleaning:
liquid acids, 33
solvent sprays, 32
steam cleaning, 32
Starters, electrical, 233, 245–51
Subcooling refrigerant:
definition, 8
effects of, 439, 441
Superheating refrigerants:
definition, 8
effects of, 141–42
Supermarkets, 273–89, 217–19
Systems, refrigeration, 170–88
automatic defrost, 173–74
automatic icemaking, 174–76
basic systems, 171–73
cascade systems, 181–83
classification, 170–71
cryogenic systems, 178–83
intercoolers, 180, 182
heat pumps, 184–86
secondary refrigerant systems, 186–88
supermarket systems, 176–78
two-stage compound systems, 179–81

Tables, refrigerant:
R-12, 5
R-22, 283

R-502, 485-86
R-717, 488-89
Temperature:
absolute zero, 1
dry bulb, 14
wet bulb, 14
Temperature controls, 223-25
Thermal expansion valves (TEV), for halocarbon refrigerants:
advantages, 117
bulb locations, 124-25
charges, 122-24
distributors, 121
electric, 120
external equalizers, 120
hunting, 124
operating pressures, 117-18
pilot operation, 119
pressure limiting, 121
selection chart, 118-19
servicing, 351-52
superheat setting, 121
Thermometers:
Celcius, 2
dial, 27
electronic, 2
Fahrenheit, 2
pocket, 27
recording, 28
Time clocks, 225
Tools:
bending, 23
caring for, 16
cutting, 22
flaring, 23, 24
swaging, 23
wrenches, 26-27
Transformers, 216-18
Troubleshooting, 358-65
correcting the problem, 360
interviewing the operator, 359
making final checks, 360
quick fix, 359
reviewing troubleshooting aids, 359
step-by-step search, 359
verifying symptoms, 359
Troubleshooting aids, 360-65
air cooled condensers, 361
basic cycle controls, 363-65
metering devices, 363
pressure controls, 364
solenoid valves, 364
strainer driers, 364
water regulating valves, 364
compressors, 362
evaporative condensers, 362
evaporators, 362
quick fix procedures, 360
water chillers, 362
water cooled condensers, 361
Truck refrigeration:
applications, 383-85
equipment, 296-98
refrigeration cycle diagrams, 299-301
wiring diagrams, 320-23
Tubing, copper:
ACR, hard drawn, 18
safe working pressures, 17-18
service tubing, 17
soft copper, 17

Valve plates, 347-48
Vapor, saturated, 8
Volumetric efficiency, 432-33

Water regulating valve, 108-11
description, 108
selection 109-10
three-way type, 110-11
Wiring diagrams:
drinking water coolers, 306
flake ice machines, 311-13
large systems, 312-13
package units, 311-12
ice cube makers, 309-10
ice dispensers, 310-11
refrigeration compressors, 315-16
supermarket systems, 317-19
compressor banks, 318
defrost systems, 319
pressure controls, solid state, 319
transport refrigeration, 320-23
connection diagram, 322
control sequence, 323
schematic diagram, 321
types, 305
walk-in coolers:
electric defrost, 307
hot gas defrost, 308-9
Wrenches:
adjustable, 27
box, 26
flare nut, 26
open end, 26
socket, 26
special service, 27